Mathematics
for Physicists

Mathematics for Physicists

Huaiyu Wang
Tsinghua University, China

Published by

World Scientific Publishing Co. Pte. Ltd.

5 Toh Tuck Link, Singapore 596224

USA office: 27 Warren Street, Suite 401-402, Hackensack, NJ 07601

UK office: 57 Shelton Street, Covent Garden, London WC2H 9HE

Library of Congress Cataloging-in-Publication Data

Names: Wang, Huaiyu, author.

Title: Mathematics for physicists / Wang Huaiyu (Tsinghua University, China).

Description: Hackensack, NJ : World Scientific, [2017] |
    Includes bibliographical references and index.

Identifiers: LCCN 2016035728| ISBN 9789813146488 (hardcover ; alk. paper) |
    ISBN 9813146486 (hardcover ; alk. paper) | ISBN 9789813148000 (pbk. ; alk. paper) |
    ISBN 9813148004 (pbk. ; alk. paper)

Subjects: LCSH: Mathematics. | Mathematical physics.

Classification: LCC QC20 .W275 2017 | DDC 510--dc23

LC record available at https://lccn.loc.gov/2016035728

British Library Cataloguing-in-Publication Data

A catalogue record for this book is available from the British Library.

Mathematics for Physicists

© Wang Huaiyu

The Work is originally published by Science Press in 2013.

This edition is published by World Scientific Publishing Company Pte Ltd by
arrangement with Science Press, Beijing, China.

All rights reserved. No reproduction and distribution without permission

Desk Editor: Christopher Teo

Typeset by Stallion Press

Email: enquiries@stallionpress.com

Printed in Singapore

Introduction

This book is mainly for graduate students in physics and engineering, and also suitable for undergraduate students at senior level and those who intend to enter the field of theoretical physics. The contents cover the mathematical knowledges of the following fields: the theories of variational method, of Hilbert space and operators, of ordinary linear differential equations, of Bessel functions, of Dirac delta function, of the Green's function in mathematical physics, of norm, of integral equations, the application of number theory in physics, the basic equations in multidimensional spaces and non-Euclid spaces. The book explains the concepts and deduces the formulas in great details. It is very learner-friendly with its content level gradually from being easy to being difficult. A great amount of exercises are beneficial to readers.

Preface

This book grew from a graduate course that I have taught at Tsinghua University. It was originally written in Chinese and published by Science Press in Beijing. World Scientific Publishing and Science Press have been very kind to jointly publish this English version. This English version has been compressed in space compared to the original Chinese version.

The aim of this book is to integrate the necessary aspects of mathematics for graduate students in physics and engineering. The undergraduate students at senior level and researchers who intend to enter the field of theoretical physics were supposed possible readers. Readers are assumed to have knowledges of linear algebra and complex analysis. Some familiarity with elementary physical knowledges could be a prerequisite for deriving full benefit from reading this book.

This book consists of ten chapters. Chapter 1 introduces the variational method. Chapter 2 introduces the theories of Hilbert space and operators. In chapter 3, the theory of ordinary linear differential equations of second order is systematically presented. The polynomial solutions are given. The method of drawing series solutions based on the complex analysis is given. The problems concerning the adjoint equations are discussed. Chapter 4 comprehensively introduces Bessel functions and their various deforms. Chapter 5 introduces Dirac delta function. Chapter 6 presents the theory of Green's function in mathematical physics. By the way, the author has written a graduate textbook systematically introducing the Green's function in condensed matter physics. Chapter 7 introduces the theory with respect to norm. Chapter 8 introduces the theory of integral equations. In these chapters, the author payed attention to the link up with the contents that readers may have learnt in undergraduate courses.

The last two chapters are an attempt to introduce some recent achievements of scientific research into the textbook while presenting mathematical basic knowledges.

Chapter 9 introduces the basis of number theory and its application in physics, material science and other scientific fields. This kind of application was initiated by a Chinese scientist, Prof. Chen Nanxian. The author thought that this ingenious method was worthy of being introduced to readers. The mathematical basis in this chapter is relatively simple, but it leads to useful and wide results. This achievement has never been introduced in textbooks except a literary work by Chen himself.

Chapter 10 introduces the fundamental equations in spaces with arbitrary dimensions. This is because in modern physics, research has been not limited to three-dimensional space and one-dimensional time, and not limited to Euclid space. The author tries hard to introduce some basic knowledge in multidimensional spaces starting from the ordinary differential equations of second order. The associated Gegenbauer equation and its solutions enable us to realize the values of the angular momentum and their projections in Euclid spaces. The pseudo spherical coordinates were introduced. Although they were employed for the discussion on Euclid spaces, they apparently were also useful to investigations on non-Euclid spaces. Plain terminologies were utilized to present the concept of metric, without resorting to symmetry or group theory. The author believed that this is a way easily grasped by readers. The work on the Klein-Gorden equation and Maxwell equation were new and interesting.

When presenting the mathematical basic theory, the logic rigor was assured without loss of understandability. The author tried hard to clearly narrate the basic concepts and the relations between them. The derivations of the formulas were given in detail as far as possible. For proofs of the theorems, when they were too long or needed knowledges that were beyond the scope of this paper, we had to omit them. The explanations of the questions asked by students in the course of my teaching have been covered.

This textbook does not include the content of group theory, for there have been textbooks specialized for group theory.

The author thought that a great amount of exercises are beneficial to readers. Most of the exercises in this book were collected from materials. A few were prepared by the author.

The author thanks Prof. Chen Nanxian for introducing his smart work to the author. The contents in Chapter 9 in this book are all from Prof. Chen's work.

The investigations of the Klein-Gorden equation and Maxwell equations in de Sitter spacetime are from the work of Prof. Zhou Bin. The author thanks him for providing his achievements and for his helpful discussions.

I acknowledge and express my deep sense of gratitude to the valuable discussions and helps from Professors Wang Chongyu, Zhou Yunsong, Xun Kun, Han Rushan, Tong Dianmin, Zheng Yujun and Yu Yabin.

I wish to express my thanks to my wife Miao Qing and my family members Miao Hui, Miao JiChun and Wang Nianci for their constant help in my work and life.

A special thank goes to Prof. Lu Xiukun, who taught me mathematics when I was a student in secondary school.

Finally, I thank editor Qian Jun for his help in publishing the original Chinese version and present English version.

The author acknowledges the National Key Research and Development Program of China under Grant No. 2016YFB0700102.

This English version has been translated from Chinese by myself, with slight updating over the original Chinese version. I shall be most grateful to those readers who are kind enough to bring to my notice any remaining mistakes, typographical or otherwise for remedial action. Please feel free to contact me.

Wang Huaiyu
Tsinghua University, Beijing, April 2016
wanghuaiyu@mail.tsinghua.edu.cn

Contents

Chapter 1

Variational Method

Variational method is an important means in mathematical physics. It is used to study the extremes of functionals that are a kind of special variables. In this chapter, some essential knowledge and its applications are presented.

1.1. Functional and Its Extremal Problems

1.1.1. *The conception of functional*

First of all, some simplest examples are given in order to introduce the concept of functionals.

Example 1. Suppose that two points $x = x_0$ and $x = x_1$ on the x axis are given, and $y = y(x)$ is a function defined on the interval $[x_0, x_1]$ and has a continuous first derivative. Then the length of the curve $y = y(x)$ is

$$l[y(x)] = \int_{x_0}^{x_1} \sqrt{1 + y'^2} \mathrm{d}x. \qquad (1.1.1)$$

As long as the functions $y(x)$ is changed, say, to be another function $y_1(x)$, the length $l[y(x)]$ will be accordingly changed to be $l[y_1(x)]$. That is to say, the variable l defined by (1.1.1) depends on the "whole function" $y(x)$.

Definition 1. A function set having some common features is defined as a **function class**.

1

For instance, a set of functions that are continuous on the interval $[x_0, x_1]$ is denoted as $C[x_0, x_1]$. A set of functions that have first continuous derivatives on the interval $[x_0, x_1]$ is called C_1 class functions on $[x_0, x_1]$, and is denoted as $C_1[x_0, x_1]$. In this analogy, a set of functions that have the n-th continuous derivatives on the interval $[x_0, x_1]$ is called C_n class functions on $[x_0, x_1]$, and is denoted as $C_n[x_0, x_1]$. If a function $y(x)$ has the n-th continuous derivative in the interval $[x_0, x_1]$, it belongs to $C_n[x_0, x_1]$, and is simply denoted as $y(x) \in C_n$. This denotation is also used for multi-variable functions. For instance, $z(x, y) \in C_2(D)$ means that the function $z(x, y)$ has the second continuous partial derivatives on domain D.

Example 2. Suppose that D is a given domain in the xy plane, and a function $z(x, y) \in C_1(D)$. Then the area of the surface corresponding to the function is

$$S[z(x, y)] = \iint\limits_{D} \sqrt{1 + z_x'^2 + z_y'^2}\,\mathrm{d}x\mathrm{d}y. \qquad (1.1.2)$$

Obviously, the variable S depends on the "whole function" $z(x, y)$.

Having given the examples above, we now introduce the concept of **functional**.

Definition 2. Let R be a number domain, and Y be a given function class denoted as $\{y(x)\}$. For every function $y(x)$ belonging to Y, there is a number $J \in R$. In this case, the variable J is defined as a **functional** of the function $y(x)$, and denote as $J = J[y(x)]$. The function class is called the **definition domain** of the functional $J[y(x)]]$, which sometimes is also called the **admissible function** of the functional. In short, a functional is a map from a function set Y to a number domain R. Each "argument" of the map is a function, and each function $y(x)$ belonging to Y is called an admissible function. It is easy for readers to give similar definition of the functionals depending on multifunctions.

According to this definition, the integral (1.1.1) and (1.1.2) are respectively the functionals at $C_1[x_0, x_1]$ and $C_1(D)$.

Example 3. The Fourier transformation

$$F[f(x)] = \int_{-\infty}^{\infty} f(x)e^{-ikx}dx$$

is a functional. It has a parameter k. Once the k value is set, for each function $f(x)$, the value of the functional is determined. The function $f(x)$ is the admissible one of the functional. Alternatively, this functional can be regarded as a function of argument k, and is denoted as $F(k)$. However, the value of $F(k)$ does not come from the definition of a function, but from the integral above which relies on the concrete form of $f(x)$.

1.1.2. *The extremes of functionals*

1. Functional extremes

The basic problem of the variational method is with respect to the functional extremes. For instance, in 1696 J. Bernoulli raised the **brachistochrone problem** as follows, which significantly promoted the variational method. It is to determine a curve passing through two fixed points A and B in a plane perpendicular to the earth's surface such that a point mass moving along the curve under the influence of gravity travels from A and B in the shortest possible time (all possible resistances such as friction are neglected).

From the viewpoint of distance, the straight line connecting A and B is the shortest one. However, in the initial period, the dropping could not gain fast speed, so that the time spent would not be the shortest.

In Fig. 1.1, A is the origin, Ax is a horizontal axis and Ay is a vertical axis downwards. Let $y = y(x)$ be a smooth curve connecting points $A(0,0)$ and $B(a,b)$. The point mass slips downwards along this curve. Since the curve is smooth, the mass does not feel a force along the tangential direction of the curve. The normal force changes the direction of the velocity but not its magnitude. Since the initial velocity is zero, the speed of the mass when it moves to a point $M(x,y)$ is

$$v = \sqrt{2gy}, \tag{1.1.3}$$

where g is the acceleration of gravity.

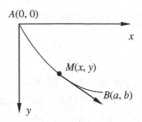

$$A(0,0)$$
$$x$$
$$M(x, y)$$
$$B(a, b)$$
$$y$$

Fig. 1.1.

Let S be the length of the curve, and $\mathrm{d}t$ the differential with respect to time. Then,

$$v = \frac{\mathrm{d}S}{\mathrm{d}t} = \frac{\sqrt{1 + y'^2}\mathrm{d}x}{\mathrm{d}t}. \tag{1.1.4}$$

Therefore,

$$\mathrm{d}t = \frac{\sqrt{1 + y'^2}\mathrm{d}x}{v} = \frac{\sqrt{1 + y'^2}}{\sqrt{2gy}}\mathrm{d}x. \tag{1.1.5}$$

Consequently, the time T needed for the mass to slip along the curve $y = y(x)$ from A to B is

$$T[y(x)] = \int_0^a \frac{\sqrt{1 + y'^2}}{\sqrt{2gy}}\mathrm{d}x. \tag{1.1.6}$$

Thus, the formal statement of the barchistochrone problem is that it is to find a curve $y = y(x)$ satisfying the boundary conditions

$$y(0) = 0, \quad y(a) = b \tag{1.1.7}$$

such that the functional $T[y(x)]$ reaches its minimum.

The functional extremes defined below are similar to those of functions. Before doing so, the concept of ε-neighborhood of curve $y = y(x)$ is defined.

Definition 3. The **ε-neighborhood** of a curve $y = y(x)$ defined on $[x_0, x_1]$ means all the possible curves $y = y_1(x)$ satisfying the condition

$$|y_1(x) - y(x)| \leq \varepsilon \tag{1.1.8}$$

within the whole interval $[x_0, x_1]$ (Fig. 1.2). It is also termed that $y_1(x)$ has an **ε-proximity of zero order** to $y(x)$. For C_1 class

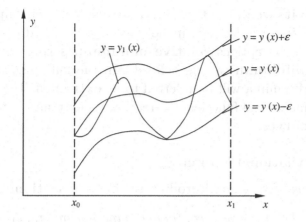

Fig. 1.2.

functions, the ε-neighborhood of $y = y(x)$ requires not only (1.1.8), but also the following inequality:

$$\left| y_1'(x) - y'(x) \right| \leq \varepsilon. \qquad (1.1.9)$$

In this case, it is said that $y_1(x)$ has an ε-**proximity of first order** to $y(x)$. Similarly, the concept of the ε-**proximity of n-th order** can be defined. Given a curve $y = y_0(x)$, the set including all the curves $y_0(x)$ that have the ε-**proximity of n-th order** is called the ε-**neighborhood of n-th order** of $y_0(x)$.

Definition 4. Let $J[y(x)]$ be a functional of a certain function class $\{y(x)\}$. Suppose that function $y_1(x)$ belongs to function class $\{y(x)\}$ which has an ε-proximity of a certain order to $y_0(x)$, where $\varepsilon > 0$. If the functional value at $y_0(x)$ is not less than that at any such $y_1(x)$, i.e.,

$$J[y_0(x)] \geq J_1[y_1(x)], \qquad (1.1.10)$$

then it is said that the functional $J[y(x)]$ reaches its **relative maximum** at $y_0(x)$. If (1.1.10) is always valid when the set $\{y(x)\}$ is the ε-neighborhood of zero order of the function $y_0(x)$, then $J[y(x)]$ is said to reach its **strong relative maximum** at $y_0(x)$, or simply **strong maximum**. If (1.1.10) is valid when the set $\{y(x)\}$ is the ε-neighborhood of first order of the function $y_0(x)$, then $J[y(x)]$ is said to reach **its weak relative maximum** at $y_0(x)$, or simply **weak maximum**. In a similar way, if the symbol $\geq$ in (1.1.10) is changed to be $\leq$, the conceptions **relative minimum, strong minimum** and **weak minimum** can be defined . In the following, all of the maxima and minima will be referred to as **extremes**. The function $y_0(x)$ that makes the functional J reaches its extremes is termed as **extremal curves**.

2. Basic variational lemma

For later usage, let us now introduce the **basic variational lemma**.

Lemma 1. *Let $f(x)$ be a continuous function on $[x_0, x_1]$. For any function $\eta(x)$ belonging to C_2 class functions and being zero at two points $x = x_0$ and $x = x_1$, the integral*

$$\int_{x_0}^{x_1} f(x)\eta(x)\,dx = 0, \qquad (1.1.11)$$

then $f(x) \equiv 0$.

Proof. We use reduction *ad absurdum*. Suppose that at point $\xi(x_0 < \xi < x_1)$, $f(\xi) \neq 0$, e.g., $f(\xi) > 0$. Since $f(x)$ is a continuous one, there must be a neighborhood of ξ, $\xi_1 < \xi < \xi_2$. In this neighborhood $f(x) > 0$. Now we construct a function as follows.

$$\eta(x) = \begin{cases} 0 & x_0 \leq x \leq \xi_1, \\ (x - \xi_1)^4(x - \xi_2)^4 & \xi_1 \leq x \leq \xi_2, \\ 0 & \xi_2 \leq x \leq x_1. \end{cases}$$

It meets all the conditions for the function $\eta(x)$ in the lemma. As a matter of fact, it is easily verified that $\eta(x_0) = \eta(x_1) = 0$, and the first and second derivatives of $(x - \xi_1)^4(x - \xi_2)^4$ with respect to x exist and are zero at $x = \xi_1$ and $x = \xi_2$. Outside the interval $[\xi_1, \xi_2]$, $\eta(x)$ is always zero. Hence the function itself and its first and second derivatives are continuous on the whole $[x_0, x_1]$. For such an $\eta(x)$,

$$\int_{x_0}^{x_1} f(x)\eta(x)dx = \int_{\xi_1}^{\xi_2} f(x)\eta(x)dx > 0, \qquad (1.1.12)$$

i.e., the integral is not zero. This contradicts to the premise (1.1.11). Hence there should be $f(x) = 0$. $\qquad\square$

Please note that the conclusion of the lemma stands under the conditions that the function $\eta(x)$ is any C_2 class one being zero at $x = x_0$ and $x = x_1$ and satisfying (1.1.11). If some of the conditions are not met, one cannot gain the conclusion $f(x) = 0$ from only (1.1.11).

Lemma 2. *Suppose a function $f(x, y)$ is continuous in region D. For any function $\eta(x, y)$ belonging to C_2 class functions and being zero at the boundary of the region, the integral*

$$\iint_D f(x, y)\eta(x, y)\,dx\,dy = 0, \qquad (1.1.13)$$

then $f(x, y) = 0$.

Proof. Suppose that at point (a, b) within the region $f(a, b) > 0$. Then it must be that in a circle K of radius ρ centered on (a, b), $f(x, y) > 0$. Now we construct a function as follows

$$\eta(x, y) = \begin{cases} 0, & (x - a)^2 + (y - b)^2 \geq \rho^2, \\ [(x - a)^2 + (y - b)^2 - \rho^2]^4, & (x - a)^2 + (y - b)^2 < \rho^2. \end{cases}$$

Obviously, this function $\eta(x, y)$ meets the conditions required in the lemma. However, we have

$$\iint_D f(x, y)\eta(x, y)\mathrm{d}x\mathrm{d}y = \iint_K f(x, y)\eta(x, y)\mathrm{d}x\mathrm{d}y > 0.$$

This contradicts to (1.1.13). The conclusion is that $f(x, y)$ should be zero everywhere on D. □

Apparently, lemmas 1 and 2 are for 1- and 2-dimensional cases, respectively. In the cases of triple and multiple integrals, there are similar lemmas.

1.2. The Variational of Functionals and the Simplest Euler Equation

1.2.1. *The variational of functionals*

Let a functional be of the following form:

$$J[y(x)] = \int_{x_0}^{x_1} F(x, y, y')\mathrm{d}x, \qquad (1.2.1)$$

where F is a continuous function of three arguments x, y, y', and has continuous partial derivatives of second order. In (1.2.1), $y(x) \in C_2$ is in turn a function of x.

Before discussing how to calculate such kind of functionals, the concept of variational of the functionals should be given, since it is of importance in searching the extremal curve of the functionals. This concept is similar to the derivative in studying the extremes of functions.

Suppose that a function $y(x)$ varies slightly, and becomes $y(x) + \delta y(x)$, where $\delta y(x)$ means a function, instead of δ multiplied by $y(x)$. The function $\delta y(x)$ is termed as the **variational** of $y(x)$. Let us see what is the increment of the functional (1.2.1) provided that both y and $y + \delta y$ are its admissible functions. Hereafter, we denote $\delta y' = (\delta y(x))'$ for the sake of convenience. The increment is defined as

$$
\begin{aligned}
\Delta J &= J[y + \delta y] - J[y] \\
&= \int_{x_0}^{x_1} [F(x, y + \delta y, y' + \delta y') - F(x, y, y')]\mathrm{d}x \\
&= \int_{x_0}^{x_1} \left[\frac{\partial F}{\partial y}\delta y + \frac{\partial F}{\partial y'}\delta y'\right]\mathrm{d}x + \int_{x_0}^{x_1} (\varepsilon_1 \delta y + \varepsilon_2 \delta y')\mathrm{d}x, \quad (1.2.2)
\end{aligned}
$$

where

$$
\lim_{\delta y \to 0, \delta y' \to 0} \varepsilon_1 = 0, \qquad \lim_{\delta y \to 0, \delta y' \to 0} \varepsilon_2 = 0. \qquad (1.2.3)
$$

The first integral in the second line of (1.2.2) is called the variational of the functional $J[y]$ at "point" $y(x)$, denoted as δJ:

$$
\delta J = \int_{x_0}^{x_1} \left(\frac{\partial F}{\partial y}\delta y + \frac{\partial F}{\partial y'}\delta y'\right)\mathrm{d}x. \qquad (1.2.4)
$$

Apparently, δJ is linear with respect to δy. Let us inspect the difference of the increment and variational, $\Delta J - \delta J$. Denoting

$$
\|\delta y\| = \max_{x_0 \le x \le x_1} \{|\delta y(x)|, |\delta y'(x)|\}, \qquad (1.2.5)
$$

we have

$$|\Delta J - \delta J| = \left| \int_{x_0}^{x_1} (\varepsilon_1 \delta y + \varepsilon_2 \varepsilon y') \mathrm{d}x \right|$$

$$\leq \max(|\varepsilon_1| + |\varepsilon_2|)(x_1 - x_0) \|\delta y\| . \qquad (1.2.6)$$

It is seen from (1.2.3) that $\Delta J - \delta J$ is the higher-order infinitesimal of $\|\delta y\|$. Thus, the variational δJ is named as the linear main part of functional increment ΔJ. This coincides with the case in differential calculus, where the differential of a function is the linear main part of the infinitesimal increment of the function.

For example, let $J[y] = \int_{x_0}^{x_1} (y^2 + y'^2) \mathrm{d}x$. Then

$$\Delta J = J[y + \delta y] - J[y]$$

$$= \int_{x_0}^{x_1} (2y\delta y + 2y'\delta y') \mathrm{d}x + \int_{x_0}^{x_1} [(\delta y)^2 + (\delta y')^2] \mathrm{d}x. \qquad (1.2.7)$$

In this equation, the first integral is linear to δy, and the second one is a higher-order infinitesimal of $\|\delta y\|$. Thus

$$\delta J = \int_{x_0}^{x_1} (2y\delta y + 2y'\delta y') \mathrm{d}x. \qquad (1.2.8)$$

Surely, this result can be achieved directly from (1.2.4).

The differential of a function $f(x)$ can be written as its derivative with respect to a parameter α:

$$\mathrm{d}f = \frac{\mathrm{d}}{\mathrm{d}\alpha} f(x + \alpha \Delta x) |_{\alpha=0} . \qquad (1.2.9)$$

In the same way, the variational δJ of a functional $J[y]$ can be written as its derivative with respect to a parameter α, namely,

$$\delta J = \frac{\mathrm{d}}{\mathrm{d}\alpha} J[y + \alpha \delta y] |_{\alpha=0} . \qquad (1.2.10)$$

As a matter of fact,

$$\frac{\mathrm{d}}{\mathrm{d}\alpha} J[y + \alpha \delta y]|_{\alpha=0} = \int_{x_0}^{x_1} \left(\frac{\partial F}{\partial y} \delta y + \frac{\partial F}{\partial y'} \delta y' \right) \mathrm{d}x = \delta J. \qquad (1.2.11)$$

The variational of a functional (1.2.4) can also be put into the following form:

$$\delta J = \int_{x_0}^{x_1} \delta F \mathrm{d}x. \qquad (1.2.12)$$

In this way, the variational of a functional is written as an integral. The integrand F is regarded as a functional depending on functions y and y' at fixed x. By comparison of (1.2.12) and (1.2.4), it is seen that

$$\delta F = \frac{\partial F}{\partial y}\delta y + \frac{\partial F}{\partial y'}\delta y'. \qquad (1.2.13)$$

This reminds us that the differential of a binary function $f = f(x, y)$ is $\mathrm{d}f = \frac{\partial f}{\partial x}\mathrm{d}x + \frac{\partial f}{\partial y}\mathrm{d}y$. Evidently, the first order variational has the same form as the first order differential. One needs only distinguish that the former is the function of a function, while the latter is the function of an argument. We use letter d to present differential and Greek letter δ to present variational. When a functional F depends on a certain function y, the partial derivative $\partial F/\partial y$ with respect to y is manipulated just as that of a function f with respect to one of its arguments x.

Next, let us see the second variational. When δJ is regarded as a functional, and δF is also a functional in the same sense as F, we have

$$\delta^2 J = \int_{x_0}^{x_1} \delta^2 F \mathrm{d}x. \qquad (1.2.14)$$

Application of (1.2.13) leads to

$$\delta^2 F = \frac{\partial \delta F}{\partial y}\delta y + \frac{\partial \delta F}{\partial y'}\delta y'$$

$$= \frac{\partial^2 F}{\partial y^2}(\delta y)^2 + 2\frac{\partial^2 F}{\partial y \partial y'}\delta y'\delta y + \frac{\partial^2 F}{\partial y'^2}(\delta y')^2.$$

This can be compared to the second differential of a binary function:

$$\mathrm{d}^2 f = \frac{\partial^2 f}{\partial x^2}(\mathrm{d}x)^2 + 2\frac{\partial^2 f}{\partial x \partial y}\mathrm{d}x\mathrm{d}y + \frac{\partial^2 f}{\partial y^2}(\mathrm{d}y)^2$$

$$= \left(\mathrm{d}x\frac{\partial}{\partial x} + \mathrm{d}y\frac{\partial}{\partial y}\right)^2 f.$$

It is seen that the manipulation of variation is just the same as that of function differentials.

Similarly, it is easy to show that the variationals of the product of two functions and of the inverse of a function are

$$\delta(F_1 F_2) = F_2\delta F_1 + F_1\delta F_2 \qquad (1.2.15)$$

and

$$\delta\left(\frac{1}{F}\right) = -\frac{1}{F^2}\delta F, \tag{1.2.16}$$

respectively.

For functionals, one has variationals with the same forms as these two equations. We show them in the following. Suppose that there are two functionals

$$J_1 = \int_{a_1}^{b_1} F_1(x_1, y_1, y_1')\mathrm{d}x_1$$

and

$$J_2 = \int_{a_2}^{b_2} F_2(x_2, y_2, y_2')\mathrm{d}x_2.$$

The variational of their product is

$$\begin{aligned}
\delta(J_1 J_2) &= \int_{a_2}^{b_2} \mathrm{d}x_2 \int_{a_1}^{b_1} \mathrm{d}x_1 \delta(F_1 F_2) \\
&= \int_{a_2}^{b_2} \mathrm{d}x_2 \int_{a_1}^{b_1} \mathrm{d}x_1 (F_2 \delta F_1 + F_1 \delta F_2) \\
&= \int_{a_2}^{b_2} F_2 \mathrm{d}x_2 \int_{a_1}^{b_1} \mathrm{d}x_1 \delta F_1 + \int_{a_2}^{b_2} \delta F_2 \mathrm{d}x_2 \int_{a_1}^{b_1} \mathrm{d}x_1 F_1 \\
&= J_2 \delta J_1 + J_1 \delta J_2. \tag{1.2.17}
\end{aligned}$$

The variational of the inverse of a functional being of the form of (1.2.1) is as follows.

$$\begin{aligned}
\delta\left(\frac{1}{J}\right) &= \frac{1}{\int_{x_0}^{x_1} F(x, y + \delta y, y' + \delta y')\mathrm{d}x} - \frac{1}{\int_{x_0}^{x_1} F(x, y, y')\mathrm{d}x} \\
&= \frac{1}{J + \delta J} - \frac{1}{J} = \frac{1}{J}\left(\frac{1}{1 + \delta J/J} - 1\right).
\end{aligned}$$

We denote

$$\|\delta y\| \equiv \max_{x_0 \le x \le x_1} \left\{|\delta y|, |\delta y'|\right\}, \quad \varepsilon_1 = \max_{x_0 \le x \le x_1} \left\{\left|\frac{\partial F}{\partial y}\right|\right\},$$

$$\varepsilon_2 = \max_{x_0 \le x \le x_1} \left\{\left|\frac{\partial F}{\partial y'}\right|\right\}.$$

Then $\delta J \leq |x_1 - x_0| (\varepsilon_1 + \varepsilon_2) \|\delta y\|$, i.e., δJ is an infinitesimal with one order higher than $\|\delta y\|$. Thus, by expanding Taylor series to the first order term, we have

$$\delta\left(\frac{1}{J}\right) = \frac{1}{J}\left[\left(1 + \frac{\delta J}{J}\right)^{-1} - 1\right] = \frac{1}{J}\left(1 - \frac{\delta J}{J} - 1\right) = -\frac{1}{J^2}\delta J$$

The conclusion is that

$$\delta\left(\frac{1}{J}\right) = -\frac{1}{J^2}\delta J. \tag{1.2.18}$$

1.2.2. *The simplest Euler equation*

1. Euler equation

Theorem 1. *Let $y(x)$ be the extremal curve of functional* (1.2.1). *Then the variational of the functional at $y = y(x)$ is*

$$\delta J = 0. \tag{1.2.19}$$

Proof. Because we are talking about the necessary condition that the extremal curve $y(x)$ must satisfy, we select a function y^* with a special form to compare $J[y]$ with $J[y^*]$. For example,

$$y^*(x) = y(x) + \alpha\delta y(x). \tag{1.2.20}$$

Here α is a small parameter, so that $y^*(x)$ is of given ε-proximity to $y(x)$, and $\delta y(x)$ is any C_2 class function.

By the premise, the function of α, $J[\alpha] = J[y^*] = J[y + \alpha\delta y]$, takes its extreme at $\alpha = 0$. Therefore, (1.2.10) leads to

$$\delta J = \frac{\mathrm{d}}{\mathrm{d}\alpha} J[y + \alpha\delta y]\,|_{\alpha=0} = 0\,. \tag{1.2.21}$$

That is to say, the variational of $J[y]$ at $y = y(x)$ is $\delta J = 0$.   □

Theorem 2. *Suppose that $y(x)$ is the extremal curve of functional* (1.2.1). *Then the function $y = y(x)$ necessarily observes the following differential equation:*

$$F_y - \frac{d}{dx}F_{y'} = 0 \tag{1.2.22}$$

or

$$F_{y'y'}y'' + F_{yy'}y' + F_{xy'} - F_y = 0, \tag{1.2.23}$$

where $F_{yy'}$ is the second partial derivative of F with respect to y and y', and so on.

Proof. Now we take a function of a more special form $y^* = y + \alpha \delta y$ so as to compare $J[y]$ with $J[y^*]$. It is required that y^* and y take the same values at points $x = x_0$ and $x = x_1$, i.e., $y^*(x_0) = y(x_0), y^*(x_1) = y(x_1)$, which results in

$$\delta y(x_0) = 0, \quad \delta y(x_1) = 0. \tag{1.2.24}$$

Now we take integration by parts for the second integrand in (1.2.4). This brings

$$\delta J = \int_{x_0}^{x_1} (F_y \delta y + F_{y'} \delta y') \mathrm{d}x = F_{y'} \delta y(x)|_{x_0}^{x_1}$$

$$+ \int_{x_0}^{x_1} (F_y - \frac{\mathrm{d}}{\mathrm{d}x} F_{y'}) \delta y \mathrm{d}x. \tag{1.2.25}$$

The first term is zero by (1.2.24). As a result,

$$\delta J - \int_{x_0}^{x_1} (F_y - \frac{\mathrm{d}}{\mathrm{d}x} F_{y'}) \delta y \mathrm{d}x. \tag{1.2.26}$$

Finally, it is known by the basic variational lemma that the extremal curve $y(x)$ obeys the differential equation (1.2.22). □

The ordinary differential equation of second order (1.2.22) or (1.2.23) is called the **Euler equation** of extremal problem of functional (1.2.1). Its general solution contains two arbitrary constants. Usually, when discussing functional extremes, there may be some additional conditions on the values of the admissible function at boundaries x_0 and x_1. For example, the two ends are fixed, which means that

$$y(x_0) = y_0, \quad y(x_1) = y_1, \tag{1.2.27}$$

where x_0, y_0, x_1, y_1 are constants. By the boundary conditions (1.2.27), one is able to determine the two constants in the solution of the Euler equation. Thus the extremal curve of functional (1.2.1) is obtained.

2. Two special cases

In this subsection the following two specials cases of Euler equation are presented.

The first case is that F is independent of y, i.e., $F = F(x, y')$. Because $F_y = 0$, Euler equation becomes

$$\frac{\mathrm{d}}{\mathrm{d}x} F_{y'} = 0. \tag{1.2.28}$$

This brings **the first integral**

$$F_{y'}(x, y') = C_1. \tag{1.2.29}$$

This is a differential equation not explicitly containing y. From this equation $y\prime$ can be solved, and then its integration brings the solution. Sometimes, properly selecting parameters is helpful.

Example 1. Find the extremal curve of functional

$$J = \int_{x_v}^{x_1} \frac{\sqrt{1 + y'^2}}{x} \mathrm{d}x$$

which satisfies the boundary conditions

$$y(x_0) = y_0, \quad y(x_1) = y_1.$$

Solution. Because F is independent of y, the first integral of the Euler equation is

$$F_{y'} = \frac{y'}{x\sqrt{1 + y'^2}} = C_1.$$

Its integration brought to

$$x^2 + (y - C_2)^2 = 1/C_1^2, \quad (C_1 \neq 0).$$

This is a set of circles centered at the y axis. The undetermined constants C_1 and C_2 can be solved from boundary conditions.

The second case is that F is independent of variable x, i.e., $F = F(y, y')$. Because $F_{xy'} = 0$, the Euler equation, with the help of (1.2.23), becomes

$$y'' F_{y'y'} + y' F_{y'y} - F_y = 0. \tag{1.2.30}$$

On the other hand,

$$\frac{d}{dx}(F - y'F_{y'}) = -y'(y''F_{y'y'} + y'F_{yy'} - F_y) = 0. \qquad (1.2.31)$$

Subsequently, (1.2.30) has the first integral:

$$F - y'F_{y'} = C_1. \qquad (1.2.32)$$

Integrating once more gives the possible extremal curves. The integral constants are determined by boundary conditions.

Example 2. In the previous section, we have restated the barchistochrone problem as to find the extremal curves of the functional

$$T[y(x)] = \int_0^a \frac{\sqrt{1 + y'^2}}{\sqrt{2gy}} dx$$

which meets boundary conditions

$$y(0) = 0, \quad y(a) = b.$$

Please find the solution.

Solution. Because F is independent of x, the Euler equation has the first integral (1.2.32), which leads to

$$\sqrt{\frac{1 + y'^2}{2gy}} - y'\frac{1}{\sqrt{2gy}}\frac{y'}{\sqrt{1 + y'^2}} = C.$$

With the denotation $C_1 = 1/2gC^2$, it is simplified to be

$$y(1 + y'^2) = C_1.$$

We solve this equation with parameter method. Let $y' = \cot(\theta/2)$. Then the equation becomes

$$y - \frac{C_1}{1 + y'^2} = C_1 \sin^2 \frac{\theta}{2} = \frac{C_1}{2}(1 - \cos\theta).$$

As a consequence,

$$dx = \frac{dy}{y'} = \frac{C_1 \sin(\theta/2)\cos(\theta/2)d\theta}{\cot(\theta/2)} = \frac{C_1}{2}(1 - \cos\theta)d\theta.$$

Integrating the last expression gives $x = \frac{C_1}{2}(\theta - \sin\theta)$. Hence, the required curve is

$$x = \frac{C_1}{2}(\theta - \sin\theta) + C_2, \quad y = \frac{C_1}{2}(1 - \cos\theta).$$

In terms of the boundary conditions $y(0) = 0$, one obtains $C_2 = 0$. This is a class of cycloids with their radius being $C_1/2$. The constant C_1 can be determined by the condition of the cycloid's value at point B. Hence, the solved barchistochrone is a cycloid through points A and B.

In the end of this section, we stress one point. Just as that the root of equation $f'(x) = 0$ is merely the necessary condition of $y = f(x)$ taking its extremes, the solutions of the Euler equation are the necessary but not sufficient condition of functional (1.2.1) taking its extremes. Nevertheless, in practical applications to the problems raised from such fields as engineering, mechanics, physics and so on, the solved extremal function $y = y(x)$ often happens to be the one that makes the functional $J[y(x)]$ take its extremes.

In the following sections, the equations obtained through the variational method are also the necessary conditions of functionals taking their extremes.

1.3.   The Cases of Multifunctions and Multivariates

1.3.1.   *Multifunctions*

When a functional depends on more than one admissible function, the corresponding Euler equation can also be achieved in a similar way as above. Here we discuss the case of two admissible functions:

$$J[y(x), z(x)] = \int_{x_0}^{x} F(x, y, y', z, z')\mathrm{d}x. \qquad (1.3.1)$$

Suppose that $y(x), z(x) \in C_2$, and there exist all possible second partial derivatives of F with respect to its two admissible functions. If functional (1.3.1) takes its extremes at $y(x)$ and $z(x)$, let us find the necessary conditions that $y(x)$ and $z(x)$ must satisfy.

The following routine is in fact the same as the case of one admissible function. This time, the comparison curves are taken as $y^* = y + \alpha\delta y$ and $z^* = z + \alpha\delta z$, where α is a small parameter and

$$\delta y(x_0) = \delta y(x_1) = \delta z(x_0) = \delta z(x_1) = 0. \qquad (1.3.2)$$

Substituting y^* and z^* into (1.3.1), we represent J as being a function of the parameter α:

$$J(\alpha) = \int_{x_0}^{x_1} F(x, y + \alpha\delta y, y' + \alpha\delta y', z + \alpha\delta z, z' + \alpha\delta z')\mathrm{d}x.$$

$J(\alpha)$ takes its extremes at $\alpha = 0$. Therefore,

$$J'(\alpha)\,|_{\alpha=0} = \int_{x_0}^{x_1} (F_y\delta y + F_{y'}\delta y' + F_z\delta z + F_{z'}\delta z')\mathrm{d}x = 0.$$

We regard

$$\delta J = \int_{x_0}^{x_1} (F_y\delta y + F_{y'}\delta y' + F_z\delta z + F_{z'}\delta z')\mathrm{d}x \qquad (1.3.3)$$

as the **variational** of functional (1.3.1). It is seen that the necessary conditions that (1.3.1) takes for its extremes at $y(x)$ and $z(x)$ are that its variational at these two functions should be zero.

Now we take integration by parts for the second and fourth terms in (1.3.3) and use (1.3.2) to get

$$\delta J = [F_{y'}\delta y + F_z\delta z]|_{x_0}^{x_1}$$

$$+ \int_{x_0}^{x_1} \left[\left(F_y - \frac{\mathrm{d}}{\mathrm{d}x}F_{y'}\right)\delta y + \left(F_z - \frac{\mathrm{d}}{\mathrm{d}x}F_{z'}\right)\delta z\right]\mathrm{d}x$$

$$= \int_{x_0}^{x_1}\left(F_y - \frac{\mathrm{d}}{\mathrm{d}x}F_{y'}\right)\delta y\mathrm{d}x + \int_{x_0}^{x_1}\left(F_z - \frac{\mathrm{d}}{\mathrm{d}x}F_{z'}\right)\delta z\mathrm{d}x = 0.$$

$$(1.3.4)$$

Specially, when $\delta z = 0$, we have

$$\int_{x_0}^{x_1}(F_y - \frac{\mathrm{d}}{\mathrm{d}x}F_{y'})\delta y\mathrm{d}x = 0.$$

That is, by the basic variational lemma, the function in the parenthesis should be zero. When $\delta z = 0$, we obtain another equation. In summary, the functions $y(x)$ and $z(x)$ that make functional (1.3.1) take its extremes should satisfy the following differential equations of second order:

$$\begin{cases} F_y - \dfrac{\mathrm{d}}{\mathrm{d}x}F_{y'} = 0, \\[2mm] F_z - \dfrac{\mathrm{d}}{\mathrm{d}x}F_{z'} = 0. \end{cases} \qquad (1.3.5)$$

These equations are combined with given boundary conditions, so that the extremal curves can be solved. For example, in an extremal problem with fixed boundaries, the conditions are

$$y(x_0) = y_0, \quad y(x_1) = y_1,$$
$$z(x_0) = z_0, \quad z(x_1) = z_1. \tag{1.3.6}$$

In the present case, one can image that x is a function of y and z, and there is a curve in a three-dimensional Cartesian system such that it in fact reflects the two functions $y(x)$ and $z(x)$.

Example 1. Find the extremal curves of the functional

$$J = \int_0^{\pi/2} (y'^2 + z'^2 + 2yz)\mathrm{d}x,$$

which satisfies the boundary conditions

$$y(0) = 0, \quad y\left(\frac{\pi}{2}\right) = 1; \quad z(0) = 0, \quad z\left(\frac{\pi}{2}\right) = -1.$$

Solution.

$$F = y'^2 + z'^2 + 2yz.$$

By (1.3.5), the corresponding Euler equations are

$$\begin{cases} y'' - z = 0, \\ z'' - y = 0. \end{cases}$$

To solve the equations, the function z is eliminated:

$$y^{(4)} - y = 0.$$

Its general solution is

$$y = C_1 e^x + C_2 e^{-x} + C_3 \cos x + C_4 \sin x.$$

By $z = y''$, we get

$$z = C_1 e^x + C_2 e^{-x} - C_3 \cos x - C_4 \sin x.$$

Now using the boundary conditions, the constants are

$$C_1 = 0, \quad C_2 = 0, \quad C_3 = 0, \quad C_4 = 1.$$

Finally, the required curves are

$$y = \sin x, \quad z = -\sin x.$$

If there are M admissible functions and one variable: $y_\alpha(x), (\alpha = 1, 2, \ldots, M)$, we use $\{y_\alpha(x)\}$ to denote the function set. The functional is expressed as

$$J[\{y_\alpha(x)\}] = \int_{x_0}^{x} F(x, \{y_\alpha(x)\}, \{y'_\alpha(x)\}) \mathrm{d}x. \qquad (1.3.7)$$

Then it is natural to extend (1.3.5) to the following M equations:

$$\frac{\partial F}{\partial y_\alpha} - \frac{\mathrm{d}}{\mathrm{d}x}\frac{\partial F}{\partial y'_\alpha} = 0, \quad (\alpha = 1, 2, \ldots, M). \qquad (1.3.8)$$

1.3.2. *Multivariates*

We turn to the cases of multivariates. For simplicity without loss of generality, we take a functional of a binary function $u(x, y)$ as an example:

$$J[u] = \iint_D F(x, y, u, u_x, u_y) \mathrm{d}x\mathrm{d}y. \qquad (1.3.9)$$

It involves double integrals. Here $u(x, y) \in C_2(D)$ and D is a given region in the xy plane. The function F is assumed to have second partial derivatives with respect to all of its argument and admissible functions.

Suppose that $u = u(x, y)$ has been the extremal surface that makes the functional (1.3.9) take its extremes. In order to find the necessary conditions it meets, we once more select comparison function as before in the following form:

$$u^* = u(x, y) + \alpha\eta(x, y),$$

where $\eta(x, y)$ is an arbitrary C_2 class function and satisfies boundary condition $\eta(x, y)|_C = 0$, and α is a small parameter. Evidently, u^* and u take the same value at the boundary. Substituting u^* into (1.3.9), we have

$$J[u^*] = \iint_D F(x, y, u + \alpha\eta, p + \alpha\eta_x, q + \alpha\eta_y) \mathrm{d}x\mathrm{d}y = J(\alpha).$$

For the sake of simplicity, here we have denoted $p = u_x$ and $q = u_y$. So, the extremal surface $u(x, y)$ should make $J'(0) = 0$. As a result,

$$J'(0) = \iint_D F_u\eta + F_p\eta_x + F_q\eta_y)\mathrm{d}x\mathrm{d}y$$

$$= \iint_D \left(F_u - \frac{\partial}{\partial x}F_p - \frac{\partial}{\partial y}F_q\right)\eta(x,y)\mathrm{d}x\mathrm{d}y$$

$$+ \iint_D \left[\frac{\partial}{\partial x}(F_p\eta) + \frac{\partial}{\partial y}(F_q\eta)\right]\mathrm{d}x\mathrm{d}y = 0. \quad (1.3.10)$$

The last line can be, by use of Green's formula, transformed to integration along the boundary line:

$$\iint_D \left[\frac{\partial}{\partial x}(F_p\eta) + \frac{\partial}{\partial y}(F_q\eta)\right]\mathrm{d}x\mathrm{d}y = \int_C (F_p\eta\mathrm{d}y - F_q\eta\mathrm{d}x). \quad (1.3.11)$$

By the boundary condition $\eta(x, y)|_C = 0$, this integration is zero. Hence, using the basic variational lemma leads to

$$F_u - \frac{\partial}{\partial x}F_p - \frac{\partial}{\partial y}F_q = 0. \quad (1.3.12)$$

This is the Euler equation that $u(x, y)$ must satisfy. It contains the partial derivatives of undetermined function, so that is a partial differential equation.

Example 2. Functional

$$J[u] = \iint_D (u_x^2 + u_y^2)\mathrm{d}x\mathrm{d}y$$

takes its extreme when its admissible function meet

$$\Delta u = \frac{\partial^2 u}{\partial x^2} + \frac{\partial^2 u}{\partial y^2} = 0.$$

This is a well-known Laplace equation. Solving $u(x, y)$ from this equation accompanied by boundary condition $u(x, y)|_C = u_0(x, y)$ is the first boundary value problem or Dirichlet problem in region D:

$$\begin{cases} \Delta u = \dfrac{\partial^2 u}{\partial x^2} + \dfrac{\partial^2 u}{\partial y^2} = 0, (x, y) \in D, \\ u(x, y)|_C = u_0(x, y). \end{cases}$$

In summary, the idea of proving the above formulas is as follows. The functional is written as a function depending on a parameter α. Then the derivative with respect to this parameter is set to be zero to obtain the extremes. One may also take more than one parameter such as α, β and so on, if necessary, and then take derivatives with respect to every parameter to obtain extremes of the functional.

If there are N variables $x_i, (i = 1, 2, \ldots, N)$, we use $\{x_i\}$ to denote them. The function is $y = y(x_1, x_2, \ldots, x_N) = y(\{x_i\})$. The functional is

$$J[y] = \int_R F\left(x_1, x_2, \ldots, x_N, y, \frac{\partial y}{\partial x_1}, \frac{\partial y}{\partial x_2}, \ldots, \frac{\partial y}{\partial x_N}\right) \mathrm{d}x_1 \mathrm{d}x_2 \cdots \mathrm{d}x_N,$$

$$(1.3.13a)$$

which is in short written as

$$J[y] = \int_R F\left(\{x_i\}, y, \left\{\frac{\partial y}{\partial x_i}\right\}\right) \mathrm{d}^N x. \qquad (1.3.13b)$$

It should not be difficult to extend (1.3.12) to be the following equations:

$$\frac{\partial F}{\partial y} - \sum_{i=1}^N \frac{\partial}{\partial x_i} \frac{\partial F}{\partial(\partial y/\partial x_i)} = 0. \qquad (1.3.14)$$

Finally, if there arc N variables $x_i, (i = 1, 2, \ldots, N)$ and M functions $y_\alpha(\{x_i\}), (\alpha = 1, 2, \ldots, M)$, the functional will be

$$J[\{y_\alpha(\{x_i\})\}] = \int_R F\left(\{x_i\}, \{y_\alpha(\{x_i\})\}, \left\{\frac{\partial y_\alpha}{\partial x_i}\right\}\right) \mathrm{d}^N x. \quad (1.3.15)$$

In this case, combination of (1.3.8) and (1.3.14) results in the following Euler equations:

$$\frac{\partial F}{\partial y_\alpha} - \sum_{i=1}^N \frac{\partial}{\partial x_i} \frac{\partial F}{\partial(\partial y_\alpha/\partial x_i)} = 0, \quad (\alpha = 1, 2, \ldots, M), \qquad (1.3.16)$$

which contain M equations.

1.4.   Functional Extremes under Certain Conditions

In many functional extremal problems, the admissible functions themselves may also be exerted some restrictions, which is similar to the conditional extreme value problems of functions. In this section, we discuss two types of conditional extreme value problems of functionals.

1.4.1.   *Isoperimetric problem*

This was originated from the following geometric problem. One is seeking a curve l through points A and B with fixed length, which makes a trapezoid with curve sides $ABCD$ take its maximum area. Suppose $y = y(x)$ is the required solution. It is easy to see that the problem is attributed to finding the extreme of a functional

$$S[y] = \int_{x_0}^{x_1} y(x) \mathrm{d}x, \qquad (1.4.1)$$

which satisfies the boundary conditions $y(x_0) = y_0, y(x_1) = y_1$ and constraint $\int_{x_0}^{x} \sqrt{1 + y'^2} \mathrm{d}x = l$. Because it is required that the perimeter of the trapezoid with curve sides $ABCD$ is fixed, this problem is called **isoperimetric problem**.

The general statement of the problem is as follows. Among all the admissible curves of the functional

$$J_1 = \int_{x_0}^{x} G(x, y, y') \mathrm{d}x = l \text{ (constant)}, \qquad (1.4.2)$$

find one which makes another functional

$$J = \int_{x_0}^{x} F(x, y, y') \mathrm{d}x \qquad (1.4.3)$$

take its extremes. This type of problem is termed as an **isoperimetric problem**. Equation (1.4.2) is termed as an **isoperimetric condition**.

Similar to the conditional extreme problem of multivariate functions, there is a theorem for the isoperimetric problem. This theorem utilizes the method of Lagrange multiplier, and makes the variational problem of conditional extremes to be one without the conditions.

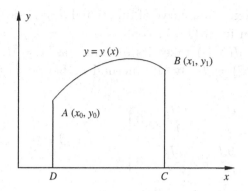

Fig. 1.3.

Theorem 1 (Lagrange). *If a curve $y = y(x)$ is the extremal one of* (1.4.3), *but not that of functional J_1, then there must be a constant λ which makes $y = y(x)$ be the extremal one of the functional*

$$\int_{x_0}^{x_1} H(x, y, y')\,dx, \qquad (1.4.4)$$

where

$$H = F + \lambda G. \qquad (1.4.5)$$

Hence, $y = y(x)$ satisfies Euler equation

$$H_y - \frac{d}{dx}H_{y'} = 0. \qquad (1.4.6)$$

Proof. We take comparison function

$$y^*(x) = y(x) + \alpha\eta_1(x) + \beta\eta_2(x), \qquad (1.4.7)$$

where α and β are small parameters, and $\eta_1(x)$ and $\eta_2(x)$ are given C_2 class functions which satisfy $\eta_1(x_0) = \eta_1(x_1) = \eta_2(x_0) = \eta_2(x_1) = 0$. Substituting $y^*(x)$ into (1.4.2) and (1.4.3), respectively, one gains

$$J(\alpha, \beta) = \int_{x_0}^{x} F(x, y + \alpha\eta_1 + \beta\eta_2, y' + \alpha\eta_1' + \beta\eta_2')dx \qquad (1.4.8)$$

and

$$J_1(\alpha, \beta) = \int_{x_0}^{x} G(x, y + \alpha\eta_1 + \beta\eta_2, y' + \alpha\eta_1' + \beta\eta_2')dx = l. \qquad (1.4.9)$$

Let $y(x)$ be the extremal curve of functional $J[y]$ under the condition (1.4.2). It is seen from (1.4.7) that as $\alpha = \beta = 0$, $y^* = y$. Thus, the binary function $J(\alpha, \beta)$ takes its extremes as $\alpha = \beta = 0$ under the condition $J_1(\alpha, \beta) = l$. By the method of Lagrange multiplier, we have

$$\left.\begin{aligned} \frac{\partial J}{\partial \alpha} + \lambda \frac{\partial J_1}{\partial \alpha} = 0 \\ \frac{\partial J}{\partial \beta} + \lambda \frac{\partial J_1}{\partial \beta} = 0 \end{aligned}\right\}, \quad (\alpha = \beta = 0), \tag{1.4.10}$$

where λ is to be determined. Substitution of J and J_1 into these two equations leads to

$$\left.\begin{aligned} \int_{x_0}^{x_1} [(F_y + \lambda G_y)\eta_1 + (F_{y'} + \lambda G_{y'})\eta_1']dx = 0 \\ \int_{x_0}^{x_1} [(F_y + \lambda G_y)\eta_2 + (F_{y'} + \lambda G_{y'})\eta_2']dx = 0 \end{aligned}\right\} \tag{1.4.11}$$

Taking integration by parts for the second terms in the two equations, and noticing $\eta_1(x_0) = \eta_1(x_1) = \eta_2(x_0) = \eta_2(x_1) = 0$, we obtain

$$\int_{x_0}^{x_1} \left[(F_y + \lambda G_y) - \frac{d}{dx}(F_{y'} + \lambda G_{y'}) \right] \eta_1 dx = 0 \tag{1.4.12}$$

and

$$\int_{x_0}^{x_1} \left[(F_y + \lambda G_y) - \frac{d}{dx}(F_{y'} + \lambda G_{y'}) \right] \eta_2 dx = 0. \tag{1.4.13}$$

By assumed condition, $y(x)$ does not make J_1 take its extremes, so that does not satisfy the equation $G_y - \frac{d}{dx}G_{y'} = 0$. Therefore, it is possible to select a $\eta_2(x)$ which yields

$$\int_{x_0}^{x_1} \left(G_y - \frac{d}{dx}G_{y'} \right) \eta_2 dx \neq 0. \tag{1.4.14}$$

Subsequently, one gains from (1.4.13) that

$$\lambda = \frac{\int_{x_0}^{x_1} \left(F_y - \frac{d}{dx}F_{y'} \right) \eta_2 dx}{\int_{x_0}^{x_1} \left(G_y - \frac{d}{dx}G_{y'} \right) \eta_2 dx}. \tag{1.4.15}$$

Because η_2 and η_1 are independent of each other, λ is independent of η_1. Similarly, λ is also independent of η_2. Therefore, λ must be a constant unrelated to both η_1 and η_2. Finally, applying the basic variational lemma to (1.4.12) leads to

$$F_y + \lambda G_y - \frac{\mathrm{d}}{\mathrm{d}x}(F_{y'} + \lambda G_{y'}) = 0. \qquad (1.4.16)$$

□

Equation (1.4.16) is an ordinary differential equation of second order containing parameter λ. Its general solution comprises three undetermined constants C_1, C_2 and λ, which could be determined by the isoperimetric condition (1.4.2) and given boundary conditions, e.g., $y(x_0) = y_0$ and $y(x_1) = y_1$.

Example 1. Find the extremal curve of the functional

$$S = \int_{x_0}^{x_1} y\mathrm{d}x, y(x_0) = y_0, y(x_1) = y_1,$$

which satisfies the isoperimetric condition

$$\int_{x_0}^{x_1} \sqrt{1 + y'^2}\mathrm{d}x = l.$$

Solution. We take an auxiliary functional

$$J = \int_{x_0}^{x_1} (y + \lambda\sqrt{1 + y'^2})\mathrm{d}x.$$

Let $H - y + \lambda\sqrt{1 + y'^2}$. Then its Euler equation is

$$H_y - \frac{\mathrm{d}}{\mathrm{d}x}H_{y'} = 0.$$

Because H does not involve x, it has the first integral

$$y + \lambda\sqrt{1 + y'^2} - \frac{\lambda y'^2}{\sqrt{1 + y'^2}} = C_1.$$

We employ parameter method to solve the differential equation. Let $y' = \tan t$. Then,

$$y - C_1 = -\lambda\cos t, \quad \mathrm{d}x = \frac{\mathrm{d}y}{y'} = \frac{\lambda\sin t\mathrm{d}t}{\tan t} = \lambda\cos t\mathrm{d}t, \quad x = \lambda\sin t + C_2.$$

Thus, the required curve meets the parameter equation

$$x - C_2 = \lambda\sin t, \quad y - C_1 = -\lambda\cos t.$$

The parameter t can be eliminated to have $(x-C_2)^2+(y-C_1)^2 = \lambda^2$. This is a class of circles. The constants C_1, C_2 and λ can be determined by the given isoperimetric condition and boundary conditions.

1.4.2. *Geodesic problem*

The problem is to find two functions $y(x)$ and $z(x)$, which make the functional

$$J = \int_{x_0}^{x_1} F(x,y,y',z,z')\mathrm{d}x \qquad (1.4.17)$$

take its extremes and satisfy an additional condition

$$G(x,y,z) = 0. \qquad (1.4.18)$$

In mechanics, this problem is the so-called constraint problem. From aspect of geometry, it is to find a curve on the surface (1.4.18) making the functional (1.4.17) take its extremes. Note that here are one argument x and two functions $y(x)$ and $z(x)$. Image that in the three-dimensional Cartesian system, there is a surface where x is a function of y and z. The required extremal curve is on the surface, and in fact contains two functions $y(x)$ and $z(x)$.

A natural idea in solving this problem is taking z as a function of x and y from (1.4.18). Then, this function is substituted into (1.4.17). As a result, we achieve an ordinary variational problem which seeks a function $y(x)$ without the additional condition. Now we follow this routine to find the equations that the extremal curves $y(x)$ and $z(x)$ must satisfy.

Suppose $G_z \neq 0$. According to the existence theorem of implicit functions, the function $z = \varphi(x,y)$ can be deduced from (1.4.18) $z = \varphi(x,y)$. It is then substituted into (1.4.17) to get

$$J = \int_{x_0}^{x_1} F(x,y,y',\varphi,\varphi_x + \varphi_y y')\mathrm{d}x. \qquad (1.4.19)$$

For the sake of simplicity, we denote $F^*(x,y,y') = F(x,y,y',\varphi,\varphi_x + \varphi_y y')$. So, $y(x)$ should meet the Euler equation

$$F_y^* - \frac{\mathrm{d}}{\mathrm{d}x}F_{y'}^* = 0. \qquad (1.4.20)$$

Because

$$F_y^* = F_y + F_z\varphi_y + F_{z'}(\varphi_{xy} + \varphi_{yy}y'),$$

$$\frac{\mathrm{d}}{\mathrm{d}x}F_{y'}^* = \frac{\mathrm{d}}{\mathrm{d}x}F_{y'} + \varphi_y\frac{\mathrm{d}}{\mathrm{d}x}F_{z'} + F_{z'}(\varphi_{xy} + \varphi_{yy}y'),$$

(1.4.21)

(1.4.20) becomes

$$F_y + \varphi_y\left(F_z - \frac{\mathrm{d}}{\mathrm{d}x}F_{z'}\right) - \frac{\mathrm{d}}{\mathrm{d}x}F_{y'} = 0. \qquad (1.4.22)$$

On the other hand, $\varphi_y = \frac{\partial z}{\partial y} = -\frac{G_y}{G_z}$. Subsituting it into (1.4.22), one obtains

$$\frac{1}{G_y}\left(F_y - \frac{\mathrm{d}}{\mathrm{d}x}F_{y'}\right) = \frac{1}{G_z}\left(F_z - \frac{\mathrm{d}}{\mathrm{d}x}F_{z'}\right). \qquad (1.4.23)$$

The two sides of the equation should always be identical. Therefore, both should equal to a common constant λ. Thus one gets

$$\begin{cases} \dfrac{\mathrm{d}}{\mathrm{d}x}F_{y'} - [F_y + \lambda(x)G_y] = 0, \\[2mm] \dfrac{\mathrm{d}}{\mathrm{d}x}F_{z'} - [F_z + \lambda(x)G_z] = 0. \end{cases} \qquad (1.4.24)$$

These are the differential equations of second order that the extremal functions $y(x)$ and $z(x)$ ought to satisfy.

In summary, we achieve the following theorem.

Theorem 2. *Suppose that the functions $y(x)$ and $z(x)$ make the functional (1.4.17) take its extremes and satisfy (1.4.18). Then there must be a proper factor $\lambda(x)$, which makes $y(x)$ and $z(x)$ satisfy the Euler equations of the functional $J^* = \int_{x_0}^{x_1} H(x, y, y', z, z')\mathrm{d}x$:*

$$\begin{cases} \dfrac{\mathrm{d}}{\mathrm{d}x}H_{y'} - H_y = 0, \\[2mm] \dfrac{\mathrm{d}}{\mathrm{d}x}H_{z'} - H_z = 0, \end{cases} \qquad (1.4.25)$$

where $H = F + \lambda(x)G$.

Note the formal difference between (1.4.24) and (1.4.25). The first term of the former is written as $H_{y'}$ instead of $F_{y'}$. They are actually *the same. The reason is that in the constraint condition* (1.4.18), *G does not contain y' and z'.*

Example 2 (geodesic problem). Find the shortest distance between two fixed points $A(x_0, y_0, z_0)$ and $B(x_1, y_1, z_1)$ on a surface $\varphi(x, y, z) = 0$.

Solution. It is known that the distance between two points is expressed by

$$l = \int_{x_0}^{x_1} \sqrt{1 + y'^2 + z'^2}\, \mathrm{d}x.$$

Hence, the problem is to find the minimum of the functional l under the condition $\varphi(x, y, z) = 0$. We take an auxiliary function

$$l^* = \int_{x_0}^{x_1} [\sqrt{1 + y'^2 + z'^2} + \lambda(x)\varphi(x, y, z)]\mathrm{d}x.$$

Its corresponding Euler equations are

$$\lambda(x)\varphi_y - \frac{\mathrm{d}}{\mathrm{d}x}\frac{y'}{\sqrt{1 + y'^2 + z'^2}} = 0,$$

$$\lambda(x)\varphi_z - \frac{\mathrm{d}}{\mathrm{d}x}\frac{z'}{\sqrt{1 + y'^2 + z'^2}} = 0.$$

From these two equations and the constraint condition $\varphi(x, y, z) = 0$, the factor $\lambda(x)$ and functions $y = y(x)$ and $z = z(x)$ can be solved. The undetermined constants in the general solutions can be found under the boundary conditions $y(x_0) = y_0, y(x_1) = y_1$ and $z(x_0) = z_0, z(x_1) = z_1$.

In mechanics, the constraint problem often manifests a more general form as follows. One wants to find the extremes of a functional containing n admissible functions

$$J[y_1, y_2, \ldots, y_n] = \int_{x_0}^{x_1} F(x; y_1, y_2, \ldots, y_n; y_1', y_2', \ldots, y_n')\mathrm{d}x \quad (1.4.26)$$

under m constraint conditions

$$\varphi_k(x, y_1, y_2, \ldots, y_n) = 0, \quad k = 1, 2, \ldots m, \quad (m < n). \quad (1.4.27)$$

This problem is attributed to solve Euler equations of the functional

$$J^* = \int_{x_0}^{x_1} \left[F + \sum_{k=1}^{m} \lambda_k(x)\varphi_k \right] \mathrm{d}x. \qquad (1.4.28)$$

The equations are

$$\frac{\mathrm{d}}{\mathrm{d}x} F_{y_i'} - \left[F_{y_i} + \sum_{k=1}^{m} \lambda_k(x)\frac{\partial \varphi_k}{\partial y_i} \right] = 0, \quad i = 1, 2, \ldots, n. \qquad (1.4.29)$$

1.5. Natural Boundary Conditions

The Euler equations that the extremal curves or surfaces should satisfy are ordinary or partial differential equations. We have shown that some boundary conditions are needed in order to explicitly determine the extremal functions. In the examples above, the boundary conditions are that the extremal functions take fixed values at the boundaries, e.g., (1.2.27). Correspondingly, the variational of the extremal functions at the boundaries are zero, see, for example, (1.2.24).

In practical applications, the boundary conditions may be beyond this kind. Even the boundaries themselves may vary. For instance, let us see the simplest functional

$$J[y] = \int_{x_0}^{x_1} F(x, y, y')\mathrm{d}x. \qquad (1.5.1)$$

Generally speaking, its lower and upper integral limits x_0, x_1 and the values of its admissible functions at the boundaries $y(x_0)$ and $y(x_1)$ may all vary. That is to say, the boundaries vary. Here we only discuss a simplest case: the integral limit x_0, x_1 in (1.5.1) remain unchanged, while the values of $y(x_0)$ and $y(x_1)$ may vary. In the aspect of geometry, the two ends of the admissible curve can vary along the lines going through $x = x_0$ and $x = x_1$ and parallel to the y axis, see Fig. 1.4.

The basic condition of taking extremes of a functional is that its variational is zero.

$$\delta J = 0. \qquad (1.5.2)$$

This is our start point in the following discussion.

From (1.2.25) it is known that

$$\delta J = F_{y'}\delta y(x)|_{x_0}^{x_1} + \int_{x_0}^{x_1}\left(F_y - \frac{\mathrm{d}}{\mathrm{d}x}F_{y'}\right)\delta y\,\mathrm{d}x = 0. \qquad (1.5.3a)$$

Please note that the Euler equations obtained before is based on the condition (1.2.24) where the variationals of the extremal curve at ends are zero. It is not so in the present case, as the curve ends may vary. Nevertheless, the admissible curves include those having fixed ends as well as those having unfixed ends. Therefore, it can be certain that as long as $y(x)$ is the extremal curve of the functional (1.5.1), it meets the Euler equation. Assuming that we have found the right extremal curve $y = y^*(x)$ with fixed ends $y(x_0)$ and $y(x_1)$, these boundary conditions can be applied to compose a new variational problem with fixed boundaries. Then obviously, $y^*(x)$ is surely the solution of the new problem. That is to say, $y^*(x)$ certainly meet the Euler equation

$$F_y - \frac{\mathrm{d}}{\mathrm{d}x}F_{y'} = 0.$$

In other words, whether the ends are fixed or not, the extremal curve always meets the Euler equation. Hence, the second term in (1.5.3a) is zero, leaving the first term. By (1.5.2), we have

$$\delta J = (F_{y'}\delta y)_{x_1} - (F_{y'}\delta y)_{x_0} = 0. \qquad (1.5.3b)$$

Because the changes of the extremal curve at the two ends $(\delta y)_{x_0}$ and $(\delta y)_{x_1}$ are independent of each other, see Fig. 1.4, it should

$$F_{y'}\big|_{x=x_0} = 0, \quad F_{y'}\big|_{x=x_1} = 0. \qquad (1.5.4)$$

These are the conditions that the function $y(x)$ should obey at ends $x = x_0$ and $x = x_1$, called **natural boundary conditions**.

A special case is that one of the ends, say the left one, is fixed, i.e.,

$$y(x_0) = y_0. \qquad (1.5.5)$$

The right end may vary along the line $x = x_1$. Then we have $\delta y(x) = 0$ and varying $\delta y(x_1)$. By (1.5.3) the natural boundary

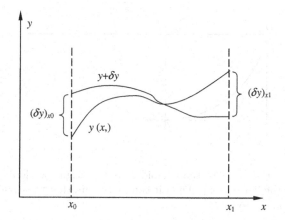

Fig. 1.4. The integral limit x_0 and x_1 remain unchanged, while the values of $y(x_0)$ and $y(x_1)$ may vary.

conditions degrade to

$$F_{y'}|_{x=x_1} = 0. \tag{1.5.6}$$

Example 1. In Subsection 1.2.2, we have obtained the general solution of the brachistochrone problem is a class of cycloids:

$$x = C_1(\theta - \sin\theta) + C_2, \quad y = C_1(1 - \cos\theta).$$

Now let the left end of the cycloid fixed, $y(0) = 0$, while the right end varies along the line $x = x_1$. Then, from $y(0) = 0$ one still gets $C_2 = 0$. The constant C_1 is to be determined by (1.5.6). Because $F = \sqrt{\frac{1+y'^2}{2gy}}$,

$$F_{y'}|_{x=x_1} = \left.\frac{y'}{\sqrt{2gy(1+y'^2)}}\right|_{x=x_1} = 0.$$

It follows that $y'(x_1) = 0$. Hence, the required cycloid should be vertical to the line $x = x_1$, see Fig. 1.5. By $y'(x_1) = 0$, one easily find that the point B corresponds to $\theta = \pi$, which results in $x_1 = C_1\pi$ and $C_1 = \frac{x_1}{\pi}$. Finally, the required curve is expressed as

$$x = \frac{x_1}{\pi}(\theta - \sin\theta), \quad y = \frac{x_1}{\pi}(1 - \cos\theta).$$

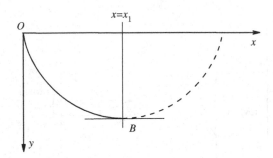

Fig. 1.5.   The initial point of the cycloid is at the origin, while its other end B may vary along the line $x = x_1$. Then it can be certain that the cycloid is vertical to the line $x = x_1$ at point B.

If both ends x_0, x_1, as well as the function values at the ends $y(x_0) = y_0$ and $y(x_1) = y_1$ can vary, i.e., the two ends (x_0, y_0) and (x_1, y_1) can vary arbitrarily, it is a more complicated case. We do not intend to introduce the derivation here, but simply present the final results. In such a case, the natural boundary conditions that the extremal curve should meet are as follows.

$$F_{y'} \mid_{x=x_0} = 0 \, , F_{y'} \mid_{x=x_1} = 0 \, , \tag{1.5.7a}$$

$$(F - y' F_{y'})_{x=x_0} = 0, (F - y' F_{y'})_{x=x_1} = 0. \tag{1.5.7b}$$

The Euler equation should be solved under these boundary conditions. In other words, the two integration constants in the general solution and the positions x_0, x_1 all should be determined by Eqs. (1.5.7).

If the two ends (x_0, y_0) and (x_1, y_1) of the admissible curve vary along two given lines $g_0(x_0)$ and $g_1(x_1)$ but not arbitrarily, i.e.,

$$y_0 = g_0(x_0), \quad y_1 = g_1(x_1), \tag{1.5.8}$$

then, one has

$$\delta y_0 = g_0'(x_0)\delta x_0, \quad \delta y_1 = g_1'(x_1)\delta x_1. \tag{1.5.9}$$

Equations (1.5.8) are just constraint conditions. In this case, the admissible function should meet the following natural boundary conditions:

$$[F + (g_0' - y')F_{y'}]_{x=x_0} = 0 \tag{1.5.10a}$$

and

$$[F + (g_1' - y')F_{y'}]_{x=x_1} = 0. \qquad (1.5.10b)$$

They are termed as transversality conditions, manifesting the relations between the derivatives of the extremal function at the ends and constraint lines $g_0(x_0)$ and $g_1(x_1)$. In solving Euler equations, Eqs. (1.5.9) and (1.5.10) are employed to determine the integral constants and the positions x_0 and x_1.

1.6. Variational Principle

The application of the variational method in physics is called the variational principle. The variational principle was firstly induced from natural phenomena, and then expressed in the form of a principle. It is usually considered as a starting point for deducing the equations that a matter should obey when it moves or exists. This principle belongs to the category of scientific hypothesis, and it manifests different forms in different scientific disciplines.

An example is the **Fermat's principle** in optics: light propagates from point A to B along such a path that it takes least time. Suppose light propagates along a path $y = y(x)$ in a plane. We denote light speed by $v(x, y)$ at point (x, y) in the media. Then, similar to the case of the brachistochrone problem, it is derived that the time needed when light propagates from $A(x_0, y_0)$ to $B(x_1, y_1)$ along path $y = y(x)$ is

$$T[y] = \int_{x_0}^{x_1} \frac{\sqrt{1 + y'^2}}{v(x, y)} dx. \qquad (1.6.1a)$$

Thus, following Fermat's principle, the problem becomes the one to find a curve which makes the functional $T[y]$ take its minimum. If we let time t be an argument, every point x, y along the path is a function of time: $x = x(t), y = y(t)$. Then (1.6.1a) is brought to be

$$T[x(t), y(t)] = \int_{t_0}^{t_1} \frac{\sqrt{\dot{x}^2 + \dot{y}^2}}{v(x, y)} dt. \qquad (1.6.1b)$$

In the following, we will introduce respectively the variational principles in classical mechanics and quantum mechanics.

1.6.1. *Variational principle of classical mechanics*

In classical mechanics, the variational principle mainly means
Hamilton's least-action principle, or simply **Hamilton's
principle**.

In a mechanical system, a **Lagrangian** is defined as its kinetic
energy T minus potential U:

$$L = T - U. \tag{1.6.2}$$

Here T and U may be functions of generalized coordinates and
momenta which vary with time. Let us construct a functional:

$$J = \int_{t_0}^{t_1} L \, \mathrm{d}t. \tag{1.6.3}$$

Hamilton's principle says that among all possible movements of the
system within a period of time (compatible to constraint conditions,
if any), the system necessarily takes one that makes the variational
of the functional J be zero:

$$\delta J = 0. \tag{1.6.4}$$

Or, the functional reaches its extremes. Usually, the extreme is a
minimum. Hence, the principle is also called the **least-action principle**. The J defined by (1.6.3) is termed as the **action**.

In the following, we take a conservative field as an example of
applying the Hamilton's principle.

Suppose that there is a given particle system. The coordinates
of the particles are denoted as $(x_i, y_i, z_i), i = 1, 2, 3 \cdots, n$. The i-th
mass point has mass m_i, and is applied a force $\boldsymbol{F}_i (i = 1, 2, \ldots n)$
which is generated by the potential $U(t, x_1, y_1, z_1, \ldots, x_n, y_n, z_n)$ of
the system:

$$F_{i,x} = -\frac{\partial U}{\partial x_i}, \quad F_{i,y} = -\frac{\partial U}{\partial y_i}, \quad F_{i,z} = -\frac{\partial U}{\partial z_i}. \tag{1.6.5}$$

The movement of the particle system is described by Newton equations. This means that the coordinates of each mass point as functions of time $x = x_i(t), y = y_i(t), z = z_i(t)$ observe the following

equations:

$$
\begin{cases}
m_i \ddot{x}_i = F_{i,x} = -\dfrac{\partial U}{\partial x_i}, \\[2mm]
m_i \ddot{y}_i = F_{i,y} = -\dfrac{\partial U}{\partial y_i}, \\[2mm]
m_i \ddot{z}_i = F_{i,z} = -\dfrac{\partial U}{\partial z_i}.
\end{cases}
\tag{1.6.6}
$$

For this system, the potential is U and its kinetic energy is

$$
T = \frac{1}{2}\sum_{i=1}^{n} m_i(\dot{x}_i^2 + \dot{y}_i^2 + \dot{z}_i^2) = T(\dot{x}_1, \dot{y}_1, \dot{z}_1, \ldots, \dot{x}_n, \dot{y}_n, \dot{z}_n).
$$

Let the initial positions $x_1(t_0), y_1(t_0), z_1(t_0), \ldots, z_n(t_0)$ and final positions $x_1(t_1), \ldots, z_n(t_1)$ be fixed. We consider the following two functionals:

$$
J_1 = \int_{t_0}^{t_1} T(\dot{x}_1, \ldots \dot{z}_n)\,dt
$$

and

$$
J_2 = \int_{t_0}^{t_1} U(t_1, x_1, \ldots, z_n)\,dt.
$$

The variational of the first functional is

$$
\delta J_1 = \int_{t_0}^{t_1} \left(\frac{\partial T}{\partial \dot{x}_1}\delta\dot{x}_1 + \cdots + \frac{\partial T}{\partial \dot{z}_n}\delta\dot{z}_n \right) dt.
$$

Integrating by parts for every term and noticing

$$
\delta x_1(t_0) = \delta x_1(t_1) = \cdots = \delta z_n(t_0) = \delta z_n(t_1) = 0,
$$

one obtains

$$
\delta J_1 = -\int_{t_0}^{t} \left[\left(\frac{d}{dt}\frac{\partial T}{\partial \dot{x}_1} \right)\delta x_1 + \cdots + \left(\frac{d}{dt}\frac{\partial T}{\partial \dot{z}_n} \right)\delta z_n \right] dt
$$

$$
= -\int_{t_0}^{t} (m_1\ddot{x}_1\delta x_1 + \cdots + m_n\ddot{z}_n\delta z_n)\,dt.
$$

For the second functional,

$$
\delta J_2 = \int_{t_0}^{t} \left(\frac{\partial U}{\partial x_1}\delta x_1 + \cdots + \frac{\partial U}{\partial z_n}\delta z_n \right) dt.
$$

It follows from Newton equations (1.6.6) that

$$\delta J_1 = \delta J_2,$$

or

$$\delta(J_1 - J_2) = 0.$$

In summary, if functions $x_1(t), y_1(t), z_1(t), \ldots, x_n(t), y_n(t), z_n(t)$ describe the movement of particle system within time period $t_0 \leq t \leq t_1$, then these functions make the functional

$$J_1 - J_2 = \int_{t_0}^{t_1} (T - U)\mathrm{d}t = \int_{t_0}^{t_1} L\mathrm{d}t \qquad (1.6.7)$$

take its extremes. In another word, the system obeys Hamilton's principle.

Reversely, once the functional of the system is written in the form of (1.6.7), the equations must be (1.6.6) that particles in the system obey, so that the functional takes its extremes. In the above sections, these equations of motion were called the Euler equations. When applied in physics, they are usually called the **Euler-Lagrange equations**, or even further simplified as the **Lagrange equations**.

Usually, kinetic energy and potential are the functions of generalized coordinates and momenta. If there are n pairs of generalized coordinates and momenta, which are represented by $\{q_i(t)\}$ and $\{\dot{q}_i(t)\}$, respectively, then according to the cases of multifunctions discussed in Subsection 1.3.1, the Lagrange equations can be written following (1.3.8) as

$$\frac{\mathrm{d}}{\mathrm{d}t}\frac{\partial L}{\partial \dot{q}_i} - \frac{\partial L}{\partial q_i} = 0, \quad (i = 1, 2, \ldots, n). \qquad (1.6.8)$$

Kinetic energy does not explicitly contain time t, and usually it is a quadratic function of generalized momenta. If potential U does not contain time t either, L will not either. In Subsection 1.2.2 we have discussed a special case that if a functional did not contain an argument, there would exist a first integral. Its integration constant E here is just the total energy of the system, and equals the kinetic energy plus potential:

$$E = T + U.$$

Therefore, that the potential does not explicitly contain time means the conservation of the system's total energy. Furthermore, if potential does not explicitly contain one generalized coordinate q_i, then by (1.6.8) one knows that the derivative of Lagrange function with respect to the generalized momentum corresponding to this coordinate is a constant:

$$\frac{\partial L}{\partial \dot{q}_i} = P_i. \tag{1.6.9}$$

This constant P_i is proportional to the generalized momentum. Hence, that the potential does not explicitly contain a coordinate q_i means the conservation of the momentum corresponding to this coordinate.

Furthermore, we assume that the movement of the particles is restricted by m constraint conditions:

$$\varphi_j(t, x_1, y_1, z_1, \ldots, x_n, y_n, z_n) = 0, \quad j = 1, 2, \ldots, m. \tag{1.6.10}$$

By utilizing Hamilton's principle, it is able to derive the equations of motion of the particle system.

The trajectory of every mass point

$$x_i = x_i(t), \quad y_i = y_i(t), \quad z_i = z_i(t)$$

should make the functional

$$J = \int_{t_0}^{t_1} (T - U)\mathrm{d}t = \int_{t_0}^{t_1} \left[\frac{1}{2}\sum_{i=1}^{n} m_i(\dot{x}_i^2 + \dot{y}_i^2 + \dot{z}_i^2) - U \right]\mathrm{d}t$$

take its extremes. Hence, this is a variational problem of functional extreme under certain conditions. Let us put down an auxiliary functional

$$J^* = \int_{t_0}^{t_1} \left[\frac{1}{2}\sum_{i=1}^{n} m_i(\dot{x}_i^2 + \dot{y}_i^2 + \dot{z}_i^2) - U + \sum_{j=1}^{m} \lambda_j(t)\varphi_j \right]\mathrm{d}x.$$

The Euler equations related to this functional

$$\left.\begin{array}{l}
m_i\ddot{x}_i = -\dfrac{\partial U}{\partial x_i} + \displaystyle\sum_{j=1}^{m}\lambda_j(t)\dfrac{\partial \varphi_j}{\partial x_i} \\[2mm]
m_i\ddot{y}_i = -\dfrac{\partial U}{\partial y_i} + \displaystyle\sum_{j=1}^{m}\lambda_j(t)\dfrac{\partial \varphi_j}{\partial y_i} \\[2mm]
m_i\ddot{z}_i = -\dfrac{\partial U}{\partial z_i} + \displaystyle\sum_{j=1}^{m}\lambda_j(t)\dfrac{\partial \varphi_j}{\partial z_i}
\end{array}\right\} \quad (i = 1, 2 \cdots n)$$

are the differential equations of the movement of this particle system.

Now we generalized Hamilton's principle to an arbitrary system. Given the Lagrangian of the system, the corresponding action is constructed by (1.6.3). Usually, the action is denoted as S:

$$S = \int_{t_0}^{t_1} L \, dt. \qquad (1.6.11)$$

The equation of motion of the system is determined by taking the action's variation to be zero, $\delta S = 0$. Most often, $\delta S = 0$ will make the action S reach its minimum. This is called the least-action principle. The function L in (1.6.11) is usually called the Lagrangian density.

In an arbitrary system, the Lagrangian may not be kinetic minus potential as in (1.6.7). We will see an example in relativistic mechanics.

Example 1. A particle is confined on the surface of a sphere with radius R, and moves by gravity. Find its equations of motion.

Solution. Since it moves on a spherical surface, it is convenient to adopt spherical polar coordinates (r, θ, φ). The sphere center taken as the origin, the kinetic energy of the particle is

$$T = \frac{1}{2}m(\dot{r}^2 + r^2\dot{\theta}^2 + r^2\sin^2\theta\,\dot{\varphi}^2),$$

and its potential is

$$U = mgz = mgr\cos\theta.$$

Subsequently, its Lagrangian is

$$L = \frac{1}{2}m(\dot{r}^2 + r^2\dot{\theta}^2 + r^2\sin^2\theta\,\dot{\varphi}^2) - mgr\cos\theta. \qquad (1.6.12)$$

It is required to be moving on the spheric surface, i.e., $r = R$, which is a constraint condition:

$$G(t, r, \theta, \varphi) = r - R = 0.$$

According to Subsection 1.4.2, we are searching the extremes of functional $\int_{t_0}^{t_1} F(t, r, \theta, \varphi) dt$, where

$$F(t, r, \theta, \varphi) = L + \lambda(t)G$$

$$= \frac{1}{2}m(\dot{r}^2 + r^2\dot{\theta}^2 + r^2\sin^2\theta\dot{\varphi}^2) - mgr\cos\theta + \lambda(t)(r - R).$$

This problem belongs to geodesic one. Please note that λ may be a function of time t.

Applying the Euler-Lagrange equations (1.4.29), one is able to put down an equation for each argument. For r,

$$m\ddot{r} - (mr\dot{\theta}^2 + mr\sin^2\theta\dot{\varphi}^2) + mg\cos\theta - \lambda(t) = 0. \qquad (1.6.13a)$$

For θ,

$$mr^2\ddot{\theta} - mr^2\sin\theta\cos\theta\dot{\varphi}^2 - mgr\sin\theta = 0. \qquad (1.6.13b)$$

Notice that the function F does not explicitly contain the angle φ. So, there is a first integral, which can be, observing (1.2.28), written as

$$mr^2\sin^2\theta\dot{\varphi} = p_\varphi. \qquad (1.6.13c)$$

Here notation p_φ is to represent this first integral. It has dimension of angular momentum, indicating that it is the generalized momentum related to the generalized coordinate φ. The constraint condition gives the fourth equation,

$$r = R. \qquad (1.6.13d)$$

From the four equations (1.6.13), one solves four undetermined functions $r(t), \theta(t), \varphi(t), \lambda(t)$.

As a matter of fact, (1.6.13d) has given the equation of motion of $r(t)$. From (1.6.13c), one gets

$$\dot{\varphi} = \frac{p_\varphi}{mR^2\sin^2\theta}.$$

The expression of function $\varphi(t)$ depends on the form of $\theta(t)$. Substituting (1.6.13c) into (1.6.13b), one obtains

$$mR^2\ddot{\theta} = \frac{p_\varphi^2}{mR^2\sin^3\theta}\cos\theta + mgR\sin\theta.$$

Multiplying $\dot\theta dt$ and then integrating on the two sides of the equation, we have

$$\frac{1}{2}mR^2\dot\theta^2 + \frac{p_\varphi^2}{2mR^2\sin^2\theta} + mgR\cos\theta = C_2. \qquad (1.6.14)$$

This equation shows that the terms at the left hand side are kinetic energies in the two angular directions and potential. Evidently, the integral constant C_2 is the total energy of the particle. As a matter of fact, in the present case, Lagrangian (1.6.12) does not explicitly contain time, so that the total energy of this system is conserved.

It is hard to achieve the analytical solution of function $\theta(t)$. Equation (1.6.13a) means that

$$\lambda(t) = -mR\dot\theta^2 - mR\sin^2\theta\dot\varphi^2 + mg\cos\theta. \qquad (1.6.15)$$

On the right hand side of this equation, the first two terms are the minus centrifugal forces of two angular motion and the third term is minus gravity. Substitution of (1.6.14) and (1.6.13c) into (1.6.15) leads to

$$\lambda(t) = \frac{p_\varphi^2}{mR^3\sin^2\theta} + 2mg\cos\theta - 2\frac{C_2}{R} - \frac{p_\varphi^2}{mR^3\sin^2\theta} + mg\cos\theta$$

$$= \frac{1}{R}(3mgR\cos\theta - 2C_2).$$

In the present case, the two integration constants manifest conservation of the total energy and one generalized momentum of the system. In any physical system, the integration constants have certain explicit physical significances.

1.6.2.   *Variational principle of quantum mechanics*

A quantum system is of a following fundamental property: if its Hamiltonian H is known and its ground state energy is E_0, then for an arbitrary function Φ satisfying required boundary conditions, there is an inequality,

$$\frac{\int \Phi^* H\Phi d\Omega}{\int \Phi^*\Phi d\Omega} \geq E_0.$$

The equal sign stands when and only when Φ is just the ground state wavefunction.

With help of this inequality, one may search a wave function that is close the real ground state as far as possible by means of the variational method. Suppose the required function is $\Phi(r, \alpha_1, \alpha_2, \alpha_3, \ldots, \alpha_l)$ which satisfies the boundary conditions, where $(\alpha_1, \alpha_2, \alpha_3, \ldots, \alpha_l)$ are l variable parameters. The average of Hamiltonian in this function is

$$E(\alpha_1, \alpha_2, \ldots, \alpha_l) = \frac{\int \Phi^* H \Phi \mathrm{d}\Omega}{\int \Phi^* \Phi \mathrm{d}\Omega} \geq E_0.$$

Choosing parameter $(\alpha_1, \alpha_2, \alpha_3, \ldots, \alpha_l)$ properly until E takes its minimum, which is close to the ground state energy E_0 as far as possible. After this is done, one gets Φ as an approximate ground state function. The parameter set should meet the following necessary conditions:

$$\frac{\partial E}{\partial \alpha_k} = 0, \quad (k = 1, 2, \ldots, l).$$

Solving the equations, one may determine the best parameter values such that the evaluated energy E being closest to E_0.

The concrete procedure is that one first suggests, based on physical consideration, a form of required wavefunction containing some parameters. Then he determines these parameter values by means of the variational method. This procedure is called Leitz variational method. It is applicable to all the problems of searching approximate extremal curves.

If one is able to put down the Lagrangian of a quantum system, or he can construct an action composed of the total energy of the system, to give the functional of wavefunctions of the system, then he will obtain proper equations by taking variational with respect to wavefunctions. We will see an example below.

1.7. The Applications of the Variational Method in Physics

As narrated in the previous section, the variational method is an important means for deriving the state of a system or the equation that the motion of the system should satisfy. In principle, the equations of motion of any system can be deduced in terms of the variational method provided that reasonable premise has been given, although actually a large amount of equations of motion were obtained by approaches other than the variational method.

The variation of physical quantities of a system depending on time t and space coordinates r is described by a set of functions $\phi_\alpha(r, t)$, where how to choose the subscript α is determined by the intrinsic properties of the system. For instances, for an electromagnetic field, α means the space components of the electric and magnetic fields. For a quantum system, α labels its eigenstates, called the quantum number. For electrons in a solid crystal, wave vector k is usually used as a quantum number to label the state of the electron system. In general, the function $\phi_\alpha(r, t)$ observes a set of equations. Since the functions represent the distribution of a physical quantity in space, they are called fields, and the differential equations they must satisfy are called field equations.

To achieve the field equations, the prerequisite is to select proper Lagrangian density L of the field. In the field of classical mechanics, it is kinetic energy minus potential $L = T - U$, a conclusion from analytical mechanics. However, only when arguments are space coordinates and momenta, the kinetic energy and potential are the real ones. If adopting generalized coordinates and momenta, one will write down generalized kinetic energy and potential. Such examples can be found in the case of electromagnetic fields. In non-classical mechanics, say, relativity and quantum mechanics, the Lagrangian is not simply kinetic energy minus potential.

Generally speaking, the Lagrangian density is of the form of

$$L = L(\{x_i\}, \{\phi_\alpha(\{x_i\})\}, \{\partial_i \phi_\alpha(\{x_i\})\}), \qquad (1.7.1)$$

where English letter i labels arguments and Greek letter α labels functions. There can be N arguments and M functions. In (1.7.1), an abbreviation $\partial_i = \frac{\partial}{\partial x_i}$ is made.

Once an action

$$S = \int_R L d^N x \qquad (1.7.2)$$

is constructed, its variational ought to be zero according to the least-action principle. This is a variational problem of both multivariates and multifunctions already introduced in Section 1.3. Following (1.3.16), the Euler-Lagrange equations are

$$\frac{\partial L}{\partial \phi_\alpha} - \sum_{i=1}^{N} \frac{\partial}{\partial x_i} \frac{\partial L}{\partial(\partial_i \phi_\alpha)} = 0, \quad (\alpha = 1, 2, \ldots, M). \tag{1.7.3}$$

These equations are just the required field equations that physical quantities must satisfy.

When suggesting a Lagrangian, the functions $\{\phi_\alpha\}$ and arguments $\{x_i\}$ should be explicitly clarified.

In previous sections, differential equations were derived by the variational process. In many cases, one may achieve differential or integral equations, or even integral differential equations, through variational method, which depends on the construction of a Lagrangian.

It should be pointed out that Lagrangian density of a system obeying variational principle is not unique. As an illustration, let us see (1.7.2). Suppose that the area of the integral is known, and the values of the functions $\{\phi_\alpha\}$ at the boundary are given. We apply to L an additional term $\sum_{j=1}^{N} \partial_j \Gamma_j(\{x_i\}, \{\phi_\alpha(\{x_i\})\})$ where Γ_j can be an arbitrary derivable function. By the least-action principle, we have

$$\delta \int_R \left[L + \sum_{j=1}^{N} \partial_j \Gamma_j(\{x_i\}, \{\phi_\alpha(\{x_i\})\}) \right] \mathrm{d}^N x$$

$$= \delta \int_R L \mathrm{d}^N x + \delta \oint \sum_{j=1}^{N} \Gamma_j \mathrm{d}l_j.$$

The last term is the line integral along the boundary line, and its result depends on the values of functions $\{\phi_\alpha\}$ at the boundary. Because the values are fixed, this term is no doubt zero. This results in

$$\delta \int_R \left[L + \sum_{j=1}^{N} \partial_j \Gamma_j \right] \mathrm{d}^N x = \delta \int_R L \mathrm{d}^N x. \tag{1.7.4}$$

It is therefore seen that the Lagrangian L and $L + \sum_{j=1}^{N} \partial_j \Gamma_j$ are equivalent, so that they describe the same one physical system.

Variational principle itself does not tell how to select Lagrangian density. Its selection is based on physical equations already known, and often is "conjectured". To testify the correctness of the derived equations is to check if they are consistent with practical physical equation, or if the calculated results by the equation agree with experiments. An example of the latter case was the discovery of Shrödinger equation.

1.7.1. *The applications in classical physics*

1. Maxwell electromagnetic field equations

From known equations of electromagnetism, the Lagrangian of electromagnetic field can be written as

$$L = -\frac{1}{4\mu} \sum_{ij} F_{ij} F_{ij} + \sum_{i} J_i A_i, \qquad (1.7.5)$$

where

$$F_{ij} = \frac{\partial A_j}{\partial x_i} - \frac{\partial A_i}{\partial x_j} \qquad (1.7.6)$$

are components of the four-dimensional antisymmetric tensor of an electromagnetic field. It is called an electromagnetic tensor. Four-dimensional vectors $\{A_i\}$ and $\{J_i\}$ are defined by

$$\{A_i\} \equiv \{A_1, A_2, A_3, i\varphi/c\} \equiv \{\boldsymbol{A}, i\varphi/c\} \qquad (1.7.7a)$$

and

$$\{J_i\} \equiv \{j_1, j_2, j_3, ic\rho\} \equiv \{\boldsymbol{j}, ic\rho\}, \qquad (1.7.7b)$$

respectively. Here, $\boldsymbol{A}$ and φ are vector and scalar potentials of electromagnetic field, and $\boldsymbol{j}$ and ρ are electric current density and charge density, respectively. The coordinates in four-dimensional time and space are

$$\{x_i\} \equiv \{x_1, x_2, x_3, ict\} \equiv \{\boldsymbol{r}, ict\}. \qquad (1.7.7c)$$

Equation (1.7.5) is invariant under Lorentz transformation. In relativistic systems, the Lorentz invariance of Lagrangian is an important foundation for finding Lagrangians of the systems.

As soon as a Lagrangian is given, one is able to derive field equations in terms of Lagrange equations. The components expressed by (1.7.6) are those of electric field $\boldsymbol{E}$ and magnetic induction $\boldsymbol{B}$. With this expression, (1.7.5) is brought to

$$L = \frac{1}{2}\varepsilon E^2 - \frac{1}{2\mu}B^2 + \boldsymbol{j} \cdot \boldsymbol{A} - \rho\varphi. \tag{1.7.8}$$

By comparison to $L = T - U$ in classical physics, one may comprehend (1.7.8) as follows: the first two terms are generalized kinetic energy and the latter two are generalized potential related to velocity. Note that they are not the energies of an electromagnetic field, because the energies are embodied in a Hamiltonian which is to be deduced from the Lagrangian. In the present case, the Lagrangian includes functions $\{A_i\}$ and arguments $\{x_i\}$. This is a variational problem of a functional with multifunctions and multivariates. The aim is to derive equations that an electromagnetic field should meet.

After (1.7.6) is substituted into (1.7.5), the Lagrangian becomes

$$L = -\frac{1}{4\mu}\left(\frac{\partial A_j}{\partial x_i} - \frac{\partial A_i}{\partial x_j}\right)\left(\frac{\partial A_j}{\partial x_i} - \frac{\partial A_i}{\partial x_j}\right) + J_i A_i$$

$$= -\frac{1}{2\mu}\frac{\partial A_j}{\partial x_i}\left(\frac{\partial A_j}{\partial x_i} - \frac{\partial A_i}{\partial x_j}\right) + J_i A_i,$$

where index exchange has been carried out to simplify the expression. The Euler-Lagrange equations appear as

$$\frac{\partial L}{\partial A_i} + \frac{1}{\mu}\sum_{j=1}^{4}\frac{\partial}{\partial x_j}\left(\frac{\partial A_i}{\partial x_j} - \frac{\partial A_j}{\partial x_i}\right) = 0.$$

For the first three generalized coordinates, we have the following results.

$$j_1 - \frac{1}{\mu}\left[\frac{\partial}{\partial x_2}B_z - \frac{\partial}{\partial x_3}B_y - \frac{1}{c^2}\frac{\partial}{\partial t}E_x\right]$$

$$= j_1 - \frac{1}{\mu}\left[(\nabla \times \boldsymbol{B})_1 + \frac{1}{c^2}\frac{\partial E_x}{\partial t}\right] = 0.$$

Therefore,

$$\frac{1}{\mu}(\nabla \times \boldsymbol{B})_1 = j_1 + \varepsilon \frac{\partial E_x}{\partial t}. \tag{1.7.9a}$$

Similarly, we have

$$\frac{1}{\mu}(\nabla \times \boldsymbol{B})_2 = j_2 + \varepsilon \frac{\partial E_y}{\partial t} \tag{1.7.9b}$$

and

$$\frac{1}{\mu}(\nabla \times \boldsymbol{B})_3 = j_3 + \varepsilon \frac{\partial E_z}{\partial t}. \tag{1.7.9c}$$

In (1.7.9), the latter two can be simply put down by taking index and coordinate rotation from the first one, i.e., taking $1 \to 2 \to 3 \to 1$ and $x \to y \to z \to x$. Combination of the three equations leads to

$$\frac{1}{\mu}\nabla \times \boldsymbol{B} = \boldsymbol{J} + \varepsilon \frac{\partial \boldsymbol{E}}{\partial t}. \tag{1.7.9}$$

For the fourth generalized coordinate,

$$\frac{\partial L}{\partial A_4} - \sum_{j=1}^{4} \frac{\partial}{\partial x_j} \frac{\partial L}{\partial (\partial A_4/\partial x_j)} = \mathrm{i}c\rho + \frac{1}{\mathrm{i}c\mu}\nabla \cdot \boldsymbol{E} = 0.$$

This results in

$$\varepsilon \nabla \cdot \boldsymbol{E} = \rho. \tag{1.7.10}$$

Equation (1.7.6) itself means the following two equations:

$$\boldsymbol{B} = \nabla \times \boldsymbol{A}, \quad \boldsymbol{E} = -\nabla\varphi - \frac{\partial \boldsymbol{A}}{\partial t}.$$

Thus, one obtains

$$\nabla \cdot \boldsymbol{B} = \nabla \cdot (\nabla \times \boldsymbol{A}) = 0 \tag{1.7.11}$$

and

$$\nabla \times \boldsymbol{E} = \nabla \times \left(-\nabla\varphi - \frac{\partial \boldsymbol{A}}{\partial t}\right) = -\nabla \times \nabla\varphi - \frac{\partial}{\partial t}\nabla \times \boldsymbol{A} = -\frac{\partial \boldsymbol{B}}{\partial t}. \tag{1.7.12}$$

In summary, (1.7.9)–(1.7.12) are well-known Maxwell equations.

This example shows that the Maxwell equations can indeed be deduced from the Lagrangian (1.7.5) by means of Euler-Lagrange equations of the variational method.

2. A charged particle in electromagnetic field

We first put down a Lagrange equation.

$$\frac{\partial L}{\partial q_\alpha} - \frac{d}{dt}\frac{\partial L}{\partial \dot{q}_\alpha} = 0. \tag{1.7.13}$$

For the movement of a charged particle, the function $\{q_\alpha\}$ in Lagrangian L is space coordinates $r(t)$ and the variable is time t. Note that the vector potential A and scalar potential φ of the field, as the ingredients of L, depend on $r(t)$. Hence there are three functions $r(t)$ and one variable t.

(1) Non-relativistic case

Lagrangian is

$$L = \frac{1}{2}mv^2 - e(\varphi - v \cdot A). \tag{1.7.14}$$

Employing (1.7.29), we get

$$\frac{\partial}{\partial x}L - \frac{d}{dt}\frac{\partial}{\partial v_x}L = -e\left(\frac{\partial}{\partial x}\varphi - \frac{\partial}{\partial x}v \cdot A\right) - \frac{d}{dt}mv_x - e\frac{d}{dt}A_x = 0,$$

where

$$v \cdot \frac{\partial}{\partial x}A - \frac{d}{dt}A_x$$

$$= v_x\frac{\partial}{\partial x}A_x + v_y\frac{\partial}{\partial x}A_y + v_z\frac{\partial}{\partial x}A_z - \frac{d}{dt}A_x$$

$$= (v \cdot \nabla)A_x - \frac{d}{dt}A_x - v_y\frac{\partial}{\partial y}A_x - v_z\frac{\partial}{\partial z}A_x + v_y\frac{\partial}{\partial x}A_y + v_z\frac{\partial}{\partial x}A_z$$

$$= -\frac{\partial}{\partial t}A_x + v_yB_z - v_zB_y = -\frac{\partial}{\partial t}A_x + (v \times B)_x.$$

The other two equations can also be given similarly. In summary, the result is

$$\frac{d}{dt}p = e(E + v \times B), \tag{1.7.15}$$

where $E = -\nabla\varphi - \frac{\partial}{\partial t}A$. This is the equation that a charged particle moving in an electromagnetic field should obey.

(2) Relativistic case

Lagrangian is

$$L = -mc^2\sqrt{1 - v^2/c^2} - e(\varphi - \boldsymbol{v} \cdot \boldsymbol{A}). \tag{1.7.16}$$

By use of (1.7.13) one obtains the equation of motion, which is of the same form as (1.7.15), with the momentum in the relativistic form. Note that in (1.7.16) the first term is neither kinetic energy nor total energy of the particle, and it is even a negative value. Obviously, Lagrangian in this case is not kinetic energy minus potential.

1.7.2.   *The applications in quantum mechanics*

We take the origination of Shrödinger equation as an example. Schrödinger in his primary article derived the essential equation of quantum mechanics indeed through the variational method. This equation was named by its inventor.

In analytical mechanics, there is an essential equation named as Hamilton-Jacobi equation

$$H\left(q, \frac{\partial S}{\partial q}\right) = E. \tag{1.7.17}$$

This demonstrates that Hamiltonian takes just the value of total energy. S is the action. Suppose that there is an undetermined quantity ψ, through which the action can be expressed in the following way:

$$S = K \ln \psi. \tag{1.7.18}$$

Or $\psi = \exp(S/K)$. Evidently, ψ should be dimensionless. K is a constant, and should takes the same dimension of action. Now the ψ is given a function to describe a microscopic particle's movement or its existence. Mathematically, it is required that in the whole space it is singly valued, finite and continuous, and has first and second order derivatives. Then one replaces S in (1.7.17) by (1.7.18) so as to get

$$H\left(q, \frac{K}{\psi}\frac{\partial \psi}{\partial q}\right) = E. \tag{1.7.19}$$

Shrödinger took into account that Hamiltonian was kinetic energy plus potential, $H = T + U$ and potential was a function of coordinates. Then he thought that kinetic energy should be a function of ψ and its derivatives. Further, he assumed that kinetic energy was a function of ψ and its second derivatives. In suggesting this, he actually had a conceptual breakthrough. In classical mechanics, kinetic energy was related to the derivatives of generalized coordinates with respect to time, while Shrödinger thought that it was related to the derivative of a function with respect to space coordinates! Thus, in a Cartesian system, (1.7.19) is expanded to be the following form:

$$\frac{K^2}{2m}\left[\left(\frac{\partial\psi}{\partial x}\right)^2 + \left(\frac{\partial\psi}{\partial y}\right)^2 + \left(\frac{\partial\psi}{\partial z}\right)^2\right] - (E - V(r))\psi^2 = 0. \quad (1.7.20)$$

Adding factor $1/2m$ was to make kinetic energy close to the form in classical mechanics. At first glance, it seems that (1.7.20) can be used to solve function ψ. However, it is a quadratic function of ψ, which would probably yield multi-values of ψ. In order to obtain a singly-valued unique solution, the variational method was resorted to.

The left hand side of (1.7.20) was regarded a function instead of zero. Further, it was thought as the Lagrangian of a microscopic particle. Then the corresponding functional was

$$J = \iiint dxdydz$$

$$\times \left[\left(\frac{\partial\psi}{\partial x}\right)^2 + \left(\frac{\partial\psi}{\partial y}\right)^2 + \left(\frac{\partial\psi}{\partial z}\right)^2 - \frac{2m}{K^2}(E - V(\mathbf{r}))\psi^2\right]. \quad (1.7.21)$$

In the Lagrangian, there were one function ψ and three arguments (x, y, z). The variational should make

$$\delta J = \delta \iiint dxdydz$$

$$\times \left[\left(\frac{\partial\psi}{\partial x}\right)^2 + \left(\frac{\partial\psi}{\partial y}\right)^2 + \left(\frac{\partial\psi}{\partial z}\right)^2 - \frac{2m}{K^2}(E - V(\mathbf{r}))\psi^2\right] = 0.$$

Note that although potential $V(\mathbf{r})$ was a function of space coordinates, its values at all the coordinates were fixed, so that remained unchanged. Hence, variational did not apply to potential $V(r)$. Following standard variational procedure, he got

$$\delta J = 2 \iiint \mathrm{d}x \mathrm{d}y \mathrm{d}z$$

$$\times \left[\psi_x \delta \psi_x + \psi_y \delta \psi_y + \psi_z \delta \psi_z - \frac{2m}{K^2}(E - V(r))\psi \delta \psi \right] = 0.$$

Integration by parts for the kinetic energy gave

$$\delta J = \int \delta \psi \frac{\partial \psi}{\partial n} \mathrm{d}s$$

$$+ \iiint \mathrm{d}x \mathrm{d}y \mathrm{d}z \left[-\psi_{xx} - \psi_{yy} - \psi_{zz} - \frac{2m}{K^2}(E - V(r))\psi \right] \delta \psi = 0.$$

The result was

$$-\nabla^2 \psi - \frac{2m}{K^2}(E - V(r))\psi = 0 \qquad (1.7.22)$$

and

$$\int \delta \psi \frac{\partial \psi}{\partial n} \mathrm{d}s = 0, \qquad (1.7.23)$$

where $\mathrm{d}s$ was an element of the surface of the integral volume.

The function ψ ought to meet Eq. (1.7.22). Compared to (1.7.20), in (1.7.23) squared terms of ψ did not appear. Then Shrödinger put the potential of a hydrogen atom into (1.7.22), and solved this differential equation under spherical polar coordinates to obtain eigenvalues of energy. The results agreed very well with the Balmer formula, and the K happened to be the Planck constant. This showed that (1.7.22) was right, and it was named as the Schrödinger equation.

The function ψ is called the **wavefunction** of a microscopic particle. The Schrödinger equation describes the movement of a single microscopic particle.

Exercises

1. Assume that $f(x)$ and $g(x)$ are functions on $C_1[a, b]$ and $\eta(x)$ is any function on $C_3[a, b]$ with $\eta(a) = 0$ and $\eta(b) = 0$. Prove the following two propositions.

(1) If $\int_a^b f(x)\eta'(x)\mathrm{d}x = 0$, then $f(x)$ is a constant on this interval.

(2) If $\int_a^b [f(x)\eta(x) + g(x)\eta'(x)]\mathrm{d}x = 0$, then $f(x) = g'(x)$ on this interval.

2. Evaluate the increments and variationals of the following functionals.

(1) $J[y] = \int_a^b \sqrt{1 + y'^2}\mathrm{d}x.$

(2) $J[y] = \int_{x_0}^{x_1} (x^2 y + y^2 + yy')\mathrm{d}x.$

3. Find the extremal curves of the following functionals with their two ends fixed.

(1) $J[y] = \int_{x_0}^{x_1} \sqrt{y(1 + y'^2)}\mathrm{d}x.$

(2) $J[y] = \int_{x_0}^{x_1} \frac{1+y^2}{y'^2}\mathrm{d}x.$

(3) $J[y] = \int_{x_0}^{x_1} (y^2 + y'^2 - 2y\sin x)\mathrm{d}x.$

(4) $J[y] = \int_{x_0}^{x_1} \sqrt{x^2 + y^2}\sqrt{1 + y'^2}\mathrm{d}x.$

4. Find the extremal curves of the following functionals with end conditions. Judge that the obtained curves make the functionals be minima by calculation of increments.

(1) $J[y] = \int_0^2 (xy' + y'^2)\mathrm{d}x, y(0) = 1, y(2) = 0.$

(2) $J[y] = \int_0^1 (12xy + yy' + y'^2)\mathrm{d}x, y(0) = 1, y(1) = 4.$

(3) $J[y] = \int_1^2 y'(1 + x^2 y')\mathrm{d}x, y(1) = 3, y(2) = 5.$

(4) $J[y] = \int_0^{\pi/4} (y'^2 - y^2 + y)\mathrm{d}x, y(0) = y(\pi/4) = 0.$

5. Show that the necessary condition for the functional $J[y] = \int_{x_0}^{x_1} F(x, y, y', y'')\mathrm{d}x$ to take its extreme is that

$$F_y - \frac{\mathrm{d}}{\mathrm{d}x}F_{y'} + \frac{\mathrm{d}^2}{\mathrm{d}x^2}F_{y''} = 0.$$

Here it is assumed that the function F has at least the third derivative with respect to its arguments, and the values of $y(x)$ and $y'(x)$ at $x = x_0$ and $x = x_1$ have already been given.

6. Find the differential equations and their solutions of the extremal curves of the following functionals.

   (1) $J[y] = \int_{x_0}^{x_1} (16y^2 - y''^2 + x^2)\,dx$.

   (2) $J[y] = \int_{x_0}^{x_1} (y'''^2 + 2xy)\,dx$.

7. Find the extremal curves of the following functionals.

   (1) $J[y, z] = \int_{x_0}^{x_1} (2yz - 2y^2 + y'^2 - z'^2)\,dx$.

   (2) $J[y, z] = \int_{x_0}^{x_1} (y'^2 + z'^2 + y'z')\,dx$.

8. Find the differential equations of the extremal curves of the following functionals.

   (1) $J[z(x, y)] = \iint\limits_{D} \left[\left(\frac{\partial z}{\partial x}\right)^2 - \left(\frac{\partial z}{\partial y}\right)^2 \right] dxdy$.

   (2) $J[u(x, y, z)] = \iiint\limits_{V} \left[\left(\frac{\partial u}{\partial x}\right)^2 + \left(\frac{\partial u}{\partial y}\right)^2 + \left(\frac{\partial u}{\partial z}\right)^2 + 2uf(x, y, z) \right]$
   $dxdydz$.

   (3) $J[u(x, y)] = \iint\limits_{D} [au_x^2 + bu_y^2 + cu^2 + 2uf(x, y)]dxdy$, where a and b are constants and $f(x, y)$ is a given function.

9. Show that under the constraint $\iiint\limits_{V} u^2(x, y, z)dxdydz = 1$, the necessary condition that makes the functional

$$J[y] = \iiint\limits_{V} F(x, y, z, u, u_x, u_y, u_z)dxdydz$$

take its extreme is that $u(x, y, z)$ satisfies the equation

$$F_u - \frac{\partial}{\partial x}F_{u_x} - \frac{\partial}{\partial y}F_{u_y} - \frac{\partial}{\partial z}F_{u_z} - \lambda u = 0,$$

where λ is a parameter.

10. Find the extremal curves of the following isoperimetric problems.

    (1) $J[y] = \int_0^1 (y'^2 + x^2)dx$, $y(0) = 0, y(1) = 0$ with isoperimetric condition $\int_0^1 y^2 dx = 2$.

(2) $J[y, z] = \int_0^1 (y'^2 + z'^2 - 4xz' - 4z)\mathrm{d}x, y(0) = z(0) = 0;$
$y(1) = z(1) = 1;$ $\int_0^1 (y'^2 - xy' - z'^2)\mathrm{d}x = 2.$

(Hint: solving the isoperimetric problems of a functional containing multi-functions is similar to solving that containing one function.)

11. Find the geodesic line on the side surface of a cylinder with radius $r = R$.

   (Hint: it is convenient using cylindrical coordinates r, φ, z.)

12. Put down the differential equation of the extremal curve of functional $J[y] = \int_0^{x_1} [p(x)y'^2 + q(x)y^2]\mathrm{d}x$ under conditions $\int_0^{x_1} r(x)y^2\mathrm{d}x = 1; y(0) = 0, y(x_1) = 0$, where $r(x), p(x), q(x)$ are three given functions.

13. Let the left end of function $y(x)$ be fixed, $y(0) = 0$, and the other end move along the line $x = \pi/4$. Under these conditions, find the extremal curve of the functional $J[y] = \int_0^{\pi/4} (y^2 - y'^2)\mathrm{d}x.$

14. A curve connecting two points on the same side of an axis. The rotation of this curve with respect to the axis forms a rotation surface. Find the function of the curve that makes the surface have the least area.

15. Given a point (x_0, y_0), e.g., $(0, 1)$, draw catenaries through this point with various parameters. These catenaries should be tangent to one envelop.

16. Find the extremal curves $y(x)$ of the functional $J[y] = \int_0^1 (y'^2 - yy' + y^2)\mathrm{d}x$ satisfying the following conditions.

   (1) $y(x)$ goes through point $P_1(0, 1)$ and satisfies natural boundary condition at $x = 1$.
   (2) $y(x)$ goes through $P_2(1, 2)$, and satisfies natural boundary condition at $x = 0$.
   (3) $y(x)$ satisfies natural boundary condition at both ends $x = 0$ and $x = 1$.

17. Show that the Lagrangian (1.7.5) becomes (1.7.8) when it is expressed by electromagnetic fields.

18. Derive the equation of motion (1.7.15) of a particle from the Lagrangian (1.7.16).

19. Let ψ and ψ^* be taken as independent field functions, and the Lagrangian of the fields is

$$L = \frac{\hbar^2}{2m}\nabla\psi \cdot \nabla\psi^* + U(r,t)\psi\psi^* - \frac{i\hbar}{2}(\dot{\psi}\psi^* - \psi\dot{\psi}^*),$$

where ∇ is gradient operator and the dot on $\dot{\psi}$ represents the derivative with respect to time, $\dot{\psi} = \frac{\partial\psi}{\partial t}$. What differential equation will be resulted in when this Lagrangian is substituted into Euler-Lagrange equation?

Chapter 2

Hilbert Space

2.1. Linear Space, Inner Product Space and Hilbert Space

2.1.1. *Linear space*

1. Distance space

Among the relationships between two elements in a set, one essential knowledge is to describe how close to each other or on the contrary how far apart they are. For example, the absolute value of the difference of two numbers x and y reflects this kind of relation. A concept of "distance" is used to define this kind of relation.

Definition 1. Let X be a non-empty set. For arbitrary two elements x and y in this set, there is associated a non-negative real number $\rho(x, y)$ defined in a certain way. If $\rho(x, y)$ satisfies the three conditions below, it is called the **distance** between the two elements, and X is called a **distance space** or **metric space** following the distance $\rho(x, y)$. Every element in X is called a **point** in this space. All of the points in X meeting $\rho(x, a) \leq r$ compose a **sphere** (**closed sphere**) centered at point a with radius r, also called the **neighborhood** of point a. The three conditions are:

(i) $\rho(x, y) \geq 0$ with the equal sign standing when and only when $x = y$, (non-negative);

(ii) $\rho(x, y) \leq \rho(x, z) + \rho(y, z)$, (triangle inequality);

(iii) $\rho(x, y) = \rho(y, x)$, (symmetry).

These conditions are called **the three axioms of distance**.

The symmetry can also be derived from two other conditions. In fact, taking $z = x$ in (ii), we have

$$\rho(x, y) \leq \rho(x, x) + \rho(y, x).$$

By non-negative (i) we know that $\rho(x, x) = 0$. This results in

$$\rho(x, y) \leq \rho(y, x).$$

Since x and y are arbitrary elements, they can be exchanged to get

$$\rho(x, y) \leq \rho(x, y).$$

Only the equal sign can let the two equations be valid simultaneously. Thus the symmetry is obtained.

The followings are distances usually encountered.

Example 1. The real number space R. For arbitrary real numbers x and y in R, one may define

$$\rho(x, y) = |x - y|$$

as the distance between x and y. It meets the three axioms of distance. Hence, R is a distance space.

The complex number space Z can also be a distance space when the distance is defined in the same way as in R.

Example 2. Let R_n be an n-dimensional Euclid space. Its point is of the form of $(x_1, x_2, \ldots, x_n)$. For two elements $x = (x_1, x_2, \ldots, x_n)$ and $y = (y_1, y_2, \ldots, y_n)$, one defines distance

$$\rho(x, y) = \left[\sum_{i=1}^{n} (x_i - y_i)^2 \right]^{1/2}.$$

Alternatively, he can also define

$$\rho(x, y) = \sum_{i=1}^{n} |x_i - y_i|$$

or

$$\rho(x, y) = \max_i |x_i - y_i|.$$

According to any one of these distances, the n-dimensional Euclid space is a distance one.

Example 3. Let $C[a, b]$ represent all continuous functions defined on the interval $[a, b]$. For two points $x(t), y(t) \in C[a, b]$, one defines a distance by

$$\rho(x, y) = \max_{a \le t \le b} |x(t) - y(t)|.$$

Then, $C[a, b]$ is a distance space.

Definition 2. If the integration of function $x(t)$ on the interval $[a, b]$ defined as

$$\int_a^b |x(t)|^p dt < \infty, \quad (p \ge 1)$$

is finite, then the function $x(t)$ is said to be **pth-order-integrable** or simply **p-integrable** on $[a, b]$. All of the p-integrable functions in $[a, b]$ compose a **pth-order-integrable space**, denoted as $L_p[a, b], (p \ge 1)$.

Example 4. Given a p-integrable space $L_p[a, b], (p \ge 1)$, for any $x(t), y(t) \in L_p[a, b]$, one defines a distance

$$\rho(x, y) = \left(\int_a^b |x(t) - y(t)|^p dt \right)^{1/p}. \tag{2.1.1}$$

Then the $L_p[a, b], (p \ge 1)$ is a distance space. Henceforth, when we mention a p-integrable space L_p, we always default $p \ge 1$.

2. Convergence and limit

Having introduced the concept of distance, we go further to introduce in distance space the concepts of convergence and limit.

Definition 3. Suppose $x, x_n (n = 1, 2, \ldots) \in X$. If as $n \to \infty$, sequence $\rho(x, x_n) \to 0$, or, given a $\varepsilon > 0$, there exists an $N > 0$

which makes $\rho(x, x_n) < \varepsilon$ when $n > N$, then the series $\{x_n\}$ is said to **converge to** x according to $\rho(x, y)$, which is denoted as

$$\lim_{n \to \infty} x_n = x, \tag{2.1.2}$$

or $x_n \to x(n \to \infty)$. In that case, $\{x_n\}$ is called **convergent point sequence** and x the **limit** of $\{x_n\}$.

By the three axioms of distance and the definition of limit, it is easily known that the limit of a convergent series in a distance space is unique.

Convergence and limit are defined with the aid of distance. Henceforth, when we mention the convergence and limit, we have implicitly defined a distance. Any pair of convergence and limit is with respect to a defined distance.

Distance $\rho(x, y)$ is a continuous function of x, y, i.e., if $x_n \to x_0$, $y_n \to y_0$, then $\rho(x_n, y_n) \to \rho(x_0, y_0)$. It is easily proved.

By the triangle inequality of distance,

$$\rho(x_n, y_n) \le \rho(x_n, x_0) + \rho(x_0, y_0) + \rho(y_n, y_0).$$

Consequently,

$$\rho(x_n, y_n) - \rho(x_0, y_0) \le \rho(x_n, x_0) + \rho(y_n, y_0).$$

As $n \to \infty$, $x_n \to x_0$, $y_n \to y_0$ and $\rho(x_n, x_0) \to 0$, $\rho(y_n, y_0) \to 0$. The limit of the right hand side is zero. Thus, $\rho(x, y)$ is continuous.

Definition 4. Let $\{x_n\}$ be a point series in distance space X. If for every $\varepsilon > 0$, there exists an $N > 0$ which makes

$$\rho(x_m, x_n) < \varepsilon, \tag{2.1.3}$$

as natural numbers $m, n > N$. Then $\{x_n\}$ is called a **fundamental sequence** or **Cauchy sequence** in X.

A convergent sequence in a distance space X is necessarily a Cauchy sequence in this space. On the contrary, a Cauchy sequence in a distance space X is not necessarily a convergent one in this space.

Point x is the limit of sequence x_n when and only when any neighborhood of x contains all the points starting from a certain index of the sequence. A point set is said to be **bounded** when it is entirely included in a sphere.

3. The completeness of a space

Definition 5. A distance space X is said to be **complete** if each Cauchy sequence in this space converges to a point in this space, called a **complete space**. Otherwise, it is **incomplete**.

This is the definition about the completeness of space. There is no other definition about the completeness. Hereafter, when we mention completeness of a space, it means a space that has been endowed a distance and this space is complete.

Example 5. Real number space R is complete after a distance $\rho(x, y) = |x - y|$ is defined.

Example 6. The set of all the rational numbers is denoted as Y. For arbitrary two elements x and y in Y, a real number $\rho(x, y) = |x - y|$ is defined as the distance between x and y. This distance obeys the three axioms of distance, so that Y is a distance space. We take a sequence $\{S_n\}$ in this space: $S_n = \sum_{m=1}^{n} \frac{1}{m!}$. It is a Cauchy sequence. Its limit is e − 1 which is not a rational number. So, the limit of the sequence is not in the rational number space Y. The conclusion is that Y is incomplete.

Definition 6. A distance space X is said to be separable if there exists a countable set $\{x_n\} \subset X$ where for every $x \in X$ there is a subset $\{x_{n_k}\}$ in $\{x_n\}$, which makes $\lim_{n_k \to \infty} \rho(x_{n_k}, x) = 0$.

This definition demonstrates that if we collect all of such subsets to construct a space, then it is also a distance space, denoted as Y. Obviously, any point in Y is also necessarily in space X. Therefore, $Y \subset X$, i.e., Y is a subspace, or subset, of X.

4. Linear space

Although the concept of distance represents a relation between elements in a space, there may be other relations between elements. The concept of distance only cannot sufficiently reveal the relations between elements. Usually, a space considered is also an algebra system, that is, there exist some algebra relations between elements. If we focus on the algebra structure of a space, which includes addition between elements and scalar multiplication between a number (scalar) and an element in the space, we can propose a concept of linear space.

Definition 7. Let X be a set of elements and K complex number domain (or a real number domain). If the conditions (i) and (ii) below are satisfied, X is called a linear space on K, or simply X is a **linear space**. Corresponding to complex or real domain K, X is called **complex linear space** or **real linear space**. In this case, X is also called a **vector space**, and the elements in this space called **vectors**.

(i) There is an addition operation in X, denoted by $+$, which makes $x + y \in X$ when $x, y \in X$, meet the following conditions:

    (a) $x + y = y + x$ (commutative law of addition);
    (b) $(x + y) + z = x + (y + z)$ (combination law of addition);
    (c) There is a **zero element** in X, denoted as θ, which makes $x + \theta = x$ for any $x \in X$;
    (d) For any $x \in X$ there exists an **inverse element** $-x \in X$ which makes $x + (-x) = \theta$.

(ii) For any element $x \in X$ and an arbitrary number $\alpha \in K$, a scalar multiplication between them is defined, denoted by αx, such that the result is still an element in this space $\alpha x \in X$. The scalar multiplication should meet the following conditions.

    (e) $1x = x$;
    (f) $(\alpha\beta)x = \alpha(\beta x)$ (commutative law of scalar multiplication);
    (g) $(\alpha+\beta)x = \alpha x + \beta x$ (distribution law of scalar multiplication);
    (h) $\alpha(x + y) = \alpha x + \alpha y$ (combination law of scalar multiplication).

Zero element θ is also called **zero vector**. The conditions (a)$\sim$(h) are termed as the **eight axioms of linear space**.

    Here we give some examples of linear spaces.

Example 7. Real vector set R_2 in a plane. i.e., $R_2 = \{(x_1, x_2)\}$ where x_1, x_2 are real numbers. The addition operation is defined as the addition of corresponding coordinate components, and the multiplication by number is defined as that both coordinate components are multiplied by a number. Specifically, for a given real number α and two vectors $x = (x_1, x_2)$, $y = (y_1, y_2)$, we define

$$x + y = (x_1 + y_1, x_2 + y_2), \quad \alpha x = (\alpha x_1, \alpha x_2).$$

The zero element in R_2 is $\theta = (0,0)$ and the inverse element of $x = (x_1, x_2)$ is $-x = (-x_1, -x_2)$. Evidently, under the definitions of addition and scalar multiplication, R_2 becomes a linear space.

Example 8. In the set $C[a, b]$ in Example 3, an element is $f(x) \in C[a, b]$. For a given real number α and two elements $f, g \in C[a, b]$, we define operators of addition and scalar multiplication as follows

$$(f + g)(x) = f(x) + g(x), \quad a \leq x \leq b,$$

$$(\alpha f)(x) = \alpha f(x), \quad a \leq x \leq b.$$

The zero element θ in $C[a, b]$ is a function on $[a, b]$ that is identical to zero: $\theta = f(x) \equiv 0$, and the inverse of an element $f(x)$ is $-f(x)$. By the properties of continuous functions, it is known that $C[a, b]$ is a real linear space.

Example 9. An nth-degree polynomial $p_n(x)$ with real number coefficients is of the form of

$$p_n(x) = a_0 + a_1 x + \cdots + a_n x^n,$$

where a_i, $(i = 1, 2, \ldots, n)$ are given real numbers. Some of them can be zero. For example, fourth-order polynomials include $1 + 3x + 5x^3$, although the highest power in it is 3. All of the nth-degree polynomials defined in $[a, b]$ is denoted as $P_n[a, b]$.

$$P_n[a, b] = a_0 + a_1 x + \cdots + a_n x^n, \quad a \leq x \leq b.$$

We define operators of addition and scalar multiplication as usual ones. Then $P_n[a, b]$ is a real linear space.

Obviously, $P_n[a, b] \subset C[a, b]$, i.e., linear space $P_n[a, b]$ is a subset of $C[a, b]$.

5. Dimensions and basis of a linear space

Definition 8. Let X be a linear space with its vector being $x_1, x_2, \ldots, x_n \in X$. If a vector $x \in X$ can be written in the following form:

$$x = a_1 x_1 + a_2 x_2 + \cdots + a_k x_k,$$

where not all of the coefficients $a_1, a_2, \ldots, a_k$ are zero, then we say that x is the **linear combination** of $x_1, x_2, \ldots, x_k$, and $k+1$ vectors $x, x_1, x_2, \ldots, x_k$ are **linearly dependent**. If when and only when $a_1 = a_2 = \cdots = a_n = 0$,

$$a_1 x_1 + a_2 x_2 + \cdots + a_n x_n = \theta,$$

then $x_1, x_2, \ldots, x_n$ are said to be **linearly independent**, otherwise they are said to be linearly dependent. If $x_1, x_2, \ldots, x_n$ are linearly independent, they are called a **linearly independent group** in space X. If arbitrary finite number of elements in $x_1, x_2, \ldots, x_n, \ldots$ are linearly independent, then $\{x_n\}$ construct a **linearly independent system** in X.

Definition 9. Let $\{x_1, x_2, \ldots, x_n\}$ be a linearly independent group in linear space X. If every nonzero element x in X is a linear combination of $\{x_1, x_2, \ldots, x_n\}$, then X is called an n-**dimensional linear space**, and the set $\{x_1, x_2, \ldots, x_n\}$ is called a **basis** in X which is **complete**. If $\{x_1, x_2, \ldots\}$ is a linearly independent system in X and every nonzero element x in X can be expressed by $x = \sum_{i=1}^{\infty} a_i x_i$, where $a_i, (i = 1, 2, \ldots)$ are real numbers, then X is said to be an **infinite-dimensional space** and $\{x_1, x_2, \ldots\}$ is a basis of it.

Please note that the two concepts of the completeness of a basis and completeness of space are different from each other. The completeness of a basis means that any vector in this space can be expressed by the linear combination of this basis.

The dimension of a space is unique, and is independent of selection of basis. Each basis in an n-dimensional linear space consists of n elements, and in an n-dimensional linear space there exists at least one basis.

Properly choosing a basis will be very helpful for practical calculation.

Example 10. Let $X = P_n[a, b]$ and choose $B_n = \{1, x, x^2, \ldots, x^n\}$. Then B_n is a basis in X, and X is an $n+1$-dimensional linear space. It is possible to choose other basis in $P_n[a, b]$. Some of them are more convenient than B_n in application.

Example 11. Let $X = C[-1, 1]$ and choose $A = \{x, |x|\}$. Then A is a linearly independent group, but it cannot be a basis in $C[-1, 1]$ because some of elements in $C[-1, 1]$ cannot be expressed by the linear combination of $x, |x|$. For example, element $\sin x \in C[-1, 1]$ cannot be expressed by $a_1 x + a_2 |x|$ no matter how we choose coefficients a_1, a_2.

Example 12. Let $X = C[0, 2\pi]$, and then $A = \{1, \sin x, \sin 2x, \ldots, \sin nx\}$ is a linearly independent group. If there exist coefficients $a_0, a_1, \ldots, a_n$ which make

$$a_0 + a_1 \sin x + a_2 \sin 2x + \cdots + a_n \sin nx \equiv 0,$$

then it is necessary that $a_1 = a_2 = \cdots = a_n = 0$. In fact, if we take $x_0, x_1, \ldots, x_n \in [0, \pi/2]$ and $x_i \neq x_j, (i, j = 0, 1, 2, \ldots, n)$ for $i \neq j$, then substituting $x_i, (i = 0, 1, 2, \ldots, n)$ into the equation above leads to homogeneous linear equations with respect to $a_0, a_1, \ldots, a_n$. It is easy to verify that the determinant factor is not zero. Therefore, $a_1 = a_2 = \cdots = a_n = 0$.

2.1.2. *Inner product space*

1. Inner product space

Definition 10. Let X be a complex linear space. A pair of vectors x and y in X, is mapped to a number in number domain F, denoted as (x, y). If (x, y) satisfies the following four conditions, it is called an **inner product** or **scalar product** of x and y.

(i) $(x, y) = (y, x)^*$, (Hermitian property);
(ii) $(x + y, z) = (x, z) + (y, z)$;

(iii) $(\alpha x, z) = \alpha^*(x, z)$;

(iv) $(x, x) \geq 0$, the equal sign is valid when and only when $x = \theta$.

In (i) the superscript * means taking complex conjugate. These four conditions are called the **four axioms of inner product**.

In real vector spaces, the superscript * conditions in (i) and (iii) does not work and can be discarded. In both real and complex inner product spaces, condition (i) means that the inner product of any vector with itself is always a real number, which guarantees that the inequality (iv) is of significance.

Definition 11. A linear space X in which an inner product on number domain K has been defined is called an **inner product space** or **quasi Hilbert space**. When K is real a (complex), the space is called a **real (complex) inner product space**, also called an **Euclid (unitary) space**.

Here are some examples of inner product.

Example 13. Let $x = (\xi_1, \xi_2, \ldots, \xi_n), y = (\eta_1, \eta_2, \ldots, \eta_n)$ be two vectors in complex n-dimensional space C^n. We define their inner product to be

$$(x, y) = \sum_{i=1}^{n} \xi_i^* \eta_i. \qquad (2.1.4)$$

This definition meets the axioms of an inner product. If ξ_i at the right hand side does not take complex conjugate, then the axioms of an inner product are not satisfied when F^n is complex.

If we define

$$(x, y) = \sum_{i=1}^{n} p_i \xi_i^* \eta_i, \qquad (2.1.5)$$

then for given positive numbers $p_i, (i = 1, 2, \ldots, n)$, the definition meets the axioms of an inner product. The numbers $p_i, (i = 1, 2, \ldots, n)$ are called **weights**. In this case, (2.1.5) defines **weighted inner product**.

Example 14. Let $x(t)$ and $y(t)$ be two vectors in polynomial space defined on $[0, 1]$. The inner product is defined by

$$(x, y) = \int_0^1 x^*(t)y(t)\mathrm{d}t. \qquad (2.1.6)$$

Alternatively, it can also be defined by

$$(x, y) = \int_0^1 x^*(t)y(t)w(t)\mathrm{d}t, \qquad (2.1.7)$$

where the function $w(t) > 0$ in the given interval and it is independent of vectors. In the latter case $w(t)$ is called the **weight function**. The inner product with the weight function is called the weighted inner product.

Example 15. Let $f(x_1, x_2, \ldots, x_n)$ and $g(x_1, x_2, \ldots, x_n)$ be two vectors in n-dimensional function space, where every argument varies within $[a, b]$. The weighted inner product is defined by

$$(f, g) = \int_a^b \mathrm{d}x_1 \int_a^b \mathrm{d}x_2 \cdots \int_a^b \mathrm{d}x_n \rho(x_1, x_2, \ldots, x_n)f^*$$
$$\times (x_1, x_2, \ldots, x_n)g(x_1, x_2, \ldots, x_n),$$

where the weight function $\rho(x_1, x_2, \ldots, x_n) > 0$.

Theorem 1. *Inner product meets* **Schwartz inequality:**

$$|(x, y)| \le \sqrt{(x, x)}\sqrt{(y, y)}. \qquad (2.1.8)$$

Proof. When $x = \theta$ or $y = \theta$, the equal sign stands naturally. When both x and y are nonzero, let $z = \dfrac{x}{\sqrt{(x,x)}}$. Then apparently $(z, z) = 1$. For a given complex number λ, we have

$$0 \le (\lambda z - y, \lambda z - y)$$
$$= [\lambda^* - (z, y)^*]\,[\lambda - (z, y)]] - |(z, y)|^2 + (y, y).$$

As a special case, let us take $\lambda = (z, y)$. Then the above inequality degrades to $|(z, y)|^2 \le (y, y)$. That is to say, $|(\frac{x}{\sqrt{(x,x)}}, y)|^2 \le (y, y)$, or $|(x, y)|^2 \le (y, y)(x, x)$. $\qquad \square$

Example 16. In the two-dimensional Euclid space, because of Schwartz inequality, one may define

$$|\cos\varphi| = \frac{|(x,y)|}{\sqrt{(x,x)}\sqrt{(y,y)}}.$$

Evidently, $|\cos\varphi| \leq 1$. The quantity φ is called the angle between vectors x and y.

Example 17. In an n-dimensional unitary space, the inner product is defined in the same way as Example 13:

$$(x,y) = \sum_{i=1}^{n} p_i \xi_i^* \eta_i.$$

Then by Schwartz inequality, we have

$$\left|\sum_{i=1}^{n} p_i \xi_i^* \eta_i\right|^2 \leq \sum_{i=1}^{n} p_i |\xi_i|^2 \sum_{i=1}^{n} p_i |\eta_i|^2.$$

This is called the **Cauchy inequality**.

Example 18. In Example 14, the inner product is defined by

$$(x,y) = \int_0^1 x^*(t)y(t)w(t)\mathrm{d}t.$$

Then Schwartz inequality results in

$$\left|\int_0^1 x^*(t)y(t)y(t)w(t)\mathrm{d}t\right|^2 \leq \int_0^1 |x(t)|^2 w(t)\mathrm{d}t \int_0^1 |y(t)|^2 w(t)\mathrm{d}t.$$

Theorem 2. *In any inner product space, the following inequality stands for arbitrary two vectors x and y:*

$$\sqrt{(x+y,x+y)} \leq \sqrt{(x,x)} + \sqrt{(y,y)}. \qquad (2.1.9)$$

This is called the **triangle inequality**.

Proof. By use of Schwartz inequality, one gets

$$(x+y,x+y) = (x,x) + (y,y) + (x,y) + (y,x)$$

$$\leq \sqrt{(x,x)}^2 + \sqrt{(y,y)}^2 + 2\sqrt{(x,x)}\sqrt{(y,y)}$$

$$= \left(\sqrt{(x,x)} + \sqrt{(y,y)}\right)^2.$$

$\square$

Theorem 3. *In any inner product space, the following equation stands for arbitrary two vectors x and y:*

$$(x + y, x + y) + (x - y, x - y) = 2[(x, x) + (y, y)]. \qquad (2.1.10)$$

This is called the **parallelogram formula.** *Its geometrical meaning is that in a parallelogram, the sum of squares of its diagonal lengths equals to the sum of squares of its four edge lengths.*

Definition 12. The square root of the inner product of a vector x with itself, $\sqrt{(x, x)}$, is called the **modulus** of the vector, denoted as $\|x\|$, i.e., its definition is

$$\|x\| \equiv \sqrt{(x, x)}. \qquad (2.1.11)$$

Modulus is necessarily non-negative, $\|x\| \geq 0$.

Definition 13. A vector with modulus 1 is called a **unit vector.**
Obviously, every vector x can become a unit one by the procedure $\frac{x}{\sqrt{(x,x)}}$. This procedure is called **normalization.**

2. Orthonormalized vector group

Definition 14. When and only when the inner product of vector x and y is zero, $(x, y) = 0$, it is said that x is **orthogonal** to y.
Because $(y, x) = (x, y)^*$, when $(x, y) = 0$, $(y, x) = 0$. The orthogonalization has such a symmetry that is similar to distance, although the definition of inner product does not has such a similarity. Specifically, zero vector is orthogonal to all vectors, $(x, \theta) = 0$.

Definition 15. If in a vector set $\{x_1, x_2, \ldots\}$, $(x_i, x_j) = 0$ for $i \neq j$, then this set is called an **orthogonal vector set.** If for all i and j,

$$(x_i, x_j) = \delta_{ij}, \qquad (2.1.12a)$$

then this set $\{x_1, x_2, \ldots\}$ is said to be **orthonormalized.** An orthonormalized vector group is also called a **unitary basis,** usually denoted by $\{e_1, e_2, \ldots\}$.

$$(e_i, e_j) = \delta_{ij}. \qquad (2.1.12b)$$

For example, the unit vectors along the three Cartesian coordinates are orthogonal to each other.

Definition 16. After choosing one orthogonal basis in an n-dimensional vector space, any vector in this space can be expressed by this basis. Then, this basis is said to be **complete**.

If a nonzero vector x can be expressed as the linear combination of an orthonormalized vector set,

$$x = \sum_{i=1}^{n} \alpha_i e_i, \qquad (2.1.13)$$

Then the combination coefficients α_i are easily solved. Making inner product to (2.1.13) with e_j under the help of (2.1.12), one gets

$$\alpha_j = (e_j, x), (j = 1, 2, \ldots, n). \qquad (2.1.14)$$

Theorem 4. *Orthonormalized vectors are linearly independent of each other.*

Proof. When the zero vector θ is expressed by the linear combination in terms of an orthonormalized basis $\{e_1, e_2, \ldots, e_n\}$, $\theta = \sum_{i=1}^{n} \alpha_i e_i$, the coefficients can be computed by (2.1.14) to be $\alpha_j = (e_j, \theta) = 0, (j = 1, 2, \ldots, n)$. All the coefficients are zero. Conversely, when all the coefficients are zero, the combination yields the zero vector. $\qquad \square$

By this theorem, an othonormalized set composed by n vectors is a basis in n-dimensional space.

Theorem 5. *Let $\{e_1, e_2, \ldots, e_m\}$ be an orthonormalized set in an m-dimensional space. One is able to express any vector x in this space in terms of the linear combination of the orthonormalized set, $x = \sum_{i=1}^{m} \alpha_i e_i$. Then if he takes the former $n(n \leq m)$ base vectors $\{e_1, e_2, \ldots, e_n\}$, he will get*

$$\|x\|^2 \geq \sum_{i=1}^{n} |\alpha_i|^2, \qquad (2.1.15)$$

where the equal sign stands when and only when $n = m$. The vector $x' = x - \sum_{i=1}^{n} \alpha_i e_i$ is orthogonal to everyone of $\{e_1, e_2, \ldots, e_n\}$. Equation (2.1.15) is called the **Bessel inequality**.

Proof.

$$0 \leq \|x'\|^2 = (x', x') = \left(x - \sum_{i=1}^{n} \alpha_i e_i, x - \sum_{i=1}^{n} \alpha_i e_i \right)$$

$$= \|x\|^2 - \sum_{i=1}^{n} |\alpha_i|^2 - \sum_{i=1}^{n} |\alpha_i|^2 + \sum_{i=1}^{n} |\alpha_i|^2$$

$$= \|x\|^2 - \sum_{i=1}^{n} |\alpha_i|^2.$$

That is (2.1.15). Furthermore,

$$(e_j, x') = \left(e_j, x - \sum_{i=1}^{n} \alpha_i e_i \right)$$

$$= (e_j, x) - \left(e_j, \sum_{i=1}^{n} \alpha_i e_i \right) = \alpha_j - \sum_{i=1}^{n} \alpha_i (e_j, e_i) = 0. \quad \square$$

It is easily obtained, by use of orthonormalized basis, that

$$(x, y) = \left(\sum_{i=1}^{n} (e_i, x) e_i, \sum_{j=1}^{n} (e_j, y) e_j \right) = \sum_{i=1}^{n} (x, e_i)(e_i, y). \quad (2.1.16)$$

This equation is called **Parseval equality**. It can be an equivalent statement of the completeness of an orthonormalized basis.

3. Metric matrix

Suppose that $\{d_1, d_2, \ldots, d_n\}$ is a basis. It can be used to express two vectors x and y, $x = \sum_{i=1}^{n} \alpha_i d_i$ and $y = \sum_{i=1}^{n} \beta_i d_i$. Now making the inner product of the two vectors, we have

$$(x, y) = \left(\sum_{i=1}^{n} \alpha_i d_i, \sum_{j=1}^{n} \beta_j d_j \right) = \sum_{i.j=1}^{n} \alpha_i^* (d_i, d_j) \beta_j$$

$$= (\alpha_1^*, \alpha_2^*, \ldots, \alpha_n^*) M \begin{pmatrix} \beta_1 \\ \beta_2 \\ \vdots \\ \beta_n \end{pmatrix} = \alpha^+ M \beta. \quad (2.1.17)$$

Here a matrix M has been defined:

$$M = \begin{pmatrix} (d_1, d_1) & (d_1, d_2) & \cdots & (d_1, d_n) \\ (d_2, d_1) & (d_2, d_2) & \cdots & (d_2, d_n) \\ \vdots & \vdots & \ddots & \vdots \\ (d_n, d_1) & (d_n, d_2) & \cdots & (d_n, d_n) \end{pmatrix}.$$

Its elements are determined by the inner products between base vectors. The matrix M is called a **metric matrix** under basis $\{d_1, d_2, \ldots, d_n\}$.

Metric matrix is of the following properties.

(i) It is easily seen $M^+ = M$, so that it is a Hermitian matrix. In real space, it is a real symmetric one.

(ii) For any $x \neq 0$,

$$(x, x) = \alpha^+ M \alpha \geq 0$$

so that metric matrix is a positive definite Hermitian matrix. In real space, it is a positive definite matrix.

(iii) When the basis is unitary, a metric matrix becomes a unit matrix, a simplest form.

Apparently, when basis is unitary, the inner product is of the simplest form. In each definition of inner products (2.1.4)–(2.1.7), an orthonormalized basis has been implicitly adopted. If the basis is not unitary, a metric matrix in the form of (2.1.17) is inevitably involved in the inner product. Nevertheless, equations (2.1.15)–(2.1.17) are independent of basis.

Hereafter the inner product will always be in the simplest forms of (2.1.4)–(2.1.7). The general form (2.1.17) will not be used. That is to say, we always implicitly adopt a unitary basis. If a basis is unitary, there is a way to transform it into a unitary one. This method is the well-known Gram-Schmidt orthonormalization method.

4. Gram-Schmidt orthonormalization method

Suppose that $\{x_1, x_2, \ldots, x_n\}$ is a linearly independent vector group in an n-dimensional space. It can be converted to be a complete orthonormalized set. The procedure is called the **Gram-Schmidt orthonormalization method.** We set the following equations:

$$y_k = x_k - \sum_{i=1}^{k-1}(e_i, x_k)e_i, \qquad e_k = \frac{y_k}{\|y_k\|}, \qquad (k = 1, 2, \ldots, n). \quad (2.1.18)$$

The built $\{e_1, e_2, \ldots, e_n\}$ is a unitary basis. If e_{k-1} has been orthogonal to every $e_j, (j < k - 1)$, then the inner product of e_k and every $e_j, (j < k)$ is

$$(e_k, e_j) = \left(\frac{y_k}{\|y_k\|}, e_j\right) = \frac{1}{\|y_k\|}\left(x_k - \sum_{i=1}^{k-1}(e_i, x_k)e_i, e_j\right)$$

$$= \frac{1}{\|y_k\|}\left[(x_k, e_j) - \sum_{i=1}^{k-1}(e_i, x_k)(e_i, e_j)\right]$$

$$= \frac{1}{\|y_k\|}\left[(x_k, e_j) - \sum_{i=1}^{k-1}(e_i, x_k)^*\delta_{ij}\right] = 0. \quad (2.1.19)$$

The conclusion is that the new basis $\{e_1, e_2, \ldots, e_n\}$ is unitary.

The above procedure and proof are summarized to be a theorem as follows.

Theorem 6. *There always exists at least one unitary basis* $\{e_1, e_2, \ldots, e_n\}$ *in an n-dimensional inner product space.*

2.1.3. *Hilbert space*

Definition 17. An inner product space is said to be a **complete inner product space**, or a **Hilbert space**, only when every Cauchy sequence converges to a limit in this space.

In an inner product space, distance can be defined, so that it is a distance space. If it is of completeness, it is called a Hilbert space. Apparently a space in which distance cannot be defined is not a Hilbert one since there is no completeness in this space.

Hereafter the letter H represents Hilbert space, or simply H space, which will be denoted by H_n if it is of n-dimensions.

Definition 18. If a complex function $f(x)$ defined on $[a, b]$ and the integration $\int_a^b |f|^2 \mathrm{d}x$ exists and is finite, this function is said to be

complex square-integrable, or simply **square integrable**, and is denoted as $L_2[a, b]$. This is in fact a special case $p = 2$ of Definition 2, explicitly mentioned here for it is used mostly.

A square-integrable function multiplied by a scalar is still square-integrable. A square-integrable function plus another is still square-integrable.

Definition 19. If a set consists of square-integrable complex functions defined on the closed interval $[a, b]$, which satisfy the following two conditions (i) and (ii), then this set constitutes a vector space, called a **square-integrable function space**, denoted as $L_2[a, b]$, or in short L_2. It is usually simply called a **function space**.

(i) Addition rule: if $f_1(x) \in L_2$ and $f_2(x) \in L_2$,

$$(f_1 + f_2)(x) \equiv f_1(x) + f_2(x) = f(x) \in L_2; \qquad (2.1.20)$$

(ii) Scalar multiplication rule: if $f(x) \in L_2, \alpha \in Z, \alpha f \equiv \alpha f(x) \in L_2$. As in (2.1.1), the distance between two vectors f and g is defined by

$$\rho(f, g) = \left[\int_a^b |f(x) - g(x)|^2 \mathrm{d}x \right]^{1/2}. \qquad (2.1.21)$$

Thus L_2 is a distance space.

Definition 20. The inner product of two functions f_1 and f_2 that belong to L_2 is defined by

$$(f_1, f_2) = \int_a^b f_1^*(x) f_2(x) \mathrm{d}x,$$

or

$$(f_1, f_2) = \int_a^b f_1^*(x) f_2(x) w(x) \mathrm{d}x,$$

where the weight function $w(x) > 0$.

A complete inner product space consisting of complex square-integrable functions with real arguments defined on a closed interval $[a, b]$ is a Hilbert space. This is a mostly useful Hilbert space.

By definition, the square-integrable also means that the modulus is finite:

$$\|f\| = \left[\int_a^b |f|^2 w(x)\mathrm{d}x \right]^{1/2} < \infty. \tag{2.1.22}$$

By the way, we mention that the square-integrability of a two-variable function $k(x, y)$ means that

$$\int_a^b w(y)\mathrm{d}y \int_a^b |k(x, y)|^2 w(x)\mathrm{d}x < \infty. \tag{2.1.23}$$

The square-integrability of an n-variable is easily defined similarly.

The inner product of any two square-integrable functions (f_1, f_2) exists. Proof:

$$\left| \int_a^b f_1^* f_2 w(x)\mathrm{d}x \right| \leq \int_a^b |f_1^* f_2| w(x)\mathrm{d}x = \int_a^b |f_1||f_2| w(x)\mathrm{d}x$$

$$\leq \frac{1}{2} \int_a^b (|f_1|^2 + |f_2|^2 w(x)\mathrm{d}x$$

$$= \frac{1}{2}(\|f_1\| + \|f_2\|) < \infty. \tag{2.1.24}$$

Null inner product $(f, f) = 0$ does not imply $f(x) = 0$ for all $x \in [a, b]$. However, it can take nonzero values at isolated points, although there may be infinite isolated points. A function that can take nonzero points is said to be an almost everywhere zero function.

Definition 21. A function $f(x)$ is said to be a **zero function** if it is zero at everywhere in $[a, b]$: $f(x) = 0, x \in [a, b]$. A function that is almost everywhere zero on interval $[a, b]$ is called a **generalized zero function**. A function is called a **nonzero function** if it is neither a zero function nor a generalized zero function. A function is called a **classical function** if its value at each isolated point is finite.

Hereafter, the functions mentioned in this book are all nonzero functions, unless specified.

For example, a function is defined on interval $[0, 1]$ such that it takes value 1 at rational positions and 0 at irrational positions. This is a generalized zero function.

If $(f, f) = 0$, f is a zero or generalized zero function.

Conversely, if a function $f(x)$ defined on $[a, b]$ is a zero or generalized zero function, then it must be $(f, f) = 0$ and

$$\int_a^b f(x)\mathrm{d}x = 0. \qquad (2.1.25)$$

The generalized zero function takes finite values at isolated positions. The case that the values of the function are infinite will be discussed in Chapter 5.

Theorem 7 (Riesz-Fisher). *Suppose that functions* $f_1(x)$, $f_2(x), \ldots$ *are components of a function space. If*

$$\lim_{n,m\to\infty} \|f_n - f_m\|^2 \equiv \lim_{n,m\to\infty} \int_a^b |f_n - f_m|^2 \mathrm{d}x = 0, \qquad (2.1.26)$$

there exists a square-integrable function $f(x)$ *to which the sequence* $\{f_n(x)\}$ *"mean" converges, i.e., there exists an* f *that makes*

$$\lim_{n\to\infty} \int_a^b |f(x) - f_n(x)|^2 \, \mathrm{d}x = 0. \qquad (2.1.27)$$

Then the square-integrable function (with finite modulus) space is complete.

According to (2.1.24), the modulus of the difference of two functions $[\int_a^b |f(x) - f_n(x)|^2 \mathrm{d}x]^{1/2}$ *is a distance. In this distance space, (2.1.26) agrees with (2.1.3). The function sequence* $f_n(x)$ *and function* $f(x)$ *are both in this distance space. Therefore, it is a complete distance space.*

2.2. Operators in Inner Product Spaces

2.2.1. *Operators and adjoint operators*

1. Operators

A function $y = f(x)$ means that for a number x in the set R of real numbers, there is associated another real number y computed by this expression. That is to say, the function reflects a rule of one-to-one correspondence from R to R. An n-argument real function is a rule of one-to-one correspondence from R^n to R. The concept of mapping is the generalization of that function. If the domain of definition and codomain are replaced by sets, one obtains the concept of mapping.

Definition 1. Let X and Y be two sets. F is a rule such that for any element $x \in X$, there is associated an element $y \in Y$. If so, F is called a **mapping** from set X to set Y, denoted as $F : X \to Y$. The element y is called the **image** of x under mapping F, denoted as $y = F(x)$, or the **value** of y at x under F. The element x is called the **preimage** of y under F. The set X is called the **domain of definition** of the mapping F, and the set Y is called the **codomain** of F. When $X = Y$, F is said to be a mapping from the set to itself.

For example, distance is a mapping from a set to R of real numbers.

When both the definition domain and codomain are the set of real numbers, a real number as a preimage is mapped to a real number, i.e., an ordinary real function.

The definition of distance is an example of the mapping of two preimages to one image.

Taking the modulus of a complex vector with dimension n is also a mapping from a preimage to an image, where the definition domain is an n-dimensional complex number set and the codomain is the set of positive real numbers.

Definition 2. Suppose that V and U are two sets on the same domain K, and T is a mapping from V to U. If for $x \in V$ and $y \in U$, there is an associated $Tx = y$, then T is called an **operator** from V to U, or a **transformation** from V to U. Let T be a mapping from V to V. T is called an **identity operator**, or a **unit operator** or **unit transformation**, if for all $x \in V$, $Tx = x$. Usually the unit transformation is denoted as $I: Ix = x$.

It should be pointed out that only when two spaces V and U are on the same domain K can the mapping be called a transformation. The general definition of mapping does not set any constraint to the definition domain of X and the codomain of Y.

Briefly, that an operator acts on a vector yields a vector. For example, derivation is an operator, which makes a function become its derivative function.

Definition 3. Let V and U be two inner product spaces on the same domain K, and T be an operator from V to U. T is called a **linear operator** or **linear transformation** from V to U if it satisfies the following two conditions:

(i) addition rule,

$$\text{for all} \quad x, y \in V, \quad T(x+y) = Tx + Ty; \qquad (2.2.1)$$

(ii) scalar multiplication rule,

$$\text{for all} \quad x \in V \quad \text{and all scalar } \alpha \in K, \quad T(\alpha x) = \alpha T x. \quad (2.2.2)$$

Example 1. An integral operator is defined by

$$Kf \equiv \int_a^b k(x, y) f(y) \mathrm{d}y$$

or

$$Kf \equiv (k, f) \equiv \int_a^b k(x, y) f(y) w(y) \mathrm{d}y.$$

The operator thus defined is linear, because

$$K(f_1 + f_2) \equiv (k, f_1 + f_2) = (k, f_1) + (k, f_2) = Kf_1 + Kf_2,$$

and

$$K(\alpha f) \equiv (k, \alpha f) = \alpha(k, f) = \alpha K f.$$

Such an operator meets the conditions (2.2.1) and (2.2.2) so that it is a **linear integral operator**. The function $k(x, y)$ in the integral is called an **integral kernel**.

Hereafter, we only discuss linear operators.

By Definition 2, an identical operator does not alter any element it acts. There may be another kind of operator: when it acts on a certain element, the element remains unchanged:

$$u = Tu. \qquad (2.2.3)$$

When this is met, the element u is called the **fixed point** of operator T. Equation (2.2.3) can be used to find a solution by means of iteration. The procedure is that in order to search for solution u, one first assumes a u_0 and uses $u_1 = Tu_0$ to get a u_1, then $u_2 = Tu_1$ to get u_2, and so on. He will get a series $u_0, u_1, \ldots, u_n, \ldots$ which gradually approaches u. Here u_0 is called an initial value. This procedure is called the **iteration method**, or the **successive approximation method**. The solution will surely be found through the successive approximation if the operator T is a contraction. Thus let us define the concept of contraction.

Definition 4. If two elements u and v in a metric space are acted on by operator T then the following inequality is satisfied:

$$\rho(Tu, Tv) \leq \lambda\rho(u, v). \tag{2.2.4}$$

The T is said to be of **Lipschitz continuity**. If $\lambda < 1$ in (2.2.4), T is called a **contraction operator**, or in short a **contraction**.

Theorem 1. *Let T be a contraction in a complete metric space X. Then (2.2.3) has one and only one solution. This solution can be found by setting any initial value through the iteration method.*

Proof. (i) First we show the uniqueness of the fixed point. Assuming that there are two fixed points subjected to this contraction denoted as u and v: $u = Tu$ and $v = Tv$, then $\rho(Tu, Tv) = \rho(u, v)$. Because T is a contraction, there is necessarily $\rho(Tu, Tv) \leq \lambda\rho(u, v)$, i.e., $\rho(u, v) \leq \lambda\rho(u, v)$, where $\lambda < 1$. Thus $\rho(u, v) = 0$, which results in $u = v$. (ii) Next we show the existence of the fixed point. Let u_0 be the initial value. Through the operation above, a series $u_0, u_1, \ldots, u_n, \ldots$ is obtained. We show that this is a Cauchy sequence. It is easy to put down

$$\rho(u_n, u_{n+1}) = \rho(Tu_{n-1}, Tu_n) \leq \lambda\rho(u_{n-1}, u_n) \leq \cdots \leq \lambda^n \rho(u_0, u_1).$$

For $k > n$, applying repeatedly the triangle inequality or distance, or by Exercise 1 of this chapter, one gets

$$\rho(u_n, u_k) \leq \rho(u_n, u_{n+1}) + \rho(u_{n+1}, u_{n+2}) + \cdots + \rho(u_{k-1}, u_k)$$
$$\leq \lambda^n \rho(u_0, u_1) + \lambda^{n+1} \rho(u_0, u_1) + \cdots + \lambda^{k-1} \rho(u_0, u_1)$$
$$\leq \rho(u_0, u_1)\frac{\lambda^n}{1 - \lambda}.$$

Since $\lambda < 1$, when $k, n \to \infty$, distance $\rho(u_n, u_k) \to 0$. Hence, $\{u_n\}$ is a Cauchy sequence, the limit of which is denoted as u. Because the space is complete, the limit is in this space. The conclusion is that $u_n \to u$. $\square$

The proof process demonstrates that the solution can be found by selecting any initial value through the iteration method. This is also called **contraction mapping**.

Please note that this theorem does not make use of the concept of inner product, but only distance. Therefore, it is valid in complete metric spaces.

2. Adjoint operators

Definition 5. The Hermitian conjugate of a linear operator T is denoted as $T^\dagger$. $T^\dagger$ is called the **adjoint operator** or **conjugate operator** of T if it, when acting on any two vectors x and y, satisfies

$$(Tx, y) = (x, T^\dagger y). \tag{2.2.5}$$

Note that there will always be a concrete form when an operator acts on a vector. For instance, in a vector space with finite dimension, a linear transformation is expressed by a matrix, and a differential operator is expressed by a differential symbol. When we say the Hermitian conjugate, we mean taking the complex conjugate of the operator's concrete form, and at the same time, its indices are taken in reverse order.

Adjoint operators have the following properties.

Theorem 2. *Let X be a Hilbert space, and A and B are linear operators from X to X. For any complex number α in the domain C of complex numbers, $\alpha \in C$, the following equations hold.*

(i) $(A + B)^\dagger = A^\dagger + B^\dagger$.
(ii) $(\alpha A)^\dagger = \alpha^* A^\dagger$.
(iii) $(A^\dagger)^\dagger = A$.
(iv) $(A \bullet B)^\dagger = B^\dagger \bullet A^\dagger$.

The symbol $\bullet$ means the complex of mappings, which will hereafter be omitted, so that $A \bullet B$ will be shortly written as AB.

By (ii), *when we mention the adjoint of a number, we mean its complex conjugate.*

Definition 6. Let A, B and C be all linear operators. If $AB = BA = I$, B is called the **inverse operator** of A, or in short **inverse**, denoted as $B = A^{-1}$. If $A^\dagger A = AA^\dagger$, A is called a **normal operator**. If $A^\dagger A = AA^\dagger = I$, A is called a **unitary operator**. B is called the **left inverse** of A if $BA = I$, and C is called the **right inverse** of A if $AC = I$.

If both the left and right inverses of an operator exists, they are certainly identical, because $B = B(AC) = (BA)C = C$. In this case, both B and C are the inverse of operator A, $B = C = A^{-1}$. Hereafter, we take into account the cases where both left and right inverses exist or both does not. In the latter case, we say that the inverse does not exist.

Example 2. An n-dimensional Euclid space E is an n-dimensional vector space on the real number domain. Each vector x in this space is a column one containing n components. The linear operator T from E to E is expressed by a transformation matrix B which is a unitary matrix, i.e., $BB^{\mathrm{T}} = B^{\mathrm{T}}B = I$, where B^{T} is the transpose of B. The transformation of vector u in this space is $v = Bu$. The transformation matrix of the adjoint operator $T^\dagger$ is B^{T}. The inner product of two vectors x and y is written as $(x, y) = x^{\mathrm{T}}y$. Then (2.2.5) in the present case becomes

$$(Bx)^{\mathrm{T}}y = x^{\mathrm{T}}B^{\mathrm{T}}y.$$

Example 3. A unitary space U is an n-dimensional vector space on the complex number domain. Given a basis in this space, the linear transformation T from U to U is expressed as a transformation matrix under the given basis, denoted by A. If T acts on a vector u in this space to produce another vector v, the transformation is

$$v = Au.$$

The transformation matrix of T's adjoint operator $T^\dagger$ is A^+ which is the Hermitian conjugate of A. The inner product of the vectors x and y in this space is written as $(x, y) = x^+y$. Then (2.2.5) in the present case becomes

$$(Ax)^+y = x^+A^+y.$$

In a unitary space, the transformation matrix associated with the linear transformation is A, the adjoint of which is its Hermitian conjugate A^+. When $AA^+ = A^+A = I$, this is a unitary transformation.

Example 4. Let us see a linear integral operator. Suppose that $k(s,t)$ is a complex function and is integrable on the interval $a \leq s$, $t \leq b$. $L_2[a,b]$ is a square-integrable space. In this space the inner product is defined by

$$(x, y) = \int_a^b x^*(x)y(t)\mathrm{d}t.$$

An integral operator T is defined by

$$(Tx)(s) = \int_a^b k(s,t)x(t)\mathrm{d}t, \quad s \in [a,b], \quad \text{for any } x \in L_2[a,b].$$

This is a linear operator from $L_2[a,b]$ to itself. Its adjoint operator $T^\dagger$ is

$$(T^\dagger y)(s) = \int_a^b k^*(t,s)y(t)\mathrm{d}t, \quad s \in [a,b], \quad \text{for any } y \in L_2[a,b].$$

Note that the integral kernel should be taken with its complex conjugate and its arguments exchanged. Then (2.2.5) in the present case becomes

$$\int_a^b \mathrm{d}s\, y(s) \int_a^b k^*(s,t)x^*(t)\mathrm{d}t = \int_a^b \mathrm{d}s\, x^*(s) \int_a^b k^*(t,s)y(t)\mathrm{d}t.$$

3. The adjoint of a differential operator

The adjoints of transformation matrices in linear algebra and integral operators have been easily given, but those of differential operators are not so. Now we show how to determine the latters by examples.

Example 5. A first-order differential operator $L = \frac{\mathrm{d}}{\mathrm{d}x}$ acting on a function $u(\mathrm{x})$ defined on $[0,1]$ will produce $Lu(x) = \frac{\mathrm{d}}{\mathrm{d}x}u(x)$. If at the boundaries the function obeys the boundary condition $u(0) = 2u(1)$, the whole problem will become

$$Lu(x) = \frac{\mathrm{d}}{\mathrm{d}x}u(x), \quad u(0) = 2u(1). \tag{2.2.6}$$

Let us find the adjoint of this differential operator.

By definition (2.2.5), it should be

$$(v, Lu) = (L^\dagger v, u).\qquad(2.2.7)$$

We take the inner product on the interval [0,1] as follows.

$$(v, Lu) = \int_0^1 v^*(x)\frac{\mathrm{d}}{\mathrm{d}x}u(x)\mathrm{d}x$$

$$= [v^*(x)u(x)]_0^1 - \int_0^1 u(x)\frac{\mathrm{d}}{\mathrm{d}x}v^*(x)\mathrm{d}x$$

$$= [v^*(x)u(x)]_0^1 + \int_0^1 \left[-\frac{\mathrm{d}}{\mathrm{d}x}v(x)\right]^* u(x)\mathrm{d}x.$$

In order to meet (2.2.7), the first term on the right hand side has to be zero. By use of the boundary condition (2.2.6), we have

$$[v^*(x)u(x)]_0^1 = u(1)v^*(1) - u(0)v^*(0) = u(1)[v^*(1) - 2v^*(0)] = 0.$$

The expression of the adjoint is

$$L^\dagger v(x) = -\frac{\mathrm{d}}{\mathrm{d}x}v(x), \quad v(1) = 2v(0).\qquad(2.2.8)$$

Note that we have the complex conjugate of the boundary condition (2.2.6). The equation and boundary condition in (2.2.8) are respectively complex conjugates of those in (2.2.6), i.e., the adjoint equation and adjoint boundary condition.

This example demonstrates that before seeking the adjoint of a differential operator, the boundary condition of the function it will act on should be given. The adjoint consequently consists of two parts: its form and the boundary condition of the function it acts on. For a differential operator, its adjoint has to relate to the boundary condition, because to find it integration by parts is employed.

Example 6. A differential operator and the boundary conditions of the function it acts on are as follows:

$$Lu(x) = \frac{\mathrm{d}^2}{\mathrm{d}x^2}u(x), \quad 0 < x < 1;$$

$$u(1) - \alpha u(0) = 0, \quad u'(1) - \beta u'(0) = 0.$$

Here we have two boundary conditions due to second-order differential. Let us find the adjoints of the operator and boundary conditions.

By definition of adjoint operator (2.2.5), we manipulate the inner product through integration by parts to transfer the differential of u to that of v.

$$\int_0^1 \mathrm{d}x v^*(x) L u(x)$$

$$= \int_0^1 \mathrm{d}x v^*(x) u''(x) = [v^*(x)u'(x)]_0^1 - [v^{*\prime}(x)u(x)]_0^1$$

$$+ \int_0^1 \mathrm{d}x u(x) v^{*\prime\prime}(x)$$

$$= v^*(1)u'(1) - v^*(0)u'(0) - [v^{*\prime}(1)u(1) - v^{*\prime}(0)u(0)]$$

$$+ \int_0^1 \mathrm{d}x u(x) v^{*\prime\prime}(x)$$

$$= [\beta v^*(1) - v^*(0)]u'(0) - [v^{*\prime}(1)\alpha - v^{*\prime}(0)]u(0)$$

$$+ \int_0^1 \mathrm{d}x u(x) v^{*\prime\prime}(x).$$

Thus we obtain

$$L^\dagger v(x) = \frac{\mathrm{d}^2}{\mathrm{d}x^2} v(x), \quad 0 < x < 1;$$

$$\beta^* v(1) - v(0) = 0, \quad v'(1) - \alpha^* v'(0) = 0.$$

In Example 5, $L = \frac{\mathrm{d}}{\mathrm{d}x}$ so that $L^\dagger = -\frac{\mathrm{d}}{\mathrm{d}x}$. We call $-\frac{\mathrm{d}}{\mathrm{d}x}$ the formal adjoint operator of $\frac{\mathrm{d}}{\mathrm{d}x}$. When we say a formal adjoint, we do not consider the interval and boundary conditions. So, a formal adjoint is not a real adjoint. Only combined with boundary conditions, can a formal adjoint be a real one. In Example 5, when $v(1) = 2v(0)$ on $[0, 1]$ is satisfied, $-\frac{\mathrm{d}}{\mathrm{d}x}$ becomes the real adjoint of $\frac{\mathrm{d}}{\mathrm{d}x}$ satisfying $u(0) = 2u(1)$.

In Example 6, the formal adjoint happens to be the same form of the operator itself: $L^\dagger = \frac{\mathrm{d}^2}{\mathrm{d}x^2} = L$.

Example 7. In general, a first-order differential operator is of the form of

$$L = q_1(x)\frac{\mathrm{d}}{\mathrm{d}x} + q_0(x). \tag{2.2.9a}$$

In the same way as Example 5, we carry out the integration by parts to get

$$(v, Lu) = \int_a^b v^* \left[\left(q_1\frac{\mathrm{d}}{\mathrm{d}x} + q_0 \right) u \right] \mathrm{d}x$$

$$= [q_1 v^* u]_a^b + \int_0^1 \left[\left(-q_1\frac{\mathrm{d}}{\mathrm{d}x} + q_0 - q_1'(x) \right) v^* \right] u \mathrm{d}x$$

$$= \int_0^1 \left[\left(-q_1^*\frac{\mathrm{d}}{\mathrm{d}x} + q_0^* - q_1^{*\prime}(x) \right) v \right]^* u(x)\mathrm{d}x + [q_1 v^* u]_a^b.$$

As a result, the formal adjoint is

$$L^\dagger = -q_1^*(x)\frac{\mathrm{d}}{\mathrm{d}x} + q_0^*(x) - q_1^{*\prime}(x). \tag{2.2.9b}$$

When it acts on a function defined on $[a, b]$, it is required to meet the boundary condition

$$[q_1(x)v^*(x)u(x)]_a^b = 0 \tag{2.2.9c}$$

for it to be an adjoint.

Example 8. In general, a second-order differential operator is of the form of

$$L = p_2(x)\frac{\mathrm{d}^2}{\mathrm{d}x^2} + p_1(x)\frac{\mathrm{d}}{\mathrm{d}x} + p_0(x). \tag{2.2.10a}$$

Taking integration by parts similar to (2.2.7), we put down its formal adjoint:

$$L^\dagger = p_2^*\frac{\mathrm{d}^2}{\mathrm{d}x^2} + (2p_2^{\prime*} - p_1^*)\frac{\mathrm{d}}{\mathrm{d}x} + p_2^{\prime\prime*} - p_1^{*\prime} + p_0^*. \tag{2.2.10b}$$

When it acts on a function defined on $[a, b]$, it is required to meet the boundary conditions

$$[p_2(u'v^* - uv^{*\prime}) + (p_1 - p_2')uv^*]_a^b = 0 \tag{2.2.10c}$$

for it to be an adjoint.

For any differential acting on a function defined on $[a, b]$, we have

$$(v, Lu) - (L^\dagger v, u) = \int_a^b dx(v^* Lu - (L^\dagger v)^* u) = [J(u, v)]_a^b.$$

$$(2.2.11)$$

If the upper limit b in this equation is regarded as a variable, we can take derivative to b to obtain

$$v^* Lu - (L^\dagger v)^* u = \frac{d}{dx} J(u, v) \qquad (2.2.12)$$

This equation is called the **Lagrange equation**. Its integral form (2.2.11) is called **Green's formula**. The function $J(u, v)$ is called the **knot** of functions u and v. The concrete expression of knot is determined by the form of differential operator L. For example, in Example 5 the knot is

$$J(u, v) = v^*(x)u(x),$$

and in Example 6 it is

$$J(u, v) = v^*(x)u'(x) - v^{*\prime}(x)u(x).$$

For the first-order differential operator in Example 7, the knot is

$$J(u, v) = q_1 u v^*. \qquad (2.2.13a)$$

Example 5 is a special case of Example 7 when $q_1 = 1$ and $q_0 = 0$. For the second-order differential operator in Example 8, the knot is

$$J(u, v) = p_2(u'v^* - uv^{*\prime}) + (p_1 - p_2')uv^*. \qquad (2.2.13b)$$

Example 6 is a special case of Example 8 when $p_2 = 1$ and $p_1 = p_0 = 0$.

2.2.2.   *Self-adjoint operators*

1. The definition and properties of self-adjoints

Definition 7. A linear operator A is said to be **self-adjoint** if it is identical to its adjoint,

$$A^\dagger = A. \qquad (2.2.14)$$

This is a **self-adjoint operator**. The self-adjoint operator in a real space is called a symmetric operator, and in a complex space called a

Hermitian conjugate operator, or in short a **Hermitian operator**. In this case,

$$(Ax, y) = (x, Ay). \tag{2.2.15}$$

Self-adjoints have the following properties.

Theorem 3. *If A and B are self-adjoint, (i) $A + B$ is also self-adjoint;*

(ii) *when and only when $AB = BA$, AB is self-adjoint;*

(iii) *when and only when α is a real number, αA is self-adjoint.*

Theorem 4. *In real spaces, if a linear operator A is symmetric, i.e., A is a symmetric operator, then*

$$(y, Ax) = (x, Ay). \tag{2.2.16}$$

Proof. Let x and y be real vectors so that

$$(y, Ax) = (Ay, x) = (x, Ay)^* = (x, Ay).$$

Note that the vectors x and y and the operator A are all real, so that (x, Ay) is real. □

Theorem 5. *If an operator A in a space is self-adjoint, for any vector x in this space, $(x, Ax) = 0 \Leftrightarrow A = 0$.*

Proof. The sufficiency is obvious: when $A = 0$, for any vector x, it is necessarily $(x, Ax) = 0$. In the following, we discuss the necessity in the two cases of real and unitary spaces.

(i) In a real space, we make the inner product of two vectors x and y:

$$(x + y, A(x + y)) = (x, Ax) + (y, Ay) + (x, Ay) + (y, Ax) = 0. \tag{2.2.17}$$

By the condition given in the theorem,

$$(x, Ay) + (y, Ax) = 0. \tag{2.2.18}$$

Because x and y are real, (2.2.16) is brought to

$$(x, Ay) = 0. \tag{2.2.19}$$

It leads to $A = 0$ since x and y are arbitrary two vectors.

(ii) In a unitary space, we also do inner product of two complex vectors x and y so as to get

$$(x, Ay) + (y, Ax) = 0. \qquad (2.2.20)$$

Although this equation is formally the same as (2.2.18), here x and y are complex. Replacement of y by iy yields

$$(x, Aiy) + (iy, Ax) = 0.$$

When the two i factors are taken out of the inner product, there will appear a minus sign in the second term. This leads to

$$(x, Ay) - (y, Ax) = 0. \qquad (2.2.21)$$

Adding (2.2.21) and (2.2.20) gives

$$(x, Ay) = 0.$$

As x and y are arbitrary, $A = 0$.   □

Theorem 6. *Let A be a linear operator in a unitary space. For any vector x in this space, the sufficient and necessary conditions for (x, Ax) to be always real are that A is Hermitian.*

Proof.   Sufficiency. If A is Hermitian, $(x, Ax) = (Ax, x) = (x, Ax)^*$. That is to say, (x, Ax) is real.

Necessity. When (x, Ax) is real, $(x, Ax) = (x, Ax)^* = (Ax, x) = (x, A^\dagger x)$. This results in $(x, (A - A^\dagger)x) = 0$. This equation shows that $A = A^\dagger$, i.e., A is Hermitian, as x is an arbitrary vector.   □

Please note that this theorem does not apply to real spaces, because in real spaces (x, Ax) is always real and the symmetry of A is unnecessary. Therefore, in real spaces, that A is symmetric is the sufficient condition of (x, Ax) being real, but not the necessary condition.

Example 9. The transformation matrices B in Example 2 and A in Example 3 are both formal matrices. They represent the transformations of formal operators, called formal transformation. They are also unitary matrices, representing the transformations of unitary operators, called unitary transformation.

Hermitian matrix $A = A^+$ and real symmetric matrix $B = B^T$ are representation matrices of self-adjoint operators in finite-dimensional spaces.

For self-adjoint operators, the inner products in Examples 2, 3 and 4 are respectively written in the following forms:

$$(Bx)^T y = x^T By,$$

$$(Ax)^+ y = x^+ Ay$$

and

$$\int_a^b \mathrm{d}s y(s) \int_a^b k^*(t,s) x^*(t) \mathrm{d}t = \int_a^b \mathrm{d}s x^*(s) \int_a^b k(s,t) y(t) \mathrm{d}t.$$

An operator L is said to be formally self-adjoint when its formal adjoint is of the same form as that of its complex conjugate, $L^\dagger = L$. A formally self-adjoint operator is not necessarily self-adjoint. Whether it is self-adjoint or not depends on boundary conditions.

A differential operator of the first order is in the form of (2.2.9a) and it adjoint in (2.2.9b). The conditions for it to be a formally self-adjoint one are that

$$q_1(x) = -q_1^*(x), q_0(x) = q_0^*(x) - q_1^{*\prime}(x). \tag{2.2.22a}$$

Let $q_1(x) = ic(x)$ and $q_0(x) = a(x) + ib(x)$, where a, b and c are all real functions. Then, we get $q_0(x) = a(x) + ib(x) = a(x) - ib(x) + ic'(x)$. The necessary conditions are that function q_1 has only imaginary part and $c'(x) = 2b(x)$. When (2.2.9c) is satisfied, the operator (2.2.9a) is a self-adjoint one. If c is a constant, then $c'(x) = b(x) = 0$ and $q_0(x)$ is a real function:

$$q_0(x) = q_0^*(x), \quad q_1 = -q_1^*. \tag{2.2.22b}$$

A differential operator of the second order is in the form of (2.2.10a) and its adjoint in (2.2.10b). The conditions for it to be a formally self-adjoint one are, by comparing (2.2.10a) and (2.2.10b), that

$$p_2(x) = p_2^*(x), \quad p_1(x) = 2p_2^{*\prime}(x) - p_1^*(x),$$
$$p_0(x) = p_2^{*\prime\prime}(x) - p_1^{*\prime}(x) + p_0^*(x).$$

It is seen that p_2 must be real. Let $p_i(x) = a_i(x) + ib_i(x), i = 0, 1$. Then we get

$$a_1(x) + ib_1(x) = 2p_2'(x) - a_1(x) + ib_1(x).$$

This requires that $p_2'(x) = a_1(x)$. It is observed from the p_0 term that

$$a_0(x) + ib_0(x) = p_2''(x) - a_1'(x) + ib_1'(x) + a_0(x) - ib_0(x)$$
$$= p_2''(x) - p_2'(x) + ib_1'(x) + a_0(x) - ib_0(x).$$

This results in $b_1'(x) = 2b_0(x)$. The simplest case is that p_2, p_1 and p_0 are all real and

$$p_2' = p_1. \tag{2.2.23a}$$

In this case the operator can be written in a compact form:

$$L = \frac{\mathrm{d}}{\mathrm{d}x}\left(p_2(x)\frac{\mathrm{d}}{\mathrm{d}x}\right) + p_0(x). \tag{2.2.23b}$$

The knot (2.2.13) of the functions u and v is simplified to be

$$J(u, v) = p_2(u'v^* - uv^{*\prime}). \tag{2.2.24}$$

If

$$\left[p_2(u'v^* - uv^{*\prime})\right]_a^b = 0, \tag{2.2.25}$$

the operator will be self-adjoint.

For example, the operator $L = \frac{\mathrm{d}}{\mathrm{d}x}$ in Example 5 is a not formally self-adjoint, but the momentum operator $p = i\frac{\mathrm{d}}{\mathrm{d}x}$ is. The second-order differential operator $L = \frac{\mathrm{d}^2}{\mathrm{d}x^2}$ is a formally self-adjoint one, $L^\dagger = L$, and will become self-adjoint when the boundary conditions in Example 6 are satisfied. Four possible cases of (2.2.11) are listed in Table 2.1.

Table 2.1. Four possible cases of (2.2.11).

	$L^\dagger \neq L$	$L^\dagger = L$
$[J(u, v)]_a^b \neq 0$	$L^\dagger$ is not the adjoint of L	L is a formally self-adjoint operator
$[J(u, v)]_a^b = 0$	$L^\dagger$ is the adjoint of L	L is a self-adjoint operator

2. Eigenvalues of operators

Definition 8. When an operator A acting on an nonzero vector x equals to a number λ multiplying this vector,

$$Ax = \lambda x, \qquad (2.2.26a)$$

this equation is called an **eigenvalue equation** or a **characteristic equation** and λ is called an **eigenvalue,** or **characteristic value,** of this operator. The vector corresponding to the eigenvalue λ is called an **eigenvector.** The eigenvalue equation of the adjoint $A^\dagger$ of an operator is

$$A^\dagger y = \gamma y, \qquad (2.2.26b)$$

which is called an **adjoint eigenvalue equation** of (2.2.26a), or in short an **adjoint equation.** The numbers γ and y are respectively called the **adjoint eigenvalue** of λ and the **adjoint eigenvector** of x.

An eigenvector is always associated with a certain eigenvalue. An eigenvalue may be associated with more than one linearly independent eigenvectors. If an eigenvalue λ is associated with k linearly independent eigenvectors,

$$Ax_i = \lambda x_i, \quad (i = 1, 2, \ldots, k), \qquad (2.2.27)$$

it is said that the **rank,** or **degeneracy,** of λ is k. Mathematicians usually use the name rank, while physicists like to use the name degeneracy. If an eigenvector is a function, it is also called an **eigenfunction.** We also say that in (2.2.27) x_i is an eigenvector belonging to the eigenvalue λ.

In a finite-dimensional linear space, a linear operator manifests as a transformation matrix. We have known how to find the eigenvalues of a matrix in the course of linear algebra. The eigenvalues of a differential operator will be discussed in Chapter 3, and those of an integral operator will be introduced in Chapter 8. No matter how the forms of the operators are of, some common properties of them can be immediately obtained by the definitions of adjoint operators and inner product.

We assume that the eigenvalues of an operator can always be arranged in order by some rules, e.g., real numbers in order from smaller ones to larger ones, such that nth eigenvalues is denoted as λ_n and its associated eigenvector denoted as x_n.

Theorem 7. *If an operator A is a normal one, the following propositions hold.*

(1) *If x is an eigenvector of A belonging to an eigenvalue λ, it is also an eigenvector of $A^\dagger$ belonging to eigenvalue λ^*, i.e.,*

$$Ax = \lambda x \Leftrightarrow A^\dagger x = \lambda^* x.$$

(2) *Eigenvectors belonging to different eigenvalues are orthogonal to each other.*

Proof.   (1) Suppose $Ax = \lambda x$ and A is normal. Then

$$(Ax, Ax) = (A^\dagger Ax, x) = (AA^\dagger x, x) = (A^\dagger x, A^\dagger x).$$

When A is normal, $A - \lambda I$ is so. Replacing A by $A - \lambda I$, we get

$$((A - \lambda I)x, (A - \lambda I)x) = ((A - \lambda I)^\dagger x, (A - \lambda I)^\dagger x).$$

So, $(A - \lambda I)x = 0 \Leftrightarrow (A^\dagger - \lambda^* I)x$, and $Ax = \lambda x \Leftrightarrow A^\dagger x = \lambda^* x$.

(2) Assume that the operator has two different eigenvalues, $Ax_1 = \lambda_1 x_1$ and $Ax_2 = \lambda_2 x_2$. In the following inner product

$$(A^\dagger x_1, x_2) = (x_1, Ax_2),$$

we substitute the eigenvalue equation at the two sides to obtain

$$(\lambda_1^* x_1, x_2) = \lambda_1(x_1, x_2) = \lambda_2(x_1, x_2).$$

So, $(\lambda_1 - \lambda_2)(x_1, x_2) = 0$. As $\lambda_1 \neq \lambda_2$, it must be $(x_1, x_2) = 0$. The conclusion is that the eigenvector belonging to different eigenvalues are orthogonal to each other.   □

If the operator A in this theorem is self-adjoint, it is necessarily $\lambda = \lambda^*$. Therefore, we acquire a corollary.

Corollary. If an operator A is self-adjoint, (i) its eigenvalues are necessarily real, and (ii) its eigenvectors belonging to different eigenvalues are orthogonal to each other.

The proof of the theorem indicates that the "adjoint" of a number is its complex conjugate, and the "adjoint" of a matrix is its complex conjugate and transpose.

In quantum mechanics, physical quantities are represented by differential operators. The eigenvalues of the operators represent measurable values of the physical quantities. Measurable quantities are real. This means that the differential operators that represent physical quantities must be self-adjoint. We have known from Table 2.1 that a necessary condition for a differential operator to de self-adjoint is that the knot when the upper and lower limits are substituted in should be zero. This condition is always satisfied in practical systems, so that usually it is not mentioned in text books of quantum mechanics.

As for the eigenvalue problems of adjoint operators, there is a general theorem as follows. Before introducing it, we make an assertion without proof that if the inner product

$$(f, y) = 0, \tag{2.2.28}$$

where f is any vector in space except y, then it must be $y = 0$, i.e., y is a zero vector.

Theorem 8. *Assume the condition (2.2.28) holds. (i) If x_n is an eigenvector belonging to eigenvalue λ_n of A, then λ_n^* is an eigenvalue of $A^\dagger$ and its associated eigenvector is denoted as y_n. That is to say, the eigenvalues of A and $A^\dagger$ are of one-to-one correspondence and are complex conjugates of each other. Accordingly, the eigenvectors and adjoint eigenvectors have also one-to-one correspondence. (ii) An eigenvector and an adjoint one are orthogonal to each other if they respectively belong to different eigenvalues*

$$(\lambda_n - \lambda_m)(y_m, x_n) = 0. \tag{2.2.29}$$

When $m \neq n$, it becomes

$$(y_m, x_n) = 0, \quad m \neq n. \tag{2.2.30}$$

An eigenvector of an operator is orthogonal to that of its adjoint operator. This feature is called **biorthogonality**. *In this case,* *whether the different eigenvectors belonging to the same operator are orthogonal to each other or not cannot be determined. This means that when $m \neq n$, (x_m, x_n) is not necessarily zero.*

3. Isometric transformation

Definition 9. A linear transformation is said to be **isometric** if it retains the modulus of vectors unchanged. Explicitly, if U is a linear operator acting on any vector x and resulting in

$$\|Ux\| = \|x\|, \tag{2.2.31}$$

then U is called an **isometric operator**.

The above definition of isometric operators are equivalent to the following two:

$$U^\dagger U = I \tag{2.2.32}$$

and

$$(Ux, Uy) = (x, y). \tag{2.2.33}$$

It is easy to prove that any two of (2.2.31)–(2.2.33) are sufficient and necessary conditions of each other.

Theorem 9. *The absolute value of an eigenvalue of an isometric operator is* 1.

Proof. If $Ux = \lambda x$, $(Ux, Ux) = (\lambda x, \lambda x) = \|\lambda\|^2 (x, x) = (x, x)$, which reads $\|\lambda\| = 1$. □

Theorem 10. (i) *In an inner product space, if an orthonormalized basis $\{e_i\}$ is acted by an isometric operator U, the resultant $\{Ue_i\}$ is still an orthonormalized basis.* (ii) *In a Hilbert space, if a complete orthonormalized basis $\{e_i\}$ is acted by an isometric operator U, the resultant $\{Ue_i\}$ is still a complete orthonormalized basis.*

Proof.   (i) Because $(e_i, e_j) = \delta_{ij}$, $(Ue_i, Ue_j) = (e_i, e_j) = \delta_{ij}$.

(ii) We only need to prove that the orthogonalized basis is still complete after the isometric transformation. To do so, we make use of Parseval equality (2.1.16), because it is an equivalent statement of the completeness of an orthonormalized basis.

$$\sum_i (x, Ue_i)(Ue_i, y) = \sum_i (U^\dagger x, e_i)(e_i, U^\dagger y) = (U^\dagger x, U^\dagger y)$$

$$= (x, UU^\dagger y) = (x, y),$$

where we have noticed that the orthogonalized basis $\{e_i\}$ is complete. This equation demonstrates that the resultant $\{Ue_i\}$ after the isometric transformation is complete. $\qquad\Box$

A simplest example of isometric transformation is the rotation of the Cartesian system in real space. This transformation converts a Cartesian system into another, while the length of any vector in this space remains unchanged. All of the absolute values of the eigenvalues of the transformation matrix are 1.

We have defined in Example 16 in Subsection 1.2 an angle between two vectors. Its cosine is expressed as

$$|\cos\varphi| = \frac{|(x, y)|}{\sqrt{(x, x)}\sqrt{(y, y)}}.$$

Obviously, after the isometric transformation, the angle between any two vectors is preserved. In this sense, it is a conformal transformation.

Definition 10. An operator A is said to be **anti-self-adjoint** if $A^\dagger = -A$.

An anti-self-adjoint transformation is obviously a formal one. The first-order differential operator d/dx in Example 5 is an anti-self-adjoint operator.

Theorem 11. *The eigenvalues of an anti-self-adjoint operator are pure imaginary numbers.*

Proof. Consider $Ax = \lambda x$. Then

$$\lambda^*(x, x) = (\lambda x, x) = (Ax, x) = (x, A^\dagger x) = (x, -Ax) = -\lambda(x, x),$$

which leads to $\lambda = -\lambda^*$. $\qquad\Box$

The momentum operator therefore must include an imaginary factor for it to possess real eigenvalues. Equation (2.2.22) meets the requirement.

Finally, we stress that the discussion with respect to operators above are valid for any linear operators including linear transformations in linear algebra, integral operators, differential operators and so on.

In Chapter 7, we will have further discussions about operators.

2.2.3.   *The alternative theorem for the solutions of linear algebraic equations*

Recall the resolution of linear algebraic equations. Having the concept of an adjoint operator, we are able to comprehensively summarize the conditions of the resolutions. The following alternative theorem clearly expresses the conditions.

Consider an inhomogeneous linear algebraic equations:

$$Ay = b. \tag{2.2.34}$$

where

$$A = \begin{pmatrix} a_{11} & a_{12} & \cdots & a_{1n} \\ a_{21} & a_{22} & \cdots & a_{2n} \\ \vdots & \vdots & \ddots & \vdots \\ a_{n1} & a_{n2} & \cdots & a_{nn} \end{pmatrix} \tag{2.2.35a}$$

and

$$y = (y_1, y_2, \ldots, y_n)^T, \quad b = (b_1, b_2, \ldots, b_n)^T. \tag{2.2.35b}$$

This is in fact a linear transformation which converts the vector y into vector b with A being the representation matrix of this transformation. If the vector b is replaced by zero, it becomes homogeneous linear equations:

$$Au = 0. \tag{2.2.36}$$

Its adjoint is achieved by taking the Hermitian conjugate of A:

$$A^\dagger v = 0. \tag{2.2.37}$$

As long as A is not self-adjoint, $A^\dagger \neq A$, then usually $u \neq v$.

We have known in linear algebra that if the determinant of the matrix A is not zero, the homogeneous equations (2.2.36) have only zero solution, and so does (2.2.37) because $\det A^\dagger = (\det A)^*$. As a result, Eqs. (2.2.34) have only one solution: $y = A^{-1}b$.

What about the case of $\det A = 0$? In this case, homogeneous equations (2.2.36), as well as its adjoint (2.2.37), have nonzero solutions. Then, do not Eqs. (2.2.34) have nonzero solutions? The answer is that it cannot be certain. If the inhomogeneous term b in (2.2.34) is orthogonal to the solution v in the adjoint equations (2.2.37):

$$(b, v) = 0, \qquad (2.2.38)$$

then Eqs. (2.2.34) may have nonzero solutions. This can be understood by the following inner product:

$$(Ay, v) - (y, A^\dagger v) = (b, v) = 0,$$

where (2.2.34) and (2.2.37) have been employed. By the definition of adjoint operators. The left hand side of this equation is necessarily zero. Thus we get (2.2.38).

Note that if (2.2.37) have k linearly independent solutions v_i, $i = 1, 2, \ldots, k$, then (2.2.38) means that b should be orthogonal to every v_i:

$$(b, v_i) = 0, \quad i = 1, 2, \ldots, k. \qquad (2.2.39)$$

The discussion above is summarized as the following theorem.

Theorem 12. *The conditions for inhomogeneous linear equations* (2.2.34) *to have solutions are that either the determinant of the coefficient matrix* A *is nonzero,* $\det A \neq 0$, *or* $\det A = 0$ *but the inhomogeneous term* b *is orthogonal to all of the linearly independent solutions of the homogeneous adjoint equations* (2.2.37).

Because the conditions are to choose one in the two cases, this theorem is also called the **alternative theorem.**

Equations (2.2.39) *are called the* **compatible condition** *for the inhomogeneous equations to have nonzero solutions.*

Up to now we discussed the alternative theorem for resolution of linear algebraic equations. There are also corresponding alternative theorems for the resolution of differential and integral equations, which will be introduced in Chapters 3 and 8, respectively.

2.3. Complete Set of Orthonormal Functions

2.3.1. *Three kinds of convergences*

The concept of the completeness of function sets depends on that of convergence. First of all, we introduce the definitions of three convergences.

Definition 1. Consider a function sequence $\{h_n(x)\}$ where the functions are defined on $[a, b]$. This sequence is said to **converge in the mean-square sense** to function $h(x)$ on $[a, b]$ If

$$\lim_{n \to \infty} \int_a^b |h(x) - h_n(x)|^2 \mathrm{d}x = 0. \qquad (2.3.1)$$

That is to say, for any given ε there is associated an integer $N(\varepsilon)$ which makes

$$\int_a^b |h(x) - h_n(x)|^2 \mathrm{d}x < \varepsilon \qquad (2.3.2)$$

when $n > N$.

Definition 2. A function sequence $\{h_n(x)\}$ is defined on $[a, b]$. This sequence is said to **converge in the pointwise sense** to function $h(x)$ on $[a, b]$. If for any $x \in [a, b]$ and $\varepsilon > 0$, there is associated an integer $N(x, \varepsilon)$ which makes

$$|h(x) - h_n(x)| < \varepsilon \qquad (2.3.3)$$

when $n > N$.

Definition 3. A function sequence $\{h_n(x)\}$ defined on $[a, b]$ is said to **converge in the uniform sense** to $h(x)$ if for any $\varepsilon > 0$, there is associated an integer $N(\varepsilon)$ independent of x, which makes

$$|h(x) - h_n(x)| < \varepsilon \qquad (2.3.4)$$

for all $x \in [a, b]$ when $n > N$.

Equation (2.3.4) is in fact the same as the definition of ε-proximity of zero order in Chapter 1.

We have known in Subsection 2.1.1 that any convergence and limit imply that a distance has been defined. Here, (2.3.2)–(2.3.4) actually have implicitly defined three distances. Each convergence corresponds to a kind of distance. The distances defined by (2.3.3) and (2.3.4) appear the same form. In spite of this they have different implications.

In the three definitions above, the function $h_n(x)$ itself can be a partial sum of another sequence: $h_n(x) = \sum_{i=1}^{n} k_i(x)$. For example, if $\lim_{n \to \infty} \int_a^b |h(x) - \sum_{i=1}^{n} k_i(x)|^2 \mathrm{d}x = 0$, the sequence $\sum_{i=1}^{\infty} k_i(x)$ converges to $h(x)$ in the mean-square sense. The integer $N(x, \varepsilon)$ depends on x in pointwise convergence, but the integer $N(\varepsilon)$ does not in uniform convergence. The latter has a stronger convergence condition than the former.

There are relationships between different convergences. The uniform convergence necessarily leads to pointwise convergence, as well as leads to mean-square convergence.

For the uniform convergence, there is the following theorem.

Theorem 1. *The function sequence $\{h_n(x)\}$ is uniformly convergent on $[a, b]$ if for any $\varepsilon > 0$, there is associated an integer $N(\varepsilon)$ which makes $|h_r(x) - h_s(x)| < \varepsilon$ for all $r > N$, $s > N$ and $x \in [a, b]$.*

Referring to (2.1.3), it is known that as long as a distance $\rho(h_r(x), h_s(x)) = |h_r(x) - h_s(x)|$ is defined, the sequence $\{h_n(x)\}$ is a Cauchy sequence and it is uniformly convergent.

If the function $h_n(x)$ itself is the partial sum of another sequence, $h_n(x) \equiv \sum_{i=1}^{n} k_i(x)$, one has

$$|h_r(x) - h_s(x)| = \left| \sum_{i=1}^{r} k_i - \sum_{i=1}^{s} k_i \right| = \left| \sum_{i=r+1}^{s} k_i(x) \right| < \varepsilon.$$

If it is uniformly convergent or pointwise convergent, we can write

$$h(x) = \lim_{n \to \infty} h_n(x) = \sum_{i=1}^{\infty} k_i(x).$$

From Subsection 2.1.1, it is seen that based on the concepts of convergence and limit, one is able to introduce the concept of space completeness. Similarly, having the concepts of convergence and limit of function sets, one is able to introduce the concept of the completeness of function sets or function spaces.

2.3.2.   *The completeness of a set of functions*

Definition 4. A function set $\{f_n\}$ defined on $[a, b]$ is said to be **orthonormalized** if

$$(f_n, f_m) \equiv \int_a^b f_n^*(x) f_m(x) w(x) \mathrm{d}x = \delta_{nm},$$

where $w(x)$ is a nonnegative **weight function** on $[a, b]$, then $\{f_n\}$ is said to be **orthonormalized with respect to weight function** $w(x)$, or in short **orthonormalized**.

Example 1. The Fourier function set defined on $[-\pi, \pi]$, $f_n(x) = \frac{e^{inx}}{\sqrt{2\pi}}$, $(n = 0, \pm 1, \pm 2, \ldots)$, is orthonormalized on $[-\pi, \pi]$. It is easy to verify that

$$(f_n, f_m) = \int_{-\pi}^{\pi} f_n^*(x) f_m(x) \mathrm{d}x$$

$$= \frac{1}{2\pi} \int_{-\pi}^{\pi} e^{i(m-n)x} \mathrm{d}x = 0, \quad \text{as } m \neq n$$

and

$$(f_n, f_n) = \int_{-\pi}^{\pi} f_n^*(x) f_n(x) \mathrm{d}x = \frac{1}{2\pi} \int_{-\pi}^{\pi} 1 \mathrm{d}x = 1.$$

Thus $(f_n, f_m) = \delta_{mn}$.

The concept of mean-square convergence can be used to define the completeness of an orthogonal function set.

Definition 5. Suppose that in a Hilbert space, there is any function $g(x)$ and an orthonormalized function set $\{f_i(x)\}$. If there is a constant series $\{a_i\}$ which makes the partial sum sequence $g_n(x) \equiv \sum_{i=1}^{n} a_i f_i(x)$ mean-square converge to $g(x)$, the set $\{f_i(x)\}$ is said to be a **complete and orthonormalized set**.

Equivalently, if the following mean-square error can be arbitrarily small:

$$\lim_{n \to \infty} \int_a^b |g - g_n|^2 \, \mathrm{d}x = \lim_{n \to \infty} \int_a^b \left| g - \sum_{i=1}^{n} a_i f_i \right|^2 \mathrm{d}x = 0,$$

the set $\{f_i\}$ is a complete orthonormalized one. It is required that the coefficients $\{a_i\}$ are independent of n.

Equation $\lim_{n\to\infty} \int_a^b |f - \sum_{i=1}^{n} a_i f_i|^2\, \mathrm{d}x = 0$ is also expressed in a symbolic way:

$$f(x) \doteq \sum_{i=1}^{\infty} a_i f_i(x). \qquad (2.3.5)$$

The point added to the equal sign means the sense of mean-square convergence.

Definition 6. Suppose that in a Hilbert space, there is an orthonormalized function set $\{f_i(x)\}$. Let $f(x)$ be any function in this space. If a sequence $\sum_{i=1}^{\infty} c_i f_i(x)$ can uniformly converge to $f(x)$, i.e.,

$$f(x) = \sum_{i=1}^{\infty} c_i f_i(x), \qquad (2.3.6)$$

the coefficients c_i are called **generalized Fourier coefficients** or **expansion coefficients**, and also called the projected components of $f(x)$ on basis functions $f_i(x)$. Equation (2.3.6) is called the **generalized Fourier expansion** of $f(x)$. The expansion coefficients are easily calculated by

$$(f_n, f) = \sum_{i=1}^{\infty} c_i(f_n, f_i) = \sum_{i=1}^{\infty} c_i \delta_{ni} = c_n.$$

Please note that when defining the completeness of function sets, the condition of mean-square convergence is required, while when defining the generalized Fourier expansion, uniform convergence is required. That is to say, the completeness only does not guarantee to implement generalized Fourier expansion. The completeness (2.3.5) of a function set $\{f_i(x)\}$ does not imply (2.3.6). Conversely, uniform convergence implies mean-square convergence, i.e., (2.3.6) guarantees (2.3.5). That any function $f(x)$ can be expanded by $\{f_i(x)\}$ indicates the completeness of $\{f_i(x)\}$.

Consider a nonnegative quantity

$$M_n = \int_a^b \left| f(x) - \sum_{i=1}^{n} a_i f_i(x) \right|^2 \mathrm{d}x \geq 0,$$

where $\{f_i(x)\}$ is an orthonormalized set and $f(x)$ is any function in Hilbert space. A question is raised of how to choose coefficients a_i for a given $n(\geq 1)$ which can make the mean-square error M_n as small as possible. To answer this question, we proceed to expand M_n as follows.

$$M_n = \left(f - \sum_{i=1}^{n} a_i f_i, f - \sum_{i=1}^{n} a_i f_i \right)$$

$$= (f, f) - \sum_{i=1}^{n} a_i c_i^* - \sum_{i=1}^{n} a_i^* c_i - \sum_{i,j=1}^{n} a_i^* a_j \delta_{ij},$$

where $c_i = (f_i, f)$. Supplementary with positive and negative $\sum_{i=1}^{n} |c_i|^2$, we obtain

$$M_n = (f, f) + \sum_{i=1}^{n} |a_i - c_i|^2 - \sum_{i=1}^{n} |c_i|^2.$$

Evidently, taking $a_i = c_i, (i = 1, 2, \ldots, n)$ will make M_n be a minimum, and

$$M_n = (f, f) - \sum_{i=1}^{n} |c_i|^2 \geq 0.$$

This means that

$$(f, f) \geq \sum_{i=1}^{n} |c_i|^2 = \sum_{i=1}^{n} |(f_i, f)|^2.$$

Denoting $s_n = \sum_{i=1}^{n} |c_i|^2$, the sequence $s_1, s_2, \ldots, s_n, \ldots$ is monotonic and is of an upper bound (f, f). Because f is mean-square-integrable, when n goes to infinity, it becomes

$$(f, f) \geq \sum_{i=1}^{\infty} |c_i|^2. \tag{2.3.7}$$

The infinite series $\sum_{i=1}^{n} |c_i|^2$ is convergent. Equation (2.3.7) is the Bessel inequality in infinite-dimensional spaces, similar to (2.1.15) in finite-dimensional spaces.

An orthonormalized function set $\{f_i\}$ is complete when and only when there exists a constant set $\{a_i\}$ which makes $\lim_{n\to\infty} M_n = 0$, as manifested by (2.3.5). If $\{f_i\}$ is complete and for any f the equal sign in (2.3.7) is valid, then $a_i = c_i, (i = 1, 2, \ldots)$ and

$$(f, f) = \sum_{i=1}^{\infty} |c_i|^2 = \sum_{i=1}^{\infty} |(f_i, f)|^2. \tag{2.3.8}$$

This equation is called the **completeness relation**.

Similar to the finite-dimensional case, in infinite-dimensional spaces, there is also a Parseval relation

$$(f, g) = \sum_{i=1}^{\infty} (f, f_i)(f_i, g). \tag{2.3.9}$$

Its proof is left for exercises.

Definition 7. An orthonormalized function set is said to be **closed** if there is no one nonzero function that is orthogonal to everyone in the set.

Theorem 2. *An orthonormalized function set in a Hilbert space is complete when and only when it is closed.*

2.3.3. *N-dimensional space and Hilbert function space*

Any vector in an N-dimensional number space C^N is a group of N numbers, while any vector in a Hilbert function space is a function $f(x)$. Both belong to linear spaces. When inner products are defined in respective spaces, both belong to inner product spaces. In both spaces, the definition and concept with respect to linear spaces discussed above apply. Most of the formulas are one-to-one correspondent, such as vector basis, expansion, inner product, orthonormalization, modulus, completeness, Gram-Schmidt orthonormalization method, Bessel inequality, Parseval equality, and so on. In Table 2.2, we list side by side the concepts and definitions in an N-dimensional number space C^N and in a Hilbert spaces.

As a matter of fact, in a Hilbert space, any function can be expanded by a complete basis set. This complete function set must be linearly independent, but may not be orthogonal. This is similar

Table 2.2. The concepts and definitions in an N-dimensional number space C^N and in a Hilbert spaces. In C^N the vectors are expressed by bold letters and the inner product is denoted by a dot $\cdot$.

	Number domain C^N	Hilbert spaces
Dimension	N	N (can be infinite)
Vector set	$\{\boldsymbol{x}_1, \boldsymbol{x}_2, \ldots, \boldsymbol{x}_m\}$	$\{\varphi_1(x), \varphi_2(x), \ldots, \varphi_m(x)\}$
Linear dependence (or independence)	For a vector group $\{\boldsymbol{x}_1, \boldsymbol{x}_2, \ldots, \boldsymbol{x}_m\}$ and $\sum_i \lambda_i \boldsymbol{x}_i = 0$, when $\lambda_1 = \lambda_2 = \cdots = \lambda_m = 0$, the vector group is linearly independent; or it is linearly dependent.	For a function group $\{\varphi_1(x), \varphi_2(x), \ldots, \varphi_m(x)\}$ and $\sum_i a_i \varphi_i(x) = 0$, when $a_1 = a_2 = \cdots = a_m = 0$, the function group is linearly independent; or it is linearly dependent.
Basis	Basis vectors $\{\boldsymbol{e}_i\}$	Basis functions $\{f_i(x)\}$
Expansion of a vector	By a vector basis $\boldsymbol{x} = \sum_i x_i \boldsymbol{e}_i$	By a basis function set $f(x) = \sum_{i=1}^{N} c_i f_i(x)$
Inner product	$\boldsymbol{x} \cdot \boldsymbol{y} = \sum_i x_i^* y_i$	$(f_1, f_2) = \int_a^b f_1^*(x) f_2(x) \mathrm{d}x$
Weighted inner product	$\boldsymbol{x} \cdot \boldsymbol{y} = \sum_i p_i x_i^* y_i \; p_i > 0,$ $(i = 1, 2, \ldots, n)$	(f_1, f_2) $= \int_a^b f_1^*(x) f_2(x) w(x) \mathrm{d}x$ $w(x) > 0$
Orthonomal basis	$\boldsymbol{e}_i \cdot \boldsymbol{e}_j = \delta_{ij}$	$\int_a^b f_n^*(x) f_m(x) w(x) \mathrm{d}x$ $= \delta_{nm}$
Expansion coefficients	$x_i = \boldsymbol{x} \cdot \boldsymbol{e}_i$	$c_i = \int_a^b f^*(x) f_i(x) w(x) \mathrm{d}x$
Square modulus	$\|\boldsymbol{x}\|^2 = \boldsymbol{x} \cdot \boldsymbol{x} = \sum_{i=1}^{N} |x_i|^2$	$\|f\|^2 = (f, f) = \sum_{i=1}^{N} |c_i|^2$
Zero vector	$x_i = 0, \; (i = 1, 2, \ldots, N)$ $\Leftrightarrow \boldsymbol{x} = 0$	$c_i = 0, \; (i = 1, 2, \ldots, N)$ $\Leftrightarrow f(x) = 0$

(*Continued*)

Table 2.2. (*Continued*)

	Number domain C^N	Hilbert spaces
Bessel inequality	$\|x\|^2 \geq \sum_{i=1}^{n} |x_i|^2$	$(f,f) \geq \sum_{i=1}^{\infty} |c_i|^2$
Parseval relation	$(x,y) = \sum_{i=1}^{n} (x,e_i)(e_i,y)$	$(f,g) = \sum_{i=1}^{\infty} (f,f_i)(f_i,g)$

to the case in space C^N where a vector can be expanded by a non-orthogonal basis. In a Hilbert space, if the complete basis set is not orthogonal, it will be comparatively more difficult to calculate the expansion coefficients. In this case, it may be better to construct an orthonormalized set by means of Gram-Schmidt process.

A vector in a Hilbert space may also be a multivariate function. It then can be expanded in the same way. For example, a binary function is expanded by

$$f(x,y) = \sum_{i,j=1}^{N} c_i d_j f_i(x) g_j(y).$$

The coefficients compose a tensor.

It is desirable in practical application to find the necessary complete function set. In Chapter 3, we will see that the eigenfunctions of ordinary differential equations of second order are able to provide the complete orthonormalized function sets.

Because Hilbert spaces are complete inner product spaces, the operator theory in inner product spaces introduced in Section 2.2 applies to Hilbert function spaces.

2.3.4. *Orthogonal polynomials*

1. Fundamental properties of orthogonal polynomials

Among almost all functions, polynomials are formally the simplest. They have plain and useful properties. Here are some theorems concerning polynomials.

Definition 8. Suppose that $\rho(x)$ is a weight function on $[a, b]$ and $\varphi_n(x)$ is an n-degree polynomial on $[a, b]$ with the coefficient of the higher power $a_n \neq 0$. The $\varphi_n(x)$ is said to be an **n-degree weighted orthogonal polynomial with weight** $\rho(x)$ if any two in the polynomial sequence $\{\varphi_n(x)\}$ are orthogonal to each other on $[a, b]$ with weight $\rho(x)$:

$$(\varphi_n, \varphi_m) = \int_a^b \varphi_n(x)\varphi_m(x)\rho(x)\mathrm{d}x = h_m \delta_{nm}. \qquad (2.3.10)$$

Theorem 3. (i) *Orthogonal polynomials $\varphi_0(x), \varphi_1(x), \ldots, \varphi_n(x)$ are linearly independent of each other;* (ii) *any n-degree polynomial $P_n(x)$ can be expressed by their linear combination;* (iii) *$\varphi_n(x)$ is orthogonal to any polynomial of degree less than n.*

The proof of this theorem is left to be an exercise.

Theorem 4. *Polynomial $\varphi_n(x)$ of degree $n(\geq 1)$ on $[a, b]$ has n different real roots, all being nondegenerate and in the interval (a, b).*

Proof. Because

$$(\varphi_n, \varphi_0) = \int_a^b \varphi_n(x)\varphi_0(x)\rho(x)\mathrm{d}x = 0, \quad n \geq 1$$

and $\rho(x) > 0$, $\varphi_n(x)$ necessarily changes its sign within (a, b). So $\varphi_n(x)$ certainly has roots in (a, b).

Suppose that the roots of $\varphi_n(x)$ in (a, b) are $x_1, x_2, \ldots, x_k$. Then $\varphi_n(x)$ can be written in the form of

$$\varphi_n(x) = (x - x_1)^{\gamma_1}(x - x_2)^{\gamma_2} \cdots (x - x_k)^{\gamma_k}.$$

When x varies across each x_i, the function must change its sign, so that $\gamma_1, \gamma_2, \ldots, \gamma_k$ must all be odd. We are to prove that there must be $k = n$, so that $\gamma_1 = \gamma_2 = \cdots = \gamma_k = 1$. Let

$$q(x) = (x - x_1)(x - x_2) \cdots (x - x_k),$$

which is a k-degree polynomial. When $k < n$, it must be

$$(\varphi_n, q) = \int_a^b \varphi_n(x)q(x)\rho(x)\mathrm{d}x = 0. \qquad (2.3.11)$$

Because $\gamma_1 = \gamma_2 = \cdots = \gamma_k = 1$ in $q(x)$, the integrand in (2.3.11) contains even powers of every $(x - x_i)$. Thus the integrand is always positive. Consequently, it must be

$$(\varphi_n, q) = \int_a^b \varphi_n(x) q(x) \rho(x) \mathrm{d}x > 0.$$

This is in contradiction to (2.3.11). Thus $k = n$. ☐

In the following two theorems, the coefficient of the highest power term of $\varphi_n(x)$ is a_n, and h_n is determined by (2.3.10).

Theorem 5. *There are recurrence relations for any three adjacent polynomials as follows:*

$$\varphi_{n+1}(x) = (\alpha_n x - \beta_n)\varphi_n(x) - \gamma_{n-1}\varphi_{n-1}(x), \qquad (2.3.12)$$

where $\alpha_n, \beta_n, \gamma_{n-1}$ are constants independent of x.

Proof. Let

$$\varphi_{n+1}(x) - \frac{a_{n+1}}{a_n} x \varphi_n(x) = \sum_{i=1}^{n} A_i \varphi_i(x).$$

Multiplying $\varphi_m(x)$ on the two sides and making weighted inner products lead to

$$(\varphi_{n+1}, \varphi_m) - \frac{a_{n+1}}{a_n}(\varphi_n, x\varphi_m) = \sum_{i=1}^{n} A_i(\varphi_i, \varphi_m).$$

As $m \leq n - 2$, the first term on the left hand side is evidently zero. The second term is also zero, because the $x\varphi_m(x)$ is a polynomial of degree less than n and is orthogonal to $\varphi_n(x)$ with the weight. Consequently, the right hand side must also be zero. Because

$$\sum_{i=1}^{n} A_i(\varphi_i, \varphi_m) = \sum_{i=1}^{n} A_i \delta_{mi}(\varphi_m, \varphi_m) = A_m h_m,$$

$A_m = 0, m \leq n - 2$. When m is respectively $m = n - 1$ and n,

$$A_{n-1}(\varphi_{n-1}, \varphi_{n-1}) = -\frac{a_{n+1}}{a_n}(\varphi_n, x\varphi_{n-1})$$

and

$$A_n(\varphi_n, \varphi_n) = -\frac{a_{n+1}}{a_n}(\varphi_n, x\varphi_n).$$

It follows that

$$\varphi_{n+1}(x) - \frac{a_{n+1}}{a_n}x\varphi_n(x) = A_{n-1}\varphi_{n-1}(x) + A_n\varphi_n(x).$$

This is (2.3.12), where

$$\alpha_n = \frac{a_{n+1}}{a_n}, \quad \gamma_{n-1} = -A_{n-1} = \frac{a_{n+1}}{a_n}\frac{(\varphi_n, x\varphi_{n-1})}{(\varphi_{n-1}, \varphi_{n-1})},$$

$$\beta_n = -A_n = \frac{a_{n+1}}{a_n}\frac{(\varphi_n, x\varphi_n)}{(\varphi_n, \varphi_n)}. \tag{2.3.13}$$

They are all constants independent of x.   □

The constant γ_{n-1} can also be recast in another form:

$$\gamma_{n-1} = \frac{a_{n+1}a_{n-1}}{a_n^2}\frac{h_n}{h_{n-1}}, \tag{2.3.14}$$

where a_{n-1} is the coefficient of the highest power term of φ_{n-1}.

Theorem 6. *Orthogonal polynomials observe the following formula:*

$$\sum_{m=1}^{n} \frac{1}{h_m}\varphi_m(x)\varphi_m(y) = \frac{a_n}{a_{n+1}h_n}\frac{\varphi_{n+1}(x)\varphi_n(y) - \varphi_n(x)\varphi_{n+1}(y)}{x - y}. \tag{2.3.15}$$

Especially, as $x = y$,

$$\sum_{m=1}^{n} \frac{1}{h_m}\varphi_m^2(x) = \frac{a_n}{a_{n+1}h_n}[\varphi'_{n+1}(x)\varphi_n(x) - \varphi'_n(x)\varphi_{n+1}(x)]. \tag{2.3.16}$$

Proof.   It follows from (2.3.12) that

$$\varphi_{m+1}(x)\varphi_m(y) = (\alpha_m x - \beta_m)\varphi_m(x)\varphi_m(y) - \gamma_{m-1}\varphi_{m-1}(x)\varphi_m(y).$$

Exchanging x and y yields

$$\varphi_{m+1}(y)\varphi_m(x) = (\alpha_m y - \beta_m)\varphi_m(y)\varphi_m(x) - \gamma_{m-1}\varphi_{m-1}(y)\varphi_m(x).$$

Subtraction of the two terms leaves

$$\alpha_m(x - y)\varphi_m(x)\varphi_m(y)$$
$$= \varphi_{m+1}(x)\varphi_m(y) - \varphi_{m+1}(y)\varphi_m(x)$$
$$\quad - \gamma_{m-1}[\varphi_m(x)\varphi_{m-1}(y) - \varphi_m(y)\varphi_{m-1}(x)].$$

Using (2.3.13) and (2.3.14) to remove α_n and γ_{n-1}, we get

$$\frac{a_{m+1}}{a_m}(x-y)\varphi_m(x)\varphi_m(y)$$

$$= \varphi_{m+1}(x)\varphi_m(y) - \varphi_{m+1}(y)\varphi_m(x)$$

$$- \frac{a_{m+1}a_{m-1}}{a_n^2}\frac{h_m}{h_{m-1}}[\varphi_m(x)\varphi_{m-1}(y) - \varphi_m(y)\varphi_{m-1}(x)].$$

After rearrangement, it becomes

$$(x-y)\frac{\varphi_m(x)\varphi_m(y)}{h_m} = \frac{a_m}{a_{m+1}h_m}[\varphi_{m+1}(x)\varphi_m(y) - \varphi_{m+1}(y)\varphi_n(x)]$$

$$- \frac{a_{m-1}}{a_m h_{m-1}}[\varphi_m(x)\varphi_{m-1}(y) - \varphi_m(y)\varphi_{m-1}(x)].$$

Let

$$q_m = \frac{a_m}{a_{m+1}h_m}[\varphi_{m+1}(x)\varphi_m(y) - \varphi_{m+1}(y)\varphi_n(x)].$$

Then

$$(x-y)\frac{\varphi_m(x)\varphi_m(y)}{h_m} = q_m - q_{m-1}.$$

As $q_{m-1} = 0$, making a summation of m from 0 to n results in (2.3.15).

As $x = y$, we have

$$\sum_{m=1}^{n}\frac{1}{h_m}\varphi_m^2(x)$$

$$= \frac{a_n}{a_{n+1}l_{n}}\lim_{y\to x}\frac{[\varphi_{n+1}(x) - \varphi_{n+1}(y)]\varphi_n(y) - [\varphi_n(x) - \varphi_n(y)]\varphi_{n+1}(y)}{x-y}$$

$$= \frac{a_n}{a_{n+1}h_n}[\varphi_{n+1}'(n)\varphi_n(x) - \varphi_n'(x)\varphi_{n+1}(x)].$$

This is just (2.3.16). □

Corollary. For all x, $\varphi_{n+1}'(n)\varphi_n(x) - \varphi_n'(x)\varphi_{n+1}(x) > 0$.

In Chapter 3, we will be able to acquire some commonly used polynomials in terms of the solutions of certain ordinary differential equations under appropriate boundary conditions.

2. Construction of orthogonal polynomials

Theorem 5 enables one to construct intentionally a polynomial set $\{\varphi_n(x)\}$ with weight $\rho(x)$ on $[a, b]$. The procedure is as follows:

$$\varphi_0(x) = 1, \tag{2.3.17a}$$

$$\varphi_1(x) = (x - x_0), \tag{2.3.17b}$$

$$\varphi_{n+1}(x) = (\alpha_n x - \beta_n)\varphi_n(x) - \gamma_{n-1}\varphi_{n-1}(x), \tag{2.3.17c}$$

$$\alpha_n = \frac{a_{n+1}}{a_n}, \quad \beta_n = \frac{a_{n+1}}{a_n}\frac{(\varphi_n, x\varphi_n)}{(\varphi_n, \varphi_n)},$$

$$\gamma_{n-1} = \frac{a_{n+1}a_{n-1}}{a_n^2}\frac{(\varphi_n, \varphi_n)}{(\varphi_{n-1}, \varphi_{n-1})}, \tag{2.3.17d}$$

where

$$(\varphi_n, \varphi_n) = \int_a^b \varphi_n(x)\varphi_n(x)\rho(x)\mathrm{d}x,$$

$$(\varphi_n, x\varphi_x) = \int_a^b \varphi_n(x)\varphi_n(x)x\rho(x)\mathrm{d}x. \tag{2.3.17e}$$

The x_0 value can be reckoned by the orthogonalization of $\varphi_0(x)$ and $\varphi_1(x)$. It follows from $(\varphi_0, \varphi_1) = \int_a^b (x - x_0)\rho(x)\mathrm{d}x = 0$ that $x_0 = \frac{\int_a^b x\rho(x)\mathrm{d}x}{\int_a^b \rho(x)\mathrm{d}x}$.

Explicitly, starting from $n = 1$ and having given the expressions of $\varphi_0(x)$ and $\varphi_1(x)$, $a_0 = a_1 = 1$. Let $a_2 = 1$. Then $\alpha_1 = 1$ from (2.3.17d), β_1 and γ_0 are also computed. Subsequently, $\varphi_2(x)$ is obtained by (2.3.17c). Next, let $a_3 = 1$. Then $\alpha_2 = 1$ from (2.3.17d), β_2 and γ_1 are also computed. Subsequently, $\varphi_3(x)$ is obtained by (2.3.17c). This process continues repeatedly. Theorem 5 has proved that an orthogonal polynomial set has necessarily a recurrence relation (2.3.12). Conversely, it is easily proven that the polynomial set constructed following (2.3.17a)–(2.3.17e) is certainly an orthogonal one.

It has been shown that for any n, $a_n = 1$. Therefore, (2.3.17c) and (2.3.17d) can be simplified to

$$\varphi_{n+1}(x) = (x - \beta_n)\varphi_n(x) - \gamma_{n-1}\varphi_{n-1}(x) \tag{2.3.18a}$$

and

$$\beta_n = \frac{(\varphi_n, x\varphi_n)}{(\varphi_n, \varphi_n)}, \quad \gamma_{n-1} = \frac{(\varphi_n, \varphi_n)}{(\varphi_{n-1}, \varphi_{n-1})}. \tag{2.3.18b}$$

2.4. Polynomial Approximation

2.4.1. *Weierstrass theorem*

Theorem 1 (Weierstrass). *If a function $f(x)$ is continuous on $[a,b]$, then there exists a sequence of polynomials $P_n(x)$ such that it uniformly converges to $f(x)$ on $[a,b]$: $\lim_{n\to\infty} P_n(x) = f(x)$. This is called the* **Weierstrass theorem.**

This theorem demonstrates that one is able to construct a polynomial sequence which can uniformly approximate to any continuous function on $[a,b]$. As a corollary, it can be proved that there certainly exists a complete orthonormalized function set on $[a,b]$. There are various ways to prove the theorem. Here we are content with a rather simple proof.

The idea of the proof is that the function $f(x)$ defined on $[a,b]$ can be transformed, by variable substitution and subtraction of some polynomials, to a new function $h(x)$ defined on interval $[0,1]$, which is of $h(0) = h(1) = 0$. Without loss of generality, one merely needs to prove that $h(x)$ satisfies the theorem on $[0,1]$. The required approximate polynomial is chosen to be the form of $P_n(x) = \int_{-1}^{1} f(x+t)\delta_n(t)\mathrm{d}t$, $0 \le x \le 1$, where $\delta_n(x)$ is a polynomial with some particular properties. Taking variable substitution in $P_n(x)$ and making use of the uniform continuity of $f(x)$, one thus proves the theorem.

Proof. Let $f(x)$ be defined on $[a,b]$. Consider a function defined by $h(z) = h\left(\frac{x-a}{b-a}\right) \equiv f(x)$. Obviously $f(a) = h(0)$ and $f(b) = h(1)$. Any point x in $[a,b]$ is mapped to a point z in $[0,1]$. Thus, if $h(z)$ can be approximated by polynomials of z, then this approximation can be transferred to the polynomial approximation of f, because any polynomial of $z = \frac{x-a}{b-a}$ is necessarily a polynomial of x.

Furthermore, we assume that $h(0) = h(1) = 0$. If not so, we define a new function $g(z) = h(z) - h(0) - z[h(1) - h(0)]$ on $[0,1]$ such that $g(0) = g(1) = 0$. Since $g(z)$ differs from $h(z)$ only by a polynomial, as long as $g(z)$ can be approximated by a polynomial $P_n(x)$, $h(0) + z[h(1) - h(0)]$ can be so by $P_n(x) + h(0) + z[h(1) - h(0)]$.

So, we assume that $f(x)$ is defined on $[0,1]$ and $f(0) = f(1) = 0$. The values of $f(x)$ outside of $[0,1]$ are irrelevant to what we want to prove, and so they are set to be zero for convenience.

Suppose that

$$P_n(x) = \int_{-1}^{1} f(x+t)\delta_n(t)\mathrm{d}t, \quad 0 \le x \le 1, \qquad (2.4.1)$$

where

$$\delta_n(t) = \begin{cases} c_n(1-t^2)^n, & 0 \le |t| \le 1. \\ 0, & |t| > 1. \end{cases} \qquad (2.4.2)$$

We intend to prove that as $n \to \infty$, $\lim_{n\to\infty} P_n(x) = f(x)$.

The function $\delta_n(x)$ defined by (2.4.2) will be discussed in Chapter 5 in detail. Here we merely introduce some of its fundamental properties. Obviously, it is an even function. When taking $c_n = \frac{(2n+1)!}{2^{2n+1}(n!)^2}$, $0 \le \delta_n(x) \le 1$ on $[-1,1]$. The function is normalized on the interval:

$$\int_{-1}^{1} \delta_n(x)\mathrm{d}x = 1. \qquad (2.4.3)$$

For any given positive small number $\gamma(0 < \gamma < 1)$,

$$\int_{\gamma}^{1} \delta_n(x)\mathrm{d}x \le \sqrt{n}(1-\gamma^2)^n, \qquad (2.4.4)$$

where $1 - \gamma^2 = z < 1$. As $n \to \infty$, z^n goes to zero more rapidly than $1/\sqrt{n}$. This is easily proved as follows. As $n \to \infty$, the ratio of z^n and $1/\sqrt{n}$ is zero nil type. Let $y = 1/z > 1$. Then

$$\lim_{n\to+\infty} z^n \sqrt{n} = \lim_{n\to+\infty} \frac{\sqrt{n}}{y^n} = \lim_{n\to+\infty} \frac{1}{2\sqrt{n}y^n \ln y} = 0, \qquad (2.4.5)$$

where both numerator and denominator are taken derivative with respect to n following l'Hospital's rule.

By assumption, outside of $[0,1]$, $f(x)$ is identical to zero. Alternatively, when $t \le -x$ or $t \ge 1 - x$, $f(x+t) \equiv 0$. Thus, (2.4.1) is recast to

$$P_n(x) = \int_{-x}^{1-x} f(x+t)\delta_n(t)\mathrm{d}t.$$

It is brought, after replacement $t \to t - x$, to

$$P_n(x) = \int_0^1 f(t)\delta_n(t - x)\mathrm{d}t = \int_0^1 f(t)c_n[1 - (t - x)^2]^n \mathrm{d}t.$$

This form demonstrates that the coefficients of the powers of x are the definite integrals of t. Therefore, $P_n(x)$ is a polynomial of x of degree $2n$.

The continuity of $f(x)$ on the closed interval $[0, 1]$ implies its uniform continuity. For any given $\varepsilon > 0$, there exists a γ which makes $|f(x + \gamma) - f(x)| < \varepsilon$ for all x in $[0, 1]$. Therefore,

$$|P_n(x) - f(x)| = \left| \int_{-1}^1 [f(x + t) - f(x)]\,\delta_n(t)\mathrm{d}t \right|$$

$$\leq \int_{-1}^1 |f(x + t) - f(x)|\,\delta_n(t)\mathrm{d}t$$

$$= \left(\int_{-1}^{-\gamma} + \int_{-\gamma}^{\gamma} + \int_{\gamma}^1 \right) |f(x + t) - f(x)|\,\delta_n(t)\mathrm{d}t,$$

where the absolute value symbol of function δ_n has been removed as it is nonnegative. Let M be the maximum of $|f(x)|$. Then

$$\int_{\gamma}^1 |f(x + t) - f(x)|\,\delta_n(t)\mathrm{d}t$$

$$\leq \int_{\gamma}^1 [|f(x + t)| + |f(x)|]\delta_n(t)\mathrm{d}t \leq 2Mn^{1/2}(1 - \gamma^2)^n.$$

Another term $\int_{-1}^{-\gamma} |f(x + t) - f(x)|\,\delta_n(t)\mathrm{d}t$ is evaluated in the same way. By the continuity of $f(x)$ it is seen that when $|t| < \gamma$, $|f(x + t) - f(x)| < \varepsilon/2$. This leads to

$$\int_{-\gamma}^{\gamma} |f(x + t) - f(x)|\,\delta_n(t)\mathrm{d}t < \frac{\varepsilon}{2} \int_{-\gamma}^{\gamma} \delta_n(t)\mathrm{d}t \leq \frac{\varepsilon}{2}.$$

In summary,

$$|P_n(x) - f(x)| < 4Mn^{1/2}(1 - \gamma^2)^n + \frac{\varepsilon}{2}.$$

From (2.4.5), as n is sufficiently large, $n^{1/2}(1 - \gamma^2)^n$ can be arbitrarily small, especially smaller than $\varepsilon/2$. Therefore, there exists an N such that as $n > N$, for a given small ε, $|P_n(x) - f(x)| < \varepsilon$. That is to say, $\lim_{n\to\infty} |P_n(x) - f(x)| = 0$, or $P_n(x)$uniformly converges to the function $f(x)$ on $[0,1]$. □

2.4.2. *Polynomial approximation*

Weierstrass theorem enables one to adopt a polynomial that is approximate to a function, and to use the polynomial as an alternative to calculate the values of the function.

Recall Taylor expansion of a function, which is equivalent to using a special polynomial to approximate the function. The comparisons of Taylor expansion and Weierstrass theorem of polynomial approximation are listed in Table 2.3.

Taylor expansion is convergent in a definite range. For polynomial approximation, Weierstrass theorem reveals that in the convergence range, the approximation certainly exists. Outside of the range, the theorem does not assert if the polynomial approximation applies or not. Therefore, in the outer range, one may still use the polynomial approximation. In the following, only the approximation in the convergence range is considered.

Taylor expansion has infinite terms. For example, the expansion of e^{-x} at point $x = 0$ is

$$e^{-x} = 1 - x + \frac{1}{2}x^2 - \frac{1}{3}x^3 + \cdots + (-1)^n \frac{x^n}{n!} + \cdots.$$

Very near $x = 0$, the series converges rapidly, and the first few terms are sufficient to calculate the values of the function. With the x

Table 2.3. The comparisons of Taylor expansion and Weierstrass theorem.

	Taylor expansion	Weierstrass theorem
The conditions that the function should meet	All-order derivatives exist, or "analytical"	The function is continuous in the closed interval
Convergence range	Radius of convergence can be either finite or infinite	Finite interval
Application range	Only applicable to convergence range	Outside of convergence range, the polynomial approximation may also applies

growing up, the series' convergence slows down. To reach the required precision, one has to take many terms in the series, which brings difficulty for practical computation.

For the polynomial approximation proved by the Weierstrass theorem, the highest power of the polynomial approaches to infinite. Of course, a polynomial including infinite terms is not practical. We hope that for a function $f(x)$, we are able to find a polynomial $P_n(x)$ with its order n being not too high such that its error with $f(x)$ at each point x may not be accurate but in the whole interval the error distribution of $f(x)$ is comparatively uniform. This involves how to measure the error. Below, the degree n of polynomial $P_n(x)$ is finite.

Definition 1. Suppose that a polynomial

$$P_n(x) = \sum_{k=0}^{n} a_k x^k$$

of degree n approximates a given function $f(x)$ on $[a, b]$. $|f(x) - P_n(x)|$ is called the **absolute error** of $f(x)$.

Apparently, absolute error varies with x in the interval. We now introduce two concepts that assess the error in the whole interval.

1. Uniform approximation

Definition 2. Suppose that a function $f(x)$ is continuous on $[a, b]$. It is said that an n-degree polynomial $P_n(x)$ **uniformly approximates** to $f(x)$ on $[a, b]$ if for any given $\varepsilon > 0$,

$$|f(x) - P_n(x)| \leq \varepsilon$$

holds everywhere in the interval $[a, b]$. (This definition is similar to that of the concept of ε-proximity in Chapter 1.) In other words, the absolute value of the error between $f(x)$ and $P_n(x)$ at any point within $[a, b]$ is less than a required precision ε. The value of ε is called **deviation**. If at $x = \xi$, the difference between $P_n(x)$ and $f(x)$ is the largest, i.e.,

$$\max_{x \in [a,b]} |f(x) - P_n(x)| = |f(x) - P_n(x)|_{x=\xi} = \varepsilon,$$

then $x = \xi$ is called the **deviation point**. It may be called the
positive deviation point or **negative deviation point** depending
on the sign of $[f(x) - P_n(x)]_{x=\xi}$.

This definition measures the uniform approximation by deviation
ε. There may be many polynomials that can uniformly approximate
$f(x)$. Nevertheless, different polynomials may have different approx-
imation speed.

Definition 3. Let n be fixed and polynomial $P_n(x)$ vary. Then

$$E_n = \min_{P_n} \left\{ \max_{x \in [a,b]} |f(x) - P_n(x)| \right\} = \max_{x \in [a,b]} |f(x) - P_n^*(x)|$$

is called the **least deviation** of $f(x)$. $P_n^*(x)$ which reaches the least
deviation is called the **optimal uniform approximation polynomial** of $f(x)$.

For the problem of optimal uniform approximation polynomial,
Chebyshev proved the following four theorems.

Theorem 2. *When n is fixed, in all of the n-degree polynomials,
there is one and only one that is the optimal uniform approximation
polynomial of function $f(x)$.*

Theorem 3. *Assume that the deviation ε is given. Among all of
the polynomials that uniformly approximate the function $f(x)$, the
optimal one has the lowest degree. The lower the deviation, the higher
the degree of the optimal uniform approximation polynomial.*

Theorem 4. *The necessary and sufficient conditions for $P_n^*(x)$ to be
the optimal uniform approximation polynomial of continuous function $f(x)$ on $[a, b]$ are that $P_n^*(x) - f(x)$ has at least $n + 2$ positive
and negative staggered deviation points $\xi_i (i = 1, 2, \ldots, n+2)$ on $[a, b]$.
These points are called* **Chebyshev staggered points.**

Theorem 4 provides theoretically an approach to construct an
optimal uniform approximation polynomial. Suppose that $f(x)$ is a
continuous function on $[a, b]$ and has an optimal uniform approximation polynomial

$$P_n^*(x) = \sum_{k=0}^{n} a_k^* x^k.$$

There are $2n + 4$ numbers to be found in determining the desired polynomial. They include $n + 1$ coefficients $a_i^*(i = 0, 1, 2, \ldots, n)$ of $P_n^*(x)$, the least deviation E_n and $n + 2$ deviation points $a \leq x_1^* < x_2^* < \cdots < x_{n+2}^* \leq b$. These numbers should satisfy the following equations:

$$\begin{cases} (f(x_k^*) - P_n^*(x_k^*))^2 = E_n^2, \\ (x_k^* - a)(x_k^* - b)(f'(x_k^*) - P_n^{*\prime}(x_k^*)) = 0, \end{cases} \quad (k = 0, 1, 2, \ldots, n+2).$$

Definition 4. The polynomial defined by

$$T_n(x) = \cos(n \cos^{-1} x), \quad (-1 \leq x \leq 1) \tag{2.4.6}$$

is called a **Chebyshev polynomial**. $T_n(x)$ is an n-degree polynomial, and its coefficient of the highest power term is 2^{n-1}.

From the definition (2.4.6), it is easily seen that $T_n(x)$ is of the following properties on the interval $[-1, 1]$.

(i) The absolute values of $T_n(x)$ are less than or equal to 1.

$$|T_n(x)| \leq 1. \tag{2.4.7}$$

(ii) $T_n(x)$ has $n + 1$ extreme points:

$$x_l = \cos \frac{l\pi}{n}, \quad (l = 0, 1, 2, \ldots, n). \tag{2.4.8}$$

The values of $T_n(x)$ at these points are

$$T_n(x_l) = T_n \left(\cos \frac{l\pi}{n} \right) = (-1)^l, \quad (l = 0, 1, 2, \ldots, n). \tag{2.4.9}$$

It is obvious that these $n + 1$ extremes are positively and negatively staggered, so that there are n zeros between them.

(iii) $T_n(x)$ is orthonormalized with respect to the weight function $w(x) = 1/\sqrt{1 - x^2}$.

$$(T_m(x), T_n(x)) = \int_{-1}^{1} \frac{T_m(x)T_n(x)}{\sqrt{1 - x^2}} dx = \delta_{mn} \tag{2.4.10}$$

(iv) Any n-degree polynomial $P_n(x) \in H_{n+1}$ can be expanded by a Chebyshev polynomial in the following form:

$$P_n(x) = \sum_{l=0}^{n} c_l T_l(x), \qquad (2.4.11)$$

where coefficients c_l are determined by the inner products

$$c_l = (P_n(x), T_l(x)) = \int_{-1}^{1} \frac{P_n(x) T_l(x)}{\sqrt{1 - x^2}} dx, \quad (l = 0, 1, 2, \ldots, n).$$

$$(2.4.12)$$

Usually, an n-degree polynomial with its coefficient of the highest power term being 1 is called a **monic polynomial**. The n-degree **Chebyshev monic polynomial** is denoted by $\tilde{T}_n(x)$, and can be easily obtained by

$$\tilde{T}_n(x) = \frac{1}{2^{n-1}} T_n(x). \qquad (2.4.13)$$

Theorem 5. *Among all of the n-degree monic polynomials defined on $[-1, 1]$, $\tilde{T}_n(x)$ has the smallest deviation with respect to a zero function, and the deviation is $\frac{1}{2^{n-1}}$. $\tilde{T}_n(x)$ is the optimal uniform approximation polynomial on this interval.*

Note the difference between Theorems 5 and 4. Theorem 4 says that $P_n^*(x) - f(x)$ should have $n + 2$ positive and negative staggered deviation points. Now $\tilde{T}_n(x)$ is an n-degree polynomial and has $n + 1$ positive and negative staggered deviation points in $[-1, 1]$.

We do not present the proofs of the above four theorems. We merely mention that to prove Theorem 5 one can construct a polynomial by $P_{n-1}^*(x) = x^n - \tilde{T}_n(x)$. Subsequently, by

$$x^n - P_{n-1}^*(x) = \tilde{T}_n(x), \qquad (2.4.14)$$

it is known that if $\tilde{T}_n(x)$ is the optimal uniform approximation polynomial of a zero function on $[-1, 1]$, then the polynomial $P_{n-1}^*(x)$ defined by (2.4.14) is the optimal uniform approximation polynomial of function x^n on $[-1, 1]$. Thus, Theorem 5 is equivalent to the following theorem.

Theorem 5'. Among all the $n-1$-degree polynomials $P_{n-1}(x)$ defined on $[-1, 1]$, the $P_{n-1}^*(x)$ defined by (2.4.14) has the smallest deviation relative to x^n, and the deviation is $\frac{1}{2^{n-1}}$. This $P_{n-1}^*(x)$ is the optimal uniform approximation polynomial of function x^n.

Proof. It follows from definition (2.4.6) and property (2.4.9) that $x^n - P_{n-1}^*(x) = \tilde{T}_n(x)$ has $n+1$ deviation points on $[-1, 1]$, and the deviations at these points are

$$\left|\tilde{T}_n(x_l)\right| = \left|\tilde{T}_n\left(\cos\frac{l\pi}{n}\right)\right| = \frac{1}{2^{n-1}}, \quad (l = 0, 1, 2, \dots, n), \quad (2.4.15)$$

which are positively and negatively staggered. $x^n - P_{n-1}^*(x)$ has $n+1$ positive and negative staggered deviation points in $[-1, 1]$. Therefore, by Theorem 4, $P_{n-1}^*(x) = x^n - \tilde{T}_n(x)$ is the optimal uniform approximation polynomial of x^n on $[-1, 1]$. □

Since now the polynomial $P_{n-1}^*(x)$ defined by (2.4.14) is of degree $n-1$, $x^n - P_{n-1}^*(x)$ should have $n+1$ positive and negative staggered deviation points. This agrees with Theorem 4.

From the above discussion, the optimal uniform approximation polynomials of nonzero and zero functions should be formulated respectively as follows.

If an n-degree polynomial $P_n(x)$ is the optimal uniform approximation polynomial of a nonzero function $f(x)$. Then $P_n(x) - f(x)$ has at least $n+2$ staggered points on the interval.

If an n-degree monic polynomial $P_n(x)$ is the optimal uniform approximation polynomial of a zero function, then $P_n(x)$ has at least $n+1$ staggered points on the interval. Please do not neglect the requirement of the monic. If the monic is removed, the optimal uniform approximation function of a zero function is just a zero function.

Because Theorems 5 and 5' are equivalent to each other, in some literature, the former form is expressed, while others use the latter form.

We now prove by reduction *ad absurdum* that the n-degree monic polynomials other than $\tilde{T}_n(x)$ on $[-1, 1]$ cannot be the optimal uniform approximation of a zero function. Suppose that there is another n-degree monic polynomials $T_n^*(x)$ that is the optimal uniform approximation of a zero function. That is say,

$$|T_n^*(x)| < \frac{1}{2^{n-1}}, \quad x \in [-1,1]. \tag{2.4.16}$$

Let

$$R(x) = \tilde{T}_n(x) - T_n^*(x). \tag{2.4.17}$$

Then $R(x)$ is an $n-1$ degree polynomial. $\tilde{T}_n(x)$ has the same deviation at all deviation points and the deviation is 2^{1-n}. Because the deviations of $T_n^*(x)$ should be less than that of $\tilde{T}_n(x)$, $R(x)$ should have the same sign as $\tilde{T}_n(x)$. $\tilde{T}_n(x)$ has $n+1$ deviation points on $[-1,1]$, as manifested by (2.4.15). So, at these $n+1$ deviation points, $R(x)$ should change its sign alternatively, and consequently, $R(x)$ has at least n zeros on $[-1,1]$, which is contradictory to the assumption that $R(x)$ is an $n-1$th degree polynomial.

In summary, $\tilde{T}_n(x)$ is the unique optimal uniform approximation polynomial on $[-1,1]$. Simultaneously, the $n-1$-degree polynomial $P_{n-1}^*(x) = x^n - \tilde{T}_n(x)$ is the unique optimal uniform approximation polynomial of function x^n on $[-1,1]$.

2. Square approximation

Definition 5. Suppose a function $f(x)$ is square integrable on $[a,b]$. If an n-degree polynomial $P_n(x)$ makes

$$\delta = \int_a^b [f(x) - P_n(x)]^2 \mathrm{d}x \tag{2.4.18}$$

the smallest, then the quantity δ is called the **mean–square error** of $P_n(x)$ and $f(x)$ on $[a,b]$. This is a way of measuring the approximation of $f(x)$ by $P_n(x)$ in terms of the value of δ. Please note that even though $P_n(x)$ is such a function that makes the mean-square error with $f(x)$ very small, it may still have very large errors with $f(x)$ at some isolated points.

Definition 6. Suppose that a given function $f(x) \in C[a,b]$ is square-integrable. Among all the n-degree polynomials, if there is one $P_n^*(x)$

that makes

$$\Delta = \int_a^b [f(x) - P_n^*(x)]^2 w(x) \mathrm{d}x \qquad (2.4.19)$$

the smallest, then $P_n^*(x)$ is said to be **the optimal square approximation (the least square approximation)** on $[a,b]$. Δ is called the **mean-square error with weight** $w(x)$. This method measures the approximation of $P_n^*(x)$ to $f(x)$ in terms of the value of Δ. The square error defined in Definition 5 is actually a special case with weight function $w(x) = 1$.

The optimal square approximation polynomial exists and is unique.

$\tilde{T}_n(x)$ can square approximate $f(x) \equiv 0$ on $[-1, 1]$. That is to say, when $P_n^*(x) = \tilde{T}_n(x)$,

$$\Delta = \int_{-1}^1 \frac{[P_n^*(x)]^2}{\sqrt{1 - x^2}} \mathrm{d}x \qquad (2.4.20)$$

is the smallest. Let $P_n^*(x)$ be an n-degree monic polynomial. It can be expanded in terms of (2.4.12):

$$P_n^*(x) = \sum_{l=0}^n a_l \tilde{T}_l(x).$$

Substituting this into (2.4.20) and using the orthonormalization of $T_n(x)$, (2.4.11), we get

$$\Delta = \sum_{l=0}^n a_l^2 \frac{1}{(2^{l-1})^2} \int_{-1}^1 \frac{[T_l(x)]^2}{\sqrt{1 - x^2}} \mathrm{d}x$$

$$= \frac{\pi}{2} \sum_{l=0}^n \frac{a_l^2}{(2^{l-1})^2} = \frac{\pi}{2} \left(\frac{a_0^2}{2^{-2}} + \frac{a_1^2}{2} + \cdots + \frac{a_n^2}{2^{2n-2}} \right).$$

Obviously, as $P_n^*(x) = \tilde{T}_n(x)$, Δ has the smallest value. In other words, when $a_0 = a_1 = \cdots = a_{n-1} = 0$ and $a_n = 1$,

$$\Delta = \int_{-1}^1 \frac{[\tilde{T}_n(x)]^2}{\sqrt{1 - x^2}} \mathrm{d}x = \frac{\pi}{2^{2n-1}}$$

is the smallest. Therefore $\tilde{T}_n(x)$ square approximates a zero function.

Here we do not assert that there is no other square approximation polynomials relative to a zero function. In fact, on $[-1, 1]$ the optimal square approximation polynomials of a zero function are Legendre polynomials which will be introduced in the next chapter.

The theory of the optimal approximation polynomials introduced in this section has important application in calculation methods.

Exercises

1. Show that distance satisfies the following inequality:

$$\rho(x, y) \leq \rho(x, z_1) + \rho(z_1, z_2) + \cdots + \rho(z_{n-1}, z_n) + \rho(z_n, y).$$

   Its geometric significance is that one side length of a polygon is less than the sum of all other side lengths of this polygon.

2. Show that a convergent sequence is necessarily a Cauchy sequence.

3. The continuous function space $C[-1, 1]$ is not complete. An example is the function sequence $\{y_n(t)\}$ defined by

$$y_n(t) = \begin{cases} 0, & -1 \leq t \leq 0 \\ nt, & 0 < t < 1/n \\ 1, & 1/n \leq t \leq 1 \end{cases}$$

   Every element in this sequence is a continuous function. Show that this is a Cauchy sequence, but converges to a discontinuous function.

4. We have explained that the rational number space is not complete by an example. Is there any other incomplete space? Give some examples.

5. Judge whether the following sets are linear spaces on real number domain in given mathematical operations or not.

   (1) All invertible matrices of n order observing the addition and scalar multiplication of matrices.

   (2) All $m \times n$ real matrices observing the addition and scalar multiplication of matrices.

(3) All $m \times n$ complex matrices observing the addition and scalar multiplication of matrices.

6. Prove the inequality of inner product: $(x - y, x - y)^{1/2} \geq (x,x)^{1/2} - (y,y)^{1/2}$.

7. Show that zero vector is orthogonal to all vectors.

8. Prove the parallelogram formula (2.1.10).

9. Suppose that $\{x_1, x_2, \ldots, x_m\}$ is an orthonormalized set in an m-dimensional inner product space. For any two vectors x and y in this space, prove Parseval equality $(x,y) = \sum_{i=1}^{m}(x, x_i)(x_i, y)$. Note that this is the case of finite-dimensional spaces.

10. In Example 12 in subsection 2.1.1, $A = \{1, \sin x, \sin 2x, \ldots, \sin nx\}$ is a linearly independent group in space $X = C[0, 2\pi]$. The inner product of two continuous functions $f_1(x)$ and $f_2(x)$ is defined by $(f_1, f_2) = \int_0^{2\pi} f_1^*(x)f_2(x)dx$. Show that this is an orthogonal set but not normalized. What is the form of its orthonormalization?

11. In the space of polynomials up to power 6, $X = P_6[-1,1]$, $A = \{1, x, x^2, x^3, x^4, x^5, x^6\}$ is a linearly independent group. The inner product of two polynomials of degree 6, $f_1(x)$ and $f_2(x)$, is defined by $(f_1, f_2) = \int_{-1}^{1} f_1^*(x)f_2(x)dx$. Show that this is not an orthonormalized set. Construct the orthonormalized set. What is the resultant polynomials? Why is it this polynomials but not others?

12. Show that trigonometric function set $1, \cos x, \sin x, \cos 2x, \sin 2x, \ldots, \cos nx, \sin nx, \ldots$ is an orthogonal set on $[-\pi, \pi]$. What is its orthonormalized form?

13. Legendre polynomials are defined by

$$P_n(x) = \frac{1}{2^n n!} \frac{d^n}{dx^n}(x^2 - 1)^n, \quad (n = 1, 2, \ldots).$$

Show that $\{P_n(x)\}$ is an orthogonal basis on $[-1, 1]$. What are the coefficients of $P_n(x)$ if it is written in the orthonormalized form?

14. Show that if among three vectors a, b, c, there is a relation

$$c = a + b$$

and a is orthogonal to b, then the relation among their moduli is

$$\|c\|^2 = \|a\|^2 + \|b\|^2$$

15. The effect of a differential operator acting on a function and the boundary condition that the function satisfies are as follows:

$$Lu(x) = \left(\frac{\mathrm{d}}{\mathrm{d}x} + 1\right) u(x), \quad 0 < x < 1; \quad u(0) - \alpha u(1) = 0.$$

Find the adjoint operator and the boundary condition of the function that it acts.

16. Prove that the formal adjoint of the differential operator (2.2.10a) is (2.2.10b) and its knot is (2.2.13b).

17. Let an operator $L = \dfrac{\mathrm{d}^2}{\mathrm{d}x^2} + 4\dfrac{\mathrm{d}}{\mathrm{d}x} - 3$ and its differential equation be $Lu(x) = 0$, $a < x < b$ with boundary conditions $4u(a) + u'(a) = 0, 4u(b) + u'(b) = 0$. Find the adjoint operator in equation $L^\dagger v(x) = 0, a < x < b$ and the boundary conditions the function $v(x)$ satisfies. What is the result of the knot $[J(u(x), v(x))]_a^b$?

18. The differential equation $u''(x) = f(x), 0 < x < 1$ has only one boundary condition $u(0) = \gamma$. Show that there are three boundary conditions for its adjoint equation. (Hint: in finding the adjoint boundary conditions, the original differential equation has to be employed.)

19. The general form of differential operator of the third order is

$$L = r_3(x)\frac{\mathrm{d}^3}{\mathrm{d}x^3} + r_2(x)\frac{\mathrm{d}^2}{\mathrm{d}x^2} + r_1(x)\frac{\mathrm{d}}{\mathrm{d}x} + r_0(x).$$

Put down its formally adjoint operator. What are the conditions for it to be self-adjoint?

20. Prove that if operators A and B are self-adjoint, then AB is self-adjoint when and only when $AB = BA$.

21. Imitating one-dimensional case, put down the adjoint of the differential operator of second order

$$L = p_2(\boldsymbol{r})\nabla^2 + p_1(\boldsymbol{r})\nabla + p_0(\boldsymbol{r})$$

in three-dimensional case. What are the conditions of its formal adjoint? Write its formal adjoint. Write the knot $J(u(\boldsymbol{r}), v(\boldsymbol{r}))$ of functions $u(\boldsymbol{r})$ and $v(\boldsymbol{r})$. What should the knot $J(u(\boldsymbol{r}), v(\boldsymbol{r}))$ meet if L is self-adjoint?

22. Show that among the three definitions of isometric transformations, (2.2.31)–(2.2.33), any two of them are sufficient and necessary conditions of each other.

23. Show that uniform convergence contains pointwise convergence, and uniform convergence contains mean-square convergence.

24. Show that the function sequence

$$f_n(x) = \frac{2\sqrt{n}}{(\pi/2)^{1/4}} n x e^{-(nx)^2}$$

converges to zero for all points x:

$$\lim_{n\to\infty} f_n(x) = 0,$$

but does not converge to zero in the mean-square sense:

$$\lim_{n\to\infty} \int_{-\infty}^{\infty} |f_n(x) - 0|^2 \mathrm{d}x = 1 \neq 0.$$

This is an instance revealing that the pointwise convergence does not contain mean-square convergence.

25. Prove Theorem 1 in Subsection 2.3.1.

26. Prove Parseval equality (2.3.9) in Hilbert spaces. The method is to apply the completeness relation (2.3.8) to function $f + \lambda g$ as well as functions f and g. Exercise 9 is the case of finite-dimensional spaces, while this is the infinite-dimensional case.

27. Prove Theorem 3 in subsection 2.3.4.

28. Prove (2.3.14).

29. Construct polynomials with weight $\rho(x) = 1$ on $[1/4, 1]$ up to $\varphi_3(x)$.

30. Construct polynomials with weight $\rho(x) = 1$ on $[0, 1]$ up to $\varphi_3(x)$.

31. Put down Chebyshev polynomials $T_n(x)$, $n = 0, 1, 2, 3, 4, 5$.

32. Show that $T_n(x)$ has $n+1$ extremes on $[-1, 1]$ as shown by (2.4.9).

33. For Chebyshev polynomials $T_n(x) = \cos(n \cos^{-1} x)$, do the following things.

   (1) Find the values at specials points: $T_n(1) = ?$ $T_n(-1) = ?$ $T_{2n+1}(0) = ?$ $T_{2n}(0) = ?$
   (2) Find their parity, the relation between $T_n(-x)$ and $T_n(x)$.
   (3) Find the zeros of $T_n(x)$.
   (4) Prove the following recursion formulas:

$$T_{n+1} - 2xT_n + T_{n-1} = 0,$$
$$(1 - x^2)T_n' = nxT_n - nT_{n+1},$$
$$2(1 - x^2)T_n' = n(T_{n-1} - T_{n+1}),$$
$$2T_nT_m = T_{n+m} - T_{n-m}, n > m.$$

   (5) Show that $T_n(x)$ satisfies the differential equation of second order: $(1 - x^2)T_n''(x) - xT_n'(x) + n^2 T_n(x) = 0$.
   (6) Show that $T_m(T_n(x)) = T_{mn}(x)$.
   (7) Show that $T_{2n}(x) = 2T_n^2(x) - 1$.
   (8) Show that $T_{2n+1}'(x) = (2n + 1)\left[2\sum_{i=0}^{n} T_{2i}(x) - 1\right]$.
   (9) Show that $T_{2n}'(x) = 4n\sum_{i=0}^{n-1} T_{2i+1}(x)$.

34. Show that $T_n(x)$'s are orthogonal to each other with weight $\rho(x) = \frac{1}{\sqrt{1-x^2}}$: $\int_{-1}^{1} \frac{1}{\sqrt{1-x^2}}T_n(x)T_m(x)\mathrm{d}x = \frac{\pi}{2}\delta_{nm}\varepsilon_m$, where $\varepsilon_0 = 2$, $\varepsilon_m = 1, (m \geq 1)$.

35. By use of the results in the last exercise and (2.3.16), show that Chebyshev polynomials follow that

$$T_{n+1}'(n)T_n(x) - T_n'(x)T_{n+1}(x) > 1.$$

36. By Theorem 5 in subsection 2.3.4, derive the recursion formulas of Chebyshev polynomials.

37. Evaluate the summation $\sum_{n=0}^{\infty} t^n e^{in\theta}$. Take the real part of the resultant to prove the formula

$$\frac{1 - xt}{1 - 2xt + t^2} = \sum_{n=0}^{\infty} T_n(x)t^n.$$

The left hand side of this equation is the generating function of Chebyshev polynomials.

38. Show that $|T_n'(x)| \leq n^2, -1 \leq x \leq 1$. When does the equal sign stand?

26. In Theorem 3.9 (Schoenberg's Theorem) determine the spline-basis-method, also polynomials.

27. Evaluate the summation $\sum_{i=0}^{n} h_i b^i$. Take the real part of the complex $\sum_{i} \dots$ with limits $n \to \infty$.

$$\sum = \frac{?}{(?)(?)\dots}$$

Between the nodes of this solution by an interpolating flux, and $p, q \dots$

28. Show that $\dots \sum_i \dots = \int \dots$ Would it work or not, find $\dots$

Chapter 3

Linear Ordinary Differential Equations of Second Order

3.1. General Theory

3.1.1. *The existence and uniqueness of solutions*

Physical problems are often reflected mathematically by differential equations, either ordinary or partial ones. In many cases, the partial differential equations are solved by means of separation of variables so as to become several ordinary differential equations. Therefore, roughly speaking, solving ordinary ones are the fundamental of solving the partial ones.

Definition 1. Equations associated with variable x, unknown function y and its derivative functions are called **ordinary differential equations**, or in short **differential equations**.

Definition 2. The highest order of derivatives of the unknown function in a differential equation is called **the order of the differential equation**.

Definition 3. In a differential equation, if the highest power of the unknown function and its derivatives is 1 and there are no products between them, this equation is called a **linear differential equation**. Otherwise it is called a **nonlinear differential equation**.

The mostly encountered in solving practical problems are ordinary differential equations of second order. Its general form is

$$y''(x) + p(x)y'(x) + q(x)y(x) = f(x), \qquad (3.1.1)$$

which is called **the standard form of ordinary differential equations of second order**. If $f(x) \equiv 0$, the equation is said to be **homogeneous**, and if not, called **inhomogeneous**. The solution of a homogeneous equation

$$y''(x) + p(x)y'(x) + q(x)y(x) = 0 \qquad (3.1.2)$$

exists. In fact, there is an existence and uniqueness theorem for the following initial value problem:

$$\begin{cases} y'' + p(x)y' + q(x)y = 0, \\ y(x_0) = \alpha, y'(x_0) = \beta, \end{cases} \qquad (3.1.3)$$

where $x \in [a, b]$ and x_0 is in $[a, b]$. This theorem asserts that the solution of problem (3.1.3) exists and is unique.

We do not intend to prove it here, but are content with introducing briefly the idea of the proof. Let $z = y'$, then (3.1.2) is recast to be the following equations:

$$\begin{cases} z' = -p(x)z - q(x)y, \\ y' = z. \end{cases}$$

This is a homogeneous linear ordinary differential equations of first order, and it is equivalent to (3.1.2). Thus, the problem turns out to be the following one:

$$\begin{cases} y_1' = a_{11}(x)y_1 + a_{12}(x)y_2, \\ y_2' = a_{21}(x)y_1 + a_{22}(x)y_2, \\ y_1(x_0) = \alpha, y_2(x_0) = \beta. \end{cases} \qquad (3.1.4)$$

The process of proving the existence and uniqueness of the solution is as follows. First, set two initial functions $y_{1,1}, y_{2,1}$ (the most convenient ones are $y_{1,1} = \alpha, y_{2,1} = \beta$), and substitute them into the right hand sides of (3.1.4). Next, integrate the equations such that two new functions $y_{1,2}, y_{2,2}$ are obtained on the left hand sides of (3.1.4). Then they are substituted into the right hand sides of (3.1.4) again, and integration once more gives new functions $y_{1,3}, y_{2,3}$, and so on. In such a way, we will obtain a function sequence $\{y_{1,n}, y_{2,n}\}$. It can be proved to converge uniformly to the solution of (3.1.4). This means that (3.1.4) has a solution. This is the existence of the solution. Because the function sequence approaches the solution step by step, this process is called the **successive approximation method**.

To prove the uniqueness of the solution, assume that two pairs of functions $\{y_1(x), y_2(x)\}$ and $\{z_1(x), z_2(x)\}$ satisfy (3.1.4) simultaneously. Then, let $u_1 = y_1 - z_1$, $u_2 = y_2 - z_2$. Consider the initial value problem:

$$\begin{cases} u_1' = a_{11}(x)u_1 + a_{12}(x)u_2, \\ u_2' = a_{21}(x)u_1 + a_{22}(x)u_2, \\ u_1(x_0) = 0, u_2(x_0) = 0. \end{cases} \tag{3.1.5}$$

Its solution must be zero, $u_1 = 0, u_2 = 0$. It is easily obtained by means of the successive approximation method with initial functions $u_{1,1} = \alpha = 0, u_{2,1} = \beta = 0$. Therefore, $y_1 \equiv z_1, y_2 \equiv z_2$. The statement of the theorem is as follows.

Theorem 1 (The existence and uniqueness of solutions). *If functions $a_{ij}(x)$ $(i, j = 1, 2)$ are continuous on $[a, b]$ which contains x_0, then the initial value problem (3.1.4) has one and only one group of solutions $y_1 - y_1(x), y_2 = y_2(x)$ on this interval.*

Regarding the proof process of the theorem, we can restate the uniqueness as follows: if two functions y_1 and y_2 both satisfy (3.1.2) and the same initial conditions, they are certainly identical.

For the homogeneous linear differential equations of higher orders, there is also the theorem of the existence and uniqueness of solutions. The idea of the proof is the same as above: the equation is transformed into differential equations of first order, and then the successive approximation method is used.

If two functions $y_1(x)$ and $y_2(x)$ are solutions of homogeneous equation (3.1.2), then their linear combination $y(x) = c_1 y_1(x) + c_2 y_2(x)$ is also a solution of (3.1.2). This is so-called the **superposition principle**.

In the following, we first introduce the solutions' structure of homogeneous equations, and then present the expression of the solution of the inhomogeneous equation.

3.1.2. *The structure of solutions of homogeneous equations*

1. Basic set of solutions

Definition 4. Consider two functions $y_1(x)$ and $y_2(x)$. If there exist two constants c_1 and c_2 not both zero, which make

$$c_1 y_1(x) + c_2 y_2(x) = 0, \quad x \in [a, b], \tag{3.1.6}$$

then $y_1(x)$ and $y_2(x)$ are **linearly dependent** on $[a, b]$. If (3.1.6) stands only when both c_1 and c_2 are zero, the two functions are **linearly independent**. For examples, $\sin x$ and $\cos x$ are linearly independent; x^2 and x^3 are linearly independent. The concept of linearly independent has been defined in Chapter 2 when we introduced Hilbert space. Here is a repeat for the specific case of the solutions of a differential equation.

By this definition, if functions $y_1(x)$ and $y_2(x)$ are linearly dependent on $[a, b]$, there must exist two constants c_1 and c_2 not both zero, which makes (3.1.6) stand on $[a, b]$. Take a derivative of (3.1.6). Then

$$c_1 y_1'(x) + c_2 y_2'(x) = 0. \tag{3.1.7}$$

Definition 5. Determinant

$$W(y_1, y_2) = \begin{vmatrix} y_1(x) & y_2(x) \\ y_1'(x) & y_2'(x) \end{vmatrix} \tag{3.1.8}$$

is called the **Wronskian** of two functions $y_1(x)$ and $y_2(x)$. The Wronskian of N functions can be defined similarly, which are composed of N functions and their first, second, up to $N - 1$th derivative functions.

If the linear equations (3.1.6) and (3.1.7) have nonzero solutions c_1 and c_2, its determinant factor, i.e., the Wronskian of $y_1(x)$ and $y_2(x)$ ought to be zero:

$$W(y_1, y_2) = 0 \qquad (3.1.9)$$

Thus, we have the following theorem.

Theorem 2. *If $y_1(x)$ and $y_2(x)$ are linearly dependent, their Wronskian is zero everywhere.*

It should be pointed out that its converse theorem is not valid. If the Wronskian of $y_1(x)$ and $y_2(x)$ are identical to zero, the functions are not necessarily linearly dependent. Here is an example.

$$y_1(x) = \begin{cases} (x-1)^2, & 0 \le x \le 1 \\ 0, & 1 \le x \le 2 \end{cases} \qquad y_2(x) = \begin{cases} 0, & 0 \le x \le 1 \\ (x-1)^2, & 1 \le x \le 2 \end{cases}$$

The two functions are linearly independent on $[0, 2]$. When the identity

$$c_1 y_1(x) + c_2 y_2(x) = 0$$

stands on $[0, 1]$, it must be $c_1 = 0$, and when it stands on $[1, 2]$, $c_2 = 0$. So, only when $c_1 = c_2 = 0$ that the identity stands on $[0, 2]$. However, their Wronskian is always zero on $[0, 2]$.

Under what condition does the converse theorem of Theorem 2 become valid? We have the following theorem.

Theorem 3. *Let $y_1(x)$ and $y_2(x)$ be the solution of (3.1.2). If their Wronskian $W(y_1, y_2)$ equals to zero at one point in the interior of (a, b), then they are linearly dependent on $[a, b]$.*

Proof. If at point x_0, $W(y_1(x_0), y_2(x_0)) = 0$, there must exist two constants c_1 and c_2 not both zero, which makes

$$c_1 y_1(x_0) + c_2 y_2(x_0) = 0 \qquad (3.1.10a)$$

and

$$c_1 y_1'(x_0) + c_2 y_2'(x_0) = 0. \qquad (3.1.10b)$$

Consider the function $y(x) = c_1 y_1(x) + c_2 y_2(x)$. It is a solution of (3.1.2) because both $y_1(x)$ and $y_2(x)$ are according to the superposition principle. On the other hand, by (3.1.10), $y(x)$ meets $y(x_0) = 0$, $y'(x_0) = 0$. That means $y(x)$ satisfies the initial value problem (3.1.5). So its solution is identical to zero on (a, b). The conclusion is that there exist c_1 and c_2 not both zero on $[a, b]$, which makes $y(x) = c_1 y_1(x) + c_2 y_2(x) = 0$.   □

Through the above proof process, we get a corollary as follows

Corollary. Let $y_1(x)$ and $y_2(x)$ be the solutions of (3.1.2). If they are linearly dependent, their $W(y_1, y_2)$ is identical to zero everywhere; if not, $W(y_1, y_2)$ is nonzero everywhere.

Suppose that $y_1(x)$ and $y_2(x)$ are the solutions of (3.1.2). Their Wronskian has an important property. Taking its derivative with respect to x,

$$\frac{\mathrm{d}}{\mathrm{d}x} W(y_1, y_2) = \frac{\mathrm{d}}{\mathrm{d}x} \begin{vmatrix} y_1 & y_2 \\ y_1' & y_2' \end{vmatrix} = \begin{vmatrix} y_1' & y_2' \\ y_1' & y_2' \end{vmatrix} + \begin{vmatrix} y_1 & y_2 \\ y_1'' & y_2'' \end{vmatrix}$$

$$= \begin{vmatrix} y_1 & y_2 \\ -py_1' - qy_1 & -py_2' - qy_2 \end{vmatrix} = \begin{vmatrix} y_1 & y_2 \\ -py_1' & -py_2' \end{vmatrix}$$

$$= -p \begin{vmatrix} y_1 & y_2 \\ y_1' & y_2' \end{vmatrix}.$$

Here to achieve the derivative of an n-order determinant, one first takes the derivatives of each row to obtain a determinant; then he sums up all of the determinants. When a multiple of one row of the determinant is added to another row, the value of the determinant remains unchanged. To reach the final result, the original differential equation has been used:

$$y_i''(x) = -p(x)y_i'(x) - q(x)y_i(x), \quad i = 1, 2.$$

Thus W observes a differential equation of first order

$$\frac{\mathrm{d}}{\mathrm{d}x} W = -p(x)W.$$

Its integration is

$$W = W_0 \exp\left[-\int_{x_0}^{x} p(t)\mathrm{d}t\right], \qquad (3.1.11)$$

where $W_0 = W(x_0)$ is the value of the Wronskian at x_0.

Equation (3.1.11) is called the **Liouville formula**, which exhibits the relationship between the coefficient function and the Wronskian of the solutions of the differential equation. This formula also indicates that under the condition that y_1 and y_2 are the solutions of (3.1.2) on $[a, b]$, then in the interior of (a, b), if W is zero at one point, it will be zero everywhere; if it is nonzero at one point, it will be nonzero everywhere.

It should be noted that (1) Theorem 3 and its corollary apply to open interval (a, b) but not include the ends. We will see from Exercise 7 that sometimes, the Wronskian of two linearly independent solutions that meet the same boundary conditions may be zero at ends of the interval. (2) Liouville formula (3.1.11) does not apply to the singular points of $p(x)$ and non-integrable points. Therefore, Theorem 3 and its corollary do not apply to such points.

Theorem 4 (Structure of the solutions of homogeneous equations). *Suppose that $y_1(x)$ and $y_2(x)$ are a pair of linearly independent solutions of (3.1.2). For any constants c_1 and c_2,*

$$y(x) = c_1 y_1(x) + c_2 y_2(x) \qquad (3.1.12)$$

is also a solution of (3.1.2). Conversely, any solution of (3.1.2) can be expressed in the form of (3.1.12).

Proof. The first part of this theorem is obvious. We prove the second part. Since $y_1(x)$ and $y_2(x)$ are linearly independent, we have $W(x_0) \neq 0$. Suppose $y(x)$ be any solution. It is certain that by $W(x_0) \neq 0$, one is able to find two constants c_1 and c_2 which make

$$c_1 y_1(x_0) + c_2 y_2(x_0) = y(x_0)$$

and

$$c_1 y_1'(x_0) + c_2 y_2'(x_0) = y'(x_0).$$

Mathematics for Physicists

That is to say, the function $y(x)$ and

$$c_1 y_1(x) + c_2 y_2(x) \tag{3.1.13}$$

are both the solutions of (3.1.2) and satisfy the same initial conditions. Therefore, according to the uniqueness theorem, it must be $y(x) = c_1 y_1(x) + c_2 y_2(x)$. □

Definition 6. Suppose that $y_1(x)$ and $y_2(x)$ are linearly independent solutions of (3.1.2). Let us take two independent constants c_1 and c_2 to constitute expression (3.1.13), and when c_1 and c_2 vary, all of possible solutions of (3.1.2) can be achieved. Expression (3.1.13) is called the **general solution** of (3.1.2). In (3.1.13) the functions $y_1(x)$ and $y_2(x)$ are called the **special solutions** of the equation. A pair of linearly independent solutions $y_1(x)$ and $y_2(x)$ are called **a basic set of solutions**.

From Theorem 4, it is known that any solution of the equation can be expressed as the linear combination of a pair of independent solutions in the form of (3.1.13). The general solution is so. Disregarding the initial conditions, the combination coefficients are arbitrary. The initial conditions determine the values of the combination coefficients. In other words, the combination that satisfy the initial conditions are definite.

Now to solve an ordinary differential equation of second order, what we should do is to find a basic set of solutions, two linearly independent solutions of (3.1.2). Yet the theorem about the structure of the solutions does not tell us if the basic set exists. This needs another theorem to affirm.

Theorem 5. *There must exist basic sets of solutions of* (3.1.2).

Proof. Let us take any two groups of constants α_1, β_1 and α_2, β_2, and they satisfy $\begin{vmatrix} \alpha_1 & \alpha_2 \\ \beta_1 & \beta_2 \end{vmatrix} \neq 0$. Then we will obtain solution y_1 and y_2 respectively from the following two initial problems

$$\begin{cases} y'' + p(x)y' + q(x)y = 0. \\ y(x_0) = \alpha_1, y'(x_0) = \beta_1, \end{cases}$$

and

$$\begin{cases} y'' + p(x)y' + q(x)y = 0, \\ y(x_0) = \alpha_2, y'(x_0) = \beta_2. \end{cases}$$

The y_1 and y_2 must be linearly independent, because at $x = x_0$,

$$W(y_1, y_2) = \begin{vmatrix} \alpha_1 & \alpha_2 \\ \beta_1 & \beta_2 \end{vmatrix} \neq 0.$$

□

This proof reveals not only there exist but also there are many basic sets of solutions, even an infinite number. But different special solutions meet different initial conditions. There is no general way of finding a pair of linearly independent solutions.

Nevertheless, if one nonzero special solution $y_1(x)$ has been found in some way, it is possible to acquire another linearly independent solution $y_2(x)$.

Let $y_1(x)$ be a nonzero solution of (3.1.2). There must be another solution $y_2(x)$ such that $y_1(x)$ and $y_2(x)$ are linearly independent. By Liouville formula,

$$y_1 y_2' - y_1' y_2 = c \exp\left[-\int p(t)\mathrm{d}t\right].$$

When divided by y_1^2, it becomes

$$\frac{y_1 y_2' - y_1' y_2}{y_1^2} = \frac{\mathrm{d}}{\mathrm{d}x}\frac{y_2}{y_1} = \frac{c}{y_1^2}\exp\left[-\int p(t)\mathrm{d}t\right].$$

After integration, we obtain

$$y_2 = y_1 \int \mathrm{d}x \frac{c}{y_1^2}\exp\left[-\int p(x)\mathrm{d}x\right] + c_1 y_1. \tag{3.1.14}$$

Thus, y_2 is achieved in terms of y_1.

Equation (3.1.14) is also called the **Liouville formula**. It presents a general way to get one special solution from another. It also exhibits such a fact that once y_2 is obtained that is linearly independent of y_1, y_2 plus a multiple of y_1 is still linearly independent of y_1. In summary, as long as a nonzero special solution is known, the general solutions of (3.1.2) can be easily obtained.

Example 1. Find the solutions of $xy'' - y' = 0$.

Solution. By checking, $y = 1$ is a special solution. Another linearly independent solution is evaluated by the Liouville formula.

$$y_2 = \int dx \exp\left[\int \frac{1}{x} dx\right] = \frac{x^2}{2}.$$

Thus, the general solutions of the equation are

$$y = c_1 + c_2 \frac{x^2}{2}.$$

2. Zeros of solutions

Theorem 6 (Sturm Separation Theorem). *If $y_1(x)$ and $y_2(x)$ are linearly independent solutions of (3.1.2), then the zeros of $y_1(x)$ are distinct from those of $y_2(x)$, and the two sequences of zeros alternate; that is, $y_1(x)$ has exactly one zero between two successive zeros of $y_2(x)$, and vice versa.*

Let

$$y(x) = u(x) \exp\left[-\frac{1}{2}\int p(x)dx\right]. \tag{3.1.15}$$

Substituting its first and second derivatives y' and y'' into (3.1.2), we have

$$\exp\left(-\frac{1}{2}\int p dx\right)$$
$$\times \left[u'' - pu' - \frac{1}{2}p'u + \frac{1}{4}p^2 u + p\left(u' - \frac{1}{2}pu\right) + qu\right] = 0.$$

After defining $Q(x) = q(x) - \frac{1}{4}p^2(x) - \frac{1}{2}p'(x)$, (3.1.2) is finally converted to be

$$u''(x) + Q(x)u(x) = 0. \tag{3.1.16}$$

Theorem 7. *Let $u(x)$ be any nonzero solution of (3.1.16). If $Q(x) < 0$, then $u(x)$ possesses at most one zero. If for all $x > 0$, $Q(x) > 0$*

and $\int_1^\infty Q(x)\mathrm{d}x = \infty$, then $u(x)$ have infinite zeros at positive x-axis.

For example, as $Q(x) = 1$ then the solution is $u(x) = \sin x$ or $\cos x$; as $Q(x) = -1$, the solution is $u(x) = \sinh x$ or $\cosh x$.

Theorem 8 (Sturm Comparison Theorem). *Let $u(x)$ and $v(x)$ be nonzero solutions of the equations $u''(x) + Q(x)u(x) = 0$ and $v''(x) + R(x)v(x) = 0$, respectively, where $Q(x) > R(x) > 0$. Then $u(x)$ has at least one zero between every two consecutive zeros of $v(x)$, unless $Q(x) \equiv R(x)$ and $u(x)$ is a constant multiple of $v(x)$.*

3.1.3. The solutions of inhomogeneous equations

1. Expression of general solution

Having learnt the structure of solutions of homogeneous equation (3.1.2), we now turn to seek for the solution of inhomogeneous equation (3.1.1). For inhomogeneous differential equation, a theorem is available.

Theorem 9. *If $w(x)$ is a special solution of (3.1.1), and it plus any solution of (3.1.2), the resultant function $u(x)+w(x)$ is also a solution of (3.1.1). Conversely, any solution $y(x)$ of (3.1.1) can be expressed in the form of*

$$y(x) = u(x) + w(x). \tag{3.1.17}$$

Proof. If $u(x)$ is a solution of (3.1.2), substitution of $u(x) + w(x)$ into (3.1.1) brings

$$(u + w)'' + p(u + w)' + q(u + w)$$
$$= u'' + pu' + qu + w'' + pw' + qw = f.$$

This reveals that $u(x) + w(x)$ is also a solution of (3.1.1).

Conversely, assume $y(x)$ is any solution of (3.1.1). Substitution of $y(x) - w(x)$ into (3.1.1) leads to

$$(y - w)'' + p(y - w)' + q(y - w)$$
$$= y'' + py' + qy - w'' - pw' - qw = f - f = 0.$$

So, function $u(x) = y(x) - w(x)$ is a solution of (3.1.2), and any solution of inhomogeneous equation can be written in the form of (3.1.17).   □

Therefore, in order to find all of the solutions of (3.1.1), one has to search a solution $w(x)$ of (3.1.2) and one takes the sum of $w(x)$ and all the solutions of (3.1.2). Let $y_1(x)$ and $y_2(x)$ be a pair of linearly independent solutions of (3.1.2). Then in expression

$$y = c_1 y_1 + c_2 y_2 + w, \tag{3.1.18}$$

one takes all possible values of c_1, c_2, so as to obtain all the solutions of (3.1.1). Equation (3.1.18) is called a **general solution** of (3.1.1), and w is called a **special solution** of (3.1.1).

2. Expression of the special solution

In the next section and Chapter 4, we will discuss the linearly independent solutions $y_1(x)$ and $y_2(x)$ for various concrete functions $p(x)$, $q(x)$. Here we assume that they have been found, which enables us to put down the special solution $w(x)$.

We make a postulation that $w(x)$ can "be linearly combined" with $y_1(x)$ and $y_2(x)$:

$$w(x) = a_1(x) y_1(x) + a_2(x) y_2(x), \tag{3.1.19}$$

where the "combination coefficients" $a_1(x), a_2(x)$ are actually functions of x. Equation (3.1.19) must satisfy (3.1.1). Since now two functions $a_1(x), a_2(x)$ are to be found, there must be two constraint conditions. To find the conditions, we take a derivative of (3.1.19),

$$w'(x) = a_1(x) y_1'(x) + a_2(x) y_2'(x) + a_1'(x) y_1(x) + a_2'(x) y_2(x).$$

We exert a constraint condition. Let

$$a_1'(x) y_1(x) + a_2'(x) y_2(x) = 0. \tag{3.1.20}$$

This brings

$$w'(x) = a_1(x) y_1'(x) + a_2(x) y_2'(x). \tag{3.1.21}$$

Again, we take a derivative of this equation and get

$$w''(x) = a_1(x)y_1''(x) + a_2(x)y_2''(x) + a_1'(x)y_1'(x) + a_2'(x)y_2'(x).$$

$$(3.1.22)$$

Substitution of (3.1.17), (3.1.19), (3.1.21) and (3.1.22) into (3.1.1) gives

$$a_1(y_1'' + py_1' + qy_1) + a_2(y_2'' + py_2' + qy_2 + a_1'y_1' + a_2'y_2') = f.$$

Since $y_1(x)$ and $y_2(x)$ are solutions of a homogeneous equation, this equation leaves

$$a_1'(x)y_1'(x) + a_2'(x)y_2'(x) = f(x). \qquad (3.1.23)$$

Equations (3.1.20) and (3.1.23) are the linear equations of solving $a_1'(x)$ and $a_2'(x)$. Since $y_1(x)$ and $y_2(x)$ are linearly independent, their Wronskian is nonzero everywhere. It is easy to get

$$a_1'(x) = -\frac{y_2(x)f(x)}{W(y_1, y_2)}, \quad a_2'(x) = \frac{y_1(x)f(x)}{W(y_1, y_2)}.$$

Then integration results are

$$a_1(x) = -\int \frac{y_2(x)f(x)}{W(y_1, y_2)}\,dx, \quad a_2(x) = \int \frac{y_1(x)f(x)}{W(y_1, y_2)}\,dx.$$

Here the integral constants need not be taken into account, since we are content with a special solution. Finally, the required special solution is

$$w(x) = -y_1(x)\int \frac{y_2(x)f(x)}{W(y_1, y_2)}\,dx + y_2(x)\int \frac{y_1(x)f(x)}{W(y_1, y_2)}\,dx. \qquad (3.1.24)$$

Example 2. Find the solutions of $xy'' - y' = x^2$ satisfying initial conditions $y(1) = 1$, $y'(1) = 1$.

Solution. The two linearly independent solutions of the homogeneous equation have been solved in Example 1, and they are 1 and $x^2/2$, respectively. Their Wronskian is

$$W(y_1, y_2) = \begin{vmatrix} y_1 & y_2 \\ y_1' & y_2' \end{vmatrix} = \begin{vmatrix} 1 & x^2/2 \\ 0 & x \end{vmatrix} = x.$$

Notice that the original equation should be recast into the form of $y'' - y'/x = x$ so as to get a correct $f(x)$. By (3.1.24), a special solution is

$$w(x) = -\int \frac{x^2 x}{2x} dx + \frac{x^2}{2} \int \frac{x}{x} dx = -\frac{x^3}{6} + \frac{x^2}{2} x = \frac{x^3}{3}.$$

Thus the general solutions of the original equation are

$$y(x) = c_1 + c_2 \frac{x^2}{2} + \frac{x^3}{3}.$$

The two constants are found to be $c_1 = \frac{2}{3}$ and $c_2 = 0$ through the initial conditions. Finally, the required solution is

$$y(x) = \frac{2}{3} + \frac{x^3}{3}.$$

Another standard method for solving inhomogeneous equation is Green's function method, which will be introduced in Chapter 6.

3.2.   Sturm-Liouville Eigenvalue Problem

3.2.1.   *The form of Sturm-Liouville equations*

The ordinary differential equation is now rewritten in a form including a parameter λ:

$$A(x)y'' + B(x)y' + [\lambda - C(x)]y = 0, \quad (a \leq x \leq b), \qquad (3.2.1)$$

where $A(x)$, $B(x)$ and $C(x)$ are real functions. Let

$$\rho(x) = \frac{1}{A(x)} \exp \int \frac{B(x)}{A(x)} dx \qquad (3.2.2)$$

and

$$p(x) = A(x)\rho(x) = \exp \int \frac{B(x)}{A(x)} dx. \qquad (3.2.3)$$

They are continuous and derivable on $[a, b]$. Their derivatives are

$$\rho'(x) = -\frac{A'(x)}{A(x)}\rho(x) + \frac{B(x)}{A(x)}\rho(x),$$

and

$$p'(x) = A'(x)\rho(x) + A(x)\rho'(x) = B(x)\rho(x),$$

respectively. Multiplied by $\rho(x)$, (3.2.1) becomes

$$\frac{\mathrm{d}}{\mathrm{d}x}\left[p(x)\frac{\mathrm{d}y}{\mathrm{d}x}\right] + [\lambda\rho(x) - q(x)]y = 0, \quad (a \leq x \leq b), \qquad (3.2.4)$$

where

$$q(x) = C(x)\rho(x) \qquad (3.2.5)$$

is a continuous function on $[a, b]$.

Definition 1. The ordinary differential equation in the form of (3.2.4) is called the **Sturm-Liouville (SL) equation**. The function $p(x)$ is called the **kernel** and $\rho(x)$ is called the **weight function** (or briefly **weight**) of orthogonal function solutions of the equation.

We define a differential operator (**Sturm-Liouville operator**) L as follows:

$$L = -A(x)\frac{\mathrm{d}^2}{\mathrm{d}x^2} - B(x)\frac{\mathrm{d}}{\mathrm{d}x} + C(x) = \frac{1}{\rho(x)}\left[-\frac{\mathrm{d}}{\mathrm{d}x}p(x)\frac{\mathrm{d}}{\mathrm{d}x} + q(x)\right].$$
$$(3.2.6)$$

Then (3.2.3) becomes a short form:

$$Ly = \lambda y, \quad (a \leq x \leq b). \qquad (3.2.7)$$

Let us define the inner product with weight $\rho(x)$:

$$(f, g) = \int_{-\infty}^{\infty} f^*(x)g(x)\rho(x)\mathrm{d}x. \qquad (3.2.8)$$

If the operator L is required to be a Hermitian one, or self-adjoint one, it should meet

$$(Lf, g) = (f, Lg). \qquad (3.2.9)$$

Because all of $A(x)$, $B(x)$ and $C(x)$ are real, it is easily seen by (2.2.23) that the operator L defined by (3.2.6) is already formally a self-adjoint. In the following, we will use the L as a self-adjoint operator. Therefore, it should satisfy boundary condition (2.2.25):

$$[p(x)(g(x)f^{*\prime}(x) - f^*(x)g'(x))]_a^b = [p(x)W(g, f^*)]_a^b = 0, \quad (3.2.10)$$

where $W(g, f^*)$ is the Wronskian of g and f^*. Note that the factor $\rho(x)$ in the denominator of (3.2.6) is the weight in the inner product as in (3.2.8) concerning S-L equations. Hereafter, S-L operator (3.2.6) is always considered a self-adjoint one satisfying (3.2.10), unless specified.

3.2.2. *The boundary conditions of Sturm-Liouville equations*

When solving the Sturm-Liouville equation, boundary conditions are inevitable. Here we present some of them usually encountered in practical problems.

1. Three kinds of homogeneous boundary conditions

The first kind of homogeneous boundary conditions is

$$y(a) = 0, \quad y(b) = 0. \quad (3.2.11)$$

The second kind of homogeneous boundary conditions is

$$y'(a) = 0, \quad y'(b) = 0. \quad (3.2.12)$$

The third kind of homogeneous boundary conditions is

$$y(a) - hy'(a) = 0, \quad y(b) + hy'(b) = 0. \quad (3.2.13)$$

This condition contains linear combinations of the values of the function and its derivative at the ends of the interval.

It is also possible that at the ends a and b different homogeneous boundary conditions are given. For example, at $x = a$ is the first kind and at $x = b$ is the second kind of homogeneous boundary conditions.

The three kinds of condition can be compactly summarized to be

$$\alpha_1 y'(a) + \alpha_2 y(a) = 0, \quad \beta_1 y'(b) + \beta_2 y(b) = 0, \quad (3.2.14a)$$

where $\alpha_1, \alpha_2, \beta_1, \beta_2$ are real constants, and

$$\alpha_1^2 + \alpha_2^2 \neq 0, \quad \beta_1^2 + \beta_2^2 \neq 0. \quad (3.2.14b)$$

This means that α_1 and α_2 are not simultaneously zero, and neither are β_1 and β_2. Note that by (3.2.13), the signs of α_1 and α_2 are contrast, and those of β_1 and β_2 are the same.

2. Periodic boundary conditions

At the two ends of the interval, the following conditions are required.

$$y(a) = y(b), \quad y'(a) = y'(b). \tag{3.2.15a}$$

In this case all three functions $p(x)$, $q(x)$, $\rho(x)$ are required to be periodic:

$$p(a) = p(b), \quad q(a) = q(b), \quad \rho(a) = \rho(b). \tag{3.2.15b}$$

3. Natural boundary conditions

Suppose that $y_1(x)$ and $y_2(x)$ are linearly independent solutions of (3.2.3). If at one end of $[a, b]$, say the left end a, $p(a) = 0$, it can be shown theoretically that the value of $y_2(x)$ at $x = a$ is unbounded provided that $\lim_{x \to a} y_1(a) = $ finite. Therefore, when $p(a) = 0$, in order to guarantee the boundedness of physical solutions, a natural boundary condition is imposed:

$$y(a) < \infty, \quad \text{as } p(a) = 0. \tag{3.2.16a}$$

In this case, since $y(a)$ is finite and $p(a) = 0$, we have

$$[p(x)y(x)y'(x)]_{x=a} = 0. \tag{3.2.16b}$$

Similarly, if $p(b) = 0$, the natural boundary condition is imposed at end b:

$$y(b) < \infty, \quad \text{as } p(b) = 0. \tag{3.2.17a}$$

In this case, since $y(a)$ is finite and $p(b) = 0$, we have

$$[p(x)y(x)y'(x)]_{x=b} = 0. \tag{3.2.17b}$$

If both $p(a) = 0$ and $p(b) = 0$, the natural boundary conditions are imposed at both ends: $y(a) < \infty$ and $y(b) < \infty$. Such boundary conditions usually appear when the values of the general solutions at the boundary are infinite.

4. General cases

The most general boundary conditions of ordinary differential equations are of the form of

$$B_1(y) = \alpha_{1,1}y(a) + \alpha_{1,2}y'(a) + \beta_{1,1}y(b) + \beta_{1,2}y'(b) = \gamma_1, \quad (3.2.18a)$$

$$B_2(y) = \alpha_{2,1}y(a) + \alpha_{2,2}y'(a) + \beta_{2,1}y(b) + \beta_{2,2}y'(b) = \gamma_2, \quad (3.2.18b)$$

where $B_1(y)$ and $B_2(y)$ are abbreviations, usually written as

$$B_1(y) = \gamma_1, \quad B_2(y) = \gamma_2. \quad (3.2.18c)$$

They are further written in shorthand as

$$B(y) = \gamma. \quad (3.2.18d)$$

This, when written in matrix form, is

$$\begin{pmatrix} \alpha_{1,1} & \alpha_{1,2} \\ \alpha_{2,1} & \alpha_{2,2} \end{pmatrix} \begin{pmatrix} y(a) \\ y'(a) \end{pmatrix} + \begin{pmatrix} \beta_{1,1} & \beta_{1,2} \\ \beta_{2,1} & \beta_{2,2} \end{pmatrix} \begin{pmatrix} y(b) \\ y'(b) \end{pmatrix} = \begin{pmatrix} \gamma_1 \\ \gamma_2 \end{pmatrix}. \quad (3.2.18e)$$

When both γ_1 and γ_2 are zero, or $\gamma = 0$, they are called **homogeneous boundary conditions**, otherwise called **inhomogeneous boundary conditions**. The three kinds of boundary conditions, periodic boundary conditions and natural boundary conditions listed above are specific cases, and also are those often met in a practical problem. In addition, there may be other conditions. For instance, the vibration of an oscillator often gives its displacement and velocity at initial time, which are initial conditions as in (3.1.3). We will vaguely say boundary condition no matter what are boundary conditions, initial conditions or others.

3.2.3. *Sturm-Liouville eigenvalue problem*

Definition 2. Equation (3.2.4) and given boundary conditions constitute a **Sturm-Liouville eigenvalue problem**, or briefly an **SL problem**. Under the given boundary conditions, the λ values that make the equation which have nonzero solutions are called **eigenvalues** of the equations. Each solution corresponding to an eigenvalue λ is called the **eigenfunction** of λ.

1. Four theorems of an SL problem

The four theorems below summarize the common features of eigenvalues and eigenfunctions of an SL problem.

Theorem 1 (Existence Theorem). *If on $[a, b]$ $p(x)$ and its derivative are continuous, and $q(x)$ is continuous or at most has first-order poles at the boundaries, then an SL problem is of infinite number of discrete real eigenvalues, which constitute a monotonically rising sequence:*

$$\lambda_1 \leq \lambda_2 \leq \lambda_3 \leq \lambda_4 \leq \cdots \leq \lambda_n \leq \lambda_{n+1} \leq \cdots . \tag{3.2.19}$$

There are infinite characteristic functions corresponding to these eigenvalues:

$$y_1(x), y_2(x), y_3(x), \ldots, \tag{3.2.20}$$

which are called an eigenfunction set. If the region is infinite and natural boundary conditions are used, the eigenvalues may be continuous.

Definition 3. The ensemble of eigenvalues $\{\lambda_n\}$ is called the **spectrum** of the eigenvalue problem. If $\{\lambda_n\}$ contains a series of discrete values, then it is called a **discrete spectrum**, and if it is continuous, called a **continuous spectrum**.

Theorem 2 (Nonnegative Theorem). *If in $(3.2.4)$ $\rho(x) > 0$, $p(x) \geq 0$, $q(x) \geq 0$, then all the eigenvalues are nonnegative:*

$$\lambda_n \geq 0, \quad n = 1, 2, 3, \ldots . \tag{3.2.21}$$

Theorem 3 (Orthogonality Theorem). *If two eigenvalues λ_m and λ_n are different, then the eigenfunctions $y_m(x)$ and $y_n(x)$ belonging to λ_m and λ_n respectively are orthogonal with weight $\rho(x)$ on $[a, b]$, i.e.,*

$$\int_a^b y_m^*(x) y_n(x) \rho(x) \mathrm{d}x = 0. \tag{3.2.22}$$

Theorem 4 (Completeness Theorem). *The eigenfunction set* $\{y_n(x)\}$ *is a complete one on* $[a, b]$. *Suppose a function* $f(x)$ *has first derivative and at least piecewise continuous second derivative on the closed interval. Then the function* $f(x)$ *can be expanded by the eigenfunction set* $\{y_n(x)\}$:

$$f(x) = \sum_{n=1}^{\infty} f_n y_n(x). \qquad (3.2.23)$$

The series on the right hand side converges uniformly to $f(x)$ *at every point. The numbers* f_n *are called expansion coefficients. They are determined by making inner products of* $f(x)$ *and eigenfunctions with weight* $\rho(x)$:

$$f_n = \frac{\int_a^b \rho(x) f(x) y_n^*(x) \mathrm{d}x}{\int_a^b \rho(x) |y_n(x)|^2 \mathrm{d}x}. \qquad (3.2.24)$$

The series on the right hand side of (3.2.23) *is also called a **generalized Fourier series**, and* f_n *called **generalized Fourier coefficients**. In* (3.2.24), *the denominator is just the square of the modulus of* $y_n(x)$, *denoted as* N_n:

$$N_n^2 = \int_a^b \rho(x) |y_n(x)|^2 \mathrm{d}x. \qquad (3.2.25)$$

N_n *is also called a **normalizing factor**. Hence,* (3.2.24) *is also recast to*

$$f_n = \frac{1}{N_n^2} \int_a^b \rho(x) f(x) y_n^*(x) \mathrm{d}x. \qquad (3.2.26)$$

If the modulus of characteristic function $y_n(x)$ equals to 1, $y_n(x)$ is called a **normalized eigenfunction**. In this specific case, (3.2.26) is simplified to be

$$f_n = \int_a^b \rho(x) f(x) y_n^*(x) \mathrm{d}x. \qquad (3.2.27)$$

Any $y_n(x)$ can be normalized if divided by its modulus.

In many cases, the condition of Theorem 4 can be relaxed. It is required that $f(x)$ and $f'(x)$ are piecewise continuous. Then the series on the right hand side of (3.2.23) converges uniformly to $f(x)$ at every point where $f(x)$ is continuous and to the value $[f(x_0 + 0^+) + f(x_0 - 0^+)]/2$ if x_0 is a point of discontinuity. In summary,

$$\sum_{n=1}^{\infty} f_n y_n(x) = \frac{1}{2}[f(x+0^+) + f(x-0^+)].$$

We will give some explicit examples in Sections 3.4 and 4.7 for various eigenfunction sets.

2. Proofs of the theorems

We intend to present the proofs of the first three theorems. As a preparation, we can prove that the following results.

$$[p(x)y(x)y'(x)]_{x=a} \geq 0, \quad [p(x)y(x)y'(x)]_{x=b} \leq 0. \quad (3.2.28)$$

$$[p(x)W(y_1, y_2)]_a^b = 0. \quad (3.2.29)$$

Let λ_1 and λ_2 be eigenvalues and their corresponding eigenfunctions be y_1 and y_2. Then we have two equations

$$\frac{\mathrm{d}}{\mathrm{d}x}\left[p(x)\frac{\mathrm{d}y_1}{\mathrm{d}x}\right] = -[\lambda_1 \rho(x) - q(x)]y_1$$

and

$$\frac{\mathrm{d}}{\mathrm{d}x}\left[p(x)\frac{\mathrm{d}y_2^*}{\mathrm{d}x}\right] = -[\lambda_2^* \rho(x) - q(x)]y_2^*,$$

respectively. Multiplying y_2^* to the former and y_1 to the latter, and then substracting them, one obtains

$$(\lambda_1 - \lambda_2^*)\rho(x)y_1 y_2^* = y_2^* \frac{\mathrm{d}}{\mathrm{d}x}\left[p(x)\frac{\mathrm{d}y_1}{\mathrm{d}x}\right] - y_1 \frac{\mathrm{d}}{\mathrm{d}x}\left[p(x)\frac{\mathrm{d}y_2^*}{\mathrm{d}x}\right]$$

$$= -\frac{\mathrm{d}}{\mathrm{d}x}[p(x)W(y_1, y_2)].$$

Integrating this equation leads to

$$(\lambda_1 - \lambda_2^*)\int_a^b \mathrm{d}x \rho(x)y_1 y_2^* = -\frac{\mathrm{d}}{\mathrm{d}x}[p(x)W(y_1, y_2)]_a^b = 0, \quad (3.2.30)$$

where (3.2.29) is used. Having these as preparation, we are now ready to prove Theorem 1–3.

Proof of Theorem 1.

(i) Every eigenvalue λ is real. In (3.2.30), let $\lambda_1 = \lambda_2$, $y_1 = y_2$. Then

$$(\lambda_1 - \lambda_1^*) \int_a^b \mathrm{d}x \rho(x) |y_1(x)|^2 = 0.$$

Because $\rho(x)$ does not change its sign on $[a, b]$, $\int_a^b \mathrm{d}x \rho(x)$ $|y_1(x)|^2 \neq 0$. The only possibility is $\lambda_1 = \lambda_1^*$. Thus every eigenvalue is real.

(ii) Eigenvalues λ are discrete. The two linearly independent solutions belonging to eigenvalue λ are denoted as $y_1(x, \lambda)$ and $y_2(x, \lambda)$, which explicitly show the dependence on the eigenvalue λ. The general solution of (3.2.4) is

$$y = A y_1(x, \lambda) + B y_2(x, \lambda),$$

where constants A and B are to be determined by boundary conditions. For the three kinds of homogeneous boundary conditions (3.2.14),

$$\alpha_1 [A y_1'(a, \lambda) + B y_2'(a, \lambda)] + \alpha_2 [A y_1(a, \lambda) + B y_2(a, \lambda)] = 0,$$
$$\beta_1 [A y_1'(b, \lambda) + B y_2'(b, \lambda)] + \beta_2 [A y_1(b, \lambda) + B y_2(b, \lambda)] = 0.$$

The condition that A and B have nonzero solutions is

$$\Delta(\lambda) = \begin{vmatrix} \alpha_1 y_1'(a, \lambda) + \alpha_2 y_1(a, \lambda) & \alpha_1 y_2'(a, \lambda) + \alpha_2 y_2(a, \lambda) \\ \beta_1 y_1'(b, \lambda) + \beta_2 y_1(b, \lambda) & \beta_1 y_2'(b, \lambda) + \beta_2 y_2(b, \lambda) \end{vmatrix} = 0.$$

$$(3.2.31)$$

For the periodic boundary condition (3.2.15a),

$$A y_1(a, \lambda) + B y_2(a, \lambda) = A y_1(b, \lambda) + B y_2(b, \lambda),$$
$$A y_1'(a, \lambda) + B y_2'(a, \lambda) = A y_1'(b, \lambda) + B y_2'(b, \lambda).$$

The condition that A and B have nonzero solutions is

$$\Delta(\lambda) = \begin{vmatrix} y_1(a, \lambda) - y_1(b, \lambda) & y_2(a, \lambda) - y_2(b, \lambda) \\ y_1'(a, \lambda) - y_1'(b, \lambda) & y_2'(a, \lambda) - y_2'(b, \lambda) \end{vmatrix} = 0. \qquad (3.2.32)$$

The eigenvalue λ can be solved from (3.2.31) and (3.2.32), i.e., λ is a root of $\Delta(\lambda)$. According to the theory of differential equations, $\Delta(\lambda)$ is analytical on the whole complex λ plane. Eigenvalue λ is known to be real, so that the zeroes of $\Delta(\lambda)$ are on the real axis. At finite places, $\Delta(\lambda)$ has merely a finite number of zeros. That is to say, in a finite interval the roots of $\Delta(\lambda) = 0$ is finite (otherwise $\Delta(\lambda) \equiv 0$). Therefore, eigenvalues are discrete. $\qquad\square$

In the case of the natural boundary conditions and finite interval $[a, b]$, eigenvalues are also discrete.

For some specific $p(x), \rho(x), q(x)$, their corresponding eigenfunctions have been obtained. Under this condition, (3.2.31) and (3.2.32) in fact present the way of solving eigenvalues. For example, the eigenvalues of Bessel equation are solved in this way, as will be seen in Subsection 4.7.2.

For the case of a natural boundary condition at infinity, it is possible for continuous eigenvalues to appear. For example, equation $y'' + \lambda^2 y = 0$ with natural boundary condition at infinity is such a case.

Proof of Theorem 2.

In (3.2.4), let $\lambda = \lambda_n$ be an eigenvalue and its eigenfunction be $y = y_n$. Multiplying y_n^* and integrating from a to b on both hand sides, one has

$$-\int_a^b dx[\lambda_n\rho(x) - q(x)]y_n y_n^*$$

$$= \int_a^b dx y_n^* \frac{d}{dx}\left[p(x)\frac{dy_n}{dx}\right] = [y_n^* p(x)y_n']_a^b - \int_a^b dx p(x)\frac{dy_n}{dx}\frac{dy_n^*}{dx},$$

where integration by parts has been carried out one time. Transposition leads to

$$\lambda_n \int_a^b dx \rho(x)|y_n|^2 = -[yp(x)_n^*(x)y_n'(x)]_{x=b} + [p(x)y_n^*(x)y_n'(x)]_{x=a}$$

$$+ \int_a^b dx p(x)\left|\frac{dy_n}{dx}\right|^2 + \int_a^b dx q(x)|y_n|^2.$$

On the right hand side of this equation, the latter two terms are nonnegative because $p(x) \geq 0$, $q(x) \geq 0$, and the former two terms are either so due to (3.2.28). Thus, $\lambda_n \int_a^b \mathrm{d}x \rho(x)|y_n|^2 \geq 0$. From $\rho(x) \geq 0$, it is known that $\lambda_n \geq 0$. $\qquad \square$

Proof of Theorem 3.

In (3.2.30), let $\lambda_1 = \lambda_m, \lambda_2^* = \lambda_n, y_1 = y_m, y_2^* = y_n^*$. Then

$$(\lambda_m - \lambda_n) \int_a^b y_m(x) y_n^*(x) \rho(x) \mathrm{d}x = 0.$$

Because $\lambda_m \neq \lambda_n$, $\int_a^b y_m(x) y_n^*(x) \rho(x) \mathrm{d}x = 0$. $\qquad \square$

SL operator is a Hermitian one. The conclusions obtained in (i) of Theorem 1 and Theorem 3 are those of self-adjoint operators. They have actually been included in the corollary of Theorem 7 in Subsection 2.2.2. Here is a restate in the case of an SL operator.

The proof of Theorem 4 is beyond the scope of this book. We show below the proof only for a special case that the eigenfunctions of SL equations are polynomials.

In view of the definition of the SL operator L (3.2.6), (3.2.1) has been rewritten in the form of (3.2.7). When we mention the eigenvalues and eigenfunctions we mean that they satisfy

$$Ly_n(x) = \lambda_n y_n(x), \quad n = 1, 2, 3, \ldots. \tag{3.2.33}$$

Eigenvalues are also called **intrinsic values**, and eigenfunctions called **intrinsic functions**. In the following (3.2.19) is briefly denoted as $\{\lambda_n\}$, and (3.2.20) denoted as $\{y_n(x)\}$.

The theorem of the existence and uniqueness of solutions presented in Subsection 3.1.1 is still valid. If one solution is found, another linearly independent solution can be calculated by the Liouville formula. The solution is the linear combination of them. The combination coefficients are determined by boundary conditions coming from a concrete physical problem, which determines the eigenvalue. In view of Theorem 4, eigenfunction set $\{y_n(x)\}$ is complete, and in the set each $y_n(x)$ is the linear combination of two linearly independent solutions.

We have proved the existence and uniqueness of solutions of differential equations under initial conditions. For the cases of boundary conditions, we have not done so yet. For an ordinary differential equation, there can be nonzero eigenfunctions under certain boundary conditions, and there may be not so under other boundary conditions.

3.3. The Polynomial Solutions of Sturm-Liouville Equations

Polynomials are functions with the simplest form. According to Weierstrass theorem, any function on a closed interval can be approximated by polynomials. Therefore, we pay special attention to polynomials. Here we discuss the cases that the solutions of SL problems arc polynomials. The form of the solutions

$$y_n(x) = Q_n(x) = \sum_{i=0}^{n} c_i x^i \tag{3.3.1}$$

are n-degree polynomials. In these cases, the SL problem is also called **SL systems of polynomials**.

3.3.1. *Possible forms of kernel and weight functions*

1. Possible parameters

That solutions must be polynomials stipulates some requirements for the kernel and weight functions. Taking $n = 0, 1, 2$ in (3.3.1) and then substituting them into (3.2.1), we get

$$[\lambda_0 - C(x)]Q_0 = [\lambda_0 - C(x)]c_0 = 0,$$

$$B(x)Q_1' + [\lambda_1 - C(x)]Q_1 = B(x)c_1 + [\lambda_1 - C(x)](c_1 x + c_0) = 0$$

and

$$A(x)Q_2'' + B(x)Q_2' + [\lambda - C(x)]Q_2$$
$$= A(x)c_2 + B(x)(2c_2 x + c_1) + [\lambda - C(x)](c_2 x^2 + c_1 x + c_0) = 0.$$

It is thus seen that the three functions $A(x)$, $B(x)$ and $C(x)$ have to be of the following forms:

$$A(x) = A_0 + A_1 x + A_2 x^2, \tag{3.3.2a}$$

$$B(x) = B_0 + B_1 x, \tag{3.3.2b}$$

$$C(x) = C_0. \tag{3.3.2c}$$

These equations reveal that all of the features of the SL systems of polynomials can be determined by six parameters $A_0, A_1, A_2, B_0, B_1, C_0$. Obviously, C_0 can be merged into λ, or set $C_0 = 0$, so that five adjustable parameters are left. Furthermore, for the SL operator L defined by (3.2.6), three among the five parameters are not independent.

(i) The operator L can be multiplied by a constant α_1, which does not change its solution Q_n but simply multiplies the eigenvalues by α_1.

(ii) When argument x is shifted by α_2, $Q_n(x)$ is simply replaced by $Q_n(x + \alpha_2)$, while the eigenvalues remain unchanged.

(iii) When argument x is multiplied by a constant α_3, $x \to \alpha_3 x$, $Q_n(x)$ is simply replaced by $Q_n(\alpha_3 x)$ while its eigenvalues remain unchanged.

In view of the features of the L operator, three parameters can be fixed. Subsequently, the totally independent parameters left becomes $5 - 3 = 2$. We will see below that how to pick the two parameters determine the eigenvalues and eigenfunctions of the SL systems.

In the condition (3.2.10), if both f and g are polynomials, it is required that as $x \to \infty$, $p(x)$ has to approach zero more rapidly than the inverse of any powers of x. Apparently, when $B(x) = 0$, $p(x)$ will be a constant, and thus cannot meet this requirement. A simplest example is the equation $y'' + y = 0$, where $B(x) = 0$. Its two linearly independent solutions are $y = \sin x$ and $y = \cos x$, which do not go to zero as $x \to \infty$. When $B(x) = $ constant, the requirement is neither met, see Exercise 9. So, $B(x)$ must possess the x term.

$A(x)$ can be a quadratic or linear function of x, and can either be a constant. We discuss the three cases below.

2. $A(x)$ is a quadratic function

In this case, it follows from (3.2.3) that

$$p(x) = \exp \int \frac{B_0 + B_1 x}{A_0 + A_1 x + A_2 x^2} \, dx. \qquad (3.3.3)$$

(1) $A(x)$ has complex roots. For the sake of convenience, set $A_2 = 1$. Let ω and ω^* be two complex roots of $A(x)$, then (see Exercise 10),

$$p(x) = [A(x)]^{B_1/2} \exp \left[\frac{B_0 + B_1 \text{Re}\omega}{|\text{Im}\omega|} \tan^{-1} \left(\frac{x - \text{Re}\omega}{|\text{Im}\omega|} \right) \right]. \qquad (3.3.4)$$

It is seen that $p(x)$ will never be zero, even if $x \to \infty$. Thus (3.2.10) is not satisfied. The case of complex roots of $A(x)$ is excluded.

(2) $A(x)$ has real roots. For convenience, set $A_2 = 1$ and $A(x) = 1 - x^2$. Then its roots are $\omega = \pm 1$. Take $B_0 = s - r$ and $B_1 = -(r + s + 2)$, so that the two parameters r and s are to be determined. We have

$$\frac{B(x)}{A(x)} = \frac{-(r + s + 2)x + s - r}{1 - x^2} = \frac{s + 1}{1 + x} - \frac{r + 1}{1 - x}. \qquad (3.3.5)$$

It is derived from (3.2.3) that

$$p(x) = (1 + x)^{s+1}(1 - x)^{r+1} \qquad (3.3.6)$$

and

$$\rho(x) = (1 + x)^s (1 - x)^r. \qquad (3.3.7)$$

As $|x| \to \infty$, $p(x)$ is unable to approach zero faster than the inverse of any power of x. Nevertheless, we can let $r > -1$, $s > -1$, and then at $x = \pm 1$, $p(x = \pm 1) = 0$. Outside of the interval, the kernel is set to be zero. Accordingly, $\rho(x)$ is also set to zero outside of the interval $[-1, 1]$. Thus the desired weight function $\rho(x)$ is as follows.

$$\rho(x) = \begin{cases} (1+x)^s(1-x)^r, & |x| \le 1, \\ 0, & |x| > 1, \end{cases} \quad (r, s > -1). \qquad (3.3.8)$$

The solution $Q_n(x)$ thus carries two parameters r and s, called a **Jacobi polynomial** (or a **Generalized ultraspheric polynomial**) with indices r and s, denoted by $P_n^{(r,s)}(x)$.

If we let $x \to 1 - 2x$, then the interval is converted to $[0, 1]$, and in the interval $A(x) = x(1-x)$ and $B(x) = -(r+s+2)x + r + 1$. Subsequently,

$$\frac{B(x)}{A(x)} = \frac{-(r+s+2)x + r + 1}{x(1-x)} = \frac{r+1}{x} - \frac{s+1}{1-x}. \qquad (3.3.9)$$

It brings

$$p(x) = x^{r+1}(1-x)^{s+1}. \qquad (3.3.10)$$

So, for $r, s > -1$, the weight is

$$\rho(x) = x^r(1-x)^s. \qquad (3.3.11)$$

The corresponding polynomials are also called Jacobi polynomials, and denoted by $J_n^{(r,s)}(x)$. This means that we have two cases: one is the weight $\rho(x) = (1+x)^s(1-x)^r$ on $[-1, 1]$ and the other is the weight $\rho(x) = x^r(1-x)^s$ on $[0, 1]$. The polynomials of the two cases differ from each other merely due to a linear transformation of the argument, both being called **Jacobi polynomials**.

Let us see some particular cases of Jacobi polynomials on $[-1, 1]$, or some particular values of r and s. In the cases of $r = s$, it is seen from (3.3.5) that $B(x) = -2(r+1)x$, i.e., $B_1 = -2(r+1)$ and $B_0 = 0$.

As long as $B_0 = 0$, the SL operator L is invariant under the argument replacement $x \to -x$. Let P be an operator which converts x to $-x$, then $PL = LP$. That the two operators exchangeable means that any characteristic function of L can also be the characteristic function of P. P is called the **parity operator**. It plays an important role in physics.

(i) When $r = s = \mu(> -1)$, $\rho(x) = (1 - x^2)^\mu$. The corresponding polynomials are called **Gegenbauer polynomials** (or **ultra-spheric polynomials**) with index μ, denoted by $G_n^\mu(x) \equiv P_n^{(\mu,\mu)}(x)$.

(ii) When $r = s = -1/2$, $\rho(x) = (1 - x^2)^{-1/2}$. We denote $T_n(x) \equiv J_n^{(-1/2,-1/2)}(x)$. If we let $x = \cos\theta$, then $\theta \in [-\pi, \pi]$ and $T_n(\cos\theta) \propto \cos n\theta$. So, on $[-1, 1]$, $T_n(x) \propto \cos(n\cos^{-1}x)$. Such polynomials are **Chebyshev polynomials**, which have been introduced in Subsection 2.4.2.

(iii) When $r = s = 0$, $\rho(x) = 1$. Then $J_n^{(0,0)}(x) = P_n(x)$ are called **Legendre polynomials**.

3. $A(x)$ is a linear function of x

Let $A(x) = x$ for convenience. Thus the two free parameters are left in $B(x)$. Then, (3.2.3) leads to

$$p(x) = \exp\left[\int \frac{B_0 + B_1 x}{x} dx\right] = \exp(B_0 \ln x + B_1 x) = x^{B_0} e^{B_1 x}.$$

$$(3.3.12)$$

If $B_1 < 0$, then $\rho(x)$ must be zero for all $x < a$ (a is to be determined) so as to be zero at infinity. If $B_0 > 0$, then at $x = 0$, $p(0) = 0$. Thus choosing $x = 0$ as the connecting point we can meet the requirement. If $B_1 > 0$, then $p(x)$ must be zero along the positive x axis. The convention is to take $B_1 < 0$. For simplicity, let $B_1 = -1$.

Then it should be taken $B_0 = s + 1, s > -1$. For positive x,

$$\rho(x > 0) = x^s e^{-x}, \qquad (3.3.13)$$

and for negative x, $\rho(x \leq 0) = 0$. The polynomials $Q_n(x)$ under this condition are called **Sonine polynomials**, denoted by $S_n^\mu(x)$. If $s = m$ is an integer, then $S_n^\mu(x)$ becomes $L_n^m(x)$, called **associated Laguerre polynomials** of m order. In particular, if $m = 0$, then $L_n^0(x) \equiv L(x)$ are called **Laguerre polynomials**.

In this case the polynomials are defined on the interval $[0, \infty)$ which is not symmetric with respect to the origin. Thus, there is no requirement of the invariance of the SL operator L under the transformation $x \rightarrow -x$. B_0 can therefore be nonzero. These equations and functions are encountered when treating problems involving radial components. For example, separation of variables of Helmholtz equations in 2-dimensional or 3-dimensional spherical coordinates will create radial equations.

4. $A\,(x)$ is a constant

Choosing $A(x) = 1$ is the most convenient. It follows from (3.2.3) that

$$p(x) = \exp \left[\int (B_0 + B_1 x)\mathrm{d}x \right]$$

$$= \mathrm{e}^{B_0 x} \mathrm{e}^{B_1 x^2/2} = \beta \exp \left[\frac{1}{2} B_1 \left(x + \frac{B_0}{B_1} \right)^2 \right]. \quad (3.3.14)$$

Choosing $B_1 < 0$ makes $p(x)$ approach zero fast as $x \rightarrow \pm\infty$. Hence the condition at infinity is always met. Let $B_1 = -2, B_0 = 0$ and $\beta = 1$. Then

$$\rho(x) = \mathrm{e}^{-x^2}. \quad (3.3.15)$$

The corresponding $Q_n(x)$ are called **Hermite polynomials**.

In this case, the interval $(-\infty, \infty)$ is symmetric with respect to the origin, and the SL operator L is invariant under transformation $x \rightarrow -x$.

The results above are summarized in Table 3.1.

Please note that the ordinary differential equation of second order (3.2.1) possesses two linearly independent solutions. When their linear combination satisfies certain boundary conditions, the parameter λ in (3.2.1) takes a series of specific values, i.e., eigenvalues. Accordingly, the solutions that meet the boundary conditions are eigenfunctions. In Table 3.1, one of the special solutions related to specific weight $\rho(x)$ is given, which is a polynomial expansion around $x = 0$. Another specific solution is not given, but the eigenvalues λ_n are listed. This is because under the present boundary conditions there is only the solution of polynomials, and the other solution does not appear (or its linear combination coefficient is zero). The present boundary condition is: the solutions is finite at the boundaries of

Table 3.1.   Complete sets of orthogonal polynomials with parameters of Sturm-Liouville systems.

$A(x)$	A_0	A_1	A_2	B_0	B_1	Weight $\rho(x)$	Interval	Name of polynomial eigenvalue λ_n
Quadratic function of x	1	0	-1	$s-r$	$-(r+s+2)$	$(1-x)^r(1+x)^s$	$[-1,1]$	Jacobi $n(n+r+s+1)$
	0	1	-1	$r+1$	$-(r+s+2)$	$x^r(1-x)^s$	$[0,1]$	Jacobi $n(n+r+s+1)$
	1	0	-1	0	$-2(\mu+1)$	$(1-x^2)^m$	$[-1,1]$	Gegenbauer $n(n+2\mu+1)$
	1	0	-1	0	-1	$(1-x^2)^{-1/2}$	$[-1,1]$	Chebyshev n^2
	1	0	-1	0	-2	1	$[-1,1]$	Legendre $n(n+1)$
Linear function of x	0	1	0	$\mu+1$	-1	$x^\mu e^{-x}$	$[0,\infty)$	Sonine n
	0	1	0	$m+1$	-1	$x^m e^{-x}$	$[0,\infty)$	Associated Laguerre $n(\geq m)$
	0	1	0	1	-1	e^{-x}	$[0,\infty)$	Laguerre n
constant	1	0	0	0	-2	e^{-x^2}	$(-\infty,\infty)$	Hermite $2m$

$r, s > -1$, m and n are natural numbers, and μ is any complex number other than negative integers.

the interval. (Please note that generally, the points in the boundary conditions may be inside the interval, not necessarily at the ends.)

3.3.2. *The expressions in series and in derivatives of the polynomials*

1. Series expressions of Sturm-Liouville polynomials

The series expressions of Sturm-Liouville polynomial are listed in Table 3.2.

In Table 3.2, the following symbols are used.

$$\left[\frac{l}{2}\right] = \begin{cases} l/2, & l \text{ is even} \\ (l-1)/2, & l \text{ is odd} \end{cases}$$

$$(\alpha)_k = \alpha(\alpha+1)\cdots(\alpha+k-1) = \frac{\Gamma(\alpha+k)}{\Gamma(\alpha)}, \quad (\alpha)_0 = 1. \quad (3.3.16a)$$

$(\alpha)_k$ is called a **Gauss symbol**. If $\alpha = n$ is an integer,

$$(n)_k = n(n+1)\cdots(n+k-1) = \frac{(n+k-1)!}{(n-1)!}. \quad (3.3.16b)$$

$\Gamma(\alpha)$ function is defined by

$$\Gamma(z) = \int_0^\infty e^{-t}t^{z-1}dt, \quad \text{Re } z > 0, \quad (3.3.17)$$

where z is a complex number.

2. Generalized Rodrigues' formula

The polynomial sets of SL system can be uniformly expressed by a generalized formula:

$$Q_n(x) = K_n \frac{1}{\rho(x)} \frac{d^n}{dx^n}[A^n(x)\rho(x)], \quad (3.3.18)$$

where the constant K_n is introduced because different polynomial functions are normalized differently. This expression declares orthogonal polynomials. To confirm this we have to show three points: one is that $Q_n(x)$ in this expression is a polynomial with the highest

Table 3.2. Series expressions of Sturm-Liouville polynomials.

Name of polynomials	Series expression
Jacobi	$P_n^{(r,s)}(x) = (r+1)_n \sum\limits_{k=0}^{n} \dfrac{(r+s+1+n)_k}{k!(n-k)!(r+1)_k} \left(\dfrac{x-1}{2}\right)^k$
Jacobi	$J_n^{(r,s)}(x) = \dfrac{(s-r-n)_n}{(s)_n} \sum\limits_{k=0}^{n} \dfrac{(-1)^k n!(r+n)_k}{k!(n-k)!(1+r-s)_k}(1-x)^k$
Gegenbauer	$G_n^{\mu}(x) = \sum\limits_{k=0}^{n} \dfrac{(n-k-1)!!}{k!(n-2k)!} \dfrac{(2\mu)_{n+k}}{(\mu+1/2)_k} \left(\dfrac{x-1}{2}\right)^k$
Chebyshev	$T_0(x) = 1,$ $T_n(x) = \dfrac{n}{2} \sum\limits_{k=0}^{[n/2]} (-1)^k \dfrac{(n-k-1)!}{k!(n-2k)!}(2x)^{n-2k}, \quad (n \geq 1)$
Legendre	$P_n(x) = \sum\limits_{k=0}^{[n/2]} \dfrac{(-1)^k(2n-2k)!}{2^n k!(n-k)!(n-2k)!} x^{n-2k}$
Sonine	$S_n^{\mu}(x) = \sum\limits_{k=0}^{n} \dfrac{(-1)^k}{k!(n-k)!} \dfrac{(\mu+1)_n}{(\mu+1)_k} x^k$
Associated Laguerre	$L_n^{(m)}(x) = \sum\limits_{k=0}^{n-m} \dfrac{(-1)^{m+k}(n!)^2}{k!(m+k)!(n-m-k)!} x^k$
Laguerre	$L_n(x) = \sum\limits_{k=0}^{n} \dfrac{(-1)^k(n!)^2}{(k!)^2(n-k)!} x^k$
Hermite	$H_n(x) = \sum\limits_{k=0}^{[n/2]} \dfrac{(-1)^k n!}{k!(n-2k)!}(2x)^{n-2k}$

degree n; the next is that the polynomials are orthogonal to each other; the third is that the $Q_n(x)$ is a solution of SL systems. The following are the proof.

(1) $\dfrac{1}{\rho(x)}\dfrac{\mathrm{d}^n}{\mathrm{d}x^n}[A^n(x)\rho(x)]$ is a polynomial.

Firstly, we show that if

$$f(x) = A^k(x)\rho(x)r_l(x), \qquad (3.3.19)$$

where $r_l(x) = \sum_{i=0}^{l} c_i x^i$ is a polynomial of x of degree l, then

$$f'(x) = A^{k-1}(x)\rho(x)s_{l+1}(x), \qquad (3.3.20\text{a})$$

where $s_{l+1}(x)$ is a polynomial of x of degree $l+1$. In view of (3.2.3) and (3.2.5),

$$\begin{aligned}
f'(x) &= [A^{k-1}(x)]'p(x)r_l(x) + A^{k-1}(x)p'(x)r_l(x) \\
&\quad + A^{k-1}(x)p(x)r_l'(x) \\
&= A^{k-1}(x)\rho(x)[(k-1)A'(x)r_l(x) + B(x)r_l(x) + A(x)r_l'(x)] \\
&= A^{k-1}(x)\rho(x)[c_l(A_2(2k+l-2)+B_1)x^{l+1} + \cdots].
\end{aligned}$$

$$(3.3.20\text{b})$$

Given $c_l \neq 0$, $B_1 \neq 0$ and $B_1 < 0$, then A_2 and B_1 have the same sign, as reflected in Table 3.1. Therefore, when $k \geq 1$, the coefficient of the $l+1$ power term in the brackets is nonzero $c_l(A_2(2k+l-2)+B_1)$. That is to say it is a polynomial of $l+1$ degrees, as manifested by (3.3.20).

Now take the derivative of $f(x) = A^n(x)\rho(x)$ n times with respect to x. then, from (3.3.19), where $r_0(x) = 1$, we have

$$\frac{\mathrm{d}^l}{\mathrm{d}x^l}[A^n(x)\rho(x)] = A^{n-l}(x)\rho(x)s_{0+l}(x) = A^{n-l}(x)\rho(x)s_n(x).$$

$$(3.3.21\text{a})$$

Especially, when $l = n$,

$$\frac{\mathrm{d}^n}{\mathrm{d}x^n}[A^n(x)\rho(x)] = \rho(x)s_n(x). \qquad (3.3.21\text{b})$$

Therefore, $\frac{1}{\rho(x)}\frac{\mathrm{d}^n}{\mathrm{d}x^n}[A^n(x)\rho(x)]$ is a polynomial of x of n degrees.

(2) The polynomials are orthogonal to each other:

$$\int_a^b Q_m^* Q_n \rho \mathrm{d}x = 0, \qquad \text{as } n \neq m. \qquad (3.3.22)$$

For simplicity without loss of generality, let $n > m$.

$$\int_a^b Q_m^* Q_n \rho \mathrm{d}x = \int_a^b Q_m^* \frac{\mathrm{d}^n}{\mathrm{d}x^n}[A^n(x)\rho(x)]\mathrm{d}x$$

$$= \left\{ Q_m^* \frac{\mathrm{d}^{n-1}}{\mathrm{d}x^{n-1}}[A^n(x)\rho(x)] \right\}_{-\infty}^{\infty}$$

$$- \int_a^b \left(\frac{\mathrm{d}}{\mathrm{d}x} Q_m^* \right) \frac{\mathrm{d}^{n-1}}{\mathrm{d}x^{n-1}}[A^n(x)\rho(x)]\mathrm{d}x$$

$$= \{Q_m^* A(x)\rho(x)s_{n-1}(x)\}_{-\infty}^{\infty}$$

$$- \int_a^b \left(\frac{\mathrm{d}}{\mathrm{d}x} Q_m^* \right) \frac{\mathrm{d}^{n-1}}{\mathrm{d}x^{n-1}}[A^n(x)\rho(x)]\mathrm{d}x,$$

where integration by parts is carried out and (3.3.21a) is utilized. Because as $x \to \infty$, $p(x) = A(x)\rho(x)$ approaches zero faster than the inverse of any power of x, the first term in the last line is zero. The second term is treated in the same way. Repeating the process m times, we get

$$\int_a^b Q_m^* Q_n \rho \mathrm{d}x$$

$$= (-1)^{m+1} \int_a^b \left(\frac{\mathrm{d}^{m+1}}{\mathrm{d}x^{m+1}} Q_m^* \right) \frac{\mathrm{d}^{n-m-1}}{\mathrm{d}x^{n-m-1}}[A^n(x)\rho(x)]\mathrm{d}x.$$

Here the integrand is zero. So, (3.3.22) is proved. Equation (3.3.22) demonstrates that $\{Q_n(x)\}$ is a linearly independent set of polynomials.

Following this result, it is easy to prove that for a polynomial r_k of m degrees,

$$\int_a^b r_k(x)Q_n(x)\rho(x)\mathrm{d}x = 0, \quad n > k. \tag{3.3.23}$$

This is because r_m can be expanded by the linearly independent set $\{Q_n(x)\}$:

$$r_k(x) = \sum_{m=0}^k g_m Q_m(x). \tag{3.3.24}$$

Then, in view of (3.3.22), we get

$$\int_a^b r_k(x)Q_n(x)\rho(x)\mathrm{d}x = \sum_{m=0}^{k} g_m \int_a^b Q_m(x)Q_n(x)\rho(x)\mathrm{d}x = 0.$$

(3) $Q_n(x) = \frac{1}{\rho(x)}\frac{\mathrm{d}^n}{\mathrm{d}x^n}[A^n(x)\rho(x)]$ is a solution of SL systems.

Let

$$R_n(x) = \frac{1}{\rho(x)}\frac{\mathrm{d}}{\mathrm{d}x}\left[p(x)\frac{\mathrm{d}Q_n(x)}{\mathrm{d}x}\right]. \qquad (3.3.25)$$

Because $Q_n(x)$ is a polynomial of degree n, $\frac{\mathrm{d}Q_n(x)}{\mathrm{d}x}$ is one with degree $n-1$. In view of (3.3.20a), $R_n(x)$ is a polynomial of degree n.

We first show that $R_n(x)$ and $Q_m(x)$ are orthogonal to each other with weight $\rho(x)$ as $m < n$:

$$I_{mn} = \int_a^b Q_m(x)R_n(x)\rho(x)\mathrm{d}x = 0, \quad (m < n). \qquad (3.3.26)$$

Integration by parts of this equation results in

$$I_{mn} = \int_a^b Q_m(x)\frac{\mathrm{d}}{\mathrm{d}x}\left[p(x)\frac{\mathrm{d}Q_n(x)}{\mathrm{d}x}\right]\mathrm{d}x$$

$$= \left[Q_m(x)p(x)\frac{\mathrm{d}Q_n(x)}{\mathrm{d}x}\right]_{-\infty}^{+\infty} - \int_a^b \frac{\mathrm{d}Q_m(x)}{\mathrm{d}x}p(x)\frac{\mathrm{d}Q_n(x)}{\mathrm{d}x}\mathrm{d}x.$$

The first term is zero again because as $x \to \infty$, $p(x)$ goes to zero faster than the inverse of any power of x. In the second term, exchange the positions of $Q_n(x)$ $Q_m(x)$ in the integrand and integrate by parts once more.

$$I_{mn} = -\int_a^b \left[p(x)\frac{\mathrm{d}Q_m(x)}{\mathrm{d}x}\right]\frac{\mathrm{d}Q_n(x)}{\mathrm{d}x}\mathrm{d}x$$

$$= -\left[Q_n(x)p(x)\frac{\mathrm{d}Q_m(x)}{\mathrm{d}x}\right]_{-\infty}^{+\infty}$$

$$+ \int_a^b Q_n(x)\rho(x)\frac{1}{\rho(x)}\frac{\mathrm{d}}{\mathrm{d}x}\left[p(x)\frac{\mathrm{d}Q_m(x)}{\mathrm{d}x}\right]\mathrm{d}x = 0.$$

The first term is again zero. In the second term, the integrand $r_m(x) = \frac{1}{\rho(x)} \frac{d}{dx}[p(x)\frac{dQ_m(x)}{dx}]$ is a polynomial of degree m. Thus, the final result is zero due to (3.3.23), as manifested by (3.3.26).

Here we rewrite (3.2.4) as

$$\frac{d}{dx}\left[p(x)\frac{dy}{dx}\right] + [\lambda\rho(x) - C(x)\rho(x)]y = 0, \qquad (3.3.27)$$

where weight $\rho(x)$ and kernel $p(x)$ are expressed by (3.2.2) and (3.2.3), respectively. In the cases of SL systems of polynomials, $A(x)$ and $B(x)$ in (3.2.1) are polynomials of degrees 2 and 1, respectively, and $C(x)$ is a constant merged into λ. Hence, (3.3.27) is simplified to be

$$\frac{d}{dx}\left[p(x)\frac{dy}{dx}\right] + \lambda\rho(x)y = 0. \qquad (3.3.28)$$

Because $R_n(x)$ is a polynomial of degree n, it is expanded by $\{Q_m(x)\}$:

$$R_n(x) = -\sum_{m=0}^{n} c_m Q_m(x). \qquad (3.3.29)$$

Making use of the orthogonality of $\{Q_m(x)\}$ with weight, see (3.3.22), the coefficients are easily evaluated:

$$c_m = -\frac{\int_a^b Q_m(x)\rho(x)R_n(x)dx}{\int_a^b Q_m^2(x)\rho(x)dx}. \qquad (3.3.30)$$

As $m < n$, $c_m = 0$ by (3.3.26). Thus the only nonzero coefficient is c_n. That is to say, there is only one term $c_n Q_n(x)$ left on the right hand side of (3.3.29):

$$R_n(x) = -c_n Q_n(x).$$

Combination with (3.3.25) leads to

$$\frac{d}{dx}\left[p(x)\frac{dQ_n(x)}{dx}\right] + c_n\rho(x)Q_n(x) = 0.$$

Table 3.3.   Rodrigues' formulas of Sturm-Liouville polynomials.

Polynomial $Q_n(x)$	$K_n \dfrac{1}{\rho(x)} \dfrac{d^n}{dx^n}[A^n(x)\rho(x)]$
Jacobi, $P_n^{(r,s)}(x)$	$\dfrac{(-1)^2}{2^n n!}(1-x)^{-r}(1+x)^{-s}\dfrac{d^n}{dx^n}[(1-x)^{n+r}(1+x)^{n+s}]$
Jacobi, $J_n^{(r,s)}(x)$	$\dfrac{x^{1-s}(1-x)^{s-r}}{(s)_n}\dfrac{d^n}{dx^n}[x^{n+s-1}(1+x)^{n+r-s}]$
Gegenbauer, $G_n^\mu(x)$	$\dfrac{(-1)^n(\mu+1)_n}{2^n n!(\mu)_n}(1-x^2)^{-\mu}\dfrac{d^n}{dx^n}(1-x^2)^{n+\mu}$
Chebyshev, $T_n(x)$	$\dfrac{(-1)^n}{(2n-1)!!}(1-x^2)^{1/2}\dfrac{d^n}{dx^n}(1-x^2)^{n-1/2}$
Legendre, $P_n(x)$	$\dfrac{(-1)^n}{2^n n!}\dfrac{d^n}{dx^n}(1-x^2)^n$
Sonine, $S_n^\mu(x)$	$\dfrac{1}{n!}x^{-\mu}e^x\dfrac{d^n}{dx^n}(x^{n+\mu}e^{-x})$
Associated Laguerre, $L_n^m(x)$	$\dfrac{(-1)^n n!}{(n-m)!}e^x\dfrac{d^{n-m}}{dx^{n-m}}(x^n e^{-x})$
Laguerre, $L_n(x)$	$e^x\dfrac{d^n}{dx^n}(x^n e^{-x})$
Hermite, $H_n(x)$	$(-1)^n e^{x^2}\dfrac{d^n}{dx^n}e^{-x^2}$

Here, $r,s > -1$, m and n are natural numbers and μ is any complex number except negative integer numbers.

This means that the polynomial (3.3.18) is a solution of an SL equation. □

Substitution of functions $A(x)$ and $\rho(x)$ in Table 3.1 into (3.3.18) gives various Rodrigues' formulas, which are listed in Table 3.3.

In complex functions there is a Cauchy integral formula,

$$f(x) = \frac{1}{2\pi i}\oint_C \frac{f(t)}{t-x}dt, \qquad (3.3.31)$$

where the point x is within the region encircled by the integral contour, and the contour takes counterclockwise. Taking derivative of (3.3.31) n times with respect to x leads to

$$f^{(n)}(x) = \frac{n!}{2\pi i} \oint_C \frac{f(t)}{(t-x)^{n+1}} dt. \qquad (3.3.32)$$

This equation is called the **Goursat formula**.

By means of the Goursat formula, a generalized Rodrigues' formula can be expressed in the form of integration:

$$Q_n(x) = K_n \frac{1}{\rho(x)} \frac{n!}{2\pi i} \oint_C \frac{A^n(t)\rho(t)}{(t-x)^{n+1}} dt. \qquad (3.3.33)$$

Equation (3.3.18) is also called the derivative expression of polynomial solution $Q_n(x)$. There are other functions other than the polynomials that may have their derivative expressions. In next section, we will introduce some of the equations and functions that are related to an SL system of polynomials. These functions are not polynomials, but can also have their derivative expressions, although not uniformly in the form of (3.3.18). Remember that an SL system of polynomials has another linearly independent solution beside the polynomial one, while (3.3.18) exhibits polynomials. Some of the derivative expressions are listed in Table 3.4.

3.3.3.   *Generating functions*

Definition 1. Suppose that a function $f(t)$ be in the neighborhood of $t = 0$ expanded as a convergent power series

$$f(t) = \sum_{n=0}^{\infty} a_n t^n.$$

Then, $f(t)$ is called the **generating function**, or **ordinary generating function**, of the series $\{a_n\}$. If a function $g(t)$ can be expanded

Table 3.4.   Derivative expressions of functions that are related to an SL system of polynomials.

$Q_n(x)$	Derivative expression
Chebyshev functions of the second kind, $U_n(x)$	$\dfrac{(-1)^{n-1}n}{(2n-1)!!}\dfrac{\mathrm{d}^{n-1}}{\mathrm{d}x^{n-1}}(1-x^2)^{n-1/2}$
Chebyshev polynomials of the second kind, $U_n^*(x)$	$\dfrac{(-1)^{n-1}(n+1)}{(2n+1)!!\sqrt{1-x^2}}\dfrac{\mathrm{d}^{n-1}}{\mathrm{d}x^{n-1}}(1-x^2)^{n+1/2}$
Associated Legendre functions of the first kind, $P_l^m(x)$	$\dfrac{(-1)^l}{2^l l!}(1-x^2)^{m/2}\dfrac{\mathrm{d}^{l+m}}{\mathrm{d}x^{l+m}}(1-x^2)^l$
Weber-Hermite functions of first kind, $\Psi_n(x)$	$(-1)^n e^{x^2/2}\dfrac{\mathrm{d}^n}{\mathrm{d}x^n}e^{-x^2}$
Weber functions (parabolic cylinder functions), $D_n(x)$	$(-1)^n e^{x^2/4}\dfrac{\mathrm{d}^n}{\mathrm{d}x^n}e^{-x^2/2}$

Here, $r, s > -1$ and m and n are natural numbers.

as a convergent power series in the following form:

$$g(t) = \sum_{n=0}^{\infty}\frac{a_n}{n!}t^n,$$

$g(t)$ is called the **exponential generating function** of the series $\{a_n\}$.

Definition 2. Suppose that $\{Q_n(x)\}$ is a function sequence. If a two-variable function $F(x,t)$ can be, in a certain region in (x,t) space, expanded by a convergent series in the following form:

$$F(x,t) = \sum_{n=0}^{\infty}Q_n(x)t^n, \qquad (3.3.34)$$

then $F(x,t)$ is called the **generating function**, or **ordinary generating function**, of the function sequence $\{Q_n(x)\}$. If a two-variable function $G(x,t)$ can be expanded by a convergent series in the form of

$$G(x,t) = \sum_{n=0}^{\infty}\frac{Q_n(x)}{n!}t^n, \qquad (3.3.35)$$

Table 3.5. Some ordinary generating functions.

Function sequence, $Q_n(x)$	Generating function, $G(x,t) = \sum\limits_{n=0}^{\infty} Q_n(x)t^n$
Jacobi polynomial, $P_n^{(r,s)}(x)$	$2^{r+s}\dfrac{(1-t+x)^{-r}(1+t+x)^{-s}}{x}$
Gegenbauer polynomial, $G_n^{\mu}(x)$	$\dfrac{2^{\mu}\Gamma(\mu+1/2)}{(1-2xt+t^2)^{\mu+1/2}\sqrt{\pi}}$
Chebyshev polynomial, $T_n(x)$	$\dfrac{1-xt}{1-2xt+t^2}$
Chebyshev function of second kind, $U_{n+1}(x)$	$\dfrac{\sqrt{1-x^2}}{1-2xt+t^2}$
Chebyshev polynomial of second kind, $U_n^*(x)$	$\dfrac{1}{1-2xt+t^2}$
Legendre polynomial, $P_n(x)$	$\dfrac{1}{\sqrt{1-2xt+t^2}}$
Associated Legendre function of first kind, $P_n^m(x)$	$\dfrac{(2m-1)!!(1-x^2)^{m/2}}{(1-2xt+t^2)^{m+1/2}}$
Sonine polynomial, $S_n^{\mu}(x)$	$\dfrac{1}{(1-t)^{\mu+1}}\exp\left(-\dfrac{xt}{1-t}\right)$

then $G(x,t)$ is called the **exponential generating function** of the function sequence $\{Q_n(x)\}$.

The generating functions are considerably helpful in deriving some recurrence relations.

In Tables 3.5 and 3.6 we list the generating functions of some functions that will be mentioned later. From the two tables, the following features are observed: taking $m = 1/2$ in Gegenbauer polynomials $G_n^m(x)$ leads to Legendre polynomials $P_n(x)$, $P_n(x) = G_n^{1/2}(x)$, which agrees with Table 3.1; taking $m = 0$ in associated Legendre function $P_n^m(x)$ results in Legendre polynomials $P_n(x)$, $P_n(x) = P_n^0(x)$; multiplying Hermite polynomials $H_n(x)$ by a factor $\exp(-x^2/2)$ yields a Weber-Hermite function of first kind $\Psi_n(x)$,

Table 3.6. Some exponential generating functions.

Function sequence, $Q_n(x)$	Generating function, $$F(x,t) = \sum_{n=0}^{\infty} \frac{Q_n(x)}{n!} t^n$$
Jacobi polynomial, $(s)_n J_n^{(r,s)}\left(\dfrac{1-x}{2}\right)$	$\dfrac{2^{r-1}(1-t+\sqrt{1-2xt+t^2})^{1-s}}{(1+t+\sqrt{1-2xt+t^2})^{r-s}}$
Laguerre polynomial, $L_n(x)$	$\dfrac{1}{1-t}\exp\left(-\dfrac{xt}{1-t}\right)$
Hermite polynomial, $H_n(x)$	$\exp(2xt-t^2)$
Weber-Hermite function of first kind, $\Psi_n(x)$	$\exp(2xt-t^2-x^2/2)$
Weber function (parabolic cylinder function), $D_n(x)$	$\exp(2xt-t^2-x^2/4)$
Bessel function of half integer order, $J_{n-1/2}(x)$	$\sqrt{\dfrac{2}{\pi x}}\cos\sqrt{x^2-2xt}$
Bessel function of half integer order, $(-1)^n J_{-n+1/2}(x)$	$\sqrt{\dfrac{2}{\pi x}}\sin\sqrt{x^2-2xt}$
Spherical Bessel function of first kind (spherical Bessel function), $j_{n-1}(x)$	$\dfrac{1}{x}\cos\sqrt{x^2-2xt}$
Spherical Bessel function of second kind (Neumann function), $y_{n-1}(x)$	$\dfrac{1}{x}\sin\sqrt{x^2-2xt}$

$\Psi_n(x) = H_n(x)\exp(-x^2/2)$; multiplying Hermite polynomials $H_n(x)$ by a factor $\exp(-x^2/4)$ leads to the Weber function (parabolic cylinder function) $D_n(x)$, $D_n(x) = H_n(x)\exp(-x^2/4)$.

Please note that $\frac{2^{r-1}(1-t+\sqrt{1-2xt+t^2})^{1-s}}{(1+t+\sqrt{1-2xt+t^2})^{r-s}}$, although called the generating function of Jacobi polynomials, in fact generates $(s)_n J_n^{(r,s)}(\frac{1-x}{2})$, which multiplies Jacobi polynomials by an additional constant factor involving n. There are also extraordinary cases such that the generating functions of function sequences are neither ordinary nor exponential ones, e.g., associated Laguerre polynomials, Legendre functions of second kind. When expanding the generating function of Bessel functions of integer orders, the index is from $-\infty$ to $+\infty$, instead of from 0 to $+\infty$, and so does the generating function of modified Bessel functions. These extraordinary cases are not listed in Tables 3.5 and 3.6.

By means of a generating function, we are able to obtain expressions of contour integrals of the polynomials.

It follows from (3.3.34) that

$$\frac{F(x,t)}{t^{m+1}} = \sum_{n=0}^{\infty} Q_n(x)t^{n-m-1}.$$

Now we take integration along a contour surrounding the origin in a complex t plane. Because the polynomial $Q_n(x)$ itself does not have poles, on the right hand side only one term with $n = m$ has a simple pole. Thus the result of the integration is

$$Q_m(x) = \frac{1}{2\pi i} \oint \frac{F(x,z)}{z^{m+1}} dz,$$

(the integral contour surrounds the origin).

Similarly, given the exponential generating function $G(x,t)$ of a certain polynomial set, we have

$$Q_m(x) = \frac{m!}{2\pi i} \oint \frac{G(x,z)}{z^{m+1}} dz,$$

(the integral contour surrounds the origin).

3.3.4. *The completeness theorem of orthogonal polynomials as Sturm-Liouville solutions*

Theorem 1. *A Hilbert function space is defined on an interval where the weight function $\rho(x) \neq 0$. In such a space an orthogonal polynomial set $\{Q_n(x)\}$ of an SL system is complete.*

This theorem is a special case of Theorem 4 introduced in Subsection 3.2.3 when the complete set consists of polynomials. All of the solution sets of SL systems are complete, and the polynomial sets are not the exceptions. But for the cases of polynomial sets, the proof of this theorem is particularly simple. The completeness of a polynomial set can be proved by its closure. Please note that in this

Hilbert space, the inner product is with weight function $\rho(x)$:

$$(f, g) \equiv \int_{-\infty}^{\infty} f^*(x)g(x)\rho(x)\mathrm{d}x.$$

The integration interval is that the weight function is nonzero on it, see Table 3.1.

In view of Theorem 2 in Section 2.3, a function set is complete if and only if it is closed. Therefore, what we need to do is to prove the closure of a polynomial set as solutions of an SL system. The closure means that there does not exist a nonzero function that is orthogonal to every function in this set. Alternatively, a function is necessarily not a nonzero function if it is orthogonal to everyone in the polynomial set.

Proof. Suppose that a function $f(x)$ is orthogonal to all $Q_n(x)$, i.e., $(f, Q_n) = 0$ for all n. Because x^m can be written as the linear combination of $Q_0, Q_1, \ldots, Q_m$: $x^m = \sum_{i=0}^{m} c_i Q_i$, then

$$(f, x^m) = 0 \quad \text{for all } m. \tag{3.3.36}$$

Now consider the following inner product

$$g(k) = (e^{ikx}, f) = \int_{-\infty}^{\infty} f(x)e^{-ikx}\rho(x)\mathrm{d}x \tag{3.3.37}$$

for all real k. Note that the upper and lower limits have been extended to infinity, which does not affect the integral because in the extended intervals the weight function is zero. Next we expand e^{ikx} by the Taylor series: $e^{-ikx} = \sum_{m=0}^{\infty} \frac{(-ik)^m}{m!} x^m$. It is immediately known by (3.3.37) that $g(k) = 0$. The $g(k)$ is just the Fourier component of function $f(x)\rho(x)$, so that $f(x)\rho(x)$ should be zero almost everywhere. However, in this Hilbert space $\rho(x) \neq 0$. It must be that $f(x)$ in this Hilbert space is zero almost everywhere. Thus, the function $f(x)$ is not a nonzero one when it is orthogonal to everyone in a polynomial set as solutions of an SL system. The conclusion is that the set $\{Q_n(x)\}$ is closed, so that it is complete. $\qquad\square$

3.3.5. *Applications in numerical integrations*

1. Integral with a weight function

The integral with weight means

$$F = \int_a^b \mathrm{d}x \rho(x) f(x), \tag{3.3.38}$$

where $\rho(x)$ is the weight function on $[a, b]$ and $f(x)$ is the integrand. If the right hand side of (3.3.38) cannot be carried out analytically, but the numerical result is desired in practical usage, numerical integration has to be done.

The way of numerical integration is selecting n points on $[a, b]$ following some rules, with each point having a weight. Suppose that at the ith point x_i the weight is w_i. Then the original integration is written as a summation in the following form:

$$\int_a^b \mathrm{d}x \rho(x) f(x) \approx \sum_{i=1}^n w_i f(x_i). \tag{3.3.39}$$

The result thus calculated by the summation in (3.3.39) is regarded as that of the integral in (3.3.38). Equation (3.3.39) is called a **mechanical integral formula**, with x_i called an **integral node** and w_i called an **integral coefficient** or **weight coefficient**.

There are different ways of selecting the positions and number of the nodes and corresponding weights. Of course the selection influences the workload and precision of the numerical calculation. We hope that for a given weight function $\rho(x)$ and a precision, the integral nodes are as few as possible. Alternatively, given the node number, the positions of the nodes and corresponding coefficients are selected such that the precision is as high as possible. In the latter case, equation (3.3.39) is called a **Gauss integration formula**. It is practical to use polynomial sets corresponding to the known weight function $\rho(x)$ under appropriate conditions.

If the integral interval $[a, b]$ and weight function $\rho(x)$ in (3.3.38) happen to be those of the polynomial set $\{\varphi_n(x)\}$ in Table 3.1, then the integrand $f(x)$ in (3.3.38) can be expanded by the corresponding polynomial set. Subsequently, given a node number, one is able to determine the positions and coefficients at these nodes. Such integral nodes and coefficients are called **Gauss nodes** and **Gauss coefficients**, respectively.

The theoretical analysis in selecting the positions of the nodes and corresponding coefficients is not introduced here. We are content with giving the final conclusions. By theoretical analysis, when the node number n is fixed, the node positions should be just at the zeros of the orthogonal polynomial of degree n $\varphi_n(x)$. It has been known from Theorem 4 in Subsection 2.3.4 that $\varphi_n(x)$ happens to have n nondegenerate roots on $[a, b]$. This condition is met naturally. The coefficients at the nodes are computed as follows:

$$w_i = \int_a^b dx \rho(x) \frac{\varphi_n(x)}{(x - x_i)\varphi_n'(x)} > 0, \quad i = 1, 2, \ldots, n. \quad (3.3.40)$$

For each explicit polynomial set, the computation formulas of the positions of the nodes and corresponding coefficients have been provided, and even their numerical computation codes have been prepared. In the following we present the conclusions for five cases.

2. Gauss-Legendre integral formula

Integral interval is $[-1, 1]$ and weight function is $\rho(x) = 1$. Gauss-Legendre integral formula is

$$\int_{-1}^1 dx f(x) = \sum_{i=1}^n w_i f(x_i). \quad (3.3.41)$$

The advantage of this formula is that the weight function being 1 means that the formula is applicable to any integrand, as long as the integral interval is finite.

If the integral interval is $[a, b]$ instead of $[-1, 1]$, the former can be converted to be the latter by

$$x = \frac{a + b}{2} + \frac{b - a}{2} t. \quad (3.3.42)$$

Thus, the integration becomes

$$\int_a^b dx f(x) = \frac{b - a}{2} \int_{-1}^1 dt f\left(\frac{a + b}{2} + \frac{b - a}{2} t\right). \quad (3.3.43)$$

This transformation does not affect the weight function since it is always 1.

3. Gauss-Chebyshev integral formula

Integral interval is $[-1, 1]$ and weight function is $\rho(x) = \frac{1}{\sqrt{1-x^2}}$. Gauss-Chebyshev integral formula is

$$\int_{-1}^{1} dx \frac{1}{\sqrt{1-x^2}} f(x) = \sum_{i=1}^{n} w_i f(x_i) = \frac{\pi}{n} \sum_{i=1}^{n} f(x_i). \qquad (3.3.44)$$

In this formula the positions of zeros are fairly simple:

$$x_i = \cos \frac{2(n-i)+1}{2n} \pi. \qquad (3.3.45)$$

The coefficients can be calculated by (3.3.40):

$$w_i = \frac{\pi}{n}, \quad i = 1, 2, \ldots, n. \qquad (3.3.46)$$

All the coefficients are the same, which is quite convenient for numerical computation.

If the weight function of an integrand $g(x)$ on $[-1, 1]$ is not $1/\sqrt{1-x^2}$, we reform the integral to be

$$\int_{-1}^{1} dx g(x) = \int_{-1}^{1} dx \frac{1}{\sqrt{1-x^2}} \sqrt{1-x^2} g(x) = \int_{-1}^{1} dx \frac{1}{\sqrt{1-x^2}} f(x). \qquad (3.3.47)$$

If the interval is $[a, b]$ instead of $[-1, 1]$, the way of (3.3.43) can be used to convert the interval, so that a Gauss-Chebyshev integral formula (3.3.47) can be employed.

4. Gauss-Laguerre integral formula

Integral interval is $[0, \infty]$ and weight function is $\rho(x) = e^{-x}$. Gauss-Laguerre integral formula is

$$\int_{0}^{\infty} dx e^{-x} f(x) = \sum_{i=1}^{n} w_i f(x_i).$$

This formula is suitable to calculate integrations on the whole positive real half axis which is an infinite interval.

5. Gauss-Hermite integral formula

Integral interval is $[-\infty, \infty]$ and weight function is $\rho(x) = e^{-x^2}$. Gauss-Hermite integral formula is

$$\int_{-\infty}^{\infty} dx e^{-x^2} f(x) = \sum_{i=1}^{n} w_i f(x_i).$$

This formula is suitable to calculate integrations on the whole real axis which is an infinite interval.

6. Gauss-Jacobi integral formula

Integral interval is $[-1, 1]$ and weight function is $\rho(x) = (1-x)^\alpha (1+x)^\beta$. Gauss-Jacobi integral formula is

$$\int_{-1}^{1} dx (1+x)^\alpha (1+x)^\beta f(x) = \sum_{i=1}^{n} w_i f(x_i).$$

This formula has two evident special cases: as $\alpha = \beta = 0$, it becomes a Gauss-Legendre integral formula; as $\alpha = \beta = -1/2$, it becomes a Gauss-Chebyshev integral formula.

3.4. Equations and Functions that Relate to the Polynomial Solutions

By the discussion above, once it is required that the solutions of SL equation (3.2.1) are polynomials, the coefficients of $A(x)$ and $B(x)$ have to be those listed in Table 3.1. Consequently, the eigenvalues and corresponding polynomial solutions can be achieved. Moreover, as we have addressed in Subsection 3.1.2, an ordinary differential equation had a pair of linearly independent special solutions. Since one of them has been a polynomial, the other should not be. In principle, the other special solution can be calculated from the polynomial one with the aid of a Liouville formula (3.1.14).

The linear combination of the two independent solutions forms the general solution of this eigenvalue. When the general solution is inserted in the boundary conditions, the combination coefficients can be acquired.

When taking derivative of SL polynomials several times, the resultants may not belong to SL systems any more, but to those sometimes called associated equations. Since the associated equations are obtained by the derivatives of SL equations, their eigenvalues can be obtained comparatively easily.

In this section, we summarized four types of typical solutions of SL equations. They are Laguerre functions, Legendre functions, Chebyshev functions and Hermite functions.

3.4.1. *Laguerre functions*

1. Sonine polynomials (generalized Laguerre polynomials)

The following equation is called a **generalized Laguerre equation**:

$$xy'' + (\mu + 1 - x)y' + \lambda y = 0, \quad (0 \le x \le \infty), \qquad (3.4.1)$$

where μ is any complex number except negative real integers. Under the boundary conditions

$$y(x = 0) < \infty, \quad y(x = +\infty) \sim x^n, \qquad (3.4.2)$$

its eigenvalues are

$$\lambda = n, \quad (n = 0, 1, 2, \ldots). \qquad (3.4.3)$$

The corresponding special solution in the form of polynomials are expressed by Rodrigues' formula:

$$S_n^\mu(x) = \frac{1}{n!} x^{-\mu} e^x \frac{d^n}{dx^n} (x^{n+\mu} e^{-x}). \qquad (3.4.4)$$

These are polynomial of degree n, called **Sonine polynomials**, or **generalized Laguerre polynomials**, denoted by $S_n^\mu(x)$. Equation (3.4.4) has been listed in Table 3.3. The series expression of Sonine polynomials has been listed in Table 3.2. Its generating function of Sonine polynomials has been listed in Table 3.5.

The square of the normalizing factor of a Sonine polynomial is

$$\int_0^\infty dx x^\mu e^{-x} |S_n^\mu(x)|^2 = \frac{\Gamma(\mu + n + 1)}{n!} \qquad (3.4.5)$$

Sonine polynomials are of the following addition formula:

$$S_n^{\mu+\nu+1}(x+y) = \sum_{k=0}^{n} S_k^{\mu}(x) S_{n-k}^{\nu}(y). \tag{3.4.6}$$

The other special solution of (3.4.1) can be calculated by the Liouville formula.

2. Laguerre polynomials

When $\mu = 0$ taken, (3.4.1) becomes

$$xy'' + (1-x)y' + \lambda y = 0, \tag{3.4.7}$$

which is called **Laguerre equation**. Under the boundary conditions (3.4.2), its eigenvalues are (3.4.3), and the special solution in the form of polynomials are called **Laguerre polynomials**, denoted by $L_n(x)$. Rodrigues' formula is

$$L_n(x) = e^x \frac{d^n}{dx^n}(x^n e^{-x}). \tag{3.4.8}$$

It presents the Laguerre polynomial of degree n. Equation (3.4.8) has been listed in Table 3.3.

The series expression of Laguerre polynomials has been listed in Table 3.2. Although a Laguerre equation (3.4.7) is obtained by taking $\mu = 0$ in the generalized Laguerre equation (3.4.1), the eigenfunctions Laguerre polynomials are not obtained by simply taking $\mu = 0$ in Sonine polynomials because of historical reason. The relation between them is

$$L_n(x) = n! S_n^0(x). \tag{3.4.9}$$

The generating function of Laguerre polynomials has been listed in Table 3.6. The factor on the right hand side of (3.4.9) makes Laguerre polynomials have an exponential generating function, while Sonine polynomials have an ordinary generating function.

There is another generating function of Laguerre polynomials:

$$J_0(2\sqrt{xt}) = e^{-t} \sum_{n=0}^{\infty} \frac{t^n}{n!^2} L_n(x). \tag{3.4.10}$$

The left hand side is a Bessel function of zero order to be introduced in Chapter 4.

By (3.4.5) and (3.4.9), the square of the normalizing factor of a Laguerre polynomial is

$$\int_0^\infty \mathrm{d}x\, x^\mu e^{-x} |L_n(x)|^2 = (n!)^2. \qquad (3.4.11)$$

The other special solution of (3.4.7) can be calculated by the Liouville formula.

3. Associated Laguerre polynomials

Taking derivative of a Laguerre equation (3.4.7) m times leads to an **associated Laguerre equation**:

$$xy'' + (m + 1 - x)y' + (\lambda - m)y = 0, \qquad (3.4.12)$$

where m is a positive integer, and $0 \leq m \leq n$. Under the boundary conditions

$$y(x = 0) < \infty, \quad y(x \to +\infty) \sim x^{n-m}, \qquad (3.4.13)$$

the eigenvalues of the equation are

$$\lambda = n, \quad (n = m, m+1, m+2, \ldots). \qquad (3.4.14)$$

The polynomial solutions can be expressed by Rodrigues' formula:

$$L_n^{(m)}(x) = (-1)^n \frac{n!}{(n-m)!} e^x \frac{\mathrm{d}^{n-m}}{\mathrm{d}x^{n-m}} (x^n e^{-x}). \qquad (3.4.15)$$

This presents a polynomial of degree $n - m$, called an **associated Laguerre polynomials**. Equation (3.4.15) has been listed in Table 3.3.

The series expression of associated Laguerre polynomials has been listed in Table 3.2.

The generating function of the associated Laguerre polynomials is

$$\frac{(-1)^m}{(1-t)^{m+1}} \exp\left(-\frac{xt}{1-t}\right) = \sum_{n=m}^\infty L_n^{(m)}(x) t^{n-m}. \qquad (3.4.16)$$

Note the abnormal lower limit of the summation. This equation was not listed in tables of generating functions.

Theorem 1. *If a function $f(x)$ is piecewise smooth on $(0, \infty)$, then*

$$\sum_{n=0}^{\infty} c_n L_n^{(m)}(x) = \frac{1}{2}[f(x+0^+) + f(x-0^+)], \qquad (3.4.17)$$

where

$$c_n = \frac{n!}{\Gamma(n+m+1)} \int_0^{\infty} x^m e^{-x} L_n^{(m)}(x) f(x) \mathrm{d}x. \qquad (3.4.18)$$

An associated Laguerre equation (3.4.12) is obtained by taking derivative of a Laguerre equation (3.4.7) m times with respect to x. Accordingly, the polynomial solutions of the former are obtained by also taking m times derivative of those of the latter with respect to x:

$$L_n^{(m)}(x) = \frac{\mathrm{d}^m}{\mathrm{d}x^m} L_n(x). \qquad (3.4.19)$$

When taking $m = 0$ in an associated Laguerre equation (3.4.12), it will degrade to a Laguerre equation (3.4.7). Accordingly, taking $m = 0$ in associated Laguerre polynomials will go back to Laguerre polynomials: $L_n^{(0)}(x) = L_n(x)$.

The relation between associated Laguerre polynomials and Sonine polynomials is

$$L_n^{(m)}(x) = (-1)^m n! S_{n-m}^m(x) \qquad (3.4.20a)$$

or

$$L_{n+m}^{(m)}(x) = (-1)^m (n+m)! S_n^m(x). \qquad (3.4.20b)$$

Each of the generalized Laguerre equation, Laguerre equation and associated Laguerre equation has another linearly independent solution, which belongs to so-called **confluent hypergeometric functions**.

Laguerre equation is encountered when solving the radial characteristic function of the hydrogen atom.

3.4.2. *Legendre functions*

1. Legendre polynomials

Legendre equation reads

$$(1 - x^2)y'' - 2xy' + \lambda y = 0. \tag{3.4.21}$$

Under boundary conditions

$$y(x = \pm 1) < \infty, \tag{3.4.22}$$

its eigenvalues are

$$\lambda = l(l + 1), \quad (l = 0, 1, 2, \ldots). \tag{3.4.23}$$

Substituting the eigenvalue expression into (3.4.21), it becomes

$$(1 - x^2)y'' - 2xy' + l(l + 1)y = 0, \tag{3.4.24}$$

which is called a **Legendre equation of order** l. The polynomial solutions can be expressed by Rodrigues' formula:

$$P_l(x) = \frac{(-1)^l}{2^l l!} \frac{d^l}{dx^l}(1 - x^2)^l, \tag{3.4.25}$$

which is a polynomial of degree l, called a **Legendre polynomial**. Equation (3.4.25) has been listed in Table 3.3.

The series expression of Legendre polynomials has been listed in Table 3.2. The generating function of Legendre polynomials has been listed in Table 3.5. Legendre polynomials $P_l(x)$ are also called **Legendre functions of the first kind**.

The square of the normalizing factor of a Legendre polynomial is

$$\int_{-1}^{1} dx |P_n(x)|^2 = \frac{2}{2n + 1}. \tag{3.4.26}$$

Theorem 2. *Suppose that a function $f(x)$ is defined on $[-1, 1]$. If the integral of $(1 - x^2)^{-1/4} f(x)$ on $[-1, 1]$ exists and absolutely converges, then*

$$\sum_{n=0}^{\infty} c_n P_n(x) = \frac{1}{2}[f(x + 0^+) + f(x - 0^+)], \quad (-1 < x < 1),$$

where

$$c_n = \frac{2n+1}{2} \int_{-1}^{1} P_n(x) f(x) \mathrm{d}x. \tag{3.4.27}$$

This expansion series is called a **Fourier-Legendre series.**

The other linearly independent solution of a Legendre equation (3.4.24) is denoted as $Q_l(x)$ and can be calculated by means of the Liouville formula:

$$Q_l(x) = P_l(x) \int \frac{\mathrm{d}x}{[P_l(x)]^2(1-x^2)}. \tag{3.4.28}$$

$Q_l(x)$'s are called **Legendre functions of the second kind.** Their values at boundaries $x = \pm 1$, $Q_l(\pm 1)$ are infinitely large. The generating function of $Q_l(x)$ is

$$\frac{1}{x-t} = \sum_{n=0}^{\infty} (2n+1) Q_n(x) P_n(t). \tag{3.4.29}$$

This expression has a feature that it includes the two linearly independent special solutions. This generating function was not listed in either of generating function tables.

The general solution of (3.4.24) should be

$$y = A P_l(x) + B Q_l(x).$$

If (3.4.24) is extended to be the following form:

$$(1-z^2) y''(z) - 2zy'(z) + \nu(\nu+1) y(z) = 0, \tag{3.4.30}$$

where ν can be any constant complex number and variable z is also complex, then (3.4.30) is called Legendre equation of order ν. The two special solutions of this equation are denoted by $P_\nu(z)$ and $Q_\nu(z)$, called Legendre functions of order ν of the first and second kinds, respectively. Their integral expressions are as follows:

$$P_\nu(z) = \frac{1}{2\pi \mathrm{i}} \oint_{C_1} \frac{(t^2-1)^\nu}{2^\nu (t-z)^{\nu+1}} \mathrm{d}t, \tag{3.4.31}$$

$$Q_\nu(z) = \frac{1}{4\mathrm{i}\sin\nu\pi} \oint_{C_2} \frac{(t^2-1)^\nu}{2^\nu (t-z)^{\nu+1}} \mathrm{d}t. \tag{3.4.32}$$

In the first integral, the contour C_1 is a counterclockwise loop in a complex t plane and cut along $(-\infty, -1)$ so that points 1 and z are in the interior of the region surrounded by C_1. In the second integral, C_2 is an ∞-shape contour in complex t plane: it first surrounds point 1 clockwise one circle, and then surrounds point -1 counterclockwise one circle. Equation (3.4.31) is called a **Schlöfli integral** of $P_\nu(z)$. If $Re(\nu + 1) > 0$, the integral contour of (3.4.32) can be deformed so as to get

$$Q_\nu(z) = \frac{1}{2^{\nu+1}} \int_{-1}^{1} \frac{(1 - t^2)^\nu}{(z - t)^{\nu+1}} \, dt. \tag{3.4.33}$$

This expression is particularly convenient when ν is an integer.

The general solution of (3.4.30) is

$$y = AP_\nu(x) + BQ_\nu(x).$$

When $\nu = l$ is a positive integer, $P_\nu(x)$ degrades to the Legendre polynomial $P_l(x)$ and $Q_\nu(x)$ to $Q_l(x)$.

2. Associated Legendre functions

Taking derivative of Legendre equation (3.4.24) m times and replacing y in (3.4.24) by $(1 - x^2)^{k/2} y^{(k)}$, we get an associated Legendre equation:

$$(1 - x^2)y'' - 2xy' + \left[l(l + 1) - \frac{m^2}{1 - x^2} \right] y = 0, \tag{3.4.34}$$

where m is an positive integer and $0 \le m \le l$. Under the boundary conditions (3.4.22), its eigenvalues are

$$l, \ (l = m, m + 1, m + 2, \ldots). \tag{3.4.35}$$

The two linearly independent special solutions are an **associated Legendre function of the first kind**

$$P_l^m(x) = (1 - x^2)^{m/2} \frac{d^m P_l(x)}{dx^m} \tag{3.4.36}$$

and an **associated Legendre function of the second kind**

$$Q_l^m(x) = (1 - x^2)^{m/2}\frac{\mathrm{d}^m Q_l(x)}{\mathrm{d}x^m}, \qquad (3.4.37)$$

respectively. At ends $x = \pm 1$, $P_l^m(x)$ satisfies the boundary conditions (3.4.22), but $Q_l^m(\pm 1)$ is infinitely large.

The derivative expression of $P_l^m(x)$ has been listed in Table 3.4. The generating function of $P_l^m(x)$ has been listed in Table 3.5.

The square of the normalizing factor of an associated Legendre polynomial is calculated by

$$\int_{-1}^{1}\mathrm{d}x|P_n^m(x)|^2 = \frac{(n+m)!}{(n-m)!}\frac{2}{2n+1}. \qquad (3.4.38)$$

Theorem 3. *Any function $f(x)$ that is continuous on $[-1,1]$ and is zero at the ends can be, in the sense of mean-square convergence, expanded by an associated Legendre functions of the first kind as*

$$f(x) = \sum_{n=m}^{\infty} c_n P_n^m(x), \qquad (3.4.39a)$$

where

$$c_n = \frac{2n+1}{2}\frac{(n-m)!}{(n+m)!}\int_{-1}^{1}P_n^m(x)f(x)\mathrm{d}x. \qquad (3.4.39b)$$

The generalized solution of (3.4.34) is

$$y = AP_l^m(x) + BQ_l^m(x). \qquad (3.4.40)$$

We stress here that although an associated Legendre equation is obtained by taking derivative of Legendre equation m times with respect to x, the solutions of the former cannot be obtained by taking derivative of the latter m, see (3.4.36) and (3.4.37).

Taking $m = 0$ in an associated Legendre equation makes it go back to a Legendre equation, and taking $m = 0$ in associated Legendre functions makes them go back to Legendre functions: $P_n^0(x) = P_n(x), Q_n^0(x) = Q_n(x).$

3. Spherical harmonics

Three-dimensional Laplace equation under spherical coordinates has the form of

$$\nabla^2 u = \frac{1}{r^2}\frac{\partial}{\partial r}\left(r^2\frac{\partial u}{\partial r}\right) + \frac{1}{r^2\sin\theta}\frac{\partial}{\partial\theta}\left(\sin\theta\frac{\partial u}{\partial\theta}\right) + \frac{1}{r^2\sin^2\theta}\frac{\partial^2 u}{\partial\varphi^2} = 0.$$

$$(3.4.41)$$

By separation of variables, let $u = R(r)\Theta(\theta)\Phi(\varphi)$ and then substitute it into (3.4.41) (3.4.40). After being multiplied by $\frac{r^2}{R\Theta\Phi}$, one gets a radial equation

$$\frac{1}{R}\frac{d}{dr}\left(r^2\frac{dR}{dr}\right) = \lambda \qquad (3.4.42)$$

and an angular equation

$$\frac{1}{\Theta\sin\theta}\frac{d}{d\theta}\left(\sin\theta\frac{d\Theta}{d\theta}\right) + \frac{1}{\Phi\sin^2\theta}\frac{d^2\Phi}{d\varphi^2} = -\lambda. \qquad (3.4.43)$$

Equations (3.4.43) is further treated by separation of variables to become

$$\frac{1}{\Phi}\frac{d^2\Phi}{d\varphi^2} = -m^2, \quad m = 0, \pm 1, \pm 2, \ldots, \qquad (3.4.44)$$

$$\frac{\sin\theta}{\Theta}\frac{d}{d\theta}\left(\sin\theta\frac{d\Theta}{d\theta}\right) + \lambda\sin^2\theta = m^2. \qquad (3.4.45)$$

In (3.4.45), let $x = \cos\theta$ and change denotation $\Theta(\theta)$ by $p(x)$. Then it is recast to be

$$(1 - x^2)\frac{d^2 p}{dx^2} - 2x\frac{dp}{dx} + \left(\lambda - \frac{m^2}{1 - x^2}\right)p = 0. \qquad (3.4.46)$$

This is just an **associated Legendre equation**. If the physical problem is independent of φ, then $\Phi(\varphi)$ is so. That means $m = 0$. In this case (3.4.46) becomes

$$(1 - x^2)\frac{d^2 p}{dx^2} - 2x\frac{dp}{dx} + \lambda p = 0. \qquad (3.4.47a)$$

This is just a **Legendre equation** (3.4.21). The eigenvalues being physical meanings of (3.4.46) and (3.4.47a) are

$$\lambda = l(l+1), \quad (l = 0,1,2,\ldots), \quad m = 0,\pm 1,\pm 2,\ldots,\pm l. \quad (3.4.47b)$$

Please note that from (3.4.44) the absolute value of the eigenvalues $|m|$ is unbounded. However, because of the constraint in (3.4.46), $|m|$ has an upper limit, as shown by (3.4.47b).

The solutions of (3.4.49) that are analytical at the boundaries are associated Legendre functions $\Theta(\theta) = P_l^m(\cos\theta)$. The solutions of (3.4.44) are $\Phi(\varphi) = e^{im\varphi}$. The products

$$Y_l^m(\theta,\varphi) = \Theta(\theta)\Phi(\varphi) = P_l^m(\cos\theta)e^{im\varphi} \quad (3.4.48)$$

are called **spherical harmonics**, which are the solutions of (3.4.43).

Theorem 4. *Any function $f(\theta,\varphi)$ that is continuous on a spherical surface can be, in the mean-square convergent sense, expanded by spherical harmonics as follows:*

$$f(\theta,\varphi) = \sum_{n=0}^{\infty}\sum_{m=0}^{n} A_{mn}Y_n^m(\theta,\varphi)$$

$$= \sum_{n=0}^{\infty}\sum_{m=0}^{n} (a_n^m \cos m\varphi + b_n^m \sin m\varphi)P_n^m(\cos\theta),$$

where

$$a_n^m = \frac{2n+1}{2\pi\varepsilon_m}\frac{(n-m)!}{(n+m)!}\int_0^\pi d\theta \int_0^{2\pi} d\varphi f(\theta,\varphi)P_n^m(\cos\theta)\cos m\varphi \sin\theta,$$

$$b_n^m = \frac{2n+1}{2\pi}\frac{(n-m)!}{(n+m)!}\int_0^\pi d\theta \int_0^{2\pi} d\varphi f(\theta,\varphi)P_n^m(\cos\theta)\sin m\varphi \sin\theta,$$

$$\varepsilon_m = \begin{cases} 2, & m = 0 \\ 1, & m \neq 0 \end{cases}.$$

The functions introduced in this subsection, such as Legendre functions including Legendre polynomials, associated Legendre functions, spherical harmonic and so on, can all be referred to as **spherical functions**.

4. The optimal square approximation

The coefficient of the highest power term of Legendre polynomial $P_n(x)$ is $\frac{2n(2n-1)\cdots(n+1)}{2^n n!} = \frac{(2n)!}{2^n(n!)^2}$. If we let $P_n(x)$ be divided by this coefficient, $P_n(x)$ becomes a **monic Legendre polynomial**, denoted as $\tilde{P}_n(x)$:

$$\tilde{P}_n(x) = \frac{2^n(n!)^2}{(2n)!} P_n(x).$$

Theorem 5. *Among all of the monic polynomials of degree n defined on $[-1,1]$, $\tilde{P}_n(x)$ has the least mean-square error relative to a zero function. Therefore, $\tilde{P}_n(x)$ is the optimal square approximation polynomial of degree n of a zero function.*

3.4.3. *Chebyshev functions*

1. Chebyshev functions

The so-called **Chebyshev equation** is

$$(1 - x^2)y'' - xy' + \lambda y = 0, \quad (-1 \le x \le 1). \tag{3.4.49}$$

This equation is easy to solve. Let

$$x = \cos\theta. \tag{3.4.50}$$

Then the equation is converted to be

$$\frac{\mathrm{d}^2 z(\theta)}{\mathrm{d}\theta^2} + \lambda z(\theta) = 0. \tag{3.4.51}$$

Obviously, an eigenvalue of the converted equation is

$$\lambda = n^2, \tag{3.4.52}$$

and its two linearly independent solutions are

$$z_1(\theta) = \cos n\theta \tag{3.4.53}$$

and

$$z_2(\theta) = \sin n\theta, \tag{3.4.54}$$

respectively.

Accordingly, the linearly independent solutions of the original equation (3.4.49) are

$$y_1 = \cos(n \cos^{-1} x) = T_n(\cos \theta) = T_n(x) \qquad (3.4.55)$$

and

$$y_2 = \sin(n \cos^{-1} x) = U_n(\cos \theta) = U_n(x). \qquad (3.4.56)$$

These two functions are of a feature that on $[-1, 1]$,

$$\max |T_n(x)| = 1, \quad \max |U_n(x)| = 1. \qquad (3.4.57)$$

That is to say, the absolute values of the functions are always less than or equal to 1.

$T_n(x)$ is a polynomial of degree n, which is just the Chebyshev polynomial already introduced in Subsection 2.4.2, also called a **Chebyshev function of the first kind**. This solution is analytical at boundaries $x = \pm 1$. From (3.4.55) it is easily seen that $T_n(1) = 1, T_n(-1) = (-1)^n$. In Subsection 2.4.2 we have already listed some of the properties of Chebyshev polynomials.

The Rodrigues' formula of Chebyshev polynomials is

$$T_n(x) = \frac{(-1)^n}{(2n-1)!!}(1-x^2)^{1/2}\frac{\mathrm{d}^n}{\mathrm{d}x^n}(1-x^2)^{n-1/2}, \qquad (3.4.58)$$

which has been listed in Table 3.3.

The series expression of Chebyshev polynomials is

$$T_0(x) = 1, \qquad (3.4.59a)$$

$$T_n(x) = \frac{n}{2}\sum_{k=0}^{[n/2]}(-1)^k\frac{(n-k-1)!}{k!(n-2k)!}(2x)^{n-2k}, \quad (n \geq 1), \quad (3.4.59b)$$

which have been listed in Table 3.2.

The generating function of Chebyshev polynomials is

$$\frac{1-xt}{1-2xt+t^2} = \sum_{n=0}^{\infty}T_n(x)t^n, \qquad (3.4.60)$$

which has been listed in Table 3.5.

Theorem 6. *If $f(x)$ is piecewise smooth on $[-1, 1]$, then*

$$\frac{c_0}{2} + \sum_{n=0}^{\infty}c_nT_n(x) = \frac{1}{2}[f(x+0^+) + f(x-0^+)], \quad (-1 \leq x \leq 1),$$

where

$$c_n = \frac{2}{\pi}\int_{-1}^{1}\frac{T_n(x)}{\sqrt{1-x^2}}f(x)\mathrm{d}x.$$

The second solution $U_n(x)$ by (3.4.56) is called a **Chebyshev function of the second kind**. It is not a polynomial. It satisfies the boundary conditions $U_n(x = \pm 1) = 0$, as can easily be seen by (3.4.56). There is a relation between $U_n(x)$ and Chebyshev polynomials:

$$U_n(x) = \frac{\sqrt{1-x^2}}{n} \frac{dT_n(x)}{dx}. \qquad (3.4.61)$$

It then brings the series expression of $U_n(x)$:

$$U_n(x) = \sqrt{1-x^2} \sum_{k=0}^{[(n-1)/2]} (-1)^k \frac{(n-k-1)!}{k!(n-2k-1)!} (2x)^{n-2k-1}.$$

$$(3.4.62)$$

The derivative expression of $U_n(x)$ has been listed in Table 3.4. The generating function of Chebyshev functions of the second kind is

$$\frac{\sqrt{1-x^2}}{1-2xt+t^2} = \sum_{n=0}^{\infty} U_{n+1}(x)t^n. \qquad (3.4.63)$$

Note that the expansion coefficients are U_{n+1} instead of the usual U_n, see Table 3.5.

The general solution of (3.4.49) is

$$y = AT_n(x) + BU_n(x). \qquad (3.4.64)$$

2. Associated Chebyshev functions

Taking derivative of a Chebyshev equation (3.4.49) m times, one obtains an **associated Chebyshev equation**:

$$(1-x^2)y'' - (2m+1)xy' + (\lambda - m^2)y = 0, \qquad (3.4.65)$$

where m is positive integers and $0 \le m \le n$. Under the boundary conditions

$$y'(x = \pm 1) < \infty, \tag{3.4.66}$$

its eigenvalues are

$$\lambda = (n + m)^2, \quad (n = 0, 1, 2, \ldots). \tag{3.4.67}$$

Both the two special solutions of the equation are called **associated Chebyshev functions**. They can be gained by taking the derivative of Chebyshev functions m times with respect to x:

$$T_n^{(m)}(x) = \frac{\mathrm{d}^m}{\mathrm{d}x^m} T_n(x) \tag{3.4.68}$$

and

$$U_n^{(m)}(x) = \frac{\mathrm{d}^m}{\mathrm{d}x^m} U_n(x). \tag{3.4.69}$$

$T_n^{(m)}(x)$ and $U_n^{(m)}(x)$ are called the first and second associated Chebyshev functions of degree n and order m, respectively. An associated Chebyshev equation (3.4.65), when $m = 0$ is taken, will go back to a Chebyshev equation (3.4.49). Accordingly, associated Chebyshev functions, when $m = 0$ is taken, will go back to Chebyshev functions:

$$T_n^{(0)}(x) = T_n(x), \quad U_n^{(0)}(x) = U_n(x).$$

The general solution of (3.4.65) is

$$z = AT_n^{(m)}(x) + BU_n^{(m)}(x).$$

The cases of $T_n^{(m)}(x)$ and $U_n^{(m)}(x)$ as $m > 1$ are seldom encountered. Let

$$\lambda = n + m.$$

Then associated Chebyshev equation becomes

$$(1 - x^2)z'' - (2m + 1)xz' + n(n + 2m)z = 0. \tag{3.4.70}$$

This form is in fact just the Gegenbauer equation shown in Table 3.1. Its general solution is

$$z = AC_n^m(x) + DU_n^m(x), \tag{3.4.71}$$

where

$$C_n^m(x) = \frac{1}{2^{m-1}(m-1)!(n+m)} \frac{\mathrm{d}^m}{\mathrm{d}x^m} T_{n+m}(x),$$

$$D_n^m(x) = \frac{1}{2^{m-1}(m-1)!(n+m)} \frac{\mathrm{d}^m}{\mathrm{d}x^m} U_{n+m}(x).$$

$C_n^m(x)$ is called a Gegenbauer polynomial of degree n and order m. A Gegenbauer equation will become a Legendre equation when m is not an integer but $m = 1/2$.

3. Chebyshev polynomials of the second kind

As $m = 1$, an associated Chebyshev equation turns out to be

$$(1 - x^2)y'' - 3xy' + (\lambda - 1)y = 0. \tag{3.4.72}$$

Under the requirement that $y(x = \pm 1)$ are analytical, the eigenvalues are

$$\lambda = (n+1)^2, \quad (n = 0, 1, 2, \ldots). \tag{3.4.73}$$

The corresponding eigenfunctions are

$$U_n^*(x) = \frac{1}{n+1} \frac{\mathrm{d}T_{n+1}(x)}{\mathrm{d}x}, \tag{3.4.74}$$

which are called **Chebyshev polynomials of the second kind**. $U_n^*(x)$ is a polynomial of degree n, because $T_{n+1}(x)$ is one of degree $n + 1$. Since $T_{n+1}(x) = \cos(n + 1)\theta$, it brings

$$U_n^*(x) = \frac{\sin(n+1)\theta}{\sin\theta}. \tag{3.4.75}$$

Having the series expression of $T_n(x)$ (3.4.58), that of $U_n^*(x)$ immediately follows:

$$U_n^*(x) = \sum_{k=0}^{[n/2]} (-1)^k \frac{(n-k)!}{k!(n-2k)!} (2x)^{n-2k}. \tag{3.4.76}$$

The derivative expression of $U_n^*(x)$ has been listed in Table 3.4. The generating function of the second kind Chebyshev polynomials has been listed in Table 3.5.

Beside (3.4.72), there are two more relations between the second kind Chebyshev polynomials $U_n^*(x)$ and the first kind ones $T_n(x)$ as follows:

$$T_n(x) = U_n^*(x) - xU_{n-1}^*(x) \qquad (3.4.77)$$

and

$$(1 - x^2)U_n^*(x) = xT_{n+1}(x) - T_{n+2}(x). \qquad (3.4.78)$$

The second kind Chebyshev polynomials $U_n^*(x)$ are related to the second kind Chebyshev functions $U_n(x)$ by

$$U_{n+1}(x) = \sqrt{1 - x^2}U_n^*(x). \qquad (3.4.79)$$

The other linearly independent solution of (3.4.70) is $\frac{\mathrm{d}U_{n+1}(x)}{\mathrm{d}x}$.

3.4.4. *Hermite functions*

In this subsection, we briefly introduce the solutions of a Hermite equation, Hermite functions, Weber-Hermite functions and Weber functions. The three kinds of functions are closely connected.

1. Hermite polynomials

A Hermite equation reads

$$y'' - 2xy' + \lambda y = 0, \quad (-\infty \le x \le \infty). \qquad (3.4.80)$$

The stipulation for boundary conditions are $y(x \to \pm\infty) \sim x^n$, i.e., as x goes to infinity, the solutions also go to infinity in the form of powers of x. This condition brings us to the eigenvalues

$$\lambda = 2n. \qquad (3.4.81)$$

The eigenfunctions are **Hermite polynomials** $H_n(x)$. Its Rodrigues' formula has been listed in Table 3.3. The series expression of Hermite polynomials has been listed in Table 3.2. The generating function of Hermite polynomials has been listed in Table 3.6.

The square of the normalizing factor of a Hermite polynomial is

$$\int_{-\infty}^{\infty} \mathrm{e}^{-x^2}|H_n(x)|^2\mathrm{d}x = \sqrt{\pi}2^n n!. \qquad (3.4.82)$$

Theorem 7. *If $f(x)$ is piecewise smooth, then*

$$\sum_{n=0}^{\infty} c_n H_n(x) = \frac{1}{2}[f(x+0^+) + f(x-0^+)], \quad (-\infty < x < \infty),$$

$$(3.4.83)$$

where

$$c_n = \frac{1}{\sqrt{\pi}2^n n!} \int_{-\infty}^{\infty} e^{-x^2} H_n(x) f(x) \mathrm{d}x.$$

Hermite polynomials obey the following addition formula.

$$H_n(x+y) = 2^{-n/2} \sum_{k=0}^{n} \frac{n!}{k!(n-k)!} H_n(\sqrt{2}x) H_{n-k}(\sqrt{2}y). \quad (3.4.84)$$

The following equation is useful in deriving the propagation function of a one-dimensional oscillator:

$$\frac{e^{x^2+y^2}}{\sqrt{1-z^2}} \exp\left(-\frac{x^2+y^2-2xyz}{1-z^2}\right) = \sum_{n=0}^{\infty} \left(\frac{z^n}{2^n n!}\right) H_n(x) H_n(y).$$

The boundary conditions can be transferred to $y(x \to \pm\infty) \sim \exp(x^2)$, i.e., as $x \to \pm\infty$, the solution goes to infinity in the way of $\exp(x^2)$. If so, then the other linearly independent solution of the Hermite equation is achieved. It is in fact calculated by the Liouville formula:

$$G_n(x) = H_n(x) \int \frac{\exp(x^2)}{[H_n(x)]^2} \mathrm{d}x. \quad (3.4.85)$$

$G_n(x)$ is called the **Hermite function of the second kind**.
   Thus the general solution of (3.4.80) is

$$y = A H_n(x) + B G_n(x).$$

If in (3.3.14) $B_1 = -1$ is taken, we will get another form of Hermite equation:

$$y'' - xy' + \lambda y = 0, \quad (-\infty \leq x \leq \infty). \quad (3.4.86)$$

If in (3.4.80) let $x = u/\sqrt{2}$, then $\frac{\mathrm{d}}{\mathrm{d}x}y = \sqrt{2}\frac{\mathrm{d}y}{\mathrm{d}u}$, $\frac{\mathrm{d}^2}{\mathrm{d}x^2}y = 2\frac{\mathrm{d}^2y}{\mathrm{d}u^2}$. Subsequently, we also get (3.4.86). The eigenvalues of this equation are

$$\lambda = n. \quad (3.4.87)$$

The corresponding characteristic functions are denoted by $He_n(x)$, which relate to Hermite functions by $He_n(x) = C(n)H_n(x/\sqrt{2})$. The constant factor is determined by normalization and its result is $C(n) = 2^{-n/2}$ which depends on n. Therefore, the explicit relation is

$$He_n(x) = 2^{-n/2}H_n(x/\sqrt{2}).\qquad(3.4.88)$$

The other linearly independent solution of (3.4.86) is

$$Ge_n(x) = 2^{-n/2}G_n(x/\sqrt{2}),\qquad(3.4.89)$$

where G_n is defined by (3.4.85).

2. Weber-Hermite functions

The first type of Weber equation reads:

$$z'' + (\lambda - x^2)z = 0, \quad (-\infty \le x \le \infty).\qquad(3.4.90)$$

The boundary conditions are $z(x = \pm 1) < \infty$. Let $z = e^{-x^2/2}y$ and substitute it into (3.4.90). Then the equation becomes

$$y'' - 2xy' + (\lambda - 1)y = 0,$$

which is just a Hermite equation, although the eigenvalues are shifted:

$$\lambda = 2n + 1, \quad (n = 0, 1, 2, \ldots).\qquad(3.4.91)$$

It has been known that a pair of linearly independent solutions of the Hermite equation (3.4.80) are the Hermite polynomial $H_n(x)$ and the second kind Hermite function $G_n(x)$. Accordingly, those of the first type of Weber equation should be $\Psi_n(x) = C_1(n)e^{-x^2/2}H_n(x)$ and $\Phi_n(x) = C_2(n)e^{-x^2/2}G_n(x)$. If all the functions have been normalized, the undetermined coefficients are 1: $C_1(n) = 1, C_2(n) = 1$. So, we have

$$\Psi_n(x) = e^{-x^2/2}H_n(x)\qquad(3.4.92)$$

and

$$\Phi_n(x) = e^{-x^2/2}G_n(x).\qquad(3.4.93)$$

$\Psi_n(x)$ and $\Phi_n(x)$ are called the **first and second types of Weber-Hermite functions**, respectively.

By means of the relationship between Weber-Hermite functions and Hermite functions, the properties of the former can be deduced from those of the latter.

For example, multiplying the generating function of Hermite polynomials $H_n(x)$ by $e^{-x^2/2}$ will immediately yield the generating function of $\Psi_n(x)$:

$$\exp(2xt - t^2 - x^2/2) = \sum_{n=0}^{\infty} \frac{\Psi_n(x)}{n!} t^n, \tag{3.4.94}$$

which has been listed in Table 3.6. Similarly, multiplying Rodrigues' formula of $H_n(x)$ by $e^{-x^2/2}$, we get the derivative expression of $\Psi_n(x)$:

$$\Psi_n(x) - (-1)^n e^{x^2/2} \frac{d^n}{dx^n} e^{-x^2}, \tag{3.4.95}$$

which has been listed in Table 3.4.

Weber-Hermite functions are of the following addition formula:

$$\Psi_n(x+y) = 2^{-n/2} e^{(x-y)^2/2} \sum_{k=0}^{n} \frac{n!}{k!(n-k)!} \Psi_n(\sqrt{2}x) \Psi_{n-k}(\sqrt{2}y).$$

3. Weber function (parabolic cylinder function)

The **second type of Weber equation** reads

$$z'' + \frac{1}{2}\left(\lambda - \frac{1}{2}x^2\right) z = 0, \quad (-\infty \le x \le \infty). \tag{3.4.96}$$

Its boundary conditions are set to be $z(x \to \pm\infty) < \infty$. Let $z = e^{-x^2/4} y$. Then (3.4.94) turns out to be

$$y'' - xy' + \frac{1}{2}(\lambda - 1)y = 0.$$

This is just (3.4.86) with eigenvalues being $(\lambda - 1)/2 = n$. That is to say,

$$\lambda = 2n + 1, \quad (n = 0, 1, 2, \ldots). \tag{3.4.97}$$

Naturally, its two linearly independent solutions are $He_n(x)$ and $Ge_n(x)$, respectively, see (3.4.88) and (3.4.89). So, a pair of linearly independent solutions of the second type of Weber equation are as follows.

The first characteristic function is

$$D_n(x) = e^{-x^2/4} He_n(x) = 2^{-n/2} e^{-x^2/4} H_n(x/\sqrt{2})$$
$$= 2^{-n/2} e^{-x^2/4} e^{(x/\sqrt{2})^2/2} \Psi_n(x/\sqrt{2}) = 2^{-n/2} \Psi_n(x/\sqrt{2}).$$
(3.4.98)

It is called a **Weber function**, or a **parabolic cylinder function**.

The second characteristic function is

$$E_n(x) = e^{-x^2/4} Ge_n(x) = 2^{-n/2} e^{-x^2/4} G_n(x/\sqrt{2}) = 2^{-n/2} \Phi_n(x/\sqrt{2}).$$
(3.4.99)

It is called the **second kind Weber function**.

The generating function of $D_n(x)$ can be immediately obtained by multiplying (3.4.94) by $\exp(-x^2/4)$:

$$\exp(2xt - t^2 - x^2/4) = \sum_{n=0}^{\infty} \frac{D_n(x)}{n!} t^n,$$
(3.4.100)

which has been listed in Table 3.6. In (3.4.95), let $x \to x/\sqrt{2}$ and multiply a factor $2^{-n/2}$. Then we get the derivative expression of $D_n(x)$:

$$D_n(x) = 2^{-n/2}(-1)^n e^{(x/\sqrt{2})^2/2} \frac{d^n}{d(x/\sqrt{2})^n} e^{-(x/\sqrt{2})^2}$$
$$= (-1)^n e^{x^2/4} \frac{d^n}{dx^n} e^{-x^2/2},$$
(3.4.101)

which has been listed in Table 3.4.

Weber function is of the following addition formula:

$$D_n(x+y) = 2^{-n/2} e^{(x-y)^2/4} \sum_{k=0}^{n} \frac{n!}{k!(n-k)!} D_n(\sqrt{2}x) D_{n-k}(\sqrt{2}y).$$

4. The solutions of Laplace equation under parabolic coordinates

Laplace equation reads

$$\nabla^2 u = \frac{\partial^2 u}{\partial x^2} + \frac{\partial^2 u}{\partial y^2} + \frac{\partial^2 u}{\partial z^2} = 0. \tag{3.4.102}$$

We now introduce parabolic coordinates ξ, η, z in the way of

$$x = \frac{1}{2}(\xi^2 - \eta^2), \quad y = \xi\eta, \quad z = z. \tag{3.4.103}$$

The transformations of the derivatives with respect to coordinates are as follows.

$$\frac{\partial u}{\partial \xi} = \xi\frac{\partial u}{\partial x} + \eta\frac{\partial u}{\partial y}, \quad \frac{\partial u}{\partial \eta} = -\eta\frac{\partial u}{\partial x} + \xi\frac{\partial u}{\partial y},$$

$$\frac{\partial^2 u}{\partial \xi^2} = 2\frac{\partial u}{\partial x} + 2\xi\eta\frac{\partial^2 u}{\partial x\partial y} + \left(\frac{\eta}{\xi} - \frac{\xi}{\eta}\right)\frac{\partial u}{\partial y} + \xi^2\frac{\partial^2 u}{\partial^2 x} + \eta^2\frac{\partial^2 u}{\partial y^2},$$

$$\frac{\partial^2 u}{\partial \eta^2} = -2\frac{\partial u}{\partial x} - 2\xi\eta\frac{\partial^2 u}{\partial x\partial y} + \left(\frac{\xi}{\eta} - \frac{\eta}{\xi}\right)\frac{\partial u}{\partial y} + \eta^2\frac{\partial^2 u}{\partial x^2} + \xi^2\frac{\partial^2 u}{\partial y^2}.$$

Then Laplace equation becomes

$$\frac{1}{\xi^2 + \eta^2}\left(\frac{\partial^2 u}{\partial \xi^2} + \frac{\partial^2 u}{\partial \eta^2}\right) + \frac{\partial^2 u}{\partial z^2} = 0.$$

It is easily seen that separation of variables is applicable. Let $u(\xi, \eta, z) = X(\xi)Y(\eta)Z(z)$. Then, the equation is decomposed to the following three equations.

$$\begin{cases} X'' + (\alpha - m^2\xi^2)X = 0, \\ Y'' - (\alpha + m^2\eta^2)Y - 0, \\ Z'' + m^2 Z = 0. \end{cases}$$

Here α and m are constants. If homogeneous boundary conditions are given, one is able to determine the value of m from the third equation. In the former two equations, let $\sqrt{2m}\xi = \zeta, \alpha = m\mu$ and $\sqrt{2m}\eta = \zeta, \alpha = -m\mu$. Then both are transformed to the same form:

$$\frac{d^2 w}{d\zeta^2} + \frac{1}{2}\left(\mu - \frac{1}{2}\zeta^2\right)w = 0,$$

where w can be either X or Y. This equation is just the second type of Weber equation. Under the conditions $w(\zeta \to \pm\infty) < \infty$, its eigenvalues are

$$\mu = 2n + 1, \quad (n = 0, 1, 2, \ldots),$$

and the corresponding eigenfunctions are

$$w = D_n(\zeta).$$

This should be the reason why $D_n(x)$ is called parabolic cylinder function.

In Table 3.7 the parities and values at special points of some functions are listed.

3.5. Complex Analysis Theory of the Ordinary Differential Equations of Second Order

When we discuss the ordinary differential equations of second order above, the argument was defaulted as real. In general, the argument should be complex, which was first noticed by Cauchy. In this section, we briefly introduce the analytical theory of the ordinary differential equations of second order. In this section, the argument is always defaulted as complex, unless specified.

3.5.1. *Solutions of homogeneous equations*

1. Fundamental theory of homogeneous linear equations

(1) Existence and uniqueness theorem of ordinary differential equations

Let us first inspect the ordinary differential equations of first order. Consider a set of equations as follows:

$$w'_s = f_s(z, w_1, w_2, \ldots, w_n), \quad (s = 1, 2, \ldots, n). \tag{3.5.1}$$

We define a column vector $w = (w_1, w_2, \ldots, w_n)^{\mathrm{T}}$, and subsequently $w' = \frac{\mathrm{d}w}{\mathrm{d}z} = (w'_1, w'_2, \ldots, w'_n)^{\mathrm{T}}$. Then (3.5.1) is written in a compact

Table 3.7. The parities and values at special points of some functions.

Function, $Q_n(x)$	Parity	Values at special points
Generalized ultraspheric Polynomial, $P_n^{(r,s)}(x)$	$P_n^{(r,s)}(-x) = (-1)^n P_n^{(s,r)}(x)$	$P_n^{(r,s)}(1) = \dfrac{(r+1)_n}{n!}$
Jacobi polynomial, $J_n^{(r,s)}(x)$		$J_n^{(r,s)}(0) = 1,$ $J_n^{(r,s)}(1) = \dfrac{(s-r-n)_n}{(s)_n}$
Gegenbauer polynomial, $G_n^m(x)$	$G_n^m(-x) = (-1)^n G_n^m(x)$	$G_n^m(1) = \dfrac{(2m)_n}{n!}, G_{2n+1}^m(0) = 0$ $G_{2n}^m(x)(0) = (-1)^n \dfrac{(m)_n}{n!}$
Chebyshev polynomial, $T_n(x)$	$T_n(-x) = (-1)^n T_n(x)$	$T_n(1) = 1$ $T_{2n+1}(0) = 0, T_{2n}(0) = (-1)^n$
Chebyshev function of second kind, $U_n(x)$	$U_n(-x) = (-1)^{n+1} U_n(x)$	$U_n(\pm 1) = 1$ $U_{2n}(0) = 0, U_{2n+1}(0) = (-1)^n$
Chebyshev polynomial of second kind, $U_n^*(x)$	$U_n^*(-x) = (-1)^n U_n^*(x)$	$U_n^*(1) = n+1$ $U_{2n}^*(0) = (-1)^n, U_{2n+1}^*(0) = 0$
Legendre polynomial, $P_n(x)$	$P_n(-x) = (-1)^n P_n(x)$	$P_n(1) = 1, P_{2n+1}(0) = 0$ $P_{2n}(0) = (-1)^n \dfrac{(2n)!}{2^{2n}(n!)^2}$
Associated Legendre function of first kind, $P_n^m(x)$	$P_n^m(-x) = (-1)^{n+m} P_n^m(x)$	$P_n^m(\pm 1) = 0$ $P_n^m(0) = 0, \quad n-m$ is odd $P_n^m(0) = (-1)^{(n-m)/2} \dfrac{(n+m-1)!!}{(n-m)!!}, \quad n-m$ is even
Sonine polynomial, $S_n^\mu(x)$		$S_n^\mu(0) = \dfrac{(\mu+1)_n}{n!}$
Associated Laguerre polynomial, $L_n^{(m)}(x)$		$L_n^{(m)}(0) = (-1)^n \dfrac{(n!)^2}{(n-m)!m!}$

(Continued)

Table 3.7.   (*Continued*)

Function, $Q_n(x)$	Parity	Values at special points
Laguerre polynomial, $L_n(x)$		$L_n(0) = n!$
Hermite polynomial, $H_n(x)$	$H_n(-x) = (-1)^n H_n(x)$	$H_{2n}(0) = (-1)^n \dfrac{(2n)!}{n!}$ $H_{2n+1}(0) = 0$
Weber-Hermite function of the first kind, $\Psi_n(x)$	$\Psi_n(-x) = (-1)^n \Psi_n(x)$	$\Psi_{2n}(0) = (-1)^n \dfrac{(2n)!}{n!}$ $\Psi_{2n+1}(0) = 0$
Weber function (parabolic cylinder function), $D_n(x)$	$D_n(-x) = (-1)^n D_n(x)$	$D_{2n}(0) = (-1)^n (2n-1)!$ $D_{2n+1}(0) = 0$
Bessel function, $J_n(x)$	$J_n(-x) = (-1)^n J_n(x)$	$J_n(0) = \begin{cases} 1, & n = 0 \\ 0, & n \geq 1 \end{cases}$
Modified Bessel function of the first kind, $I_n(x)$	$I_n(-x) = (-1)^n I_n(x)$	$I_n(0) = \begin{cases} 1, & n = 0 \\ 0, & n \geq 1 \end{cases}$
Spherical Bessel function of the first kind (spherical Bessel function), $j_n(x)$	$j_n(-x) = (-1)^n j_n(x)$	$j_n(0) = \begin{cases} 1, & n = 0 \\ 0, & n \geq 1 \end{cases}$
Spherical Bessel function of the second kind (Neumann function), $y_n(x)$	$y_n(-x) = (-1)^{n+1} y_n(x)$	$y_n(0) \sim x^{-(n+1)} \to \infty$

form:

$$w' = f(z, w). \tag{3.5.2}$$

The corresponding initial value problem is:

$$\begin{cases} w' = f(z, w), \\ w(z_0) = w_0. \end{cases} \tag{3.5.3}$$

We will be asked to find an analytical function $\varphi(z)$ defined on a certain domain R including a given point z_0 in complex plane C, which satisfies

$$\varphi(z_0) = w_0, \quad \varphi'(z) = f(z, \varphi), \quad z \in R.$$

Theorem 1 (Cauchy Theorem). *If a function $f_s(z, w_1, w_2, \ldots, w_n)$, $(s = 1, 2, \ldots, n)$ is analytical on*

$$|z - z_0| < r, \quad |w_s - w_s(z_0)| < \rho, \quad (s = 1, 2, \ldots, n)$$

and satisfies

$$|f_s(z, w_1, w_2, \ldots, w_n)| \leq M, \quad (s = 1, 2, \ldots, n),$$

where $r > 0, \rho > 0, M > 0$ and they are all constants, then there exists a unique solution for initial value problem (3.5.3) at least on the domain

$$|z - z_0| < r(1 - e^{-\rho/(n+1)Mr}).$$

Mathematically, when discussing a homogeneous equation of higher order, it is considerably convenient to convert it to a homogeneous equations of first order. Specifically, consider the following initial value problem

$$\begin{cases} y^{(n)}(z) + \alpha_1(z)y^{(n-1)}(z) + \cdots + \alpha_{n-1}(z)y'(z) + \alpha_n y(z) = 0, \\ y(z_0) = \gamma_0, y'(z_0) = \gamma_1, y''(z_0) = \gamma_2, \ldots, y^{(n-1)}(z_0) = \gamma_{n-1}. \end{cases}$$

$$(3.5.4)$$

After defining

$$w_1 = y, w_1' = y' = w_2, w_2' = w_1'' = y'' = w_3, \ldots, w_n' = w_1^{(n)} = y^{(n)},$$

$$(3.5.5)$$

(3.5.4) becomes the following linear equations:

$$\begin{cases} w_1' = w_2, \\ w_2' = w_3, \\ \quad \vdots \\ w_{n-1}' = w_n, \\ w_n' = -\alpha_1 w_n - \cdots - \alpha_{n-1} w_2 - \alpha_n w_1. \end{cases}$$

$$(3.5.6)$$

This set of equations (3.5.6) can be concisely written in the form of

$$w' = A(z)w, \tag{3.5.7}$$

where $A(z) = (a_{ij}(z))$ is a single-valued and analytical $n \times n$ matrix function on a simply connected domain $G \subset C$.

The form of (3.5.7) is more or less an extended form of (3.5.6). On the one hand, this compact form has universality to a certain extent. On the other hand, this extended form is convenient to discuss, since some conclusions of linear algebra can be utilized.

In Subsection 3.1.1, an ordinary differential equation of second order was converted to a linear differential equations of first order so as to discuss the existence and uniqueness of its solutions. This process also applies to complex functions.

Equation (3.5.7) is only a special case of 3.5.2), where $f(z, w) = A(z)w$. Therefore, the existence and uniqueness of the solutions of (3.5.7) are obvious. Compared to (3.5.2), (3.5.7) has a simpler form and is an ordinary differential equation of first order. Because of this reason, the existence and uniqueness of its solutions can be expressed more explicitly as follows.

Theorem 2. *If $A(z)$ is single-valued and analytical on a simply connected domain $G \subset C$, $z_0 \in G$ and $w_0 \in C^n$, then there exists a single-valued and analytical solution $w(z)$ on G for the initial value problem*

$$\begin{cases} w' = A(z)w, \\ w(z_0) = w_0. \end{cases} \tag{3.5.8}$$

(2) Basic solution matrix

All of the solutions of (3.5.7) compose an n-dimensional complex linear space. In this solution space, there necessarily exist n linearly independent special solutions $w_1(z), w_2(z), \ldots, w_n(z)$, and the general solution w can always be expressed by the linear combination of the n special solutions:

$$w(z) = c_1 w_1(z) + c_2 w_2(z) + \cdots + c_n w_n(z), \quad (c_i \in C).$$

Suppose that $w_{(1)}(z), w_{(2)}(z), \ldots, w_{(n)}(z)$ (the subscripts represent column index) are arbitrary n special solutions of (3.5.7). They are denoted by

$$W(z) = (w_{(1)}(z), w_{(2)}(z), \ldots, w_{(n)}(z)). \tag{3.5.9a}$$

We say that (3.5.9a) composes a **solution matrix** of (3.5.7).

The n solutions in (3.5.9a) may be either **linearly dependent** or **linearly independent**. If they are linearly independent, they are denoted by

$$W(z) = (w^{(0)}(z), w^{(1)}(z), \ldots, w^{(n-1)}(z)). \tag{3.5.9b}$$

Here the superscripts represent column index, not the derivatives of the vectors. We say that (3.5.9b) composes a **basic solution matrix** of (3.5.7). Both the solution matrix and basic solution matrix satisfy the equation $W'(z) = A(z)W(z)$.

The sufficient and necessary conditions for a solution matrix to be a basic one are that its determinant $\Phi(z) = \det W(z)$ is not zero at a given point $z_0 \in G$. $\Phi(z) = \det W(z)$ is just the Wronskian of the solution of (3.5.4), which satisfies

$$\Phi'(z) = \mathrm{tr}(A(z))\Phi(z).$$

Theorem 3. *If $W(z)$ is one basic solution matrix of (3.5.7), then its general solution matrix of V can be expressed by*

$$V(z) = W(z)C, \tag{3.5.10}$$

where C is a nonsingular matrix.

2. Isolated singularities and their classification

(1) Isolated singularities bring out multi-valued functions

Up to now, it has been implicitly assumed that $A(z) = (a_{ij}(z))$ in (3.5.7) is a single-valued and analytical $n \times n$ matrix function on a simply connected domain $G \subset C$. Now we turn to discussion of cases with isolated singularities.

Definition 1. In equations

$$w' = A(z)w, \qquad (3.5.11)$$

if $A(z) = (a_{ij}(z))$ is single-valued and analytical on a domain $0 < |z - z_0| < r$ (r is a given positive real number) except point z_0, which means that at least one element $a_{ij}(z)$ of $A(z)$ is not analytical at point z_0, then the point z_0 is called an **isolated singularity** of $A(z)$, also called an **isolated singularity** of (3.5.11).

Although Eqs. (3.5.11) and (3.5.7) appear as the same form, their coefficient matrices $A(z)$ have different analytical properties. Below, when we mention (3.5.7), its $A(z)$ is single-valued and analytical on a simply connected domain $G \subset C$; when we mention (3.5.11), its $A(z)$ is not analytical at z_0.

Theorem 2 does not apply to the case when there is an isolated singularity. Because $0 < |z - z_0| < r$ is not a simply connected domain, Eqs. (3.5.11) may have solutions which are multivalued functions.

Example 1. Consider a differential equation of $n = 1$:

$$w' = \frac{\alpha}{z}w, \qquad (3.5.12)$$

where α is a constant. The whole complex plane except the point $z = 0$ is denoted by $C - \{0\}$ or $K = C - \{0\}$. The function $A(z) = \frac{\alpha}{z}$ is singly-valued and analytical on $C - \{0\}$, while $z = 0$ is its isolated singularity. There is a basic solution $w = z^\alpha$ where $z^\alpha \neq 0$. It is multivalued as $\alpha = \frac{1}{2}$, because $w = \sqrt{z}$ is a multivalued function.

Example 2. Consider differential equations of $n = 2$:

$$\begin{pmatrix} w_1' \\ w_2' \end{pmatrix} = \begin{pmatrix} 0 & 1/z \\ 0 & 0 \end{pmatrix} \begin{pmatrix} w_1 \\ w_2 \end{pmatrix}.$$

Here the matrix $A(z) = \begin{pmatrix} 0 & 1/z \\ 0 & 0 \end{pmatrix}$ is singly-valued and analytical on $C - \{0\}$, and $z = 0$ is its isolated singularity. Hence, these equations have a multi-valued basic solution matrix

$$W = \begin{pmatrix} \text{Ln}\, z & 1 \\ 1 & 0 \end{pmatrix}$$

where the first special solution is multi-valued and the second is singly-valued. This is because

$$\text{Ln}\, z = \ln |z| + i \arg z + i2k\pi \qquad (3.5.13)$$

is a multi-valued function.

These two examples indicate that in the cases where there exist isolated singularities, there will appear multi-valued functions in the special solutions of the equation.

(2) Solutions of equations when there are isolated singularities

Consider simple equations

$$w' = \frac{A}{z} w, \qquad (3.5.14)$$

where $A = (a_{ij})$ is an $n \times n$ constant matrix. These are called **Cauchy equations**. The Examples 1 and 2 are their two specific cases. Obviously, $z = 0$ is an isolated singularity of (3.5.14). The process of solving the equations reflect the main idea of dealing with the equations with isolated singularities.

The coefficient matrix of (3.5.14) is A/z and it is singly-valued and analytical on $K = C - \{0\}$. In order to seek its solutions, let us make variable transformation

$$z = e^s \qquad (3.5.15)$$

and the function turns to $w(z) = u(s)$. Subsequently, (3.5.14) becomes

$$u' = A u. \qquad (3.5.16)$$

This is a constant coefficient linear equation, and has basic solution matrix $U = e^{As} = \sum_{k=0}^{\infty} \frac{A^k}{k!} s^k$, which is singly-valued and analytical on the whole complex plane. Accordingly, the original equation (3.5.14) also has a basic solution matrix $W(z) = U(\text{Ln}\, z) = e^{A\text{Ln}\, z} = \sum_{k=0}^{\infty} \frac{A^k}{k!} (\text{Ln}\, z)^k$, which is analytical on $K = C - \{0\}$, but not singly-valued.

By Theorem 3, the general solution matrix of (3.5.16) is $u = e^{As} C = \sum_{k=0}^{\infty} \frac{A^k}{k!} s^k C$, where C is any constant nonsingular matrix.

Consequently, the general solution matrix of (3.5.14) is $w(z) = e^{A \operatorname{Ln} z} C = \sum_{k=0}^{\infty} \frac{A^k}{k!} (\operatorname{Ln} z)^k C$. This case demonstrates that the variable transformation (3.5.15) enables one to turn equations with some isolated singularities to those on a simply connected domain, so as to obtain multi-valued solutions in terms of singly-valued solutions.

If (3.5.11) has at most one isolated singularity $z_0 \in C$, then its solutions near z_0 are given by the following theorem.

Theorem 4. *Suppose that $A(z)$ is singly-valued and analytical on $0 < |z - z_0| < r$ with r being a certain positive number, then there necessarily exists a basic solution matrix $W(z)$ of (3.5.11). It is of a form of*

$$W(z) = V(z)(z - z_0)^B, \quad 0 < |z - z_0| < r, \tag{3.5.17}$$

where $V(z)$ is an $n \times n$ function matrix singly-valued and analytical on $0 < |z - z_0| < r$ and B is an $n \times n$ constant matrix.

The idea of proving this theorem is as follows. Without loss of generality, assume $z_0 = 0$. Then make variable transformation as (3.5.15). So, (3.5.11) is turned into the form of

$$u'(s) = e^s A(e^s) u(s). \tag{3.5.18}$$

The coefficient matrix $e^s A(e^s)$ is singly-valued and analytical on the domain $\operatorname{Re} s < \ln r$. Therefore, (3.5.18) and (3.5.7) hold the same properties. A basic solution matrix of (3.5.18) is denoted by $U(s)$. Note that the coefficient matrix $e^s A(e^s)$ has a period $2\pi i$. By Theorem 3, $U(s)C$ is also a solution of (3.5.18). Hence, there exists a nonsingular constant matrix C satisfying $U(s + 2\pi i) = U(s)C$. We here cite without proof a lemma in complex analysis: for any nonsingular $n \times n$ matrix, there must be an $n \times n$ matrix B which makes $e^B = A$. In terms of this lemma, one is able to rewrite C into the form of $C = e^{2\pi i B}$. Let $T(s) = U(s)e^{-Bs}$. Then $T(s)$ has a period $2\pi i \cdot U(s) = T(s)e^{Bs}$. The solution of the original equation (3.5.11) is $W(z) = U(\operatorname{Ln} z) = T(\operatorname{Ln} z)e^{B \operatorname{Ln} z} = T(\operatorname{Ln} z)z^B = V(z)z^B$, which is indeed in the form of (3.5.17).

Under the premise of Theorem 4, any basic solution matrix $W(z)$ of (3.5.11) is of the form of (3.5.17). Now, if C is a nonsingular constant matrix, $U(s)C$ is a solution of (3.5.18), and $W(z)C$ is surely that of (3.5.11).

$$W(z)C = V(z)z^B C = V(z)\, CC^{-1}z^B C = V_1(z)z^{B_1},$$

where $V_1(z) = V(z)\, C, B_1 = C^{-1}BC$.

The domain $0 < |z - z_0| < r$ is denoted by $K_r^{z_0}$, and the domain $0 < |z| < r$ denoted by K_r^0. If in K_r^0 one cuts along the negative real axis from $z = 0$, he will get a singly connected domain, denoted as K_r^-. This is a set consisting of all the z points inside the circular domain $|z| < r$ except those with $\operatorname{Re} z \leq 0, \operatorname{Im} z = 0$.

Theorem 5. *Suppose that $A(z)$ is singly-valued and analytical in the domain K_r^0, and it has a pole of $m\,(\geq 1)$ order at $z = 0$. Then for any singly-valued and analytical solution $w(z)$ of (3.5.11) on K_r^-, there must be two positive constant a and b, which make*

$$|w(z)| = \begin{cases} a|z|^{-b}, & m = 1 \\ a\exp(b|z|^{1-m}), & m > 1 \end{cases} \quad z \in K_{r/2}^-. \qquad (3.5.19)$$

(3) Classification of singularities, Fuchs-type equations

(i) Finite singularities

Definition 2. Suppose that $A(z)$ in (3.5.11) is a singly-valued and analytical $n \times n$ matrix function on $0 < |z - z_0| < r$. Point z_0 is said to be an **ordinary point** of $A(z)$ if it is analytical at this point. Point z_0 is said to be a **removable singularity** of $A(z)$ if it is nonanalytic at this point but $\lim_{z \to z_0} A(z)$ exists. For example, $\lim_{z \to 0} \frac{\sin z}{z}$ exists so that $z_0 = 0$ is a removable singularity of $\frac{\sin z}{z}$. Ordinary points and removable singularities are all called **regular points** of (3.5.11).

Suppose that a function $A(z)$ is expanded by a Laurent series as follows:

$$A(z) = \sum_{i=N}^{M} a_i(z - z_0)^i, \quad |z - z_0| > r,$$

where the lower limit N can be either negatively finite or negatively infinite, and the upper limit M can be either positively finite or positively infinite. If the last negative power coefficient which does not vanish is a_{-N}, then z_0 is said to be a **pole of order** N. If the series of negative powers in the Laurent expansion does not terminate, i.e., $N = -\infty$, the point z_0 is called an **essential singularity** of $A(z)$.

If z_0 is a pole of order one of $A(z)$, z_0 is also called a **simple pole**, or the **first kind (weak) singularity** of $A(z)$.

If z_0 is neither a regular point of $A(z)$ nor its first kind singularity, then z_0 is called the **second kind (strong) singularity** of $A(z)$. That means that this kind of singularities are poles of order two or more.

For the first kind singularities, the conclusion of Theorem 4 can be strengthened. This yields the following theorem.

Theorem 6. *Suppose that $A(z)$ in* (3.5.11) *is singly-valued and analytical in the domain* $0 < |z - z_0| < r$, *and z_0 is a first kind singularity of* (3.5.11). *Then every basic solution matrix $W(z)$ of* (3.5.11) *is of the form of*

$$W(z) = V(z)(z - z_0)^B, \qquad (3.5.20)$$

where $V(z)$ is an $n \times n$ function matrix singly-valued and analytical on $|z - z_0| < r$ and B is an $n \times n$ constant matrix.

The difference between Theorem 6 and Theorem 4 is that the domain $0 < |z - z_0| < r$ in (3.5.17) is relaxed to be $|z - z_0| < r$ in (3.5.20).

(ii) Infinite singularity

Consider again Eqs. (3.5.11). If $A(z)$ is singly-valued and analytical on domain $|z| > r$, then the infinite point ∞ is also an isolated singularity of (3.5.11). In order to conduct classification of an ∞ point, we make transformation $\xi = \frac{1}{z}$ and $w(z) = u(\xi)$ so as to convert (3.5.11) to be a form of

$$u'(\xi) = -\frac{1}{\xi^2} A\left(\frac{1}{\xi}\right) u(\xi), \quad 0 < |\xi| < \frac{1}{r}. \qquad (3.5.21)$$

In this case, the singularity properties of $\xi = 0$ are just those of $z = \infty$. For example, when $\xi = 0$ is the regular point, $z = \infty$ is also the regular point, and so on.

If the matrix function $A(\frac{1}{\xi})$ in (3.5.21) is expanded to be a Laurent series:

$$A\left(\frac{1}{\xi}\right) = \sum_{k=-\infty}^{\infty} B_k \xi^k, \quad 0 < |\xi| < \frac{1}{r},$$

then by the definition above, the ∞ point will be a (i) regular point if $B_k = 0$ for all $k \leq 1$, (ii) the first kind singularity if $B_k = 0$ for $k \leq 0$ but $B_1 \neq 0$, (iii) the second kind singularity if some $B_k \neq 0$ for $k \leq -2$.

Theorem 7. *Suppose that $A(z)$ in (3.5.11) is singly-valued and analytical in the domain $|z| > r$, and z_0 is a first kind singularity of (3.5.11). Then the sufficient and necessary conditions for $z = \infty$ to be the first kind singularity of (3.5.11) are that $z - \infty$ is a simple pole of $A(z)$ or $A(z)$ can be expanded into*

$$A(z) = \frac{B_1}{z} + \frac{B_2}{z^2} + \frac{B_3}{z^3} + \cdots, \quad |z| > r, \qquad (3.5.22)$$

and $B_1 \neq 0$. The sufficient and necessary conditions for $z = \infty$ to be a regular point are that $A(z)$ can be expanded in the form of

$$A(z) = \frac{B_2}{z^2} + \frac{B_3}{z^3} + \cdots, \quad |z| > r. \qquad (3.5.23)$$

In these two equations, $B_i (i = 1, 2, 3, \ldots)$ are constant matrices.

Besides, every basic solution matrix $W(z)$ of (3.5.11) is of the form of

$$W(z) = \begin{cases} U(z)z^B, & z = \infty \text{ is first kind singularity,} \\ U(z), & z = \infty \text{ is regular point,} \end{cases} \quad |z| > r,$$

where $U(z)$ is a singly-valued and analytical matrix function on $|z| > r$ (including $z = \infty$) and B is a constant matrix.

(iii) Fuchs-type equations

Definition 3. The equations

$$w' = A(z)w \tag{3.5.24}$$

are said to be **Fuchs-type equations** if all the points on complex plane C are regular points except a finite number of first kind singularities (including ∞).

Theorem 8. *The sufficient and necessary conditions for equations (3.5.24) to be of Fuchs-type are that the coefficient matrix is of the form of*

$$A(z) = \sum_{j=1}^{k} \frac{1}{z - z_j} R_j, \tag{3.5.25}$$

where $z_1, z_2, \ldots, z_k$ are k first singularities different from each other and R_j is a nonzero constant matrix.

By this theorem, if equations (3.5.24) are of Fuchs-type, i.e., $A(z)$ holds the form of (3.5.25), taking the limit of $zA(z)$ as $z \to \infty$, one sees that when $\sum_{j=1}^{k} R_j = 0$ (or $\neq 0$) the infinite point is a regular point (or first kind singularity) of (3.5.24).

The solutions of Fuchs-type equations can be searched by means of power series method.

3. Power series solutions

Consider equations

$$w' = A(z)w, \tag{3.5.26}$$

where $A(z)$ is an $n \times n$ matrix function singly-valued and analytical on $0 < |z - z_0| < r$ and z_0 is a first kind singularity. That is to say, (3.5.26) is a special case of Fuchs-type equations where there is only one simple pole.

$$A(z) = \frac{1}{z - z_0} \sum_{k=0}^{\infty} A_k (z - z_0)^k, \quad 0 < |z - z_0| < r. \tag{3.5.27a}$$

For the sake of simplicity, let $z_0 = 0$. Then $A(z)$ is written as

$$A(z) = \frac{1}{z} \sum_{k=0}^{\infty} A_k z^k, \quad 0 < |z| < r, \qquad (3.5.27\text{b})$$

where $A_k (i = 0, 1, 2, \dots)$ are $n \times n$ constant matrices and $A_0 \neq 0$.

In order to find the analytical solution $w(z)$ of (3.5.26) near $z = 0$, we let

$$w(z) = \sum_{k=0}^{\infty} w_k z^k, \qquad (3.5.28)$$

where w_k are $n \times n$ constant matrices. Substituting (3.5.28) into (3.5.26) and comparing coefficients of z^k yield

$$k w_k = \sum_{i=0}^{k} A_{k-i} w_i, \quad (k = 0, 1, 2, \dots). \qquad (3.5.29)$$

Explicitly, the series equations are listed as follows:

$$\begin{cases} -A_0 w_0 = 0, \\ (I - A_0) w_1 = A_1 w_0, \\ \quad \vdots \\ (kI - A_0) w_k = A_k w_0 + A_{k-1} w_1 + \cdots + A_1 w_{k-1}, \\ \quad \vdots \end{cases} \qquad (3.5.30)$$

Here I is the notation of unit matrix.

The coefficient sequence in (3.5.28) is denoted by $(w_k)_0^{\infty}$, and any such a sequence which satisfies (3.5.29) or (3.5.30) in called a **formal solution** of (3.5.26).

Theorem 9 (Theorem of power series expansion of solutions). *For equations* (3.5.26), *we assume that*

(i) *the matrix series $A(z)$ in* (3.5.27) *are convergent on $0 < |z| < r$;*
(ii) *$(w_k)_0^{\infty}$ is a formal solution of* (3.5.26).

Then the power series (3.5.28) *generated by* $(w_k)_0^\infty$ *is certainly convergent on* $|z| < r$, *and thus gives a singly-valued and analytical solution of* (3.5.26).

This theorem has the following three significant corollaries.

Corollary 1. *Suppose that the series* (3.5.27) *is convergent on* $0 < |z| < r$, $\lambda = 0$ *is a characteristic root of* A_0 *and for all natural numbers* k *matrix* $kI - A_0$ *is nonsingular. Then the eigenvector corresponding to the eigenvalue* $\lambda = 0$ *of* A_0 *is not zero,* $w_0 \neq 0$, *and there exists a singly-valued and analytical solution of* (3.5.26) *on* $|z| < r$, *which is of the form of* (3.5.28) *with its coefficients* $w_k (k \geq 1)$ *determined by* (3.5.30).

This corollary substantially means that w_0 *corresponding to* $\lambda = 0$ *is the nonzero solution of the first equation of* (3.5.30). *Once* w_0 *is solved, one can successively solve* w_1, w_2, w_3 *etc., the condition being that all the* $kI - A_0$ *in* (3.5.30) *are nonsingular.*

Corollary 2. *Under the conditions of corollary 1, if the root* $\lambda = 0$ *of* A_0 *has* p *linearly independent eigenvectors, then there must exist* p *linearly independent singly-valued and analytical solutions of* (3.5.26) *on* $|z| < r$, *the forms of which are* (3.5.28).

This corollary actually means that if the first equation of (3.5.30) *is of* p *linearly independent nonzero eigenvectors, then starting from each of them, say, the* mth *eigenvector* $w_{m,0}$, *one is able to solve all* $w_{m,k}$ *successively from* (3.5.30), *which constitute a solution set denoted by*

$$w_m(z) = \sum_{k=0}^{\infty} w_{m,k} z^k, \quad (m = 1, 2, \ldots, p).$$

These p *solutions are linearly independent of each other. If not so, let* $z = 0$ *and then* $w_m(0) = w_{m,0}$, $(m = 1, 2, \ldots, p)$. *These* p $w_m(0)$ *are linearly independent of each other, and so do* p *eigenvectors* $w_{m,0}$, $(m = 1, 2, \ldots, p)$, *which contradicts to the premise.*

Corollary 3. *Suppose that the series* (3.5.27) *is convergent on* $0 < |z| < r$, *a certain complex number* λ *is a characteristic root of* A_0

and for all natural numbers k, $\lambda + k$ are not the roots of A_0. Then the eigenvector w_0 corresponding to the eigenvalue λ of A_0, and there exists a singly-valued and analytical solution of (3.5.26) on $|z| < r$, which is of the form of

$$w(z) = z^\lambda \sum_{k=0}^{\infty} w_k z^k. \tag{3.5.31}$$

The coefficients $w_k (k \geq 1)$ are determined, after (3.5.31) is substituted into (3.5.26), by equations

$$((\lambda + k)I - A_0)w_k = \sum_{i=0}^{k-1} A_{k-i}w_i, \quad (k = 0, 1, 2, \ldots). \tag{3.5.32}$$

If the root λ of A_0 has p linearly independent eigenvectors, then there exist p linearly independent singly-valued and analytical solutions of (3.5.26) on $|z| < r$, which are of the form of (3.5.31).

Corollaries 1 and 2 are the cases of roots $\lambda = 0$, and Corollary 3 is the case of the root $\lambda \neq 0$ of A_0. The latter can be understood as follows. After the transformation $w = z^\lambda u$, the equation that u satisfies will hold the same features as (3.5.26) that w satisfies. Theorem 9 and its first two corollaries apply to solution u, the prerequisite being that λ is the root of A_0. As $\lambda = 0$, this goes back to the cases of Corollaries 1 and 2.

4. General solutions

The discussion above provided a solution of (3.5.26) under the condition (3.5.27), the form of which was a power series (3.5.28). This is a special solution. It has been mentioned that isolated singularities might lead to multi-valued solutions. For the case of a coefficient matrix with the form of (3.5.27), making transformation (3.5.15) can yield multi-valued solutions which are called **general solutions**. We proceed to find them. When making transformation

$$z = e^s, \quad w(z) = u(s), \tag{3.5.33}$$

(3.5.26) becomes

$$u'(s) = e^s A(e^s)u = \sum_{k=0}^{\infty} A_k e^{ks} u. \tag{3.5.34}$$

There is a theorem supplying the power series of general solutions, which is not introduced here. We merely recount how to construct the solutions.

The process of the construction of solutions of (3.5.34) is similar to that in (3.5.28). Nevertheless, we start from the very beginning (3.5.31). Let the solutions of (3.5.34) be of the form of

$$u(s) = e^{\lambda s} \sum_{k=0}^{\infty} u_k(s) e^{ks}. \tag{3.5.35}$$

It is substituted into (3.5.34) to get

$$e^{\lambda s} \sum_{k=0}^{\infty} [u_k' + (\lambda + k)u_k]e^{ks} = \sum_{k=0}^{\infty} A_k e^{ks} e^{\lambda s} \sum_{l=0}^{\infty} u_l(s)e^{ls}. \tag{3.5.36}$$

Comparing the coefficients of e^{ks}, one gets the following series of equations.

$$u_0' + (\lambda I - A_0)u_0 = 0$$

$$u_1' + ((\lambda + 1)I - A_0)u_1 = A_1 u_0$$

$$\cdots\cdots$$

$$u_k' + ((\lambda + k)I - A_0)u_k = A_k u_0 + A_{k-1}u_1 + \cdots + A_1 u_{k-1}$$

$$\cdots\cdots \tag{3.5.37}$$

Consider the first equation. Once a root λ of A_0 and its corresponding eigenvectors are figured out, $u_0(s)$ is a vector polynomial of s. (This is because the solution of $y' = A_0 y$ is $y = e^{\lambda s}u_0(s)$, where $u_0(s)$ is a vector polynomial of s. We do not intend to introduce the related theory.) Then with the help of this $u_0(s)$ and starting from the second equation of (3.5.37), one is able to successively solve $u_1(s), u_2(s), \ldots$, all of which are vector polynomials of s.

Assume that A_0 has l eigenvalues $\lambda_1, \lambda_2, \ldots, \lambda_l$ different from each other, and each λ_j has m_j linearly independent eigenvectors. Obviously, there should be $m_1 + m_2 + \cdots + m_l = n$. The eigenvector $u_0^{(ij)}(s)$ belonging to the ith eigenvector of eigenvalue λ_j can be worked out by the first equation in (3.5.37). Then by following equations, $u_k^{(ij)}(s)$, $(k = 1, 2, \ldots)$ can be solve successively. In this way, n groups of solutions $\{u_k^{(ij)}(s))_0^{(ij)}, (k = 1, 2, \ldots)$ are solved. When they are substituted into (3.5.35), the solutions of (3.5.34) are achieved as

$$u^{(ij)}(s) = e^{\lambda_j s} \sum_{k=0}^{\infty} u_k^{(ij)}(s) e^{ks}, \quad (i = 1, 2, \ldots, m_l; \ j = 1, 2, \ldots, l).$$

(3.5.38)

There are n solutions, and all of them are singly-valued and analytical. It is easily shown that they are linearly independent of each other, so that they constitute a basic solution group of (3.5.34).

Then replacing back the variable by (3.5.33), we acquire the basic solution group in the following form:

$$w^{(ij)}(z) = z^{\lambda_j} \sum_{k=0}^{\infty} u_k^{(ij)}(\mathrm{Ln}\, z) z^k, \quad (i = 1, 2, \ldots, m_l; \ j = 1, 2, \ldots, l).$$

(3.5.39)

After rearranged in the powers of $\mathrm{Ln}\, z$, it becomes

$$w^{(ij)}(z) = z^{\lambda_j} \{ h_0^{(ij)}(z) + (\mathrm{Ln}\, z) h_1^{(ij)}(z) + \cdots + (\mathrm{Ln}\, z)^q h_q^{(ij)}(z) \},$$
$$\times (i = 1, 2, \ldots, m_l; \ j = 1, 2, \ldots, l), \qquad (3.5.40)$$

where $h_k^{(ij)}(z)$ are vector functions singly-valued and analytical on $|z| < r$. As for how many terms which include Ln factors there are in (3.5.40), it has to be determined by the Jordan canonical form of A_0, and it can be certain that $q < n$.

5. The case of two equations

As an example, let us consider the following equations with $n = 2$:

$$w' = A(z)w, \qquad (3.5.41)$$

where $A(z)$ is a 2×2 matrix function singly-valued and analytical on $0 < |z| < r$ and $z = 0$ is a first kind singularity. The explicit form of the coefficient matrix is expressed by

$$A(z) = \frac{1}{z} \sum_{k=0}^{\infty} A_k z^k, \quad 0 < |z| < r. \qquad (3.5.42)$$

Assume that λ and μ are two roots of A_0 and $\mathrm{Re}\,\lambda \leq \mathrm{Re}\,\mu$.

Observing Corollary 3 of Theorem 9, these equations have a solution in the form of

$$w_1(z) = z^\mu h(z) \tag{3.5.43}$$

corresponding to root μ, where $h(z)$ is a two-dimensional vector function singly-valued and analytical on $|z| < r$. The other linearly independent solution can be expressed by

$$w_2(z) = z^\lambda [h_1(z) + h_2(z)\text{Ln } z] \tag{3.5.44a}$$

where $h_1(z), h_2(z)$ are two-dimensional vector functions singly-valued and analytical on $|z| < r$. If the Ln z term does not appear, the solution becomes

$$w_2(z) = z^\lambda h_1(z). \tag{3.5.44b}$$

In summary, one solution is (3.5.43), and the other linearly independent solution is either (3.5.44a) or (3.5.44b).

We narrate how to gain the two solutions in more detail as follows.

Since now $n = 2$, the equations must have two linearly independent solutions.

For the first eigenvalue μ of A_0, one assumes that the first solution w_1 is of the form of (3.5.31). Then following the procedure (3.5.32), a set of coefficients $(w_{\mu,k})_0^\infty$ can be solved. Thus, the first solution (3.5.43) is obtained. For the second eigenvalue λ of A_0, we have to distinguish the following cases.

(1) $\lambda \neq \mu$

There are following two cases.

   (i) $\lambda - \mu$ is not an integer.

   The second solution w_2 can still be assumed in the form of (3.5.31). Then following the procedure (3.5.32), a set of coefficients $(w_{\lambda,k})_0^\infty$ are obtained. Thus the second solution (3.5.44b) is obtained.

   (ii) $\lambda - \mu = m$ is a positive integer.

   If w_2 is still assumed to have the form of (3.5.31) and we derive the coefficients following the procedure (3.5.32), the process will stop at $k = m$. Therefore, the second solution cannot be in the form of (3.5.31). In this case, one has to follow the procedure of general solutions, i.e., (3.5.33)–(3.5.37), to find the second solution w_2. There may appear Ln z in the w_2 in the form of (3.5.44a).

(2) $\lambda = \mu$

Although there is only one eigenvalue of A_0 there are still two cases to be distinguished.

(i) There are two linearly independent eigenvectors w_0's belonging to the eigenvalue of A_0.

In this case, both solutions can be assumed to be the form of (3.5.31). Starting from the two different w_0's one observes the procedure (3.5.32) to derive two sets of coefficients $(w_{\lambda,k})_0^\infty$. Thus the obtained two linearly independent solutions are (3.5.43) and (3.5.44b), where $\lambda = \mu$.

(ii) There is only one eigenvector w_0 belonging to the eigenvalue of A_0.

In this case, it is impossible to find the second solution following the procedure (3.5.31) and (3.5.32). Instead, the second solution w_2 is to be found through the procedure of general solutions, i.e., (3.5.33)–(3.5.37). There may appear Ln z in the w_2 in the form of (3.5.44a).

The discussion above is summarized as follows.

If $\lambda - \mu$ is not an integer or $\lambda = \mu$ and A_0 has two linearly independent eigenvectors, then there will be no Ln term appearing in w_2, i.e., $h_2(z) = 0$ in (3.5.44a).

If $\lambda - \mu$ is an integer or $\lambda = \mu$ and A_0 has solely one eigenvector, then it is possible that a logarithm term appears in w_2: $h_2(z) \neq 0$ in (3.5.44a).

Finally, we retell the two simplest cases intuitively seen from (3.5.37). If in series (3.5.42), there is only one term A_0, then it can be seen from (3.5.37) that all u_k but u_0 are zero. In this case, the solution (3.5.39) has only the first term but not a series form. If in (3.5.42), A_0 term is absent but there are nonzero A_i, $i \neq 0$, then the solution (3.5.39) will be a series, although $\lambda = 0$.

3.5.2.   *Ordinary differential equations of second order*

In this subsection, we will apply the theory of linear differential equations introduced in the previous subsection to ordinary differential equations of second order. The standard of the equation is

$$w''(z) + a(z)w'(z) + b(z)w(z) = 0, \qquad (3.5.45)$$

where $a(z)$ and $b(z)$ are singly-valued and analytical functions on $0 < |z - z_0| < r$.

1. The structure of solutions of ordinary differential equations of second order

In Section 3.1 we have presented the existence and uniqueness theorem and solution structure of homogeneous differential equations with real functions. All the conceptions, theorems and formulas from the beginning of that section to (3.1.14) are applicable to the case of complex functions, such as the proof of the existence and uniqueness theorem by successive approximation method, basic solution group, general solutions, Wronskian and its properties and so on. One simply replaces real functions $y(x)$ by complex ones $w(z)$. The only exceptions are two Liouville formulas (3.1.11) and (3.1.14). In the two formulas, integrations are carried out. For real functions, the integral paths are unambiguous, while for complex functions, integration paths can be any one connecting two given end points but not pass the singularities of the integrands.

The two Liouville formulas are repeated explicitly as follows. Suppose that $w_1(z)$ and $w_2(z)$ are two linearly independent solutions of (3.5.45). Then their Wronskian is

$$\Delta(z) = w_1 w_2' - w_2 w_1'. \qquad (3.5.46)$$

It relates the coefficient $a(z)$ in (3.5.45) by

$$\Delta(z) = \Delta(z_0) \exp\left[-\int_{z_0}^{z} a(\zeta)d\zeta\right]. \qquad (3.5.47)$$

Given one special solution $w_1(z)$, the other $w_2(z)$ can be solved by the Liouville formula

$$w_2 = w_1 \int d\zeta \frac{c}{w_1^2} \exp\left[-\int a(\zeta)d\zeta\right] + c_1 w_1. \qquad (3.5.48)$$

In these two formulas, the integral paths should not pass the singularities of the integrands.

Beside Liouville formula, there are also other ways to find one special solution from another.

2. Classification of singularities

Definition 4. Point z_0 is called a **regular point** of (3.5.45) if it is an **ordinary point** or **removable singularity** of $a(z)$ and $b(z)$.

Suppose that z_0 is not a regular point of (3.5.45). The point z_0 is called a **first kind singularity** of (3.5.45) if it is a pole of $a(z)$ of order not greater than one, and a pole of $b(z)$ of order not greater than 2.

Point z_0 is called **second kind singularity** of (3.5.45) if it is neither a regular point nor a first kind singularity.

If z_0 is the simple pole of $a(z)$ and the second order singularity of $b(z)$, then the functions can be written as

$$a(z) = \frac{c(z)}{z - z_0}, \quad b(z) = \frac{d(z)}{(z - z_0)^2}, \quad 0 < |z - z_0| < r, \qquad (3.5.49)$$

where $c(z)$ and $d(z)$ are analytical functions on $|z - z_0| < r$. The transformation

$$u_1 = w, \quad u_2 = (z - z_0)w'$$

makes (3.5.45) to be the following differential equations of the first order

$$u' = Au, \quad u = \begin{pmatrix} u_1 \\ u_2 \end{pmatrix} = \begin{pmatrix} w \\ (z - z_0)w' \end{pmatrix}, \qquad (3.5.50)$$

where

$$A = \frac{1}{z - z_0} \begin{pmatrix} 0 & 1 \\ -d(z) & 1 - c(z) \end{pmatrix}. \tag{3.5.51}$$

Therefore, the second order singularity $z = z_0$ of $b(z)$ is actually the simple pole of (3.5.53).

For the infinite point, refer to that discussed previously. Consider that $a(z)$ and $b(z)$ in (3.5.45) are singly-valued and analytical on $|z| > R$. Then the ∞ point is an isolated singularity of the equation. In order to distinguish this singularity, make transformation $\xi = 1/z$ and $w(z) = u(\xi)$. Then (3.5.45) is converted to

$$u''(\xi) + a^*(\xi)u'(\xi) + b^*(\xi)u(\xi) = 0, \tag{3.5.52}$$

where

$$a^*(\xi) = \frac{2}{\xi} - \frac{1}{\xi^2}a\left(\frac{1}{\xi}\right), \quad b^*(\xi) = \frac{1}{\xi^4}b\left(\frac{1}{\xi}\right). \tag{3.5.53}$$

They are singly-valued and analytical on $0 < |\xi| < \frac{1}{R}$.

So, if $\xi = 0$ is a regular point, first kind or second kind singularity of (3.5.49), then the ∞ point is correspondingly the same type of point of (3.5.45).

A theorem follows the definition.

Theorem 10. *The sufficient and necessary conditions for the ∞ point to be the first kind point of* (3.5.45) *are that $z = \infty$ is a pole of $a(z)$ of order not greater than 1 and a pole of $b(z)$ of order not greater than 2. That is to say, near $z = \infty$ there are expansions*

$$a(z) = \frac{a_1}{z} + \frac{a_2}{z^2} + \frac{a_3}{z^3} + \cdots \tag{3.5.54a}$$

and

$$b(z) = \frac{b_2}{z^2} + \frac{b_3}{z^3} + \cdots. \tag{3.5.54b}$$

Particularly, the sufficient and necessary conditions for $z = \infty$ to be a regular one of (3.5.45) *are that in* (3.5.54) *$a_1 = 2, b_2 = b_3 = 0$.*

Example 3. In equation

$$(z+2)z^2 w''(z) + (z+2)w'(z) - 4zw(z) = 0,$$

because

$$a(z) = \frac{1}{z^2}, \quad b(z) = \frac{-4}{z(z+2)},$$

$z = 0$ is a second kind singularity, $z = -2$ and $z = \infty$ are first kind singularities, and all remaining points in the whole complex plane are regular ones.

Example 4. In equation

$$(\sin z)w''(z) - zw'(z) + (e^z - 1)w(z) = 0$$

$$a(z) = -\frac{z}{\sin z}, \quad b(z) = \frac{e^z - 1}{\sin z}.$$

Then $z = 0$ is a regular point. All $z = k\pi\,(k = \pm1, \pm2, \ldots)$ are first kind singularities because they are zeros of $\sin z$. It is seen that $z = \infty$ is the limit of series singularities, so that it is not isolated and do not belong to any kind of singularities defined above.

3. Series expansion of solutions

Suppose that in (3.5.45), $a(z)$ and $b(z)$ are singly-valued and analytical functions on $0 < |z - z_0| < r$ and $z = z_0$ is a regular point or first kind singularity of the equations. Then the equation can be transformed to be (3.5.53). We expand the matrix A by the Laurent series such that (3.5.53) becomes

$$u' = \frac{1}{z - z_0} \sum_{m=0}^{\infty} A_m (z - z_0)^m u,$$

where all A_m are 2×2 constant matrices computed by

$$A_m = \frac{1}{m!}\frac{d^m}{dz^m}\begin{pmatrix} 0 & 1 \\ -d(z) & 1 - c(z) \end{pmatrix}_{z=z_0}.$$

Now the form of the equation can be compared to (3.5.26) and (3.5.27).

Especially,

$$A_0 = \begin{pmatrix} 0 & 1 \\ -b_0 & 1 - a_0 \end{pmatrix}, \qquad (3.5.55)$$

where $a_0 = c(z_0), b_0 = d(z_0)$. The simple way of evaluating them is

$$a_0 = c(z_0) = \lim_{z \to z_0} (z - z_0)a(z), \quad b_0 = \lim_{z \to z_0} (z - z_0)^2 b(z).$$

The eigenvalue equation of A_0 is $\det(A_0 - \lambda I) = 0$:

$$\lambda(\lambda - 1) + a_0\lambda + b_0 = 0. \qquad (3.5.56)$$

This equation is called the **indicial equation** of (3.5.53) with respect to singularity z_0. The roots of this indicial equation determine the structure of the solutions of (3.5.53), as well as that of (3.5.45). Due to this reason, the roots of (3.5.56) are called the **indices** of equation (3.5.45) with respect to z_0.

Suppose the two roots of (3.5.55) be λ and μ and $\operatorname{Re}\lambda \leq \operatorname{Re}\mu$. Observing the discussion previously, the basic solutions of (3.5.45) must take the forms of

$$u_1(z) = (z - z_0)^\mu h(z - z_0) \qquad (3.5.57a)$$

and

$$u_2(z) = (z - z_0)^\lambda [h_1(z - z_0) + h_2(z - z_0)\operatorname{Ln}(z - z_0)], \quad (3.5.57b)$$

where $h(z - z_0)$, $h_1(z - z_0)$, $h_2(z - z_0)$ are singly-valued and analytical functions on $|z - z_0| < r$. They can be solved by means of recurrence equations (3.5.37). Sometimes, directly comparing the coefficients of power series may be more convenient.

The simplest case is that there is only one term A_0 in A. Then the solution is not a power series. First, evaluate the two roots λ and μ of A_0 from (3.5.56) and assume $\operatorname{Re}\lambda \leq \operatorname{Re}\mu$. Then denote the eigenvectors corresponding to μ and λ by c_1 and c_2, respectively. The first special solution is

$$u_1(z) = (z - z_0)^\mu. \qquad (3.5.58a)$$

Referring to (3.5.57a), here is $h(z) = 1$. If $c_1 \neq c_2$, the second special solution is

$$u_2(z) = (z - z_0)^\lambda \ln(z - z_0). \qquad (3.5.58b)$$

Referring to (3.5.57b), here is $h_1 = 0$, $h_2 = 1$.

In general, put $w(z) = (z - z_0)^\mu h(z - z_0)$ into (3.5.45) and notice (3.5.52). Then, one obtains the equation for solving $h(z - z_0)$:

$$(z - z_0)^2 h''$$
$$+[2\mu + c(z)](z - z_0)h' + [\mu(\mu - 1) + c(z)\mu + d(z)]h = 0.$$
$$(3.5.59)$$

Next, letting $h(z - z_0) = \sum_{i=0}^\infty \beta_i(z - z_0)^i$ in (3.5.59), one can work out $h(z - z_0)$.

If $c(z)$ and $d(z)$, just as a_0 and b_0, happen to meet the indicial equation (3.5.56), then (3.5.59) will be simplified to be

$$(z - z_0)h'' + [2\mu + c(z)]h' = 0.$$

In this case, an obvious solution is h being a constant. This is manifested in (3.5.58a).

As long as one special solution is found, the other can be achieved in principle by the Liouville formula.

4. An alternative way of finding the other independent solution

If the coefficient matrix A in (3.5.53) is of the series form of (3.5.54), evaluation of the other independent solution by means of the Liouville formula will sometimes be rather difficult. Here an alternative method is presented, which is sometimes practical. Consider the equation with standard form:

$$A(z)w''(z) + B(z)w'(z) + C(z)w(z) = 0. \qquad (3.5.60)$$

Suppose that its one special solution w_1 has been known. Assume that the other solution is of the following form:

$$w_2(z) = u(z)w_1(z) + v(z), \qquad (3.5.61)$$

where functions u and v are to be determined. Substituting this form into (3.5.60) and subtracting (3.5.60) result in

$$[A(z)u'' + B(z)u']w_1 + A(z)v'' + B(z)v' + C(z)v + A(z)2u'w_1' = 0.$$
$$\text{(3.5.62a)}$$

It is then, following the way of (3.2.7), written in the form of

$$\frac{d}{dz}\left[p(z)\frac{d}{dz}u\right] + \left[\frac{d}{dz}p(z)\frac{d}{dz} - q(z)\right]v + A(z)\rho(z)2u'w_1' = 0.$$
$$\text{(3.5.62b)}$$

Let the first term be zero:

$$[p(z)u'(z)]' = 0. \qquad\qquad (3.5.63a)$$

Then, integration produces

$$p(z)u'(z) = c_1, \qquad\qquad (3.5.63b)$$

where c_1 is an integral constant. So, the function $u(z)$ is obtained:

$$u(z) = c_1 \int \frac{dz}{p(z)}. \qquad\qquad (3.5.63c)$$

From (3.5.62), the equation that v should satisfy is

$$A(z)v'' + B(z)v' + C(z)v = -2c_1 w_1'. \qquad (3.5.64)$$

This equation seems not simpler than the original one (3.5.60), even plus an inhomogeneous term. However, utilizing some properties of known w_1 will facilitate the resolution of (3.5.64). For instance, if w_1 is a polynomial and its properties have been known such as the generating function, recurrence relationships and so on, it will be easy to solve (3.5.64). One example is the Legendre equation, where the second independent solution is of the form of (3.5.61).

5. Infinite point

If the infinite point ∞ is at most the first kind singularity of (3.5.45), then the indicial equation of (3.5.49) with respect to $\xi = 0$ is that of (3.5.45) with respect to ∞; the roots of this indicial equation are called the indices of (3.5.45) with respect to the ∞ point. We now present explicitly this indicial equation.

After transformation $\xi = 1/z$ and $w(z) = u(\xi)$, (3.5.45) is converted to be

$$u''(\xi) + \frac{1}{\xi}\left[2 - \frac{1}{\xi}a\left(\frac{1}{\xi}\right)\right]u'(\xi) + \frac{1}{\xi^4}b\left(\frac{1}{\xi}\right)u(\xi) = 0.$$

If $\xi = 0$ is at most the first kind singularity of this equation, then the matrix A_0 should be

$$A_0 = \begin{pmatrix} 0 & 1 \\ -z^2 b(z) & -1 + za(z) \end{pmatrix}_{z=\infty}.$$

The indicial equation is

$$\lambda(\lambda + 1) - (za(z))_{z=\infty}\lambda + (z^2 b(z))_{z=\infty} = 0. \qquad (3.5.65)$$

As long as the roots of (3.5.65) are figured out, one is able to work out a basic solution set of (3.5.64) near the ∞ point.

6. Polynomial solutions of ordinary differential equations of second order

The series expressions of polynomial solutions in Table 3.2 are still valid after the real argument x is replaced by complex z. Consequently, all the formulas in Tables 3.2 to 3.6 are still valid after the real argument x is replaced by complex z.

For example, for Chebyshev equation and its solutions the Chebyshev functions introduced in Subsection 3.4.3, one merely needs to modify (3.4.49) to (3.4.79) as follows: replace real function $y(x)$ by complex $w(z)$; replace real number θ by complex ξ; get rid of the constraint $(-1 \leq x \leq 1)$ in (3.4.49) and all boundaries; drop (3.4.57); change $\lambda = n^2$ in (3.4.52) to be $\lambda = \alpha^2$ where α is a complex number. Then all the formulas will be applicable in the present subsection. Readers easily reform (3.4.49) to (3.4.79) to the forms of complex functions.

When we discuss complex analysis of ordinary differential equations, the boundaries are not involved, so that eigenvalues are not concerned. We merely pay attention to two linearly independent special solutions.

In the first four sections of the present chapter, the cases of real argument are discussed. It is easy to judge that for each case in Table 3.1, the roots of indicial equation are identical. Therefore, near $x = 0$, one special solution is polynomial and the other necessarily includes a factor $\ln x$. This is why under the condition that the solutions should be finite on the interval, the polynomial solution appears but the other does not. The eigenvalues are consequently determined by the polynomial solution, as given in Table 3.1.

3.6.   Non-Self-Adjoint Ordinary Differential Equations of Second Order

3.6.1.   *Adjoint equations of ordinary differential equations*

In the last section, the argument is complex, and so are functions. In this section, we consider the case where argument is real but functions can be complex.

If the operator in an equation is replaced by its adjoint, one gets the adjoint equation. In the case of linear algebra equations, taking the Hermitian conjugate of the coefficient matrix leads to the adjoint equations, which was discussed in Subsection 2.2.3.

Ordinary differential equation and its boundary conditions constitute a boundary value problem. Accordingly, adjoint equation and the adjoint boundary conditions constitute an adjoint boundary value problem. In order to put down an adjoint boundary value problem, one has to not only replace the differential operator in the equation by the adjoint differential operator, but also put down the corresponding boundary conditions by means of (2.2.11).

In equation

$$Lu(x) - \lambda u(x) = 0 \tag{3.6.1}$$

if the operator L is substituted by its adjoint $L^\dagger$, one gets

$$L^\dagger v(x) - \gamma v(x) = 0 \tag{3.6.2}$$

Please note that the solutions v of (3.6.2) are usually different from the solutions u of (3.6.1). The solution v is often written by $u^\dagger$ to indicate its identity as a solution of the adjoint equation, and called the adjoint function of u. We stress once more that a differential equation is necessarily accompanied by boundary conditions. The boundary conditions of the adjoint equation are obtained by

$$(v, Lu) - (L^\dagger v, u) = [J(u, v)]_a^b = 0. \qquad (3.6.3)$$

If the solutions of a boundary value problem is identical to those of the adjoint boundary value problem, this boundary value problem is called self-adjoint. This requires two conditions satisfied: the differential operator is formally adjoint and the boundary conditions of the original problem and of the adjoint problem are the same. The problems considered in the first four sections of this chapter are all self-adjoint boundary value problems.

3.6.2. *Sturm-Liouville operator*

Consider the SL operator of ordinary differential equations of second order:

$$L = -A(x)\frac{\mathrm{d}^2}{\mathrm{d}x^2} - B(x)\frac{\mathrm{d}}{\mathrm{d}x} + C(x) = \frac{1}{\rho(x)}\left[-\frac{\mathrm{d}}{\mathrm{d}x}p(x)\frac{\mathrm{d}}{\mathrm{d}x} + q(x)\right],$$
$$(3.6.4)$$

where the functions can all be complex and the coefficients in the boundary conditions are all complex. Let us inspect its adjoint.

Theorem 1. *Suppose that the SL operator in the form of (3.6.4) is self-adjoint. Then the operator and its self-adjoint*

$$L^\dagger = \frac{1}{\rho^*}\left(-p^*\frac{\mathrm{d}^2}{\mathrm{d}x^2} - p'^*\frac{\mathrm{d}}{\mathrm{d}x} + q^*\right) = L^* \qquad (3.6.5)$$

have the same weight function $\rho(x)$. The eigenvalue equation

$$L(x)u_n(x) = \lambda_n u_n(x) \qquad (3.6.6)$$

and its adjoint

$$L^\dagger(x)v_n(x) = \gamma_n v_n(x) \qquad (3.6.7)$$

*have the following relations: their eigenvalues are complex conjugates
of each other,*

$$\gamma_n = \lambda_n^* \qquad\qquad (3.6.8)$$

and so are their eigenfunctions,

$$v_n(x) = u_n^*(x). \qquad\qquad (3.6.9)$$

Proof.   We first show that on the left hand side of (3.6.3), the
weight functions of the two inner products are the same.

$$
\begin{aligned}
(v, Lu) &= \int_a^b \rho(x) v^*(x) \frac{1}{\rho(x)} \left[-\frac{\mathrm{d}}{\mathrm{d}x} p(x) \frac{\mathrm{d}}{\mathrm{d}x} + q(x) \right] u(x) \mathrm{d}x \\
&= \int_a^b \left[\left(-p \frac{\mathrm{d}^2}{\mathrm{d}x^2} - p' \frac{\mathrm{d}}{\mathrm{d}x} + q \right) v^* \right] u \mathrm{d}x + [J(u,v)]_a^b \\
&= \int_a^b \rho \left[\frac{1}{\rho^*} \left(-p^* \frac{\mathrm{d}^2}{\mathrm{d}x^2} - p'^* \frac{\mathrm{d}}{\mathrm{d}x} + q^* \right) v \right]^* u \mathrm{d}x + [J(u,v)]_a^b \\
&= (L^+ v, u) + [J(u,v)]_a^b. \qquad\qquad (3.6.10)
\end{aligned}
$$

It is seen that (v, Lu) and $(L^\dagger v, u)$ share the same weigh function.
Next, taking complex conjugates of (3.6.6) yields

$$L^*(x) u_n^*(x) = \lambda_n^* u_n^*(x). \qquad\qquad (3.6.11)$$

Comparing it with the adjoint equation (3.6.7), we get (3.6.8) and
(3.6.9) since the operator is self-adjoint.   □

The eigenvalues of the eigenvalue equation and its adjoint equation are complex conjugates of each other. This property has been
proved by Theorem 8 in Subsection 2.2.2. The present theorem discloses that in the case of an SL operator, their eigenfunctions are
complex conjugates of each other.

The boundary conditions accompanying (3.6.7) are to be obtained
by combination of those of (3.6.6) and the constraint (3.6.3). Here
we show a simple case often encountered.

Theorem 2. *Suppose that in (3.6.4)*

$$p(x) = 1, \qquad\qquad (3.6.12)$$

and the solutions of (3.6.6) *satisfy the following homogeneous boundary conditions*:

$$\begin{cases} \alpha_1 u(a) + \alpha_2 u'(a) = 0, \\ \beta_1 u(b) + \beta_2 u'(b) = 0. \end{cases} \tag{3.6.13}$$

Then the adjoint boundary conditions of the adjoint equation (3.6.7) *are as follows*:

$$\begin{cases} \alpha_1^* v(a) + \alpha_2^* v'(a) = 0, \\ \beta_1^* v(b) + \beta_2^* v'(b) = 0. \end{cases} \tag{3.6.14}$$

That is to say, the coefficients are simply the complex conjugates of those in (3.6.13).

Proof. Making use of (2.2.25), we have

$$
\begin{aligned}
\left[J(u,v) \right]_a^b &= u'(b)v^*(b) - u(b)v^{*\prime}(b) - u'(a)v^*(a) + u(a)v^{*\prime}(a) \\
&= u'(b)v^*(b) + \beta_2 u'(b)v^{*\prime}(b)/\beta_1 \\
&\quad + \alpha_1 u(a)v^*(a)/\alpha_2 + u(a)v^{*\prime}(a) \\
&= [\beta_1 v^*(b) + \beta_2 v^{*\prime}(b)]u'(b)/\beta_1 \\
&\quad + [\alpha_1 v^*(a) + \alpha_2 v^{*\prime}(a)]u(a)/\alpha_2 = 0,
\end{aligned}
$$

where (3.6.13) has been employed. The last line reflects (3.6.14). $\square$

In this case, if these coefficients are real, and the weight ρ and kernel q in (3.6.4) are both real, then the operator is self-adjoint.

Example 1. Consider the following boundary value problem:

$$u''(x) + \lambda u(x) = 0, \quad 0 \le x \le 1,$$

$$u'(0) - \alpha u(0) = 0, \quad u'(1) - \beta u(1) = 0, \tag{3.6.15}$$

where coefficients α and β are complex. Find its eigenvalues and corresponding eigenfunctions, and write down its adjoint boundary value problem.

Solution. For convenience, let $\lambda = \mu^2$. The solution of the equation is

$$u(x) = A \sin \mu x + B \cos \mu x.$$

The two coefficients are determined by the boundary conditions.

$$A\mu - \alpha B = 0.$$

$$A\mu \cos \mu - B\mu \sin \mu - \beta A \sin \mu - \beta B \cos \mu = 0.$$

The general eigenvalues obtained by these two equations are complicated. We here consider a simple case, where $\alpha = \beta \neq 0$. Then the eigenvalues satisfy

$$(\mu + \alpha^2/\mu) \sin \mu = 0.$$

Either of the two factors should be zero. If $\sin \mu_n = 0$, the eigenvalues are

$$\mu_n = n\pi, \quad n = 1, 2, 3, \ldots \qquad (3.6.16a)$$

and the corresponding eigenfunctions are

$$u_n(x) = \sqrt{\frac{2}{n^2\pi^2 + \alpha^2}} (\alpha \sin \mu_n x + n\pi \cos \mu_n x) \qquad (3.6.16b)$$

which have been normalized. If $\mu + \alpha^2/\mu = 0$, there is only one eigenvalue,

$$\mu = i\alpha \qquad (3.6.17a)$$

and the corresponding normalized eigenfunction is

$$u_\alpha(x) = \sqrt{\frac{2\alpha}{e^{2\alpha} - 1}} e^{\alpha x}, \qquad (3.6.17b)$$

which is labeled by a subscript α. The whole eigenfunctions are (3.6.16b) plus (3.6.17b). The last eigenfunction seems isolated, but is of physical significance. By the boundary conditions, the values of function at the ends of the interval are not zero. The function (3.6.17b) is one exponentially increasing or decreasing starting from the ends. This is a function closely related to boundaries.

Now we turn to put down the adjoint boundary value problem of (3.6.15). The differential operator and boundary conditions meet the prerequisites of Theorem 2. So, the adjoint problem is easily written as

$$v''(x) + \gamma v(x) = 0, \quad 0 \leq x \leq 1, \quad v'(0) - \alpha^* v(0) = 0,$$
$$v'(1) - \beta^* v(1) = 0. \tag{3.6.18}$$

The eigenvalues and eigenfunctions of the adjoint problem are just complex conjugates of (3.6.16)–(3.6.17). The following biorthogonality is easily testified:

$$(v_n, u_m) = \int_0^1 v_n^*(x) u_m(x) \mathrm{d}x = \delta_{nm}. \tag{3.6.19}$$

It can also be checked that

$$(u_n, u_m) = \int_0^1 u_n^*(x) u_m(x) \mathrm{d}x \neq 0, \quad n \neq m. \tag{3.6.20}$$

Because eigenfunctions u_α and u_n belong to different eigenvalues, there is a biorthogonality between them:

$$(v_n, u_\alpha) = (v_\alpha, u_n) = 0. \tag{3.6.21}$$

3.6.3. *Complete set of non-self-adjoint ordinary differential equations of second order*

The example above shows the eigenvalues and eigenfunctions of a nonself-adjoint problem. Is such a set of eigenfunctions a complete one? In general, one is unable to get a conclusion. However, in some particular cases, this conclusion can be addressed. The following theorem lists these particular cases.

Theorem 3. *Consider a boundary value problem as follows:*

$$L(x)u(x) + \lambda u(x) = 0, \quad a \leq x \leq b,$$
$$a_1 u'(a) + b_1 u'(b) + a_0 u(a) + b_0 u(b) = 0,$$
$$c_1 u'(a) + d_1 u'(b) + c_0 u(a) + d_0 u(b) = 0. \tag{3.6.22}$$

Here $L(x)$ is an SL operator (3.6.4), functions $A(x), B(x), C(x)$ are piecewise continuous and the coefficients in boundary conditions are complex. Suppose that the adjoint of $L(x)$ exists and every eigenvalue of $L(x)$ is nondegenerate, i.e., each eigenvalue corresponds to one eigenfunction. If these conditions are provided, and one of the following three conditions is satisfied:

$$a_1 d_1 - b_1 c_1 \neq 0, \tag{3.6.23}$$

$$a_1 d_1 - b_1 c_1 = 0, \quad |a_1| + |b_1| > 0,$$

$$2(a_1 c_0 + b_1 d_0) \neq \pm(b_1 c_0 + a_1 d_0) \neq 0, \tag{3.6.24}$$

$$a_1 = b_1 = c_1 = d_1 = 0, \quad a_0 d_0 - b_0 c_0 \neq 0 \tag{3.6.25}$$

then the eigenfunction set of $L(x)$ is a complete set on complex square-integrable space $L_2[a, b]$, and any function $f(x)$ on $L_2[a, b]$ can be expanded in the form of:

$$f(x) = \sum_{n=1}^{\infty} (v_n, f) u_n(x),$$

where $v_n(x)$'s are the eigenfunctions of the adjoint problem. The eigenvalues κ_n of $L^\dagger(x)$ and μ_n of $L(x)$ are one-to-one correspondent and complex conjugates of each other: $\kappa_n = \mu_n^*$.

We point out that the conditions in Theorem 3 are sufficient but not necessary. The boundary conditions are homogeneous.

The boundary conditions in Example 1 meet (3.6.24), so that the set of eigenfunctions in that problem is complete. A function $f(x)$ can be expanded by this set as follows:

$$f(x) = \sum_{n=1}^{\infty} (v_n, f) u_n(x) + (v_\alpha, f) u_\alpha(x),$$

where all the eigenfunctions have been solved in Example 1.

The nonself-adjoint boundary value problem have been encountered in dealing with the electromagnetic field problems. In this kind of problems, the argument is space coordinates, and permittivity and permeability are complex and piecewise continuous. When complex coefficients appear in differential operator of second order, the operator is not self-adjoint anymore.

3.7. The Conditions under Which Inhomogeneous Equations have Solutions

In this section, we consider the case where argument is real. Up to now in this chapter beginning with Section 3.2, we have discussed how to find the solutions of homogeneous equations. Let us investigate the case of inhomogeneous equations. Let L be a differential operator of second order. Its boundary value problem is of the following form:

$$\begin{cases} Ly(x) = f(x), & a < x < b, \\ B(y) = \gamma. \end{cases} \tag{3.7.1}$$

The boundary condition $B(y) = \gamma$ is the shorthand of two equations:

$$\begin{cases} \alpha_{1,1}y(a) + \alpha_{1,2}y'(a) + \beta_{1,1}y(b) + \beta_{1,2}y'(b) = \gamma_1, \\ \alpha_{2,1}y(a) + \alpha_{2,2}y'(a) + \beta_{2,1}y(b) + \beta_{2,2}y'(b) = \gamma_2, \end{cases} \tag{3.7.2}$$

see (3.2.18). These are inhomogeneous boundary conditions. The homogeneous ones are denoted by $B(y) = 0$. The parameter λ appearing such as (3.2.1) has been in (3.7.1) merged into operator L. Equation (3.7.1) is an inhomogeneous one.

We separate (3.7.1) into the following two boundary value problems:

$$\begin{cases} Lw(x) = 0, & a < x < b, \\ B(w) = \gamma \end{cases} \tag{3.7.3}$$

and

$$\begin{cases} Ls(x) = f(x), & a < x < b, \\ B(s) = 0. \end{cases} \tag{3.7.4}$$

The former is of homogeneous equation and inhomogeneous boundary conditions, and the latter is of inhomogeneous equation and homogeneous boundary conditions. As long as the solutions $w(x)$ and $s(x)$ of these two problems are found, the solution of (3.7.1) is naturally

$$y(x) = w(x) + s(x). \tag{3.7.5}$$

The problem (3.7.3) contains homogeneous equation and inhomogeneous boundary conditions, which has been discussed in Sections 3.2 to 3.5 and considered being figured out.

What we have to inspect is whether (3.7.4) has solutions or not. To this aim, we put down its corresponding homogeneous boundary value problem:

$$
\begin{cases} Lu(x) = 0, & a < x < b, \\ B(u) = 0 \end{cases}
\tag{3.7.6}
$$

and its adjoint problem:

$$
\begin{cases} L^\dagger v(x) = 0, & a < x < b, \\ B^\dagger(v) = 0. \end{cases}
\tag{3.7.7}
$$

The adjoint homogeneous boundary condition $B^\dagger(v) = 0$ is obtained through (3.6.3).

As has been mentioned in the end of Subsection 3.2.3, we are unable to know in advance whether the problem (3.7.6) has solutions or not.

What we can be certain is that if (3.7.4) has solutions can be judged according to if the adjoint problem (3.7.7) has solutions. This is similar to the case of the alternative theorem of linear algebra equations presented in Subsection 2.2.3.

Theorem 1. (i) *If homogeneous boundary value problem (3.7.6) has no nonzero solutions, then neither does its adjoint problem (3.7.7), and the problem (3.7.4) has unique nonzero solution. (ii) If (3.7.6) has nonzero solutions, then so does its adjoint problem (3.7.7). In this case, their solutions are not necessarily the same, because the forms of their operators and boundary conditions are not the same, e.g., Example 1 in Subsection 2.7.2. Furthermore, the condition for (3.7.4) to have nonzero solutions is that*

$$
(f, v) = 0.
\tag{3.7.8}
$$

This means that the inhomogeneous term in $f(x)$ (3.7.4) is orthogonal to the solutions $v(x)$ of (3.7.7). If there are k solutions of

(3.7.7): $v_1(x), v_2(x), \ldots, v_k(x)$, then $f(x)$ has to be orthogonal to every one of these functions.

Condition (3.7.8) is easily tested from the following inner product:

$$(Ls, v) - (s, L^\dagger v) = (f, v) = 0. \tag{3.7.9}$$

According to the definition of adjoint operators, it results in (3.7.8).

This theorem is **the alternative theorem** for the solutions of linear differential equations. It reveals that the condition for (3.7.4) to have solutions is either one of the following two: one is that (3.7.7) has no nonzero solutions, and the other is that (3.7.7) has nonzero solutions and meanwhile (3.7.8) is satisfied.

Equation (3.7.8) is also called **the compatibility condition** for problem (3.7.4) to have nonzero solutions. Note that this is for homogeneous boundary conditions. For inhomogeneous boundary conditions, there is no requirement of the compatibility condition, since this theorem does not involve (3.7.3).

Here we merely furnish a criterion to judge if an inhomogeneous equation has solutions. As for how to find the solutions, a formula was provided in Subsection 3.1.3. We will investigate this question comprehensively in Chapter 6 by means of Green's function method.

Example 1. Consider a boundary value problem

$$-u''(x) - \pi^2 u(x) = f(x), \quad 0 < x < 1,$$
$$u(1) + u(0) = 0, \quad u'(1) + u'(0) = 0.$$

Please give the condition that this problem has solutions.

Solution. Firstly, we should write down its adjoint homogeneous boundary value problem. The adjoint boundary conditions are derived by (3.6.3), where the knot was (2.2.13b).

$$[J(u, v)]_0^1 = -[u'v^* - uv^{*\prime}]_0^1$$
$$= -[u'(1)v^*(1) - u(1)v^{*\prime}(1) - u'(0)v^*(0) + u(0)v^{*\prime}(0)]$$
$$= u'(0)[v^*(1) + v^*(0)] - u(0)[v^{*\prime}(1) - v^{*\prime}(0)] = 0.$$

So, the adjoint problem is

$$-v''(x) - \pi^2 v(x) = 0, \quad 0 < x < 1,$$
$$v(1) + v(0) = 0, \quad v'(1) + v'(0) = 0.$$

The general solution of the differential equation is $v(x) = A \sin \pi x + B \cos \pi x$, which always meets the boundary conditions. The general solution is the linear combination of the following two linearly independent special solutions: $v_1(x) = \sin \pi x$. And $v_2(x) = \cos \pi x$. Consequently, there are two conditions for the original problem to have solutions: $\int_0^1 f(x) \sin \pi x \mathrm{d}x = 0$, and $\int_0^1 f(x) \cos \pi x \mathrm{d}x = 0$.

Finally, let us inspect what kind of boundary value problems are self-adjoint, which were defaulted in previous sections. For an SL operator with real functions and three kinds of boundary conditions listed in Subsection 3.2.3 with real coefficients, the boundary value problem is self-adjoint. The homogeneous and periodic boundary conditions can be merged to the uniform condition (3.7.2) with homogeneous form $B(u) = 0$. Note that if the boundary conditions are inhomogeneous, $B(u) = \gamma \neq 0$, then the boundary value problem is not self-adjoint.

For a self-adjoint problem, from the self-adjoint feature of the differential operator, one has

$$Lu^* = (Lu)^*, \quad B(u^*) = [B(u)]^*. \tag{3.7.10}$$

That is to say, the self-adjoint operator acting on the complex conjugate of a function is identical to the complex conjugate of the self-adjoint operator acting on the function. The boundary conditions that the complex conjugate of the function satisfies is identical to the complex conjugate of the boundary conditions that the function satisfies.

Example 2. Consider the following boundary value problem of differential equation of first order:

$$\left[q_1(x) \frac{\mathrm{d}}{\mathrm{d}x} + q_0(x) \right] y(x) = f(x), \quad a < x < b, \quad \alpha y(a) + \beta y(b) = 0. \tag{3.7.11}$$

What are the adjoint boundary conditions when $f(x) = 0$? What is the condition for the problem to have solutions when $f(x) \neq 0$?

Solution. The homogeneous boundary value problem is

$$\left[q_1(x) \frac{\mathrm{d}}{\mathrm{d}x} + q_0(x) \right] u(x) = 0, \quad a < x < b, \quad \alpha u(a) + \beta u(b) = 0,$$

$$(3.7.12)$$

and its adjoint problem is

$$\left[-q_1^*(x) \frac{\mathrm{d}}{\mathrm{d}x} + q_0^*(x) \right] v(x) = 0, \quad a < x < b. \quad (3.7.13a)$$

The formally self-adjoint of a differential operator of first order was given by (2.2.22): $q_0 = q_0^*$, $q_1 - -q_1^*$. Now we seek for the boundary conditions that $v(x)$ should satisfy. Observing (2.2.13a), the knot is

$$[J(u, v)]_a^b = q_1(b)u(b)v^*(b) - q_1(a)u(a)v^*(a) = 0.$$

This, combined with boundary condition, leads to

$$q_1(b)\alpha v^*(b) + q_1(a)\beta v^*(a) = 0.$$

Take its complex conjugate and note that q_1 must be a complex number, see (2.2.22a). Then

$$q_1(b)\alpha^* v(b) + q_1(a)\beta^* v(a) = 0. \quad (3.7.13b)$$

If the homogeneous boundary value problem is self-adjoint, this boundary condition is the same as that in (3.7.12). This requires that $q_1(x)$ is a constant, in accordance with (2.2.22b), and coefficients α and β are complex conjugates of each other. By Theorem 1, the condition for (3.7.11) to have solution is that the solution of (3.7.13) is orthogonal to the function $f(x)$: $\int_a^b v^*(x)f(x)\mathrm{d}x = 0$.

It can be shown that if the boundary condition in (3.7.12) is inhomogeneous, $\alpha u(a) + \beta u(b) = \gamma$, the boundary value problem will not be self-adjoint.

Exercises

1. Why do physical problems often appear in mathematics as differential equations of second order?

2. Similarly to the case of two functions, we discuss the Wronskian of three functions. Suppose that there are three functions $y_1(x), y_2(x), y_3(x)$ and they have derivatives up to the second order. Their Wronskian is defined by

$$\Delta(y_1, y_2, y_3) = \begin{vmatrix} y_1(x) & y_2(x) & y_3(x) \\ y_1'(x) & y_2'(x) & y_3'(x) \\ y_1''(x) & y_2''(x) & y_3''(x) \end{vmatrix}.$$

   Show that if y_1, y_2, y_3 are linearly dependent of each other, $\Delta(y_1, y_2, y_3)$ is zero everywhere. Similar conclusion can be deduced for the case of n functions.

3. $y''(x) + \lambda^2 y(x) = 0$ has two linearly independent solutions. Show that one of them can be calculated from another by Liouville formula.

4. Show that the solutions of Bessel equation $x^2 y'' + xy' + (x^2 - \nu^2)y = 0$ and spheric Bessel equation $x^2 y'' + 2xy' + [x^2 - l(l + 1)]y = 0$ are not polynomials.

5. Show that if in (3.3.3) $B(x)$ is a constant and $A(x)$ is a quadratic polynomial, then as $x \to \infty$, $p(x)$ does not go to zero faster than the inverse of any powers of x.

6. Obtain the result (3.3.4) from (3.3.3).

7. Calculate the weight coefficients (3.3.46) of Gauss-Chebyshev integral formula by (3.3.40).

8. Show, by means of mathematical induction, the associated Laguerre equation $xz'' + (m+1-x)z' + (\nu - m)z = 0$ is generated from Laguerre equation $xy'' + (1-x)y' + \nu y = 0$, where $z = y^{(m)}$.

9. If $(1 - x^2)^n$ is a solution of an ordinary differential equation of second order, write this differential equation. Is this Legendre

equation? From this equation, show that Legendre polynomi-
als $P_n(x) = \frac{d^n}{dx^n}(1-x^2)^n = [(1-x^2)^n]^{(n)}$ satisfy Legendre
equation. Further, show that the associated Legendre functions
$(1-x^2)^{m/2}\frac{d^m}{dx^m}P_n(x) = (1-x^2)^{m/2}[P_n(x)]^{(m)}$ are the solutions
of Legendre equation

$$(1-x^2)\frac{d^2}{dx^2}y - 2x\frac{d}{dx}y + \left[n(n+1) - \frac{m^2}{1-x^2}\right]y = 0.$$

10. Deduce the parity of Legendre polynomials $P_n(x)$. Calculate the
    values of $P_n(1)$ and $P_n(-1)$ from the generating function of
    $P_n(x)$.

11. Show, by the generating function of Legendre polynomials $P_n(x)$,
    the following recursion relations.

    $$(n+1)P_{n+1} - (2n+1)xP_n + nP_{n-1} = 0.$$

    $$nP_n - xP_n' + P_{n-1}' = 0.$$

    $$nP_{n-1} - P_n' + xP_{n-1}' = 0.$$

    $$(2n+1)P_n - P_{n+1}' + P_{n-1}' = 0.$$

12. Show that for Legendre polynomials

    $$P_n(x), \quad \int_{-1}^{1} x^n P_n(x)dx = \frac{2^{n+1}(n!)^2}{(2n+1)!}.$$

13. Expand two functions

    $$f_1(x) = \begin{cases} -1 & -1 \le x < 0 \\ +1 & 0 < x \le 1 \end{cases}$$

    and $f_2(x) = |x|, -1 \le x \le 1$ by Legendre polynomials. Evaluate
    the expansion coefficients.

14. Expand function

    $$f(x) = \begin{cases} 0 & -1 \le x < \alpha \\ 1/2 & x = \alpha \\ 1 & \alpha < x \le 1 \end{cases}$$

by Legendre polynomials, and evaluate the expansion coefficients.

15. Prove (3.4.28).

16. The first five Legendre polynomials are

$$P_0(x) = 1, \quad P_1(x) = x, \quad P_2(x) = \frac{1}{2}(3x^2 - 1),$$

$$P_3(x) = \frac{1}{2}(5x^3 - 3x), \quad P_4(x) = \frac{1}{8}(35x^4 - 30x^2 + 3).$$

(1) Show that the function

$$\tanh^{-1} x = \frac{1}{2} \ln \left(\frac{1+x}{1-x} \right)$$

is the solution of Legendre equation of zero order. Can this
solution be obtained by Liouville formula?

(2) When $l = 0, 1, 2$, derive the expressions of Legendre functions
of the second kind $Q_0(x), Q_1(x), Q_2(x)$ by Liouville formula.

(3) Let $x = \cos\theta$. Put down the expressions $Q_0(\cos\theta), Q_1(\cos\theta)$,
$Q_2(\cos\theta)$ as functions of θ.

(4) Write the following Legendre functions of the first and second kinds: $P_1^1(x), P_2^1(x), P_2^2(x), P_3^1(x), P_3^2(x), P_3^3(x), Q_1^1(x)$,
$Q_2^1(x), Q_2^2(x)$ by (3.4.36) and (3.4.37).

17. By use of mathematical induction, obtain the associated Chebyshev equation $(1 - x^2)z'' - (2m + 1)xz' + (\nu^2 - m^2)z = 0$ from
Chebyshev equation $(1 - x^2)y'' - xy' + \nu^2 y = 0$, where $z = y^{(m)}$.

18. Prove the expansion formula of Chebyshev polynomials:

$$\frac{1 - t^2}{1 - 2xt + t^2} = T_0(x) + 2\sum_{n=1}^{\infty} T_n(x)t^n.$$

19. Show that the relation between Chebyshev polynomials $T_n(x)$
and Chebyshev functions of second kind $U_n(x)$ is $U_n(x) = \frac{\sqrt{1-x^2}}{n} \frac{dT_n(x)}{dx}$. Using this formula and combining the results of
Exercise 33 in Chapter 2, do the following work.

(1) Calculate the values at special points: $U_n(1) = ?$ $U_n(-1) = ?$
$U_{2n+1}(0) = ?$ $U_{2n}(0) = ?$

(2) Find the parity of $U_n(x)$, i.e., the relation between $U_n(-x)$ and $U_n(x)$.

(3) Prove the following recursion formulas.

$$U_{n+1} - 2xU_n + U_{n-1} = 0.$$

$$(1 - x^2)U_n' = nxU_n - nU_{n+1}.$$

$$2(1 - x^2)U_n' = n(U_{n-1} - U_{n+1}).$$

$$(1 - x^2)U_n' = nU_{n-1} - nxU_n.$$

$$(1 - x^2)U_n''(x) - xU_n'(x) + n^2 U_n(x) = 0.$$

(4) Show that $U_n(x)$'s are orthogonal to each other with weight $\rho(x) = \frac{1}{\sqrt{1-x^2}}$: $\int_{-1}^{1} \frac{1}{\sqrt{1-x^2}} U_n(x)U_m(x)\mathrm{d}x = \frac{\pi}{2}\delta_{nm}$.

20. The relation between Chebyshev polynomials $T_n(x)$ and those of the second kind $U_n^*(x)$ is $U_n^*(x) = \frac{1}{n+1}\frac{\mathrm{d}T_{n+1}(x)}{\mathrm{d}x}$. Using this formula and combining the results of Exercise 33 in Chapter 2, do the following work.

(1) Calculate the values at special points: $U_n^*(1) =?$ $U_n^*(-1) =?$ $U_{2n+1}^*(0) =?$ $U_{2n}^*(0) =?$

(2) Find the parity of $U_n^*(x)$, i.e., the relation between $U_n^*(-x)$ and $U_n^*(x)$.

(3) Prove the following recursion formulas.

$$U_{n+1}^* - 2xU_n^* + U_{n-1}^* = 0.$$

$$(1 - x^2)U_n^{*\prime} = (n + 2)xU_n^* - (n + 1)U_{n+1}^*.$$

$$2(1 - x^2)U_n^{*\prime} = (n + 2)U_{n-1}^* - nU_{n+1}^*.$$

$$(1 - x^2)U_n^{*\prime} = (n + 1)U_{n-1}^* - nxU_n^*.$$

$$(1 - x^2)U_n^{*\prime\prime}(x) - 3xU_n^{*\prime}(x) + n(n + 2)U_n^*(x) = 0.$$

(4) Show that $U_n^*(x)$'s are orthogonal to each other with weight $\rho(x) = \sqrt{1 - x^2}$: $\int_{-1}^{1} \sqrt{1 - x^2}U_n^*(x)U_m^*(x)\mathrm{d}x = \frac{\pi}{2}\delta_{nm}$.

21. Prove (3.4.77) and (3.4.78).

22. What transformation can convert equation $y'' - B_1 xy' + \lambda y = 0$ to be a Hermite equation? What are the eigenvalues and eigenfunctions of this equation? Show that the nomalization coefficient of Hermite polynomial $He_n(x)$ is $C_n = 2^{n/2}$.

23. Prove the expansion of Hermite polynomials $H_n(x)$ by its generating function. Derive derivative expression of $H_n(x)$ by its generating function.

24. Find the parity of Hermite polynomials $H_n(x)$.

25. By the generating function of Hermite polynomials $H_n(x)$, do the following work.

    (1) Prove the following recursion relations.

    $$H_{n+1} - 2xH_n + 2nH_{n-1} = 0.$$

    $$H'_n = 2nH_{n-1}.$$

    $$H_n - 2xH_{n-1} + H'_{n-1} = 0.$$

    $$xH'_n - nH'_{n-1} - nH_n = 0.$$

    (2) Calculate the values at $x = 0$: $H_{2n}(0) = ?$, $H_{2n+1}(0) = ?$

26. Prove the integral expression of Hermite polynomials:

    $$H_n(x) = \frac{2^n}{\sqrt{\pi}} \int_{-\infty}^{\infty} (x + it)^n e^{-t^2} dt.$$

    Deduce the series expression (3.4.82) of Hermite polynomials by this integral expression.

27. For equation

    $$\frac{1}{\xi^2 + \eta^2} \left(\frac{\partial^2 u}{\partial \xi^2} + \frac{\partial^2 u}{\partial \eta^2} \right) + \frac{\partial^2 u}{\partial z^2} = 0,$$

    use separation of variables $u(\xi, \eta, z) = X(\xi)Y(\eta)Z(z)$ to derive equations

    $$\begin{cases} X'' + (\alpha - m^2 \xi^2)X = 0, \\ Y'' - (\alpha + m^2 \eta^2)Y = 0, \\ Z'' + m^2 Z = 0. \end{cases}$$

28. Make use of Theorem 5 in Subsection 2.3.4 to derive the recursion formulas among adjacent three polynomials of the following polynomials.

(1) Legendre polynomials.
(2) Hermite polynomials.

29. Put down the expressions of hyperbolic Chebyshev polynomials $T_0^\times(p), T_1^\times(p), T_2^\times(p), T_3^\times(p), T_4^\times(p), T_5^\times(p), T_6^\times(p)$.

30. Show that Chebyshev polynomials can be written in the following form:

$$T_n(x) = \frac{1}{2}\left[\left(x + \sqrt{x^2-1}\right)^2 + \left(x - \sqrt{x^2-1}\right)^2\right].$$

The function is defined on $-\infty < x < \infty$. When $|x| \le 1$ it goes back to triangle Chebyshev polynomials. When $|x| \ge 1$ it becomes hyperbolic Chebyshev polynomials. What is the form of the Chebyshev functions of second kind $U_n(x)$ accordingly?

31. The two independent solutions of the associated Chebyshev equation $(1 - z^2)w'' - (2m + 1)zw' + (\alpha^2 - m^2)w = 0$ are $T_\alpha^{(m)}(z) = \frac{d^m}{dz^m}T_\alpha(z)$ and $U_\alpha^{(m)}(z) = \frac{d^m}{dz^m}U_\alpha(z)$, respectively. Show that one of them can be evaluated by the other by the Liouville formula.

32. Show that the coefficient of the highest power term of the Chebyshev functions of second kind $U_n(x)$ is 2^n.

33. Prove the relation between the Chebyshev polynomials and Chebyshev functions of the second kind: $\frac{1}{2} + \sum_{k=1}^{n} T_{2k}(x) = \frac{1}{2}U_{2k}(x)$. (Hint: $\sin A \cos B = \frac{1}{2}[\sin(A - B) + \sin(A + B)]$.)

34. The following differential equation is one established by the author in his work:

$$f''(u) + \frac{V(u)}{W(u)}\frac{[W(u)]^2 - 2(R^2 - 1)}{[W(u)]^2 + 4(R^2 - 1)}f'(u)$$

$$- \frac{[W(u)]^2}{[W(u)]^2 + 4(R^2 - 1)}S(S + 1)f(u) = 0,$$

where $V(u) = (Q + 1)e^{u/2} + (Q - 1)e^{-u/2}$ and $W(u) = (Q + 1)e^{u/2} - (Q - 1)e^{-u/2}$. Here, Q, R and S are three parameters. The initial conditions are that $f(0) = 1$ and

$$f'(0) = \frac{(n+1)(d_n + 1/d_n)}{2R(d_n - 1/d_n)} - \frac{Q}{2R}$$

where $d_n = \sqrt{\frac{Q+R}{Q-R}}^{n+1}$. In order to find the solution, one has to make a transformation $p = \frac{V(u)}{2\sqrt{Q^2 - R^2}}$. What will the equation turn to? Under what condition for the S value does the transformed equation has solutions? Please give the expression of $f(u)$, and put down its first and second derivatives $f'(u)$ and $f''(u)$.

35. Find singularities of the following equations and point out their types.
   (1) $z^2(z-1)w''(z) - 2(z-1)w'(z) + 3zw(z) = 0$.
   (2) $z(3z+1)w''(z) - (z+1)w'(z) + 2w(z) = 0$.

36. Show that the sufficient and necessary conditions for $z = \infty$ to be the regular point of (3.5.45) are that there is the expansion (3.5.54) around $z = \infty$, where $a_1 = 2, b_2 = b_3 = 0$.

37. There is an ordinary differential equation of second order:

$$w''(z) + \frac{\alpha}{z^2} w(z) = 0.$$

Convert it to be ordinary differential equations of first order. Find a basic set of solutions of the equations. What is the values of α at which all of the solutions are rational functions?

38. Find the series solutions near $z = 0$ of the following differential equations.
   (1) $w''(z) + bz^2 w(z) = 0$.
   (2) $w''(z) + zw'(z) + w(z) = 0$.
   (3) $2z^2 w''(z) - (z + z^2)w'(z) + w(z) = 0$.
   (4) $4z^2 w''(z) + 4z^2 w'(z) + w(z) = 0$.
   (5) $4z^2 w''(z) + 4zw'(z) - w(z) = 0$.
   (6) $z^2 w''(z) + (2z + z^2)w'(z) + (z - 2)w(z) = 0$.

39. Find the series solutions around $z = \infty$ of the following differential equations.

    (1) $z^4 w''(z) - (z - 2z^2)w'(z) + w(z) = 0.$
    (2) $(z^2 - 1)w''(z) + 2zw'(z) + w(z) = 0.$

40. Show that a Fuchs-type equation of second order with at most one singularity at $z = z_0$ is of the form of

$$w''(z) + \frac{r}{z - z_0}w'(z) + \frac{s}{(z - z_0)^2}w(z) = 0,$$

    where r and s are complex constants. Find a basic set of solutions of it. What are the features of the solutions in the cases of $r = s = 0$ and $r = 2, s = 0$?

41. Find a basic set of solutions of confluent hypergeometric equation $zw''(z) + (\gamma - z)w'(z) - \alpha w(z) = 0$ near $z = 0$, where γ and α are complex constants.

42. Show that the orthonormalized eigenfunctions corresponding to eigenvalues (3.6.16a) and (3.6.17a) are (3.6.16b) and (3.6.17b). Find the eigenfunctions of the boundary problem (3.6.18), and prove (3.6.19)–(3.6.21).

43. There is a boundary value problem as follows:

$$u'(x) = f(x), \quad 0 < x < 1; \quad u(0) = 0, \quad u(1) = 0 \qquad (1)$$

    This is a differential equation of first order, but has two boundary conditions. The boundary conditions are too many. Nevertheless, this does not means that the boundary value problem does not have a solution. Apply the alternative theorem of the solutions of differential equations. Put down the adjoint problem of (1) and find its solutions. Then by the compatibility condition (3.7.8), write the condition for (1) to have solutions. Choose a function $f(x)$ that satisfies the compatibility condition and find the solution of (1).

44. Put down the conditions for the following boundary value problem to have solutions:

$$-u''(x) = f(x), \quad 0 < x < 1; \quad u(1) - u(0) = 0, \quad u'(1) - u'(0) = 0.$$

45. Suppose that the kernel of the differential operator of second order (3.6.4) is

$$p(x) = 1$$

and the boundary conditions are

$$\begin{cases} \alpha_1 u(a) + \beta_1 u(b) = 0, \\ \alpha_2 u'(a) + \beta_2 u'(b) = 0. \end{cases}$$

Then what boundary conditions are satisfied by the function v in the adjoint problem? Is the boundary value problem self-adjoint if the coefficients in the boundary conditions are real numbers?

46. Show that if the boundary conditions in (3.7.12) are inhomogeneous: $\alpha u(a) + \beta u(b) = \gamma$, then this boundary value problem is not self-adjoint.

Appendix 3A Generalization of Sturm-Liouville Theorem to Dirac Equation

It goes without saying that Sturm-Liouville theory of ordinary differential equations of second order has been successfully applied to physics. In the case of non-relativistic quantum mechanics, the state or movement of a microscopic particle obeys Schrödinger equation. If the potential is independent of time, the time factor of the wave function can be separated, and the space coordinate factor satisfies the stationary Schrödinger equation. This is an ordinary differential equation of second order, and its eigenvalues are the energies the system may have. The corresponding eigenfunctions are the states the system may exists. The third Sturm-Liouville theorem tells us that the eigenenergy has a lower limit. In a quantum system, the energy has indeed a minimum. The state corresponding to the minimum energy is called ground state. The states with higher energies are called excitation states. When the system is not disturbed, the particle is in its ground state. When an external field is applied, the particle may transit from the ground state to an excitation state. In the course of transition, the total energy is conserved.

However, in the case of relativistic quantum mechanics, new features appear. A relativistic particle obeys Dirac equation of relativistic quantum mechanics. Solving Dirac equation, one obtains the energy expressed by $E = \pm c\sqrt{p^2 + m^2 c^2}$, where m and p are the particle's rest mass and momentum, respectively, and c is light speed. It is seen that there are two branches of energies: positive and negative ones. The positive branch has a lower limit but not the upper limit, which agrees with the Sturm-Liouville theory. However, the negative branch has no lower limit, which is not consistent with the Sturm-Liouville theory. The appearance of the branch without a lower limit arises from the fact that the form of Dirac equation is different from that of Sturm-Liouville equations. Therefore, it was desirable to extend the Sturm-Liouville theory to one that can be applied to Dirac equation. C. N. Yang did this work (Yang C N, Generalization of Sturm-Liouville Theory to a System of Ordinary Differential Equations with Dirac Type Spectrum. Commun. Math. Phys., 1987, **112**: 205–216). He noticed that in Dirac equation, derivations are written in a matrix form. Consequently, the solution functions have to comprise at least two components. He proposed four theorems applicable to Dirac equation. His theory was parallel to Sturm-Liouville theory. Here we do not intend to introduce his theory. Readers are encouraged to read his article.

Chapter 4

Bessel Functions

In Chapter 3 we introduced complex analysis theory of ordinary differential equations of second order. The main point was the process of how to construct solutions. In this chapter, we will apply the process to solve Bessel equations. The same process can be employed to deal with other differential equations such as Legendre equations, Hermite equations etc.

In this chapter, the argument is represented by z if it is a complex number, and by x if it is real.

4.1. Bessel Equation

4.1.1. *Bessel equation and its solutions*

A Bessel equation is

$$z^2 w'' + z w' + (z^2 - \nu^2) w = 0, \tag{4.1.1}$$

where ν is a complex number and $\operatorname{Re} \nu \geq 0$. This equation has eigenvalues ν^2, and accordingly has two linearly independent solutions. By the method introduced in Section 3.5, it is easily distinguished that $z = 0$ is a singularity of first kind, and $z = \infty$ is a singularity of second kind. Equation (4.1.1) is called a Bessel equation of order ν. We are seeking for the expression of its solutions near $z = 0$.

Recall (3.5.45), (3.5.49) and (3.5.51). We can put down for (4.1.1) $a_0 = 1$, $b_0 = -\nu^2$. The indicial equation, therefore, with respect to

$z = 0$ can be obtained by means of (3.5.56):

$$P_\nu(\lambda) = \lambda^2 - \nu^2 = 0. \tag{4.1.2}$$

Its two roots are $\lambda_1 = \nu$ and $\lambda_2 = -\nu$, where λ_1 and λ_2 are the two indices with respect to $z = 0$ of (4.1.1). If the eigenvectors are denoted by $\varphi = \binom{a_1}{a_2}$, then the eigenvalue equation $A_0\varphi = \lambda\varphi$ is solved to obtain $\frac{a_2}{a_1} = \lambda$. Where λ is one of λ_1 and λ_2. This equation shows that when λ_1 and λ_2 are different, they correspond to different eigenvectors. As $\nu = 0$, the two eigenvalues are identical. Then there must be a special solution with a logarithm factor. Another case should be noticed: if the difference of the two eigenvalues is any integer, it is probable that one special solution includes a logarithm factor. We stress here that when (4.1.1) is reformed to be the form of (3.5.50), the coefficient matrix A is a matrix series, not only just one term A_0. This is because the coefficient of the w term in (4.1.1) is a series, although it contains merely two terms. Thus the series method has to be resorted to.

We now search for a special solution on $0 < |z| < \infty$ following (3.5.57a). It is of the form of

$$w(z) = z^\lambda h(z). \tag{4.1.3a}$$

When this is substituted into (4.1.1), the equation that $h(z)$ satisfies can be obtained observing (3.5.59):

$$zh'' + (2\lambda + 1)h' + zh = 0, \tag{4.1.3b}$$

where $\lambda = \nu$ is utilized. At this stage we assume $h(z)$ to be a series near $z = 0$:

$$h(z) = \sum_{k=0}^{\infty} c_k z^k. \tag{4.1.3c}$$

Then it follows from (4.1.3b), and by comparing coefficients with the same power, we get

$$\sum_{k=0}^{\infty} c_k k(k + 2\lambda)z^k + \sum_{k=2}^{\infty} c_{k-2}z^k = 0.$$

Explicitly,

$$0(0 + 2\lambda)c_0 = 0, \quad (1 + 2\lambda)c_1 = 0, \dots, \tag{4.1.4}$$

$$k(k + 2\lambda)c_k + c_{k-2} = 0, \quad k \geq 2. \tag{4.1.5}$$

We first consider the case of $\lambda \neq 1/2$, i.e.,

$$\nu \neq \pm 1/2. \tag{4.1.6}$$

(The case of $\nu = \pm 1/2$ will be included in case (2) below.) It can be certain by (4.1.4) that $c_1 = 0$. Then from (4.1.5) we immediately get

$$c_1 = c_3 = c_5 = \cdots = 0. \tag{4.1.7}$$

The coefficient c_0 can be nonzero. We set

$$c_0 = 1. \tag{4.1.8}$$

As $\lambda = \lambda_1 = \nu$, (4.1.5) is written as $k(k + 2\nu)c_k + c_{k-2} = 0$, $k \geq 2$. Thus we get a recurrence relation:

$$c_k = -\frac{c_{k-2}}{k(k + 2\nu)} = 0, \quad k \geq 0. \tag{4.1.9}$$

Starting from (4.1.8) and in terms of

$$c_{2m} = -\frac{c_{2(m-1)}}{4m(m + \nu)} = 0, \quad m \geq 1, \tag{4.1.10}$$

the series of coefficients can all be calculated:

$$c_{2m} = -\frac{(-1)^m}{4^m m!(\nu + 1)_m}, \quad m \geq 1. \tag{4.1.11}$$

Here $(\nu + 1)_m$ is a Gauss symbol, the definition of which is (3.3.16). Substituting the coefficients (4.1.11) into (4.1.3), we achieve the expression of the solution:

$$w_\nu(z) = \sum_{m=0}^{\infty} \frac{(-1)^m}{4^m m!(\nu + 1)_m} z^{\nu+2m}, \quad 0 < |z| < \infty. \tag{4.1.12}$$

Next, we seek for another linearly independent solution. Considering the discussions presented in Subsection 3.5.1, we have to distinguish four cases of $\lambda = \lambda_2 = -\nu$.

(1) $\lambda_1 - \lambda_2 = 2\nu$ is not an integer

In this case, when k is not an integer, $P_\nu(\lambda + k) \neq 0$, and the derivation from (4.1.7) to (4.1.12) applies also. So, one simply replaces ν by $-\nu$ in these equations to gain another solution:

$$w_{-\nu}(z) = \sum_{m=0}^{\infty} \frac{(-1)^m}{4^m m!(-\nu+1)_m} z^{-\nu+2m}, \quad 0 < |z| < \infty. \quad (4.1.13)$$

The two solutions $w_{-\nu}$ and w_ν are linearly independent of each other, because they are series beginning with different powers of z. As a result, the linear combination $a_1 w_\nu + a_2 w_{-\nu}$ is zero only when $a_1 = a_2 = 0$.

(2) $\lambda_1 - \lambda_2 = 2\nu = 2n + 1$, $(n \geq 0)$, i.e., ν is positive half integer

In (4.1.5),

$$k(k - 2n - 1) = \begin{cases} \neq 0, & k \neq 2n + 1 \\ = 0, & k = 2n + 1 \end{cases} \quad (4.1.14)$$

So, when $k < 2n + 1$, (4.1.9) leads to

$$k(k - 2n - 1)c_k + c_{k-2} = 0. \quad (4.1.15)$$

It follows that

$$c_1 = c_3 = c_5 = \cdots = c_{2n-1} = 0. \quad (4.1.16)$$

On the other hand, when $k = 2n+1$, (4.1.9) gives $0c_{2n+1}+c_{2n-1} = 0$. Taking $c_{2n+1} = 0$ yields

$$c_1 = c_3 = c_5 = \cdots = 0. \quad (4.1.17)$$

Anyhow, the coefficients with odd indices are all zero.

For even k, the derivation from (4.1.7) to (4.1.12) is still valid. Thus the two linearly independent solutions are obtained easily:

$$w_{n+1/2}(z) = \sum_{m=0}^{\infty} \frac{(-1)^m}{4^m m!(n+1/2+1)_m} z^{n+1/2+2m}, \quad 0 < |z| < \infty$$

$$(4.1.18a)$$

and

$$w_{-n-1/2}(z) = \sum_{m=0}^{\infty} \frac{(-1)^m}{4^m m! (-n-1/2+1)_m} z^{-n-1/2+2m},$$

$$0 < |z| < \infty \qquad (4.1.18b)$$

(3) $\nu = n$ is integer

In this case, $\lambda_1 = n > 0$, $\lambda_2 = -n$. One special solution again comes from (4.1.12) and has the form of

$$w_1(z) = \sum_{m=0}^{\infty} \frac{(-1)^m}{4^m m! (n+1)_m} z^{n+2m}, \qquad 0 < |z| < \infty \qquad (4.1.19)$$

According to the discussion of (3.5.44), another special solution has to comprise a logarithm factor. To find this solution, the following transformation is made:

$$z = e^s, \qquad w(z) = u(s). \qquad (4.1.20)$$

Then (4.1.1) is converted to

$$u''(s) + (e^{2s} - n^2) u(s) = 0. \qquad (4.1.21)$$

Assume that its solution is of the form of

$$u(s) = e^{\lambda_2 s} \sum_{k=0}^{\infty} u_k(s) e^{ks}, \qquad \text{where } u_k(s) \text{ is linear of } s. \qquad (4.1.22)$$

Since $u_k(s)$ is linear, its second derivative is zero. Substituting (4.1.22) into (4.1.21) leads to

$$\sum_{k=0}^{\infty} \left[2(\lambda_2 + k) u_k'(s) + k(2\lambda_2 + k) u_k(s) \right] e^{ks} + \sum_{k=2}^{\infty} u_{k-2}(s) e^{ks} = 0.$$

where we have made use of $\lambda_2^2 - n^2 = 0$. The coefficient relations are as follows:

$$2\lambda_2 u_0' = 0,$$

$$2(\lambda_2 + 1) u_1' + (2\lambda_2 + 1) u_1 = 0, \qquad (4.1.23a)$$

$$2(\lambda_2 + k) u_k' + k(2\lambda_2 + k) u_k + u_{k-2} = 0, \qquad k \geq 2.$$

Remember that $\lambda_2 = -n$. So,

$$-2nu_0' = 0,$$

$$2(1-n)u_1' + (1-2n)u_1 = 0, \tag{4.1.23b}$$

$$2(k-n)u_k' + k(k-2n)u_k + u_{k-2} = 0, \quad k \geq 2.$$

It is thus $u_1 = 0$, and subsequently,

$$u_1 = u_3 = u_5 = \cdots = 0 \tag{4.1.24}$$

The first coefficient u_0 must be a constant. Let $u_0 = 1$. Then as $k = 2$, $2(2-n)u_2' + 2(2-2n)u_2 + u_0 = 0$. To meet this equation, u_2 must not depend on s, but only a constant. Similarly, $u_k(0 < k = 2m < 2n)$ are all constants. They are explicitly reckoned as follows:

$$2m(2m-2n)u_{2m} + u_{2(m-1)} = 0, \quad (0 < m < n),$$

$$u_{2m} = \frac{1}{4^m m!(n-1)(n-2)\cdots(n-m)}$$

$$= \frac{(n-m-1)!}{4^m m!(n-1)!}, \quad (0 < m < n). \tag{4.1.25}$$

As $k = 2n$, we have $2nu_{2n}' + 0u_{2n} + u_{2n-2} = 0$. It leads to

$$u_{2n} = \alpha_0(s + \beta_0), \tag{4.1.26}$$

where

$$\alpha_0 = -\frac{u_{2n-2}}{2n} = -\frac{1}{2n}\frac{1}{4^{n-1}(n-1)!}\frac{(-1)^{n-1}}{(1-n)_{n-1}} = -\frac{2}{4^n n!(n-1)!} \tag{4.1.27}$$

and β_0 is an arbitrary constant.

As $k = 2n+2$, $2(n+2)u_{2n+2}' + 4nu_{2n+2} + u_{2n} = 0$. When $k \geq 2n+2m$, $(m \geq 1)$, we assume $u_{2n+2m} = \alpha_m(s + \beta_m)$, where α_m and β_m are constants to be determined.

When this is substituted into (4.1.23b), we obtain

$$2(n+2m)\alpha_m + 4m(n+m)\alpha_m(s+\beta_m) + \alpha_{m-1}(s+\beta_{m-1}) = 0.$$

The relationships of the coefficients of the one and zero powers of s are as follows:

$$4m(n+m)\alpha_m + \alpha_{m-1} = 0,$$

$$2(n+2m)\alpha_m + 4m(n+m)\alpha_m\beta_m + \alpha_{m-1}\beta_{m-1} = 0.$$

Form the first equation, we get

$$\alpha_m = -\frac{\alpha_{m-1}}{4m(n+m)} = \cdots = \frac{(-1)^m \alpha_0}{4^m m!(n+1)_m}. \tag{4.1.28}$$

This is then used in the second equation:

$$2(n+2m)\alpha_m + 4m(n+m)\alpha_m\beta_m - 4m(n+m)\alpha_m\beta_{m-1} = 0,$$

$$\beta_m = \beta_{m-1} - \frac{n+2m}{2m(n+m)} = \beta_{m-1} - \frac{1}{2}\left(\frac{1}{m} + \frac{1}{n+m}\right)$$

$$= \beta_0 - \frac{1}{2}\sum_{j=1}^{m}\left(\frac{1}{j} + \frac{1}{n+j}\right).$$

Let

$$H_m = \sum_{j=1}^{m}\frac{1}{j} = 1 + \frac{1}{2} + \frac{1}{3} + \cdots + \frac{1}{m} \tag{4.1.29}$$

and adopt

$$\beta_0 = -\frac{1}{2}(H_n - \gamma) = -\frac{1}{2}\psi(n), \tag{4.1.30}$$

where γ is the Euler constant. The function $\psi(n)$ is introduced. Its definition is $\psi(z) = \frac{d\ln\Gamma(z)}{dz} = \frac{\Gamma'(z)}{\Gamma(z)}$ where $\Gamma(z) = \int_0^\infty e^{-t}t^{z-1}dt, \operatorname{Re} z > 0$. Then the coefficient β_m can be expressed by the function $\psi(n)$ as

$$\beta_m = -\frac{1}{2}(H_m + H_{m+n} - 2\gamma) = -\frac{1}{2}[\psi(m+1) + \psi(m+n+1)]. \tag{4.1.31}$$

In summary, as $k = 2m < 2n$, we have (4.1.25); as $k \geq 2n + 2m$, $(m \geq 0)$, we have (4.1.26) to (4.1.31). These are substituted into (4.1.22) to gain

$$u(s) = e^{\lambda_2 s}\sum_{m=0}^{n-1} u_{2m}e^{2ms} + e^{\lambda_2 s}\sum_{m=0}^{\infty}\alpha_m(s+\beta_m)e^{(2n+2m)s}. \tag{4.1.32}$$

Now we are at the stage to replace $z = e^s$ back. Finally, the second special solution is obtained:

$$w_2(z) = z^{-n} \sum_{m=0}^{n-1} u_{2m} z^{2m} + z^{-n} \sum_{m=0}^{\infty} \alpha_m (\operatorname{Ln} z + \beta_m) z^{2n+2m}$$

$$= \operatorname{Ln} z \sum_{m=0}^{\infty} \alpha_m z^{n+2m} + \sum_{m=0}^{\infty} \alpha_m \beta_m z^{n+2m} + \sum_{m=0}^{n-1} u_{2m} z^{2m-n}$$

$$= \alpha_0 \operatorname{Ln} z \sum_{m=0}^{\infty} \frac{(-1)^m}{4^m m!(n+1)_m} z^{n+2m}$$

$$- \frac{\alpha_0}{2} \sum_{m=0}^{\infty} \frac{(-1)^m [\psi(m+1) + \psi(m+n+1)]}{4^m m!(n+1)_m} z^{n+2m}$$

$$+ \sum_{m=0}^{n-1} \frac{(n-m-1)!}{4^m m!(n-1)!} z^{2m-n}, \tag{4.1.33}$$

where α_0 is expressed by (4.1.27).

The two solutions (4.1.19) and (4.1.33) are linearly independent of each other. They constitute a basic solutions of Eq. (4.1.1) when $\nu^2 = n^2 \neq 0$.

(4) $\nu = 0$

In this case $\lambda_1 = \lambda_2 = 0$, which can be considered as a special case $n = 0$ of (3). Once letting $n = 0$ in (4.1.23b), all the following equations are valid. Accordingly, there is only the second term left while the first term is excluded in (4.1.32). Hence, the two special solutions are deduced from (4.1.19) and (4.1.33) to be

$$w_1(z) = \sum_{m=0}^{\infty} \frac{(-1)^m}{4^m m!(1)_m} z^{2m} = \sum_{m=0}^{\infty} \frac{(-1)^m}{4^m (m!)^2} z^{2m}, \quad 0 < |z| < \infty \tag{4.1.34}$$

and

$$w_2(z) = \alpha_0 \operatorname{Ln} z \sum_{m=0}^{\infty} \frac{(-1)^m}{4^m m!(1)_0} z^{2m}$$

$$- \frac{\alpha_0}{2} \sum_{m=0}^{\infty} \frac{(-1)^m [\psi(m+1) + \psi(m+1)]}{4^m m!(1)_0} z^{2m}$$

$$= \alpha_0 \text{Ln} \, z \sum_{m=0}^{\infty} \frac{(-1)^m}{m!} \left(\frac{z}{2}\right)^{2m} - \alpha_0 \sum_{m=0}^{\infty} \frac{(-1)^m \psi(m+1)}{m!} \left(\frac{z}{2}\right)^{2m}.$$

$$(4.1.35)$$

In conclusion, the eigenvalues of Bessel equation is ν^2, and its two linearly independent solutions correspond respectively to indices ν and $-\nu$. When ν is not integer, solutions (4.1.12) and (4.1.13) follow (3.5.43) and (3.5.44), or (3.5.57) with $h_2(z) = 0$. When ν is integer, solutions (4.1.19) and (4.1.33) also follow (3.5.43) and (3.5.44), or (3.5.57) but with $h_2(z) \neq 0$. Equations (4.1.34) and (4.1.35) are the special case of (4.1.19) and (4.1.33) at $\nu = 0$.

4.1.2.   *Bessel functions of the first and second kinds*

Now we rewrite the basic solutions of a Bessel equation as follows.

1. ν is not negative integer

The first special solution (4.1.12) stands for any ν. In convention, it is divided by $2^\nu \Gamma(\nu + 1)$ and then denoted as J_ν.

$$J_\nu(z) = \frac{w_\nu(z)}{2^\nu \Gamma(\nu + 1)}$$

$$= \sum_{m=0}^{\infty} \frac{(-1)^m}{m! \Gamma(m + \nu + 1)} \left(\frac{z}{2}\right)^{2m+\nu}, \quad 0 < |z| < \infty,$$

where Γ function was defined by (3.3.17). Another linearly independent special solution is

$$J_{-\nu}(z) = \frac{w_{-\nu}(z)}{2^{-\nu} \Gamma(-\nu + 1)} = \sum_{m=0}^{\infty} \frac{(-1)^m}{m! \Gamma(m - \nu + 1)} \left(\frac{z}{2}\right)^{2m-\nu},$$

$$0 < |z| < \infty.$$

These two solution can be uniformly written as

$$J_{\pm\nu}(z) = \sum_{m=0}^{\infty} \frac{(-1)^m}{m! \Gamma(m \pm \nu + 1)} \left(\frac{z}{2}\right)^{2m\pm\nu}. \qquad (4.1.36)$$

$J_\nu(z)$ is called a **Bessel function of the first kind with order ν** (ν is not negative integer), or simply a **Bessel function**. Note that in (4.1.36) the summation covers infinite terms, so that it belongs to special functions, not just a polynomial.

2. $\nu = n$ is positive integer

In this case (4.1.36) is still valid. Since $\Gamma(m+n+1) = (m+n)!$, this solution becomes

$$J_n(z) = \sum_{m=0}^{\infty} \frac{(-1)^m}{m!(m+n)!} \left(\frac{z}{2}\right)^{2m+n}. \tag{4.1.37}$$

$J_n(z)$ (n is positive integers) is called a Bessel function with integer order. We emphasize that the $J_n(z)$ is linearly dependent on $J_{-n}(z)$. It will be proved later that the relationship between them is

$$J_{-n}(z) = (-1)^n J_n(z). \tag{4.1.38}$$

Another special solution of a Bessel equation is of the form of (4.1.33). Conventionally, (4.1.33) and (4.1.37) are linearly combined, and the resultant is denoted as Y_n.

$$Y_n(z) = -\frac{2^n(n-1)!}{\pi} w_2(z) - \frac{2\ln 2}{\pi} J_n(z). \tag{4.1.39}$$

Let us check in detail the expression of $w_2(z)$, i.e., (4.1.33). As n is positive integer,

$$(n+1)_k = \frac{(n+k)!}{n!},$$

the right hand side of (4.1.33) can be written as

$$\sum_{m=0}^{\infty} \frac{(-1)^m}{4^m m!(n+1)_m} z^{n+2m} = 2^n n! \sum_{m=0}^{\infty} \frac{(-1)^m}{m!(n+m)!} \left(\frac{z}{2}\right)^{n+2m}$$

$$= 2^n n! J_n(z),$$

where (4.1.37) is employed. Thus, (4.1.33) becomes

$$w_2(z) = \alpha_0 2^n n! J_n(z) \mathrm{Ln}\, z$$

$$- \frac{\alpha_0}{2} 2^n n! \sum_{m=0}^{\infty} \frac{(-1)^m [\psi(m+1) + \psi(m+n+1)]}{m!(n+m)!} \left(\frac{z}{2}\right)^{n+2m}$$

$$+ \frac{1}{2^n} \sum_{m=0}^{n-1} \frac{(n-m-1)!}{m!(n-1)!} \left(\frac{z}{2}\right) z^{2m-n}.$$

After this equation and (4.1.27) are substituted into (4.1.39), we have

$$Y_n(z) = -\frac{2^n(n-1)!}{\pi}\left\{-\frac{2}{4^n n!(n-1)!}2^n n! J_n(z)\text{Ln}\,z\right.$$

$$+\frac{2^n n!}{4^n n!(n-1)!}\sum_{m=0}^{\infty}\frac{(-1)^m[\psi(m+1)+\psi(m+n+1)]}{m!(n+m)!}$$

$$\left.\times\left(\frac{z}{2}\right)^{n+2m}+\frac{1}{2^n}\sum_{m=0}^{n-1}\frac{(n-m-1)!}{m!(n-1)!}\left(\frac{z}{2}\right)^{2m-n}\right\}-\frac{2\ln 2}{\pi}J_n(z)$$

$$=\frac{2}{\pi}J_n(z)\text{Ln}\frac{z}{2}-\frac{1}{\pi}\sum_{m=0}^{\infty}\frac{(-1)^m[\psi(m+1)+\psi(m+n+1)]}{m!(n+m)!}$$

$$\times\left(\frac{z}{2}\right)^{n+2m}-\frac{1}{\pi}\sum_{m=0}^{n-1}\frac{(n-m-1)!}{m!}\left(\frac{z}{2}\right)^{2m-n}. \qquad (4.1.40)$$

$Y_n(z)$ is called a **Bessel function of the second kind** or **Neumann function**. In literature it is also denoted by $N_n(z)$.

A Neumann function is often expressed in a simple form:

$$Y_n(z) = \lim_{\nu\to n}\frac{J_\nu(z)\cos\nu\pi - J_{-\nu}(z)}{\sin\nu\pi}. \qquad (4.1.41)$$

The right hand side of this equation is, according to (4.1.38), zero nil type, but its limit exists. In can be shown that (4.1.41) and (4.1.40) are identical.

When ν is not real integer, the denominator in (4.1.41) is not zero. So, the limit can be removed:

$$Y_\nu(z) = \frac{J_\nu(z)\cos\nu\pi - J_{-\nu}(z)}{\sin\nu\pi}. \qquad (4.1.42)$$

Here $J_\nu(z)$ and $J_{-\nu}(z)$ are linearly independent of each other. Therefore, the resultant $Y_\nu(z)$ is also linearly independent of $J_\nu(z)$. The general solution of a Bessel equation can be written as the linear combination of $J_\nu(z)$ and $J_{-\nu}(z)$, as well as that of $J_\nu(z)$ and $Y_\nu(z)$.

$$w(z) = AJ_\nu(z) + BY_\nu(z) \qquad (4.1.43)$$

Usually, the general solution is expressed in this way. As $\nu = n$ is integer, (4.1.37) becomes the linear combination of J_n and Y_n of (4.1.41).

4.2. Fundamental Properties of Bessel Functions

In the definition of a Bessel function of order ν, ν can be any complex number. In practical cases usually encountered, ν is real. Hereafter we default that ν is real.

It is easily obtained by the definition of a Bessel function that

$$J_\nu(-z) = \sum_{m=0}^{\infty} \frac{(-1)^m}{m!\,\Gamma(m+\nu+1)} \left(-\frac{z}{2}\right)^{2m+\nu} = (-1)^\nu J_\nu(z). \quad (4.2.1)$$

If the argument $z = x$ is real, then both functions $J_\nu(x)$ and $Y_\nu(x)$ are real.

4.2.1. *Recurrence relations of Bessel functions*

1. Basic recursion formulas

$$\left(z^\nu J_\nu\right)' = z^\nu J_{\nu-1}. \qquad\qquad\qquad\qquad\qquad (4.2.2)$$

$$\left(z^{-\nu} J_\nu\right)' = -z^{-\nu} J_{\nu+1}. \qquad\qquad\qquad\qquad (4.2.3)$$

$$J_\nu = \frac{z}{2\nu}\left(J_{\nu-1} + J_{\nu+1}\right). \qquad\qquad\qquad\quad (4.2.4)$$

$$J_\nu' = \frac{1}{2}\left(J_{\nu-1} - J_{\nu+1}\right)$$

$$= J_{\nu-1}(z) - \frac{\nu}{z}J_\nu(z) = \frac{\nu}{z}J_\nu(z) - J_{\nu+1}(z). \quad (4.2.5)$$

These four equations are collectively referred to as recursion formulas. Among them, the two, (4.2.2) and (4.2.3), with derivatives can also be regarded as differential relations. Now we proceed to prove them.

Proof.   Multiply z^ν to the expression of J_ν, and then take derivative with respect to z. The result is

$$\frac{d}{dz}(z^\nu J_\nu) = \frac{d}{dz}\left[\sum_{k=0}^{\infty} \frac{(-1)^k}{k!\,\Gamma(k+1+\nu)} \left(\frac{1}{2}\right)^{2k+\nu} z^{2k+2\nu}\right]$$

$$= \sum_{k=0}^{\infty} \frac{(-1)^k(2k+2\nu)}{k!\,\Gamma(k+1+\nu)} \left(\frac{1}{2}\right)^{2k+\nu} z^{2k+2\nu-1} = z^\nu J_{\nu-1}.$$

This is (4.2.2). Similarly,

$$\frac{d}{dz}(z^{-\nu}J_\nu) = \frac{d}{dz}\left[\sum_{k=0}^{\infty} \frac{(-1)^k}{k!\Gamma(k+1+\nu)}\left(\frac{1}{2}\right)^{2k+\nu} z^{2k}\right]$$

$$= -\sum_{k=1}^{\infty} \frac{z^{-\nu}(-1)^{k-1}}{(k-1)!\Gamma(k+1+\nu)}\left(\frac{z}{2}\right)^{2k+\nu-1}.$$

Let $n = k - 1$. Then

$$\frac{d}{dz}(z^{-\nu}J_\nu) = -\sum_{n=0}^{\infty} \frac{z^{-\nu}(-1)^n}{\Gamma(n+1)\Gamma(n+2+\nu)}\left(\frac{z}{2}\right)^{2n+1+\nu}$$

$$= -z^{-\nu}J_{\nu+1},$$

which is (4.2.3). The two equations

$$(z^\nu J_\nu)' = \nu z^{\nu-1}J_\nu + z^\nu J_\nu' = z^\nu J_{\nu-1} \qquad (4.2.6a)$$

and

$$(z^{-\nu}J_\nu)' = -\nu z^{-\nu-1}J_\nu + z^{-\nu}J_\nu' = -z^{-\nu}J_{\nu+1} \qquad (4.2.6b)$$

are rearranged to be

$$\nu J_\nu + z J_\nu' = z J_{\nu-1} \qquad (4.2.7a)$$

and

$$-\nu J_\nu + z J_\nu' = -z J_{\nu+1}. \qquad (4.2.7b)$$

Subtraction of these two equations produces $2\nu J_\nu = z(J_{\nu-1} + J_{\nu+1})$. This is (4.2.4).

The summation of the two equations of (4.2.7) yields $2J_\nu' = J_{\nu-1} - J_{\nu+1}$, which is the first equal sign in (4.2.5). Application of (4.2.4) leads to the latter two equal signs in (4.2.5).

By the two differential relations (4.2.2) and (4.2.3), one is able to evaluate Bessel functions of order $\nu - 1$ or $\nu + 1$ in terms of the function of order ν. Especially, as $\nu = 0$, we have

$$J_0'(z) = J_{-1}(z) = -J_1(z).$$

It can be thus asserted that the extreme points of $J_0(z)$ are zeros of $J_1(z)$.

By the other two formulas (4.2.4) and (4.2.5), one is able to use two Bessel functions with adjacent orders to evaluate the function with one more order. □

2. Further recursion formulas

$$\left(\frac{1}{z}\frac{d}{dz}\right)^m (z^\nu J_\nu) = z^{\nu-m} J_{\nu-m}. \qquad (4.2.8a)$$

$$\left(\frac{1}{z}\frac{d}{dz}\right)^m (z^{-\nu} J_\nu) = (-1)^m z^{-\nu-m} J_{\nu+m}. \qquad (4.2.8b)$$

Proof. It follows from (4.2.2) that

$$\frac{1}{z}\frac{d}{dz}(z^\nu J_\nu) = z^{\nu-1} J_{\nu-1}. \qquad (4.2.9)$$

This is the case of $m = 1$ in (4.2.8a). Now regarding $\frac{1}{z}\frac{d}{dz}$ as an operator and employing (4.2.9) repeatedly, we get

$$\left(\frac{1}{z}\frac{d}{dz}\right)^2 (z^\nu J_\nu) = \left(\frac{1}{z}\frac{d}{dz}\right)\left(\frac{1}{z}\frac{d}{dz}\right)(z^\nu J_\nu)$$

$$= \left(\frac{1}{z}\frac{d}{dz}\right)(z^{\nu-1} J_{\nu-1}) = z^{\nu-2} J_{\nu-2}. \qquad (4.2.10)$$

It is seen that the result of the operator $\frac{1}{z}\frac{d}{dz}$ acting on $z^\nu J_\nu$ each time is to lower the order ν by 1. Thus (4.2.8a) is obtained. Equation (4.2.8b) is proved similarly by use of (4.2.3). □

3. Recursion formulas for Bessel functions of the second kind

In (4.2.2)–(4.2.5), J represents Bessel functions of the first kind. If J in these equations is replaced by Y, the formulas remain valid for Bessel functions of the second kind. They are as follows.

$$\begin{cases} (z^\nu Y_\nu)' = z^\nu Y_{\nu-1}. \\ (z^{-\nu} Y_\nu)' = -z^{-\nu} Y_{\nu+1}. \\ Y_{\nu-1} + Y_{\nu+1} = \frac{2\nu}{z} Y_\nu. \\ Y_{\nu-1} - Y_{\nu+1} = 2Y_\nu'. \end{cases} \qquad (4.2.11)$$

The first two equations can be derived by definition (4.1.42) and formulas (4.2.2) and (4.2.3). For example,

$$[z^\nu Y_\nu(z)]' = \frac{[z^\nu J_\nu(z)]' \cos \nu\pi - [z^{-(-\nu)} J_{-\nu}(z)]'}{\sin \nu\pi}$$

$$= z^\nu \frac{J_{\nu-1}(z) \cos \nu\pi + J_{-(\nu-1)}(z)}{\sin \nu\pi} = z^\nu Y_{\nu-1}(z).$$

The latter two equations in (4.2.11) can be derived by the first two.

4.2.2. *Asymptotic formulas of Bessel functions*

When $z \to \infty$ and ν is fixed,

$$J_\nu(z \to \infty) \sim \sqrt{\frac{2}{\pi z}} \cos \left(z - \frac{\nu\pi}{2} - \frac{\pi}{4} \right), \quad (-\pi < \arg z < \pi)$$

$$(4.2.12)$$

and

$$Y_\nu(z \to \infty) \sim \sqrt{\frac{2}{\pi z}} \sin \left(z - \frac{\nu\pi}{2} - \frac{\pi}{4} \right), \quad (-\pi < \arg z < \pi).$$

$$(4.2.13)$$

When z is fixed and $|\nu|$ is very large, the asymptotic expression of $J_\nu(z)$ is:

$$J_{\nu\to\infty}(z) \approx \exp \left[v + v \ln \frac{z}{2} - \left(v + \frac{1}{2} \right) \ln v \right]$$

$$\times \left[\frac{1}{\sqrt{2\pi}} + \frac{c_1}{v} + \frac{c_2}{v^2} + \cdots \right]. \qquad (4.2.14)$$

As $z \to 0$,

$$J_\nu(z \to 0) \sim \frac{1}{\Gamma(\nu + 1)} \left(\frac{z}{2} \right)^\nu. \qquad (4.2.15a)$$

For real ν and $\nu > 0$, we have

$$J_\nu(z \to 0) \sim \frac{1}{\Gamma(\nu + 1)} \left(\frac{z}{2} \right)^\nu \to 0 \qquad (4.2.15b)$$

and

$$J_{-\nu}(z \to 0) \sim \frac{1}{\Gamma(-\nu + 1)} \left(\frac{z}{2} \right)^{-\nu} \to \infty. \qquad (4.2.15c)$$

4.2.3. *Zeros of Bessel functions*

1. The properties of zeros of Bessel functions

(1) For any given real ν, $J_\nu(z)$ has infinite number of zeros, i.e., at
these real z points, $J_\nu(z) = 0$.

(2) As ν is real and greater than 0,

$$J_\nu(0) = 0. \tag{4.2.16}$$

This can easily be seen from the expression (4.1.36).

(3) All of the zeros of $J_\nu(z)$, except $z = 0$ if it is a zero, are of first
order.

In fact, if $z_0 \neq 0$ is a zero of $J_\nu(z)$ with second or higher order,
then $J_\nu(z_0) = 0$ and $J'_\nu(z_0) = 0$. Recall that $J_\nu(z)$ satisfies a
homogeneous linear differential equation. So, by the theorem of
the existence and uniqueness of solutions of homogeneous differential equations introduced in Subsection 3.1.1, it must follow
that $J_\nu(z) \equiv 0$.

(4) As $\nu > -1$, all the zeros of $J_\nu(z)$ are real. This means that the
z points that make $J_\nu(z) = 0$ are on the real axis. Please note
the discrepancy with (1) where there might be real as well as
complex zeros.

(5) If $J_\nu(z) = 0$, then $J_\nu(-z) = 0$.

This indicates that the zeros of $J_\nu(z)$ symmetrically distribute
with respect to the origin.

(6) The zeros of $J_\nu(z)$ and of $J_{\nu+1}(z)$ are not the same.

In fact, from (4.2.6b), we have $J'_\nu(z) - \nu z^{-1} J_\nu(z) = -J_{\nu+1}(z)$.
If at $z_0 J_\nu(z_0) = J_{\nu+1}(z_0) = 0$, there must be $J'_\nu(z_0) = 0$. This
indicates that z_0 is a zero of $J_\nu(z)$ with order 2, a conclusion
contrary to the assertion in (3).

The following three conclusions are for the cases of $\nu > 0$. We
have known from (4) and (5) that the zeros are on the real axis
and distribute symmetrically with respect to origin. Therefore,
it is enough to consider the zeros on the positive real axis, $x > 0$,
which are called **positive zeros**.

Before addressing the following three conclusions, let us recall
the **Rolle theorem** in differential calculus: if an $f(x)$ is continuous on $[a, b]$, is derivable on (a, b) and $f(a) = f(b)$, there must

exist a $\xi \in (a, b)$ such that $f'(\xi) = 0$. In geometric viewpoint, this means that the tangent $y = f(x)$ at point $C(\xi, f(\xi))$ is horizontal.

(7) Between the adjacent two zeros of $J_\nu(x)$, there is one and only one zero of $J_{\nu-1}(x)$ and $J_{\nu+1}(x)$, and vice versa.

(8) The smallest positive zero of $J_\nu(x)$ is closer to the origin than that of $J_{\nu+1}(x)$.

(9) Equation $J_\nu'(x) = 0$ has an infinite number of real roots.

It is known that there are an infinite number of real zeros of $J_\nu(x)$. By Rolle theorem, there must be a point a between any adjacent two zeros such that $J_\nu'(a) = 0$. More generally, it can be shown that equation

$$J_\nu(x) + h J_\nu'(x) = 0, \quad (h \text{ is constant})$$

has infinite real roots.

2. Proof of property (4)

Here we intend to prove the property (4) above. Before doing so let us make some preparation. Replacement of complex z by real x in (4.1.1) leads to

$$\frac{\mathrm{d}}{\mathrm{d}x}\left[x J_\nu'(x)\right] + \left(x - \frac{\nu^2}{x}\right) J_\nu(x) = 0. \qquad (4.2.17)$$

Then x is replaced by $z_1 x$, where z_1 is a complex constant.

$$\frac{\mathrm{d}}{\mathrm{d}x}\left[z_1 x J_\nu'(z_1 x)\right] + \left(z_1^2 x - \frac{\nu^2}{x}\right) J_\nu(z_1 x) = 0.$$

Similarly, for another complex z_2, we have

$$\frac{\mathrm{d}}{\mathrm{d}x}\left[z_2 x J_\nu'(z_2 x)\right] + \left(z_2^2 x - \frac{\nu^2}{x}\right) J_\nu(z_2 x) = 0.$$

The two equations are respectively multiplied by $J_\nu(z_2 x)$ and $J_\nu(z_1 x)$, and then one subtracts the other. After that, integration on any interval $[a_1, a_2]$ yields

$$(z_1^2 - z_2^2) \int_0^a x J_\nu(z_1 x) J_\nu(z_2 x) \mathrm{d}x$$

$$= -a \left[J_\nu(z_2 x) z_1 J_\nu'(z_1 x) - J_\nu(z_1 x) z_2 J_\nu'(z_2 x)\right]_{a_1}^{a_2}.$$

$$(4.2.18a)$$

In the case that $a_1 = 0$ and $a_2 = a$, we have

$$(z_1^2 - z_2^2) \int_0^a x \, J_\nu(z_1 x) J_\nu(z_2 x) \mathrm{d}x$$

$$= -a \left[J_\nu(z_2 x) z_1 J_\nu'(z_1 x) - J_\nu(z_1 x) z_2 J_\nu'(z_2 x) \right]_{x=a}.$$

$$(4.2.18\mathrm{b})$$

We begin to prove by reduction *ad absurdum*. Two cases are discussed.

(1) If $z_1 = b + \mathrm{i}c$, $(b \neq 0, c \neq 0)$ is a zero of $J_\nu(z)$, then since the coefficients in (4.1.36) are all real, $z_2 = z_1^* = b - \mathrm{i}c$ is also a zero of $J_\nu(z)$. Substitute z_1 and z_2 into (4.2.18) and let $a = 1$. Since $J_\nu(z_1) = J_\nu(z_2) = 0$, the right hand side is zero. As $z_1 \neq z_2$, $z_1^2 \neq z_2^2$. So, $\int_0^1 x \, J_\nu(z_1 x) J_\nu(z_2 x) \mathrm{d}x = 0$. By assumption, z_1 and z_2, as well as $z_1 x$ and $z_2 x$, are complex conjugates of each other, and so are $J_\nu(z_1 x)$ and $J_\nu(z_2 x)$. Thus, we get $\int_0^1 x |J_\nu(z_1 x)|^2 \mathrm{d}x = 0$. However, the integrand is positive and continuous on $[0, 1]$, so that the integral cannot be zero. The conclusion is that it is impossible for $z_1 = b + \mathrm{i}c$, $(b \neq 0, c \neq 0)$ to be a zero of $J_\nu(z)$.

(2) If $z_1 = \mathrm{i}c$, $(c \neq 0)$ and $z_2 = -\mathrm{i}c$ are zeros of $J_\nu(z)$, then $z_1^2 = z_2^2$. The proof above is not valid. However, by definition,

$$J_\nu(\mathrm{i}c) = (\mathrm{i}c)^\nu \sum_{m=0}^{\infty} \frac{c^{2m}}{m! \Gamma(m + \nu + 1) 2^{2m+\nu}}.$$

Since $\nu > -1$, $m + \nu + 1 > 0$, $(m = 0, 1, 2, \ldots)$. So, $\Gamma(m + \nu + 1) > 0$. The right hand side of the equation does not vanish, which is contrary to $J_\nu(\mathrm{i}c) = 0$.

The final conclusion is that as $\nu > -1$, all zeros of $J_\nu(z)$ are real.

4.2.4.   *Wronskian*

By (3.5.47), the Wronskian of the basic solutions $J_{\pm \nu}(z)$ can be calculated.

$$J_\nu J_{-\nu}' - J_{-\nu} J_\nu' = c \exp\left[-\int \frac{1}{z} \mathrm{d}z\right] = c \exp\left[-\ln z\right] = \frac{c}{z}.$$

To determine the constant c, we only need to know the coefficient of the z^{-1} in $J_\nu J'_{-\nu} - J_{-\nu} J'_\nu$. We put down the expressions of Bessel functions and of their derivative, and then distinguish the terms with powers of z in $J_\nu J'_{-\nu}$ and $J_{-\nu} J'_\nu$. Apparently, $m = 0$ corresponds to the required term.

$$z \left[\frac{1}{\Gamma(\nu+1)} \left(\frac{z}{2}\right)^\nu \frac{1}{2} \frac{(-\nu)}{\Gamma(-\nu+1)} \left(\frac{z}{2}\right)^{-\nu-1} \right.$$

$$\left. - \frac{1}{\Gamma(-\nu+1)} \left(\frac{z}{2}\right)^{-\nu} \frac{1}{2} \frac{\nu}{\Gamma(\nu+1)} \left(\frac{z}{2}\right)^{\nu-1} \right]$$

$$= -\frac{2\nu}{\Gamma(\nu+1)\Gamma(-\nu+1)} = -\frac{2}{\Gamma(\nu)\Gamma(-\nu+1)}.$$

Making use of the formula of Γ function

$$\Gamma(\nu)\Gamma(-\nu+1) = \frac{\pi}{\sin\nu\pi},$$

we obtain

$$J_\nu J'_{-\nu} - J_{-\nu} J'_\nu = -\frac{2}{\pi z} \sin\nu\pi. \qquad (4.2.19)$$

Here caution is needed for ν is not an integer.

If we take $J_\nu(z)$ and $Y_\nu(z)$ as basic solutions, ν can also be integers without problem. By the definition of $Y_v(z)$,

$$J_\nu Y'_\nu - Y_\nu J'_\nu = \frac{1}{\sin\nu\pi} \left[J_\nu(J'_\nu \cos\nu\pi - J'_{-\nu}) - (J_\nu \cos\nu\pi - J_{-\nu})J'_\nu \right]$$

$$= -\frac{1}{\sin\nu\pi}(J_\nu J'_{-\nu} - J_{-\nu} J'_\nu) = \frac{2}{\pi z}. \qquad (4.2.20)$$

One can further prove that

$$\begin{vmatrix} J_\nu(z) & J_{\nu-1}(z) \\ Y_\nu(z) & Y_{\nu-1}(z) \end{vmatrix} = \begin{vmatrix} J_\nu(z) & J'_\nu(z) \\ Y_\nu(z) & Y'_\nu(z) \end{vmatrix} = \frac{2}{\pi z}.$$

Especially, as $\nu = 1$,

$$J_1(z)Y_0(z) - J_0(z)Y_1(z) = \frac{2}{\pi z}.$$

4.3.   Bessel Functions of Integer Orders

Taking $\nu = n$ to be integers in the expression of J_ν, we gain Bessel functions of integer orders.

$$J_n(z) = \left(\frac{z}{2}\right)^n \sum_{k=0}^{\infty} \frac{(-1)^k}{k!\Gamma(n+k+1)} \left(\frac{z}{2}\right)^{2k}. \qquad (4.3.1)$$

Let us prove (4.1.38). Replace n by $-n$ in (4.3.1)

$$J_{-n}(z) = \left(\frac{z}{2}\right)^{-n} \sum_{k=0}^{n-1} \frac{(-1)^k}{k!\Gamma(-n+k+1)} \left(\frac{z}{2}\right)^{2k}$$

$$+ \left(\frac{z}{2}\right)^{-n} \sum_{k=n}^{\infty} \frac{(-1)^k}{k!\Gamma(-n+k+1)} \left(\frac{z}{2}\right)^{2k}.$$

The first term is zero, since as $k < n$, $\Gamma(-n+k+1) = \infty$. In the second term, let $-n+k = l$. Comparing it with (4.3.1) results in

$$J_{-n}(z) = (-1)^n J_n(z). \qquad (4.3.2)$$

From this equation and (4.1.41), we further get

$$Y_{-n}(z) = (-1)^n Y_n(z). \qquad (4.3.3)$$

In (4.2.3), set $\nu = 0$. Then

$$J_0'(z) = -J_1(z). \qquad (4.3.4)$$

In (4.2.2), set $\nu = 1$. Then

$$(zJ_1(z))' = zJ_0(z). \qquad (4.3.5)$$

If $z = x$ is real, this equation is integrated to be

$$xJ_1(x) = \int_0^x \xi J_0(\xi)\mathrm{d}\xi. \qquad (4.3.6)$$

4.3.1.   *Parity and the values at certain points*

1. Parity

It follows from (4.2.1) that

$$J_n(-z) = (-1)^n J_n(z). \qquad (4.3.7)$$

This means that the parity of $J_n(z)$ of integer orders depends on if the integer n is odd or even. It is seen that as n is odd (even), $J_n(z)$ contains odd (even) powers of z.

2. The value as $z \to 0$

$$J_0(0) = 1; \quad J_n(0) = 0, \quad n \geq 1. \tag{4.3.8a}$$

By (4.1.40), as $n = 0$,

$$Y_0(z \to 0) \sim \frac{2}{\pi} \ln \frac{z}{2} \to -\infty. \tag{4.3.8b}$$

As $n > 0$,

$$Y_n(z \to 0) \sim -\frac{(n-1)!}{\pi} \left(\frac{2}{z}\right)^n. \tag{4.3.8c}$$

4.3.2.   *Generating function of Bessel functions of integer orders*

1. Generating function

The generating function of $J_n(z)$ of integer order is $e^{z(t-1/t)/2}$:

$$e^{z(t-1/t)/2} = \sum_{n=-\infty}^{\infty} J_n(z)t^n = \cdots + J_{-1}(z)t^{-1} + J_0(z)t^0 + J_1(z)t$$

$$+ J_2(z)t^2 + \cdots. \tag{4.3.9}$$

We prove this equation. Making use of the power series of the exponential function, we write down the Taylor expansions of functions $e^{zt/2}$ and $e^{-z/2t}$. Multiplying these two equations produces

$$e^{zt/2}e^{-z/2t} = \sum_{l=0}^{\infty} \frac{1}{l!} \left(\frac{zt}{2}\right)^l \sum_{m=0}^{\infty} \frac{1}{m!} \left(-\frac{z}{2t}\right)^m$$

$$= \sum_{m=0}^{\infty} \sum_{l=0}^{\infty} \frac{(-1)^m}{m!l!} \left(\frac{z}{2}\right)^{l+m} t^{l-m},$$

where in the last step the terms are rearranged according to powers of t. Let $l - m = n$ and the summation of l is transferred to that of

n. Note that both l and m vary from 0 to ∞, while n has to vary from $-\infty$ to ∞. (If n varies from $-m$ to ∞, the terms from $-\infty$ to $-(m+1)$ will be missed.)

$$e^{zt/2}e^{-z/2t} = \sum_{n=-\infty}^{\infty} \sum_{m=0}^{\infty} \frac{(-1)^m}{m!(n+m)!} \left(\frac{z}{2}\right)^{n+2m} t^n = \sum_{n=-\infty}^{\infty} J_n(z)\, t^n.$$

Because the summation in (4.3.9) covers from $-\infty$ and $+\infty$, this generating function was not listed in Table 3.5.

Letting $t = 1$ in (4.3.9) gives

$$\sum_{n=-\infty}^{\infty} J_n(z) = 1. \qquad (4.3.10a)$$

Employing (4.3.2) we further get

$$J_0(z) + 2\sum_{n=1}^{\infty} J_{2n}(z) = 1. \qquad (4.3.10b)$$

Replacement of t by $-t$ in (4.3.9) gives

$$e^{-z(t-1/t)/2} = \sum_{n=-\infty}^{\infty} J_n(z)(-1)^n t^n.$$

If this equation is multiplied with (4.3.9),

$$1 = \sum_{n=-\infty}^{\infty} J_n(z)t^n \sum_{m=-\infty}^{\infty} J_m(z)(-1)^m t^m$$

$$= \sum_{n=-\infty}^{\infty} t^n \sum_{m=-\infty}^{\infty} (-1)^m J_m(z) J_{n-m}(z) \qquad (4.3.11)$$

Comparing the constant terms on the two sides of (4.3.11) and making use of (4.3.7), we get

$$\sum_{n=-\infty}^{\infty} J_n^2(z) = J_0^2(z) + 2\sum_{n=1}^{\infty} J_n^2(z) = 1.$$

This equation reveals that if $z = x$ is real, then

$$|J_0(x)| \le 1, \quad |J_n(x)| \le \frac{1}{\sqrt{2}}, \quad (n = 1, 2, \dots).$$

Comparison of the coefficients of nonzero powers of t in (4.3.11) leads to

$$\sum_{m=-\infty}^{\infty} (-1)^m J_n(z) J_{n-m}(z) = 0.$$

2. Addition formula

The addition formula of Bessel functions is

$$J_n(z_1 + z_2) = \sum_{m=-\infty}^{\infty} J_{n-m}(z_1) J_m(z_2). \qquad (4.3.12a)$$

It is proved by use of the generating function.

$$\exp\left[(z_1 + z_2)\frac{1}{2}\left(t - \frac{1}{t}\right)\right]$$

$$= \exp\left[z_1\frac{1}{2}\left(t - \frac{1}{t}\right)\right] \exp\left[z_2\frac{1}{2}\left(t - \frac{1}{t}\right)\right].$$

The two sides of this equation are expanded by Bessel functions:

$$\sum_{n=-\infty}^{\infty} J_n(z_1 + z_2)t^n = \sum_{m=-\infty}^{\infty} J_m(z_1)t^m \sum_{n=-\infty}^{\infty} J_n(z_2)t^n.$$

Taking the coefficients of the same power of t on both sides, we obtain (4.3.12a). Particularly, as $n = 0$,

$$J_0(z_1 + z_2) = \sum_{m=-\infty}^{\infty} J_{-m}(z_1) J_m(z_2)$$

$$= \sum_{m=-\infty}^{\infty} (-1)^m J_m(z_1) J_m(z_2). \qquad (4.3.12b)$$

Let $z_1 = z_2$. The following formula is called doubleness formula:

$$J_0(2z) = \sum_{m=-\infty}^{\infty} (-1)^m [J_m(z)]^2$$

$$= [J_0(z)]^2 + 2 \sum_{m=1}^{\infty} (-1)^m [J_m(z)]^2. \qquad (4.3.13)$$

There is another important addition formula. Suppose that there are two position vectors r_1 and r_2, both starting from the origin and ending at points P_1 and P_2, respectively. The angel between them is θ. Then the distance between points P_1 and P_2 is $R = |r_1 - r_2| = \sqrt{r_1^2 + r_2^2 - 2r_1r_2\cos\theta}$. We have

$$J_0(kR) = \sum_{m=-\infty}^{\infty} J_m(kr_1)J_m(kr_2)e^{im\theta}$$

$$= J_0(kr_1)J_0(kr_2) + 2\sum_{m=1}^{\infty} (-1)^m J_m(kr_1)J_m(kr_2)\cos m\theta.$$

This equation holds even when r_1, r_2 and θ are complex numbers.

3. Integral expressions

The contour integral expression of $J_n(z)$ of integer orders can be derived from their generating function as follows:

$$J_m(z) = \frac{1}{2\pi i}\oint \frac{e^{z\zeta/2 - z/2\zeta}}{\zeta^{m+1}}d\zeta, \qquad (4.3.14)$$

where the contour includes the origin. Let $t = e^{i\varphi}$ in (4.3.9). We have

$$\exp\left[z(e^{i\varphi} - e^{-i\varphi})/2\right] = e^{iz\sin\varphi} = \sum_{n=-\infty}^{\infty} J_n(z)e^{in\varphi}. \qquad (4.3.15)$$

Multiply $e^{-im\varphi}$ and then integrate with respect to φ on $[-\pi, \pi]$. We achieve

$$J_n(z) = \frac{1}{2\pi}\int_{-\pi}^{\pi} e^{i(z\sin\varphi - n\varphi)}d\varphi. \qquad (4.3.16)$$

Using Euler formula $e^{iz} = \cos z + i\sin z$, we have

$$J_n(z) = \frac{1}{2\pi}\int_{-\pi}^{\pi} \left[\cos(z\sin\varphi - n\varphi) + i\sin(z\sin\varphi - n\varphi)\right]d\varphi.$$

$$(4.3.17)$$

The second term in the integrand is an odd function of φ so that its integration on $[-\pi, \pi]$ vanishes. The first term is an even function of φ. So, the integral expression is

$$J_n(z) = \frac{1}{\pi}\int_0^{\pi} \cos(z\sin\varphi - n\varphi)d\varphi. \qquad (4.3.18)$$

This is called the **Bessel expression**.

4. Relations between Bessel functions and trigonometric functions

Taking $n = 0$ in (4.3.18) leads to

$$J_0(z) = \frac{1}{\pi} \int_0^\pi \cos(z \sin \varphi) d\varphi. \tag{4.3.19}$$

By (4.3.15) and (4.1.38), we get

$$e^{iz \sin \varphi} = J_0(z) + \sum_{n=1}^\infty (J_n(z)e^{in\varphi} + (-1)^n J_n(z)e^{-in\varphi})$$

$$= J_0(z) + \sum_{n=1}^\infty J_{2n}(z)(e^{i2n\varphi} + e^{-i2n\varphi})$$

$$+ \sum_{n=0}^\infty J_{2n+1}(z)(e^{i(2n+1)\varphi} - e^{-i(2n+1)\varphi})$$

$$= J_0(z) + \sum_{n=1}^\infty J_{2n}(z)2\cos 2n\varphi + \sum_{n=0}^\infty J_{2n+1}(z)2i\sin(2n+1)\varphi \tag{4.3.20a}$$

and

$$e^{iz \cos \varphi} = e^{iz \sin(\pi/2 - \varphi)}$$

$$= J_0(z) + \sum_{n=1}^\infty \left(J_n(z)e^{in(\pi/2-\varphi)} + J_{-n}(z)e^{-in(\pi/2-\varphi)} \right)$$

$$= J_0(z) + \sum_{n=1}^\infty \left(J_n(z)i^n e^{-in\varphi} + J_{-n}(z)(-1)^n e^{in\varphi} \right)$$

$$= J_0(z) + \sum_{n=1}^\infty i^n J_n(z)2\cos n\varphi. \tag{4.3.20b}$$

In these two equations, let $z \to -z$. Then,

$$e^{-iz \cos \varphi} = J_0(z) + \sum_{n=1}^\infty (-i)^n J_n(z)2\cos n\varphi \tag{4.3.21a}$$

and

$$e^{-iz\sin\varphi} = J_0(z) + \sum_{n=1}^{\infty} J_{2n}(z)2\cos 2n\varphi$$

$$- \sum_{n=0}^{\infty} J_{2n+1}(z)2i\sin(2n+1)\varphi. \qquad (4.3.21b)$$

These generate the following four formulas.

$$\cos(z\cos\varphi) = \frac{1}{2}(e^{iz\cos\varphi} + e^{-iz\cos\varphi})$$

$$= J_0(z) + \sum_{n=1}^{\infty} i^n\left[1+(-1)^n\right] J_n(z)\cos n\varphi$$

$$= J_0(z) + 2\sum_{n=1}^{\infty} (-1)^n J_{2n}(z)\cos 2n\varphi. \qquad (4.3.22a)$$

$$\sin(z\cos\varphi) = \frac{1}{2i}(e^{iz\cos\varphi} - e^{-iz\cos\varphi})$$

$$= 2\sum_{n=0}^{\infty} (-1)^n J_{2n+1}(z)\cos(2n+1)\varphi. \qquad (4.3.22b)$$

$$\cos(z\sin\varphi) = \frac{1}{2}(e^{iz\sin\varphi} + e^{-iz\sin\varphi})$$

$$= J_0(z) + 2\sum_{n=1}^{\infty} J_{2n}(z)\cos 2n\varphi. \qquad (4.3.23a)$$

$$\sin(z\sin\varphi) = \frac{1}{2i}(e^{iz\sin\varphi} - e^{-iz\sin\varphi})$$

$$= 2\sum_{n=0}^{\infty} J_{2n+1}(z)\sin(2n+1)\varphi. \qquad (4.3.23b)$$

As $\varphi = 0$, we get

$$\cos z = J_0(z) + 2\sum_{n=1}^{\infty} (-1)^n J_{2n}(z) \qquad (4.3.24a)$$

and

$$\sin z = 2\sum_{n=0}^{\infty} (-1)^n J_{2n+1}(z). \qquad (4.3.24b)$$

These are the important formulas connecting Bessel functions of integer orders and trigonometric functions.

Physically, trigonometric functions represent plane waves. Bessel functions represent the solutions of radial Helmholtz equations in cylindrical coordinates, so that they represent cylindrical waves. The relationships between these two kinds of functions mean that plane waves can be expanded by cylindrical waves. Such expansions are very useful in studying scattering of electromagnetic waves.

The Bessel function of zero order can be the generating function of Laguerre polynomials:

$$J_0(2\sqrt{xt}) = e^{-t} \sum_{n=0}^{\infty} \frac{t^n}{n!^2} L_n(x).$$

4.4. Bessel Functions of Half-Integer Orders

Bessel functions with half-integer orders, $\nu = n + 1/2$ where n is integer, can be simplified to be elementary functions. This can be viewed from two aspects.

On one hand, for a Bessel equation of $\nu = 1/2$,

$$z^2 w'' + z w' + \left(z^2 - \frac{1}{4}\right) w = 0, \qquad (4.4.1)$$

we let $w(z) = u(z)/\sqrt{z}$ and insert it into (4.4.1). This results in $u''(z) + u(z) = 0$. Two special solutions of this equation are $u(z) = \sin z$ and $u(z) = \cos z$. Thus, the two independent solutions of (4.4.1) are $u(z) = \frac{\sin z}{\sqrt{z}}$ and $u(z) = \frac{\cos z}{\sqrt{z}}$.

On the other hand, recall the definition of Bessel functions (4.1.36),

$$J_{1/2}(z) = \sum_{m=0}^{\infty} \frac{(-1)^m}{m!\,\Gamma(m + 1/2 + 1)} \left(\frac{z}{2}\right)^{2m+1/2}. \qquad (4.4.2)$$

The properties of Γ functions are $\Gamma(k + \nu + 1) = (k + \nu)(k + \nu - 1)\cdots(1 + \nu)\Gamma(1 + \nu)$ and $\Gamma(1/2) = \sqrt{\pi}$. The summation in (4.4.2) can be carried out:

$$J_{1/2}(z) = \left(\frac{z}{2}\right)^{1/2} \sum_{m=0}^{\infty} \frac{(-1)^m z^{2m}/2^{2m}}{m!\,\left(m + \frac{1}{2}\right)\left(m - \frac{1}{2}\right)\left(m - \frac{3}{2}\right)\cdots 3 \cdot 1\Gamma(1/2)}$$

$$= \frac{\sqrt{z}}{\sqrt{2\pi}} \sum_{m=0}^{\infty} \frac{(-1)^m z^{2m}}{(2m)!!(2m+1)!!}$$

$$= \sqrt{\frac{2}{\pi z}} \sum_{m=0}^{\infty} \frac{(-1)^m z^{2m+1}}{(2m+1)!} = \sqrt{\frac{2}{\pi z}} \sin z. \tag{4.4.3}$$

Similarly,

$$J_{-1/2}(z) = \sum_{m=0}^{\infty} \frac{(-1)^m}{m!\Gamma(m+1/2)} \left(\frac{z}{2}\right)^{2m-1/2}$$

$$= \left(\frac{z}{2}\right)^{-1/2} \sum_{m=0}^{\infty} \frac{(-1)^m z^{2m}/2^{2m}}{m! \left(m - \frac{1}{2}\right)\left(m - \frac{3}{2}\right) \cdots 3 \cdot 1\Gamma(1/2)}$$

$$= \frac{\sqrt{2}}{\sqrt{\pi z}} \sum_{k=0}^{\infty} \frac{(-1)^m z^{2m}}{(2m)!!(2m-1)!!} = \sqrt{\frac{2}{\pi z}} \sum_{k=0}^{\infty} \frac{(-1)^m}{(2m)!} z^{2m}$$

$$= \sqrt{\frac{2}{\pi z}} \cos z. \tag{4.4.4}$$

From the recursion formulas of Bessel functions, it is easy to derive expressions of $J_{n\pm1/2}$ with $n \geq 1$.

As $|z| < 1$, the following approximation expressions are useful.

$$J_{3/2}(z) \approx -\sqrt{\frac{2}{\pi z}} \cos z, \quad J_{-3/2}(z) \approx -\sqrt{\frac{2}{\pi z}} \sin z,$$

$$J_{5/2}(z) \approx -\sqrt{\frac{2}{\pi z}} \sin z, \quad J_{-5/2}(z) \approx -\sqrt{\frac{2}{\pi z}} \cos z.$$

By employing the recursion formulas of Bessel functions, we get the differential expressions as follows:

$$J_{n+1/2} = (-1)^n z^{n+1/2} \left(\frac{1}{z}\frac{d}{dz}\right)^n \left(z^{-1/2} J_{1/2}\right)$$

$$= (-1)^n \sqrt{\frac{2}{\pi z}} z^{n+1} \left(\frac{1}{z}\frac{d}{dz}\right)^n \frac{\sin z}{z} \tag{4.4.5a}$$

and

$$J_{-n-1/2} = \sqrt{\frac{2}{\pi z}} z^{n+1} \left(\frac{1}{z}\frac{d}{dz}\right)^n \frac{\cos z}{z}. \tag{4.4.5b}$$

From the definition of Bessel functions of the second kind (4.1.42), the relations between $Y_{n+1/2}$ and $J_{n+1/2}$ are

$$Y_{n+1/2}(z) = \frac{J_{n+1/2}(z)\cos(\pi(n+1/2)) - J_{-n-1/2}(z)}{\sin(\pi(n+1/2))}$$

$$= (-1)^{n+1} J_{-n-1/2}(z) \tag{4.4.6a}$$

and

$$Y_{-n-1/2}(z) = (-1)^n J_{n+1/2}(z). \tag{4.4.6b}$$

The generating functions of the Bessel functions of half-integer orders have been listed in Table 3.6.

The other formulas concerning Bessel functions of half-integer orders can be obtained by letting ν be half-integers in the formulas of Bessel functions.

It is easily seen that as $z \to \infty$, every Bessel function of half-integer order decays in an oscillating way. The asymptotic formulas are as follows.

$$J_{\nu+1/2}(z \to \infty) \sim \sqrt{\frac{2}{\pi z}} \cos\left(z - \frac{\nu+1}{2}\pi\right). \tag{4.4.7a}$$

$$Y_{\nu+1/2}(z \to \infty) \sim \sqrt{\frac{2}{\pi z}} \sin\left(z - \frac{\nu+1}{2}\pi\right). \tag{4.4.7b}$$

4.5. Bessel Functions of the Third Kind and Spherical Bessel Functions

4.5.1. *Bessel functions of the third kind*

1. Definition

Bessel functions of the third kind are defined by

$$H_\nu^{(1)}(z) = J_\nu(z) + iY_\nu(z) \tag{4.5.1a}$$

and

$$H_\nu^{(2)}(z) = J_\nu(z) - iY_\nu(z). \tag{4.5.1b}$$

They are also called **Hankel functions of the first and second kind**, respectively.

Because a Hankel function is the linear combination of the linearly independent solutions $J_v(z)$ and $Y_v(z)$ of Eq. (4.1.1), $H_v^{(1)}(z)$ and $H_v^{(2)}(z)$ are also two independent solutions of Bessel equation (4.1.1).

Since Hankel functions are the linear superposition of $J_v(z)$ and $Y_v(z)$, they obey the same recursion formulas as those of $J_v(z)$ and $Y_v(z)$. If in (4.2.2)–(4.2.5), J is replaced by $H_v^{(1)}$ or $H_v^{(2)}$, the formulas remain valid. Explicitly,

$$\begin{cases} \left[z^v H_v^{(i)}(z)\right]' = z^v H_{v-1}^{(i)}(z), \\[2mm] \left[z^{-v} H_v^{(i)}(z)\right]' = -z^{-v} H_{v+1}^{(i)}(z), \\[2mm] z\left[H_{v-1}^{(i)}(z) + H_{v+1}^{(i)}(z)\right] = 2v H_v^{(i)}(z), \\[2mm] H_{v-1}^{(i)}(z) - H_{v+1}^{(i)}(z) = 2[H_v^{(i)}(z)]'. \end{cases} \qquad (4.5.2)$$

In these equations, $i = 1, 2$.

Usually, Bessel functions of the first, second and third kinds are uniformly called **cylindrical functions**. Cylindrical functions have the same recursion formulas (4.2.2)–(4.2.5). Conversely, the functions satisfying recursion formulas (4.2.2)–(4.2.5) can be defined as cylindrical functions.

When the expression of $Y_v(z)$ (4.1.42) is substituted into (4.5.1), we get

$$H_v^{(1)}(z) = J_v(z) + i\frac{J_v(z)\cos v\pi - J_{-v}(z)}{\sin v\pi}$$

$$= \frac{1}{i\sin v\pi}\left[J_{-v}(z) - e^{-iv\pi} J_v(z)\right] \qquad (4.5.3a)$$

and

$$H_v^{(2)}(z) = J_v(z) - i\frac{J_v(z)\cos v\pi - J_{-v}(z)}{\sin v\pi}$$

$$= \frac{1}{i\sin v\pi}\left[e^{iv\pi} J_v(z) - J_{-v}(z)\right]. \qquad (4.5.3b)$$

Then two important formulas follow:

$$H_{-v}^{(1)}(z) = \frac{J_v(z) - e^{iv\pi} J_{-v}(z)}{-i\sin v\pi}$$

$$= \frac{e^{iv\pi}}{i\sin v\pi}\left[J_{-v}(z) - e^{-iv\pi} J_v(z)\right] = e^{iv\pi} H_v^{(1)}(z) \qquad (4.5.4a)$$

and

$$H_{-\nu}^{(2)}(z) = \frac{e^{-i\nu\pi} J_{-\nu}(z) - J_\nu(z)}{-i\sin\nu\pi} = e^{-i\nu\pi} H_\nu^{(2)}(z). \quad (4.5.4b)$$

2. Hankel functions of integer orders

As $\nu = n$ is integer, Hankel functions are

$$H_n^{(1)}(z) = J_n(z) + iY_n(z) \quad (4.5.5a)$$

and

$$H_n^{(2)}(z) = J_n(z) - iY_n(z). \quad (4.5.5b)$$

When $n = 0$, there are integral expressions for Hankel functions as $z = x, |\text{Re}\,\nu| < 1$:

$$H_\nu^{(1)}(x) = \frac{2e^{-i\nu\pi/2}}{\pi i} \int_0^\infty e^{ix\cosh t} \cosh(\nu t)dt$$

and

$$H_\nu^{(2)}(x) = -\frac{2e^{i\nu\pi/2}}{\pi i} \int_0^\infty e^{-ix\cosh t} \cosh(\nu t)dt.$$

From these two expressions one immediately sees that

$$H_0^{(1)}(-x) = -H_0^{(2)}(x). \quad (4.5.6)$$

This also valid for complex argument:

$$H_0^{(1)}(-z) = -H_0^{(2)}(z)$$

3. Hankel functions of half-integer orders

When ν is half-integer, Hankel functions degrade to elementary ones. For instance, let $\nu = 1/2$ in (4.5.1). Then

$$H_{1/2}^{(1)}(z) = J_{1/2}(z) + iY_{1/2}(z)$$

$$= \sqrt{\frac{2}{\pi z}}\sin z - i\sqrt{\frac{2}{\pi z}}\cos z = -i\sqrt{\frac{2}{\pi z}}e^{iz} \quad (4.5.7a)$$

and

$$H^{(2)}_{1/2}(z) = J_{1/2}(z) - iY_{1/2}(z)$$

$$= \sqrt{\frac{2}{\pi z}} \sin z + i\sqrt{\frac{2}{\pi z}} \cos z = i\sqrt{\frac{2}{\pi z}} e^{-iz}. \qquad (4.5.7b)$$

By use of (4.4.4) and (4.4.6), $Y_{-1/2}(z) = -J_{1/2}(z)$, we have

$$H^{(1)}_{-1/2}(z) = J_{-1/2}(z) + iY_{-1/2}(z)$$

$$= \sqrt{\frac{2}{\pi z}} \cos z - i\sqrt{\frac{2}{\pi z}} \sin z = \sqrt{\frac{2}{\pi z}} e^{-iz}$$

and

$$H^{(2)}_{-1/2}(z) = J_{-1/2}(z) - iY_{-1/2}(z)$$

$$= \sqrt{\frac{2}{\pi z}} \cos z + i\sqrt{\frac{2}{\pi z}} \sin z = \sqrt{\frac{2}{\pi z}} e^{iz}.$$

So, $H^{(1)}_{-1/2}(z) = -iH^{(2)}_{1/2}(z)$ and $H^{(2)}_{-1/2}(z) = iH^{(1)}_{1/2}(z)$.

By (4.4.6), Hankel functions of half-integer orders can be expressed by Bessel functions of half-integer orders:

$$H^{(1)}_{n+1/2}(z) = J_{n+1/2}(z) + i(-1)^{n+1} J_{-(n+1/2)}(z) \qquad (4.5.8a)$$

and

$$H^{(2)}_{n+1/2}(z) = J_{n+1/2}(z) + i(-1)^{n} J_{-(n+1/2)}(z). \qquad (4.5.8b)$$

Further, by (4.4.5), the differential expressions of Hankel function of half-integer orders can be put down:

$$H^{(1)}_{n+1/2}(z) = (-1)^{n}\sqrt{\frac{2}{\pi z}} z^{n+1} \left(\frac{1}{z}\frac{d}{dz}\right)^{n} \frac{\sin z}{z}$$

$$+ i(-1)^{n+1}\sqrt{\frac{2}{\pi z}} z^{n+1} \left(\frac{1}{z}\frac{d}{dz}\right)^{n} \frac{\cos z}{z}$$

$$= i(-1)^{n+1}\sqrt{\frac{2}{\pi z}} z^{n+1} \left(\frac{1}{z}\frac{d}{dz}\right)^{n} \frac{e^{iz}}{z} \qquad (4.5.9a)$$

and

$$H^{(2)}_{n+1/2}(z) = (-1)^n \sqrt{\frac{2}{\pi z}} z^{n+1} \left(\frac{1}{z}\frac{d}{dz}\right)^n \frac{\sin z}{z}$$

$$-i(-1)^{n+1} \sqrt{\frac{2}{\pi z}} z^{n+1} \left(\frac{1}{z}\frac{d}{dz}\right)^n \frac{\cos z}{z}$$

$$= i(-1)^n \sqrt{\frac{2}{\pi z}} z^{n+1} \left(\frac{1}{z}\frac{d}{dz}\right)^n \frac{e^{-iz}}{z}. \qquad (4.5.9b)$$

4. Asymptotic formulas of Hankel functions

The asymptotic formulas come from the definition of Hankel functions (4.5.1) and the asymptotic forms of Bessel functions of the first and second kinds.

As $z \to \infty$ and $\nu \neq 0$, from (4.2.12) we get

$$H^{(1)}_\nu(z \to \infty) \sim \sqrt{\frac{2}{\pi z}} e^{i\left(z - \frac{\nu\pi}{2} - \frac{\pi}{4}\right)} \qquad (4.5.10a)$$

and from (4.2.13) we get

$$H^{(2)}_\nu(z \to \infty) \sim \sqrt{\frac{2}{\pi z}} e^{-i\left(z - \frac{\nu\pi}{2} - \frac{\pi}{4}\right)}. \qquad (4.5.10b)$$

Obviously, when $z = x$ is real and goes to infinity, $H^{(1)}_\nu(x \to \infty)$ and $H^{(2)}_\nu(x \to \infty)$ behave as two decaying plane waves propagating along opposite directions.

As $\nu = n > 0$ is integer and $z \to 0$, the following asymptotic formulas can be derived from (4.3.8):

$$H^{(1)}_n(z \to 0) \sim -i\frac{(n-1)!}{\pi}\left(\frac{2}{z}\right)^n \qquad (4.5.10c)$$

and

$$H^{(2)}_n(z \to 0) \sim i\frac{(n-1)!}{\pi}\left(\frac{2}{z}\right)^n. \qquad (4.5.10d)$$

The case of $\nu = 0$ is special:

$$H^{(1)}_0(z \to 0) \sim i\frac{2}{\pi}\ln\frac{z}{2} \qquad (4.5.10e)$$

and

$$H_0^{(2)}(z \to 0) \sim -i\frac{2}{\pi}\ln\frac{z}{2}. \qquad (4.5.10f)$$

When $z = x$ is real, $H_0^{(1)}(x \to 0) \sim i\frac{2}{\pi}\ln\frac{|x|}{2}$ and $H_0^{(2)}(x \to 0) \sim -i\frac{2}{\pi}\ln\frac{|x|}{2}$, agreeing to (4.5.6).

4.5.2. *Spherical Bessel functions*

1. Definition

The following equation is called **spherical Bessel equation of ν order**:

$$z^2 w'' + 2zw' + (z^2 - \nu(\nu+1))w = 0. \qquad (4.5.11)$$

It seems slightly different from Bessel equation. However, the transformation $w(z) = u(z)/\sqrt{z}$ can reform the equation to be

$$z^2 u'' + zu' + \left(z^2 - \left(\nu+\frac{1}{2}\right)^2\right)u(z) = 0. \qquad (4.5.12)$$

This is a Bessel equation of order $\nu+1/2$. Its basic solutions are Bessel functions of the first and second kinds $J_{\nu+1/2}(z)$ and $Y_{\nu+1/2}(z)$. So, the basic solutions of (4.5.11) are $\frac{1}{\sqrt{z}}J_{\nu+1/2}(z)$ and $\frac{1}{\sqrt{z}}Y_{\nu+1/2}(z)$.

Divide Bessel functions of the first, second and third kinds by $\sqrt{z}$. The resultants are uniformly called **spherical Bessel functions**. Explicitly, their denotations and corresponding names are as follows.

$$j_\nu(z) = \sqrt{\frac{\pi}{2z}}J_{\nu+1/2}(z), \qquad (4.5.13a)$$

named as **spherical Bessel function of order ν**.

$$y_\nu(z) = \sqrt{\frac{\pi}{2z}}Y_{\nu+1/2}(z), \qquad (4.5.13b)$$

named as **spherical Neumann function of order ν**.

$$h_\nu^{(1)}(z) = \sqrt{\frac{\pi}{2z}}H_{\nu+1/2}^{(1)}(z), \qquad (4.5.13c)$$

named as **spherical Hankel function of the first kind of order** ν.

$$h_\nu^{(2)}(z) = \sqrt{\frac{\pi}{2z}} H_{\nu+1/2}^{(2)}(z), \qquad (4.5.13d)$$

named as **spherical Hankel function of the second kind of order** ν.

All of them are solutions of spherical Bessel equation (4.5.11).

By the definitions, the relations between spherical Bessel functions are

$$h_\nu^{(1)}(z) = j_\nu(z) + iy_\nu(z) \qquad (4.5.14a)$$

and

$$h_\nu^{(2)}(z) = j_\nu(z) - iy_\nu(z). \qquad (4.5.14b)$$

Any two functions chosen from (4.5.13) are linearly independent of each other. Hence, the general solution of spherical Bessel equation can be any of the following:

$$w(z) = Aj_\nu(z) + By_\nu(z) \qquad (4.5.15)$$

or

$$w(z) = Aj_\nu(z) + Bh_\nu^{(1)}(z) \qquad (4.5.16)$$

or

$$w(z) = Ah_\nu^{(1)}(z) + Bh_\nu^{(2)}(z) \qquad (4.5.17)$$

A spherical Bessel equation is the radial equation in spherical coordinates, so that spherical Bessel functions physically represent spherical waves.

Since the spherical functions are expressed in terms of the three kinds of Bessel functions, the properties of the former can be deduced from those of the latter. In the following, we present some of the properties.

2. Recursion formulas of spherical Bessel functions

The recursion formulas of spherical functions come from those of Bessel functions. Let ψ_ν denote any of the four functions $j_\nu(z)$, $n_\nu(z)$,

$h_\nu^{(1)}(z)$ and $h_\nu^{(2)}(z)$ defined in (4.5.13). The recursion formulas are as follows.

$$(2\nu + 1)\psi_\nu = z(\psi_{\nu+1} + \psi_{\nu-1}). \tag{4.5.18}$$

$$(2\nu + 1)\psi_\nu' = \nu\psi_{\nu+1} - (\nu + 1)\psi_{\nu+1}. \tag{4.5.19}$$

$$\psi_\nu' = \frac{\nu}{z}\psi_\nu - \psi_{\nu+1}. \tag{4.5.20}$$

$$\psi_\nu' = \psi_{\nu-1} - \frac{\nu + 1}{z}\psi_\nu. \tag{4.5.21}$$

For example, in the recursion formula of Bessel functions $(z^{-\nu}J_\nu)' = -z^{-\nu}J_{\nu+1}$, replacing ν by $\nu + 1/2$ will lead to $(z^{-\nu-1/2}J_{\nu+1/2})' = -z^{-\nu-1/2}J_{\nu+1+1/2}$, so that $\frac{1}{z}\frac{d}{dz}\frac{j_\nu(z)}{z^\nu} = -\frac{j_{\nu+1}(z)}{z^{\nu+1}}$. This becomes the recursion formula of spherical Bessel functions. Similarly, one can get that of spherical Neumann functions: $\frac{n_{\nu+1}(z)}{z^{\nu+1}} = -\frac{1}{z}\frac{d}{dz}[\frac{n_\nu(z)}{z^\nu}]$. Both of them belong to (4.5.20).

3. Expressions of spherical Bessel functions of integer orders

It has been known from Section 4.4 that when $\nu = n$, the three kinds of Bessel functions of half-integer orders degraded to elementary functions. Consequently, spherical Bessel functions can also be expressed by elementary functions. Starting from (4.4.5), we derive the differential expressions of the spherical Bessel and Neumann functions of positive integer orders.

$$j_n(z) = \sqrt{\frac{\pi}{2z}}J_{n+1/2} = \sqrt{\frac{\pi}{2z}}(-1)^n\sqrt{\frac{2}{\pi z}}z^{n+1}\left(\frac{1}{z}\frac{d}{dz}\right)^n\frac{\sin z}{z}$$

$$= (-1)^n z^n\left(\frac{1}{z}\frac{d}{dz}\right)^n\frac{\sin z}{z}. \tag{4.5.22a}$$

$$y_n(z) = \sqrt{\frac{\pi}{2z}}Y_{n+1/2}(z) = \sqrt{\frac{\pi}{2z}}(-1)^{n+1}J_{-n-1/2}$$

$$= \sqrt{\frac{\pi}{2z}}(-1)^{n+1}\sqrt{\frac{2}{\pi z}}z^{n+1}\left(\frac{1}{z}\frac{d}{dz}\right)^n\frac{\cos z}{z}$$

$$= (-1)^{n+1}z^n\left(\frac{1}{z}\frac{d}{dz}\right)^n\frac{\cos z}{z}. \tag{4.5.22b}$$

Starting from (4.5.9), we derive the differential expressions of the spherical Hankel functions of positive integer orders.

$$h_n^{(1)}(z) = \sqrt{\frac{\pi}{2z}} H_{n+1/2}^{(1)}(z) = \sqrt{\frac{\pi}{2z}} i(-1)^{n+1} \sqrt{\frac{2}{\pi z}} z^{n+1} \left(\frac{1}{z}\frac{d}{dz}\right)^n \frac{e^{iz}}{z}$$

$$= i(-1)^{n+1} z^n \left(\frac{1}{z}\frac{d}{dz}\right)^n \frac{e^{iz}}{z}. \tag{4.5.23a}$$

$$h_n^{(2)}(z) = \sqrt{\frac{\pi}{2z}} H_{n+1/2}^{(2)}(z) = \sqrt{\frac{\pi}{2z}} i(-1)^n \sqrt{\frac{2}{\pi z}} z^{n+1} \left(\frac{1}{z}\frac{d}{dz}\right)^n \frac{e^{-iz}}{z}$$

$$= i(-1)^n z^n \left(\frac{1}{z}\frac{d}{dz}\right)^n \frac{e^{-iz}}{z}. \tag{4.5.23b}$$

The simplest cases $n = 0$ of the above formulas are listed explicitly as follows.

$$j_0(z) = \sqrt{\frac{\pi}{2z}} J_{1/2}(z) = \frac{\sin z}{z}. \tag{4.5.24a}$$

$$y_0(z) = \sqrt{\frac{\pi}{2z}} Y_{1/2}(z) = -\frac{\cos z}{z}. \tag{4.5.24b}$$

$$h_0^{(1)}(z) = -\frac{i}{z} e^{iz}. \tag{4.5.24c}$$

$$h_0^{(2)}(z) = \frac{i}{z} e^{-iz}. \tag{4.5.24d}$$

For spherical Bessel functions of negative integer orders, their expressions can be derived from the formulas previously given. From (4.4.5b), we have

$$j_{-n}(z) = \sqrt{\frac{\pi}{2z}} J_{-n+1/2}(z) = \sqrt{\frac{\pi}{2z}} \sqrt{\frac{2}{\pi z}} z^n \left(\frac{1}{z}\frac{d}{dz}\right)^{n-1} \frac{\cos z}{z}$$

$$= z^{n-1} \left(\frac{1}{z}\frac{d}{dz}\right)^{n-1} \frac{\cos z}{z} = (-1)^n y_{n-1}(z), \tag{4.5.25a}$$

where the last equal sign comes from (4.5.22b). From (4.4.6b), we have

$$y_{-n}(z) = \sqrt{\frac{\pi}{2z}} Y_{-n+1/2}(z) = \sqrt{\frac{\pi}{2z}} (-1)^{n-1} J_{n-1+1/2}(z)$$

$$= \sqrt{\frac{\pi}{2z}} (-1)^{n-1} (-1)^{n-1} \sqrt{\frac{2}{\pi z}} z^n \left(\frac{1}{z}\frac{d}{dz}\right)^{n-1} \frac{\sin z}{z}$$

$$= z^{n-1} \left(\frac{1}{z}\frac{d}{dz}\right)^{n-1} \frac{\sin z}{z} = (-1)^{n-1} j_{n-1}(z), \quad (4.5.25b)$$

where the last equal sign comes from (4.5.22a). From (4.5.9), we have

$$h^{(1)}_{-n}(z) = \sqrt{\frac{\pi}{2z}} H^{(1)}_{-n+1/2}(z) = \sqrt{\frac{\pi}{2z}} H^{(1)}_{-(n-1/2)}(z)$$

$$= \sqrt{\frac{\pi}{2z}} e^{i(n-1/2)\pi} H^{(1)}_{n-1/2}(z)$$

$$= z^{n-1} \left(\frac{1}{z}\frac{d}{dz}\right)^{n-1} \frac{e^{iz}}{z} = i(-1)^{n-1} h^{(1)}_{n-1}(z) \quad (4.5.25c)$$

and

$$h^{(2)}_{-n}(z) = \sqrt{\frac{\pi}{2z}} H^{(2)}_{-n+1/2}(z) = \sqrt{\frac{\pi}{2z}} H^{(2)}_{-(n-1/2)}(z)$$

$$= \sqrt{\frac{\pi}{2z}} e^{i(n-1/2)\pi} H^{(2)}_{n-1/2}(z)$$

$$= z^n \left(\frac{1}{z}\frac{d}{dz}\right)^{n-1} \frac{e^{-iz}}{z} = i(-1)^{n-1} h^{(2)}_{n-1}(z). \quad (4.5.25d)$$

An example is

$$j_{-1}(z) = \sqrt{\frac{\pi}{2z}} J_{-1/2}(z) = \frac{\cos z}{z}.$$

4. Parity of spherical Bessel functions of integer orders

$$j_n(-z) = (-1)^n(-z)^n \left(\frac{1}{(-z)}\frac{d}{d(-z)}\right)^n \frac{\sin(-z)}{(-z)}$$

$$= (-1)^n j_n(z). \quad (4.5.26a)$$

$$y_n(-z) = (-1)^{n+1}(-z)^n \left(\frac{1}{(-z)} \frac{d}{d(-z)} \right)^n \frac{\cos(-z)}{(-z)}$$

$$= (-1)^{n+1} y_n(z). \tag{4.5.26b}$$

5. The generating functions of spherical Bessel functions of integer orders

Multiply $\sqrt{\frac{\pi}{2z}}$ to the generating functions of Bessel functions of half-integer orders (4.4.7), and the generating functions of spherical Bessel and Neumann functions of half-integer orders are obtained.

$$\frac{1}{z} \cos \sqrt{z^2 - 2zt} = \sum_{m=0}^{\infty} \frac{1}{m!} \sqrt{\frac{\pi}{2z}} J_{m-1/2}(z) t^m$$

$$= \sum_{m=0}^{\infty} \frac{1}{m!} j_{m-1}(z) t^m. \tag{4.5.27a}$$

$$\frac{1}{z} \sin \sqrt{z^2 - 2zt} = \sum_{m=0}^{\infty} \frac{1}{m!} \sqrt{\frac{\pi}{2z}} (-1)^m J_{-m+1/2}(z) t^m$$

$$= \sum_{m=0}^{\infty} \frac{1}{m!} \sqrt{\frac{\pi}{2z}} Y_{m-1/2}(z) t^m$$

$$= \sum_{m=0}^{\infty} \frac{1}{m!} y_{m-1}(z) t^m. \tag{4.5.27b}$$

Here (4.4.6a) has been used. These two generating functions have been listed in Table 3.6, but please note that in the expansion series, the indices of the spherical functions are $m-1$.

6. Values of special points of spherical Bessel functions of integer orders

$$j_0(0) = 1, \quad j_n(0) = 0, \quad n > 1. \tag{4.5.28}$$

$$y_n(z \to \infty) \sim \frac{1}{z^{n+1}}, \quad n \geq 0. \tag{4.5.29}$$

Because $j_0(z) = \frac{\sin z}{z}$, when $z \to 0$, $j_0(z \to 0) \to 1$. Form the series expression of $j_n(z)$, as well as that of $\sqrt{\frac{\pi}{2z}} J_{n+1/2}(z)$, we get $j_n(0) = 0$, $n > 1$. By the way, $y_0(z) = \frac{\cos z}{z}$ and $y_0(z \to 0) \sim \frac{1}{z}$. The behavior of $y_n(z \to \infty)$ can be derived from the series expression of $\sqrt{\frac{\pi}{2z}} Y_{n+1/2}(z)$.

7. Asymptotic formulas of spherical Bessel functions of integer orders

Multiplying $\sqrt{\frac{\pi}{2z}}$ to (4.4.8) and (4.5.10), we get the asymptotic formulas of spherical Bessel functions as $z \to \infty$.

$$j_\nu(z \to \infty) \sim \frac{1}{z} \cos\left(z - \frac{\nu+1}{2}\pi\right). \tag{4.5.30a}$$

$$y_\nu(z \to \infty) \sim \frac{1}{z} \sin\left(z - \frac{\nu+1}{2}\pi\right). \tag{4.5.30b}$$

$$h_\nu^{(1)}(z \to \infty) \sim \frac{1}{z} e^{i(z-\nu\pi/2-\pi/4)}, \quad (-\pi < \arg z < 2\pi).$$
$$\tag{4.5.30c}$$

$$h_\nu^{(2)}(z \to \infty) \sim \frac{1}{z} e^{-i(z-\nu\pi/2-\pi/4)}, \quad (-2\pi < \arg z < \pi).$$
$$\tag{4.5.30d}$$

8. Some series expansions using spherical Bessel functions of integer orders

A plane wave can be expanded by spherical Bessel functions as follows:

$$e^{iz\cos\theta} = \sum_{n=0}^{\infty} (2n+1) i^n j_n(z) P_n(\cos\theta), \tag{4.5.31}$$

where $P_n(\cos\theta)$'s are Legendre polynomials.

If there are two position vectors $\boldsymbol{r}$ and $\boldsymbol{r}'$, the absolute value of their difference is

$$|\boldsymbol{r} - \boldsymbol{r}'| = \sqrt{r^2 + r'^2 - 2rr'\cos\theta},$$

where θ is the angle between $\boldsymbol{r}$ and $\boldsymbol{r}'$. A Bessel function with the argument $|\boldsymbol{r}-\boldsymbol{r}'|$ can be expanded by Bessel functions with arguments

r and $\mathbf{r}'$, respectively. These expansion series are also called addition theorems. One of them is as follows:

$$H_0^{(2)}\left(k|\mathbf{r}-\mathbf{r}'|\right) = \sum_{m=-\infty}^{\infty} e^{\mathrm{i}m(\theta-\theta')} \begin{cases} J_m(kr)H_m^{(2)}(kr'), & r < r', \\ J_m(kr')H_m^{(2)}(kr), & r > r'. \end{cases}$$

$$(4.5.32)$$

The following four formulas are uniformly called addition theorems of spherical Bessel functions.

The expansion of spherical Bessel function of zero order:

$$j_0\left(k|\mathbf{r}-\mathbf{r}'|\right) = \frac{\sin\left(k|\mathbf{r}-\mathbf{r}'|\right)}{k|\mathbf{r}-\mathbf{r}'|}$$

$$= \sum_{n=0}^{\infty} (2n+1)j_n(kr)j_n(kr')P_n(\cos\theta). \qquad (4.5.33a)$$

The expansion of spherical Neumann function of zero order:

$$y_0\left(k|\mathbf{r}-\mathbf{r}'|\right) = \frac{\cos(k|\mathbf{r}-\mathbf{r}'|)}{k|\mathbf{r}-\mathbf{r}'|}$$

$$= \sum_{n=0}^{\infty} (2n+1)y_n(kr)y_n(kr')P_n(\cos\theta). \qquad (4.5.33b)$$

The expansion of zero order spherical Hankel function of the first kind:

$$h_0^{(1)}\left(k|\mathbf{r}-\mathbf{r}'|\right) = \sum_{n=0}^{\infty} (2n+1)P_n(\cos\theta) \begin{cases} j_n(kr)h_n^{(1)}(kr'), & r < r', \\ j_n(kr')h_n^{(1)}(kr), & r > r'. \end{cases}$$

$$(4.5.33c)$$

The expansion of zero order spherical Hankel function of the second kind:

$$h_0^{(2)}\left(k|\mathbf{r}-\mathbf{r}'|\right) = \sum_{n=0}^{\infty} (2n+1)P_n(\cos\theta) \begin{cases} j_n(kr)h_n^{(2)}(kr'), & r < r', \\ j_n(kr')h_n^{(2)}(kr), & r > r'. \end{cases}$$

$$(4.5.33d)$$

All of these expansions by spherical Bessel functions have practical usage in electromagnetic theory.

4.6.   Modified Bessel Functions

4.6.1.   *Modified Bessel functions of the first and second kinds*

1. Definition of modified Bessel functions

In the Bessel equation $z^2 w''(z) + z w'(z) + (z^2 - \nu^2) w(z) = 0$ we replace z by iz. Then the equation becomes

$$z^2 w'' + z w' - (z^2 + \nu^2) w = 0. \qquad (4.6.1)$$

This is called a **modified Bessel equation**. Here iz is called imaginary argument, but note that z is still a complex. Obviously, $J_\nu(iz)$ is a solution of (4.6.1).

$$J_\nu(iz) = i^\nu \left(\frac{z}{2}\right)^\nu \sum_{m=0}^{\infty} \frac{(-1)^m}{m!\,\Gamma(m+1+\nu)} \left(\frac{iz}{2}\right)^{2m}. \qquad (4.6.2)$$

Now we define

$$I_\nu(z) = i^{-\nu} J_\nu(iz) = \left(\frac{z}{2}\right)^\nu \sum_{m=0}^{\infty} \frac{1}{m!\,\Gamma(m+1+\nu)} \left(\frac{z}{2}\right)^{2m}. \qquad (4.6.3)$$

It is called a **modified Bessel function of the first kind**. For example, its zero order form is

$$I_0(z) = J_0(iz) = 1 + \left(\frac{z}{2}\right)^2 + \frac{1}{(2!)^2}\left(\frac{z}{2}\right)^4 + \frac{1}{(3!)^2}\left(\frac{z}{2}\right)^6 + \cdots$$

$$+ \frac{1}{(k!)^2}\left(\frac{z}{2}\right)^{2k} + \cdots$$

From (4.2.1) it is easy to get

$$I_\nu(-z) = i^{-\nu} J_\nu(-iz) = i^{-\nu}(-1)^\nu J_\nu(iz) = (-1)^\nu I_\nu(z). \qquad (4.6.4)$$

As long as the argument z is replaced by iz in formulas of Bessel functions, one will get corresponding formulas for modified Bessel functions of the first kind.

In the following we discuss the solutions of (4.6.1) in the cases ν is and is not integer.

(1) When ν is not integer, $I_{\pm \nu}(z)$ is the two linearly independent solutions of (4.6.1). Hence, the general solution is

$$w(x) = A I_\nu(z) + B I_{-\nu}(z), \qquad (4.6.5)$$

where A and B and coefficients to be determined by boundary conditions.

Analogous to Bessel functions, we define

$$K_\nu(z) = \frac{\pi}{2} \frac{I_{-\nu}(z) - I_\nu(z)}{\sin \nu \pi}, \qquad (4.6.6)$$

which is called a **modified Bessel function of the second kind**. Apparently, $K_\nu(z)$ and $I_\nu(z)$ are linearly independent of each other. Thus, the general solutions of (4.6.1) can be alternatively written as

$$w(z) = A I_\nu(z) + B K_\nu(z). \qquad (4.6.7)$$

If $z = x$ is real, $I_\nu(x)$ and $K_\nu(x)$ are real.

(2) When $\nu = n$ is integer, it follows from the definition (4.6.3) and $J_{-n}(z) = (-1)^n J_n(z)$ that

$$I_{-n}(z) = \mathrm{i}^n J_{-n}(\mathrm{i}z) = \mathrm{i}^n (-1)^n J_n(\mathrm{i}z) = \mathrm{i}^{2n}(-1)^n I_n(z) = I_n(z). \qquad (4.6.8)$$

$I_n(z)$ linearly depends on $I_{-n}(z)$, so that they cannot be used to constitute the general solution of (4.6.1). One has to find another special solution that is linearly independent of it.

Recall the definition of $K_\nu(z)$, (4.6.6). When $\nu \to n$ becomes integer, the right hand side of (4.6.6) is a zero nil type due to (4.6.8). We have to evaluate its limit of $\nu \to n$ by use of L'Hospital's rule.

$$
\begin{aligned}
K_n(z) &= \lim_{\nu \to n} \frac{\pi}{2} \frac{1}{\pi \cos \nu \pi} \left[\frac{\partial I_{-\nu}(z)}{\partial \nu} - \frac{\partial I_\nu(z)}{\partial \nu} \right] \\
&= \frac{(-1)^n}{2} \left[\frac{\partial \mathrm{i}^\nu J_{-\nu}(\mathrm{i}z)}{\partial \nu} - \frac{\partial \mathrm{i}^{-\nu} J_\nu(\mathrm{i}z)}{\partial \nu} \right]_{\nu=n} \\
&= \frac{(-1)^n}{2} \left[\mathrm{i}^\nu J_{-\nu}(\mathrm{i}z) \ln \mathrm{i} + \mathrm{i}^\nu \frac{\partial J_{-\nu}(\mathrm{i}z)}{\partial \nu} \right. \\
&\qquad \left. + \mathrm{i}^{-\nu} J_\nu(\mathrm{i}z) \ln \mathrm{i} - \mathrm{i}^{-\nu} \frac{\partial J_\nu(\mathrm{i}z)}{\partial \nu} \right]_{\nu=n} \\
&= \frac{(-1)^n}{2} \left[(I_{-\nu}(z) + I_\nu(z)) \ln \mathrm{i} + \mathrm{i}^\nu \frac{\partial J_{-\nu}(\mathrm{i}z)}{\partial \nu} - \mathrm{i}^{-\nu} \frac{\partial J_\nu(\mathrm{i}z)}{\partial \nu} \right]_{\nu=n}.
\end{aligned}
$$

$$(4.6.9)$$

Using (4.6.3) and (4.6.8), we get

$$K_n(z) = \frac{(-1)^n}{2}\left[-2I_n(z)\ln\frac{z}{2} + \sum_{m=0}^{\infty}\frac{\psi(m+1)+\psi(m+n+1)}{(m+n)!m!}\right.$$

$$\left.\times\left(\frac{z}{2}\right)^{2m+n} + (-1)^n\sum_{m=0}^{n-1}\frac{(-1)^m(n-m-1)!}{m!}\left(\frac{z}{2}\right)^{2m-n}\right]$$

$$= (-1)^{n+1}I_n(z)\ln\frac{z}{2}$$

$$+\frac{(-1)^n}{2}\sum_{m=0}^{\infty}\frac{\psi(m+1)+\psi(m+n+1)}{(m+n)!m!}\left(\frac{z}{2}\right)^{2m+n}$$

$$+\frac{1}{2}\sum_{m=0}^{n-1}\frac{(-1)^m(n-m-1)!}{m!}\left(\frac{z}{2}\right)^{2m-n}. \tag{4.6.10}$$

Equation (4.1.30) enables us to reform the expression to be

$$K_n(z) = (-1)^{n+1}I_n(z)\left(\ln\frac{z}{2}+\gamma\right)$$

$$+\frac{(-1)^n}{2}\sum_{m=0}^{\infty}\frac{H_n+H_{m+n}}{(m+n)!m!}\left(\frac{z}{2}\right)^{2m+n}$$

$$+\frac{1}{2}\sum_{m=0}^{n-1}\frac{(-1)^m(n-m-1)!}{m!}\left(\frac{z}{2}\right)^{2m-n}. \tag{4.6.11}$$

As $n = 0$, the last term will vanish. The simplest case is the zero order:

$$K_0(z) = -\left(\ln\frac{z}{2}+\gamma\right)I_0(z) + \sum_{k=0}^{\infty}\frac{1}{(k!)^2}\left(\sum_{m=1}^{k}\frac{1}{m}\right)\left(\frac{z}{2}\right)^{2k}. \tag{4.6.12}$$

Now, $K_n(z)$ is linearly independent of $I_n(z)$. Therefore, as $\nu = n$ the general solution of (4.6.1) can be written as

$$w(z) = AI_n(z) + BK_n(z).$$

In summary, the definition (4.6.6) is valid for any ν no matter whether it is an integer or not. Subsequently, the general solution of modified Bessel equation (4.6.1) can always be expressed in the form of (4.6.7).

From (4.6.8) we get

$$K_{-\nu}(z) = \frac{\pi}{2} \frac{I_\nu(z) - I_{-\nu}(z)}{\sin(-\nu)\pi} = K_\nu(z). \qquad (4.6.13)$$

As $\nu = n$ are integers, there is the following parity:

$$K_n(-z) = \frac{\pi}{2} \lim_{\nu \to n} \frac{I_{-\nu}(-z) - I_\nu(-z)}{\sin \nu\pi} = (-1)^n K_n(z). \qquad (4.6.14)$$

$I_\nu(z)$ and $K_\nu(z)$ have no real zeros.

2. Relations between modified Bessel functions of the second kind and Hankel functions

Substitution of (4.1.42) into (4.5.1) leads to

$$H_\nu^{(1)}(z) = J_\nu(z) + i \frac{J_\nu(z)\cos\nu\pi - J_{-\nu}(z)}{\sin\nu\pi}$$

$$= \frac{i}{\sin\nu\pi} \left[e^{-i\nu\pi} J_\nu(z) - J_{-\nu}(z) \right] \qquad (4.6.15a)$$

and

$$H_\nu^{(2)}(z) = J_\nu(z) - i \frac{J_\nu(z)\cos\nu\pi - J_{-\nu}(z)}{\sin\nu\pi}$$

$$= \frac{-i}{\sin\nu\pi} \left[e^{i\nu\pi} J_\nu(z) - J_{-\nu}(z) \right]. \qquad (4.6.15b)$$

When z is replaced by iz, due to (4.6.3), $I_\nu(z) = i^{-\nu} J_\nu(iz) = e^{-i\nu\pi/2} J_\nu(iz)$. This results in

$$H_\nu^{(1)}(iz) = \frac{i}{\sin\nu\pi} \left[e^{-i\nu\pi} J_\nu(iz) - J_{-\nu}(iz) \right]$$

$$= \frac{ie^{-i\nu\pi/2}}{\sin\nu\pi} \left[I_\nu(z) - I_{-\nu}(z) \right] \qquad (4.6.16a)$$

and

$$H_\nu^{(2)}(-\mathrm{i}z) = \frac{-\mathrm{i}}{\sin \nu\pi}\left[e^{\mathrm{i}\nu\pi}J_\nu(-\mathrm{i}z) - J_{-\nu}(-\mathrm{i}z)\right]$$

$$= \frac{-\mathrm{i}}{\sin \nu\pi}[J_\nu(\mathrm{i}z) - e^{\mathrm{i}\nu\pi}J_{-\nu}(\mathrm{i}z)]$$

$$= \frac{-\mathrm{i}e^{\mathrm{i}\nu\pi/2}}{\sin \nu\pi}[I_\nu(z) - I_{-\nu}(z)]. \qquad (4.6.16b)$$

The latter equation has used (4.2.1), $J_\nu(-z) = (-1)^\nu J_\nu(z) = e^{-\mathrm{i}\nu\pi}J_\nu(z)$.

Now, making use of (4.6.6), we get

$$H_\nu^{(1)}(\mathrm{i}z) = -\frac{2}{\pi}\mathrm{i}e^{-\mathrm{i}\nu\pi/2}K_\nu(z) \qquad (4.6.17a)$$

and

$$H_\nu^{(2)}(-\mathrm{i}z) = \frac{2}{\pi}\mathrm{i}e^{\mathrm{i}\nu\pi/2}K_\nu(z). \qquad (4.6.17b)$$

3. Recursion formulas

$I_\nu(z)$ and $K_\nu(z)$ satisfy the following recursion formulas.

$$\begin{cases} [z^\nu I_\nu(z)]' = z^\nu I_{\nu-1}(z). \\ [z^{-\nu}I_\nu(z)]' = z^{-\nu}I_{\nu+1}(z). \\ [z^\nu K_\nu(z)]' = -z^\nu K_{\nu-1}(z). \\ [z^{-\nu}K_\nu(z)]' = -z^{-\nu}K_{\nu+1}(z). \end{cases} \qquad (4.6.18)$$

Readers can easily prove these recursion formulas in terms of those of Bessel functions. The simplest cases are as follows.

$$\begin{cases} [zI_1(z)]' = zI_0(z). \\ I_0'(z) = I_1(z). \\ [zK_1(z)]' = -zK_0(z). \\ K_0'(z) = -K_1(z). \end{cases} \qquad (4.6.19)$$

From (4.6.18) it is derived that $I_\nu(z)$ and $K_\nu(z)$ satisfy the following recursion formulas.

$$\begin{cases} z[I_{\nu-1}(z) - I_{\nu+1}(z)] = 2\nu I_\nu(z). \\ z[K_{\nu-1}(z) - K_{\nu+1}(z)] = -2\nu I_\nu(z). \\ I_{\nu-1}(z) + I_{\nu+1}(z) = 2I'_\nu(z). \\ K_{\nu-1}(z) + K_{\nu+1}(z) = -2K'_\nu(z). \end{cases} \qquad (4.6.20)$$

Using (4.6.18) repeatedly, one obtains the following recursion formulas.

$$\begin{cases} \left(\frac{1}{z}\frac{d}{dz}\right)^m (z^\nu I_\nu) = z^{\nu-m} I_{\nu-m}(z). \\[2mm] \left(\frac{1}{z}\frac{d}{dz}\right)^m (z^{-\nu} I_\nu) = z^{-(\nu+m)} I_{\nu+m}(z). \\[2mm] \left(\frac{1}{z}\frac{d}{dz}\right)^m (z^\nu K_\nu) = (-1)^m z^{\nu-m} K_{\nu-m}(z). \\[2mm] \left(\frac{1}{z}\frac{d}{dz}\right)^m (z^{-\nu} K_\nu) = (-1)^m z^{-(\nu+m)} K_{\nu+m}(z). \end{cases}$$

4.6.2. *Modified Bessel functions of integer orders*

1. Generating function

In the generating function of Bessel functions $\exp\left(\frac{z}{2}\left(t - \frac{1}{t}\right)\right) = \sum_{n=-\infty}^\infty J_n(z)t^n$, let $z \to iz$, $t \to -it$. Then we get

$$\exp\left(\frac{iz}{2}\left(\frac{t}{i} - \frac{i}{t}\right)\right) = \exp\left(\frac{z}{2}\left(t + \frac{1}{t}\right)\right)$$

$$= \sum_{n=-\infty}^\infty J_n(iz)\left(\frac{t}{i}\right)^n = \sum_{n=-\infty}^\infty I_n(z)t^n.$$

$$(4.6.21)$$

So, the generating function of $I_n(z)$ is $\exp\left(\frac{z}{2}\left(t + \frac{1}{t}\right)\right)$. Because the summation covers from $-\infty$ to $+\infty$, this generating function was not listed in Table 3.5.

2. Addition formulas

We use the addition formula of Bessel functions to get

$$I_n(z_1 + z_2) = i^{-n} J_n(iz_1 + iz_2) = i^{-n} \sum_{m=-\infty}^{\infty} J_{n-m}(iz_1) J_m(iz_2)$$

$$= i^{-n} \sum_{m=-\infty}^{\infty} i^{n-m} I_{n-m}(z_1) i^m I_m(z_2)$$

$$= \sum_{m=-\infty}^{\infty} I_{n-m}(z_1) I_m(z_2). \tag{4.6.22a}$$

In the case of zero order,

$$I_0(z_1 + z_2) = \sum_{m=-\infty}^{\infty} I_{-m}(z_1) I_m(z_2) = \sum_{m=-\infty}^{\infty} I_m(z_1) I_m(z_2)$$

$$= I_0(z_1) I_0(z_2) + 2 \sum_{m=1}^{\infty} I_m(z_1) I_m(z_2). \tag{4.6.22b}$$

Let $z_1 = z_2$. Then the doubleness formula follows:

$$I_0(2z) = \sum_{m=-\infty}^{\infty} [I_m(z)]^2 = [I_0(z)]^2 + 2 \sum_{m=1}^{\infty} [I_m(z)]^2. \tag{4.6.23}$$

4.7. Bessel Functions with Real Arguments

4.7.1. *Eigenvalue problem of Bessel equation*

In treating practical physical problems, a Bessel equation with a real argument is usually encountered.

$$x^2 y'' + xy' + (x^2 - \nu^2)y = 0, \tag{4.7.1}$$

where ν is real. In this section, all the variables and functions are real.

The Bessel equation with real argument can be accompanied by boundary conditions. Therefore, the problems of eigenvalues and eigenfunctions emerge. Further, for Bessel functions with a real argument, the topics such as orthonormalization and generalized Fourier expansion should be investigated.

Often, the Bessel equation derived from a physical model is of the following form:

$$x^2 y'' + x y' + \left(\lambda x^2 - v^2\right) y = 0 \qquad (4.7.2)$$

or

$$\frac{1}{x} \frac{d}{dx}\left(x \frac{dy}{dx}\right) + \left(\lambda - \frac{v^2}{x^2}\right) y = 0, \quad 0 \le x \le R. \qquad (4.7.3)$$

Please note that here a parameter λ is added. If $\lambda = 1$ the equation will go back to (4.7.1). In (4.7.3) the interval of the equation is given: $a = 0$, $b = R$. In (4.7.3), $p(x) = x$, $\rho(x) = x$, $q(x) = v^2/x$. Because now $p(a) = p(0) = 0$, natural boundary condition has to be applied at end point $x = 0$:

$$|y(0)| < \infty. \qquad (4.7.4)$$

This was discussed in Subsection 3.2.2.

At another end point, we assume that the desired solution satisfies one of the following three homogeneous boundary conditions.

$$y(R) = 0. \qquad (4.7.5a)$$

$$y'(R) = 0. \qquad (4.7.5b)$$

$$\alpha y'(R) + \beta y(R) = 0. \qquad (4.7.5c)$$

They can also be written in one equation:

$$\alpha y'(R) + \beta y(R) = 0, \qquad (4.7.5d)$$

where α and β are nonnegative real numbers and at least one of them is nonzero.

Because the kernel $p(x) = x$ is not a periodical function, it is impossible to have a periodic boundary condition.

For convenience, we denote

$$\lambda = k^2 \qquad (4.7.6)$$

and take positive k values.

The general solution of (4.7.3) is

$$y(x) = A J_v(kx) + B Y_v(kx). \qquad (4.7.7)$$

Since from (4.1.40) $Y_\nu(kx \to 0) \to \infty$, to satisfy the natural boundary condition (4.7.4), it must be $B = 0$. So,

$$y(x) = AJ_\nu(kx). \tag{4.7.8}$$

A Bessel equation is a special case of a Sturm-Liouville equation, so that the theory of Sturm-Liouville eigenvalue problem can be applied. By Theorem 1 in Subsection 3.2.3, it is known that there are infinite discrete real eigenvalues which compose a monotone increasing sequence:

$$\lambda_1 \leq \lambda_2 \leq \lambda_3 \leq \lambda_4 \leq \cdots \leq \lambda_n \leq \lambda_{n+1} \leq \cdots .$$

Accordingly, there are infinite eigenfunctions

$$J_\nu(k_1 x), J_\nu(k_2 x), J_\nu(k_3 x), \ldots,$$

which compose an eigenfunction set.

The eigenvalues k_n are determined by boundary condition (4.7.5).

$$\alpha k_n J_\nu'(k_n R) + \beta J_\nu(k_n R) = 0. \tag{4.7.9}$$

Solving the roots of this equation, one can achieve infinite number of eigenvalues k_n. They are arranged in an increasing order:

$$0 < k_1 < k_2 < \cdots < k_n < \cdots .$$

We have ambiguously called λ_n or k_n eigenvalues. Anyway, they are of one-to-one correspondence.

The expressions of Bessel functions are in general complicated. Therefore, it is hard to solve analytically the roots from (4.7.9), but often this transcendental equation is solved numerically to get the roots.

After the eigenvalues $\lambda_n = k_n^2$ are calculated, the solution (4.7.8) is written in the following form:

$$y_n(x) = A_n J_\nu(k_n x). \tag{4.7.10}$$

Equation (4.7.10) is a simple case where the second term of (4.7.7) vanishes due to $p(a) = 0$. In the case that $p(a)$ is not zero, we have discussed the process of solving eigenvalues after Theorem 1 in Subsection 3.2.3. The equation to find the eigenvalues was (3.2.31) or (3.2.32).

4.7.2. *Properties of eigenfunctions*

1. Orthogonality of Bessel functions

According to Sturm-Liouville theory, the eigenfunctions $J_\nu(k_n x)$ corresponding to different eigenvalues are orthogonal to each other with weight $\rho(x) = x$ on $[0, R]$:

$$\int_0^R J_\nu(k_n x) J_\nu(k_m x) x \mathrm{d}x = 0, \quad (n \neq m). \qquad (4.7.11)$$

Now we make use of (4.2.18) and let $z_1 = k_n, z_2 = k_m, a = R$ in this equation. Then we have

$$\int_0^R x J_\nu(k_n x) J_\nu(k_m x) \mathrm{d}x$$

$$= -\frac{R}{k_n^2 - k_m^2} \left[J_\nu(k_m x) k_n J_\nu'(k_n x) - J_\nu(k_n x) k_m J_\nu'(k_m x) \right]_{x=R}.$$

$$(4.7.12)$$

We inspect the value of the right hand side under three homogeneous boundary conditions (4.7.5).

In the case of (4.7.5a), we have $J_\nu(k_n R) = J_\nu(k_m R) = 0$. In the case of (4.7.5b), we have $J_\nu'(k_n R) = J_\nu'(k_m R) = 0$. In the case of (4.7.5c), we have $\alpha k_n J_\nu'(k_n R) + \beta J_\nu(k_n R) = 0$ and $\alpha k_m J_\nu'(k_m R) + \beta J_\nu(k_m R) = 0$. These two equations appear as the linear algebraic equations for solving α and β. The α and β must have nonzero solution so that their coefficient determinant must be zero: $J_\nu(k_m R) k_n J_\nu'(k_n R) - J_\nu(k_n R) k_m J_\nu'(k_m R) = 0$. In summary, under three kinds of boundary conditions, the right hand side of (4.7.12) is always zero. Thus (4.7.12) is proved.

The present case discusses Bessel functions. This is in fact a special case of Theorem 3 in Subsection 3.2.3.

2. Modulus of Bessel functions

The square of the modulus of $J_\nu(k_n x)$ is evaluated by

$$N_n^2 = \int_0^R J_\nu^2(k_n x) x \mathrm{d}x. \qquad (4.7.13)$$

We emphasize that the eigenvalues and modulus are with respect to Bessel functions of order ν. Better written explicitly as $N_{\nu,n}^2 = \int_0^R J_\nu^2(k_{\nu,n}x)\,x\,\mathrm{d}x$. In this section we omit the index ν.

Now that $J_\nu(k_nx)$ is the solution of Bessel equation (4.7.3),

$$\frac{\mathrm{d}}{\mathrm{d}x}\left(x\frac{\mathrm{d}}{\mathrm{d}x}J_\nu(k_nx)\right) + \left(k_n^2x - \frac{\nu^2}{x}\right)J_\nu(k_nxt) = 0.$$

Multiply this equation by $x\frac{\mathrm{d}}{\mathrm{d}x}J_\nu(k_nx)$ and then integrating with respect to x from 0 to R lead to

$$\frac{1}{2}\left[x\frac{\mathrm{d}}{\mathrm{d}x}J_\nu(k_nx)\right]^2\Big|_0^R + \int_0^R (k_n^2x^2 - \nu^2)J_\nu(k_nx)\frac{\mathrm{d}}{\mathrm{d}x}J_\nu(k_nx)\,\mathrm{d}x$$

$$= \frac{1}{2}\left[R\frac{\mathrm{d}}{\mathrm{d}R}J_\nu(k_nR)\right]^2 + \frac{1}{2}\int_0^R (k_n^2x^2 - \nu^2)\frac{\mathrm{d}}{\mathrm{d}x}J_\nu^2(k_nx)\,\mathrm{d}x = 0.$$

Using integration by parts for the second term, we have

$$\frac{1}{2}\left[R\frac{\mathrm{d}}{\mathrm{d}R}J_\nu(k_nR)\right]^2 + \frac{1}{2}(k_n^2x^2 - \nu^2)J_\nu^2(k_nx)|_0^R - k_n^2\int_0^R xJ_\nu^2(k_nx)\,\mathrm{d}x$$

$$= \frac{1}{2}\left[R\frac{\mathrm{d}}{\mathrm{d}R}J_\nu(k_nR)\right]^2 + \frac{1}{2}(k_n^2R^2 - \nu^2)J_\nu^2(k_nR)$$

$$+ \frac{1}{2}\nu^2 J_\nu^2(0) - k_n^2 N_n^2 = 0.$$

Suppose that ν is a real number greater than 0. Then $J_\nu(0) = 0$ and $\nu J_\nu(0) = 0$ as $\nu = 0$. Thus, when $\nu \geq 0$,

$$N_n^2 = \frac{R^2}{2k_n^2}\left[\frac{\mathrm{d}}{\mathrm{d}R}J_\nu(k_nR)\right]^2 + \frac{1}{2}\left(R^2 - \frac{\nu^2}{k_n^2}\right)J_\nu^2(k_nR)$$

$$= \frac{R^2}{2}\left[J_\nu'(k_nR)\right]^2 + \frac{1}{2}\left(R^2 - \frac{\nu^2}{k_n^2}\right)J_\nu^2(k_nR). \qquad (4.7.14)$$

The further computation should take into account boundary conditions. In the following we do calculation for each case of (4.7.5).

(1) The first kind of homogeneous boundary condition is $J_\nu(k_nR) = 0$. Then, (4.7.14) is simplified to be $N_n^2 = \frac{R^2}{2}[J_\nu'(k_nR)]^2$.

To eliminate the derivative, we make use of recursion formula $J'_\nu(x) = \frac{\nu}{x} J_\nu(x) - J_{\nu+1}(x)$.

$$N_n^2 = \frac{R^2}{2} \left[\frac{\nu}{k_n R} J_\nu(k_n R) - J_{\nu+1}(k_n R) \right]^2 = \frac{R^2}{2} [J_{\nu+1}(k_n R)]^2.$$

(4.7.15)

The simplest case is when $\nu = 0$,

$$N_{0,n}^2 = \frac{R^2}{2} [J_1(k_{0,n} R)]^2.$$

(4.7.16)

(2) The second kind of homogeneous boundary condition is $J'_\nu(k_n R) = 0$. Then, (4.7.14) becomes

$$N_n^2 = \frac{1}{2} \left(R^2 - \frac{\nu^2}{k_n^2} \right) J_\nu^2 (k_n R).$$

(4.7.17)

Especially, as $\nu = 0$,

$$N_{0,n}^2 = \frac{1}{2} R^2 J_0^2 (k_{0,n} R).$$

(4.7.18)

(3) The third kind of homogeneous boundary condition is $\alpha k_n J'_\nu(k_n R) + \beta J_\nu(k_n R) = 0$. This condition means that $J'_\nu(k_n R) = -\frac{\beta}{\alpha k_n} J_\nu(k_n R)$. Substituting it into (4.7.14), we get

$$N_n^2 = \frac{R^2}{2} \left[\frac{\beta}{\alpha k_n} J_\nu(k_n R) \right]^2 + \frac{1}{2} \left(R^2 - \frac{\nu^2}{k_n^2} \right) J_\nu^2(k_n R)$$

$$= \frac{1}{2} \left[\left(\frac{\beta^2}{2\alpha^2 k_n^2} + 1 \right) R^2 - \frac{\nu^2}{k_n^2} \right] J_\nu^2(k_n R).$$

(4.7.19)

When $\nu = 0$,

$$N_{0,n}^2 = \frac{1}{2} \left(\frac{\beta^2}{\alpha^2 k_{0,n}^2} + 1 \right) R^2 J_0^2 (k_{0,n} R).$$

(4.7.20)

3. Fourier-Bessel series

Following Sturm-Liouville theory, the eigenfunction set $\{J_\nu(k_n x)\}$ is an orthonormalized one. Any function $f(x)$ defined on $(0, R)$ can be expanded by this set:

$$f(x) = \sum_{n=1}^{\infty} c_n J_\nu(k_n x), \tag{4.7.21}$$

where the coefficients are calculated by

$$c_n = \frac{1}{N_n^2} \int_0^R f(x) J_\nu(k_n x) x \mathrm{d}x. \tag{4.7.22}$$

The expansion (4.7.21) is called **Fourier–Bessel series**. The convergence of this series has been guaranteed by Theorem 4 in Subsection 3.2.3. Below is a theorem more widely used.

Theorem 1. *Suppose a function $f(x)$ is piecewise smooth on $(0, R)$ and integral $\int_0^R \sqrt{x}|f(x)|\mathrm{d}x$ is finite. Then its Fourier-Bessel series (4.7.21) converges to $\frac{1}{2}[f(x+0^+) + f(x-0^+)]$, where k_n is the root of $J_\nu(k_n R) = 0$.*

Please note that this theorem is valid for the first kind of homogeneous boundary Condition (4.7.5a), which is the case mostly encountered.

Example 1. Assume $\mu_m, (m = 1, 2, 3, \ldots)$ are positive zeros of $J_0(x)$. Expand function $f(x) = 1, (0 \leq x \leq 1)$ to be a Fourier-Bessel series by means of $J_0(\mu_m x)$.

Solution. In present case, $R = 1$, $\nu = 0$ in Eq. (4.7.3). By (4.7.21), it is assumed that

$$1 = \sum_{m=1}^{\infty} c_m J_0(\mu_m x),$$

where the coefficient c_m is evaluated by (4.7.22):

$$c_m = \frac{1}{N_m^2} \int_0^1 J_0(\mu_m x) x \mathrm{d}x. \tag{4.7.23}$$

Because of the first kind of homogeneous boundary condition, N_m^2 is from (4.7.16):

$$N_m^2 = \frac{1}{2}[J_1(\mu_m)]^2. \tag{4.7.24}$$

Using recursion formula $\frac{d}{dx}[xJ_1(x)] = xJ_0(x)$, we obtain

$$\int_0^1 xJ_0(\mu_m x)\mathrm{d}x = \left[\frac{xJ_1(\mu_m x)}{\mu_m}\right]_{x=0}^{x=1} = \frac{1}{\mu_m}J_1(\mu_m). \qquad (4.7.25)$$

Substituting (4.7.25) and (4.7.24) into (4.7.23), we get the coefficients

$$c_m = \frac{2}{[J_1(\mu_m)]^2}\int_0^1 1\cdot J_0\left(\mu_m x\right)x\mathrm{d}x = \frac{2}{\mu_m J_1(\mu_m)}.$$

So the function $f(x) = 1, (0 \le x \le 1)$ is expanded by

$$1 = \sum_{m=1}^{\infty}\frac{2}{\mu_m J_1(\mu_m)}J_0(\mu_m x), \quad (0 \le x \le 1).$$

4.7.3. *Eigenvalue problem of spherical Bessel equation*

Suppose a Sturm-Liouville equation is

$$\frac{\mathrm{d}}{\mathrm{d}x}\left(r^2\frac{\mathrm{d}R}{\mathrm{d}r}\right) + (k^2 r^2 - l(l+1))R = 0, \quad 0 \le r \le r_0. \qquad (4.7.26)$$

Its solution satisfies the natural boundary condition at $r = 0$ and the three kinds of homogeneous boundary conditions at $r = r_0$. Equation (4.7.26) is recast to be

$$r^2\frac{\mathrm{d}^2 R}{\mathrm{d}r^2} + 2r\frac{\mathrm{d}R}{\mathrm{d}r} + \left(k^2 r^2 - l(l+1)\right)R = 0.$$

Let $x = kr$. It becomes spherical Bessel equation:

$$x^2\frac{\mathrm{d}^2 y(x)}{\mathrm{d}x^2} + 2x\frac{\mathrm{d}y(x)}{\mathrm{d}x} + \left(x^2 - l(l+1)\right)y(x) = 0,$$

where $y(x) = R\left(\frac{x}{k}\right)$. The solution of (4.7.26) is $R(r) = Aj_l(kr) + By_l(kr)$. Because $y_l(kr \to 0) \to \infty$, in order for the solution to be significant physically, the natural boundary condition $R(0) < \infty$ is applied. Therefore, $B = 0$. The solution is $R(r) = Aj_l(kr)$. Then, the boundary condition at $r = r_0$ is applied to calculate the eigenvalues k_n^2. The corresponding eigenfunction set $\{j_l(k_n r)\}$ is the complete orthonormalized set with weight $\rho(r) = r^2$ on $(0, r_0)$. The square of the modulus of $j_l(k_n r)$ is calculated by

$$N_n^2 = \int_0^{r_0} j_l^2(k_n r) r^2 \mathrm{d}r = \int_0^{r_0} \frac{\pi}{2k_n r} J_{l+1/2}^2(k_n r) r^2 \mathrm{d}r$$

$$= \frac{\pi}{2k_n} \int_0^{r_0} J_{l+1/2}^2(k_n r) r \mathrm{d}r.$$

A function $f(r)$ that has second derivative, finite at $r = 0$ and satisfies homogeneous boundary condition at $r = r_0$, can be expanded by the spherical Bessel functions:

$$f(r) = \sum_{m=1}^{\infty} c_m j_l(k_m r),$$

where the coefficients are computed by

$$c_n = \frac{1}{N_n^2} \int_0^{r_0} f(r) j_l(k_n r) r^2 \mathrm{d}r.$$

Bessel functions with real argument in the other forms, such as modified Bessel functions, are easily discussed in the similar way.

Exercises

1. Imitating the process in Subsection 4.1.1, find the solutions of Legendre equation by means of power series method.

2. Multiply Bessel equation of order ν by J_ν' and then carry out integration. Show that $\int_0^z u J_\nu^2(u) \mathrm{d}u = \frac{1}{2} z^2 \left[J_\nu^2(z) + J_{\nu+1}^2(z) \right] - \nu z J_\nu(z) J_{\nu+1}(z)$.

3. Show that $\begin{vmatrix} J_\nu(z) & J_{\nu-1}(z) \\ Y_\nu(z) & Y_{\nu-1}(z) \end{vmatrix} = \begin{vmatrix} J_\nu(z) & J_\nu'(z) \\ Y_\nu(z) & Y_\nu'(z) \end{vmatrix} = \frac{2}{\pi z}$.

4. Carry out integration $\int \frac{\mathrm{d}z}{z[J_\nu(z)]^2}$ by use of Wronskian (4.2.19) or (4.2.20).

5. Making use of the result of the last exercise, do the following integrations. (1) $\int \frac{\mathrm{d}z}{z J_\nu(z) Y_\nu(z)}$. (2) $\int \frac{\mathrm{d}z}{z[Y_\nu(z)]^2}$. (3) $\int \frac{\mathrm{d}z}{z[J_\nu^2(z) + Y_\nu^2(z)]}$.

6. Prove the following equalities. (1) $J_2 = J_0'' - z^{-1}J_0'$. (2) $2J_0'' = J_2 - J_0$. (3) $4J_n'' = J_{n-2} - 2J_n + J_{n+2}$. (4) $z^2 J_n'' = (n^2 - n - z^2)J_n + zJ_{n+1}$. (Hint: put the expansion of the generating function of the Bessel functions with integer orders; then take the derivation with respect to z or t two times).

7. Show that $\int z J_0(z)dz = z J_1(z) + C$ and $\int z^n J_0(z)dz = z^n J_1(z) + (n-1)z^{n-1}J_0(z) - (n-1)^2 \int z^{n-2}J_0(z)dz$, $n \geq 2$. Calculate $\int z^3 J_0(z)dz$ and $\int z^4 J_0(z)dz$.

8. Prove the following indefinite integrals.

   (1) $\int J_1(z)dz = -J_0(z) + C$.

   (2) $\int z J_1(z)dz = -z J_0(z) + \int J_0(z)dz$.

   (3) $\int z^2 J_1(z)dz = 2z J_1(z) - z^2 J_0(z) + C$.

   (4) $\int \frac{1}{z^2} J_2(z)dz = -\frac{1}{3}(\frac{2}{z^2} - 1)J_1(z) + \frac{1}{3z}J_0(z) + \frac{1}{3}\int J_0(z)dz$.

   (5) $\int z^2 J_2(z)dz = -z^2 J_1(z) - 3z J_0(z) + 3 \int J_0(z)dz$.

   (6) $\int z^3 J_3(z)dz = -z^3 J_2(z) + 5 \int z^2 J_2(z)dz$. Then, continue to do the integral to get the final result.

   (7) Calculate $\int J_3(z)dz$. (Hint: use $z^{-2}J_3 = -(z^{-2}J_2)'$ and $z^{-1}J_2 = -(z^{-1}J_1)'$ and do integration by parts.)

9. Show the following equalities.
$$Y_{-\nu}(z) = \sin \nu\pi J_\nu(z) + \cos \nu\pi Y_\nu(z).$$
$$Y_\nu(ze^{im\pi}) = e^{-im\pi}Y_\nu(z) + 2i \sin m\nu\pi \cot \nu\pi J_\nu(z).$$
$$Y_{-\nu}(ze^{im\pi}) = e^{-im\pi}Y_{-\nu}(z) + 2i \sin m\nu\pi \csc \nu\pi J_\nu(z).$$

   In these equations, m's are even integers.

10. Show that (1) $z = 2\sum_{n=0}^{\infty}(2n+1)J_{2n+1}(z)$ and (3) $z^2 = 2\sum_{n=0}^{\infty}(2n)^2 J_{2n}(z)$. Following the same routines, expand z^m by Bessel functions of integer orders. Note that the cases of m being even and odd are treated separately.

11. Show that (1) $z \cos z = 2\sum_{n=1}^{\infty}(-1)^n(2n+1)^2 J_{2n+1}(z)$ and (2) $z \sin z = 2\sum_{n=0}^{\infty}(-1)^{n+1}(2n)^2 J_{2n}(z)$.

12. Show that $[1 + (-1)^n] J_n(z) = \frac{2}{\pi} \int_0^\pi \cos(z \sin \varphi) \cos(n\varphi) d\varphi$ and

$$[1 - (-1)^n] J_n(z) = \frac{2}{\pi} \int_0^\pi \sin(z \sin \varphi) \sin(n\varphi) d\varphi.$$

Show further that $J_{2n}(z) = \frac{1}{\pi} \int_0^{2\pi} \cos(z \sin \varphi) \cos(2n\varphi) d\varphi$ and

$$J_{2n+1}(z) = \frac{1}{\pi} \int_0^{2\pi} \sin(z \sin \varphi) \sin((2n+1)\varphi) d\varphi$$

13. Show, as $z = x$ is real, (1) $\sum_{n=0}^\infty J_{2n+1}(x) = \frac{1}{2} \int_0^x J_0(t) dt$ and (2) $\sum_{n=0}^\infty J_{\nu+2n+1}(x) = \frac{1}{2} \int_0^x J_\nu(t) dt$ when ν is a real number greater than 0.

14. Show, by use of recursion formulas, that (1) $J_\nu(z) = \frac{2\nu-2}{z} J_{\nu-1}(z) - J_{\nu-2}(z)$ and (2) $J_3(z) + 3J_0'(z) + 4J_0'''(z) = 0$.

15. Show that (1) $\int J_0(z) \cos z \, dz = z J_0(z) \cos z + z J_1(z) \sin z + C$ and (2) $\int J_0(z) \sin z \, dz = z J_0(z) \sin z - z J_1(z) \cos z + C$. (Hint: take integraton with respect to z by parts, and then make use of the reccurence relations of Bessel functions).

16. Show that $J_n(2\sqrt{z}) = (-1)^n z^{n/2} \frac{d^n}{dz^n} J_0(2\sqrt{z})$.

17. Prove that $\int_0^\pi \cos(z \cos \varphi) d\varphi = \int_0^\pi \cos(z \sin \varphi) d\varphi$ (Hint: make use of (4.3.22a) and (4.3.23a)).

18. Make use of (4.3.23) to put down the integral expression of $J_n(z)$. Then use (4.3.18) to show that $\int_0^\pi \sin(z \sin \varphi) \sin(2m\varphi) d\varphi = 0$ and $\int_0^\pi \cos(z \sin \varphi) \cos(2m+1)\varphi d\varphi = 0$.

19. Show that $J_0(z) = \frac{1}{\pi} \int_{-1}^1 \frac{\cos(zt)}{\sqrt{1-t^2}} dt$.

20. Write the expression of $J_{\pm 7/2}(z)$. What is its asymptotic expression as $|z| < 1$?

21. Calculate the Wronskian $J_\nu(x) H_{\nu-1}^{(2)}(x) - J_{\nu-1}(x) H_\nu^{(2)}(x)$.

22. Show that $I_\nu(z e^{im\pi}) = e^{im\nu\pi} I_\nu(z)$ and $K_\nu(z e^{im\pi}) = e^{-im\nu\pi} K_\nu(z) - i\pi \frac{\sin m\nu\pi}{\sin \nu\pi} I_\nu(z)$.

23. Find the integral expression of $I_n(z)$ in terms of the result of Exercise 12.

24. Show that $h_n^{(1)}(-z) = (-1)^n h_n^{(2)}(z)$ and $h_n^{(2)}(-z) = (-1)^n h_n^{(1)}(z)$.

25. Show that $H_0^{(1,2)}(iz) = I_0(z) \mp i\frac{2}{\pi} K_0(z)$.

26. The generating function of modified Bessel function of integer orders is $\exp(\frac{z}{2}(t + \frac{1}{t})) = \sum_{n=-\infty}^{\infty} I_n(z) t^n$. Show that (1) $e^{z \cos \varphi} = I_0(z) + \sum_{n=1}^{\infty} I_n(z) 2 \cos n\varphi$ and (2) $e^{-z \sin \varphi} = I_0(z) + \sum_{n=1}^{\infty} (-1)^n I_{2n}(z) 2 \cos 2n\varphi + \sum_{n=0}^{\infty} (-1)^n I_{2n+1}(z) 2 \sin(2n+1)\varphi$.

27. Prove the following formulas.

   (1) $\cosh(z \cos \varphi) = I_0(z) + 2 \sum_{n=1}^{\infty} I_{2n}(z) \cos 2n\varphi$.

   (2) $\sinh(z \cos \varphi) = 2 \sum_{n=1}^{\infty} I_{2n+1}(z) \cos(2n + 1)\varphi$.

28. Show that $I_0(z) = \frac{1}{\pi} \int_0^\pi \cosh(z \cos \varphi) d\varphi = \frac{1}{\pi} \int_0^\pi \cosh(z \sin \varphi) d\varphi$.

29. Show that $\cosh z = I_0(z) + 2 \sum_{n=1}^{\infty} (-1)^n I_{2n}(z)$ and $\sinh z = 2 \sum_{n=0}^{\infty} I_{2n+1}(z)$.

30. Show that (1) $I_{n+1/2}(z) = \sqrt{\frac{2}{\pi i z}} z^{n+1} \left(\frac{1}{z}\frac{d}{dz}\right)^n \frac{\sinh z}{z}$ and (2) $K_{n+1/2}(z) = (-1)^n \sqrt{\frac{\pi}{2z}} z^{n+1} \left(\frac{1}{z}\frac{d}{dz}\right)^n \frac{e^{-z}}{z}$.

31. Show that $I_0(z) = \frac{1}{\pi} \int_{-1}^1 \frac{e^{-zt}}{\sqrt{1-t^2}} dt$.

32. Show that (1) $I_{1/2}(z) = \sqrt{\frac{2}{\pi z}} \sinh z$, $I_{-1/2}(z) = \sqrt{\frac{2}{\pi z}} \cosh z$ and (2) $K_{1/2}(z) = K_{-1/2}(z) = \sqrt{\frac{\pi}{2z}} e^{-z}$.

33. Prove the following equations in the case of n being integers.

   (1) $\frac{1}{z^n} J_n(z) = (-2)^n \frac{d^n}{d(z^2)^n} J_0(z)$.

   (2) $\frac{1}{z^n} H_n(z) = (-2)^n \frac{d^n}{d(z^2)^n} H_0(z)$.

   (3) $\frac{1}{z^{n+1/2}} J_{n+1/2}(z) = (-2)^n \sqrt{\frac{2}{\pi}} \frac{d^n}{d(z^2)^n} \left(\frac{\sin z}{z}\right)$.

   (4) $\frac{1}{z^{n+1/2}} H_{n+1/2}^{(1)}(z) = -i(-2)^n \sqrt{\frac{2}{\pi}} \frac{d^n}{d(z^2)^n} \left(\frac{e^{iz}}{z}\right)$.

34. Show that when m is an integer,

   (1) $\int_0^{2\pi} d\varphi e^{\pm im\varphi} e^{iz \cos(\varphi - \varphi')} = 2\pi i^{|m|} J_{|m|}(z) e^{\pm im\varphi'}$,

(2) $\int_0^{2\pi} d\varphi e^{\pm im\varphi} \cos(\varphi - \varphi') e^{iz\cos(\varphi-\varphi')} = -i2\pi i^{|m|} J'_{|m|}(z) e^{\pm im\varphi'}$

and

(3) $\int_0^{2\pi} d\varphi e^{\pm im\varphi} \sin(\varphi-\varphi') e^{iz\cos(\varphi-\varphi')} = \pm \frac{m}{z} 2\pi i^{|m|} J_{|m|}(z) e^{\pm im\varphi'}$.

35. Bessel equation is (4.1.1) and spherical Bessel equation is (4.5.11). Their difference is that the coefficients of the first derivative terms are 1 and 2, respectively. The latter can be transferred to the former through a transformation $w(z) = u(z)/\sqrt{z}$, see (4.5.12), and the solutions are Bessel functions of half-integer orders. Assume that the coefficient of the first derivative term is an arbitrary positive integer n. That is to say, the equation becomes

$$z^2 w''(z) + nzw'(z) + (z^2 - \gamma)w(z) = 0.$$

Please find a transformation which can convert the equation to be Bessel equation. What are the solutions of this equation? Obviously, this equation naturally involves the case of $n = 2$, the spherical Bessel equation. Give the results for special cases $n = 0$ and $n = -1$.

36. Here is a differential equation $w''(z) + ae^{mz}w(z) = 0$. Use transformation $u = e^{mz/2}$ to convert it to be a spherical Bessel equation. Find the general solution of the latter, and then that of the former.

37. Here is a differential equation $w''(z) + k^2 z^2 w(z) = 0$. Use transformation $f(z) = w(z)/\sqrt{z}$ with $u = z^2$ to convert it to be a spherical Bessel equation. Find the general solution of the latter, and then that of the former.

38. Convert the following equations to be Bessel equations in terms of given transformations, and find their solutions. In the course of transformation, if there appear multi-valued functions, observe the rule about the singly-valued branch.

(1) $zw''(z) + w'(z) + \frac{1}{4}w(z) = 0$, $u = \sqrt{z}$.

(2) $z^2 w''(z) + zw'(z) + 4(z^4 - k^2)w(z) = 0$, $u = z^2$.

(3) $z^2 w''(z) + z w'(z) - (z + \frac{m^2}{4}) w(z) = 0$, $u = 2iz$.

(4) $z w''(z) - w'(z) + z w(z) = 0$, $w(z) = z f(z)$.

(5) $z w''(z) + (1 + 2n) w'(z) + z w(z) = 0$, $w(z) = z^{-n} f(z)$.

(6) $z^2 w''(z) + (z - 2z^2 \tan z) w'(z) - (m^2 + z \tan z) w(z) = 0$,
    $w(z) = \frac{1}{\cos z} f(z)$.

(7) $z^2 w''(z) + (z + 2z^2 \tan z) w'(z) - (m^2 - z \cot z) w(z) = 0$,
    $w(z) = \frac{1}{\sin z} f(z)$.

(8) $w''(z) + k^2 z w(z) = 0$, $w(z) = \sqrt{z} f(z)$, $u = \frac{2k}{3} z^{3/2}$.

(9) $z^2 w''(z) + \frac{1}{4}(z + \frac{3}{4}) w(z) = 0$, $w(z) = \sqrt{z} f(z)$, $u = \sqrt{z}$.

(10) $z^2 w''(z) - 3z w'(z) + 4(z^4 - 3) w(z) = 0$, $w(z) = z^2 f(z)$,
    $u = z^2$.

39. Let k_n be the positive roots of $J_0(2k_n) = 0$. Expand the function

$$f(x) = \begin{cases} 1, & 0 < x < 1 \\ 1/2, & x = 1 \\ 0, & 1 < x < 2 \end{cases}$$

in a Fourier series in Bessel functions $J_0(k_n x)$.

40. Let k_n be the positive roots of $J_1(k_n) = 0$. Expand the function $f(x) = x, (0 < x < 1)$ in a Fourier series in Bessel functions $J_1(k_n x)$.

41. Let k_n be the positive roots of $J_1(k_n) = 0$. Expand the function $f(x) = x^3, (0 < x < 1)$ in a Fourier series in Bessel functions $J_1(k_\mu x)$.

42. Let $k_n, (n = 1, 2, 3, \ldots)$ be positive zero of $J_0(x)$. Expand the function $f(x) = 1 - x^2, (0 \le x \le 1)$ in a Fourier series in Bessel functions $J_0(k_n x)$.

43. If $f(x) = \sum_{n=1}^{\infty} A_n J_0(k_n x)$, where $J_0(k_n) = 0, (n = 1, 2, 3, \ldots)$, show that $\int_0^1 x f^2(x) dx = \frac{1}{2} \sum_{n=1}^{\infty} A_n^2 J_1^2(k_n)$.

44. Combining $1 = \sum_{n=1}^{\infty} \frac{2}{k_n J_1(k_n)} J_0(k_n x)$ and the result of the last exercise, prove that $\sum_{n=1}^{\infty} \frac{1}{k_n^2} = \frac{1}{4}$, where $J_0(k_n) = 0, (n =$

$1, 2, 3, \ldots$). This result is a summation relationship involving all the real zeros of Bessel function of zero order. Following the routines of this and the last exercises, find the summation relationship involving all the real zeros of Bessel function of νth order with $\nu > -1$.

45. There is an infinitely long cylinder with radius R. Its side holds temperature u_0. Inside the cylinder, the initial temperature is $0°C$. Find the temperature distribution inside the cylinder.

46. There is a semicircle film with radius R. Its edges are fixed. Find its characteristic vibration modes.

47. There is a cylinder with radius R and height L. A steady and uniformly distributed heat flux with intensity q enters the cylinder through its side. Its top and bottom surfaces hold temperature $0°C$. Find the stable distribution of temperature inside the cylinder. This is a boundary value problem in polar coordinates as follows.

$$
\begin{cases}
\left[\dfrac{\partial^2}{\partial r^2} + \dfrac{1}{r}\dfrac{\partial}{\partial r} + \dfrac{1}{r^2}\dfrac{\partial^2}{\partial \varphi^2} + \dfrac{\partial^2}{\partial z^2} \right] u(r, \varphi, z) = 0, \\
\quad (r \le R, 0 < z < L), \\
\dfrac{\partial}{\partial r} u(r, \varphi, z)|_{r=R} = q, \quad u(r, \varphi, 0) = 0, \quad u(r, \varphi, L) = 0.
\end{cases}
$$

The differential equation is Laplace equation that temperature distribution must satisfy. The first boundary condition is the existence of steady heat flux through the side of the cylinder.

48. There is a cylinder with radius R and height h. The temperature of its surfaces holds $0°C$ except its top one. The top surface holds temperature $f(r)$. Find the temperature distribution inside the cylinder. This is the following boundary value problem.

$$
\begin{cases}
u_{rr}(r, z) + \dfrac{1}{r} u_r(r, z) + u_{zz}(r, z) = 0, \\
u(r, z)|_{r=0} < \infty, \quad [u_r(r, z) + k(r, z)]\,|_{r=R} = 0, \\
u(r, z)|_{z=h} = f(r), \quad u(r, z)|_{z=0} = 0.
\end{cases}
$$

49. Solve the following problem.

$$
\begin{cases}
u_{tt}(t,r) + 2hu_t(t,r) = a^2 \left[u_{rr}(t,r) + \dfrac{1}{r} u_r(t,r) \right], \\
u(t,r)|_{r=0} < \infty, \quad u(t,r)|_{r=l} = 0, \\
u(t,r)|_{t=0} = \varphi(r), \quad u_t(t,r)|_{t=0} = 0.
\end{cases}
$$

50. Solve the following problem.

$$
\begin{cases}
u_{rr}(r,z) + \dfrac{1}{r} u_r(r,z) + u_{zz}(r,z) = 0, \\
u(r,z)|_{r=0} < \infty, \quad u(r,z)|_{r=R} = f(z), \\
u(r,z)|_{z=h} = 0, \quad u(r,z)|_{z=0} = 0.
\end{cases}
$$

Find the result when $f(z) = f_0$ is a constant.

Chapter 5

The Dirac Delta Function

5.1. Definition and Properties of the Delta Function

5.1.1. *Definition of the delta function*

Dirac delta function is defined as follows.

(i)
$$\delta(x) = \begin{cases} +\infty, & x = 0, \\ 0, & x \neq 0. \end{cases} \tag{5.1.1a}$$

$$\int_{-\infty}^{\infty} \delta(x)\mathrm{d}x = 1. \tag{5.1.1b}$$

Equation (5.1.1b) is often written in another form:

$$\int_{a}^{b} \delta(x)\mathrm{d}x = \begin{cases} 1, & 0 \in (a,b), \\ 0, & 0 \notin (a,b). \end{cases} \tag{5.1.1c}$$

(ii) $\displaystyle\int_{-\infty}^{\infty} f(x)\delta(x)\mathrm{d}x = f(0).$ \hfill (5.1.2)

(iii) $\displaystyle\int_{-\infty}^{\infty} f(x')\delta(x' - x)\mathrm{d}x' = f(x).$ \hfill (5.1.3)

A Dirac delta function has a singularity. In (i) and (ii), the singularity is at $x = 0$, while in (iii) is in general an arbitrary point x. Equation (5.1.1b) can be regarded as the special case of (ii) when $f(x) = 1$. Equation (5.1.3) is also considered as the sampling property of Dirac delta function. Equations (5.1.1a) and (5.1.1b) are the definitions primarily suggested by Dirac.

According to classical theory of real functions, since the delta function $\delta(x)$ is nonzero only at one isolated point $x = 0$, it accords the definition of zero function, and then its integration on a (finite or infinite) interval should be trivial, see (2.1.25). Therefore, (5.1.1a) and (5.1.1b) seem in conflict to each other. This contradiction arises from that the generalized zero functions were defined such that their values were finite at isolated points, i.e. they belong to classical functions defined in Subsection 2.1.3, while (5.1.1a) tells us that the values at isolated points can be infinite. This means that Dirac delta function is not a classical function in the sense of what we have known previously, and it cannot be classified and analyzed according to ordinary calculus. However, a Dirac delta function can mirror many objective phenomena that classical functions cannot. For example, the electric current behavior of an electric circuit containing a source and a capacitor but not resister behaviors as the delta function at the time the circuit is turned on.

The delta function manifests its physical significance in more examples: a finite quantity of mass or electric charge is placed on a geometric point with vanishing volume; finite quantity of heat is transferred to a point of a rod; an impulsive force acts on an end of a rod so that the rod acquires a finite impulse instantaneously; and so on.

In order to reasonably explain the singularities appearing in reality and to deal with the related physical problems, it is necessary to extend the concept of functions. This promotes the generation of the concept of generalized functions.

5.1.2. *The delta function is a generalized function*

First of all, the delta function is regarded as a functional on a function space.

By the sampling property (5.1.3) it is seen that the delta function manifests its value only when it acts on some function. This is actually a kind of functional. The concept of a functional has been defined in Chapter 1. Now a functional δ on a continuous function space Φ is defined as follows:

$$\delta[\varphi(x)] = \varphi(0). \tag{5.1.4}$$

This is a linear functional:

$$\delta[\alpha\varphi_1(x) + \beta\varphi_2(x)] = \alpha\delta[\varphi_1(x)] + \beta\delta[\varphi_2(x)]. \tag{5.1.5}$$

This is easily proved: $\delta[\alpha\varphi_1(x) + \beta\varphi_2(x)] = \alpha\varphi_1(0) + \beta\varphi_2(0) = \alpha\delta[\varphi_1(x)] + \beta\delta[\varphi_2(x)]$.

Definition 1. If a functional δ on a function space Φ is linear observing (5.1.5), then it is called **linear functional** on space Φ. A linear functional on space Φ is also called a **generalized function** on space Φ.

Consider a generalized function f of integral type:

$$(f, \varphi) = \int_{-\infty}^{\infty} f(x)\varphi(x)\mathrm{d}x, \tag{5.1.6}$$

where f is a given function, $\varphi(x)$ belongs to the space Φ considered. A further postulation is that the integral on the right hand side exists. In the form of (5.1.6), the symbol (f, φ) is of the same meaning as the functional $f[\varphi(x)]$. For instance,

$$(\delta(x), \varphi(x)) = \varphi(0), \tag{5.1.7}$$

which is the same as (5.1.4). Obviously, the general function defined by (5.1.6) is linear satisfying (5.1.5). Therefore, we are able to make (f, φ) and $f[\varphi(x)]$ equal:

$$f[\varphi(x)] = (f, \varphi) = \int_{-\infty}^{\infty} f(x)\varphi(x)\mathrm{d}x. \tag{5.1.8}$$

In (5.1.6), the $f(x)$ on the right hand side is a function while the f on the left hand side is a functional. Different functions $f(x)$ lead to different generalized function f. In view of this, the generalized function of integral type f and the function $f(x)$ generating the f are considered the same thing. In this sense, the concept of generalized function extends to that of a function. Or, the $f(x)$ as a functional is also a generalized function.

Next, the delta function is regarded as a generalize function. As a generalized function, the delta function $\delta(x)$ itself can be dealt with as an ordinary function. Its argument is the same as that of its permissible function φ. In one-dimensional space, it is defined by (5.1.1a), although it is discontinuous at one point and its value at this point is infinite.

5.1.3. *The Fourier and Laplace transformations of the delta function*

The delta function is zero on the whole real axis except the singularity $x = 0$. So, $\delta(x)$ behaves as an ordinary function almost everywhere. The operations on any ordinary function can also be applied to the delta function. Two examples are the Fourier and Laplace transformations.

By the definition of the Fourier transformation, we can put down

$$F[\delta(t)] = \int_{-\infty}^{\infty} \delta(t)e^{i\omega t}dt = 1. \qquad (5.1.9)$$

Consequently, its inverse transformation retrieves the delta function itself.

$$\delta(t) = \frac{1}{2\pi} \int_{-\infty}^{\infty} e^{-i\omega t}d\omega. \qquad (5.1.10)$$

This equation reveals that $\delta(x)$ can be expressed in an integral form. By the way, here we present its another integral expression. Consider integral $\int_{-\infty}^{\infty} \frac{1}{\omega'-\omega} \frac{1}{\omega'-\omega_0}d\omega'$. As $\omega \neq \omega_0$, it equals to $\frac{1}{\omega-\omega_0} \int_{-\infty}^{\infty} (\frac{1}{\omega'-\omega} - \frac{1}{\omega'-\omega_0})d\omega'$, which is zero either because the two integrals are equal, which leads to trivial result, or because $\int_{-\infty}^{\infty} \frac{d\omega'}{\omega'-\omega} = \int_{-\infty}^{\infty} \frac{d\omega'}{\omega'} = 0$ where the integrand is an odd function. As $\omega = \omega_0$, $\int_{-\infty}^{\infty} \frac{d\omega'}{(\omega'-\omega)^2} = \int_{-\infty}^{\infty} \frac{d\omega'}{\omega'^2} = 2\int_0^{\infty} \frac{d\omega'}{\omega'^2} = \frac{2}{\omega'}|_0^{\infty} = \infty$. In summary, we have

$$\delta(\omega - \omega_0) = \int_{-\infty}^{\infty} \frac{1}{\omega' - \omega} \frac{1}{\omega' - \omega_0}d\omega'. \qquad (5.1.11)$$

This equation is useful in dealing with some physical problems.

As for the Laplace transformation of the delta function, we should be very cautious. The equation ought to be written as

$$L[\delta(t)] = \int_{-0^+}^{\infty} \delta(t)e^{-pt}dt = 1, \qquad (5.1.12a)$$

or

$$L[\delta(t)] = L[\delta(t - 0^+)] = \int_0^{\infty} \delta(t - 0^+)e^{-pt}dt = 1. \qquad (5.1.12b)$$

Anyway, the singularity of the delta function has to be definitely within the integral interval. Consequently, its inverse Laplace transformation is

$$L^{-1}[1] = \delta(t - 0^+). \qquad (5.1.12c)$$

If the integral (5.1.12a) is written as $L[\delta(t)] = \int_{0^+}^{\infty} \delta(t)e^{-pt}dt = 0$, (5.1.12c) will be unable to be its inverse transformation.

By the way, we mention that $\int_0^{\infty} \delta(t)dt = 1/2$.

5.1.4. *Derivative and integration of generalized functions*

In order to guarantee the existence of the integration in (5.1.6), it is necessary to exert some conditions on the functions $\varphi(x)$ and $f(x)$. Since $f(x)$ is the function we want to manipulate, we hope the conditions on it are as weak as possible. The stronger the conditions exerted on the function $\varphi(x)$, the weaker the conditions on $f(x)$. Usually, the space Φ is taken as a set of all of the functions $\varphi(x)$ that are derivable infinite times and are nonzero at most in a finite interval. Such a function space is denoted by **space K**.

We turn to investigate the derivative of the general function. Assume that the function $f(x)$ is derivable. Then the generalized function f' determined by $f'(x)$ should be expressed by

$$(f'(x), \varphi(x)) = \int_{-\infty}^{\infty} f'(x)\varphi(x)dx$$

$$= f(x)\varphi(x)|_{-\infty}^{\infty} - \int_{-\infty}^{\infty} f(x)\varphi'(x)dx = -(f(x), \varphi'(x)),$$

$$(5.1.13)$$

where $\varphi(x) \in K$ so that $\varphi(\pm\infty) = 0$. This equation can be defined as the derivative of the generalized function. Let $f(x)$ be a given generalized function. Since $\varphi(x) \in K$, then $\varphi'(x) \in K$. Hence, the functional $(f(x), \varphi'(x))$ is of significance. The derivative f' of the generalized function $f(x)$ is defined by

$$(f'(x), \varphi(x)) = -(f(x), \varphi'(x)), \quad \varphi(x) \in K. \tag{5.1.14}$$

The above equation is for a one-variable case. It is easily extended to multi-variable cases. For example, for a three-variable case, the function space K consists all of the three-variable functions $\varphi(x, y, z)$ that are derivable with respect to every variable infinite times and are nonzero at most in a finite interval. The derivatives of the generalized function $f(x, y, z)$ can be defined similarly to (5.1.13). For example,

$$\left(\frac{\partial f}{\partial x}, \varphi\right) = -\left(f, \frac{\partial}{\partial x}\varphi\right), \quad \left(\frac{\partial^2 f}{\partial y^2}, \varphi\right)$$

$$= -\left(\frac{\partial f}{\partial y}, \frac{\partial}{\partial y}\varphi\right) = \left(f, \frac{\partial^2}{\partial y^2}\varphi\right), \dots; \quad \varphi \in K.$$

Particularly, the derivative of the delta function is still a generalized function:

$$(\delta'(x), \varphi(x)) = -(\delta(x), \varphi'(x)) = -\varphi'(0).$$

Or in integral form:

$$\int_{-\infty}^{\infty} \delta'(x)\varphi(x)\mathrm{d}x = -\varphi'(0).$$

The derivative $\delta'(x)$ is of definite physical meaning: the charge density of a dipole in electrostatics is expressed by $\delta'(x)$.

Subsequently, the second derivative of the delta function is

$$(\delta''(x), \varphi(x)) = -(\delta'(x), \varphi'(x)) = (\delta(x), \varphi''(x)) = \varphi''(0).$$

In general,

$$(\delta^{(n)}(x), \varphi(x)) = (-1)^n(\delta(x), \varphi^{(n)}(x)) = (-1)^n\varphi^{(n)}(0).$$

With the help of the definition of the generalized functions, one is able to understand the meaning of the following differential equation:

$y' = \delta(x)$. Both sides of this equation are considered to be generalized functions. Then, for any $\varphi(x) \in K$, we have $(y', \varphi(x)) = (\delta(x), \varphi(x)) = \varphi(0)$. What satisfies the above differential equation is a so-called unit step function, or Heaviside function, defined as

$$\theta(x) = \begin{cases} 0, & x < 0, \\ 1/2, & x = 0, \\ 1, & x > 0. \end{cases}$$

It is easily verified by the definition of the derivative of generalized functions:

$$(\theta'(x), \varphi(x)) = -(\theta(x), \varphi'(x)) = -\int_{-0+}^{\infty} \varphi'(x)\mathrm{d}x = \varphi(0).$$

Consequently, we know that the integration of the delta function is

$$\int_{-\infty}^{x} \delta(t)\mathrm{d}t = \theta(x) \quad \text{or} \quad \int_{-\infty}^{x} \delta(t - x')\mathrm{d}t = \theta(x - x').$$

Almost every textbook takes the derivative of the unit step function as

$$\frac{\mathrm{d}}{\mathrm{d}x}\theta(x) = \delta(x). \tag{5.1.15}$$

Here the author of this book would like to stress that there is an infinitesimal small number in addition to the delta function. This is closely related to the fact that the two functions $\theta(x)$ and $\delta(x)$ are discontinuous at the origin. To clearly recognize this result, we write the Fourier transformation of the unit step function:

$$\theta(t - t') = \frac{-1}{2\pi i} \int_{-\infty}^{\infty} \frac{e^{-i\varepsilon(t - t')}}{\varepsilon + i0+}\mathrm{d}\varepsilon. \tag{5.1.16a}$$

This equation can be proved as follows: as $t - t' > 0$, a semi-infinite circle in the lower half-plane is supplemented to constitute a closed integral path in the complex ε plane; as $t - t' < 0$, a semi-infinite circle in the upper half-plane is supplemented to constitute a closed path. From (5.1.16a) one acquires the rigorous derivative relation:

$$\frac{\partial}{\partial t}\theta(t - t') = \delta(t - t') - 0+\theta(t - t'). \tag{5.1.16b}$$

In most cases, the infinitesimally small number on the right hand side of (5.1.16b) can be discarded such that (5.1.15) is retrieved. However, in some cases the right hand side of (5.1.16b) is to be moved to a denominator, this infinitesimally small number must be retained, because it represents the position of the pole off the real axis. This is similar to the case of (5.1.16a) where the infinitesimally small number in the denominator should not be removed. This feature is of important application in physics.

A more general version compared to (5.1.15) is that $\frac{\mathrm{d}}{\mathrm{d}x}(\theta(x) + c) = \delta(x)$, where the constant c is to be determined by the concrete problem.

The following identity is quite useful:

$$x\delta(x) = 0. \tag{5.1.17}$$

This is because when $x \neq 0$, the left hand side is obviously zero, while at $x = 0$, it is also zero, which can be testified by multiplying any function $\varphi(x) \in K$ and then integrating.

In fact, it is easy to show that $x^\alpha \delta(x) = 0$, $\alpha > 0$. So, no matter how small α is, as long as it is positive and finite, when $x \to 0$, $\delta(x)$ always approaches the infinity more slowly than $x^{-\alpha}$.

Now we make the derivative: $[x\delta(x)]' = \delta(x) + x\delta'(x) = 0$. Thus,

$$x\delta'(x) = -\delta(x). \tag{5.1.18}$$

This is a useful formula involving the derivative of the delta function.

5.1.5. *Complex argument in the delta function*

In the beginning of this chapter, the definite of the delta function

$$\int_{-\infty}^{\infty} f(x)\delta(x - a)\mathrm{d}x = f(a) \tag{5.1.19}$$

mainly embodies its effect on a function $f(x)$. Here the constant a is real. What about if $a = z$ is complex? In other words, what is the result of the integral

$$\int_{-\infty}^{\infty} f(x)\delta(x - z)\mathrm{d}x? \tag{5.1.20}$$

It is well known that if a function $f(x)$ is derivable infinite times, it can be expanded by Taylor series:

$$f(x + a) = \sum_{n=0}^{\infty} \frac{a^n}{n!} f^{(n)}(x).$$

This expansion is valid for the case when a is complex. In Exercise 3 in this chapter, the delta function can also be expanded by Taylor series as an ordinary function

$$\delta(x + a) = \sum_{n=0}^{\infty} \frac{a^n}{n!} \delta^{(n)}(x). \tag{5.1.21}$$

This is generalized to the case when $a = z$ is complex. Then this series is inserted into (5.1.20) to achieve

$$\int_{-\infty}^{\infty} f(x)\delta(x - z)\mathrm{d}x = \int_{-\infty}^{\infty} f(x) \sum_{n=0}^{\infty} \frac{(-z)^n}{n!} \delta^{(n)}(x)\mathrm{d}x$$

$$= \sum_{n=0}^{\infty} \frac{(-z)^n}{n!} (-1)^n \int_{-\infty}^{\infty} f^{(n)}(x)\delta(x)\mathrm{d}x$$

$$= \sum_{n=0}^{\infty} \frac{z^n}{n!} f^{(n)}(0) = f(z), \tag{5.1.22}$$

where the nth term has taken integration by parts n times. The final result is exactly the same as (5.1.19) where a is real. Therefore, we are able to put down the following Fourier transformation

$$F[\delta(t - z)] = \int_{-\infty}^{\infty} \delta(t - z)\mathrm{e}^{\mathrm{i}\omega t}\mathrm{d}t = \mathrm{e}^{\mathrm{i}\omega z} \tag{5.1.23}$$

and its inverse transformation

$$\delta(t - z) = \frac{1}{2\pi} \int_{-\infty}^{\infty} \mathrm{e}^{-\mathrm{i}\omega(t-z)}\mathrm{d}\omega, \tag{5.1.24}$$

where z is complex.

According to (5.1.1c), only when the singularity is on the integral path, the integral is nonzero. This is the case where the singularity is on the real axis. If the point z is not on the real axis, the definitions

(5.1.22)–(5.1.24) are still valid. This is the formula given in mathematical handbooks. In this case it can be understood that the integral
path in (5.1.22) is from $x = -\infty$ to $x = \infty$. However, the path is not
along the real axis, but is distorted to deviate from the real axis and
to enter the complex plane such that the singularity is on the path.

5.2.   The Delta Function as Weak Convergence
Limits of Ordinary Functions

1. Some examples of weak convergences of ordinary functions

Equation (5.1.1) is the primary definition of the delta function. But
this form is inconvenient for proving formulas, evaluating derivatives,
dealing with experimental data and so on. It is thus often to write
the delta function as a limit of an ordinary function depending on a
real parameter when the parameter approaches a certain value.

Let $\delta_\alpha(x)$ be a function with an index α, and stipulate its following
properties.

(i) $\displaystyle\lim_{\alpha \to \alpha_0} \delta_\alpha(x) = 0, \quad$ for all $x \neq 0$.   (5.2.1)

(ii) For any real α,

$$\int_{-\infty}^{\infty} \delta_\alpha(x)\mathrm{d}x = 1. \tag{5.2.2}$$

(iii) $\displaystyle\lim_{\alpha \to \alpha_0} \int_{-\infty}^{\infty} f(x)\delta_\alpha(x)\mathrm{d}x = f(0).$ (5.2.3a)

In (5.2.3a) the limit and integral are exchangeable. This equation is combined with (5.2.2) to become the following form:

$$\lim_{\alpha \to \alpha_0} \left| \int_{-\infty}^{\infty} f(x)\delta_\alpha(x)\mathrm{d}x - f(0) \right|$$

$$= \lim_{\alpha \to \alpha_0} \left| \int_{-\infty}^{\infty} \delta_\alpha(x)[f(x) - f(0)]\mathrm{d}x \right| = 0. \tag{5.2.3b}$$

This property is also equivalent to

(iv) $\displaystyle\lim_{\alpha \to \alpha_0} \delta_\alpha(x) = +\infty, \quad$ as $x = 0$.   (5.2.4)

When $\delta_\alpha(x)$ function satisfies (i) and (ii) and one of (iii) and (iv), we denote

$$\lim_{\alpha \to \alpha_0} \delta_\alpha(x) = \delta(x). \tag{5.2.5}$$

In this case the delta function is called the **limit of weak convergence** of $\delta_\alpha(x)$. The weak convergence means that convergence is implemented in some way, not necessarily in a specific form such as uniform, pointwise, mean-square convergence and so on. In other words, as long as (5.2.3b) is met, it is a weak convergence.

In the following, we list several $\delta_\alpha(x)$ functions commonly used. (see Fig. 5.1)

(1) $\delta_c(x) \equiv \begin{cases} 1/c, & |x| \le c/2 \\ 0, & |x| > c/2 \end{cases}, \quad c \to 0^+.$ (5.2.6)

(2) $\delta_\alpha(x) = \dfrac{1}{\alpha\sqrt{\pi}} e^{-x^2/\alpha^2}, \quad \alpha \to 0^+,$ (Gauss pulse, see Fig. 5.2).

(5.2.7a)

$\delta_t(x) = \dfrac{1}{2a\sqrt{\pi t}} \exp\left[-\dfrac{x^2}{4a^2 t}\right], \quad t \to 0^+,$ (heat conduction pulse).

(5.2.7b)

(3) $\delta_\varepsilon(x) \equiv \dfrac{1}{\pi} \dfrac{\varepsilon}{\varepsilon^2 + x^2}, \quad \varepsilon \to 0^+.$ (5.2.8)

(4) $\delta_\alpha(x) \equiv \dfrac{\sin(x/\alpha)}{\pi x}, \quad \alpha \to 0^+,$ (5.2.9a)

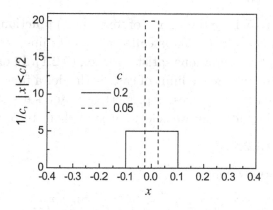

Fig. 5.1. $\delta_a(x)$ function in the form of (5.2.6).

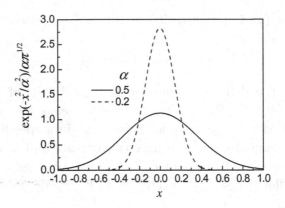

Fig. 5.2. $\delta_a(x)$ function in the form of (5.2.7).

or

$$\delta_N(x) \equiv \frac{\sin Nx}{\pi x}, \quad N \to +\infty, \text{ (sampling pulse, see Fig. 5.4).}$$
$$(5.2.9b)$$

(5) $\delta_k(x) \equiv \dfrac{1 - \cos kx}{2\pi kx^2} = \dfrac{\sin^2(kx/2)}{\pi kx^2}, \quad k \to +\infty.$ $\qquad (5.2.10)$

(6) $\delta_n(x) \equiv \begin{cases} c_n(1 - x^2)^n, & 0 \le |x| \le 1, \quad n = 1, 2, 3, \ldots, \\ 0, & |x| > 1. \end{cases}$ $\quad (5.2.11)$

(7) $\dfrac{1}{2\pi} \dfrac{1 - r^2}{1 - 2r\cos(\theta - \varphi) + r^2}, \quad r \to 1.$ $\qquad (5.2.12)$

The Figs. 5.1 to 5.4 are the curves of some $\delta_n(x)$ functions at different parameter values. Figure 5.3 is called a Lorentz line.

All of the above functions satisfy (i), i.e., (5.2.1). In order to prove that the weak convergence limit of each is the delta function, we have to show that each one possesses property (ii) and satisfies one of (iii) and (iv). In the following we select three of them to prove.

2. Proof of (5.2.9b)

For any $N > 0$, we have

$$\int_{-\infty}^{\infty} dx \delta_N(x) = \frac{1}{\pi} \int_{-\infty}^{\infty} dx \frac{\sin Nx}{x} = 1,$$

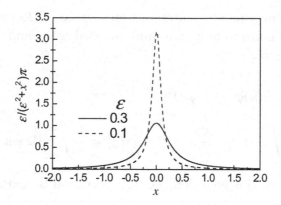

Fig. 5.3. $\delta_a(x)$ function in the form of (5.2.8).

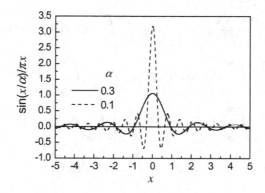

Fig. 5.4. $\delta_a(x)$ function in the form of (5.2.9b).

which satisfies (ii). Then, for any $\varphi(x) \in K$, let

$$\psi(x) = \begin{cases} \dfrac{\varphi(x) - \varphi(0)}{x}, & x \neq 0 \\ \varphi'(0), & x = 0 \end{cases}$$

and

$$\psi'(x) = \begin{cases} \dfrac{x\varphi'(x) - \varphi(x) + \varphi(0)}{x^2}, & x \neq 0 \\ \varphi''(0), & x = 0 \end{cases}.$$

These two functions are continuous on $(-\infty, \infty)$. Because the functions $\varphi(x)$ is nonzero only in a finite interval, $\psi(x)$ and $\psi'(x)$ are so, $\psi(x), \psi'(x) \in \mathbf{K}$.

$$\int_{-\infty}^{\infty} \mathrm{d}x \delta_N(x) \varphi(x) - \varphi(0)$$

$$= \int_{-\infty}^{\infty} \mathrm{d}x \frac{\sin Nx}{\pi x} [\varphi(x) - \varphi(0)] = \frac{1}{\pi} \int_{-\infty}^{\infty} \mathrm{d}x \sin Nx \psi(x).$$

On the right of this equation, we do integration by parts.

$$\int_{-\infty}^{\infty} \mathrm{d}x \sin Nx \psi(x)$$

$$= -\frac{\cos Nx}{N} \psi(x) \Big|_{-\infty}^{\infty} + \frac{1}{N} \int_{-\infty}^{\infty} \mathrm{d}x \cos Nx \psi'(x)$$

$$= \frac{1}{N} \int_{-\infty}^{\infty} \mathrm{d}x \cos Nx \psi'(x) \le \frac{1}{N} \int_{-\infty}^{\infty} \mathrm{d}x \left| \cos Nx \psi'(x) \right|$$

$$\le \frac{1}{N} \int_{-\infty}^{\infty} \mathrm{d}x \left| \psi'(x) \right| = \frac{1}{N} \left[\int_{-\infty}^{-c} \mathrm{d}x \left| \psi'(x) \right| \right.$$

$$\left. + \int_{-c}^{c} \mathrm{d}x \left| \psi'(x) \right| + \int_{c}^{\infty} \mathrm{d}x \left| \psi'(x) \right| \right].$$

Here c is a finite positive number. The functions $\psi(x)$ and $\psi'(x)$ are continuous on the whole real axis. Thus, $\psi'(x)$ is finite on the whole real axis. This guarantees

$$\int_{-c}^{c} \mathrm{d}x \left| \psi'(x) \right| \le \alpha.$$

The estimation of another term is

$$\int_{c}^{\infty} \mathrm{d}x \left| \psi'(x) \right| = \int_{c}^{\infty} \mathrm{d}x \left| \frac{x\varphi'(x) - \varphi(x) + \varphi(0)}{x^2} \right|$$

$$\le \int_{c}^{\infty} \mathrm{d}x \left| \frac{x\varphi'(x) - \varphi(x)}{x^2} \right| + |\varphi(0)| \int_{c}^{\infty} \mathrm{d}x \frac{1}{x^2} \le \beta.$$

Similarly, $\int_{-\infty}^{-c} \mathrm{d}x |\psi'(x)| \leq \gamma$. Here α, β and γ are all finite numbers. In summary, we get

$$\lim_{N\to+\infty} \left[\int_{-\infty}^{\infty} \mathrm{d}x \delta_N(x)\varphi(x) - \varphi(0) \right]$$

$$= \lim_{N\to+\infty} \frac{1}{N} \int_{-\infty}^{\infty} \mathrm{d}x \cos Nx \psi'(x) \leq \lim_{N\to+\infty} \frac{1}{N}(\alpha + \beta + \gamma) = 0,$$

which satisfies (iii). In conclusion,

$$\lim_{N\to+\infty} \int_{-\infty}^{\infty} \mathrm{d}x \delta_N(x)\varphi(x) = \varphi(0)$$

Recall the inverse Fourier transformation of the delta function (5.1.10). This integration does not exist in the sense of ordinary integration. It must be understood as a weak convergence as follows:

$$\frac{1}{2\pi} \int_{-\infty}^{\infty} e^{-i\omega t} \mathrm{d}\omega = \lim_{N\to+\infty} \frac{1}{2\pi} \int_{-N}^{N} e^{-i\omega t} \mathrm{d}\omega$$

$$= \lim_{N\to+\infty} \frac{1}{2\pi} \frac{e^{iNt} - e^{-iNt}}{it} = \lim_{N\to+\infty} \frac{\sin Nt}{\pi t}.$$

$$(5.2.13)$$

3. Proof of (5.2.11)

This function has been defined by (2.4.2). To meet the condition

$$\int_{-1}^{1} \delta_n(x)\mathrm{d}x = 1,$$

the coefficient c_n, called the normalized coefficient, should be reckoned. To do so, we make transformation $x = \sin\theta$.

$$\frac{1}{c_n} = \int_{-1}^{1} (1 - x^2)^n \mathrm{d}x$$

$$= 2 \int_0^1 (1 - x^2)^n \mathrm{d}x = 2 \int_0^{\pi/2} \cos^{2n+1}\theta \mathrm{d}\theta = \frac{2^{2n+1}(n!)^2}{(2n+1)!}.$$

Thus, when $c_n = \frac{(2n+1)!}{2^{2n+1}(n!)^2}$ is taken, property (ii) is met. As n is sufficiently large, we use Stirling formula $n! = \sqrt{2n\pi}(n/e)^n$. It is judged that

$$c_n = \frac{2n+1}{2^{2n+1}} \frac{\sqrt{4n\pi}(2n/e)^{2n}}{2n\pi(n/e)^{2n}} = \frac{2n+1}{2} \frac{1}{\sqrt{n\pi}} < \sqrt{n}. \qquad (5.2.14)$$

The final inequality can be obtained by $(\sqrt{\pi} - 1)n > 1/2$. Equation (5.2.14) reveals that c_n increases with n increasing more slowly than $\sqrt{n}$.

Now we prove that for any given positive number γ less than 1 $(0 < \gamma < 1)$,

$$\lim_{n \to +\infty} \int_\gamma^1 \delta_n(x)\mathrm{d}x = 0. \qquad (5.2.15)$$

By (5.2.14), we have

$$\int_\gamma^1 \delta_n(x)\mathrm{d}x = c_n \int_\gamma^1 (1 - x^2)^n \mathrm{d}x$$

$$\leq \sqrt{n} \int_\gamma^1 (1 - \gamma^2)^n \mathrm{d}x \leq \sqrt{n}(1 - \gamma^2)^n.$$

It was already shown in (2.4.5) that

$$\lim_{n \to +\infty} \sqrt{n}(1 - \gamma^2)^n = 0.$$

So, (5.2.15) is proved. It follows then

$$\lim_{n \to +\infty} \int_{-1}^1 \delta_n(x)\mathrm{d}x$$

$$= \lim_{n \to +\infty} \lim_{\gamma \to +0} \left[\int_{-1}^{-\gamma} \delta_n(x)\mathrm{d}x + \int_{-\gamma}^{\gamma} \delta_n(x)\mathrm{d}x + \int_{\gamma}^1 \delta_n(x)\mathrm{d}x \right]$$

$$= \lim_{n \to +\infty} \lim_{\gamma \to +0} \int_{-\gamma}^{\gamma} \delta_n(x)\mathrm{d}x = 1,$$

which agrees with (5.1.1c).

With these results, we obtain

$$\lim_{n \to +\infty} \left| \int_{-1}^1 \varphi(x)\delta_n(x)\mathrm{d}x - \varphi(0) \right|$$

$$= \left| \int_{-1}^1 [\varphi(x) - \varphi(0)]\delta_n(x)\mathrm{d}x \right|$$

$$\leq \lim_{n\to+\infty} \left\{ \left| \int_{-1}^{-\gamma} [\varphi(x) - \varphi(0)]\delta_n(x)\mathrm{d}x \right| \right.$$

$$+ \left| \int_{-\gamma}^{\gamma} [\varphi(x) - \varphi(0)]\delta_n(x)\mathrm{d}x \right|$$

$$+ \left. \left| \int_{\gamma}^{1} [\varphi(x) - \varphi(0)]\delta_n(x)\mathrm{d}x \right| \right\}$$

$$\leq \lim_{n\to+\infty} \lim_{\gamma\to+0} \max |\varphi(x) - \varphi(0)|$$

$$\times \left\{ \left| \int_{-1}^{-\gamma} \delta_n(x)\mathrm{d}x \right| + \left| \int_{\gamma}^{1} \delta_n(x)\mathrm{d}x \right| \right\}$$

$$+ \lim_{n\to+\infty} \lim_{\gamma\to+0} \left| \int_{-\gamma}^{\gamma} [\varphi(x) - \varphi(0)]\delta_n(x)\mathrm{d}x \right|.$$

Here $\max |\varphi(x)|$ is a finite number. In this equation, the first term is zero due to (5.2.15), and the last term is also zero because as $\gamma \to 0$ the integral interval goes to zero but the integrand is finite.

$$\lim_{n\to+\infty} \left| \int_{-1}^{1} \varphi(x)\delta_n(x)\mathrm{d}x - \varphi(0) \right|$$

$$\leq \lim_{n\to+\infty} \lim_{\gamma\to+0} \left| \int_{-\gamma}^{\gamma} [\varphi(x) - \varphi(0)]\delta_n(x)\mathrm{d}x \right| = 0.$$

This confirms that the $\delta_n(x)$ function defined by (5.2.11) meets property (iii). Our conclusion is that

$$\lim_{n\to\infty} \delta_n(x) = \delta(x).$$

In Chapter 2, we have used this $\delta_n(x)$ function to prove Weierstrass theorem about polynomial approximation.

4. Examples

The various weak convergences of the delta function listed above not only manifests different mathematical forms, but also have their respective mathematical or physical applications. For instances, the Fourier transformation of the delta function can be recognized as the limit of (5.2.13). The polynomial form (5.2.11) can be used to prove the theorem of polynomial approximation. Below we give two examples which demonstrate that some weak convergences of the delta function embody specific physical meanings.

Example 1.

$$\lim_{\eta \to 0^+} \frac{1}{x \pm i\eta} = \frac{1}{x \pm i0^+} = P\frac{1}{x} \mp i\pi\delta(x). \qquad (5.2.16)$$

Proof.

$$\frac{1}{x \pm i\eta} = \frac{x \mp i\eta}{x^2 + \eta^2} = \frac{x}{x^2 + \eta^2} \mp i\pi\frac{\eta}{\pi(x^2 + \eta^2)} \to P\frac{1}{x} \mp i\pi\delta(x).$$

Note that as $x = 0$, there is no real part in this equation. The imaginary part is the weak convergence of the delta function, see (5.2.8). On one hand, this form of weak convergence furnishes a convenient method for calculating the value of the delta function for practical use. On the other hand, it is a way of calculating densities of states in physics.

In physics, the densities of states at energy E is defined as the number of energy states within unit energy interval at energy E. Its expression is

$$\rho(E) = \sum_n \delta(E - E_n), \qquad (5.2.17)$$

where the summation covers all possible quantum states. For example, for electrons in a solid crystal, quantum states are labeled by wavevectors. Then, the summation covers the first Brillouin zone in wavevector space, which can be written in an integral form.

$$\rho(E) = \int d\mathbf{k}\delta(E - E_n(\mathbf{k})).$$

It is impractical to carry out numerical computation using the primary form of the delta function. In this case, the delta function is replaced by the Lorentz line (5.2.8):

$$\rho(E) = \int d\mathbf{k}\frac{1}{\pi}\frac{\varepsilon}{(E - E_n(\mathbf{k}))^2 + \varepsilon^2}. \qquad (5.2.18a)$$

In computation, an appropriate small number ε is chosen. The densities of states thus computed are a smooth curve. If there is a remarkable peak, its shape is a Lorentz line as in Fig. 5.3. This is in agreement with experimental results. For example, the peaks of Mössbauer spectra can be fitted by Lorentz lines quite well.

Why do the peaks appear as Lorentz line shapes? The reason is that energy levels have finite life. If an energy level is a real number, its peak in densities of states appears in the shape of the delta function. Accordingly, the time factor in a wavefunction is a plane wave, indicating that the lifetime of this level is infinitely long. In a system consisting of many particles, there exist interactions between particles. The interactions cause the lifetime of an energy level to be finite. Accordingly, the time factor in the wavefunction exponentially decays. That is to say, the energy has an imaginary part. The Lorentz line reflects that the energy has an imaginary part. Equation (5.2.18a) is recast to be

$$\rho(E) = -\int d\mathbf{k} \frac{1}{\pi} \text{Im} \frac{1}{E - (E_n(\mathbf{k}) - i\varepsilon)}. \qquad (5.2.18b)$$

The appearance of the imaginary part of the energy makes the peak appear a finite height and a width.

In summary, the general definition of densities of states should be

$$\rho(E) = -\frac{1}{\pi} \text{Im} \sum_n \frac{1}{E - E_n}, \qquad (5.2.18c)$$

where energies E_n are complex. When the energies are real, their imaginary parts become infinitely small and we retrieve the special case (5.2.17). $\qquad \square$

Example 2. Consider the following equation:

$$\left(\frac{i}{c} \frac{\partial}{\partial t} + \frac{\partial^2}{\partial x^2} \right) g(x, x'; t, t') = \delta(x - x') \delta(t - t'). \qquad (5.2.19)$$

When $t - t' > 0$, its solution is

$$g(x, x'; t, t') = -\frac{ic}{\sqrt{4\pi i c(t - t')}} e^{i(x - x')^2 / 4c(t - t')}. \qquad (5.2.20)$$

This is of the form of (5.2.7b), indicating that as $t \to t'$, there is a wave packet at x' with the shape of a delta function. Please note that as $t - t' > 0$, (5.2.19) is actually the Schrödinger equation. This example demonstrates how a wave packet having delta shape at initial time and obeying Schrödinger equation evolves with time. When $t - t' > 0$, as time passes the peak height lowers continuously. Eventually, the packet will become a plane wave in infinite space. Since it is a Schrödinger equation, the solution has the physical meaning of probability amplitude. The square of its absolute value is probability density.

The constant c here does not have the meaning of velocity, while it reflects the rate of the broadening of the packet width or of the lowering of its peak height. The larger the c value, the faster the peak height lowers and the packet width broadens.

We have implicitly assumed that the velocity of the packet is zero, and it broadens with the position of its center remaining unmoved. If at initial time, its velocity is u, then at later time the wave packet evolves as described above, and meanwhile moves with velocity u.

5.3. The Delta Function in Multidimensional Spaces

5.3.1. *Cartesian coordinate system*

Assume that at point $r_0 = (x_0, y_0, z_0)$ there is a particle with unit mass, and no mass elsewhere. Then the density of the mass distribution in space is expressed by

$$\delta(r - r_0) = \delta(x - x_0)\delta(y - y_0)\delta(z - z_0). \qquad (5.3.1)$$

The particle with finite mass occupies a point with vanishing volume. The total mass in a volume is expressed in the following integration:

$$\iiint_\Omega \delta(r - r_0)\mathrm{d}V = \iiint_\Omega \delta(x - x_0)\delta(y - y_0)\delta(z - z_0)\mathrm{d}x\mathrm{d}y\mathrm{d}z$$

$$= \begin{cases} 0, & r_0 \notin \Omega \\ 1, & r_0 \in \Omega \end{cases} \qquad (5.3.2)$$

In general, the factor 1 is replaced by a function $f(r)$:

$$\iiint_\Omega f(r)\delta(r - r_0)\mathrm{d}x\mathrm{d}y\mathrm{d}z = \begin{cases} 0, & r_0 \notin \Omega, \\ f(r_0), & r_0 \in \Omega. \end{cases} \qquad (5.3.3)$$

Please note that conceptually (5.3.1) does not imply the meaning of factor decomposition with respect to the three Cartesian coordinates. It is understood by its effect.

$$\int_{-\infty}^{\infty} f(\boldsymbol{r})\delta(\boldsymbol{r} - \boldsymbol{r}_0)\mathrm{d}V$$

$$= \int_{-\infty}^{\infty} f(x,y,z)\delta(x - x_0)\delta(y - y_0)\delta(z - z_0)\mathrm{d}x\mathrm{d}y\mathrm{d}z$$

$$= \int_{-\infty}^{\infty} f(x_0,y,z)\delta(y - y_0)\delta(z - z_0)\mathrm{d}y\mathrm{d}z$$

$$= \int_{-\infty}^{\infty} f(x_0,y_0,z)\delta(z - z_0)\mathrm{d}z = f(x_0,y_0,z_0) = f(\boldsymbol{r}_0)$$

The effect on a function $f(\boldsymbol{r})$ is the same as (5.3.3).

When the mass point is just at the origin, $\boldsymbol{r}_0 = (0,0,0)$, one simply replaces $\boldsymbol{r}_0$ by $(0,0,0)$ in (5.3.1)–(5.3.3). In physics, dimension should be considered. From (5.3.2) and (5.3.3), it is known that the delta function in (5.3.1) is of dimension of $1/L^3$.

In a two-dimensional or n-dimensional space, the delta function can be defined in a way similar to (5.3.1).

The delta function has two properties: sifting property and similarity transformation. A position vector in an n-dimensional space R^n is denoted by $\boldsymbol{r}$. The sifting property means that

$$\int_{R^n} f(\boldsymbol{r})\delta(\boldsymbol{r} - \boldsymbol{r}_0)\mathrm{d}\Omega = \int_{R^n} f(\boldsymbol{r} + \boldsymbol{r}_0)\delta(\boldsymbol{r})\mathrm{d}\Omega = f(\boldsymbol{r}_0). \qquad (5.3.4)$$

The similarity transformation is

$$\int_{R^n} f(\boldsymbol{r})\delta(a\boldsymbol{r})\mathrm{d}\Omega = \int_{R^n} f\left(\frac{\boldsymbol{r}}{a}\right)\frac{1}{|a|^n}\delta(\boldsymbol{r})\mathrm{d}\Omega. \qquad (5.3.5)$$

5.3.2. The transform from Cartesian coordinates to curvilinear coordinates

In resolving the equations in mathematical physics, proper curvilinear coordinates should adopted. An appropriate form in each curvilinear coordinates, such as (5.3.1), is desired for the sake of convenient computation. This concerns the transformation of the delta function from Cartesian to general curvilinear coordinates.

1. Two-dimensional case

Let the curvilinear coordinates be (q_1, q_2). The relations between these and Cartesian coordinates of a position vector $\boldsymbol{r} = (q_1, q_2) = (x, y)$ are

$$x = x(q_1, q_2), \quad y = y(q_1, q_2).$$

Their Jacobi determinant is

$$J = \frac{\partial(x, y)}{\partial(q_1, q_2)} = \begin{vmatrix} \partial x/\partial q_1 & \partial y/\partial q_1 \\ \partial x/\partial q_2 & \partial y/\partial q_2 \end{vmatrix}.$$

Accordingly,

$$dS = dx dy = |J| dq_1 dq_2.$$

There is an integral in Cartesian coordinates:

$$I = \int_{R^2} \varphi(\boldsymbol{r}) \delta(x - x') \delta(y - y') dx dy.$$

We hope that in the curvilinear coordinates the integral is of the following form:

$$I = \int_{R^2} \varphi(\boldsymbol{r}) \delta(q_1 - q_1') \delta(q_2 - q_2') dq_1 dq_2.$$

It is recast to

$$I = \int_{R^2} \varphi(\boldsymbol{r}) \frac{\delta(q_1 - q_1') \delta(q_2 - q_2')}{|J|} dS.$$

It is seen that the necessary condition for the integral to exist is that $J \neq 0$. Comparing the integrals in Cartesian and curvilinear coordinates, we have

$$\delta(x - x') \delta(y - y') = \frac{1}{|J|} \delta(q_1 - q_1') \delta(q_2 - q_2'). \tag{5.3.6}$$

2. Three-dimensional case

Let the curvilinear coordinates be (q_1, q_2, q_3). The relations between these and Cartesian coordinates of a position vector $\boldsymbol{r} = (q_1, q_2, q_3) = (x, y, z)$ are

$$x = x(q_1, q_2, q_3), \quad y = y(q_1, q_2, q_3), \quad z = z(q_1, q_2, q_3).$$

Their Jacobi determinant is

$$J = \frac{\partial(x, y, z)}{\partial(q_1, q_2, q_3)}.$$

Accordingly,

$$dV = dxdydz = |J|dq_1 dq_2 dq_3$$

There is an integral in Cartesian coordinates:

$$I = \int_{R^3} \varphi(\boldsymbol{r})\delta(x - x')\delta(y - y')\delta(y - z')dxdydz.$$

We hope that in the curvilinear coordinates the integral is of the following form:

$$I = \int_{R^3} \varphi(\boldsymbol{r})\delta(q_1 - q_1')\delta(q_2 - q_2')\delta(q_3 - q_3')dq_1 dq_2 dq_3.$$

It is recast to

$$I = \int_{R^3} \varphi(\boldsymbol{r})\frac{\delta(q_1 - q_1')\delta(q_2 - q_2')\delta(q_3 - q_3')}{|J|}dV.$$

Again, the necessary condition is that $J \neq 0$. Comparing the integrals in Cartesian and curvilinear coordinates, we have

$$\delta(x - x')\delta(y - y')\delta(z - z') = \frac{1}{|J|}\delta(q_1 - q_1')\delta(q_2 - q_2')\delta(q_3 - q_3').$$

$$(5.3.7)$$

Please note the necessary condition that the Jacobi determinant is nonzero. It is under this condition that (5.3.6) and (5.3.7) holds. We should be very cautious when J equals to zero. Below we discuss some explicit curvilinear coordinates.

3. The expression of the delta function in polar coordinates

The relations between Cartesian coordinates (x, y) and polar coordinates (r, φ) are $x = r \cos \varphi$, $y = r \sin \varphi$. When $J = \frac{\partial(x,y)}{\partial(r,\varphi)} = r \neq 0$, the two coordinates (x, y) and (r, φ) are in one-to-one correspondence. By $dS = dxdy = rdr\,d\varphi$ and (5.3.6), it is known that

$$\delta(\boldsymbol{r} - \boldsymbol{r}_0) = \frac{\delta(r - r_0)\delta(\varphi - \varphi_0)}{|J|}$$

$$= \frac{\delta(r - r_0)\delta(\varphi - \varphi_0)}{r}, \quad r_0 > 0. \qquad (5.3.8)$$

This equation is valid for any point $\boldsymbol{r}$, including $r = 0$, as long as $r_0 \neq 0$. The position of source (r_0, φ_0) is crucial. When $r_0 = 0$, the relations between $x_0 = 0$, $y_0 = 0$ and $r = 0$, $\varphi_0 = 0$ are not the same as (5.3.6). In this case, $J = 0$ so that (5.3.8) is invalid. Some modification to the right hand side of (5.3.8) is necessary.

Now we assume that the mass uniformly distributes on a circle with radius $r = r_0$. The angular uniform distribution means that a factor $1/2\pi$ ought to replace the factor $\delta(\varphi - \varphi_0)$ in (5.3.8) which reflects the distribution of the mass centralized at an angle φ_0. Then let $r_0 \to 0^+$. So, (5.3.8) becomes

$$\frac{\delta(r - r_0)}{r}\frac{1}{2\pi} = \frac{1}{2\pi r}\delta(r - r_0).$$

Then, let $r_0 \to 0^+$ to obtain

$$\delta(\boldsymbol{r}) = \frac{1}{2\pi r}\delta(r - 0^+) = \frac{1}{2\pi r}\delta(r). \qquad (5.3.9)$$

This is the polar coordinate expression of the delta function $\delta(\boldsymbol{r})$ when $r_0 = 0$. As a verification, let us see the following integrations.

$$\iint \delta(\boldsymbol{r})dS = \iint \delta(x)\delta(y)dxdy = 1.$$

$$\iint \delta(\boldsymbol{r})dS = \iint \frac{1}{2\pi r}\delta(r)rdrd\varphi = 1.$$

The results are the same.

4. The expression of the delta function in spherical coordinates

In three-dimensional space,

$$dV = dxdydz = |J|drd\varphi d\theta = r^2 \sin\theta drd\varphi d\theta.$$

From (5.3.7) we have

$$\delta(\boldsymbol{r} - \boldsymbol{r}_0) = \frac{1}{|J|}\delta(r - r_0)\delta(\varphi - \varphi_0)\delta(\theta - \theta_0)$$

$$= \frac{1}{r^2 \sin\theta}\delta(r - r_0)\delta(\varphi - \varphi_0)\delta(\theta - \theta_0),$$

$$(r_0 > 0, 0 \le \varphi \le 2\pi, 0 \le \theta \le \pi). \qquad (5.3.10)$$

When $r_0 = 0$, similar to the case of polar coordinates, assume the mass distributes uniformly on a spheric surface with radius $r = r_0$. The solid angle of a sphere is 4π. The uniform distribution in the angulars φ and θ means that a factor $1/4\pi$ ought to replace the factor $\frac{1}{\sin\theta}\delta(\varphi - \varphi_0)\delta(\theta - \theta_0)$ in (5.3.10) which reflects the distribution of a unit mass centralized at an angle (φ_0, θ_0). Then let $r_0 \to 0^+$. So, (5.3.10) becomes

$$\delta(\boldsymbol{r}) = \frac{1}{4\pi r^2}\delta(r). \qquad (5.3.11)$$

In cylindrical coordinates, the expressions of the delta function are as follows.

$$\delta(\boldsymbol{r} - \boldsymbol{r}_0) = \frac{1}{r}\delta(r - r_0)\delta(\varphi - \varphi_0)\delta(z - z_0), \quad r_0 > 0. \qquad (5.3.12)$$

$$\delta(\boldsymbol{r}) = \frac{1}{2\pi r}\delta(r)\delta(z). \qquad (5.3.13)$$

5.4. Generalized Fourier Series Expansion of the Delta Function

Although the delta function owns a singularity, it is piecewise smooth. So, it can be expanded by some commonly used complete function sets. A characteristic function set $\{y_n(x)\}$ of Sturm-Liouville equation composes a complete set on $[a, b]$. Thus, it is natural to employ them to expand the delta function, as long as the fixed point x_0 in $\delta(x - x_0)$ is within the interval, $x_0 \in [a, b]$.

Assume that the characteristic functions $\{y_n(x)\}$ of SL equation

$$\frac{\mathrm{d}}{\mathrm{d}x}\left[p(x)\frac{\mathrm{d}y}{\mathrm{d}x}\right] + [\lambda\rho(x) - q(x)]y = 0, \quad (a \le x \le b)$$

are orthonormalized on $[a, b]$ with weight function $\rho(x)$:

$$\int_a^b y_m^*(x)y_n(x)\rho(x)\mathrm{d}x = \delta_{mn}. \tag{5.4.1}$$

Then,

$$\delta(x - x_0) = \sum_n \sqrt{\rho(x_0)\rho(x)}y_n^*(x_0)y_n(x). \tag{5.4.2}$$

Proof.   Suppose that $\delta(x - x_0) = \sum_n c_n y_n(x)$. On both sides multiply $y_m^*(x)\rho(x)$ and take integration on $[a, b]$. Using (5.4.1), we get $c_n = \rho(x_0)y_n^*(x_0)$. Therefore,

$$\delta(x - x_0) = \sum_n \rho(x_0)y_n^*(x_0)y_n(x). \tag{5.4.3}$$

Because $\delta(x - x_0)$ is invariant under exchange of x and x_0, the expansion should also be so under the exchange. For this reason multiply $\sqrt{\frac{\rho(x)}{\rho(x_0)}}$ to both sides of (5.4.3) such that (5.4.2) is achieved.   □

We stress here that each eigenvalue of the ordinary differential equation of second order has two linearly independent special solutions. In (5.4.3) the solution $y_n(x)$ is the linear combination of these two special solutions, with the combination coefficients determined by boundary conditions.

In quantum mechanics, (5.4.2) is called the completeness relation of characteristic functions. This is a fundamental and significant relation in quantum mechanics.

Example 1. The orthonormalized eigenfunction set of Hermite equation is $\frac{1}{\pi^{1/4}\sqrt{2^n n!}}H_n(x), n = 0, 1, 2, \ldots$, where $H_n(x)$'s are Hermite polynomials. If the solutions satisfying the boundary conditions only contain $H_n(x)$ but not the other special solution (the second kind Hermite function), then,

$$\delta(x - x_0) = \sum_{n=0}^{\infty} \frac{1}{\sqrt{\pi} 2^n n!} e^{-(x^2 + x_0^2)/2} H_n(x_0) H_n(x).$$

Example 2. The orthonormalized eigenfunction set of Legendre equation is $\sqrt{\frac{2l+1}{2}} P_l(x), l = 0, 1, 2, \ldots$, where $P_l(x)$'s are Legendre polynomials. If the solutions satisfying the boundary conditions only contain $P_l(x)$ but not the other special solution (the second kind Legendre function), then,

$$\delta(x - x_0) = \sum_{l=0}^{\infty} \frac{2l + 1}{2} P_l(x_0) P_l(x).$$

This can be reformed to be

$$\delta(\cos \theta - \cos \theta_0) = \sum_{l=0}^{\infty} \frac{2l + 1}{2} P_l(\cos \theta_0) P_l(\cos \theta).$$

As $\theta_0 = 0$, then by $P_l(1) = 1$ (see Table 3.7), we get

$$\delta(\cos \theta - 1) = \sum_{l=0}^{\infty} \frac{2l + 1}{2} P_l(\cos \theta).$$

Example 3. The orthonormalized eigenfunctions of $\Phi''(\varphi) + m^2 \Phi(\varphi) = 0$, $0 \le \varphi \le 2\pi$ are $\frac{1}{\sqrt{2\pi}} e^{im\varphi}, m = 0, \pm 1, \pm 2, \ldots$, which satisfy the periodic boundary condition. Then,

$$\delta(\varphi - \varphi_0) = \frac{1}{2\pi} \sum_{m=-\infty}^{\infty} e^{im(\varphi - \varphi_0)}.$$

If the summation over m is replaced by integration, the limit form will be (5.1.10).

Example 4. Expand $\delta(x - x')$ on the interval $\alpha \le x, x' \le 2\pi - \alpha$.

Solution. First of all, we must know that what is the complete function set on this interval. The method is to artificially establish an ordinary differential equation of second order with certain boundary conditions. Then its solutions satisfying this boundary value problem are the desired complete functions. In the present case, we establish the following boundary value problem:

$$y''(x) + \nu^2 y(x) = 0, \quad x \in [\alpha, 2\pi - \alpha], \quad y(\alpha) = y(2\pi - \alpha) = 0.$$

Since the function to be expanded is zero at the boundaries, the first homogeneous boundary conditions are set in this problem. The general solution of the equation is $y(x) = \sin\nu(x + \beta)$. The two constants ν and β in the phase $\nu(x + \beta)$ are solved by the boundary conditions. Applying the conditions in the general solution, we get $\sin\nu(\alpha+\beta) = 0, \sin\nu(2\pi-\alpha+\beta) = 0$. It follows that $\nu(\alpha+\beta) = m\pi$, $\nu(2\pi - \alpha + \beta) = n\pi$ or $2\nu(\pi + \beta) = (n+m)\pi, \nu(2\pi - 2\alpha) = (n - m)\pi$. Thus, we obtain

$$\nu = \frac{(n - m)\pi}{2(\pi - \alpha)}, \quad \beta = \frac{(n + m)\pi}{2\nu} - \pi = \frac{(n + m)(\pi - \alpha)}{n - m} - \pi = \frac{m\pi}{\nu} - \alpha.$$

The phase is

$$\nu(x + \beta) = \nu\left(x + \frac{m\pi}{\nu} - \alpha\right) = \nu(x - \alpha) + m\pi$$

and the solution of the equation is

$$y(x) = \sin[\nu(x - \alpha) + m\pi] = (-1)^m \sin\frac{n\pi(x - \alpha)}{2(\pi - \alpha)},$$

where n is integer. It is easily verified that the functions are orthogonal to each other.

$$(-1)^{m+n} \int_\alpha^{2\pi-\alpha} \sin\frac{m\pi}{2(\pi - \alpha)}(x - \alpha) \sin\frac{n\pi}{2(\pi - \alpha)}(x - \alpha)\mathrm{d}x = 0.$$

We normalize the functions:

$$\int_\alpha^{2\pi-\alpha} \sin^2\frac{m\pi}{2(\pi - \alpha)}(x - \alpha)\mathrm{d}x$$

$$= \frac{1}{2} \int_\alpha^{2\pi-\alpha}\left[1 - \cos\frac{m\pi(x - \alpha)}{\pi - \alpha}\right]\mathrm{d}x = \pi - \alpha.$$

So, the orthonormalized function set is $\{\frac{1}{\sqrt{\pi-\alpha}}\sin\frac{n\pi(x-\alpha)}{2(\pi-\alpha)}\}$. The values of the eigenfunctions at the end points of the interval $x \in [-\alpha, 2\pi - \alpha]$ are zero. Any function defined on this interval and vanishing at the two ends can be expanded by this complete set.

Applying (5.4.2), we obtain the required expansion:

$$\delta(x - x') = \frac{1}{\pi - \alpha}\sum_{n=1}^{\infty}\sin\frac{n\pi(x' - \alpha)}{2(\pi - \alpha)}\sin\frac{n\pi(x - \alpha)}{2(\pi - \alpha)}. \qquad (5.4.4)$$

Because the delta function $\delta(x - x')$ is zero at the end points of the interval, the complete set in Example 4 is selected such that the eigenfunctions are zero at the ends. Actually, it is not necessary to expand the delta function by the eigenfunctions the values of which at ends of the interval are zero.

Up to now, the expansion makes use of the completeness of the eigenfunction of an SL equation. For a non-self-adjoint SL system, its solutions compose a complete set as long as the conditions listed in Theorem 3 in Subsection 3.7.3 are satisfied, and the delta function can be expanded by this set. The expansion is

$$\delta(x - x_0) = \sum_n \rho(x_0)v_n^*(x_0)u_n(x), \qquad (5.4.5)$$

where u_n and v_n are respectively the eigenfunctions of the boundary value problem and its adjoint problem. For example, the delta function is expanded by the eigenfunction set of Example 1 in Subsection 3.6.2 as follows:

$$\delta(x - x_0) = \sum_{n=1}^{\infty}\frac{2}{n^2\pi^2 + \alpha^2}(\alpha\sin\mu_n x + n\pi\cos\mu_n x)$$

$$\times (\alpha\sin\mu_n x_0 + n\pi\cos\mu_n x_0) + \frac{2\alpha}{e^{2\alpha} - 1}e^{\alpha x}e^{\alpha x_0}. \qquad (5.4.6)$$

Exercises

1. Prove (5.1.16).

2. Prove that $\frac{d^2}{dx^2}|x| = 2\delta(x)$. (Hint: $|x| = x\theta(x) - x\theta(-x)$.)

3. Show that

$$\delta(x + a) = \sum_{n=0}^{\infty} \frac{a^n}{n!} \delta^{(n)}(x).$$

This demonstrates that the delta function can be expanded by Taylor series like an ordinary function.

4. Show that $x^2 \delta'(x) = 0$.

5. Show that $\delta^{(n)}(x)$ is even when n is an even integer and is odd when n is an odd integer.

6. Prove that

$$\delta[g(x)] = \sum_k \frac{\delta(x - x_k)}{|g'(x_k)|},$$

where x_k's are all one-order zeros of $g(x)$. Then, prove the following special cases. (1) $\delta(ax) = \frac{\delta(x)}{|a|}$. (2) $\delta(x^2 - a^2) = \frac{\delta(x-a)+\delta(x+a)}{2|a|}$. (3) $g(x) = (x - a)(x - b)$, $g'(x = a, b) = \pm(a - b)$. (4) $\delta(e^x - 1) = \delta(x)$. (5) $\delta(\sin x) = \sum_{k=-\infty}^{\infty} \delta(x - k\pi)$.

7. Make use of the result in the last exercise to do the following integrals. (1) $\int_{-\infty}^{\infty} dx \delta(x^2 - 5x + 6)(3x^2 - 7x + 2)$. (2) $\int_{-\infty}^{\infty} dx \delta(x^2 - \pi^2) \cos x$. (3) $\int_{1/2}^{\infty} dx \delta(\sin \pi x)(\frac{2}{3})^x$.

8. Find the result of $\frac{d}{dx}\theta(x^2 - 1)$.

9. (1) Show that if there is a dipole in space, its charge density $\rho(r)$ is expressed by the derivative of the delta function; (2) what charge density does $\rho(x) = \frac{d}{dx^2}\delta(x^2 - 1)$ represent?

10. We have proved by (5.2.13) that $\delta(t) = \frac{1}{2\pi}\int_{-\infty}^{\infty} e^{-i\omega t} d\omega$ can be regarded as a weak convergence of $\delta_N(x) \equiv \frac{\sin Nx}{\pi x}$. Show that it can also be regarded as the weak convergence of $\delta_\varepsilon(x) \equiv \frac{1}{\pi}\frac{\varepsilon}{\varepsilon^2 + x^2}$ in (5.2.8).

11. Evaluate Laplace transformations: $L[\delta(x - y)]$ and $L[\delta^{(n)}(x)]$.

12. Show that

$$\lim_{\varepsilon \to 0} \frac{e^{-|x|/\varepsilon}}{2\varepsilon} = \delta(x).$$

This form of weak convergence is also called exponential pulse.

13. Show that the limit of

$$\delta_\varepsilon(x) \equiv \frac{1}{\pi} \frac{\varepsilon}{\varepsilon^2 + x^2}$$

at $\varepsilon \to 0$ is the delta function.

14. Show that the weak convergence of (5.2.12) is $\delta(\theta - \varphi)$, i.e.,

$$\lim_{r \to 1} \frac{1}{2\pi} \frac{1 - r^2}{1 - 2r \cos(\theta - \varphi) + r^2} = \delta(\theta - \varphi)$$

15. Assume

$$f_\alpha(x) = \frac{\sin \alpha}{2\pi(\cosh x + \cos \alpha)}.$$

Show that

$$\lim_{\alpha \to \pi^-} \int_a^b f_\alpha(x)dx = \begin{cases} 1, & 0 \in (a, b), \\ 0, & 0 \notin (a, b). \end{cases}$$

Therefore,

$$\lim_{\alpha \to \pi^-} \frac{\sin \alpha}{2\pi(\cosh x + \cos \alpha)} = \delta(x).$$

16. Show that

(1) $\lim\limits_{n \to \infty} \frac{1}{2}[1 + \tanh(nx)] = \theta(x)$ and

(2) $\frac{1}{2} \lim\limits_{\beta \to \infty} \frac{\beta}{1 + \tanh(\beta x)} = \delta(x)$.

17. Show that

$$\lim_{n \to \infty} \frac{2n^3 x}{\sqrt{\pi}} e^{-n^2 x^2} = -\delta'(x).$$

18. Evaluate the integral $\int_0^\infty \cos ax \cos bx dx$.

19. Evaluate the following integrals.

(1) $I = \int_{-\infty}^\infty e^{at} \sin bt \delta^{(n)}(t)dt$, $n = 0, 1, 2$.

(2) $\int_{-\infty}^\infty (\cos t + \sin t)\delta^{(n)}(t^3 + t^2 + t)dt$, $n = 0, 1$. (Note that all the singularities should be covered.)

20. Suppose that the function $f(x)$ is discontinuous at $x = a$. Its left and right side limits are denoted by $f(a - 0^+)$ and $f(a + 0^+)$, respectively. Show, by making use of (5.2.6), that $\int_{-\infty}^{\infty} f(x)\delta(x - a)\mathrm{d}x = \frac{1}{2}[f(a - 0^+) + f(a + 0^+)]$.

21. The function $\sqrt{\delta(x)}$ is the square root of the delta function $\delta(x)$. By the definition of $\delta(x)$ function (5.1.1a), it must be

$$\sqrt{\delta(x)} = \begin{cases} 0, & x \neq 0, \\ \infty, & x = 0. \end{cases}$$

   Using (5.2.6), show that $\int_{-\infty}^{\infty} \sqrt{\delta(x)}\mathrm{d}x = 0$. Can this be proven by the other forms of weak limits (5.2.7)–(5.2.11)?

22. Using (5.2.6), show that $\delta(x^2) = \frac{\delta(x)}{2|x|}$. The function on the right hand side has been encountered in physical problems.

23. In spherical coordinates $\mathbf{r} = (r, \theta, \varphi)$, a unit mass is at the point r_0 at the positive $z$axis. Put down the expression of the delta function $\delta(\mathbf{r} - \mathbf{r}_0)$. (Hint: the mass is distributed uniformly on the circle with $r = r_0, \theta = \theta_0 \neq 0$.)

24. In cylindrical coordinates $\mathbf{r} = (r, \varphi, z)$, suppose a unit mass is positioned at (1) $\mathbf{r}_0 = (0, 0, z_0)$ and (2) $\mathbf{r}_0 = (0, 0, 0)$. Put down the expressions of the delta function $\delta(\mathbf{r} - \mathbf{r}_0)$.

25. Show that function $f(x) = \sum_{n=-\infty}^{\infty} \delta(x - 2n\pi)$ has a period 2π. Find its Fourier transformation. Verify that the resultant series weakly converges to the delta function $\delta(x)$. (Hint: use formula $1 + 2\cos x + 2\cos 2x + \cdots + 2\cos nx = \frac{\sin[(n+1/2)x]}{\sin(x/2)}$.)

26. Find the Fourier transformation of the function

$$\varphi(t) = \sum_{n=1}^{\infty} \frac{1}{n!}\delta'(t - n).$$

27. Show that $\frac{\mathrm{d}}{\mathrm{d}x}\delta(f(x)) = f'(x)\delta'(f(x))$ and $\delta(f(x)) + f(x)\delta'(f(x)) = 0$, where $\delta'(f(x)) = \frac{\mathrm{d}}{\mathrm{d}f(x)}\delta(f(x))$, and prove that function $\phi(x, y) = \delta(x^2 - y^2)$ is the solution of equation $x\frac{\partial\phi}{\partial x} + y\frac{\partial\phi}{\partial y} + 2\phi = 0$.

28. Show that $\int_{-\infty}^{x} \theta(y-t)\mathrm{d}y = (x-t)\theta(x-t)$.

29. Show that $\lim_{\varepsilon \to \infty} \varepsilon e^{-\varepsilon(\alpha^* - \alpha'^*)(\alpha - \alpha')} = \pi\delta(\alpha^* - \alpha'^*)\delta(\alpha - \alpha')$. (William H. Louisell, Quantum Statistical Properties of Radiation, John Wiley 1973.)

Chapter 6

Green's Function

6.1. Fundamental Theory of Green's Function

6.1.1. *Definition of Green's function*

Definition 1. Let $L(x)$ be an operator. Then function $G(x, x')$ is called **Green's function** if it satisfies the equation

$$[\lambda - L(x)]G(x, x') = \delta(x - x') \qquad (6.1.1a)$$

or

$$[\lambda - L(x)]G(x, x') = \delta(x - x')/\rho(x). \qquad (6.1.1b)$$

Some explanations about the two equations are necessary. The argument $x(x')$ may represent one of the following cases. It can be time t: $x = t$; it can be space coordinates r: $x = r$ where the space can be one-, two- or three-dimensional; it can contain both space and time: $x = (r, t)$. If x does not include time t, it is called a **time-independent Green's function**, otherwise it will be called a **time-dependent Green's function**.

The mathematical theory of the delta function on the right hand side of (6.1.1) has been introduced in Chapter 5. When x represents time t, it is $\delta(t - t')$; when $x = r$, it is $\delta(r - r')$; when $x = (r, t)$, it is $\delta(r - r')\delta(t - t')$. Equation (5.2.19) gave an example, and its solution was (5.2.20). In Chapter 5, we have also introduced the expressions of $\delta(r - r')$ in various coordinates.

The parameter λ on the left hand side is in general complex. Green's function depends on the parameter λ, but it is not explicitly shown in $G(x, x')$. $L(x)$ is usually a differential operator: when $x = t$, it involves derivative with respect to time; when $x = \boldsymbol{r}$, it involves derivatives with respect to space coordinates; when $x = (\boldsymbol{r}, t)$, it involves derivatives with respect to time and space coordinates, e.g., (5.2.19).

The mostly encountered $L(x)$ is a differential operator of second order. If it is written in the form of an SL operator, there is a weight function $\rho(x)$, which plays a role in orthonormalization of eigenfunctions of the operator $L(x)$. In this case, we use (6.1.1b) to find Green's function. In general, (6.1.1a) is used, which implies that $\rho(x) = 1$. Please note that when the orthonormalization of eigenfunctions is not concerned, the weight $\rho(x)$ in the inner product can be chosen arbitrarily, as long as it does not change its sign on the interval.

The following formulas are derived from (6.1.1b). They will naturally go to those derived from (6.1.1a) by letting $\rho(x) = 1$, the condition being that the eigenfunctions are not concerned.

Since $L(x)$ is now a differential operator, $G(x, x')$ satisfies a differential equation. We emphasized that the right hand side of (6.1.1) is a specific function, the delta function.

Now that this is a differential equation, in the case of x being space coordinates, boundary conditions are considered, and in the case of x being time t, initial conditions are considered. Hereafter, both boundary and initial conditions are roughly referred to as boundary conditions. When solving Green's function, boundary conditions should be given at the same time. If not so, the space is infinite in every direction, and so is time.

The Green's function in an infinite space is called the **basic solution of Green's function**.

In this chapter, we consider the case of self-adjoint differential operators except the last section. The argument is real, the operator is formally self-adjoint and the coefficients in boundary conditions are all real. The boundary conditions are in most cases homogeneous.

Mathematician Green found the Green's function in his first paper in 1828, although the name Green's function was later suggested by Riemann. However, after about 100 years, the delta function was suggested by famous physician Dirac. Before the appearance of the delta function, the mathematical theory about Green's function had been consummate. That is to say, mathematicians have been able to establish all the formulas concerning Green's function, without resorting to the delta function, and they would rather do so. After all, the delta function is a generalized function, which involves problems related to the stringency in mathematics. It is interesting that (6.1.1) relates directly the two functions. The merit of this definition is that it reveals the physical meaning of Green's function, and it is quite convenience in proving and deriving some formulas. This is what physicists like.

6.1.2. *Properties of Green's function*

1. The role of Green's function

Consider an inhomogeneous differential equation

$$[\lambda - L(x)]\psi(x) = f(x), \tag{6.1.2}$$

which satisfies certain given boundary conditions. Our task is to find its solution ψ. It is assumed that under certain conditions, the eigenfunction $\varphi_n(x)$ corresponding to the eigenvalue λ_n of the operator $L(x)$ is solved,

$$L(x)\varphi_n(x) = \lambda_n \varphi_n(x) \tag{6.1.3}$$

then from the Theorem 9 in Subsection 3.1.3 we can put down

$$\psi(x) = \varphi_n(x) + \xi(x), \tag{6.1.4}$$

where $\xi(x)$ is a special solution satisfying the inhomogeneous equation

$$[\lambda - L(x)]\xi(x) = f(x). \tag{6.1.5}$$

Let us see the following equation.

$$(G(\lambda^*), (\lambda - L)\xi) - ((\lambda^* - L)G(\lambda^*), \xi) = [J(\xi, G)]_a^b. \tag{6.1.6}$$

The two terms on the left hand side are respectively as follows.

$$(G(\lambda^*), (\lambda - L)\xi) = \int_V \rho(x)G^*(x, x'; \lambda^*)\left[\lambda - L(x)\right]\xi(x)\mathrm{d}x$$

$$= \int_a^b \rho(x)G^*(x, x'; \lambda^*)f(x)\mathrm{d}x \qquad (6.1.7)$$

and

$$((\lambda^* - L)G(\lambda^*), \xi) = \int_V \rho(x)\{[\lambda^* - L(x)]G(x, x'; \lambda^*)\}^*\xi(x')\mathrm{d}x$$

$$= \xi(x') = \psi(x') - \varphi_n(x'). \qquad (6.1.8)$$

The knot on the right hand side of (6.1.6) is expressed by (3.2.10). Thus, we obtain

$$\psi(x') = \varphi(x') + \int_a^b \rho(x)G^*(x, x'; \lambda^*)f(x)\mathrm{d}x$$

$$+ \left[p(x)\left\{\xi(x)\frac{\partial}{\partial x}G^*(x, x'; \lambda^*) - \xi'(x)G^*(x, x'; \lambda^*)\right\}\right]_a^b.$$

$$(6.1.9)$$

Now we exchange x and x', and use the symmetry of the Green's function,

$$G^*(x, x'; \lambda^*) = G(x', x; \lambda) \qquad (6.1.10)$$

which will be proven later. After the manipulation, the general expression of the solution of Eq. (6.1.2) is achieved.

$$\psi(x') = \varphi_n(x') + \int_a^b \rho(x)G^*(x, x'; \lambda^*)f(x)\mathrm{d}x$$

$$+ \left[p(x)\left\{\xi(x)\frac{\partial}{\partial x}G^*(x, x'; \lambda^*) - \xi'(x)G^*(x, x'; \lambda^*)\right\}\right]_a^b.$$

$$(6.1.11)$$

However, this general form is actually not seen. We have to present the explicit form distinguishing if the operator $L(x)$ in (6.1.2) is self-adjoint or not.

(1) The $L(x)$ in (6.1.2) is self-adjoint.

In this case, the knot on the right hand side of (6.1.6) is zero. If λ happens to be an eigenvalue λ_n, then (6.1.11) is simplified to be

$$\psi(x) = \varphi_n(x) + \int \rho(x')G(x, x'; \lambda_n)f(x')\mathrm{d}x'. \qquad (6.1.12a)$$

This solution is easily proved to satisfy (6.1.2) by use of (6.1.4) and (6.1.1). If λ is not an eigenvalue, then (6.1.11) is simplified to be

$$\psi(x) = \int \rho(x')G(x, x'; \lambda)f(x')\mathrm{d}x'. \qquad (6.1.12b)$$

Only the special solution is left.

In Subsection 3.1.3 we have mentioned that the solution of an ordinary inhomogeneous differential equation was its special solution plus the general solution of the homogeneous equation. The first term in (6.1.12) is just the general solution of homogeneous equation, and the second term is just a special solution of the inhomogeneous equation. In view of this, Green's function is mainly used to find a special solution of an inhomogeneous differential equation. In Chapter 3, we have introduced a general theory about the solutions of homogeneous differential equations of second order. The present chapter provides a standard method to find a special solution of inhomogeneous equations. The key is to find the Green's function of (6.1.1). As soon as it is found, the special solution of (6.1.3) is expressed in the form of the last term in (6.1.12).

Recall that in Chapter 3 a formula (3.1.24) was given for the special solution of inhomogeneous equation. Comparatively, the merit of (6.1.4) is that the Green's function $G(x, x')$ is of clear physical significance, and easily applied to higher-dimensional cases while (3.1.24) is applicable to the case of one argument.

For the differential equation (6.1.2) or (6.1.4), there is a corresponding Green's function satisfying (6.1.1). Whether the boundary conditions of the three equations are the same or not depends on concrete physical problems. As long as Eq. (6.1.1) is given certain boundary conditions, it is called the boundary value problem that the Green's function should meet.

(2) The $L(x)$ in (6.1.2) is not self-adjoint.

In this case, the knot on the right hand side of (6.1.6) is not zero, but usually, there is no solution for the corresponding homogeneous equation (6.1.4). Then (6.1.11) degrades to be

$$\psi(x) = \int_a^b \rho(x')G(x, x'; \lambda)f(x')dx$$

$$+ \left[p(x') \left\{ \xi(x') \frac{\partial}{\partial x'} G(x, x'; \lambda) - G(x, x'; \lambda)\xi'(x') \right\} \right]_a^b.$$

$$(6.1.13)$$

The second term takes into account the influence of the boundary. When the boundary is complex, the solution is of the form of (6.1.13).

In the following, we always consider the case (1) unless specified.

2. The existence, uniqueness and symmetry of Green's function

(1) The existence and uniqueness of Green's function

Suppose that there is an ordinary differential operator of order n:

$$L(x) = \sum_{i=0}^{n} p_i(x) \frac{d^i}{dx^i}, \qquad (6.1.14)$$

where $p_i(x)$, $i = 0, 1, 2, \ldots, n$ are continuous functions. Consider the following boundary value problem:

$$L(x)y(x) = \sum_{i=0}^{n} p_i(x) \frac{d^i}{dx^i} y(x) = 0, \quad a < x < b, \qquad (6.1.15a)$$

$$\sum_{i=1}^{n} \left[\alpha_{k,i} y^{(i-1)}(a) + \beta_{k,i} y^{(i-1)}(b) \right] = 0, \quad k = 1, 2, \ldots, n, \quad (6.1.15b)$$

where equation and boundary conditions are all homogeneous.

Correspondingly, the Green's function should satisfy the following boundary value problem:

$$L(x)G(x, x') = \delta(x - x'), \quad a < x, \ x' < b, \qquad (6.1.16a)$$

$$\sum_{i=0}^{n-1} \left\{ \alpha_{k,i} \left[\frac{d^i}{dx^i} G(x, x') \right]_{x=a} + \beta_{k,i} \left[\frac{d^i}{dx^i} G(x, x') \right]_{x=b} \right\} = 0,$$

$$k = 1, 2, \ldots, n. \qquad (6.1.16b)$$

The features of this Green's function is that at $x = x'$, $\frac{\mathrm{d}^{n-1}}{\mathrm{d}x^{n-1}} G(x, x')$ is discontinuous but other derivatives are continuous.

Theorem 1. *If the homogeneous boundary value problem* (6.1.15) *does not have nonzero solution, then* (6.1.16) *has one and only one Green's function solution.*

(2) Symmetry of Green's function

In the case of $\rho(x') = 1$, only the delta function is left on the right hand side of (6.1.1) and it depends on $x - x'$. If the interval is unbounded,

$$\frac{\mathrm{d}}{\mathrm{d}x} = \frac{\mathrm{d}}{\mathrm{d}(x - x')}. \tag{6.1.17}$$

Then every term in (6.1.14) is a function of $x - x'$ or a constant. It is seen that the Green's function must be of the following form:

$$G(x, x') = G(x - x'). \tag{6.1.18}$$

If $x = t$, $G(t, t') = G(t - t')$ represents translational invariance with respect to time; if $x = r$, $G(r, r') = G(r - r')$, represents the translational invariance with respect to space.

In many cases, Green's function is symmetric with respect to transposed arguments:

$$G(x, x') = G(x', x). \tag{6.1.19}$$

Note that if the interval is not unbounded, (6.1.17) does not stand, and subsequently Green's function will do not appear in the form of (6.1.18).

In the one-dimensional case, we explicitly write the infinite interval:

$$-\infty < x, \quad x' < \infty. \tag{6.1.20}$$

For the cases of finite intervals, we will show in Subsection 6.4.1 that if $L(x)$ is a Sturm-Liouville operator, then $G^*(x_1, x_2; \lambda) = G(x_2, x_1; \lambda^*)$.

Hereafter we mainly discuss the Green's functions of differential equations of second order.

6.1.3. *Methods of obtaining Green's function*

There have been several standard methods to find Green's functions. Here we introduce the eigenfunction method, piecewise expression method and Fourier transformation method. The subsequent sections show various examples of application of the three methods. Each method applies under certain conditions.

1. Eigenfunction method

Given the Hamiltonian H of a system of quantum mechanics, its eigenvalues and corresponding eigenfunctions can be in principle solved. Mathematically, this corresponds to a boundary value problem: one solves the eigenvalues and eigenfunctions of a Sturm-Liouville differential equation under certain boundary conditions.

Let the eigenvalues of $L(x)$ be denoted by λ_n and their eigenfunctions be denoted by $\varphi_n(x)$. Then,

$$L(x)\varphi_n(x) = \lambda_n\varphi_n(x), \tag{6.1.21}$$

where the eigenfunction set $\{\varphi_n(x)\}$ obeys homogeneous boundary conditions and the eigenfunctions belonging to different eigenvalues are orthogonal to each other. We assume that every eigenfunction $\varphi_n(x)$ has been normalized. If Green's function obeys homogeneous boundary conditions, then it can be expressed by this set of eigenfunctions as follows:

$$G(x, x') = \sum_n c_n(x')\varphi_n(x). \tag{6.1.22}$$

We are to find the coefficients $c_n(x')$. Let the operator $\lambda - L(x)$ act on (6.1.22). The result is

$$\delta(x - x') = \rho(x) \sum_n c_n(x') \left[\lambda - L(x)\right] \varphi_n(x)$$

$$= \rho(x) \sum_n c_n(x')(\lambda - \lambda_n)\varphi_n(x). \tag{6.1.23}$$

We must distinguish if the operator $L(x)$ is self-adjoint or not.

(1) The $L(x)$ is self-adjoint.

In this case, multiplying $\varphi_m^*(x)$ to (6.1.23) and integrating with respect to x, we get

$$\int \mathrm{d}x \varphi_m^*(x)\delta(x-x')$$

$$= \sum_n c_n(x')(\lambda - \lambda_n)\int \mathrm{d}x \rho(x)\varphi_m^*(x)\varphi_n(x). \quad (6.1.24)$$

The functions $\{\varphi_n(x)\}$ have orthonormalization. Thus, the coefficients are

$$c_m(x') = \frac{\varphi_m^*(x')}{\lambda - \lambda_m}. \quad (6.1.25)$$

Consequently, the expression of Green's function is

$$G(x, x', \lambda) = \sum_n \frac{\varphi_n^*(x')\varphi_n(x)}{\lambda - \lambda_n}. \quad (6.1.26)$$

Here the parameter λ is explicitly shown. If the eigenvalue spectrum is continuous, the summation in (6.1.26) is replaced by an integral. Equation (6.1.26) indicates that the parameter λ should not be equal to the eigenvalues λ_n solved from (6.1.21). The eigenvalues are simple poles of Green's function so that it cannot be defined at these points.

If $\lambda = \lambda_m$ happen to be one of the spectrum $\{\lambda_n\}$, there are two possible cases: (i) if $\{\lambda_n\}$ is a discrete spectrum, a modified Green's function can be defined, which will be introduced in Subsection 6.4.3; (ii) if $\{\lambda_n\}$ is a continuous spectrum, we let the parameter λ approach to the real axis from either the upper or lower complex plane, so as to define two side limits of the Green's function at λ_n. Both side limits are of physical significance, which will be introduced when discussing the basic solutions.

The subscript n of λ_n not only means one index. It can also mean two or three indices in the two- or three-dimensional cases. The summation in (6.1.26) involves all the indices.

(2) The $L(x)$ is not self-adjoint.

In this case, the adjoint equation of (6.1.21) is solved,

$$L^\dagger(x)\zeta_n(x) = \gamma_n \zeta_n(x). \quad (6.1.27)$$

The solutions $\{\zeta_n(x)\}$ are assumed to be a complete set. Then, by orthogonality, multiplying $\zeta_m^*(x)$ (6.1.23) and integrating with respect to x, we get

$$\int \mathrm{d}x\zeta_m^*(x)\delta(x-x')$$

$$= \sum_n c_n(x')(\lambda-\lambda_n)\int \mathrm{d}x\rho(x)\zeta_m^*(x)\varphi_n(x)$$

Thus, the coefficients are

$$c_m(x') = \frac{\zeta_m^*(x')}{\lambda-\lambda_m}. \tag{6.1.28}$$

Consequently, the expression of Green's function is

$$G(x,x',\lambda) = \sum_n \frac{\zeta_n^*(x')\varphi_n(x)}{\lambda-\lambda_n}. \tag{6.1.29}$$

2. Piecewise expression method

This method merely applies to a one-dimensional case. By inspection of (6.1.1) we see that it is zero at both $x>x'$ and $x<x'$ regions. So, in these two regions the equation is simplified to be a homogeneous equation

$$[\lambda-L(x)]\varphi(x) = 0. \tag{6.1.30}$$

It is comparatively easier solved. The boundary conditions will be taken into account later.

Suppose that $L(x)$ is an SL operator,

$$L(x) = -A(x)\frac{\mathrm{d}^2}{\mathrm{d}x^2} - B(x)\frac{\mathrm{d}}{\mathrm{d}x} + C(x)$$

$$= \frac{1}{\rho(x)}\left[-\frac{\mathrm{d}}{\mathrm{d}x}p(x)\frac{\mathrm{d}}{\mathrm{d}x} + q(x)\right]. \tag{6.1.31}$$

Then, (6.1.1) becomes

$$\left[\lambda\rho(x) + \frac{\mathrm{d}}{\mathrm{d}x}p(x)\frac{\mathrm{d}}{\mathrm{d}x} - q(x)\right]G(x,x') = \delta(x-x'). \tag{6.1.32a}$$

The corresponding homogeneous equation is

$$\left[\lambda + \frac{1}{\rho(x)}\frac{\mathrm{d}}{\mathrm{d}x}p(x)\frac{\mathrm{d}}{\mathrm{d}x} - \frac{q(x)}{\rho(x)}\right]\varphi(x) = 0. \tag{6.1.32b}$$

We have known in Chapter 3 that as soon as the two linearly independent special solutions $\psi_1(x)$ and $\psi_2(x)$ of (6.1.32b) are solved, the Green's function in each region can be expressed by the linear combination of them. Thus, we can write

$$G(x, x') = \begin{cases} c_1(x')\psi_1(x) + c_2(x')\psi_2(x), & x > x'. \\ d_1(x')\psi_1(x) + d_2(x')\psi_2(x), & x < x'. \end{cases} \quad (6.1.33)$$

The Green's function is required to be continuous at $x = x'$, so that

$$\left[c_1(x') - d_1(x')\right]\psi_1(x') + \left[c_2(x') - d_2(x')\right]\psi_2(x') = 0. \quad (6.1.34)$$

Now we integrate both sides of (6.1.32a) from $x = x' - 0^+$ to $x = x' + 0^+$. Because the integral region is infinitely small, the integrals of continuous functions vanish. We get

$$\int_{x-0^+}^{x+0^+} dx \left[\lambda\rho(x) + \frac{d}{dx}p(x)\frac{d}{dx} - q(x)\right]G(x, x')$$

$$= p(x')\left[\frac{d}{dx}G(x, x')\right]_{x-0^+}^{x+0^+} = 1.$$

The Green's function at the two sides of $x = x'$ has been known as (6.1.34). When this expression is used, we have

$$\rho(x')A(x')\left[(c_1(x') - d_1(x'))\psi_1'(x) + (c_2(x') - d_2(x'))\psi_2'(x)\right] = 1. \quad (6.1.35a)$$

In the case of ordinary differential equations of second order with constant coefficients, i.e., $A(x)$, $B(x)$ and $C(x)$ are constants, we use (6.1.1a). Then

$$\int_{x-0^+}^{x+0^+} dx \left(A\frac{d^2}{dx^2} + B\frac{d}{dx} + C\right)G(x, x')$$

$$= A\left[\frac{d}{dx}G(x, x')\right]_{x-0^+}^{x+0^+} = 1.$$

When this is applied to (6.1.33), we have

$$A[(c_1(x') - d_1(x'))\psi_1'(x) + (c_2(x') - d_2(x'))\psi_2'(x)] = 1. \quad (6.1.35b)$$

Solving (6.1.35) and (6.1.34) simultaneously, we obtain

$$c_1(x') - d_1(x') = -\frac{\psi_2(x')}{\rho(x')A(x')W(x')}, \quad c_2(x') - d_2(x')$$

$$= \frac{\psi_1(x')}{\rho(x')A(x')W(x')}, \qquad (6.1.36)$$

where

$$W(x') = \psi_1(x')\psi_2'(x) - \psi_2(x')\psi_1'(x) \qquad (6.1.37)$$

is Wronskian. Since ψ_1 and ψ_2 are linearly independent of each other, their Wronskian is nonzero. We denote

$$A_W(x') = \rho(x')A(x')W(y_1(x'), y_2(x')) \qquad (6.1.38a)$$

in the case of (6.1.35a) and

$$A_W(x') = AW(y_1(x'), y_2(x')) \qquad (6.1.38b)$$

in the case of (6.1.35b).

After (6.1.37) is substituted into (6.1.33), there are yet two quantities undetermined. They are to be determined by the boundary conditions. Usually, the Green's function is written in one of the following forms.

(i)

$$G(x, x') = \begin{cases} \left[d_1(x') - \dfrac{\psi_2(x')}{A_W(x')}\right]\psi_1(x) + c_2(x')\psi_2(x), & x > x'. \\[3mm] d_1(x')\psi_1(x) + \left[c_2(x') - \dfrac{\psi_1(x')}{A_W(x')}\right]\psi_2(x), & x < x'. \end{cases}$$

$$(6.1.39a)$$

Particularly, if $d_1(x') = c_2(x') = 0$, (6.1.39a) is simplified to be

$$G(x, x') = -\frac{1}{A_W(x')}[\psi_2(x')\psi_1(x)\theta(x - x')$$

$$+\psi_1(x')\psi_2(x)\theta(x' - x)]. \qquad (6.1.39b)$$

(ii)

$$G(x, x') = \begin{cases} \left[d_1(x') - \dfrac{\psi_2(x')}{A_W(x')}\right]\psi_1(x) \\[3mm] \quad + \left[d_2(x') + \dfrac{\psi_1(x')}{A_W(x')}\right]\psi_2(x), & x > x'. \\[3mm] d_1(x')\psi_1(x) + d_2(x')\psi_2(x), & x < x'. \end{cases} \quad (6.1.40a)$$

The two quantities $d_1(x'), d_2(x')$ are to be determined by boundary conditions. A special case is that $d_1(x') = d_2(x') = 0$. Then (6.1.40b) is simplified to be

$$G(x, x') = -\frac{1}{A_W(x')}[\psi_2(x')\psi_1(x) - \psi_1(x')\psi_2(x)]\theta(x - x').$$

$$(6.1.40b)$$

(iii)

$$G(x, x') = \begin{cases} c_1(x')\psi_1(x) + c_2(x')\psi_2(x), & x > x'. \\ \left[c_1(x') + \dfrac{\psi_2(x')}{A_W(x')}\right]\psi_1(x) \\ \quad + \left[c_2(x') - \dfrac{\psi_1(x')}{A_W(x')}\right]\psi_2(x), & x < x'. \end{cases}$$

$$(6.1.41a)$$

The two quantities $c_1(x'), c_2(x')$ are to be determined by boundary conditions. A special case is that $c_1(x') = c_2(x') = 0$. Then (6.1.41b) is simplified to be

$$G(x, x') = \frac{1}{A_W(x')}[\psi_2(x')\psi_1(x) - \psi_1(x')\psi_2(x)]\theta(x' - x).$$

$$(6.1.41b)$$

Comparison of (6.1.26) and (6.1.39)–(6.1.41) shows the following features.

(1) $G(x, x')$ depends on parameter λ, so that (6.1.39)–(6.1.41), each Wronskian contains the parameter λ.
(2) The piecewise expression method is mainly used to the argument x being one-dimensional case. This is embodied in that integration from the left to right sides of $x = x'$ has been carried out, as (6.1.35). In contrast, the eigenfunction method does not have constraint of dimensions, that is to say, x can be multivariables.
(3) The eigenfunction method is based on the existence of eigenvalues and eigenfunctions. This method will fail if the boundary value problem does not have nonzero eigenfunctions.

In Section 6.5, we will introduce how to employ the piecewise expression method in the two- and three-dimensional cases.

3. The Fourier transformation method

If $L(x)$ is a differential operator with constant coefficients and the interval is unbounded as (6.1.20), the Fourier transformation can be applied to solve Green's function.

For example, $L(x)$ is a differential operator of second order with constant coefficients as follows:

$$L(x) = A\frac{\mathrm{d}^2}{\mathrm{d}x^2} + 2B\frac{\mathrm{d}}{\mathrm{d}x} + C. \tag{6.1.42}$$

Suppose that the argument x is time t. Following (6.1.1a), the differential equation that Green's function should satisfy is

$$\left(A\frac{\mathrm{d}^2}{\mathrm{d}t^2} + 2B\frac{\mathrm{d}}{\mathrm{d}t} + C\right)G(t,t') = \delta(t-t'). \tag{6.1.43}$$

By (6.1.17), Green's function is of translational invariance as the form of (6.1.18).

$$\left(A\frac{\mathrm{d}^2}{\mathrm{d}t^2} + 2B\frac{\mathrm{d}}{\mathrm{d}t} + C\right)G(t-t') = \delta(t-t'). \tag{6.1.44}$$

Now we make the Fourier transformation of Green's function:

$$G(t-t') = \frac{1}{2\pi}\int_{-\infty}^{\infty} G(\omega)\mathrm{e}^{-\mathrm{i}\omega(t-t')}\mathrm{d}\omega. \tag{6.1.45}$$

This equation, together with the Fourier transformation of $\delta(t-t')$ (5.1.10), is substituted into (6.1.44).

$$(-A\omega^2 - 2\mathrm{i}B\omega + C)G(\omega) = 1 \tag{6.1.46}$$

Then the Fourier component of Green's function is solved:

$$G(\omega) = \frac{1}{-A\omega^2 - 2\mathrm{i}B\omega + C}. \tag{6.1.47}$$

Its inverse Fourier transformation is

$$G(t-t') = \frac{1}{2\pi}\int_{-\infty}^{\infty} \frac{\mathrm{e}^{-\mathrm{i}\omega(t-t')}}{-A\omega^2 - 2\mathrm{i}B\omega + C}\mathrm{d}\omega. \tag{6.1.48}$$

The denominator of (6.1.47) is just the root discriminant of the operator (6.1.42). It has two roots. They are denoted by ω_1 and ω_2 with $\omega_1 \neq \omega_2$. Then,

$$-A\omega^2 - 2\mathrm{i}B\omega + C = -A(\omega - \omega_1)(\omega - \omega_2), \qquad (6.1.49)$$

where

$$\omega_{1,2} = \frac{1}{A}\left(-\mathrm{i}B \pm \sqrt{AC - B^2}\right). \qquad (6.1.50)$$

So, (6.1.47) can be written as

$$G(\omega) = \frac{1}{\omega_2 - \omega_1}\left(\frac{1}{\omega - \omega_1} - \frac{1}{\omega - \omega_2}\right). \qquad (6.1.51)$$

It is substituted into (6.1.48) to get

$$G(t - t') = -\frac{1}{4\pi\sqrt{AC - B^2}}\int_{-\infty}^{\infty}\left(\frac{\mathrm{e}^{-\mathrm{i}\omega(t-t')}}{\omega - \omega_1} - \frac{\mathrm{e}^{-\mathrm{i}\omega(t-t')}}{\omega - \omega_2}\right)\mathrm{d}\omega.$$

$$(6.1.52)$$

The integrand contains two terms, each having a simple pole. For each term, one can supplement an infinite semicircle in the upper or lower complex ω plane, depending on where the pole ω_1 or ω_2 is positioned, to form a closed integral path, and then use residue theorem to carry out the integration.

Besides, in electrostatics, the image method is a simple and direct way of solving Green's function of electrostatic field in some regions with special shapes. This method can be considered as derived from the basic solution of Green's function, and will be introduced in Section 6.6. For Laplace equation in two-dimensional special regions, conformal transformation in complex functions can be used to solve Green's function. This method is derived from the image method, which will not be discussed in this book.

The different methods introduced above have their own scopes of application. The eigenfunction method has almost no constraint, and can be employed in all the cases as long as the eigenfunctions exist. However, if a system has no nonzero eigenfunctions, other methods have to be used. The piecewise expression method can merely be used in a one-dimensional case. The Fourier transformation method is applicable for the cases of differential operators with constant coefficients and unbounded interval.

4. Procedure of solving inhomogeneous equations

We have mentioned that the main purpose of Green's function is to find the solution of an inhomogeneous equation. Now the general procedure of solving inhomogeneous equation (6.1.2) is outlined as follows.

(1) Solve the corresponding homogeneous equation (6.1.4) to obtain eigenvalues and eigenfunctions.
(2) Solve Green's function from (6.1.1). In this course either the eigenfunction method or others may be used. In the former case the eigenvalues and eigenfunctions solved in (1) are utilized.
(3) The general solution of (6.1.2) is achieved by means of formula (6.1.12).

6.1.4. *Physical meaning of Green's function*

1. The meaning of the distribution of a field generated from a point source

It is well known that the distribution of electrostatic potential $\varphi(\boldsymbol{r})$ in vacuum observes the Poisson equation:

$$\nabla^2\varphi(\boldsymbol{r}) = \rho(\boldsymbol{r})/\varepsilon_0, \tag{6.1.53}$$

where $\rho(\boldsymbol{r})$ is the distribution of charges in space and ε_0 is permittivity of vacuum. If there is only one point charge q at position $\boldsymbol{r} = \boldsymbol{r}_0$, the Poisson equation becomes

$$\nabla^2 G(\boldsymbol{r}, \boldsymbol{r}') = q\delta(\boldsymbol{r} - \boldsymbol{r}')/\varepsilon_0. \tag{6.1.54}$$

Formally, the $\varepsilon_0\nabla^2/q$ here corresponds to $\lambda - L(x)$ in (6.1.1). The Green's function $G(\boldsymbol{r}, \boldsymbol{r}')$ solved from (6.1.38) represents the space distribution of electrostatic potential generated by the point charge at $\boldsymbol{r} = \boldsymbol{r}_0$, so that the point $\boldsymbol{r}_0$ is a "field source". If $G(\boldsymbol{r}, \boldsymbol{r}') = G(\boldsymbol{r}', \boldsymbol{r})$, Green's function is symmetric with respect to the exchange of a field point and field source.

This conception can be generalized. Green's function $G(x, x')$ represents the space distribution of field at point x' generated by a "point source". The distribution law is determined by the operator $L(x)$. In the case of electrostatics, the distribution law of electrostatic potential is determined by Laplace operator ∇^2.

2. The meaning of response function

A system may be affected by an external field which is embodied in the term $f(x)$ on the right hand side of (6.1.2). Then the second term in the solution (6.1.12) reflects the behavior of the system at point x after it is acted on by the field. This reflects the response of the system to the external action. Therefore, Green's function is of the physical meaning of a system's **response function**. The first term in (6.1.12) is irrelative to the external action. Only when an external action appears, Green's function plays its role.

If the argument x is time t, Green's function $G(t,t')$ reflects the behavior of the system at time t when the system evolve starting form time t'. In physics, Green's function can also be called propagation function or propagator. In this case, Green's function reflects the effect at time t after it is acted on by the external field $f(t')$ at time t'. By (6.1.4), the behavior of the system at time t is determined by the total effect of $f(t')$ at all time t'.

(1) If Green's function is of the form of (6.1.40c), then only when $t > t'$ the external field shows its effect. Such a Green's function can be called a **retarded Green's function**. In Section 6.3 we will show an example.

(2) If Green's function is of the form of (6.1.41c), then only when $t < t'$ the external field shows its effect. Such a Green's function can be called an **advanced Green's function**. In classical physics, the advance Green's function seems a violation of common sense. Nevertheless, in quantum mechanics, the advanced Green's function may exist, and has definite physical meaning: this is a kind of course eliminating particles or generating anti-particles.

(3) If Green's function is of the form of (6.1.39a), the first and second terms play roles at $t > t'$ and $t < t'$, respectively. Thus both retarded and advanced effect can occur. Such a Green's function may be called **causal Green's function**.

6.2.   The Basic Solution of Laplace Operator

In this section, we consider the simplest case: $L(x)$ in (6.1.1) is Laplace operator

$$L(r) = -\nabla_r^2. \tag{6.2.1}$$

Now the weight $\rho(x) = 1$. Equation

$$\nabla_r^2 \psi(r) = 0 \tag{6.2.2}$$

is called **Laplace equation**, or **harmonic equation**. The function satisfying this equation is called a **harmonic function**. In (6.2.2), the parameter λ has been set to zero. If it is not zero, the equation

$$(\lambda + \nabla_r^2)\psi(r) = 0 \tag{6.2.3}$$

is called a **Helmholtz equation**. Often, in a Helmholtz equation the parameter is written as $\lambda = k^2$.

Corresponding to a Helmholtz equation, its Green's function should satisfy the following equation:

$$(\lambda + \nabla_r^2)G(r, r'; \lambda) = \delta(r - r'). \tag{6.2.4}$$

Once this Green's function is found, letting $\lambda \to 0$ leads to that of a Laplace equation. Equation (6.2.4) is solved in the whole space. Its boundary condition is that the values of Green's function goes to zero at infinite boundaries.

We use eigenfunction method. First of all, the eigenfunctions φ_n of Laplace operator have to be solved.

$$-\nabla_r^2 \varphi_n(r) = \lambda_n \varphi_n(r). \tag{6.2.5}$$

This is equivalent to solving a Helmholtz equation. The eigenfunction of (6.2.5) are

$$\varphi_k(r) = \frac{1}{\sqrt{2\pi}^d} e^{ik\cdot r}, \tag{6.2.6}$$

where $d = 1, 2, 3$ corresponds to one-, two- and three-dimensional cases, respectively. The eigenvalues are

$$\lambda_n = k^2. \tag{6.2.7}$$

Now the vector k replaces n to be the index labeling eigenvalues. By directly using (6.1.14), Green's function can be evaluated. Before doing that, we consider how to do summation.

Equation (6.2.7) shows that the eigenvalue spectrum is continuous, from $0 \sim +\infty$, and there is no discrete eigenvalues. Every eigenvalue corresponds to $\pm k$, while each k corresponds to an eigenfunction. Physically, this is called double degeneration. Please note that the summation in (6.1.14) should cover all possible degenerate states. In the present case, the summation over n in (6.1.14) is replaced by that over k. The vector k varies continuously in the whole k space, so that the summation has to be replaced by integration. In the two- and three-dimensional spaces, k has two and three components, respectively. The integration in two- or three-dimensional space is carried out. Based on these discussions, (6.1.14) is converted to be the following form:

$$G(r, r'; \lambda) = \sum_k \frac{\varphi_k^*(r')\varphi_k(r)}{\lambda - k^2} = \int \frac{dk}{(2\pi)^d} \frac{e^{ik \cdot (r-r')}}{\lambda - k^2}, \qquad (6.2.8)$$

where shorthand $dk = d^d k$ is used. In the following, the cases of three-, two- and one-dimensional spaces are evaluated explicitly.

In general, λ is complex, and its square root is denoted as

$$\sqrt{\lambda} = p + iq. \qquad (6.2.9)$$

In the discussions below, we assume $\text{Im}\lambda > 0$ and $p > 0$. Other cases can be discussed similarly.

6.2.1. *Three-dimensional space*

In (6.2.8), $d = 3$. Let $\rho = r - r'$ and the angle between ρ and k be θ. The direct calculation of (6.2.8) results in

$$G(r, r'; \lambda) = \frac{1}{8\pi^3} \int_0^\infty \frac{2\pi k^2 dk}{\lambda - k^2} \int_0^\pi \sin\theta d\theta e^{ik\rho\cos\theta}$$

$$= \frac{1}{4i\pi^2\rho} \int_{-\infty}^\infty \frac{ke^{ik\rho}}{\lambda - k^2} dk,$$

where the second term in the second equal sign has been made by transformation $k \to -k$ to become integration on the interval $(-\infty, 0)$. The integrand then is decomposed into two terms.

$$G(r, r'; \lambda) = \frac{1}{8i\pi^2\rho} \int_{-\infty}^\infty \left(\frac{e^{ik\rho}}{k - \sqrt{\lambda}} + \frac{e^{ik\rho}}{k + \sqrt{\lambda}} \right) dk. \qquad (6.2.10)$$

The integral path is the real axis in the complex k plane. We make use of the residue theorem. The exponential factors in the two term of the integrand indicate that an infinite semicircle in the upper half plane can be supplemented to compose a closed integral path. Since $\sqrt{\lambda}$ is as (6.2.9), the simple poles in the two terms in (6.2.10) are positioned at upper and lower half plane, respectively. Therefore, only one term the pole of which is in the upper half plane contributes to the integral.

If $q > 0$, the pole $\sqrt{\lambda}$ of the first term in (6.2.10) is in the upper half plane, while that of the second term in the lower half plane. The result is

$$G(\boldsymbol{r}, \boldsymbol{r}'; \lambda) = -\frac{e^{i\sqrt{\lambda}|\boldsymbol{r}-\boldsymbol{r}'|}}{4\pi|\boldsymbol{r}-\boldsymbol{r}'|} = -\frac{e^{i(p+iq)|\boldsymbol{r}-\boldsymbol{r}'|}}{4\pi|\boldsymbol{r}-\boldsymbol{r}'|}$$

$$= -\frac{i\sqrt{\lambda}}{4\pi}h_0^{(1)}\left(\sqrt{\lambda}|\boldsymbol{r}-\boldsymbol{r}'|\right), \quad q > 0. \quad (6.2.11a)$$

The last equal sign is from (4.5.24c), which is the first kind spherical Hankel function of zero order. If $q < 0$, the pole $\sqrt{\lambda}$ of the second term in (6.2.10) is in the upper half plane, and the result is

$$G(\boldsymbol{r}, \boldsymbol{r}'; \lambda) = -\frac{e^{-i\sqrt{\lambda}|\boldsymbol{r}-\boldsymbol{r}'|}}{4\pi|\boldsymbol{r}-\boldsymbol{r}'|} = -\frac{e^{-i(p+iq)|\boldsymbol{r}-\boldsymbol{r}'|}}{4\pi|\boldsymbol{r}-\boldsymbol{r}'|}$$

$$= \frac{i\sqrt{\lambda}}{4\pi}h_0^{(2)}\left(\sqrt{\lambda}|\boldsymbol{r}-\boldsymbol{r}'|\right), \quad q < 0. \quad (6.2.11b)$$

The final result is the second kind spherical Hankel function of zero order, see (4.5.24d).

If $\lambda < 0$ is a negative real number, then let $p = 0$ in (6.2.11a) or (6.2.11b). We obtain

$$G(\boldsymbol{r}, \boldsymbol{r}'; -q^2) = -\frac{e^{-q|\boldsymbol{r}-\boldsymbol{r}'|}}{4\pi|\boldsymbol{r}-\boldsymbol{r}'|}$$

$$= \frac{q}{4\pi}h_0^{(1)}(iq|\boldsymbol{r}-\boldsymbol{r}'|), \quad q \geq 0. \quad (6.2.12)$$

Particularly, as $q = 0$, i.e., $\lambda = 0$,

$$G(\boldsymbol{r}, \boldsymbol{r}'; 0) = -\frac{1}{4\pi|\boldsymbol{r}-\boldsymbol{r}'|}. \quad (6.2.13)$$

This is just the solution of three-dimensional Laplace equation

$$\nabla_r^2 G(\boldsymbol{r},\boldsymbol{r}';0) = \delta(\boldsymbol{r}-\boldsymbol{r}'). \tag{6.2.14}$$

The physical meaning of the Green's function satisfying (6.2.14) is the electrostatic potential generated by a point charge positioned at $\boldsymbol{r}'$.

In the case that $\lambda > 0$ is a positive real number, i.e., within the spectrum of the operator $L(\boldsymbol{r}) = -\nabla_r^2$, the poles in the integrand in (6.2.10) are just at the real axis. Therefore, the integral cannot be done, and the Green's function cannot be defined. In this case, we make use of (6.2.11) to define two side limits. In (6.2.11), let the imaginary part q be infinitely small. Then we will get

$$G^+(\boldsymbol{r},\boldsymbol{r}';p^2) = -\frac{e^{ip|\boldsymbol{r}-\boldsymbol{r}'|}}{4\pi|\boldsymbol{r}-\boldsymbol{r}'|}$$

$$= -\frac{ip}{4\pi}h_0^{(1)}(p|\boldsymbol{r}-\boldsymbol{r}'|), \quad p>0 \tag{6.2.15a}$$

and

$$G^-(\boldsymbol{r},\boldsymbol{r}';p^2) = -\frac{e^{-ip|\boldsymbol{r}-\boldsymbol{r}'|}}{4\pi|\boldsymbol{r}-\boldsymbol{r}'|}$$

$$= \frac{ip}{4\pi}h_0^{(2)}(p|\boldsymbol{r}-\boldsymbol{r}'|), \quad p>0, \tag{6.2.15b}$$

respectively. G^+ (G^-) is the side limit when the parameter λ approaches the real axis from the upper (lower) half complex k plane. Therefore, as λ is a positive real number, we can choose one of these two expressions. Which one should be chosen then? It depends on the physical problem. Equation (6.2.15a) means a spherical wave propagating from the center outward, or outward wave, while (6.2.15b) a spherical wave propagating inward, or inward wave.

6.2.2. *Two-dimensional space*

In (6.2.8) $d = 2$. Let $\boldsymbol{\rho} = \boldsymbol{r} - \boldsymbol{r}'$, and the angle between $\boldsymbol{\rho}$ and $\boldsymbol{k}$ be θ. The direct calculation of (6.2.8) results in

$$G(\boldsymbol{r},\boldsymbol{r}';\lambda) = \int \frac{d\boldsymbol{k}}{(2\pi)^2}\frac{e^{i\boldsymbol{k}\cdot(\boldsymbol{r}-\boldsymbol{r}')}}{\lambda-k^2} = \int \frac{k d\theta dk}{(2\pi)^2}\frac{e^{ik\rho\cos\theta}}{\lambda-k^2}. \tag{6.2.16}$$

The plane wave factor $e^{\mathrm{i}k\rho\cos\theta}$ can be expanded by Bessel functions as (4.3.20b). Then, (6.2.16) becomes

$$G(\boldsymbol{r}, \boldsymbol{r}'; \lambda) = \frac{1}{4\pi^2} \int_0^\infty \frac{k\,\mathrm{d}k}{\lambda - k^2} \int_0^{2\pi} \mathrm{d}\theta \left[J_0(k\rho) + 2\sum_{n=1}^\infty \mathrm{i}^n J_n(k\rho) \cos n\theta \right]$$

$$= \frac{1}{2\pi} \int_0^\infty \mathrm{d}k \frac{k J_0(k\rho)}{\lambda - k^2}$$

$$= \frac{1}{4\pi} \int_0^\infty \mathrm{d}k \frac{k \left[H_0^{(1)}(k\rho) + H_0^{(2)}(k\rho) \right]}{\lambda - k^2}. \tag{6.2.17}$$

After integration with respect to angle, only the term of Bessel function of zero order is left. This term is in turn expressed by Bessel functions of the third kind as (4.5.1), so that (6.2.17) reaches the last form. By (4.5.6), we have

$$H_0^{(1)}(-z) = -H_0^{(2)}(z). \tag{6.2.18}$$

This helps us to merge the two terms in (6.2.17) into one, and meanwhile the integral interval is extended to the whole real axis.

$$G(\boldsymbol{r}, \boldsymbol{r}'; \lambda) = \frac{1}{4\pi} \int_{-\infty}^\infty \mathrm{d}k \frac{k H_0^{(1)}(k\rho)}{\lambda - k^2}$$

$$= -\frac{1}{8\pi} \int_{-\infty}^\infty \left(\frac{H_0^{(1)}(k\rho)}{k - \sqrt{\lambda}} + \frac{H_0^{(1)}(k\rho)}{k + \sqrt{\lambda}} \right) \mathrm{d}k. \tag{6.2.19}$$

Similar to the three-dimensional case, the residue theorem is used to do the integration. Form the asymptotic behavior of a Hankel function as its argument goes to infinite,

$$H_\nu^{(1)}(k\rho \to \infty) = \sqrt{\frac{2}{\pi k\rho}} \exp\left[\mathrm{i}\left(k\rho - \frac{\nu\pi}{2} - \frac{\pi}{4} \right) \right] \tag{6.2.20a}$$

and

$$H_\nu^{(2)}(k\rho \to \infty) = \sqrt{\frac{2}{\pi k\rho}} \exp\left[-\mathrm{i}\left(k\rho - \frac{\nu\pi}{2} - \frac{\pi}{4} \right) \right], \tag{6.2.20b}$$

it is seen that for the two terms in (6.2.19) one has to supplement an infinite semicircle in the upper half plane to compose a closed integral path. Again, since $\sqrt{\lambda}$ is as (6.2.9), the simple poles in the two terms in (6.2.19) are positioned at the upper and lower half plane, respectively. Therefore, only one term the pole of which is in the upper half plane contributes to the integral.

If $q > 0$, the pole $\sqrt{\lambda}$ of the first term in (6.2.10) is in the upper half plane, while that of the second term is in the lower half plane. Thus, the result is

$$G(\boldsymbol{r}, \boldsymbol{r}'; \lambda) = -\frac{1}{8\pi} \int_{-\infty}^{\infty} \frac{H_0^{(1)}(k\rho)}{k - \sqrt{\lambda}} \mathrm{d}k$$

$$= -\frac{i}{4} H_0^{(1)}((p + iq)\rho), \quad q > 0. \qquad (6.2.21\mathrm{a})$$

If $q < 0$, the pole $\sqrt{\lambda}$ of the second term in (6.2.19) is in the upper half plane, and the result is

$$G(\boldsymbol{r}, \boldsymbol{r}'; \lambda) = -\frac{1}{8\pi} \int_{-\infty}^{\infty} \frac{H_0^{(1)}(k\rho)}{k + \sqrt{\lambda}} \mathrm{d}k$$

$$= -\frac{i}{4} H_0^{(1)}(-(p + iq)\rho), \quad q < 0. \qquad (6.2.21\mathrm{b})$$

If $\lambda < 0$ is a negative real number, then let $p = 0$ in (6.2.21a) or (6.2.21b). We obtain

$$G(\boldsymbol{r}, \boldsymbol{r}'; -q^2) = -\frac{i}{4} H_0^{(1)}(iq\rho) = \frac{1}{2\pi} K_0(q\rho), \quad q \geq 0. \qquad (6.2.22)$$

Here we have made use of formula $H_\nu^{(1)}(iz) = -\frac{2}{\pi} i e^{-i\nu\pi/2} K_\nu(z)$, and K_0 is a modified second kind Bessel function of zero order.

Particularly, as $q = 0$, i.e., $\lambda = 0$, with the asymptotic expression of Hankel function $H_0^{(1)}(z \to 0) \sim i \frac{2}{\pi} \ln \frac{z}{2}$, we obtain

$$G(\boldsymbol{r}, \boldsymbol{r}'; \lambda \to 0) \to -\frac{i}{4} i \frac{2}{\pi} \ln \frac{\sqrt{\lambda}|\boldsymbol{r} - \boldsymbol{r}'|}{2}$$

$$= \frac{1}{2\pi} \left(\ln |\boldsymbol{r} - \boldsymbol{r}'| + \ln \frac{\sqrt{\lambda}}{2} \right).$$

Although the second term goes to infinity as $\lambda \to 0$, it is a constant in the whole space.

Therefore, as $\lambda = 0$, the Green's function of a two-dimensional Laplace operator is:

$$G(\boldsymbol{r}, \boldsymbol{r}'; 0) = \frac{1}{2\pi} \ln |\boldsymbol{r} - \boldsymbol{r}'| + \text{const.} \qquad (6.2.23)$$

Please note the reason of the appearance of the infinite constant: it arises from that as $\lambda = 0$, the pole in the integrand in (6.2.19) is not of first order.

Equation (6.2.23) is just the solution of two-dimensional Laplace equation $\nabla_r^2 G(\boldsymbol{r}, \boldsymbol{r}'; 0) = \delta(\boldsymbol{r} - \boldsymbol{r}')$. The physical meaning of the Green's function is the electrostatic potential generated by a point charge positioned at $\boldsymbol{r}'$ in a plane.

In the case that $\lambda > 0$ is a positive real number, i.e., within the spectrum of the operator $L(\boldsymbol{r}) = -\nabla_r^2$, the poles in the integrand in (6.2.19) are just at the real axis. Therefore, the integral cannot be done, and the Green's function cannot be defined. In this case, we make use of (6.2.21) to define two side limits. In (6.2.21), let the imaginary part q be infinitesimally small. Then we will get

$$G^+(\boldsymbol{r}, \boldsymbol{r}'; p^2) = -\frac{i}{4} H_0^{(1)}(p\rho), \quad p > 0 \qquad (6.2.24a)$$

and

$$G^-(\boldsymbol{r}, \boldsymbol{r}'; p^2) = -\frac{i}{4} H_0^{(1)}(-p\rho) = \frac{i}{4} H_0^{(2)}(p\rho), \quad p > 0 \qquad (6.2.24b)$$

respectively. G^+ (G^-) is the side limit when the parameter λ approaches the real axis from the upper (lower) half complex k plane. From (6.2.20) it is known that the asymptotic behavior of the first kind Hankel function means a cylindrical wave propagating from the center outward, while that of the second kind Hankel function means a cylindrical wave propagating inward.

Considering the lengths of $\boldsymbol{r}, \boldsymbol{r}'$ and the angle θ between them, the argument of (6.2.21) is written more explicitly:

$$G(\boldsymbol{r}, \boldsymbol{r}'; \lambda) = -\frac{i}{4} H_0^{(1)} \left(\sqrt{\lambda} |\boldsymbol{r} - \boldsymbol{r}'| \right)$$

$$= \frac{i}{4} H_0^{(1)} \left(\sqrt{\lambda} \sqrt{r^2 - 2rr' \cos\theta + r'^2} \right). \qquad (6.2.25)$$

6.2.3. *One-dimensional space*

In (6.2.8) $d = 1$. The integration in (6.2.8) appears in the following simple form.

$$G(x, x'; \lambda) = \frac{1}{2\pi} \int_{-\infty}^{\infty} dk \frac{e^{ik(x-x')}}{\lambda - k^2}$$

$$= -\frac{1}{4\pi\sqrt{\lambda}} \int_{-\infty}^{\infty} dk \left(\frac{e^{ik(x-x')}}{k - \sqrt{\lambda}} - \frac{e^{ik(x-x')}}{k + \sqrt{\lambda}} \right). \qquad (6.2.26)$$

It is easy to carry out the integration by the residue theorem. $\sqrt{\lambda}$ is as (6.2.9).

If $q > 0$, the simple pole of the first term is in the upper half plane, while that of the second term is in the lower half plane. When $x - x' > 0$, one has to supplement an infinite semicircle in the upper half plane to compose the closed path, and thus the first term has contribution; when $x - x' < 0$, one has to supplement an infinite semicircle in the lower half plane to compose the closed path, thus the second term has contribution. The result is

$$G(x, x'; \lambda) = \frac{e^{i(p+iq)|x-x'|}}{2i\sqrt{\lambda}}, \qquad q > 0. \qquad (6.2.27a)$$

If $q < 0$, the integral result is

$$G(x, x'; \lambda) = -\frac{e^{-i(p+iq)|x-x'|}}{2i\sqrt{\lambda}}, \qquad q < 0. \qquad (6.2.27b)$$

If $\lambda < 0$ is a negative real number, then let $p = 0$ in the above two equations.

$$G(x, x'; -q^2) = -\frac{1}{2q}e^{-q|x-x'|}, \qquad q > 0. \qquad (6.2.28)$$

Particularly, as $q = 0$, i.e., $\lambda = 0$, we focus on the dependence of the Green's function on the space coordinates. We expand the exponential function to the first order term for small q:

$$G(x, x', q \to 0) = -\frac{1}{2q} \left[(1 - q|x - x'|] = \frac{1}{2}|x - x'| - \frac{1}{2q}.$$

So, the result is

$$G(x, x'; 0) = \frac{1}{2}|x - x'| + \text{const.} \qquad (6.2.29)$$

This is the Green's function of a one-dimensional Laplace equation. Similar to the two-dimensional case, the second term in (6.2.29) is a constant in coordinate space, although it goes to infinity as $\lambda \to 0$. The reason for the emergence of this infinity is that $\lambda = 0$ is not a simple pole of the integrand in (6.2.26).

In the case that $\lambda > 0$ is a positive real number, i.e., within the spectrum of the operator $L(x) = -\frac{d^2}{dx^2}$, the poles in the integrand in (6.2.26) are just at the real axis. Therefore, the integral cannot be done, and the Green's function cannot be defined. In this case, we make use of (6.2.27) to define two side limits. In (6.2.27), let the imaginary part q be infinitely small. Then we will get

$$G^{\pm}(x, x'; p^2) = \pm \frac{e^{\pm ip|x-x'|}}{2ip}, \qquad p > 0 \qquad (6.2.30)$$

G^+ (G^-) is the side limit when the parameter λ approaches the real axis from the upper (lower) half complex k plane. From (6.2.30) it is known that G^+ represents a wave propagating from the origin to the two sides along the real axis, while G^- represents a wave propagating in the opposite direction of G^+.

From the above results, we have seen that in a uniform and isotropic space, $G(\boldsymbol{r}, \boldsymbol{r}'; \lambda)$ is a function of the absolute value of difference of coordinates $|\boldsymbol{\rho}| = |\boldsymbol{r} - \boldsymbol{r}'|$. Firstly, because of the uniform, adding an arbitrary vector $\boldsymbol{A}$ to both $\boldsymbol{r}$ and $\boldsymbol{r}'$ should not change the solution: $G(\boldsymbol{r} + \boldsymbol{A}, \boldsymbol{r}' + \boldsymbol{A}; z) = G(\boldsymbol{r}, \boldsymbol{r}'; z)$. Thus G has to be the function of $\boldsymbol{\rho} = \boldsymbol{r} - \boldsymbol{r}'$. Secondly, because of isotropy, along any direction of $\boldsymbol{\rho}$, the form of G should remain unchanged. Therefore, it should be $G(\boldsymbol{r}, \boldsymbol{r}'; z) = G(\rho; z)$.

In the end of this section, we point out that (6.2.8) is actually the Fourier transformation of the basic solution. In the passing three subsections we calculated the expressions of basic solutions in three-, two- and one-dimensional coordinate spaces, respectively. The expression of the basic solution in wave vector k space is simply the Fourier component in (6.2.8):

$$G(\boldsymbol{k}; \lambda) = -\frac{1}{k^2 - \lambda + i\varepsilon}. \qquad (6.2.31)$$

Here we add an infinitesimally small imaginary part in the denominator, stressing that the parameter λ is not on the real axis.

6.3. Green's Function of a Damped Oscillator

The Cauchy problem of a forced oscillator in classical mechanics is as follows:

$$\left(\frac{d^2}{dt^2} + 2\gamma \frac{d}{dt} + \omega_0^2 \right) \psi(t) = F(t), \quad t > t_0. \tag{6.3.1a}$$

$$\gamma > 0, \quad \omega_0 > \gamma, \quad \psi(t)|_{t=t_0} = x_0; \quad \psi'(t)|_{t=t_0} = y_0. \tag{6.3.1b}$$

This is an equation with one variable time t. The initial time is t_0. The first order term shows that the oscillator is damping, and γ is called the damping coefficient. In this operator, the weight is $\rho(t) = e^{2\gamma t}$ and the kernel is $p(t) = e^{2\gamma t}$. We solve this problem following the procedure listed in the end of Subsection 6.1.3.

6.3.1. *Solution of homogeneous equation*

First of all, we solve the following problem with homogeneous differential equation.

$$-L(t)\varphi(t) = 0, \tag{6.3.2a}$$

$$\gamma > 0, \quad \omega_0 > \gamma, \quad \varphi(t)|_{t=t_0} = x_0; \quad \varphi'(t)|_{t=t_0} = y_0. \tag{6.3.2b}$$

Here the operator is

$$L(t) = -\left(\frac{d^2}{dt^2} + 2\gamma \frac{d}{dt} + \omega_0^2 \right). \tag{6.3.3}$$

This is an operator with constant coefficients. Let $\psi(t) = e^{-i\omega t}$ and substitute it into (6.3.2a). Then we get the characteristic equation

$$-\omega^2 - 2i\gamma\omega + \omega_0^2 = 0. \tag{6.3.4}$$

Its two roots are

$$\omega_{1,2} = -i\gamma \pm \sqrt{\omega_0^2 - \gamma^2} = -i\gamma \pm \alpha, \tag{6.3.5}$$

where we have let $\alpha = \sqrt{\omega_0^2 - \gamma^2}$. So, the two linearly independent solutions are

$$\psi_1(t) = e^{-i\omega_1 t}, \quad \psi_2(t) = e^{-i\omega_2 t}. \tag{6.3.6}$$

Their Wronskian is

$$W(\psi_1, \psi_2) = \begin{vmatrix} e^{-i\omega_1 t} & e^{-i\omega_2 t} \\ -i\omega_1 e^{-i\omega_1 t} & -i\omega_2 e^{-i\omega_2 t} \end{vmatrix} = 2i\alpha e^{-2\gamma t}.$$

In (6.3.3) the kernel is $A(t) = 1$, so that

$$A_W(t) = \rho(t)A(t)W(\psi_1(t), \psi_2(t)) = 2i\alpha e^{-2\gamma t}. \tag{6.3.7}$$

Here (6.1.38a) is used.

6.3.2.   Obtaining Green's function

Green's function $G(t, t')$ satisfies the equation (6.1.1a):

$$-L(t)G(t, t') = \delta(t - t'), \quad t, t' > t_0. \tag{6.3.8a}$$

The operator $L(t)$ is (6.3.3). The solution $G(t, t')$ of this equation is called the Green's function of a Cauchy problem. Because the inhomogeneous part in initial conditions have been merged into (6.3.2b), the Green's function satisfies homogeneous initial conditions:

$$G(t_0, t') = 0, \quad \left[\frac{\mathrm{d}}{\mathrm{d}t}G(t, t')\right]_{t=t_0} = 0. \tag{6.3.8b}$$

This form may also be called a single point boundary condition.

We use the piecewise expression method to solve this one-dimensional Green's function. According to (6.1.24b) the expression is

$$G(t, t') = d_1(t')\psi_1(t) + d_2(t')\psi_2(t)$$

$$- \frac{\psi_2(t')\psi_1(t) - \psi_1(t')\psi_2(t)}{A_W(t')}\theta(t - t'). \tag{6.3.9}$$

By the conditions (6.3.8b) we get $d_1(t') = d_2(t') = 0$. So,

$$G(t, t') = \frac{e^{-i\omega_2 t' - i\omega_1 t} - e^{-i\omega_1 t' - i\omega_2 t}}{2i\alpha e^{-2\gamma t'}}\theta(t - t')$$

$$= -\frac{e^{-\gamma(t-t')}}{\alpha}\sin\left[\alpha(t - t')\right]\theta(t - t'). \tag{6.3.10}$$

Finally, the required Green's function is

$$G(t, t') = \frac{e^{-\gamma(t-t')}}{\sqrt{\omega_0^2 - \gamma^2}}\sin\left[\sqrt{\omega_0^2 - \gamma^2}(t - t')\right]\theta(t - t'). \tag{6.3.11}$$

This solution includes a factor $\theta(t - t')$, meaning a retarded Green's function, which is obtained by the initial conditions (6.3.8b).

This discloses that once the oscillator system starts to move at time t_0, it evolves toward the future direction. Note that Green's function (6.3.11) is a function of $t - t'$.

In this case, the Green's function is solved by piecewise expression method.

6.3.3. *Generalized solution of the equation*

Having the preparations above, we are now able to put down the general solution of (6.3.1). The first term in (6.1.4) is the solution of homogeneous equation (6.3.2a), which should be the linear combination of the two independent solutions in (6.3.6). Therefore, by (6.1.4), we write the general solution of (6.3.1) as follows:

$$x(t) = b_1 e^{-i\omega_1 t} + b_2 e^{-i\omega_2 t}$$
$$+ \int_{t_0}^{\infty} \frac{e^{-\gamma(t-t')}}{\sqrt{\omega_0^2 - \gamma^2}} \sin\left[\sqrt{\omega_0^2 - \gamma^2}(t - t')\right] \theta(t - t')F(t')dt'.$$

$$(6.3.12)$$

The coefficients b_1, b_2 are to be determined by the initial conditions (6.3.1b). We mention again that the last term in (6.3.12) is a special solution of (6.3.1).

As $t < t'$, the term of the special solution is zero. Thus,

$$x(t < t') = b_1 e^{-i\omega_1 t} + b_2 e^{-i\omega_2 t}. \qquad (6.3.13)$$

The function $F(t)$ on the right hand side of (6.3.1) represents the affection of an external field. Before this action is exerted, the system moves following the intrinsic mode described by the homogeneous equation (6.3.2a). After the field acts on the system, response of the system emerges. The term of special solution in (6.3.12) reflects this response.

Now we suppose a particular form of the external field:

$$F(t) = F_0 e^{-\beta t}\theta(t).$$

The initial time is $t_0 = 0$, and $x(t)|_{t=0} = 0$; $x'(t)|_{t=0} = 0$. In this case, the two coefficients in (6.3.12) are $b_1 = b_2 = 0$. The remaining special solution in (6.3.12) is

$$x(t) = \int_0^\infty \frac{e^{-\gamma(t-t')}}{\sqrt{\omega_0^2 - \gamma^2}} \sin\left[\sqrt{\omega_0^2 - \gamma^2}(t - t')\right] \theta(t - t')e^{-\beta t'} dt'$$

$$= \frac{F_0}{\sqrt{\omega_0^2 - \gamma^2}} \frac{\sin\left[\sqrt{\omega_0^2 - \gamma^2}t - \phi\right]}{\sqrt{\omega_0^2 + \beta^2 - 2\beta\gamma}} e^{-\gamma t} + \frac{F_0 e^{-\beta t}}{\omega_0^2 + \beta^2 - 2\beta\gamma},$$

$$(6.3.14)$$

where the angle ϕ is calculated by

$$\tan\phi = \frac{\sqrt{\omega_0^2 - \gamma^2}}{\beta - \gamma}.$$

Both terms in (6.3.14) decay with time, but with different reasons. The decay of the first term is due to the fact that the oscillator itself is damped, and that of the second term is because the external field is exponentially decaying.

6.3.4.   *The case without damping*

In this case, let $\gamma \to 0$. Green's function (6.3.11) is simplified to be

$$G(t, t') = \frac{\sin[\omega_0(t - t')]}{\omega_0}\theta(t - t').   (6.3.15)$$

The movement of the oscillator (6.3.14) becomes

$$x(t) = \frac{F_0 \sin(\omega_0 t - \phi)}{\omega_0\sqrt{\omega_0^2 + \beta^2}} + \frac{F_0 e^{-\beta t}}{\omega_0^2 + \beta^2}.$$

After a sufficiently long time, the second term disappears because the external field does, and the pure harmonic vibration $x(t) = \frac{F_0 \sin(\omega_0 t - \phi)}{\omega_0\sqrt{\omega_0^2 + \beta^2}}$ is left. After all, a nonzero field acts in a finite period of time. This action causes two effects: one is the change of the amplitude of the vibration to be $\frac{F_0}{\omega_0\sqrt{\omega_0^2 + \beta^2}}$, which depends on the strength and decaying speed of the field; the other is the phase shift of the vibration, $\phi = \tan^{-1}(\omega_0/\beta)$.

It should be mentioned that as $\gamma = 0$, the roots from (6.3.5) become $\omega_{1,2} = \pm\omega_0$ and they are real numbers. We let $\gamma \to 0$ from $\gamma > 0$, i.e., the root goes to the real axis from the lower half complex ω plane. In such a way, we gain the retarded Green's function.

6.3.5. *The influence of boundary conditions*

We consider the case without damping. Let $\gamma = 0$ in (6.3.3)–(6.3.9). Then the eigenvalues $\omega_{1,2} = \pm\omega_0$ are real, and the eigenfunctions in (6.3.6) are simplified to be

$$\psi_1(t) = e^{i\omega_0 t}, \quad \psi_2(t) = e^{-i\omega_0 t}. \tag{6.3.16}$$

Alternatively, the eigenfunction can be chosen as

$$\psi_1(t) = \sin\omega_0 t, \quad \psi_2(t) = \cos\omega_0 t. \tag{6.3.17}$$

Their Wronskian is

$$W(\psi_1, \psi_2) = \begin{vmatrix} \sin\omega_0 t & \cos\omega_0 t \\ \omega_0\cos\omega_0 t & -\omega_0\sin\omega_0 t \end{vmatrix} = -\omega_0 \tag{6.3.18}$$

and

$$A_W(t') = A(t')W(\psi_1(t'), \psi_2(t')) = -\omega_0. \tag{6.3.19}$$

In this case, because there is no damping, the vibration does not go to zero as time goes, so that we keep both $t > t'$ and $t < t'$ terms in Green's function. We still use the form of (6.1.24b):

$$
\begin{aligned}
G(t, t') &= d_1(t')\psi_1(t) + d_2(t')\psi_2(t) \\
&\quad - \frac{\psi_2(t')\psi_1(t) - \psi_1(t')\psi_2(t)}{A_W(t')}\theta(t - t') \\
&= d_1(t')\sin\omega_0 t + d_2(t')\cos\omega_0 t \\
&\quad + \frac{\sin[\omega_0(t - t')]}{\omega_0}\theta(t - t'),
\end{aligned} \tag{6.3.20}
$$

where $d_1(t'), d_2(t')$ are to be determined by the initial conditions.

If the initial conditions are (6.3.8b), we get $d_1(t') = d_2(t') = 0$, and the Green's function is just (6.3.15).

Suppose the boundary conditions are as follows:

$$G(t_0, t') = G(0, t') = 0, \quad G(t_1, t') = G(1, t') = 0. \tag{6.3.21}$$

These are also called two-point boundary conditions, which indicates that we investigate the vibration of the oscillator in time period $0 \le t$,

$t' \leq 1$. At $t = 0$, the initial time, only the $t < t'$ part in (6.3.20) is left, so that

$$d_2(t') = 0. \tag{6.3.22a}$$

At $t = 1$, the final point of time, only the $t > t'$ part in (6.3.20) can be left, so that

$$d_1(t') \sin \omega_0 + \frac{\sin[\omega_0(1 - t')]}{\omega_0} = 0. \tag{6.3.22b}$$

In summary, the required Green's function is

$$G(t, t') = -\frac{\sin[\omega_0(1 - t')]}{\omega_0 \sin \omega_0} \sin \omega_0 t + \frac{\sin[\omega_0(t - t')]}{\omega_0} \theta(t - t'). \tag{6.3.23}$$

It is seen that different boundary conditions lead to different forms of the Green's function.

A special case of (6.3.23) is that when $\omega_0 \to 0$,

$$G(t, t') = -(1 - t')t + (t - t')\theta(t - t'). \tag{6.3.24}$$

6.4.   Green's Function of Ordinary Differential Equations of Second Order

In this section we consider self-adjoint SL operator

$$L(x) = -A(x)\frac{\mathrm{d}^2}{\mathrm{d}x^2} - B(x)\frac{\mathrm{d}}{\mathrm{d}x} + C(x)$$

$$= \frac{1}{\rho(x)} \left[-\frac{\mathrm{d}}{\mathrm{d}x} p(x)\frac{\mathrm{d}}{\mathrm{d}x} + q(x) \right], \tag{6.4.1a}$$

and

$$p(x) > 0, \quad q(x) \geq 0, \quad \rho(x) \geq 0 \tag{6.4.1b}$$

on $[a, b]$. We consider a two-point value problem: the interval $[a, b]$ is finite and there are definite boundary conditions at the ends.

Below we will frequently use homogeneous equations and homogeneous boundary conditions. The homogeneous boundary value problem is written as

$$[\lambda - L(x)]\varphi(x) = 0, \qquad (6.4.2a)$$

$$B(\varphi) = 0. \qquad (6.4.2b)$$

The boundary condition is the shorthand of the following equations:

$$B_1(y) = \alpha_{1,1}y(a) + \alpha_{1,2}y'(a) + \beta_{1,1}y(b) + \beta_{1,2}y'(b). \qquad (6.4.3a)$$

$$B_2(y) = \alpha_{2,1}y(a) + \alpha_{2,2}y'(a) + \beta_{2,1}y(b) + \beta_{2,2}y'(b). \qquad (6.4.3b)$$

As for the discussions of boundary conditions, please refer to (3.2.18) and Section 3.8. Note that this is a general form. When the coefficients in (6.4.3) take certain values, the boundary conditions become self-adjoint, see, for examples, Theorem 2 in Section 3.6 and Exercise 44 in Chapter 3.

The homogeneous boundary value problem (6.4.2) may have or may not have solutions.

In the following we will investigate the solutions of inhomogeneous boundary value problems in various cases, which are closely related to the corresponding homogeneous boundary value problems expressed by (6.4.2). Before doing so, we discuss the symmetry of Green's function.

6.4.1. *The symmetry of Green's function*

The Green's function $G(x, x'; \lambda)$ of a formally self-adjoint SL operator L satisfies the following equation:

$$(\lambda - L)G(x, x'; \lambda) = \delta(x - x')/\rho(x). \qquad (6.4.4)$$

Here the parameter λ is explicitly written in the Green's function. Suppose that $g(x, x_1; \lambda)$ satisfies

$$(\lambda - L)g(x, x_1; \lambda) = \delta(x - x_1)/\rho(x). \qquad (6.4.5a)$$

Then,

$$(\lambda^* - L)g(x, x_2; \lambda^*) = \delta(x - x_2)/\rho(x). \qquad (6.4.5b)$$

The solutions $g(x, x_1; \lambda)$ and $g(x, x_2; \lambda^*)$ satisfy the same boundary conditions. By the definition of adjoint operators, we have

$$(g\,(x, x_2; \lambda^*), (\lambda - L)g(x, x_1; \lambda\,))$$

$$= ((\,\lambda^* - L)g(x, x_2; \lambda^*), g(x, x_1; \lambda\,)). \qquad (6.4.6)$$

The results of the two sides are as follows.

$$(g(x, x_2; \lambda^*), (\lambda - L)g(x, x_1; \lambda))$$

$$= \int_a^b \mathrm{d}x \rho(x) g^*(x, x_2; \lambda^*)(\lambda - L)g(x, x_1; \lambda) = g^*(x_1, x_2; \lambda^*).$$

$$((\lambda^* - L)g(x, x_2; \lambda^*), g(x, x_1; \lambda))$$

$$= \int_a^b \mathrm{d}x \rho(x)[(\lambda^* - L)g(x, x_2; \lambda^*)]^* g(x, x_1; \lambda) = g(x_2, x_1; \lambda).$$

Thus, the symmetry of Green's function is of the form of

$$g^*(x_1, x_2; \lambda) = g(x_2, x_1; \lambda^*). \tag{6.4.7}$$

When Green's function is solved by the eigenfunction method and λ is not an eigenvalue, its expression is

$$G(x, x') = \sum_i \frac{\varphi_i^*(x')\varphi_i(x)}{\lambda - \lambda_i}.$$

This expression also manifests the symmetry (6.4.7).

6.4.2. *Solutions of boundary value problem of ordinary differential equations of second order*

1. Boundary value problems

Consider the following boundary value problem:

$$[\lambda - L(x)]\psi(x) = f(x), \tag{6.4.8a}$$

$$B(\psi) = \gamma. \tag{6.4.8b}$$

If $f(x) = 0$, this is a boundary value problem of a homogeneous equation, or an Eigenvalue problem. The so-called an eigenvalue problem means that the value of parameter λ must be eigenvalues in order for the boundary value problem to have nonzero solutions, and the solutions are eigenfunctions.

The explicit form of the eigenvalue problem is as follows:

$$[\lambda_m - L(x)]\varphi_m(x) = 0, \tag{6.4.9a}$$

$$B(\varphi_m) = \gamma, \tag{6.4.9b}$$

where λ_m is an eigenvalue and $\varphi_m(x)$ is a corresponding eigenfunction.

2. Parameter λ is not an eigenvalue

If the parameter λ is not an eigenvalue of (6.4.9), then there is no solution of the following boundary value problem:

$$[\lambda - L(x)]\varphi(x) = 0, \quad \lambda \neq \lambda_m, \tag{6.4.10a}$$

$$B(\varphi) = \gamma. \tag{6.4.10b}$$

Before solving (6.4.8) in this case, we solve the corresponding Green's function $G(x, x'; \lambda)$ which should satisfy the following boundary value problem:

$$[\lambda - L(x)]G(x, x'; \lambda) = \delta(x - x')/\rho(x), \tag{6.4.11a}$$

$$B(G) = \sigma, \tag{6.4.11b}$$

where

$$\sigma = \gamma \left[\int_a^b \rho(x') f(x') dx' \right]^{-1} \tag{6.4.12}$$

with the condition being $\int_a^b \rho(x') f(x') dx' \neq 0$. After the $G(x, x'; \lambda)$ is solved, the solution of (6.4.8) is expressed by

$$\psi(x) = \int_a^b G(x, x'; \lambda) \rho(x') f(x') dx'. \tag{6.4.13}$$

It is easily verified that this solution satisfies (6.4.8) by use of (6.4.11).

3. Parameter λ is an eigenvalue

If the parameter λ happens to be an eigenvalue of (6.4.9), $\lambda = \lambda_m$, then we divide the solution of (6.4.8) into two parts:

$$\psi(x) = \varphi_m(x) + \xi(x). \tag{6.4.14}$$

Here $\varphi_m(x)$ satisfies (6.4.9) and $\xi(x)$ satisfies the following boundary value problem:

$$[\lambda_m - L(x)]\xi(x) = f(x), \tag{6.4.15a}$$

$$B(\xi) = 0. \tag{6.4.15b}$$

In order to solve (6.4.15), we have to find its Green's function $G(x, x'; \lambda_m)$ which obeys the following boundary value problem:

$$[\lambda_m - L(x)]G(x, x'; \lambda_m) = \delta(x - x')/\rho(x), \tag{6.4.16a}$$

$$B(G) = 0. \tag{6.4.16b}$$

Once the $G(x, x'; \lambda_m)$ is solved, $\xi(x)$ is expressed by

$$\xi(x) = \int_a^b G(x, x'; \lambda_m)\rho(x')f(x')\mathrm{d}x'. \qquad (6.4.17)$$

Therefore, the solution of (6.4.8) is written as

$$\psi(x) = \varphi_m(x) + \int_a^b G(x, x'; \lambda_m)\rho(x')f(x')\mathrm{d}x', \qquad (6.4.18)$$

where the two terms meet boundary value problem (6.4.9) and (6.4.15), respectively.

In this case, let us see how to use the eigenfunction method to solve the Green's function $G(x, x'; \lambda_m)$. The $G(x, x'; \lambda_m)$ should obey the homogeneous boundary condition in (6.4.21b), so that it should be expanded by the eigenfunctions which follow the homogeneous boundary conditions, instead of by the eigenfunctions solved from (6.4.9) where the boundary condition is inhomogeneous. For this reason, we solve the eigenvalue problem corresponding to (6.4.16) as follows:

$$[\kappa_n - L(x)]\chi_n(x) = 0, \qquad (6.4.19a)$$

$$B(\chi_n) = 0, \qquad (6.4.19b)$$

After the $\chi_n(x)$'s are solved, the $G(x, x'; \lambda_m)$ is expressed by

$$G(x, x'; \lambda_m) = \sum_n \frac{\chi_n^*(x')\chi_n(x)}{\lambda_m - \kappa_n}. \qquad (6.4.20)$$

We stress that the Green's function of (6.4.16) ought to be expanded by the eigenfunctions of (6.4.19) because both obey homogeneous boundary conditions.

If the eigenvalue problem has no nonzero solution, then one has to use the piecewise expression method to determine Green's function.

We should be cautious when the parameter λ happens to be an eigenvalue. In the case of the homogeneous boundary conditions, (6.4.20) will not be valid, and one has to solve the modified Green's function which will be introduced below. In the case of natural boundary conditions, it is required that the γ in (6.4.9b) is finite but not definite, so that it may also be zero. Then the solutions of (6.4.19) and (6.4.9) may be the same. In this case the expression

(6.4.20) may not be valid either, and the modified Green's function is needed.

6.4.3. *Modified Green's function*

1. Half-homogeneous boundary problems

In (6.4.8), if at least one of γ_1, γ_2 is nonzero, the boundary conditions are inhomogeneous. If boundary conditions are homogeneous, we rewrite the boundary value problem as follows:

$$[\lambda - L(x)]\psi(x) = f(x), \qquad (6.4.21a)$$

$$B(\psi) = 0. \qquad (6.4.21b)$$

It may be called a **half-homogeneous boundary problem**, because the equation is inhomogeneous while the boundary conditions are homogeneous. The corresponding homogeneous boundary value problem is

$$[\lambda_m - L(x)]\varphi_m(x) = 0, \qquad (6.4.22a)$$

$$B(\varphi_m) = 0, \qquad (6.4.22b)$$

where λ_m is an eigenvalue and $\varphi_m(x)$ its corresponding eigenfunction.

(1) If the parameter λ in (6.4.21) is not an eigenvalue of (6.4.22), the procedure of solving solutions of (6.4.21) observes (6.4.11) to (6.4.13), as long as one lets $\gamma_{1,2} = 0$ in (6.4.11) and (6.4.12). The solution is (6.4.18).

(2) If the parameter λ in (6.4.21) is just an eigenvalue of (6.4.22), $\lambda = \lambda_m$, (6.4.21) becomes

$$[\lambda_m - L(x)]\psi(x) = f(x), \qquad (6.4.23a)$$

$$B(\psi) = 0. \qquad (6.4.23b)$$

Let us see the condition under which (6.4.23) has solutions. Now the operator L is self-adjoint, so that $(\lambda_m - L)^\dagger = (\lambda_m - L)$.

$$(\varphi_m, (\lambda_m - L)\psi) - ((\lambda_m - L)^\dagger \varphi_m, \psi) = (\varphi_m, f). \qquad (6.4.24)$$

The left hand side is zero, and the right hand side must be so. Thus we get

$$(\varphi_m, f) = \int_a^b \rho(x)\varphi_m^*(x)f(x)\mathrm{d}x = 0. \qquad (6.4.25)$$

This is the required condition. That is to say, the inhomogeneous term $f(x)$ must be orthogonal to the eigenfunction $\varphi_m(x)$ belonging to the eigenvalue λ_m. Equation (6.4.25) is called **the compatibility condition** of boundary value problem (6.4.23). This is substantially the same as the compatibility (3.7.8) in the alternative theorem in Section 3.7. In that theorem, the condition for inhomogeneous equations to have nonzero solutions is that the inhomogeneous term $f(x)$ must be orthogonal to the solutions of adjoint homogeneous boundary value problem. Here the operator has been self-adjoint, so that the adjoint problem is identical to (6.4.22).

If the condition in (6.4.23) is changed to be inhomogeneous, there is no need of this compatibility condition.

In order to find the solution of (6.4.23), we solve the corresponding Green's function of the following boundary value problem:

$$[\lambda_m - L(x)]G(x, x') = \delta(x - x')/\rho(x), \qquad (6.4.26a)$$

$$B(G) = 0. \qquad (6.4.26b)$$

We hope that after the Green's function is found, the solution (6.4.23) can be expressed in the form of (6.4.18). However, the Green's function satisfying (6.4.26) does not exist. The proof is as follows.

Since λ_m is an eigenvalue of (6.4.23), we replace $\psi(x)$ and $f(x)$ in (6.4.22a) by $G(x, x')$ and $\delta(x - x')/\rho(x)$ in (6.4.26a), respectively. Then (6.4.25) requires that

$$\int_a^b \varphi_m(x)\delta(x - x')\mathrm{d}x = \varphi_m(x') = 0. \qquad (6.4.27)$$

This is a contradiction, because $\varphi_m(x)$ is certainly nonzero.

2. Modified Green's function

Why does the above contradiction appear? The result (6.4.27) comes from (6.4.22) and (6.4.26). The former is the prerequisition. Thus the contradiction arises from the latter. Under the prerequisition (6.4.27), the corresponding Green's function does not satisfy (6.4.26), or the Green's function defined by (6.4.26) does not exist. We have to provide another equation to find a proper Green's function to achieve the solution of (6.4.23).

Suppose that the required Green's function satisfies the following equation:

$$[\lambda_m - L(x)]G(x, x') = \delta(x - x')/\rho(x)$$

$$+c(x')\varphi_m(x), \quad a \leq x, \; x' \leq b, \qquad (6.4.28a)$$

$$B(G) = 0. \qquad (6.4.28b)$$

Compared to (6.4.26), the boundary condition in (6.4.28) remains unchanged but the equation has an additional term proportional to the normalized eigenfunction $\varphi_m(x)$. The coefficient $c(x')$ is to be determined. To do so, let us inspect the following result:

$$(\varphi_m, (\lambda_m - L)G) - ((\lambda_m - L)^\dagger \varphi_m, G)$$

$$= (\varphi_m, \delta(x - x')/\rho + c(x')\varphi_m)$$

$$= (\varphi_m, \delta(x - x')/\rho) + c(x') = 0$$

It follows that

$$c(x') = -\int_a^b \varphi_m^*(x)\delta(x - x')\mathrm{d}x = -\varphi_m^*(x'). \qquad (6.4.29)$$

Thus, $G(x, x')$ satisfies

$$[\lambda_m - L(x)]G(x, x') = \delta(x - x')/\rho(x)$$

$$-\varphi_m^*(x')\varphi_m(x), \quad a \leq x, \; x' \leq b, \qquad (6.4.30a)$$

$$B(G) = 0. \qquad (6.4.30b)$$

This solved $G(x, x')$ is called a **modified Green's function**.

It can be shown that the modified Green's function is also of the symmetry (6.4.7).

3. Expression of solution

We are now ready to write the solution of (6.4.28). Consider the integrals

$$(\psi, (\lambda_m - L)G) - ((\lambda_m - L)\psi, G) = 0.$$

The left hand side is

$$(\psi_m, \delta(x - x')/\rho - \varphi_m^*(x')\varphi_m) - (f, G)$$

$$= \psi^*(x') - \varphi_m^*(x')(\psi, \varphi_m) - (f, G) = 0,$$

where $(\psi, \varphi_m) = c$ is a constant and $G(x, x')$ is the modified Green's function. We obtain $\psi^*(x') = c\varphi_m^*(x') + \int_a^b \rho(x)G(x, x')f^*(x)\mathrm{d}x$ and its complex conjugate equation can be recast, by use of the symmetry (6.4.7) of the modified Green's function, as

$$\psi(x) = c\varphi_m(x) + \int_a^b G(x, x')\rho(x')f(x')\mathrm{d}x'. \qquad (6.4.31\mathrm{a})$$

Let operator $[\lambda_m - L(x)]$ act on both sides.

$$[\lambda_m - L(x)]\psi(x)$$

$$= \int_a^b \left[\delta(x - x')/\rho(x) - \varphi_m^*(x')\varphi_m(x)\right] \rho(x')f(x')\mathrm{d}x'$$

$$= f(x) - \varphi_m(x) \int_a^b \rho(x')\varphi_m^*(x')f(x')\mathrm{d}x'.$$

To meet (6.4.28a), the second term on the right hand side must be zero:

$$\int_a^b \rho(x')\varphi_m^*(x')f(x')\mathrm{d}x' = 0. \qquad (6.4.31\mathrm{b})$$

This is just the compatibility condition (6.4.25) for (6.4.23) to have solutions.

In summary, when $f(x)$ in (6.4.23a) satisfies the compatibility condition (6.4.31b), the boundary value problem (6.4.23) has a solution (6.4.31a), where $G(x, x')$ is the modified Green's function satisfying (6.4.30).

Please note the discrepancy between (6.4.31a) and (6.4.18). The eigenfunction $\varphi_m(x)$ in (6.4.31) obey homogeneous boundary conditions, so that it can be multiplied by a constant, while the $\varphi_m(x)$ in (6.4.18) obeys inhomogeneous boundary conditions, so that no such constant factor is allowed.

4. Solving modified Green's function

The eigenfunction method and piecewise expression method can be used to solve the modified Green's function defined by (6.4.30).

(1) The eigenfunction method

Suppose that the eigenvalues $\{\lambda_n\}$ and corresponding orthonormalized eigenfunction set $\{\varphi_n(x)\}$ of (4.4.27) have been solved. The modified Green's function can be expanded by this set:

$$G(x, x') = \sum_n c_n(x')\varphi_n(x). \qquad (6.4.32)$$

It is substituted into (6.4.30a) to yield

$$\sum_{n(n\neq m)} c_n(x')(\lambda_m - \lambda_n)\varphi_n(x) = \delta(x - x')/\rho(x) - \varphi_m^*(x')\varphi_m(x).$$

Multiplying $\rho(x)\varphi_l^*(x)$ and then taking integrals on both sides, we get

$$\int_a^b \rho(x)\varphi_l^*(x)\mathrm{d}x \sum_{n(n\neq m)} c_n(x')(\lambda_m - \lambda_n)\varphi_n(x)$$

$$= \int_a^b \varphi_l^*(x)\mathrm{d}x \left[\delta(x - x') - \rho(x)\varphi_m^*(x')\varphi_m(x)\right].$$

Since $\{\varphi_n(x)\}$ are orthonormalized, we have $c_n(x') = \frac{\varphi_n^*(x')}{\lambda_m - \lambda_n}$, $n \neq m$. It is substituted back to (6.4.32). Then, the modified Green's function is

$$G(x, x') = c_m(x')\varphi_m(x) + \sum_{n(n\neq m)} \frac{\varphi_n^*(x')\varphi_n(x)}{\lambda_m - \lambda_n}. \qquad (6.4.33)$$

Here $c_m(x')$ can be an arbitrary one, so that it is still undetermined yet. To determine it, we recall the symmetry of the modified Green's function. This symmetry requires that

$$c_m(x')\varphi_m(x) + \sum_{i=0, i\neq m}^{\infty} \frac{\varphi_i^*(x')\varphi_i(x)}{\lambda_m - \lambda_i}$$

$$= c_m^*(x)\varphi_m^*(x') + \sum_{i=0, i\neq m}^{\infty} \frac{\varphi_i^*(x')\varphi_i(x)}{\lambda_m - \lambda_i}.$$

This leads to

$$c_m(x')\varphi_m(x) = c_m^*(x)\varphi_m^*(x'). \qquad (6.4.34)$$

We multiply $\rho(x)\varphi_m^*(x)$ and then integrate with respect to x on $[a, b]$ on both sides so as to get $c_m(x') = \varphi_m^*(x') \int_a^b \mathrm{d}x \rho(x)\varphi_m^*(x)c_m^*(x)$. The right hand side is a number, denoted by d: $c_m(x') = d\varphi_m^*(x')$. Apparently, any nonzero real number d meets (6.4.34). Usually it is chosen as $d = 1$. So, $c_m(x') = \varphi_m^*(x')$. Finally, the required modified Green's function is

$$G(x, x') = \varphi_m^*(x')\varphi_m(x) + \sum_{i=0, i \neq m}^{\infty} \frac{\varphi_i^*(x')\varphi_i(x)}{\lambda_m - \lambda_i}. \qquad (6.4.35)$$

This expression is of the symmetry (6.4.7). Furthermore, because the parameter λ here is an eigenvalue λ_m which is real, the symmetry is simplified to be

$$G(x, x'; \lambda_m) = G^*(x', x; \lambda_m). \qquad (6.4.36)$$

Compared with the expression of the ordinary Green's function (6.1.14), the term containing the eigenvalue λ_m is separated in (6.4.35) so as to avoid infinity.

(2) The piecewise expression method

At both $x > x'$ and $x < x'$, Eq. (6.4.30) is reduced to be

$$[\lambda_m - L(x)]\,\zeta(x, x') = -\varphi_m(x')\varphi_m(x). \qquad (6.4.37)$$

This is still an inhomogeneous one. Its corresponding homogeneous equation

$$[\lambda_m - L(x)]\zeta(x, x') = 0 \qquad (6.4.38)$$

should have two linearly independent solutions, denoted by $\zeta_1(x)$ and $\zeta_2(x)$, respectively. Besides, (6.4.37) has a special solution, denoted by $\zeta_0(x, x')$. Therefore, at both regions $x > x'$ and $x < x'$, the modified Green's function should be the superposition of the two independent solutions of (6.4.38) and the special solution of (6.4.37).

$$G(x, x') = \begin{cases} a_1(x')\zeta_1(x) + a_2(x')\zeta_2(x) + \zeta_0(x, x'), & x > x'. \\ b_1(x')\zeta_1(x) + b_2(x')\zeta_2(x) + \zeta_0(x, x'), & x < x'. \end{cases}$$
$$(6.4.39)$$

Compared to (6.1.13), (6.4.39) has an additional special solution which is the same in both regions.

It is required that the modified Green's function is continuous at $x = x'$:

$$\left[a_1(x') - b_1(x')\right]\zeta_1(x') + [a_2(x') - b_2(x')]\zeta_2(x') = 0. \qquad (6.4.40)$$

We integrate equation (6.4.30) from $x = x' - 0^+$ to $x = x' + 0^+$. Because the integral interval is infinitesimally small, the integrations of continuous functions are zero. So,

$$p_2(x')[(a_1(x') - b_1(x'))\zeta_1'(x) + (a_2(x') - b_2(x'))\zeta_2'(x)] = 1.$$

$$(6.4.41)$$

Equations (6.4.40,41) and (6.1.19,20) are solved simultaneously. We find

$$a_1(x') - b_1(x') = -\frac{\zeta_2(x')}{p_2(x')W(x')}, \quad a_2(x') - b_2(x') = \frac{\zeta_1(x')}{p_2(x')W(x')},$$

$$(6.4.42)$$

where

$$W(x') = \zeta_1(x')\zeta_2'(x) - \zeta_2(x')\zeta_1'(x) \qquad (6.4.43)$$

is the Wronskian of $\zeta_1(x)$ and $\zeta_2(x)$. The two equations in (6.4.42) plus the two boundary conditions in (6.4.30b) can determine all the four quantities $a_1(x'), b_1(x'), a_2(x'), b_2(x')$. Similarly for (6.1.23)–(6.1.25), the modified Green's function can also be written in those three forms.

The modified Green's function is introduced in a special case. It has not so strong physical significance as an ordinary Green's function.

The two methods, eigenfunction method and piecewise expression method, are in principle equivalent to each other. In using eigenfunction method, we have seen that the symmetry condition (6.4.7) had to be employed to determine the last quantity in modified Green's function. The symmetry condition is necessarily employed in the piecewise expression method too. That is to say, the four conditions of (6.4.42) and (6.4.30b) can merely determine three of $a_1(x'), b_1(x'), a_2(x'), b_2(x')$. We will see an example below.

Up to now, when we solve ordinary and modified Green's functions, both the eigenfunction method and piecewise expression method can apply. If there exists an eigenfunction set in the region, these two methods should be equivalent to each other. This may be proved. However, the author of this book has not seen such a proof yet. In some cases, it is indeed that if the results obtained by the piecewise expression method are expanded by the eigenfunctions, one achieves what is obtained by the eigenfunction method.

6.4.4.   *Examples of solving boundary value problem of ordinary differential equations of second order*

Example 1. Find the solution of the following boundary value problem.

$$-\frac{d^2}{dx^2}\psi(x) = f(x), \quad 0 < x < 1, \tag{6.4.44a}$$

$$\left.\frac{d\psi(x)}{dx}\right|_{x=0} = \left.\frac{d\psi(x)}{dx}\right|_{x=1} = 0. \tag{6.4.44b}$$

Solution. In the operator of (6.4.44a), $\rho(x) = 1$ and $q(x) = 1$. The parameter $\lambda = 0$. The boundary conditions are homogeneous. The corresponding homogeneous boundary value problem is

$$\left(\lambda_m + \frac{d^2}{dx^2}\right)\varphi_m(x) = 0, \quad 0 < x < 1, \tag{6.4.45a}$$

$$\left.\frac{d\varphi_m(x)}{dx}\right|_{x=0} = \left.\frac{d\varphi_m(x)}{dx}\right|_{x=1} = 0. \tag{6.4.45b}$$

It has an eigenvalue $\lambda_0 = 0$, and its two independent special solutions are 1 and x, respectively. The general solution belonging to this eigenvalue is $\varphi_0(x) = a + bx$. When it is substituted into (6.4.45b), we get $b = 0$, so that the function is a constant. The normalization results in

$$\varphi_0(x) = 1. \tag{6.4.46}$$

When $\lambda_m \neq 0$, the corresponding two independent solutions are $\cos\sqrt{\lambda_m}x$ and $\sin\sqrt{\lambda_m}x$. After the boundary conditions (6.4.45b) are applied, we solve the following eigenfunction and eigenvalues:

$$\varphi_0(x) = 1, \quad \varphi_n(x) = \sqrt{2}\cos n\pi x; \quad \lambda_n = n^2\pi^2, \quad n = 0, 1, 2, 3, \ldots. \tag{6.4.47}$$

Now in (6.4.44) the parameter $\lambda = 0$ is just an eigenvalue, and the boundary conditions are homogeneous. This is just the case as

(6.4.23). Two requirements immediately follow. One is the compatibility condition (6.4.25) has to be met, and the other is to find its modified Green's function.

The compatibility conditions is $\int_0^1 f(x)\,dx = 0$. We assume that it is satisfied. Below we solve the modified Green's function.

According to (6.4.30), the modified Green's function obeys the following boundary value problem:

$$-\frac{d^2}{dx^2}G(x, x') = \delta(x - x') - 1, \quad 0 < x < 1, \tag{6.4.48a}$$

$$\left.\frac{dG(x, x')}{dx}\right|_{x=0} = \left.\frac{dG(x, x')}{dx}\right|_{x=1} = 0. \tag{6.4.48b}$$

(1) The eigenfunction method

Observing (6.4.35), we immediately put down the expression of the modified Green's function:

$$G(x, x') = 1 + \frac{2}{\pi^2}\sum_{n=1}^{\infty}\frac{1}{n^2}\cos n\pi x' \cos n\pi x. \tag{6.4.49}$$

(2) The piecewise expression method

As $x \neq x'$, Eq. (6.4.48a) has a special solution

$$\zeta_0(x, x') = \frac{x^2}{2}. \tag{6.4.50}$$

The homogeneous equation corresponding to (6.4.48a) is

$$-\frac{d^2}{dx^2}\zeta(x) = 0. \tag{6.4.51}$$

Its two independent solutions are

$$\zeta_1(x) = 1, \quad \zeta_2(x) = x. \tag{6.4.52}$$

Therefore, following (6.4.39), the modified Green's function is

$$G(x, x') = \left[c_1(x') + c_2(x')x + \frac{x^2}{2}\right]\theta(x - x')$$

$$+ \left[d_1(x') + d_2(x')x + \frac{x^2}{2}\right]\theta(x' - x). \tag{6.4.53}$$

$$0 < x, x' < 1.$$

By (6.1.19) and (6.1.20), we get

$$c_1(x') - d_1(x') + [c_2(x') - d_2(x')]x' = 0 \qquad (6.4.54a)$$

and

$$-c_2(x') + d_2(x') = 1. \qquad (6.4.54b)$$

The boundary conditions (6.4.48b) lead to

$$\left.\frac{dG(x, x')}{dx}\right|_{x=0} = d_2(x') = 0 \qquad (6.4.54c)$$

and

$$\left.\frac{dG(x, x')}{dx}\right|_{x=1} = c_2(x') + 1 = 0. \qquad (6.4.54d)$$

From (6.4.54b)–(6.4.54d), we obtain $c_1(x') = d_1(x') + x'$, and thus the modified Green's function is

$$G(x, x') = \left[c_1(x') - x + \frac{x^2}{2}\right]\theta(x - x')$$

$$+ \left[c_1(x') - x' + \frac{x^2}{2}\right]\theta(x' - x). \qquad (6.4.55a)$$

Here we see that there is still an unknown $c_1(x')$. This case has been pointed out in the end of the last subsection. We have to resort to the symmetry (6.4.7) to determine $c_1(x')$. Exchanging x and x' in (6.4.55a), we have

$$G(x', x) = \left[c_1(x) - x' + \frac{x'^2}{2}\right]\theta(x' - x)$$

$$+ \left[c_1(x) - x + \frac{x'^2}{2}\right]\theta(x - x'). \qquad (6.4.55b)$$

It is required that $G(x', x) = G(x, x')$. Then, $c_1(x') + \frac{x^2}{2} = c_1(x) + \frac{x'^2}{2}$. This means that $c_1(x) = \frac{x^2}{2}$. Finally, we acquire the modified Green's function:

$$G(x, x') = \left(-x + \frac{x^2 + x'^2}{2}\right)\theta(x - x')$$

$$+ \left(-x' + \frac{x^2 + x'^2}{2}\right)\theta(x' - x). \qquad (6.4.56)$$

Subsequently, the solution of (6.4.44) is

$$\psi(x) = c + \int_0^1 G(x, x') f(x') dx', \quad 0 < x < 1, \tag{6.4.57}$$

where c is any constant.

Example 2. Solve the following boundary value problem.

$$\left[-\frac{d}{dx} \left(x \frac{d}{dx} \right) + \frac{n^2}{x} \right] \psi(x) = f(x), \quad 0 < x < 1, \tag{6.4.58a}$$

$$\psi(x = 0) < \infty, \quad \psi(x = 1) = 0, \tag{6.4.58b}$$

where n is an integer.

Solution. In the operator (6.4.58), $\rho(x) = 1$, $p(x) = x$, and $q(x) = n^2/x$. The parameter $\lambda = 0$. The corresponding eigenvalue problem is

$$\left[-\frac{d}{dx} \left(x \frac{d}{dx} \right) + \frac{n^2}{x} \right] \varphi_m(x) = \lambda_m \varphi_m(x), \quad -1 < x < 1, \tag{6.4.59a}$$

$$\varphi(x = 0) < \infty, \quad \varphi(x = 1) = 0. \tag{6.4.59b}$$

The problem with homogeneous equation corresponding to (6.4.58a) is

$$\left[-\frac{d}{dx} \left(x \frac{d}{dx} \right) + \frac{n^2}{x} \right] \varphi(x) = 0, \quad -1 < x < 1, \tag{6.4.60a}$$

$$\varphi_m(x = 0) < \infty, \quad \varphi_m(x = 1) = 0. \tag{6.4.60b}$$

The corresponding problem for solving Green's function is

$$\left[-\frac{d}{dx} \left(x \frac{d}{dx} \right) + \frac{n^2}{x} \right] G(x, x') = \delta(x - x'), \quad 0 < x, \quad x' < 1,$$

$$\tag{6.4.61a}$$

$$G(x = 0) < \infty, \quad G(x = 1) = 0. \tag{6.4.61b}$$

Here the boundary conditions are inhomogeneous, so that the modified Green's function is not concerned.

The eigenvalues and eigenfunctions of (6.4.60) are unknown. So, we resort to the piecewise expression method to find the Green's function.

When $x \neq x'$, (6.4.58a) becomes

$$\left(\frac{\mathrm{d}^2}{\mathrm{d}x^2} + \frac{1}{x} \frac{\mathrm{d}}{\mathrm{d}x} - \frac{n^2}{x^2} \right) \varphi(x) = 0. \tag{6.4.62}$$

The point $x = 0$ is a first kind of singularity. As $a_0 = 1$, $b_0 = -n^2$, the indicial equation is $\lambda(\lambda - 1) + \lambda - n^2 = \lambda^2 - n^2 = 0$. Its two roots are $\lambda = \pm n$. Assume the corresponding eigenvectors are $\binom{a_1}{a_2}$. Then, it is easily solved that $\frac{a_2}{a_1} = \pm n$. As $n \neq 0$, the two eigenvectors are linearly independent of each other. If (6.4.62) is recast in the form of (3.5.48), one finds that the coefficient matrix solely contains one term A_0. There is no need to use the series method. The two independent eigenfunctions for $n \neq 0$ are $\varphi_1(x) = x^n$ and $\varphi_2(x) = x^{-n}$. As $n = 0$, the two roots of the indicial equation are the same, so that one of the eigenfunctions should contain a logarithm factor. It is easy to get the two independent special solutions which are $\varphi_1(x) = 1$ and $\varphi_2(x) = \ln x$.

So, when $n = 0, 1, 2, \ldots$, the general solution of the homogeneous equation is respectively

$$\{a_0 + b_0 \ln x, a_1 x + b_1 x^{-1}, a_2 x^2 + b_2 x^{-2}, \ldots, a_n x^n + b_n x^{-n}, \ldots\}.$$

Now we solve the Green's function in the cases of $n = 0$ and $n \neq 0$.

(1) When $n = 0$, Green's function should be

$$G(x, x') = \left[c_1(x') + c_2(x') \ln x \right] \theta(x - x')$$
$$+ \left[d_1(x') + d_2(x') \ln x \right] \theta(x' - x).$$

By the boundary conditions (6.4.61b), we obtain $c_1(x') = 0$, $d_2(x') = 0$.

$$G(x, x') = c_2(x') \ln x \theta(x - x') + d_1(x') \theta(x' - x).$$

Applying the boundary conditions of the Green's function at $x = x'$, we get the Green's function for $n = 0$:

$$G(x, x') = -\ln x \theta(x - x') - \ln x' \theta(x' - x).$$

The general solution of Eq. (6.4.58a) is

$$\psi(x) = a_1 + a_2 \ln x + \int_0^1 G(x, x') f(x') \mathrm{d}x', \quad 0 < x < 1.$$

The linear combination coefficients are determined by the boundary conditions (6.4.58b) to be $a_1 = a_2 = 0$. Finally, the solution of problem (6.4.58) for $n = 0$ is

$$\psi(x) = \int_0^1 G(x, x') f(x') \mathrm{d}x', \quad 0 < x < 1.$$

(2) When $n \neq 0$, the Green's function is

$$G(x, x') = [c_1(x')x^n + c_2(x')x^{-n}]\theta(x - x')$$
$$+ [d_1(x')x^n + d_2(x')x^{-n}]\theta(x' - x).$$

By conditions (6.4.61b), it is solved that $c_1(x') = -c_2(x'), d_2(x') = 0$. Then, we apply the conditions at $x = x'$ to get the Green's function:

$$G(x, x') = -\frac{x'^n}{2n}(x^n - x^{-n})\theta(x - x')$$

$$-\frac{x'^n - x'^{-n}}{2n}x^n\theta(x' - x).$$

The general solution of (6.4.58) is

$$\psi(x) = b_1 x^n + b_2 x^{-n} + \int_0^1 G(x, x') f(x') \mathrm{d}x', \quad 0 < x < 1.$$

The coefficients are determined by the conditions (6.4.58b) to be $b_1 = b_2 = 0$. So, the solution of the problem (6.4.58) for $n \neq 0$ is

$$\psi(x) = \int_0^1 G(x, x') f(x') \mathrm{d}x', \quad 0 < x < 1,$$

Because we are not solving modified Green's function, we are able to determine all the quantities without resorting to the symmetry $G(x, x') = G^*(x', x)$.

Suppose that a term λy is added to the equation (6.4.58a),

$$\frac{1}{x}\frac{\mathrm{d}}{\mathrm{d}x}\left(x\frac{\mathrm{d}y}{\mathrm{d}x}\right) + \left(\lambda - \frac{n^2}{x^2}\right)y = -\frac{1}{x}f(x).$$

Then, the corresponding homogeneous equation will be

$$\frac{1}{x}\frac{\mathrm{d}}{\mathrm{d}x}\left(x\frac{\mathrm{d}y}{\mathrm{d}x}\right) + \left(\lambda - \frac{n^2}{x^2}\right)y = 0.$$

This is a Bessel equation of order n. Its eigenvalues and eigenfunctions have been known, and $\lambda = 0$ is not an eigenvalue.

Bessel eigenvalue equation is

$$\frac{1}{x}\frac{\mathrm{d}}{\mathrm{d}x}\left(x\frac{\mathrm{d}\varphi_m}{\mathrm{d}x}\right) + \left(\lambda_m - \frac{n^2}{x^2}\right)\varphi_m = 0,$$

$$\varphi_m(x=0) < \infty, \quad \varphi_m(x=1) = 0.$$

The normalized eigenfunction set is $\{\varphi_m\} = \{N_m J_n(k_m x)\}$, where J_n is Bessel function of order n, N_m is the normalization coefficient and $k_m = \sqrt{\lambda_m}$. Then, the required Green's function can be expanded by this eigenfunction set. Observing (6.1.14), we get

$$G(x, x') = -\sum_{i=1}^{\infty} \frac{N_i^2}{\lambda_i} J_n(k_i x') J_n(k_i x)$$

In this way, Green's function is achieved quickly. This is a special case that boundary conditions are not homogeneous, but the eigenfunction method happens to be applicable.

From the examples, it is seen that if eigenfunctions have been solved, it is quite easy to use the eigenfunction method, while if the eigenfunctions are not available, the piecewise expression method has to be resorted to.

6.5. Green's Function in Multi-dimensional Spaces

6.5.1. *Ordinary differential equations of second order and Green's function*

1. Ordinary differential equations of second order

In this chapter, the multi-dimensions mean two- or three-dimensions. In resolving partial differential equations in mathematical physics, the problems in multi-dimensions are usually dealt with by separation of variables.

The region where the differential equation is to be solved is denoted by V, and its surface denoted by S. For a two-dimensional space, V is actually a region in the plane, and S is its border. When we say volume integral $\int_V \mathrm{d}V$ and surface integral $\oint_S \mathrm{d}S$ in a two dimensional space, they correspond to integrations in a plane region and its border line, respectively. The theory in this section applies to both two- and three dimensional spaces.

In two- and three dimensional spaces, we consider the differential operators up to the second order, and they are presented in Sturm-Liouville form:

$$L = \frac{1}{\rho(r)} \left(-\nabla \cdot [p(r)\nabla] + q(r) \right). \tag{6.5.1a}$$

It is required that on V the three functions $p(r), q(r), \rho(r)$ are real and continuous and satisfy

$$p(r) \geq 0, \quad q(r) \geq 0, \quad \rho(r) > 0. \tag{6.5.1b}$$

The function $\rho(r)$ is the weight. In the following, the inner products are all with respect to the weight $\rho(r)$, unless specified. So, the inner product of two functions is expressed by

$$(f, g) = \int_V \rho(r) f^*(r) g(r) dV.$$

As long as $\rho(r), p(r), q(r)$ are real, the operator L defined by (6.5.1) is formally self-adjoint. This is easily shown as follows.

$$(v, Lu) = \int_V \rho v^* Lu dV = \int_V v^* \left[-\nabla \cdot (p\nabla) + q \right] u dV$$

$$= \int_V u[-\nabla \cdot (p\nabla) + q] v^* dV + \int_V \nabla \cdot (up\nabla v^* - v^* p \nabla u) dV$$

$$= (Lv, u) + \oint_S [p(u\nabla v^* - v^* \nabla u)] \cdot dS. \tag{6.5.2}$$

In the multi-dimensional spaces, the knot is defined by

$$J = \oint_S (up\nabla v - vp\nabla u) \cdot dS.$$

If this term vanishes, the operator $L(r)$ becomes self-adjoint.

Having cleared the form of $L(r)$, we address the boundary value problem of differential equation of second order. It is of the following form:

$$[\lambda - L(r)]\psi(r) = f(r), \quad (r, r' \in V), \tag{6.5.3a}$$

$$\left[\alpha \psi(r) + \beta \frac{\partial \psi(r)}{\partial n} \right]_S = u(r). \tag{6.5.3b}$$

When $f(x) = 0$, it is a boundary value problem of homogeneous equation. This is just an eigenvalue problem, and the parameter λ and solutions must be respectively eigenvalues and eigenfunctions.

The eigenvalue problem is explicitly written as follows.

$$[\lambda_m - L(\boldsymbol{r})]\varphi_m(\boldsymbol{r}) = 0, \quad (\boldsymbol{r}, \boldsymbol{r}' \in V) \tag{6.5.4a}$$

$$\left[\alpha\varphi_m(\boldsymbol{r}) + \beta\frac{\partial\varphi_m(\boldsymbol{r})}{\partial n}\right]_S = v(\boldsymbol{r}) \tag{6.5.4b}$$

where λ_m is an eigenvalue and its corresponding eigenfunction is $\varphi_m(x)$.

2. Green's function

The Green's function corresponding to (6.5.3) satisfies the following boundary value problem:

$$[\lambda - L(\boldsymbol{r})]G(\boldsymbol{r}, \boldsymbol{r}') = \delta(\boldsymbol{r} - \boldsymbol{r}')/\rho(\boldsymbol{r}), \quad (\boldsymbol{r}, \boldsymbol{r}' \in V), \tag{6.5.5a}$$

$$\left[\alpha G(\boldsymbol{r}, \boldsymbol{r}') + \beta\frac{\partial G(\boldsymbol{r}, \boldsymbol{r}')}{\partial n}\right]_S = w(\boldsymbol{r}). \tag{6.5.5b}$$

In most cases, the boundary condition is homogeneous, $w(\boldsymbol{r}) = 0$.

The Green's function is of the symmetry

$$G^*(\boldsymbol{r}_1, \boldsymbol{r}_2; \lambda) = G(\boldsymbol{r}_2, \boldsymbol{r}_1; \lambda^*). \tag{6.5.6}$$

It can be verified in a way similar to that in Subsection 6.4.1.

In multi-dimensional spaces usually two methods are used to solve Green's function: the eigenfunction method and the piecewise expression method. The latter means that the piecewise expression method is used in one of the coordinates and the eigenfunction method is used in other coordinates.

Then, which coordinate should be chosen to apply the piecewise expression method? Here are some experience for reference.

For a three-dimensional system being of spherical symmetry, spherical coordinates have to be employed. In this case, the eigenfunctions in the angular directions are easy to solve, so that the piecewise expression method is applied in the radial coordinate. Similarly, for a two-dimensional system with circular symmetry, polar coordinates are adopted and the piecewise expression method is used in the radial direction.

For a two-dimensional system, after the separation of variables, the eigenfunctions of this system are in fact the products of those of the two coordinate directions. If in one coordinate there is no nonzero eigenfunction, then the piecewise expression method is applied to this direction. Similarly, in a three-dimensional system, the piecewise expression method is applied to the coordinate along which there is no eigenfunctions.

If in a system with Cartesian coordinates, the region in one directions is infinite and those in others are finite, then the piecewise expression method is applied to the direction with infinite range.

These experiences are reflected in the following examples.

3. The expression of the solution of the differential equation

The procedure of solving the boundary value problem (6.5.3) is similar to that in one-dimensional case, see Section 6.4, and we do not repeat here. We furnish here a general expression of the solution of (6.5.3a). Its derivation is almost the same as Eqs. (6.1.2)–(6.1.13). Let

$$\psi(\boldsymbol{r}) = \varphi_m(\boldsymbol{r}) + \xi(\boldsymbol{r}) \tag{6.5.7}$$

where $\varphi_m(\boldsymbol{r})$ is the solution satisfying the homogeneous equation, i.e., it meets the boundary value problem (6.5.4), and is a special solution of (6.5.3a):

$$[\lambda - L(\boldsymbol{r})]\xi(\boldsymbol{r}) = f(\boldsymbol{r}). \tag{6.5.8}$$

Then, starting from

$$(G(\lambda^*), (\lambda - L)\xi) - ((\lambda^* - L)G(\lambda^*), \xi)$$
$$= \oint_S [p(\boldsymbol{r})(\xi(\boldsymbol{r})\nabla G^*(\boldsymbol{r}, \boldsymbol{r}'; \lambda^*) - G^*(\boldsymbol{r}, \boldsymbol{r}'; \lambda^*)\nabla\xi(\boldsymbol{r}))] \cdot d\boldsymbol{S}, \tag{6.5.9}$$

we obtain the general expression of the solution

$$\psi(\boldsymbol{r}) = \varphi_m(\boldsymbol{r}) + \int_{V'} \rho(\boldsymbol{r}')G(\boldsymbol{r}, \boldsymbol{r}'; \lambda)f(\boldsymbol{r}')d\boldsymbol{r}'$$
$$- \oint_{S'} [p(\boldsymbol{r}')(\psi(\boldsymbol{r}')\nabla_{\boldsymbol{r}'}G(\boldsymbol{r}, \boldsymbol{r}'; \lambda) - G(\boldsymbol{r}, \boldsymbol{r}'; \lambda)\nabla\psi(\boldsymbol{r}'))] \cdot d\boldsymbol{S}', \tag{6.5.10}$$

where the symmetry (6.5.6) is employed. Actually, the expression form is usually not seen. We have to present the explicit form distinguishing if the operator $L(\boldsymbol{r})$ in (6.5.4) is self-adjoint or not.

(1) The operator $L(\boldsymbol{r})$ is self-adjoint.

In this case, the knot on the right hand side of (6.5.9) is zero. When the parameter λ happens to be an eigenvalue λ_m, (6.5.10) becomes

$$\psi(\boldsymbol{r}) = \varphi_m(\boldsymbol{r}) + \int_{V'} \rho(\boldsymbol{r}')G(\boldsymbol{r},\boldsymbol{r}';\lambda_m)f(\boldsymbol{r}')\mathrm{d}\boldsymbol{r}', \qquad (6.5.11\text{a})$$

which mimics (6.1.12a) or (6.4.23). If the parameter λ is not an eigenvalue, (6.5.10) degrades to be

$$\psi(\boldsymbol{r}) = \int_{V'} \rho(\boldsymbol{r}')G(\boldsymbol{r},\boldsymbol{r}';\lambda)f(\boldsymbol{r}')\mathrm{d}\boldsymbol{r}'. \qquad (6.5.11\text{b})$$

(2) The operator $L(\boldsymbol{r})$ is non-self-adjoint.

In this case, there is no solution for (6.5.4), and (6.5.7) becomes

$$\psi(\boldsymbol{r}) = \int_{V'} \rho(\boldsymbol{r}')G(\boldsymbol{r},\boldsymbol{r}';\lambda)f(\boldsymbol{r}')\mathrm{d}\boldsymbol{r}'$$
$$- \oint_{S'} \left[p(\boldsymbol{r}') \left(\psi(\boldsymbol{r}')\frac{\partial}{\partial n'}G(\boldsymbol{r},\boldsymbol{r}';\lambda) - G(\boldsymbol{r},\boldsymbol{r}';\lambda)\frac{\partial}{\partial n'}\psi(\boldsymbol{r}') \right) \right] \cdot \mathrm{d}\boldsymbol{S}'.$$
$$(6.5.12)$$

The second term takes into account the influence of the boundary. When the boundary is complex, the solution is of the form of (6.5.12).

In the following, we always consider the case (1) unless specified.

4. Contributions from infinite surfaces

In order to evaluate the second term of (6.5.12), one has to know the boundary conditions: the values of the unknown function $\psi(\boldsymbol{r}')$ and Green's function $G(\boldsymbol{r},\boldsymbol{r}';\lambda)$ and their normal derivatives at the boundaries.

Sometimes, the integral domain can be divided into more than one region, each having respective boundary conditions. Then the surface integral in (6.5.12) is also divided into more than one term. For example, if the surface is divided into two parts, they are denoted as S_1 and S_2, respectively. Then, (6.5.12) can be written in the form of

$$\psi(\boldsymbol{r}) = \int_{V'} \rho(\boldsymbol{r}')f(\boldsymbol{r}')G(\boldsymbol{r}',\boldsymbol{r})\mathrm{d}\boldsymbol{r}' + \left(\int_{S_1'}\mathrm{d}S_1' + \int_{S_2'}\mathrm{d}S_2'\right)p(\boldsymbol{r}')$$

$$\times \left[\psi(\boldsymbol{r}')\frac{\partial}{\partial n'}G(\boldsymbol{r}',\boldsymbol{r}) - G(\boldsymbol{r}',\boldsymbol{r})\frac{\partial}{\partial n'}\psi(\boldsymbol{r}')\right]. \qquad (6.5.13)$$

Here we discuss a simplest case: the differential operator is the Laplace operator

$$L(\boldsymbol{r}) = -\nabla^2$$

and the surface is at infinity, denoted by S_∞. In this case the integral on the surface at infinity vanishes:

$$\int_{S_\infty} \mathrm{d}S_\infty \left[\psi(\boldsymbol{r}')\frac{\partial}{\partial n'}G(\boldsymbol{r}',\boldsymbol{r}) - G(\boldsymbol{r}',\boldsymbol{r})\frac{\partial}{\partial n'}\psi(\boldsymbol{r}')\right] = 0. \qquad (6.5.14)$$

The proof is as follows. For the Laplace operator, the weight $\rho(\boldsymbol{r}) = 1$ and kernel $p(\boldsymbol{r}) = 1$. The point $\boldsymbol{r}'$ in (6.5.14) is at infinity, and distance between $\boldsymbol{r}$ and $\boldsymbol{r}'$ is infinite. The functions $G(\boldsymbol{r}',\boldsymbol{r})$ and $\psi(\boldsymbol{r}')$ in (6.5.14) are those in an infinite space.

In an infinite space, the Green's function at r generated by the source at point $\boldsymbol{r}'$ has been given by (6.2.11) and (6.2.15):

$$G(\boldsymbol{r}',\boldsymbol{r};k) = -\frac{e^{ik|\boldsymbol{r}'-\boldsymbol{r}|}}{4\pi|\boldsymbol{r}'-\boldsymbol{r}|}; \quad \mathrm{Im}\,k \geq 0, \qquad (6.5.15)$$

where $k = \sqrt{\lambda}$. This reflects that a point source generates a spherical wave in a three dimensional space.

The function $\psi(\boldsymbol{r})$ satisfies the equation $(\lambda + \nabla^2)\psi(\boldsymbol{r}) = f(\boldsymbol{r})$. The first term of (6.5.14) shows that the solution $\psi(\boldsymbol{r})$ is mainly generated by the source $f(\boldsymbol{r}')$, while the $f(\boldsymbol{r}')$ distributes in a finite region. As long as point r is sufficiently apart from $\boldsymbol{r}'$, this region can be regarded as a point from the aspect of point $\boldsymbol{r}$. Thus, at places far enough, $\psi(\boldsymbol{r})$ can be written in the form of a spherical wave

$$\psi(\boldsymbol{r}) \sim g(\theta,\varphi)\frac{e^{ik_f r}}{r}, \qquad (6.5.16)$$

where $g(\theta, \varphi)$ is an angular distribution factor that is related to the distribution of the source $f(\boldsymbol{r})$. The length of the wavevector $\boldsymbol{k}_f$ is the same as k: $k_f = k$.

With this knowledge, we are able to reckon $G(\boldsymbol{r}', \boldsymbol{r})$ and $\psi(\boldsymbol{r}')$ and their normal derivatives at the surface of the integral (6.5.14). Then it follows that

$$\psi(\boldsymbol{r}')\frac{\partial}{\partial n'}G(\boldsymbol{r}', \boldsymbol{r}) - G(\boldsymbol{r}', \boldsymbol{r})\frac{\partial}{\partial n'}\psi(\boldsymbol{r}')$$

$$= g(\theta', \varphi')\frac{\mathrm{e}^{\mathrm{i}k_f r'}\mathrm{e}^{\mathrm{i}k|r'-r|}}{4\pi}\left(\frac{1}{r'|r'-r|^2} - \frac{1}{r'^2|r'-r|}\right).$$

$$(6.5.17)$$

Since the surface considered is at infinity, it can be regarded as a part of a spherical surface with radius R, and then let R go to infinity. Thus, as $r' = R$ in (6.5.17) is sufficiently large,

$$\psi(\boldsymbol{r}')\frac{\partial}{\partial n'}G(\boldsymbol{r}', \boldsymbol{r}) - G(\boldsymbol{r}', \boldsymbol{r})\frac{\partial}{\partial n'}\psi(\boldsymbol{r}') \sim g(\theta', \varphi')\frac{\mathrm{e}^{\mathrm{i}k_f r'}\mathrm{e}^{\mathrm{i}k|r'-r|}}{4\pi R^3}.$$

$$(6.5.18)$$

The area integral is $\mathrm{d}S = R^2 \sin\theta\mathrm{d}\theta\mathrm{d}\varphi$. Therefore, on this surface, we have

$$\int_{S_\infty}\left[\psi(\boldsymbol{r}')\frac{\partial}{\partial r'}G(\boldsymbol{r}', \boldsymbol{r}) - G(\boldsymbol{r}', \boldsymbol{r})\frac{\partial}{\partial r'}\psi(\boldsymbol{r}')\right]_{r'=R}\mathrm{d}S_\infty \sim \frac{1}{R} \to 0.$$

The conclusion is that for the Laplace operator, the integration (6.5.14) vanishes.

6.5.2.   *Examples in two-dimensional space*

1. Helmholtz equation in a rectangular region

Find the Green's function satisfying

$$\left(\nabla^2 + k^2\right)G\left(\boldsymbol{r}, \boldsymbol{r}'\right) = \delta(\boldsymbol{r} - \boldsymbol{r}'), \quad (0 \le x, x' \le a; 0 \le y, y' \le b),$$

$$(6.5.19a)$$

$$G(\boldsymbol{r}, \boldsymbol{r}')|_{x=0} = G(\boldsymbol{r}, \boldsymbol{r}')|_{x=a} = 0, \quad G(\boldsymbol{r}, \boldsymbol{r}')|_{y=0} = G(\boldsymbol{r}, \boldsymbol{r}')|_{y=b} = 0.$$

$$(6.5.19b)$$

In this case, the parameter $\lambda = k^2$ and operator $L = -\nabla^2$. The corresponding boundary value problem is

$$(\nabla^2 + k^2)\psi(\boldsymbol{r}) = f(\boldsymbol{r}), \quad (0 \le x \le a; 0 \le y \le b),$$

$$\psi(\boldsymbol{r})|_{x=0} = \psi(\boldsymbol{r})|_{x=a} = 0, \quad \psi(\boldsymbol{r})|_{y=0} = \psi(\boldsymbol{r})|_{y=b} = 0.$$

(1) The eigenfunction method

First of all, let us solve the solution of eigenvalue problem

$$\left(\nabla^2 + k_i^2\right)\varphi_i(\boldsymbol{r}) = 0, \quad (0 \le x \le a; 0 \le y \le b),$$

$$\varphi_i(\boldsymbol{r})|_{x=0} = \varphi_i(\boldsymbol{r})|_{x=a} = 0, \quad \varphi_i(\boldsymbol{r})|_{y=0} = \varphi_i(\boldsymbol{r})|_{y=b} = 0.$$

By separation of variables, let

$$\varphi_i(\boldsymbol{r}) = X(x)Y(y) \tag{6.5.20}$$

so as to get

$$\frac{X''(x)}{X(x)} + \frac{Y''(y)}{Y(y)} + k^2 = 0.$$

Subsequently, the characteristic equations and boundary conditions in the two directions are as follows.

$$X''(x) + X(x)k_x^2 = 0. \tag{6.5.21a}$$

$$X(0) = X(a) = 0. \tag{6.5.21b}$$

$$Y''(y) + k_y^2 Y(y) = 0. \tag{6.5.21c}$$

$$Y(0) = Y(b) = 0. \tag{6.5.21d}$$

The boundary conditions (6.5.21b) and (6.5.21d) are derived from (6.5.19b). The normalized characteristic functions and eigenvalues are solved.

$$X(x) = \sqrt{\frac{2}{a}} \sin k_x x, \quad Y(y) = \sqrt{\frac{2}{b}} \sin k_y y, \quad k_x = \frac{n\pi}{a}, \quad k_y = \frac{m\pi}{b},$$
$$\tag{6.5.22}$$

where m and n are integers. The eigenfunctions and eigenvalues of the system are thus

$$\varphi_i(\boldsymbol{r}) = \varphi_{n,m}(\boldsymbol{r}) = \sqrt{\frac{4}{ab}} \sin\frac{n\pi}{a}x \sin\frac{m\pi}{b}y, \quad k_i^2 = k_x^2 + k_y^2.$$
$$\tag{6.5.23}$$

Please note that there are double indices due to the two-dimensional system.

(i) The parameter k^2 is not an eigenvalue

The Green's function is

$$G(r, r') = \sum_i \frac{\varphi_i(r')\varphi_i(r)}{k^2 - k_i^2} = \sum_{n,m=1}^{\infty} \frac{\varphi_{n,m}(r')\varphi_{n,m}(r)}{k^2 - k_{n,m}^2}$$

$$= \frac{4}{ab} \sum_{n,m=1}^{\infty} \frac{1}{k^2 - [(n\pi/a)^2 + (m\pi/b)^2]}$$

$$\times \sin\frac{n\pi}{a}x' \sin\frac{m\pi}{b}y' \sin\frac{n\pi}{a}x \sin\frac{m\pi}{b}y. \qquad (6.5.24)$$

(ii) The parameter k^2 equals to an eigenvalue

Suppose that

$$k^2 = \left(\frac{n_1\pi}{a}\right)^2 + \left(\frac{m_1\pi}{b}\right)^2. \qquad (6.5.25)$$

In this case, because the boundary conditions are homogeneous, the modified Green's function is desired. It meets the equation

$$\left(\nabla^2 + \left(\frac{n_1\pi}{a}\right)^2 + \left(\frac{m_1\pi}{b}\right)^2\right) G(r, r') = \delta(x - x')\delta(y - y')$$

$$- \frac{4}{ab} \sin\frac{n_1\pi}{a}x' \sin\frac{m_1\pi}{b}y' \sin\frac{n_1\pi}{a}x \sin\frac{m_1\pi}{b}y. \qquad (6.5.26)$$

On the right hand side, the additional term is just the eigenfunction belonging to the eigenvalue (6.5.25). Applying (6.4.35), we obtain

$$G(r, r') = \frac{4}{ab} \sin\frac{n_1\pi}{a}x' \sin\frac{m_1\pi}{b}y' \sin\frac{n_1\pi}{a}x \sin\frac{m_1\pi}{b}y$$

$$+ \frac{4ab}{\pi^2} \sum_{n=1,n\neq n_1}^{\infty} \sum_{m=1,m\neq m_1}^{\infty} \frac{\sin\frac{n\pi}{a}x' \sin\frac{m\pi}{b}y' \sin\frac{n\pi}{a}x \sin\frac{m\pi}{b}y}{b^2(n_1^2 - n^2) + a^2(m_1^2 - m^2)}.$$

$$(6.5.27)$$

(2) The piecewise expression method

In the present case, along each of the Cartesian coordinates, the interval is finite and there are eigenfunctions, as manifested by (6.5.22).

Hence along either direction the piecewise expression method can be applied. Here we choose the x direction to do so. When $x \neq x'$, (6.5.19a) becomes

$$(\nabla^2 + k^2)G(\boldsymbol{r}, \boldsymbol{r}') = 0, \quad (0 \leq x \neq x' \leq a; 0 \leq y, y' \leq b). \quad (6.5.28)$$

Neglecting the variable $\boldsymbol{r}'$ temporarily, the equation is simplified to be

$$(\nabla^2 + k^2)\psi(\boldsymbol{r}) = 0.$$

By separation of variables, let $\psi(\boldsymbol{r}) = X(x)Y(y)$. Then the equation with respect to y is just (6.5.21), and its normalized eigenfunctions and eigenvalues are listed in (6.5.22).

The delta function $\delta(y-y')$ can be expanded by this eigenfunction set:

$$\delta(y - y') = \frac{2}{b} \sum_m \sin \frac{m\pi}{b} y' \sin \frac{m\pi}{b} y. \quad (6.5.29)$$

Because $\{\sin \frac{m\pi}{b} y\}$ is a complete set on this interval, the Green's function can be expanded by it:

$$G(\boldsymbol{r}, \boldsymbol{r}') = \frac{2}{b} \sum_m g_m(x, x') \sin \frac{m\pi}{b} y' \sin \frac{m\pi}{b} y. \quad (6.5.30)$$

(i) The parameter k^2 is not an eigenvalue

Substituting (6.5.29) and (6.5.30) into (6.5.19a) and taking derivatives, we get

$$\frac{2}{b} \sum_m \left[\frac{\partial^2}{\partial x^2} g_m(x, x') + k_{xm}^2 g_m(x, x') \right] \sin \frac{m\pi}{b} y' \sin \frac{m\pi}{b} y$$

$$= \delta(x - x') \frac{2}{b} \sum_m \sin \frac{m\pi}{b} y' \sin \frac{m\pi}{b} y.$$

This leads to the following equation:

$$\frac{\partial^2}{\partial x^2} g_m(x, x') + k_{xm}^2 g_m(x, x') = \delta(x - x'). \quad (6.5.31a)$$

This is the equation that the Green's function along the x direction $g(x, x')$ should satisfy. Its boundary conditions are derived from

(6.5.19b):

$$g_m(0, x') = g_m(a, x') = 0. \tag{6.5.31b}$$

Now, we use the piecewise expression method to solve (6.5.31). The result is

$$g_m(x, x') = -\frac{1}{k_{xm} \sin k_{xm} a}[\sin k_{xm}(x' - a) \sin k_{xm} x \theta(x - x')$$

$$+ \sin k_{xm} x' \sin k_{xm}(x - a)\theta(x' - x)]. \tag{6.5.32}$$

This is substituted into (6.5.30) to get the Green's function:

$$G(\boldsymbol{r}, \boldsymbol{r}') = -\frac{2}{b} \sum_m \sin \frac{m\pi}{b} y' \sin \frac{m\pi}{b} y \frac{1}{k_{xm} a \sin k_{xm} a}$$

$$\times [\sin k_{xm}(x' - a) \sin k_{xm} x \theta(x - x')$$

$$+ \sin k_{xm} x' \sin k_{xm}(x - a)\theta(x' - x)], \tag{6.5.33}$$

where

$$k_{xm}^2 = k^2 - \left(\frac{m\pi}{b}\right)^2. \tag{6.5.34}$$

This expression is valid when $k^2 \neq \left(\frac{n\pi}{a}\right)^2 + \left(\frac{m\pi}{b}\right)^2$ is not an eigenvalue.

(ii) The parameter k^2 equals to an eigenvalue

When $k^2 = \left(\frac{n_1\pi}{a}\right)^2 + \left(\frac{m_1\pi}{b}\right)^2$ is just an eigenvalue as (6.5.25), one has to solve the modified Green's function. It should satisfy Eq. (6.5.26) and the boundary conditions (6.5.19b). Assume that the modified Green's function can also be expanded in the way of (6.5.30). Then its expression and (6.5.29) are substituted into (6.5.26).

$$\left(\nabla^2 + \left(\frac{n_1\pi}{a}\right)^2 + \left(\frac{m_1\pi}{b}\right)^2\right) \frac{2}{b} \sum_m g_m(x, x') \sin \frac{m\pi}{b} y' \sin \frac{m\pi}{b} y$$

$$= \frac{2}{b} \sum_m \left(\frac{\partial^2}{\partial x^2} + k_{1m}^2\right) g_m(x, x') \sin \frac{m\pi}{b} y' \sin \frac{m\pi}{b} y$$

$$= \delta(x - x') \frac{2}{b} \sum_m \sin \frac{m\pi}{b} y' \sin \frac{m\pi}{b} y$$

$$- \frac{4}{ab} \sin \frac{n_1\pi}{a} x' \sin \frac{m_1\pi}{b} y' \sin \frac{n_1\pi}{a} x \sin \frac{m_1\pi}{b} y, \tag{6.5.35}$$

where

$$k_{1m}^2 = \left(\frac{n_1\pi}{a}\right)^2 + \left(\frac{m_1\pi}{b}\right)^2 - \left(\frac{m\pi}{b}\right)^2. \tag{6.5.36}$$

Multiplying $\frac{2}{b}\sin\frac{n\pi}{b}y'\sin\frac{n\pi}{b}y$ on (6.5.35) and integrating over y and y', we get

$$\left(\frac{\partial^2}{\partial x^2} + k_{1n}^2\right)g_n(x,x') = \delta(x-x') - \frac{2}{a}\sin\frac{n_1\pi}{a}x'\sin\frac{n_1\pi}{a}x\delta_{nm_1}. \tag{6.5.37}$$

We are at the stage to solve (6.5.37). When $n \neq m_1$,

$$\left(\frac{\partial^2}{\partial x^2} + k_{1n}^2\right)g_n(x,x') = \delta(x-x'), \quad n \neq m_1.$$

The equation is the same as (6.5.31a), and its solution follows (6.5.32):

$$g_m(x,x') = -\frac{1}{k_{1m}\sin k_{1m}a}\left[\sin k_{1m}(x'-a)\sin k_{1m}x\theta(x-x')\right.$$
$$\left. + \sin k_{1m}x'\sin k_{1m}(x-a)\theta(x'-x)\right]. \tag{6.5.38a}$$

When $n = m_1$, (6.5.37) becomes

$$\left(\frac{\partial^2}{\partial x^2} + k_{1m_1}^2\right)g_{m_1}(x,x') = \delta(x-x') - \frac{2}{a}\sin\frac{n_1\pi}{a}x'\sin\frac{n_1\pi}{a}x,$$

where $k_{1m_1}^2 = \left(\frac{n_1\pi}{a}\right)^2 = k_{n_1}^2$, see (6.5.36). This equation is solved following the routine of (6.4.39)–(6.4.43). As $x \neq x'$, the equation becomes

$$\left(\frac{\partial^2}{\partial x^2} + k_{n_1}^2\right)g_{m_1}(x,x') = -\frac{2}{a}\sin k_{n_1}x'\sin k_{n_1}x.$$

The two independent solutions of the homogeneous equation are

$$\zeta_1(x) = \sin k_{n_1}x, \quad \zeta_2(x) = \cos k_{n_1}x.$$

Their Wronskian is $W = -k_{n_1}$. The special solution of the inhomogeneous equation is constructed by formula (3.1.24). Let

$\zeta_0(x, x') = u(x) \sin k_{n_1} x'$. Then the expression of $u(x)$ is

$$u(x) = -\frac{2}{a} \left[\sin k_{n_1} x \int \frac{\cos k_{n_1} x}{k_{n_1}} \sin k_{n_1} x dx \right.$$

$$\left. + \cos k_{n_1} x \int \frac{\sin k_{n_1} x}{k_{n_1}} \sin k_{n_1} x dx \right]$$

$$= \frac{1}{ak_{n_1}^2} (-\sin k_{n_1} x + k_{n_1} x \cos k_{n_1} x).$$

Its values at the boundaries are $u(0) = 0$, $u(a) = (-1)^{n_1}/k_{n_1}$. Thus, we get

$$\zeta_0(x, x') = \frac{\sin k_{n_1} x'}{ak_{n_1}^2} (-\sin k_{n_1} x + k_{n_1} x \cos k_{n_1} x).$$

The required modified Green's function is expressed following (6.4.39):

$$g_{m_1}(x, x') = \begin{cases} a_1(x')\zeta_1(x) + a_2(x')\zeta_2(x) + \zeta_0(x, x'), & x > x'. \\ b_1(x')\zeta_1(x) + b_2(x')\zeta_2(x) + \zeta_0(x, x'), & x < x'. \end{cases}$$

Applying the boundary conditions and the conditions at $x = x'$, we obtain the modified Green's function:

$$g_{m_1}(x, x') = \begin{cases} a_1(x') \sin k_{n_1} x - \dfrac{1}{k_{n_1}} \sin k_{n_1} x' \cos k_{n_1} x \\ \quad + \zeta_0(x, x'), & x > x'. \\ \left[a_1(x') - \dfrac{\cos k_{n_1} x'}{k_{n_1}} \right] \sin k_{n_1} x + \zeta_0(x, x'), & x < x'. \end{cases}$$

There is still a quantity $a_1(x')$ to be determined. This needs the symmetry of the modified Green's function. Let $g_{m_1}(x', x) = g_{m_1}(x, x')$. This leads to

$$a_1(x') \sin k_{n_1} x + u(x) \sin k_{n_1} x' = a_1(x) \sin k_{n_1} x' + u(x') \sin k_{n_1} x.$$

Obviously, it should be that $a_1(x) = u(x)$. Therefore, we achieve the modified Green's function along the x direction:

$$g_{m_1}(x, x') = \begin{cases} u(x') \sin k_{n_1} x + u(x) \sin k_{n_1} x' \\ \quad - \dfrac{1}{k_{n_1}} \sin k_{n_1} x' \cos k_{n_1} x, \quad x > x'. \\[2mm] u(x') \sin k_{n_1} x + u(x) \sin k_{n_1} x' \\ \quad - \dfrac{\cos k_{n_1} x'}{k_{n_1}} \sin k_{n_1} x, \quad x < x'. \end{cases} \quad (6.5.38b)$$

Finally, we get the modified Green's function (6.5.14) as $k^2 = \left(\frac{n_1 \pi}{a}\right)^2 + \left(\frac{m_1 \pi}{b}\right)^2$:

$$G(\boldsymbol{r}, \boldsymbol{r}') = \frac{2}{b} \sum_m g_m(x, x') \sin \frac{m\pi}{b} y' \sin \frac{m\pi}{b} y, \quad (6.5.39)$$

where as $m \neq m_1$, $g_m(x, x')$ is (6.5.38a), and $g_{m_1}(x, x')$ is (6.5.38b).

In summary, when the parameter λ is not an eigenvalue, the Green's function can be expanded by (6.5.30), and the components along the x direction $g_m(x, x')$ are expressed by (6.5.32). When λ is an eigenvalue as (6.5.25), the modified Green's function can still be expanded by (6.5.30), and $g_m(x, x')$ expressed by (6.5.38a) is of the same form as (6.5.32). As $m = m_1$, $g_{m_1}(x, x')$ is (6.5.38b).

2. Helmholtz equation in a circle

Find the Green's function satisfying

$$(\nabla^2 + k^2)G(\boldsymbol{r}, \boldsymbol{r}') = \delta(\boldsymbol{r} - \boldsymbol{r}'); \quad r, r' < a, \quad (6.5.40a)$$

$$G(\boldsymbol{r}, \boldsymbol{r}')|_{r=0} < \infty, \quad G(\boldsymbol{r}, \boldsymbol{r}')|_{r=a} = 0. \quad (6.5.40b)$$

The corresponding boundary value problem is

$$(\nabla^2 + k^2)\psi(\boldsymbol{r}) = f(\boldsymbol{r}), \quad \psi(\boldsymbol{r})|_{r=0} < \infty, \quad \psi(\boldsymbol{r})|_{r=a} = 0.$$

Apparently, polar coordinates are convenient. Laplace operator and the delta function are expressed by

$$\nabla^2 = \frac{1}{r} \frac{\partial}{\partial r} \left(r \frac{\partial}{\partial r} \right) + \frac{1}{r^2} \frac{\partial^2}{\partial^2 \theta} \quad (6.5.41)$$

Mathematics for Physicists

and

$$\delta(\boldsymbol{r} - \boldsymbol{r}') = \frac{\delta(r - r')\delta(\theta - \theta')}{r}, \tag{6.5.42}$$

respectively.

(1) The eigenfunction method

The eigenfunction method is applicable to the present boundary conditions. First of all, let us solve the solution of eigenvalue problem

$$(\nabla^2 + k_i^2)\varphi_i(\boldsymbol{r}) = 0, \quad \varphi_i(\boldsymbol{r})|_{r=0} < \infty, \quad \varphi_i(\boldsymbol{r})|_{r=a} = 0.$$

In polar coordinates, the equation becomes

$$\frac{1}{r}\frac{\partial}{\partial r}\left(r\frac{\partial}{\partial r}\right)\varphi_m(\boldsymbol{r}) + \frac{1}{r^2}\frac{\partial^2}{\partial^2\theta}\varphi_m(\boldsymbol{r}) + k_m^2\varphi_m(\boldsymbol{r}) = 0. \tag{6.5.43}$$

By separation of variables, let $\varphi_i(\boldsymbol{r}) = R(r)\Theta(\theta)$. We have

$$\frac{r}{R(r)}\frac{\partial}{\partial r}\left(r\frac{\partial R(r)}{\partial r}\right) + k_i^2 r^2 = -\frac{1}{\Theta(\theta)}\frac{\partial^2\Theta(\theta)}{\partial^2\theta} = \gamma.$$

The angular factor satisfies the equation

$$\Theta''(\theta) + \gamma\Theta(\theta) = 0. \tag{6.5.44}$$

The functions should be singly-valued. From the periodical condition, we get the solutions is

$$\Theta(\theta) = e^{im\theta}, \tag{6.5.45}$$

where $\gamma = m^2$ with m being integers.

The radial equation and its boundary condition are

$$r\frac{\partial}{\partial r}\left(r\frac{\partial R(r)}{\partial r}\right) + (k_i^2 r^2 - m^2)R(r) = 0, \quad R(0) < \infty, \quad R(a) = 0. \tag{6.5.46a}$$

Let $kr = \rho$. Then,

$$\rho^2\frac{\partial^2 R(\rho)}{\partial^2\rho} + \rho\frac{\partial R(\rho)}{\partial\rho} + (\rho^2 - m^2)R(\rho) = 0. \tag{6.5.46b}$$

This is Bessel equation of order m. Its solution is the combination of the first and second kinds Bessel functions.

$$R(kr) = AJ_m(kr) + BY_m(kr). \tag{6.5.47}$$

Application of the boundary conditions leads to

$$B = 0, \quad J_m(k_{m,n}a) = 0. \tag{6.5.48}$$

Let $N_{m,n}$ denote the normalization coefficient of $J_m(k_{m,n}r)$ on $[0, a]$:

$$N_{m,n}^2 = \int_0^a J_m^2(k_{m,n}r)r\mathrm{d}r.$$

It is the first kind of boundary condition, so that (4.7.15) should be used:

$$N_{m,n}^2 = \frac{a^2}{2}\left[J_m'(k_{m,n}a)\right]^2 = \frac{a^2}{2}\left[J_{m+1}(k_{m,n}a)\right]^2. \tag{6.5.49}$$

Therefore, the normalized eigenfunctions are

$$\varphi_i(\boldsymbol{r}) = \varphi_{m,n}(\boldsymbol{r}) = \frac{J_m(k_{m,n}r)}{N_{m,n}}\frac{\mathrm{e}^{im\theta}}{\sqrt{2\pi}}. \tag{6.5.50}$$

The eigenvalues are determined by (6.5.48). There are two subscripts attached to eigenfunctions and eigenvalues.

(i) The parameter k is not an eigenvalue

The Green's function is

$$G(\boldsymbol{r}, \boldsymbol{r}') = \sum_{n=1}^{\infty}\sum_{m=-\infty}^{\infty} \frac{\varphi_{m,n}^*(\boldsymbol{r}')\varphi_{m,n}(\boldsymbol{r})}{k^2 - k_{m,n}^2}$$

$$= \frac{1}{2\pi}\sum_{n=1}^{\infty}\sum_{m=-\infty}^{\infty} \frac{J_m(k_{m,n}r')J_m(k_{m,n}r)}{(k^2 - k_{m,n}^2)N_{m,n}^2}\mathrm{e}^{im(\theta-\theta')}.$$

$$\tag{6.5.51}$$

(ii) The parameter k equals to an eigenvalue

Considering that the boundary conditions (6.5.40b) are inhomogeneous, the first thought is to follow the way described in Subsection 6.4.2 to solve the solution of the problem (6.4.24) with homogeneous boundary conditions. However, if the boundary conditions in (6.5.40b) are changed to be homogeneous, we find that the radial equation (6.5.46a) has no nonzero functions at $m = 0$, and as $m \neq 0$, the eigenfunctions obtained under the homogeneous boundary conditions are the same as those obtained under inhomogeneous ones. This is a special case. We have to solve the modified Green's function under inhomogeneous boundary conditions (6.5.40b).

Since the eigenfunctions are (6.5.50), when $k^2 = k_{m1,n1}^2$, it is easy to put down the modified Green's function:

$$G(r, r') = \frac{J_{m1}(k_{m1,n1}r')}{2\pi N_{m1,n1}^2} J_{m1}(k_{m1,n1}r)e^{im_1(\theta-\theta')}$$

$$+ \frac{1}{2\pi} \sum_{n=1,n\neq n_1}^{\infty} \sum_{m=-\infty,m\neq m_1}^{\infty} \frac{J_m(k_{m,n}r')J_m(k_{m,n}r)}{(k_{n1,m1}^2 - k_{n,m}^2)N_{m,n}^2}e^{im(\theta-\theta')}.$$

$$(6.5.52)$$

(2) The piecewise expression method

In polar coordinates, (6.5.40a) appears as

$$\left(\frac{1}{r}\frac{\partial}{\partial r}\left(r\frac{\partial}{\partial r}\right) + \frac{1}{r^2}\frac{\partial^2}{\partial^2\theta} + k^2\right)G(r, r') = \frac{\delta(r-r')\delta(\theta-\theta')}{r}.$$

$$(6.5.53)$$

We apply the piecewise expression method along the radial direction. As $r \neq r'$,

$$\left(\frac{1}{r}\frac{\partial}{\partial r}\left(r\frac{\partial}{\partial r}\right) + \frac{1}{r^2}\frac{\partial^2}{\partial^2\theta} + k^2\right)G(r, r') = 0;$$

$$(r \neq r' < a; \ 0 \leq \theta < 2\pi).$$

Neglecting the variable r' temporarily, the equation is simplified to be

$$\left(\frac{1}{r}\frac{\partial}{\partial r}\left(r\frac{\partial}{\partial r}\right) + \frac{1}{r^2}\frac{\partial^2}{\partial^2\theta} + k^2\right)\psi(r) = 0,$$

which is of the same form of (6.5.43). After separation of variable, the angular normalized eigenfunctions and eigenvalues are in (6.5.45). So, the delta function $\delta(\theta-\theta')$ can be expanded:

$$\delta(\theta-\theta') = \sum_{m=-\infty}^{\infty} e^{im(\theta-\theta')}.$$

$$(6.5.54)$$

The Green's function is expanded in the following way:

$$G(r, r') = \sum_{m=-\infty}^{\infty} g_m(r, r')e^{im(\theta-\theta')}$$

$$(6.5.55)$$

where the radial factor $g_m(r, r')$ is called radial Green's function. These two expansions are substituted into (6.5.53) to get

$$\sum_{m=-\infty}^{\infty} \left[\frac{1}{r} \frac{\partial}{\partial r} \left(r \frac{\partial}{\partial r} \right) + \left(k^2 - \frac{m^2}{r^2} \right) \right] g_m(r, r') e^{im(\theta - \theta')}$$

$$= \frac{\delta(r - r')}{r} \sum_{m=-\infty}^{\infty} e^{im(\theta - \theta')}.$$

This leads to the following radial equation that the radial Green's function should satisfy:

$$\left[\frac{1}{r} \frac{\partial}{\partial r} \left(r \frac{\partial}{\partial r} \right) + \left(k^2 - \frac{m^2}{r^2} \right) \right] g_m(r, r') = \frac{\delta(r - r')}{r}. \tag{6.5.56}$$

Its boundary conditions are derived from (6.5.40b):

$$g_m(0, r') < \infty, \quad g_m(a, r') = 0. \tag{6.5.57}$$

From (6.5.56) the jump condition of the radial Green's function is

$$\left[r \frac{\partial}{\partial r} g_m(r, r') \right]_{r = r' + 0^+} - \left[r \frac{\partial}{\partial r} g_m(r, r') \right]_{r = r' + 0^+} = 1. \tag{6.5.58}$$

Now we apply the piecewise expression method to solve (6.5.56). As $r \neq r'$,

$$\left[\frac{1}{r} \frac{\partial}{\partial r} \left(r \frac{\partial}{\partial r} \right) + \left(k^2 - \frac{m^2}{r^2} \right) \right] g_m(r, r') = 0, \tag{6.5.59}$$

which is a Bessel equation of order m. Its general solution is the linear combination of the first and second kinds of Bessel functions. So the radial Green's function is expressed by

$$g_m(r, r') = \begin{cases} a_1(r') J_m(kr) + a_2(r') Y_m(kr), & r < r'. \\ \\ b_1(r') J_m(kr) + b_2(r') Y_m(kr), & r > r'. \end{cases} \tag{6.5.60}$$

By the conditions at $r \neq r'$, it is shown that

$$c_1(r') = a_1(r') - b_1(r') = -\frac{\pi k}{2} Y_m(kr'),$$

$$c_2(r') = a_2(r') - b_2(r') = \frac{\pi k}{2} J_m(kr'). \tag{6.5.61}$$

When the boundary conditions (6.5.57) are applied to (6.5.60), we achieve

$$a_2(r') = 0, \quad b_1(r')J_m(ka) + b_2(r')Y_m(ka) = 0. \qquad (6.5.62)$$

Combination of (6.5.61) and (6.5.62) give the expressions of $b_1(r')$, $b_2(r')$ and $a_1(r')$. Finally, the radial Green's function is

$$g_m(r,r') = \frac{\pi k}{2} \left[\frac{Y_m(ka)J_m(kr') - J_m(ka)Y_m(kr')}{J_m(ka)} J_m(kr)\theta(r'-r) \right.$$
$$\left. + \frac{Y_m(ka)J_m(kr) - J_m(ka)Y_m(kr)}{J_m(ka)} J_m(kr')\theta(r-r') \right]. \qquad (6.5.63)$$

This is substituted into (6.5.55) to get the total Green's function:

$$G(\boldsymbol{r},\boldsymbol{r}') = \frac{\pi k}{2} \sum_{m=-\infty}^{\infty} e^{im(\theta-\theta')}$$
$$\times \left[\frac{Y_m(ka)J_m(kr') - J_m(ka)Y_m(kr')}{J_m(ka)} J_m(kr)\theta(r'-r) \right.$$
$$\left. + \frac{Y_m(ka)J_m(kr) - J_m(ka)Y_m(kr)}{J_m(ka)} J_m(kr')\theta(r-r') \right]. \qquad (6.5.64)$$

This expression is not valid when k meets $J_m(ka) = 0$, i.e., k is an eigenvalue. In that case, the modified Green's function is appealing.

3. Scattering of electromagnetic wave by a perfect metal wedge

Consider that in the xy plane, there is a wedge made of perfect metal with its tip at the origin, see Fig. 6.1. An electromagnetic plane wave propagates from far away and is scattered by the wedge. The half-angle of the wedge is α, and the bisector of the angle is along the x axis. We consider the electric field component with its polarization along the z axis. Because the wedge is made of perfect metal, the electric field at the wedge surfaces should be zero by the boundary conditions of an electromagnetic field.

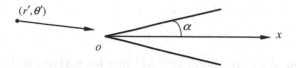

Fig. 6.1. A two-dimensional wedge made of perfect metal lies in the xy plane. Its tip is at the origin, its half-angle is α and its bisector of the angle is along the x axis. An electromagnetic plane wave propagates from far away and is scattered by this wedge.

The idea of solving this problem is that first we consider the field distribution generated by a point source at (r', θ'). Then let $r' \to \infty$ to move the source to infinity. This distribution naturally involves the scattering by the wedge.

This is the following boundary value problem for Green's function.

$$(\nabla^2 + k^2)G(\boldsymbol{r}, \boldsymbol{r}') = \delta(\boldsymbol{r} - \boldsymbol{r}'); \quad \alpha < \theta < 2\pi - \alpha, \qquad (6.5.65a)$$

$$G(\boldsymbol{r}, \boldsymbol{r}')|_{r=0} < \infty, \quad G(\boldsymbol{r}, \boldsymbol{r}')|_{\theta=\alpha, 2\pi-\alpha} = 0. \qquad (6.5.65b)$$

In polar coordinates, the equation is expressed by

$$\left(\frac{1}{r}\frac{\partial}{\partial r}\left(r\frac{\partial}{\partial r} \right) + \frac{1}{r^2}\frac{\partial^2}{\partial \theta^2} + k^2 \right) G\left(\boldsymbol{r}, \boldsymbol{r}' \right)$$

$$= \frac{\delta(r - r')\delta(\theta - \theta')}{r}; \quad \alpha < \theta < 2\pi - \alpha. \qquad (6.5.66)$$

We apply the piecewise expression method along the radial direction.

(1) Solving the angular eigenfunctions

As $r \neq r'$, (6.5.66) becomes

$$\left(\frac{1}{r}\frac{\partial}{\partial r}\left(r\frac{\partial}{\partial r} \right) + \frac{1}{r^2}\frac{\partial^2}{\partial \theta^2} + k^2 \right) G\left(\boldsymbol{r}, \boldsymbol{r}' \right) = 0;$$

$$\alpha < \theta < 2\pi - \alpha, \quad r \neq r'.$$

Neglecting the variable $\boldsymbol{r}'$ temporarily, the equation is simplified to be

$$\left(\frac{1}{r}\frac{\partial}{\partial r}\left(r\frac{\partial}{\partial r} \right) + \frac{1}{r^2}\frac{\partial^2}{\partial \theta^2} + k^2 \right) \psi(\boldsymbol{r}) = 0.$$

By separation of variables, the angular eigenvalue equation is

$$\Theta''(\theta) + \gamma\Theta(\theta) = 0 \qquad (6.5.67a)$$

and its boundary conditions are

$$\Theta(\alpha) = \Theta(2\pi - \alpha) = 0. \qquad (6.5.67b)$$

The solution of the problem (6.5.67) has been given in Example 4 in Section 5.4. The orthonormalized eigenfunction set is

$$\left\{ \frac{1}{\sqrt{\pi - \alpha}} \sin \frac{n\pi(\theta - \alpha)}{2(\pi - \alpha)} \right\}, \qquad (6.5.68)$$

where n is integer. The eigenvalues are

$$\gamma_n = \nu^2, \qquad (6.5.69)$$

where

$$\nu = \frac{n\pi}{2(\pi - \alpha)}, \quad n = 1, 2, 3, \ldots. \qquad (6.5.70)$$

So, the angular delta function can be expanded as follows:

$$\delta(\theta - \theta') = \frac{1}{\pi - \alpha} \sum_{\nu}^{\infty} \sin \nu(\theta' - \alpha) \sin \nu(\theta - \alpha). \qquad (6.5.71)$$

This is just (5.4.4).

The Green's function is expanded by this eigenfunction set:

$$G(\boldsymbol{r}, \boldsymbol{r}') = \sum_{\nu}^{\infty} g_\nu(r, r') \frac{1}{\pi - \alpha} \sin \nu(\theta' - \alpha) \sin \nu(\theta - \alpha). \qquad (6.5.72)$$

(2) The radial Green's function

The above two equations are substituted into (6.5.66). Then,

$$\sum_{\nu}^{\infty} \left(\frac{\partial}{\partial r} \left(r \frac{\partial}{\partial r} \right) - \frac{1}{r}\nu^2 + rk^2 \right) g_\nu(r, r') \sin \nu(\theta' - \alpha) \sin \nu(\theta - \alpha)$$

$$= \delta(r - r') \sum_{\nu}^{\infty} \sin \nu(\theta' - \alpha) \sin \nu(\theta - \alpha). \qquad (6.5.73)$$

This leads to

$$\left(\frac{\partial}{\partial r} \left(r \frac{\partial}{\partial r} \right) - \frac{1}{r}\nu^2 + rk^2 \right) g_\nu(r, r') = \delta(r - r'). \qquad (6.5.74)$$

This is the equation that the radial Green's function $g_\nu(r, r')$ should satisfy. We apply the piecewise expression method.

As $r \neq r'$, $\left[r \frac{\partial}{\partial r} \left(r \frac{\partial}{\partial r} \right) + \left(k^2 r^2 - \nu^2 \right) \right] g_\nu(r, r') = 0$. Let $kr = \rho$. The equation becomes $\left[\rho^2 \frac{\partial^2}{\partial^2 \rho} + \rho \frac{\partial}{\partial \rho} + (\rho^2 - \nu^2) \right] g_\nu(r, r') = 0$. This Bessel equation is of order ν. Since now the argument r can be infinite, we are unable to set a boundary condition at finite r. Therefore, there is no eigenfunction set in this case. We have to resort to the piecewise method.

Its general solution can be a linear combination of Bessel functions of either the first and second kinds or the first and third kinds. It is found that the combination of the first kind of Bessel and the second kind of Hankel functions are convenient. Please note that here the index ν does not mean integers.

$$g_\nu(r, r') = \begin{cases} a_1(r') J_\nu(kr) + a_2(r') H_\nu^{(2)}(kr), & r < r'. \\ b_1(r') J_\nu(kr) + b_2(r') H_\nu^{(2)}(kr), & r > r'. \end{cases} \tag{6.5.75}$$

The boundary condition $g_\nu(0, r') < \infty$ leads to $a_2(r') = 0$. The boundary condition at infinity is not given. But we can employ the symmetry of the Green's function. It is seen that this can be met when we take $b_1(r') = 0$ and $a_1(r') = H_\nu^{(2)}(kr')$, the radial Green's function is

$$g_\nu(r, r') = \frac{i\pi k}{2} \begin{cases} H_\nu^{(2)}(kr') J_\nu(kr), & r < r'. \\ J_\nu(kr') H_\nu^{(2)}(kr), & r > r'. \end{cases} \tag{6.5.76}$$

(3) Green's function

The Green's function is obtained by inserting (6.5.76) into (6.5.72):

$$G(\boldsymbol{r}, \boldsymbol{r}') = \frac{i\pi k}{2(\pi - \alpha)} \sum_\nu^\infty \sin \nu(\theta' - \alpha)$$

$$\times \sin \nu(\theta - \alpha) \begin{cases} H_\nu^{(2)}(kr') J_\nu(kr), & r < r'. \\ J_\nu(kr') H_\nu^{(2)}(kr), & r > r'. \end{cases} \tag{6.5.77}$$

At last, we explain why the first kind Bessel and second kind Hankel functions are selected to combine the solution in (6.5.74). When $r' \to \infty$, the source goes to infinity, so that the Green's function is in an infinitely large space. The corresponding solution of the two-dimensional basic solution has been obtained as

$$G^+(r, r'; k^2) = -\frac{i}{4} H_0^{(1)}(k|r - r'|). \tag{6.5.78a}$$

$$G^-(r, r'; k^2) = \frac{i}{4} H_0^{(2)}(k|r - r'|). \tag{6.5.78b}$$

G^+ is a divergent wave outward from the center, and G^- an aggregate wave inward to the center. Since now $r' \to \infty$, the source is at infinity, we should take G^-.

The asymptotic formula of a Hankel function of the second kind is

$$H_\nu^{(2)}(k\rho \to \infty) = \sqrt{\frac{2}{\pi k \rho}} \exp\left[-i\left(k\rho - \frac{\nu\pi}{2} - \frac{\pi}{4}\right)\right]. \tag{6.5.79}$$

As $r' \to \infty$, $|r - r'| \to r' - r\cos(\theta' - \theta)$. So,

$$H_0^{(2)}(k\rho \to \infty) \to \sqrt{\frac{2i}{\pi k r'}} e^{-ik[r' - r\cos(\theta' - \theta)]}$$

$$= \sqrt{\frac{2i}{\pi k r'}} e^{-ikr'} e^{ikr\cos(\theta' - \theta)}.$$

This means that the asymptotic behavior of the Green's function should be

$$G^-(r, r' \to \infty) = \frac{i}{4}\sqrt{\frac{2i}{\pi k r'}} e^{-ikr'} e^{ikr\cos(\theta' - \theta)} \tag{6.5.80}$$

Now we turn to consider the $r < r'$ part of the Green's function $g_\nu(r, r')$.

$$G(r, r') = \frac{i\pi k}{2(\pi - \alpha)} \sum_\nu^\infty \sin\nu(\theta' - \alpha)$$

$$\times \sin\nu(\theta - \alpha) H_\nu^{(2)}(kr') J_\nu(kr), \quad r < r'. \tag{6.5.81}$$

When $r' \to \infty$, the asymptotic behavior of (6.5.79) is

$$H_\nu^{(2)}(kr' \to \infty) = \sqrt{\frac{2i}{\pi k r'}} i^\nu e^{-ikr'}.$$

It is substituted into (6.5.81) to get

$$G(r, r' \to \infty) = \frac{i\pi k}{2(\pi - \alpha)} \sqrt{\frac{2i}{\pi k r'}} e^{-ikr'}$$

$$\times \sum_{\nu}^{\infty} i^{\nu} \sin \nu(\theta' - \alpha) \sin \nu(\theta - \alpha) J_{\nu}(kr).$$

$$(6.5.82)$$

Equation (6.5.82) should equal to (6.5.81). Apparently, their behavior as $r' \to \infty$ are the same, both being $e^{-ikr'}/\sqrt{r'}$. This is the reason we choose the Hankel function of the second kind in (6.5.75). If a Bessel function of the second kind was selected, the asymptotic behavior could not match (6.5.80), which would cause is difficulty to find the form of the Green's function.

From the equality of (6.5.80) and (6.5.82), we obtain the following expansion:

$$e^{ikr \cos(\theta' - \theta)} = \frac{2\pi k}{\pi - \alpha} \sum_{\nu}^{\infty} i^{\nu} \sin \nu(\theta' - \alpha) \sin \nu(\theta - \alpha) J_{\nu}(kr),$$

$$(6.5.83)$$

where ν values are in (6.5.70). This is the expansion of a plane wave at the region outside of the wedge.

(4) Scattering of electromagnetic wave by the half-infinite plane of a perfect metal

If the angle of the wedge in Fig. 6.1 becomes zero, this does not mean the disappearance of the wedge, but it is degraded to be a half-infinite plane lying in the positive x axis. This is because we used the boundary condition (6.5.67b) but not the periodical boundary condition. In (6.5.77) and (6.5.70), let $\alpha = 0$, we get

$$G(r, r') = \frac{ik}{2} \sum_{n=1}^{\infty} \sin\left(\frac{n}{2}\theta'\right) \sin\left(\frac{n}{2}\theta\right) \begin{cases} H_{\nu}^{(2)}(kr') J_{\nu}(kr), & r < r'. \\ J_{\nu}(kr') H_{\nu}^{(2)}(kr), & r > r'. \end{cases}$$

$$(6.5.84)$$

This is the Green's function at the whole two-dimensional plane except the positive x axis.

6.5.3.   *Examples in three-dimensional space*

The problem of Laplace equation in a sphere is as follows.

$$\nabla^2 G(\boldsymbol{r}, \boldsymbol{r}') = \delta(\boldsymbol{r} - \boldsymbol{r}'); \quad r, r < a. \tag{6.5.85a}$$

$$G(\boldsymbol{r}, \boldsymbol{r}')|_{r=0} < \infty, \quad G(\boldsymbol{r}, \boldsymbol{r}')|_{r=a} = 0. \tag{6.5.85b}$$

In spherical coordinates, the operator and delta function are in the forms of

$$\nabla^2 = \frac{1}{r^2} \frac{\partial}{\partial r} \left(r^2 \frac{\partial}{\partial r} \right) + \frac{1}{r^2 \sin\theta} \frac{\partial}{\partial\theta} \left(\sin\theta \frac{\partial}{\partial\theta} \right) + \frac{1}{r^2 \sin^2\theta} \frac{\partial^2}{\partial\varphi^2}$$

and

$$\delta(\boldsymbol{r} - \boldsymbol{r}') = \frac{\delta(r - r')\delta(\varphi - \varphi')\,\delta(\theta - \theta')}{r^2 \sin\theta}$$

respectively.

In order to solve the Laplace equation, we let $\psi(r) = R(r)\Theta(\theta)\Phi(\varphi)$. Then, the equation

$$\frac{1}{R} \frac{\mathrm{d}}{\mathrm{d}r} \left(r^2 \frac{\mathrm{d}R}{\mathrm{d}r} \right) + \frac{1}{\Theta \sin\theta} \frac{\mathrm{d}}{\mathrm{d}\theta} \left(\sin\theta \frac{\mathrm{d}\Theta}{\mathrm{d}\theta} \right) + \frac{1}{\Phi \sin^2\theta} \frac{\mathrm{d}^2\Phi}{\mathrm{d}\varphi^2} = 0$$

is separated into two equations:

$$\frac{1}{R} \frac{\mathrm{d}}{\mathrm{d}r} \left(r^2 \frac{\mathrm{d}R}{\mathrm{d}r} \right) = \gamma \tag{6.5.86}$$

and

$$\frac{1}{\Theta \sin\theta} \frac{\mathrm{d}}{\mathrm{d}\theta} \left(\sin\theta \frac{\mathrm{d}\Theta}{\mathrm{d}\theta} \right) + \frac{1}{\Phi \sin^2\theta} \frac{\mathrm{d}^2\Phi}{\mathrm{d}\varphi^2} = -\gamma. \tag{6.5.87}$$

Equation (6.5.87) is just (3.4.43) and its solutions are spherical harmonics

$$Y_l^m(\theta, \varphi) = \Theta(\theta)\Phi(\varphi) = P_l^m(\cos\theta)e^{im\varphi}, \tag{6.5.88}$$

which are products of associated Legendre functions $\Theta(\theta) = P_l^m(\cos\theta)$ and $\Phi(\varphi) = e^{im\varphi}$. The corresponding eigenvalues are

$$\gamma = l(l + 1), \quad (l = 0, 1, 2, \ldots), \quad m = 0, \pm 1, \pm 2, \ldots, \pm l.$$

With these eigenvalues, the radial equation (6.5.86) becomes

$$r^2 \frac{\mathrm{d}^2 R}{\mathrm{d}r^2} + 2r \frac{\mathrm{d}R}{\mathrm{d}r} - l(l + 1)R = 0.$$

Its boundary conditions are derived from (6.5.85b):

$$R(0) < \infty, \quad R(a) = 0, \quad G(\boldsymbol{r}, \boldsymbol{r}')|_{r=a} = 0.$$

The general solution of the radial equation is $R(r) = Ar^l + \frac{B}{r^{l+1}}$. By the boundary conditions it is known that $A = 0$, $B = 0$. Therefore, there is no nonzero solutions for this problem. This is because the parameter λ in the Bessel equation is now zero which is not an eigenvalue. We have to resort to the piecewise expression method. Since there are eigenfunctions in the radial direction, it is natural to implement the piecewise expression method in this direction.

$$(\nabla^2 + k^2)G(\boldsymbol{r}, \boldsymbol{r}') = \frac{\delta(r - r')\delta(\varphi - \varphi')\delta(\theta - \theta')}{r^2 \sin\theta},$$

$$(r, r' < a, \; 0 \le \varphi \le 2\pi, \; 0 \le \theta \le \pi).$$

$$(6.5.89)$$

As $r \neq r'$, it becomes

$$(\nabla^2 + k^2)G(\boldsymbol{r}, \boldsymbol{r}') = 0, \quad (r \neq r' < a, \; 0 \le \varphi \le 2\pi, \; 0 \le \theta \le \pi).$$

After separation of variables, the angular equation is still (6.5.87) and its normalized eigenfunctions are spherical harmonics (6.5.88). So, the angular delta function can be expanded by

$$\delta(\theta - \theta')\delta(\varphi - \varphi') = \sum_{l=0}^{\infty} \sum_{m=-l}^{l} \sin\theta Y_l^{m*}(\theta', \varphi')Y_l^m(\theta, \varphi). \quad (6.5.90)$$

Note that there is a weight factor $\sin\theta$, as shown by (5.4.3). The Green's function is also expanded by the eigenfunction set:

$$G(\boldsymbol{r}, \boldsymbol{r}') = \sum_{l=0}^{\infty} \sum_{m=-l}^{l} Y_l^{m*}(\theta', \varphi')Y_l^m(\theta, \varphi)g(r, r'). \quad (6.5.91)$$

Note that here the radial Green's function should be denoted by $g_{lm}(r, r')$ but we omitted the indices for convenience. Equations (6.5.90) and (6.5.91) are substituted into (6.5.89):

$$\sum_{l=0}^{\infty} \sum_{m=-l}^{l} Y_l^{m*}(\theta', \varphi')Y_l^m(\theta, \varphi)\left[\frac{\partial}{\partial r}\left(r^2\frac{\partial}{\partial r}\right) - l(l+1)\right]g(r, r')$$

$$= \delta(r - r')\sum_{l=0}^{\infty} \sum_{m=-l}^{l} Y_l^{m*}(\theta', \varphi')Y_l^m(\theta, \varphi).$$

This leads to

$$\left[\frac{\partial}{\partial r}\left(r^2\frac{\partial}{\partial r}\right) - l(l+1)\right]g(r,r') = \delta(r-r').$$

This is the equation that $g(r,r')$ must meet. Its boundary conditions are derived from (6.5.85b):

$$g(0,r') < \infty, \quad g(a,r') = 0. \tag{6.5.92}$$

The jump condition of the $g(r,r')$ is

$$\left[r^2\frac{\partial}{\partial r}g(r,r')\right]_{r=r'+0^+} - \left[r^2\frac{\partial}{\partial r}g(r,r')\right]_{r=r'-0^+} = 1.$$

The piecewise expression of the radial Green's function is

$$g(r,r') = \begin{cases} a_1(r')r^l + a_2(r')r^{-l-1}, & r < r'. \\\\ b_1(r')r^l + b_2(r')r^{-l-1}, & r > r'. \end{cases}$$

By the conditions at $r = r'$ and the boundary conditions, we achieve the radial Green's function:

$$g(r,r') = \frac{r'^{-2l-1} - a^{-2l-1}}{2l+1}r''^l r^l\theta(r'-r)$$

$$+ \frac{r^{-2l-1} - a^{-2l-1}}{2l+1}r''^l r^l\theta(r-r'). \tag{6.5.93}$$

Substituted into (6.5.91), the total Green's function is obtained:

$$G(\boldsymbol{r},\boldsymbol{r}') = \sum_{l=0}^{\infty}\sum_{m=-l}^{l} Y_l^{m*}(\theta',\varphi')Y_l^{m}(\theta,\varphi)$$

$$\times\left[\frac{r'^{-2l-1} - a^{-2l-1}}{2l+1}r''^l r^l\theta(r'-r)\right.$$

$$\left. + \frac{r^{-2l-1} - a^{-2l-1}}{2l+1}r''^l r^l\theta(r-r')\right]. \tag{6.5.94}$$

In this problem, there are no eigenvalues in the radial direction, but there are eigenfunctions in the other two directions.

6.6. Green's Function of Ordinary Differential Equation of First Order

For a self-adjoint operator, it can be proven in the way described in Subsection 6.4.1 that the Green's function is of the symmetry (6.4.7).

6.6.1. *Boundary value problem of inhomogeneous equations*

The general boundary value problem with a parameter λ is as follows.

$$\left[\lambda - p(x) \frac{\mathrm{d}}{\mathrm{d}x} - q(x) \right] \psi(x) = f(x); \quad a \leq x \leq b, \tag{6.6.1a}$$

$$\alpha \psi(a) + \beta \psi(b) = \gamma. \tag{6.6.1b}$$

It is assumed that $p(x) \neq 0$ on $[a, b]$.

Similar to the case of differential equation of second order, the problem (6.6.1) is divided into two parts:

$$\psi(x) = \varphi(x) + \xi(x), \tag{6.6.2}$$

where the function $\varphi(x)$ satisfies the homogeneous equation with inhomogeneous boundary condition and $\xi(x)$ satisfies the inhomogeneous equation with homogeneous boundary condition.

6.6.2. *Boundary value problem of homogeneous equations*

The function $\varphi(x)$ satisfies the following problem.

$$\left[\lambda - p(x) \frac{\mathrm{d}}{\mathrm{d}x} - q(x) \right] \varphi(x) = 0; \quad a \leq x \leq b, \tag{6.6.3a}$$

$$\alpha \varphi(a) + \beta \varphi(b) = \gamma. \tag{6.6.3b}$$

The general solution of (6.6.3a) is easily obtained. Its form can be written as

$$\varphi(x) = C \exp \left[\int_a^x \frac{\lambda - q(x_1)}{p(x_1)} \mathrm{d}x_1 \right]. \tag{6.6.4a}$$

Or

$$\varphi(x) = D \exp \left[-\int_x^b \frac{\lambda - q(x_1)}{p(x_1)} \mathrm{d}x_1 \right]. \tag{6.6.4b}$$

In the first form, the boundary condition requires that

$$\alpha C + \beta C \exp\left[\int_a^b \frac{\lambda - q(x_1)}{p(x_1)}dx_1\right] = \gamma.$$

Thus the constant C is

$$C = \gamma\left\{\alpha + \beta \exp\left[\int_a^b \frac{\lambda - q(x_1)}{p(x_1)}dx_1\right]\right\}^{-1}.$$

As $\gamma = 0$, $C = 0$. This means that the homogeneous equation of the first order has nonzero solution only when the boundary condition is inhomogeneous. The solution of the problem (6.6.3) is

$$\varphi(x) = \gamma\left\{\alpha + \beta \exp\left[\int_a^b \frac{\lambda - q(x_1)}{p(x_1)}dx_1\right]\right\}^{-1}$$

$$\times \exp\left[\int_a^x \frac{\lambda - q(x_2)}{p(x_2)}dx_2\right]. \qquad (6.6.5)$$

A simple case is that when $p(x) = -i$, and $q(x) = 0$, $\varphi(x) = \frac{\gamma e^{i\lambda(x-a)}}{\alpha+\beta e^{i\lambda(b-a)}}$.

6.6.3.   *Inhomogeneous equations and Green's function*

The function $\xi(x)$ satisfies the following problem.

$$\left[\lambda - p(x)\frac{d}{dx} - q(x)\right]\xi(x) = f(x); \quad a \le x \le b, \qquad (6.6.6a)$$

$$\alpha\xi(a) + \beta\xi(b) = 0. \qquad (6.6.6b)$$

Since the equation is inhomogeneous, we have to find the Green's function of the following problem.

$$\left[\lambda - p(x)\frac{d}{dx} - q(x)\right]G(x,x') = \delta(x-x'); \quad a \le x, \ x' \le b, \qquad (6.6.7a)$$

$$\alpha G(a,x') + \beta G(b,x') = 0. \qquad (6.6.7b)$$

Once the Green's function is solved,

$$\xi(x) = \int_a^b G(x,x')f(x')dx'. \qquad (6.6.8)$$

We use the piecewise expression method, but now the condition at $x = x'$ is that

$$-G(x' + 0^+, x') + G(x' - 0^+, x') = \frac{1}{p(x)}. \tag{6.6.9}$$

The Green's function itself is discontinuous at $x = x'$.

As $x \neq x'$, the equation

$$\left[\lambda - p(x)\frac{\mathrm{d}}{\mathrm{d}x} - q(x) \right] G(x, x') = 0.$$

has a solution the same as (6.6.4).

$$G(x, x') = \begin{cases} A(x') \exp\left[\int_a^x \dfrac{\lambda - q(y)}{p(y)} \mathrm{d}y \right], & x < x' \\[3mm] B(x') \exp\left[-\int_x^b \dfrac{\lambda - q(y)}{p(y)} \mathrm{d}y \right], & x > x' \end{cases}$$

By the jump condition (6.6.9), we get

$$A(x') \exp\left[\int_a^{x'} \frac{\lambda - q(y)}{p(y)} \mathrm{d}y \right] - B(x')$$

$$\exp\left[-\int_{x'}^b \frac{\lambda - q(y)}{p(y)} \mathrm{d}y \right] = \frac{1}{p(x')}.$$

The boundary condition (6.6.7b) requires that

$$\alpha A(x') + \beta B(x') = 0.$$

From these two equations it is found that

$$\Delta = \alpha \exp\left[-\int_{x'}^b \frac{\lambda - q(y)}{p(y)} \mathrm{d}y \right] + \beta \exp\left[\int_a^{x'} \frac{\lambda - q(y)}{p(y)} \mathrm{d}y \right]$$

and

$$A(x') = \frac{\beta}{\Delta p(x')}, \qquad B(x') = -\frac{\alpha}{\Delta p(x')}.$$

So, the Green's function is

$$G(x, x') = \frac{1}{\Delta p(x')} \begin{cases} \beta \exp\left[\int_a^x \dfrac{\lambda - q(y)}{p(y)} \mathrm{d}y \right], & x < x' \\[3mm] -\alpha \exp\left[-\int_x^b \dfrac{\lambda - q(y)}{p(y)} \mathrm{d}y \right], & x > x' \end{cases} \tag{6.6.10}$$

Let us consider the special case that the operator in (6.6.6a) is self-adjoint. There are two necessary conditions for the operator to be self-adjoint. One has been seen from (2.2.22) that a necessary condition for the operator to be self-adjoint is that $p(x)$ has merely an imaginary part. Here we simply choose

$$p(x) = -i, \quad q(x) = 0. \tag{6.6.11}$$

The other necessary condition comes from the following equation.

$$
\begin{aligned}
(g\,(x, x_2; \lambda^*), (\lambda - L)g(x, x_1; \lambda)) & \\
- ((\lambda^* - L)g(x, x_2; \lambda^*), g(x, x_1; \lambda)) & \\
= (g(x, x_2; \lambda^*), Lg(x, x_1; \lambda)) - (Lg(x, x_2; \lambda^*), g(x, x_1; \lambda)) & \\
= - [p(x)g^*(x, x_2; \lambda^*)g(x, x_1; \lambda)]_{x=a}^{x=b} &
\end{aligned}
\tag{6.6.12}
$$

The two Green's functions in (6.6.12) satisfy respectively equations

$$(\lambda - L)g(x, x_1; \lambda) = \delta(x - x_1) \tag{6.6.13a}$$

and

$$(\lambda^* - L)g(x, x_2; \lambda^*) = \delta(x - x_2). \tag{6.6.13b}$$

The last equal mark in (6.6.12) originates from (2.2.9). We substitute the boundary condition (6.6.7b) and (6.6.13) into (6.6.12) to get

$$
\begin{aligned}
-g^*(x_1, x_2; \lambda^*) + g(x_2, x_1; \lambda) & \\
= (p(b)|\alpha|^2 - p(a)|\beta|^2)g^*(b, x_2; \lambda^*)g(b, x_1; \lambda)/|\alpha|^2. &
\end{aligned}
\tag{6.6.14}
$$

If $p(x)$ is a constant and

$$\beta = \alpha^*, \tag{6.6.15}$$

there will be the symmetry:

$$g^*(x_1, x_2; \lambda^*) = g(x_2, x_1; \lambda) \tag{6.6.16}$$

Now substituting (6.6.15) and (6.6.11) into (6.6.10), we obtain

$$G(x, x'; \lambda) = ie^{i\lambda x} \frac{\alpha^* e^{-i\lambda a}\theta(x' - x) - \alpha e^{-i\lambda b}\theta(x - x')}{e^{i\lambda x'}(\alpha^* e^{-i\lambda a} + \alpha e^{-i\lambda b})}. \tag{6.6.17}$$

It is easily verified that (6.6.17) is of the symmetry (6.1.16).

6.6.4. *General solutions of boundary value problem*

By (6.6.2), the solution of the problem (6.6.1) is

$$\psi(x) = \varphi(x) + \int_a^b G(x, x') f(x') \mathrm{d}x', \qquad (6.6.18)$$

where $\varphi(x)$ and $G(x, x')$ are expressed by (6.6.5) and (6.6.10), respectively.

Equation (6.6.18) is the standard form of the solution of the problem (6.6.1). Suppose that (6.6.1) is extended to the following form:

$$\left[\lambda - \frac{\mathrm{d}}{\mathrm{d}x} - q(x) \right] y(x) = f(x, y(x)); \quad a \le x \le b, \qquad (6.6.19a)$$

$$\alpha y(a) + \beta y(b) = \gamma, \qquad (6.6.19b)$$

where the inhomogeneous term contains the unknown function $y(x)$. Then, the solution is still (6.6.18):

$$y(x) = \varphi(x) + \int_a^b G(x, x') f(x', y(x')) \mathrm{d}x'. \qquad (6.6.20)$$

Nevertheless, the integrand in (6.6.20) contains the unknown $y(x)$, so that this is an integral equation.

6.7. Green's Function of Non-Self-Adjoint Equations

6.7.1. *Adjoint Green's function*

The Theorem 1 in Section 6.1 shows that for homogeneous boundary conditions Green's function always exists and is unique, independent of whether the operator is self-adjoint or not. Therefore, Green's function can always be found under homogeneous boundary conditions. In searching for the Green's function, the piecewise expression method is always applicable.

The prerequisite of using the eigenfunction method is the existence of a complete set of eigenfunctions of an ordinary differential equation of second order.

For an equation with a non-self-adjoint operator, as long as the boundary condition is one of the three cases listed in Theorem 3 in Section 3.7, the solutions compose a complete set in the given interval. Thus, any function continuous on this interval can be expanded by this set. Consequently, Green's function can also be expanded by this set.

Let the eigenvalue equation be

$$L(x)u_n(x) = \lambda_n u_n(x), \tag{6.7.1}$$

and its adjoint equation be

$$L^+(x)v_n(x) = \lambda_n^* v_n(x). \tag{6.7.2}$$

According to Theorem 8 in Section 2.2, the eigenvalues of these equations are one-to-one correspondent and are complex conjugates of each other. An eigenvector and an adjoint one are orthogonal to each other if they respectively belong to different eigenvalues:

$$(v_m, u_n) = \delta_{mn}. \tag{6.7.3}$$

We assume that all the eigenvectors have been normalized. If the differential operator is of Sturm-Liouville type, then $v_n(x) = u_n^*(x)$, see Theorem 1 in Section 3.7.

Now we expand Green's function by eigenfunctions:

$$G(x, x') = \sum_n c_n(x')u_n(x). \tag{6.7.4}$$

In order to reckon the expansion coefficients $c_n(x')$, let operator $\lambda - L(x)$ act on the both sides:

$$\delta(x - x') = \rho(x) \sum_n c_n(x')[\lambda - L(x)]u_n(x)$$

$$= \rho(x) \sum_n c_n(x')(\lambda - \lambda_n)u_n(x).$$

Then $v_m^*(x)$ is multiplied to the equation and integration is carried out with respect to x.

$$\int dx v_m^*(x)\delta(x - x') = \sum_n c_n(x')(\lambda - \lambda_n) \int dx \rho(x)v_m^*(x)u_n(x).$$

Using biorthogonality (6.7.3), we obtain the expression of the coefficients:

$$c_m(x') = \frac{v_m^*(x')}{\lambda - \lambda_m}.$$

Green's function is

$$G(x, x', \lambda) = \sum_n \frac{v_n^*(x')u_n(x)}{\lambda - \lambda_n}. \tag{6.7.5}$$

To find the symmetry of the Green's function, we observe the way in Subsection 6.4.1. Green's function G and its adjoint Q respectively satisfy equations

$$(\lambda - L)G(x, x_1) = \delta(x - x_1)/\rho(x) \qquad (6.7.6)$$

and

$$(\lambda^* - L^+)Q(x, x_2; \lambda^*) = \delta(x - x_2)/\rho(x). \qquad (6.7.7)$$

For a SL operator, the weight function remains unchanged for the adjoint operator, see Theorem 1 in Section 3.6. By (2.2.11), we have

$$(Q(\lambda^*), (\lambda - L^+)G(\lambda)) - ((\lambda - L^+)Q(\lambda^*), G(\lambda))$$
$$= [J(G, Q)]_a^b = 0. \qquad (6.7.8)$$

This equation helps to derive the boundary conditions of Q from those of Green's function G.

Following the way in Subsection 6.4.1, we are able to achieve

$$Q^*(x_1, x_2; \lambda) = G(x_2, x_1; \lambda^*). \qquad (6.7.9)$$

This equation can be compared to (6.4.7).

Thus, the adjoint Green's function reads

$$Q(x, x', \lambda) = G^*(x', x, \lambda^*) = \sum_n \frac{v_n(x)u_n^*(x')}{\lambda - \lambda_n^*}. \qquad (6.7.10)$$

For example, for Example 1 in Subsection 3.6.2, the Green's function is expanded by the eigenfunctions as follows:

$$G(x, x', \lambda) = \frac{1}{\lambda + \alpha^2} \frac{2\alpha}{e^{2\alpha} - 1} e^{\alpha x} e^{\alpha x'}$$

$$+ \sum_{n=1}^{\infty} \frac{1}{\lambda - n^2\pi^2} \frac{2}{n^2\pi^2 + \alpha^2}$$

$$\times (\alpha \sin \mu_n x + n\pi \cos \mu_n x)(\alpha \sin \mu_n x' + n\pi \cos \mu_n x').$$

6.7.2. *Solutions of inhomogeneous equations*

Suppose that there is an inhomogeneous equation

$$(\lambda - L(x)]\psi(x) = f(x), \quad a < x < b. \qquad (6.7.11)$$

Its solution can be given by means of its Green's function. We follow the procedure (6.1.6)–(6.1.12).

$$(Q(\lambda^*), (\lambda - L)\psi) - ((\lambda^* - L^+)Q(\lambda^*), \psi) = [J(\psi, G)]_a^b. \quad (6.7.12)$$

The two inner products on the left hand side are

$$(Q(\lambda^*), (\lambda - L)\psi) = \int_V \rho(x)Q^*(x, x'; \lambda^*)[\lambda - L(x)]\psi(x)\mathrm{d}x$$

$$= \int_a^b \rho(x)Q^*(x, x'; \lambda^*)f(x)\mathrm{d}x \quad (6.7.13)$$

and

$$((\lambda^* - L^+)Q(\lambda^*), \psi)$$

$$= \int_V \rho(x)\left\{\left[\lambda^* - L^+(x)\right]Q(x, x'; \lambda^*)\right\}^*\psi(x')]\mathrm{d}x = \psi(x'),$$

$$(6.7.14)$$

where (6.7.11) and (6.7.7) have been employed. The knot is (3.2.10). Hence, the expression of the solution (6.7.11) is

$$\psi(x') = \int_a^b \rho(x)Q^*(x, x'; \lambda^*)f(x)\mathrm{d}x$$

$$+ \left[p(x)\left\{\psi(x)\frac{\partial}{\partial x}Q^*(x, x'; \lambda^*) - \psi'(x)Q^*(x, x'; \lambda^*)\right\}\right]_a^b.$$

$$(6.7.15)$$

Exchanging x, x' and using the symmetry (6.7.10), we get

$$\psi(x) = \int_a^b \rho(x')G(x, x'; \lambda)f(x')\mathrm{d}x'$$

$$+ \left[p(x')\left\{\psi(x')\frac{\partial}{\partial x'}G(x, x'; \lambda) - \psi'(x')G(x, x'; \lambda)\right\}\right]_a^b.$$

$$(6.7.16)$$

Equation (6.7.15) is expressed by the adjoint Green's function, and (6.7.16) is by Green's function. The latter is in fact of the same form as (6.1.13).

Since we have (6.7.16), it seems that the expression (6.7.15) by adjoint Green's function is not necessary. Usually it is indeed so. However, there may also be some cases where the expression (6.7.15) is inevitably used. For example, in (6.7.16), the Green's function and its derivatives at the ends are needed. If of the boundary conditions, only one is given, say the value at $x = a$, and another condition is absent, then (6.7.16) cannot be used. An example is Exercise 18 in Chapter 2. In such a case, the boundary conditions of the adjoint Green's function can be derived by (6.7.8), and then (6.7.15) is employed to evaluate the solutions.

Exercises

1. (1) Show that the solution of following problem is just (6.3.24).
$$\frac{d^2}{dx^2}G(x, x') = \delta(x - x'), \quad 0 < t, t' < 1;$$
$$G(0, x') = G(1, x') = 0.$$

   (2) Find the solution of
$$\left(\frac{d^2}{dx^2} + \lambda\right)G(x, x') = \delta(x - x'), \quad 0 < t, t' < 1;$$
$$G(0, x') = G(1, x') = 0.$$
   As $\lambda = 0$ the solution degrades to that of (1).

2. For a differential operator of third order
$$L(x) = \frac{d^3}{dx^3} + f_2(x)\frac{d^2}{dx^2} + f_1(x)\frac{d}{dx} + f_0(x),$$
   use the piecewise expression method to find its Green's function. From the result, conjecture the expression of Green's function of a differential operator of nth order.

3. Find the Green's function of the following problem.
$$\left(\frac{d^2}{dx^2} + k^2\right)G(x, x') = \delta(x - x'), \quad 0 < x, x' < \infty;$$
$$G(x = 0) = 1, \quad G(x \to \infty) \sim e^{ikx}.$$

   This is the Green's function of a one-dimensional Helmholtz equation that the voltage (or electric current) in a half-infinitely long conductor should satisfy.

4. Find the Green's functions of the following problems.

(1) $\left(\dfrac{d^2}{dx^2} + \lambda^2\right) G\left(x, x'\right) = \delta(x - x'), \quad 0 < x, x' < \pi;$

$$\left[\dfrac{d}{dx}G(x, x')\right]_{x=0} = 0, \quad \left[\dfrac{d}{dx}G(x, x')\right]_{x=\pi} = 0.$$

(2) $\dfrac{d^2}{dx^2}G(x, x') = \delta(x - x'), 0 < x, x' < 1;$

$$G(0, x') + G(1, x') = 0,$$

$$\left[\dfrac{d}{dx}G(x, x')\right]_{x=0} + \left[\dfrac{d}{dx}G(x, x')\right]_{x=1} = 0.$$

(3) $\left(\dfrac{d^2}{dx^2} + 1\right) G(x, x') = \delta(x - x'), \quad 0 < x, x' < 1;$

$$G(0, x') = G(1, x'), \quad \left[\dfrac{d}{dx}G(x, x')\right]_{x=0} = \left[\dfrac{d}{dx}G(x, x')\right]_{x=1}.$$

(4) $\left(\dfrac{d^2}{dx^2} - k^2\right) G(x, x') = \delta(x - x'), \quad 0 < x, x' < 1;$

$$G(0, x') = \left[\dfrac{d}{dx}G(x, x')\right]_{x=0}, \quad G(1, x') = \left[\dfrac{d}{dx}G(x, x')\right]_{x=1}.$$

(5) $\dfrac{d^3}{dx^3}G(x, x') = \delta(x - x'), \quad 0 < x, x' < 1;$

$$G(0, x') = G(1, x') = 0,$$

$$\left[\dfrac{d}{dx}G(x, x')\right]_{x=0} = \left[\dfrac{d}{dx}G(x, x')\right]_{x=1}.$$

5. Find the Green's functions of the following problems.

(1) $\left(\dfrac{d^2}{dx^2} - 1\right) G(x, x') = \delta(x - x'), \quad G(x \to \infty, x') < \infty.$

(2) $\left(\dfrac{d^2}{dx^2} - 1\right) G(x, x') = \delta(x - x'), \quad G(-a, x') = G(a, x') = 0.$
As $a \to \infty$, the solution degrades to that of (1).

(3) $\left(\dfrac{d^2}{dx^2} + \lambda^2\right) G(x, x') = \delta(x - x'), \quad G(a, x') = G(x, b) = 0,$

$a \le x, x' \le b.$

(4) $\dfrac{d}{dx}\left[(1 - x^2)\dfrac{d}{dx}G(x, x')\right] = \delta(x - x'), \quad G(0, x') = 0,$

$\left[(1 - x^2)\dfrac{d}{dx}G(x, x')\right]_{x \to 1^-} \to 1.$

6. Prove the contour integral formula of Green's function:

$$\frac{1}{2\pi i} \oint_C d\lambda G(x, x', \lambda) = \delta(x - x'),$$

where the contour path C contains all of the simple poles of Green's function $G(x, x', \lambda)$.

7. Substitute (6.3.12) into (6.3.1) to show that the solution satisfies the differential equation.

8. For a damping oscillator, solve its Green's function by the Fourier transformation method.

9. By (6.1.26), it seems that at the eigenvalues, the Green's function is of singularities of first order. Actually, it is not necessarily so. Consider the following boundary value problem.

$$\left(\frac{d^2}{dx^2} + \lambda^2\right) y(x) = 0, \quad y(0) = 0, \quad y'(0) + y'(\pi) = 0.$$

Find its eigenvalues and eigenfunctions. Find the Green's function satisfying the same boundary conditions. In this case, the Green's function is of singularities of second order at its eigenvalues. Nevertheless, the Green's functions discussed in this chapter have only singularities of first order.

10. In Section 6.2, we have solved the basic solutions of the Green's function for Laplace operator.

(1) Please do the same work by piecewise expression method.
(2) In spherical coordinates, the radial equations of the Green's functions in one-, two- and three-dimensional space can be uniformly obtained. Observing this method, find the basic solution in four-dimensional space.

11. There is a δ potential in the origin. Find the Green's function in an infinite one-dimensional space. That is to say, find the solution of the following equation:

$$\left(z + \frac{d^2}{dx^2} - V_0 \delta(x) \right) G_0(x, x'; z) = \delta(x - x').$$

The Green's function should be finite as x goes to infinite. Then take $V_0 \to \infty$ to get the Green's function in a semi infinite one-dimensional space.

12. Find the one-dimensional Green's function of the following equation:

$$(z - H)G(x, x'; z) = I\delta(x - x'),$$

where $H = \begin{pmatrix} mc^2 & -i\hbar cd/dx \\ -i\hbar cd/dx & -mc^2 \end{pmatrix}$. This operator is the one-dimensional Dirac relativistic Hamiltonian.

13. The equation (6.4.28) that the modified Green's function $g(x, x'; \lambda_m)$ must satisfy does not guarantee the symmetry (6.4.7). Some conditions are needed for $g(x, x'; \lambda_m)$ to be of (6.4.7). One condition is that $\int_a^b dx \rho(x) g^*(x, x'; \lambda_m^*) \varphi_m(x) = c\varphi_m(x')$, where c is a real number. Show that under this condition, the modified Green's function is of the symmetry of (6.4.7). Equation (6.4.35) meets this condition. This is a sufficient condition.

14. Solve (6.5.31) to obtain the expression of the Green's function (6.5.32).

15. Equations (6.5.32) and (6.5.24) are obtained by the piecewise expression method and eigenfunction method, respective. Expand (6.5.32) by eigenfunctions. Is the result the same as (6.5.24)?

16. Expand the radial Green's function in (6.5.64) by the eigenfunctions. Is the result the same as (6.5.51)?

17. Show that when eigenfunctions exist, the eigenfunction method and the piecewise expression method for solving Green's function are equivalent to each other, i.e., one of the expressions can be

derived from the other. We have seen two specific examples in Exercises 15 and 16.

18. For the problem (6.5.40), is it possible using the piecewise expression method along the angular direction? If it is, do it.

19. For problem (6.5.40), find its solution of the modified Green's function by the piecewise expression method when the parameter k is just an eigenvalue.

20. Find the Green's function of the following problem by means of the piecewise expression method.

$$\nabla^2 G(\boldsymbol{r}, \boldsymbol{r}') = \delta(\boldsymbol{r} - \boldsymbol{r}'); r, r' < a; \quad G(\boldsymbol{r}, \boldsymbol{r}')|_{r=0} < \infty,$$

$$G(\boldsymbol{r}, \boldsymbol{r}')|_{r=a} = 0.$$

21. Find the Green's function of the following problem in a rectangular.

$$(\nabla^2 + k^2)G(\boldsymbol{r}, \boldsymbol{r}') = \delta(\boldsymbol{r} - \boldsymbol{r}'),$$

$$(0 \le x, x' \le a; 0 \le y, y' \le b);$$

$$\left[\frac{\partial}{\partial x} G(\boldsymbol{r}, \boldsymbol{r}')\right]\Bigg|_{x=0} = \left[\frac{\partial}{\partial x} G(\boldsymbol{r}, \boldsymbol{r}')\right]\Bigg|_{x=a} = 0, \quad \left[\frac{\partial}{\partial y} G(\boldsymbol{r}, \boldsymbol{r}')\right]\Bigg|_{y=0}$$

$$= \left[\frac{\partial}{\partial y} G(\boldsymbol{r}, \boldsymbol{r}')\right]\Bigg|_{y=b} = 0.$$

22. Find the Green's function in an annular region.

$$\nabla^2 G(\boldsymbol{r}, \boldsymbol{r}') = \delta(\boldsymbol{r} - \boldsymbol{r}'); a < r, r < b;$$

$$G(\boldsymbol{r}, \boldsymbol{r}')|_{r=a} = G(\boldsymbol{r}, \boldsymbol{r}')|_{r=b} = 0.$$

23. We have figured out the scattering problem by a perfect metal wedge and obtained the solution (6.5.77). The boundary conditions were that at the border of the wedge the Green's function should be zero. This in fact represented the distribution of electric field in the direction parallel to the wedge plane. Now the problem is revised to become the following form:

$$(\nabla^2 + k^2)G(\boldsymbol{r}, \boldsymbol{r}') = \delta(\boldsymbol{r} - \boldsymbol{r}'); \quad \alpha < \theta < 2\pi - \alpha;$$

$$G(\boldsymbol{r}, \boldsymbol{r}')|_{r=0} < \infty, \quad \left[\frac{\partial}{\partial \theta} G(\boldsymbol{r}, \boldsymbol{r}')\right]_{\theta=\alpha, 2\pi-\alpha} = 0.$$

This in fact represents the distribution of electric field in the direction perpendicular to the wedge plane. Find this Green's function.

24. In an RC circuit, the electromotance is E, the voltage across the capacitor C is u and the current is i. The E, resistance R and capacitance C are constants. The voltage u, current I and the charge on the plate of the capacitor q depend on time. In this circuit we have the following basic relations: circuit voltage law $-E + iR + u = 0$; $i = \frac{dq}{dt}$; $u = \frac{q}{C}$. From these three relations we obtain $R\frac{dq}{dt} + \frac{q}{C} - E = 0$. Suppose that as $t = 0$, $q = 0$ Solve this equation in the way in Section 6.6.

25. Find the Green's function that satisfy the following boundary value problem.

$$\left(i\frac{d}{dx} - \lambda\right) G(x, x') = \delta(x - x'), \quad G(0, x') = G(1, x') = 0.$$

26. For the differential equation $(\partial_t + i\omega)G_\omega(t - t') = \delta(t - t')$, we may have the following two kinds of boundary conditions.

    (1) The solution is denoted by $G_\omega^p(t-t')$ which obeys the boundary condition

    $$G_\omega^p(0) = G_\omega^p(T).$$

    (2) The solution is denoted by $G_\omega^a(t-t')$ which obeys the boundary condition

    $$G_\omega^a(0) = -G_\omega^a(T).$$

    Please find the two solutions. The condition in (1) is usually called the periodic boundary condition. Does solution $G_\omega^p(t-t')$ is a periodic function with periodicity T and why? This point seemed misunderstood in the following book: Hagen Kleinert, Path integral in quantum mechanics, statistics and polymer physics, World Scientific Publishing Co. Pte. Ltd., 1990.

Chapter 7

Norm

For a real or complex number, its modulus represents its dimension. The concept of the modulus is useful in mathematical analysis. In Hilbert spaces, the concept of the inner product has been defined. The inner product of an element with itself is just the square of its modulus. Thus, the modulus of the element characterizes the dimension of the element. However, the definition of the inner product cannot be applied to all elements. In spite of this fact, the concept of modulus can still be generalized. For any vector, matrix and even operator, it is possible to introduce a scalar which in a sense represents its "dimension". This is the concept of norm. The inner product is related to angle, while the norm is related to the dimension, independent of angle. In short, norm is a kind of dimension, but actually has a much wider significance. This chapter is intended to introduce the norms of vectors, matrices and operators. The vector and matrix norms play an important role in matrix analysis. With the help of the concept of norm, it is convenient to conduct theoretical analysis of various operators.

7.1. Banach Space

7.1.1. *Banach space*

First of all, let us recall the concept of vector modulus.

Example 1. The modulus of a planar vector $x = ai + bj$ is defined by $\sqrt{a^2 + b^2}$. The modulus of a vector x, denoted by $\|x\|$ possesses the following three properties.

(1) If $x \neq 0$, then $\|x\| > 0$; when and only when $x = 0$, $\|x\| = 0$.
(2) $\|kx\| = |k|\,\|x\|$ for any real k.
(3) For any planer vectors x and y, the following triangle inequality stands:

$$\|x + y\| \leq \|x\| + \|y\| .$$

Example 2. By linear algebra, the modulus of a vector in n-dimensional Euclid space is defined by $\|x\| = (x, x)^{1/2}$. The zero vector is denoted by θ. The modulus thus defined is of the following three properties.

(1) If $x \neq 0$, then $\|x\| > 0$; when and only when $x = 0$, $\|\theta\| = 0$.
(2) For any real k and any vector x, $\|kx\| = |k|\,\|x\|$.
(3) For any vectors x and y, $\|x + y\| \leq \|x\| + \|y\|$.

For a general linear space, a scalar (or a function) called norm is introduced, which is of the above three properties and describes the dimension of a vector.

Definition 1. Suppose that K is a domain of real or complex numbers and V is a vector space on K. For any element in V, a positive real number is assigned and denoted by $\|f\|$. This scalar is called a **norm** if it satisfies the following conditions.

(1) Triangle inequality: for all

$$f, g \in V, \quad \|f + g\| \leq \|f\| + \|g\| . \tag{7.1.1}$$

(2) Nonnegative: when and only when

$$f \equiv 0, \quad \|f\| = 0. \tag{7.1.2}$$

(3) Homogeneity: for any scalar

$$\alpha \text{ in } K, \|\alpha f\| = |\alpha| \cdot \|f\|. \tag{7.1.3}$$

The three conditions are called the **three axioms of norm** which are also three properties that the norm possesses. A space where the norm can be defined is called a **normed space**. A vector space where the norm can be defined is called a **normed vector space**. When

$K = R$ or C, the space V is called a **normed real vector space** or a **normed complex vector space**.

Please note that these conditions are necessary conditions that a norm should satisfy. As long as a scalar is defined in some way and satisfies these conditions, it can be called a norm.

Among the three concepts of distance, norm and inner product, the epitaxial of distance is the largest, that of the norm is the second, and that of inner product is the least.

The relation between distance and norm is as follows. Distance can be defined for any set of elements. In a linear space, the norm can be defined. Once the norm is defined, distance can be defined accordingly, but not vice versa.

There is a way to define a norm based on distance: if the distance $\rho(f, g)$ between elements f and g is defined, then it is possible in some cases to define the norm of the element f by

$$\|f\| = \rho(f, \theta). \tag{7.1.4}$$

This definition shows that a necessary condition to define norm is that there exists a zero vector. Sometimes but not always, if the norm $\|f\|$ is defined, the distance can be defined by

$$\rho(f, g) = \|f - g\|. \tag{7.1.5}$$

The relation between norm and inner product is as follows. The norm can always be defined in a linear space, but the inner product is not necessarily so. Once the norm is defined, an inner product can be defined accordingly, but not vice versa.

There is a way to define a norm based on inner product: if the inner product (f, g) between elements f and g is defined, then the norm of the element f can be defined by

$$\|f\| = \sqrt{(f, f)}. \tag{7.1.6}$$

A way to define inner product based on norm is as follows. If the norm $\|f\|$ of an element f is defined and follows the parallelogram rule, then the inner product between two elements f and g can be defined accordingly. In a real inner product space, it is defined by

$$(f, g) = \frac{1}{4}(\|f + g\|^2 - \|f - g\|^2). \tag{7.1.7}$$

In a complex inner product space, it is defined by

$$(f,g) = \frac{1}{4}(\|f+g\|^2 - \|f-g\|^2 + i\|if+g\|^2 - i\|if-g\|^2). \quad (7.1.8)$$

Definition 2. Let V be a normed vector space on number domain K. The distance $\rho(f,g)$ between two elements f and g in V can be defined as the norm of the difference between them by (7.1.5). If V is complete with respect to the distance $\rho(f,g)$, then V is called a **Banach space** on K. V is called **real Banach space** or **complex Banach space** if K is R or C.

In short, a complete normed vector space is a **Banach space**.

Please note that there can be no inner product in a Banach space. Thus, the epitaxial of a Banach space is larger than a Hilbert space. As long as an inner product is defined, a norm can be defined without problem. Therefore, any Hilbert space is a normed one.

When the distance is defined after the norm is in a space, then the completeness of the space with respect to the distance is equivalent to that with respect to the norm. In literature, a Hilbert space is sometimes defined as a complete one with respect to a norm. An equivalent definition of Hilbert space is the complete normed vector space where an inner product has been defined.

Example 3. In the vector space consisting of rational numbers, select a series $S_n = \sum_{m=1}^{n} \frac{1}{m^2}$. It is a Cauchy series and can be normed. The absolute value of each element is defined as its norm, and the norm thus defined satisfies the three axioms of norm. The norm of the difference between two elements is defined as the distance between them. Then the limit of the Cauchy series exists, but the limit $\lim_{n\to\infty} S_n = \lim_{n\to\infty} \sum_{m=1}^{n} \frac{1}{m^2} \to \frac{\pi^2}{6}$ is an irrational number which is not within the rational number space. Therefore, the rational number space can be normed but is not complete, so that it is not a Banach space.

Example 4. All the smooth functions defined on $[a,b]$ compose a space, denoted as $C_1[a,b]$. A smooth function means that at least its first derivative exists. A definition of the norm of the elements in this space is that

$$\|f(x)\| = \max_{x \in [a,b]} |f(x)|. \tag{7.1.9}$$

It is easily verified that the norm thus defined meets the three axioms of a norm. Hence, this is a normed space. Since the convergent sequence of smooth and continuous functions on a closed interval is uniformly convergent, its limit function is still smooth and continuous. That is to say, $C_1[a,b]$ is a Banach space.

Please note that a space $C[a,b]$ consisting of continuous (but not smooth) functions defined on $[a,b]$ is not complete. We had an example in Exercise 3 in Chapter 2. So, such a space is not a Banach one.

Example 5. All the vectors consisting of n real numbers compose a vector space R^n. The norm of a vector $x = (x_1, x_2, \ldots, x_n) \in R^n$ is defined by

$$\|x\| = \left(\sum_{i=1}^{n} x_i^2 \right)^{1/2}.$$

All the vectors consisting of n complex numbers compose a vector space C^n. The norm of a vector $z = (z_1, z_2, \ldots, z_n) \in C^n$ is defined by $\|z\| = \left(\sum_{i=1}^{n} |z_i|^2 \right)^{1/2}$. Thus, R^n is a real Banach space and C^n is a complex Banach space.

Before further discussing norm, we introduce two inequalities.

7.1.2. *Hölder inequality*

Theorem 1. *Let*

$$p > 1, \quad q = \frac{p}{p-1}. \tag{7.1.10}$$

Then

$$\sum_{k=1}^{n} a_k b_k \leq \left(\sum_{k=1}^{n} a_k^p \right)^{1/p} \left(\sum_{k=1}^{n} b_k^q \right)^{1/q}, \tag{7.1.11}$$

where $a_k, b_k \geq 0$. Equation (7.1.11) is called the **Hölder inequality**.

Equation (7.1.10) means that

$$\frac{1}{p} + \frac{1}{q} = 1. \tag{7.1.12}$$

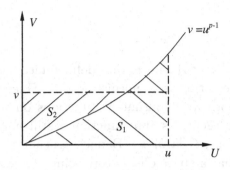

Fig. 7.1.

In this case, the number q is called the dual coefficient of p, and vice versa. Hereafter in this section, p and q follow (7.1.10).

Proof. Firstly, let us prove that if u and v are nonnegative, then

$$uv \leq \frac{u^p}{p} + \frac{v^q}{q}. \qquad (7.1.13)$$

Consider a function $v = u^{p-1}$. Without loss of generality, its curve is plotted in Fig. 7.1. We define two integrations: $S_1 = \int_0^u U^{p-1} dU = \frac{1}{p} u^p$ and $S_2 = \int_0^v V^{\frac{1}{p-1}} dV = \int_0^v V^{q-1} dV = \frac{1}{q} v^q$. It is seen from Fig. 7.1 that $S_1 + S_2 \geq uv$, where the equal sign stands when $v = u^{p-1}$. So, $uv \leq \frac{u^p}{p} + \frac{v^q}{q}$. This is just (7.1.13).

Secondly, let u and v be components u_k and v_k of two vectors, respectively, and the components satisfy

$$\sum_{k=1}^n u_k^p = 1, \quad \sum_{k=1}^n v_k^q = 1. \qquad (7.1.14)$$

Then, we take summation $\sum_{k=1}^n$ on both sides of (7.1.13) to obtain

$$\sum_{k=1}^n u_k v_k \leq \frac{1}{p} + \frac{1}{q} = 1. \qquad (7.1.15)$$

Let $u_k = \dfrac{a_k}{\left(\sum_{j=1}^n a_j^p\right)^{1/p}}$, $v_k = \dfrac{b_k}{\left(\sum_{j=1}^n b_j^q\right)^{1/q}}$. The components u_k and v_k thus defined satisfy (7.1.14). Consequently, it follows from (7.1.15) that $\sum_{k=1}^{n} \dfrac{a_k}{\left(\sum_{j=1}^{n} a_j^p\right)^{1/p}} \dfrac{b_k}{\left(\sum_{j=1}^{n} b_j^q\right)^{1/q}} \leq 1$. This results in the Hölder inequality (7.1.11). □

A special case is $p = q = 2$. Then we have

$$\sum_{k=1}^{n} a_k b_k \leq \sqrt{\sum_{k=1}^{n} a_k^2} \sqrt{\sum_{k=1}^{n} b_k^2}. \tag{7.1.16}$$

This is just the Cauchy inequality in Example 17 in Subsection 2.1.2.

The summations in (7.1.14) and (7.1.15) can be changed to be integrations. If

$$\int u^p(x)\mathrm{d}x = 1, \quad \int v^q(x)\mathrm{d}x = 1 \tag{7.1.17}$$

then taking integral $\int \mathrm{d}x$ on both sides of (7.1.12), we get

$$\int u(x)v(x)\mathrm{d}x \leq 1. \tag{7.1.18}$$

Let

$$u(x) = \frac{f(x)}{\left(\int f^p(x)\mathrm{d}x\right)^{1/p}}, \quad v(x) = \frac{g(x)}{\left(\int g^q(x)\mathrm{d}x\right)^{1/q}},$$

where $f(x) > 0, g(x) > 0$. Thus defined u and v satisfy (7.1.17). Then it follows from (7.1.18) that

$$\int \mathrm{d}x \frac{f(x)}{\left(\int f^p(x)\mathrm{d}x\right)^{1/p}} \frac{g(x)}{\left(\int g^q(x)\mathrm{d}x\right)^{1/q}} \leq 1.$$

This leads to the Hölder inequality in integral form:

$$\int f(x)g(x)\mathrm{d}x \leq \left(\int f^p(x)\mathrm{d}x\right)^{1/p} \left(\int g^q(x)\mathrm{d}x\right)^{1/q}. \tag{7.1.19}$$

In an exercise of this chapter, the weighted Hölder inequality is to be proved.

We point out here that the inequalities above are valid for $p > 1$, but not for $p = 1$.

7.1.3. *Minkowski inequality*

Theorem 2. *If $p \geq 1$, then*

$$\left(\sum_{i=1}^{n} |a_i + b_i|^p \right)^{1/p} \leq \left(\sum_{i=1}^{n} |a_i|^p \right)^{1/p} + \left(\sum_{i=1}^{n} |b_i|^p \right)^{1/p}. \qquad (7.1.20)$$

This is called the **Minkowski inequality**.

Proof.   For $p = 1$, the inequality is obvious. Let us prove the case of $p > 1$. We start with the equation

$$\sum_{i=1}^{n} |a_i + b_i|^p = \sum_{i=1}^{n} |a_i + b_i| \, |a_i + b_i|^{p-1}.$$

By (7.1.12), the right hand side becomes

$$\sum_{i=1}^{n} |a_i + b_i| \, |a_i + b_i|^{p/q} \leq \sum_{i=1}^{n} |a_i| \, |a_i + b_i|^{p/q} + \sum_{i=1}^{n} |b_i| \, |a_i + b_i|^{p/q}.$$

By the Hölder inequality, we have

$$\sum_{i=1}^{n} |a_i + b_i|^p$$

$$\leq \left(\sum_{i=1}^{n} |a_i|^p \right)^{1/p} \left(\sum_{i=1}^{n} |a_i + b_i|^p \right)^{1/q}$$

$$+ \left(\sum_{i=1}^{n} |b_i|^p \right)^{1/p} \left(\sum_{i=1}^{n} |a_i + b_i|^p \right)^{1/q}$$

$$= \left[\left(\sum_{i=1}^{n} |a_i|^p \right)^{1/p} + \left(\sum_{i=1}^{n} |b_i|^p \right)^{1/p} \right] \left(\sum_{i=1}^{n} |a_i + b_i|^p \right)^{1/q}.$$

This equation is divided by $\left(\sum_{i=1}^{n} |a_i + b_i|^p \right)^{1/q}$ to get (7.1.20).   □

Minkowski inequality in integral form reads

$$\left(\int |f(x) + g(x)|^p \mathrm{d}x \right)^{1/p}$$

$$\leq \left(\int |f(x)|^p \mathrm{d}x \right)^{1/p} + \left(\int |g(x)|^p \mathrm{d}x \right)^{1/p}. \qquad (7.1.21)$$

In an exercise of this chapter, the weighted Minkowski inequality is to be proved.

The Hölder inequality and Minkowski inequality have some generalizations.

Example 6. In a p-integrable space L_p, every function $f(x)$ satisfies condition

$$\int_D |f(x)|^p dx < \infty,$$

where D is a certain subset of the real axis (not necessarily finite). A kind of norm in this space is defined by

$$\|f\| = \left[\int_D |f(x)|^p dx \right]^{1/p}. \qquad (7.1.22)$$

It is easily verified that the norm thus defined satisfies the three axioms of norm. Therefore, L_p is a Banach space.

7.2. Vector Norms

7.2.1. *Vector norms*

The form of Minkowski inequality is the same as that of the triangle inequality of a norm. Hence, a concept of p-norm is developed from the Minkowski inequality.

Definition 1. Assume $x = (x_1, x_2, \ldots, x_n)$. For any $p \geq 1$, the scalar

$$\left(\sum_{i=1}^n |x_i|^p \right)^{1/p} \qquad (7.2.1)$$

is called the **p-norm** of the vector x, and is denoted by $\|x\|_p$.

It is obvious that $\|x\|_p$ is nonnegative and is of homogeneity. By (7.1.20), $\|x + y\|_p \leq \|x\|_p + \|y\|_p$, which is triangle inequality. So, $\|x\|_p$ satisfies the three axioms of norm.

The following three p-norms are usually used.

(1) When $p = 1$, the 1-norm $\|x\|_1$ is

$$\|x\|_1 = |x_1| + |x_2| + \cdots + |x_n|.$$

(2) When $p = 2$, the 2-norm $\|\boldsymbol{x}\|_2$ is what has been defined in Example 5 in Subsection 7.1.1, which is also called the Euclid norm.

(3) When $p = \infty$, the ∞-norm $\|\boldsymbol{x}\|_\infty$ is

$$\|\boldsymbol{x}\|_\infty = \lim_{p \to \infty} \|\boldsymbol{x}\|_p = \lim_{p \to \infty} (|x_1|^p + |x_2|^p + \cdots + |x_n|^p)^{1/p}.$$

Proposition 1. $\|\boldsymbol{x}\|_\infty = \max_i |x_i|$.

Proof. Let $\alpha = \max_i |x_i|$ and $\beta_i = \left|\frac{x_i}{\alpha}\right| \leq 1$, $(i = 1, 2, \ldots, n)$. Then,

$$\|\boldsymbol{x}\|_p = \left(\sum_{i=1}^n |x_i|^p\right)^{1/p} = \left(\sum_{i=1}^n (\beta_i \alpha)^p\right)^{1/p} = \alpha \left(\sum_{i=1}^n \beta_i^p\right)^{1/p}.$$

At least one of $\beta_1, \beta_2, \ldots, \beta_n$ is equal to 1. So, $1 \leq (\sum_{i=1}^n \beta_i^p)^{1/p} \leq n^{1/p}$. Because $\lim_{p \to \infty} n^{1/p} = 1$, we have $\lim_{p \to \infty} (\sum_{i=1}^n \beta_i^p)^{1/p} = 1$. In conclusion,

$$\|\boldsymbol{x}\|_\infty = \lim_{p \to \infty} \|\boldsymbol{x}\|_p = \alpha = \max_i |x_i|. \qquad \square$$

Briefly, the proof is as follows:

$$\|\boldsymbol{x}\|_\infty = \lim_{p \to \infty} \|\boldsymbol{x}\|_p = \lim_{p \to \infty} \left[\alpha^p \left(\frac{|x_1|^p}{\alpha^p} + \frac{|x_2|^p}{\alpha^p} + \cdots + \frac{|x_n|^p}{\alpha^p}\right)\right]^{1/p}$$

$$= \alpha \lim_{p \to \infty} \left(\frac{|x_1|^p}{\alpha^p} + \frac{|x_2|^p}{\alpha^p} + \cdots + \frac{|x_n|^p}{\alpha^p}\right)^{1/p}$$

$$= \alpha(0 + 0 + \cdots + 1 + \cdots + 0)^{1/p} = \alpha.$$

Please note that even if more than one component of $\boldsymbol{x} = (x_1, x_2, \ldots, x_n)$ satisfy $\alpha = \max_i |x_i|$, the ∞-norm is still $\|\boldsymbol{x}\|_\infty = \alpha$.

It is seen that $\|\boldsymbol{x}\|_\infty \leq \|\boldsymbol{x}\|_2 \leq \|\boldsymbol{x}\|_1$. One thus can reasonably conjecture that the larger the p, the smaller the value of the p-norm.

Similarly for the case of finite-dimensional vector spaces, the p-norm of a continuous function space can be defined in the same way. The concept of the p-integrable space $L_p[a, b]$, $(1 \leq p < \infty)$ has been defined in Definition 2 in Subsection 2.1.1. In such a space, every element $f(x)$ satisfies

$$\int_a^b |f(x)|^p \mathrm{d}x < \infty. \tag{7.2.2}$$

Definition 2. Assume that $f(x) \in C[a, b]$ and satisfies (7.2.2). Its norm defined by

$$\|f\|_p = \left[\int_a^b |f(x)|^p \mathrm{d}x \right]^{1/p} < \infty \qquad (7.2.3)$$

is called the **p-norm** of a function, also called its **Hölder norm**. $\|f\|_\infty = \max_{x \in [a,b]} |f(x)|$ is called the **uniform norm** or **Chebyshev norm** on $C[a, b]$.

Obviously, the uniform norm corresponds to the ∞-norm defined in finite-dimensional vector spaces. It is also called ∞-**norm** of function $f(x)$, which is a special case of (7.2.3) where $p = \infty$.

The other two special cases of (7.2.3) are 1-norm $\|f\|_1 = \int_a^b |f(x)| \mathrm{d}x$ and 2-norm $\|f\|_2 = [\int_a^b |f(x)|^2 \mathrm{d}x]^{1/2}$.

It is easy to verify that the p-norm defined by (7.2.3) satisfies the three axioms of norm. A p-integrable space $L_p[a, b]$, $(1 \le p < \infty)$ is complete, so that is a Banach space.

The norm $\|x\|_a$ is a function of arguments $x_1, x_2, \ldots, x_n$, denoted by

$$\|x\|_a = \varphi(x_1, x_2, \ldots, x_n). \qquad (7.2.4)$$

Theorem 1. *The norm* $\|x\|_a = \varphi(x_1, x_2, \ldots, x_n)$ *of a vector* $x = (x_1, x_2, \ldots, x_n)$ *is a continuous function of its arguments* $x_1, x_2, \ldots, x_n$.

Proof. Let $e_1, e_2, \ldots, e_n$ be a basis set of space V. Then for any $x \in V$, we have

$$x = x_1 e_1 + x_2 e_2 + \cdots + x_n e_n \qquad (7.2.5)$$

and $x' = x_1' e_1 + x_2' e_2 + \cdots + x_n' e_n$. The norm of the vector x' is

$$\|x'\|_a = \varphi(x_1', x_2', \ldots, x_n').$$

Then

$$\left| \varphi(x_1', x_2', \ldots, x_n') - \varphi(x_1, x_2, \ldots, x_n) \right|$$
$$= \left| \|x'\|_a - \|x\|_a \right| \le \|x' - x\|_a$$
$$= \left\| (x_1' - x_1)e_1 + (x_2' - x_2)e_2 + \cdots + (x_n' - x_n)e_n \right\|_a$$
$$\le \left| x_1' - x_1 \right| \|e_1\|_a + \left| x_2' - x_2 \right| \|e_2\|_a + \cdots + \left| x_n' - x_n \right| \|e_n\|_a.$$

All $\|e_i\|_a$ are constants. As all $|x_i' - x_i|$ are sufficiently small,

$$\left|\varphi(x_1', x_2', \ldots, x_n') - \varphi(x_1, x_2, \ldots, x_n)\right| < \delta.$$

This shows that $\varphi(x_1, x_2, \ldots, x_n)$ is a continuous function of $x_1, x_2, \ldots, x_n$. $\square$

Finally, we present the forms of weighted norms: $\|x\|_p = (\sum_{i=1}^n w_i |x_i|^p)^{1/p}$ and $\|f\|_p = [\int_a^b \rho(x)|f(x)|^p dx]^{1/p}$, where the weights are positive.

7.2.2.   *Equivalence between vector norms*

In a linear space, various norms can be defined. The norms defined in different manner have different values. For examples, for a vector $x = (1, 1, \ldots, 1)$, we have

$$\|x\|_2 = \sqrt{n}, \quad \|x\|_1 = n, \quad \|x\|_\infty = 1.$$

Although different norms have different values, they have relations with each other. For instance, the different norms show consistency in view of the convergence of a vector sequence. This property is called the **equivalence of norms**.

Theorem 2. *Suppose that V is an n-dimensional linear space and $\|x\|_a$ and $\|x\|_b$ are two kinds of norms (not limited to p-norms). There are necessarily positive numbers c_1 and c_2, for any $x \in V$,*

$$c_1\|x\|_b \leq \|x\|_a \leq c_2\|x\|_b. \tag{7.2.6}$$

Proof.   First let use show that the relation (7.2.6) is of transitivity. If two kinds of $\|x\|_a$ and $\|x\|_b$ have a relation to a fixed norm $\|x\|_2$ in the form of (7.2.6). That is to say, there exist c_1', c_2', c_1'' and c_2'' such that $c_1' \|x\|_2 \leq \|x\|_a \leq c_2'\|x\|_2$ and $c_1''\|x\|_b \leq \|x\|_2 \leq c_2''\|x\|_b$. Then, $c_1'c_1''\|x\|_b \leq \|x\|_a \leq c_2'c_2''\|x\|_b$. It is sufficient to prove (7.2.6) in the case of $b = 2$.

A vector x can be expressed in the form of (7.2.5), and its norm is expressed by (7.2.4). It has been proved by Theorem 1 that $\varphi(x_1, x_2, \ldots, x_n)$ is a continuous function of its arguments $x_1, x_2, \ldots, x_n$. According to the properties of continuous functions, on a bounded closed set

$$|x_1|^2 + |x_2|^2 + \cdots + |x_n|^2 = 1, \qquad (7.2.7)$$

function $\varphi(x_1, x_2, \ldots, x_n)$ can reach its maximum M and its minimum m. Since at least one of x_i in (7.2.5) is not zero, denote $m > 0$. Let $d = \sqrt{\sum |x_i|^2}$. Such a d is just a 2-norm. The components of vector $\boldsymbol{y} = \frac{x_1}{d}\boldsymbol{e}_1 + \frac{x_2}{d}\boldsymbol{e}_2 + \cdots + \frac{x_n}{d}\boldsymbol{e}_n$ satisfy $\left|\frac{x_1}{d}\right|^2 + \left|\frac{x_2}{d}\right|^2 + \cdots + \left|\frac{x_n}{d}\right|^2 = 1$. Then, we have $0 < m \le \|\boldsymbol{y}\|_a = \varphi\left(\frac{x_1}{d}, \frac{x_2}{d}, \ldots, \frac{x_n}{d}\right) \le M$. But $\boldsymbol{y} = \boldsymbol{x}/d$, so $md \le \|\boldsymbol{x}\|_a \le Md$. That is say, $m\|\boldsymbol{x}\|_2 \le \|\boldsymbol{x}\|_a \le M\|\boldsymbol{x}\|_2$.

□

Definition 3. Equation (7.2.6) is called the **equivalence relation**. The vector norms that satisfy (7.2.6) are **equivalent**.

From the equivalence of vector norms, it is deduced that there is similar relation between the distances of elements in a set defined in Chapter 2.

7.3. Matrix Norms

7.3.1. *Matrix norms*

In this section, the concept of vector norms is generalized to $m \times n$ matrices.

1. Induced matrix norms

Definition 1. Let $A \in F^{m \times n}$ (F is a real or complex number domain, R or C) be an $m \times n$ matrix. A real-value function of A is defined on $F^{m \times n}$ observing a rule, and denoted as $\|A\|$, such that the function obeys the following three axioms.

(i) If $A \ne 0$, then $\|A\| > 0$, and $\|0\| = 0$, nonnegative.
(ii) For any scalar k, $\|kA\| = |k|\|A\|$, homogeneity.
(iii) For any $A, B \in F^{m \times n}$, $\|A + B\| \le \|A\| + \|B\|$, triangle inequality.

Then $\|A\|$ is called the **norm** of the matrix A.

In the theory and application of matrices, the multiplication of matrices often plays a very important role. Therefore, in discussing the matrix norms, if a vector is generated by a matrix acting on another vector, coordination should be considered among the norms of the matrix and two vectors. This promotes the following definition.

Definition 2. (Norm compatibility) Suppose that on $F^{m\times n}$, $F^{n\times p}$ and $F^{m\times p}$, three norms $\|\cdot\|_{m\times n}$, $\|\cdot\|_{n\times p}$ and $\|\cdot\|_{m\times p}$ are respectively defined, and for any matrices $A \in F^{m\times n}$ and $B \in F^{n\times p}$, there exists a relation

$$\|AB\|_{m\times p} \le \|A\|_{m\times n}\|B\|_{n\times p}. \tag{7.3.1}$$

Then the norms $\|\cdot\|_{m\times n}$, $\|\cdot\|_{n\times p}$ and $\|\cdot\|_{m\times p}$ are said to be **compatible**.

If $p = 1$, B is a column vector with n components. In this case, (7.3.1) is written as $\|Ax\|_{m\times 1} \le \|A\|_{m\times n}\|x\|_{n\times 1}$. Alternatively, $\|Ax\|_\alpha \le \|A\|_{m\times n}\|x\|_\beta$, where $\|\cdot\|_\alpha$ and $\|\cdot\|_\beta$ are two norms in R^m and R^n, respectively.

To avoid difficulty in matrix analysis, we tacitly admit that the matrix norm discussed are always compatible.

In the following, we will define some kind of matrix norm. The prerequisition is that for vectors $x \in R^m$, $y \in R^n$, their norms $\|x\|_\alpha$ and $\|y\|_\beta$ in corresponding m- and n-dimensional spaces have been well defined.

Definition 3. Let $A \in F^{m\times n}$. We define the norm of matrix A in the following way:

$$\|A\|_\beta = \max_{x\neq 0} \frac{\|Ax\|_\beta}{\|x\|_\beta} = \max_{\|x\|_\beta=1} \|Ax\|_\beta. \tag{7.3.2}$$

This means that the norm of A is defined as the largest one among all vector norms $\|Ax\|_\beta$ while keeping $\|x\|_\beta = 1$. The matrix norm $\|A\|_\beta$ defined by (7.3.2) is called the **induced matrix norm** or **subordinate matrix norm**, since each matrix norm $\|\cdot\|_\beta$ is induced from an associate vector norm $\|\cdot\|_\beta$.

The matrix norm defined by (7.3.2) meets the three axioms of norm and compatibility of norms. In fact, it is easily seen that $\|\theta\| = 0$ and $\|A\| > 0$ for $A \neq 0$.

$$\|kA\| = \max_{\|x\|_\beta=1} \|kAx\|_\beta = |k| \max_{\|x\|_\beta=1} \|Ax\|_\beta = |k|\,\|A\|.$$

For $A, B \in F^{m \times n}$,

$$\|A + B\| = \max_{\|\boldsymbol{x}\|_\beta = 1} \|(A + B)\boldsymbol{x}\|_\beta \leq \max_{\|\boldsymbol{x}\|_\beta = 1} \|A\boldsymbol{x}\|_\beta$$

$$+ \max_{\|\boldsymbol{x}\|_\beta = 1} \|B\boldsymbol{x}\|_\beta = \|A\| + \|B\|.$$

In general,

$$\|A\boldsymbol{x}\|_\beta \leq \|A\| \|\boldsymbol{x}\|_\alpha.$$

That is to say, the matrix norm $\|A\|$ and vector norm $\|\boldsymbol{x}\|_\alpha$ and $\|\boldsymbol{y}\|_\beta$ are compatible.

When $\|\boldsymbol{x}\|_\beta$ is taken as $\|\boldsymbol{x}\|_1$, $\|\boldsymbol{x}\|_2$ and $\|\boldsymbol{x}\|_\infty$, we achieve three matrix norms, denoted by $\|A\|_1$, $\|A\|_2$ and $\|A\|_\infty$, respectively.

Theorem 1. *Let $A \in F^{m \times n}$. Then the following equations hold.*

(i) $\|A\|_1 = \max_j \left(\sum_{i=1}^{m} |a_{ij}| \right)$, $(j = 1, 2, \ldots, n)$.   (7.3.3)

(ii) $\|A\|_2 = \max_j \left(\lambda_j(A^+A) \right)^{1/2}$,   (7.3.4)

    where $\lambda_j(A^+A)$ represents the j-th eigenvalue of matrix A^+A.

(iii) $\|A\|_\infty = \max_i \left(\sum_{j=1}^{n} |a_{ij}| \right)$, $(i = 1, 2, \ldots, m)$.   (7.3.5)

Proof. (i) Let $w = \max_j \left(\sum_{i=1}^{m} |a_{ij}| \right)$, $(j = 1, 2, \ldots, n)$. Assume that $\boldsymbol{x} = (x_1, x_2, \ldots, x_n)^T$ and $\|\boldsymbol{x}\|_1 = 1$. For matrix $A = [\alpha_1, \alpha_2, \ldots, \alpha_n]$, we have

$$w = \max_j \|\alpha_j\|_1.$$

Then,

$$\|A\boldsymbol{x}\|_1 = \|x_1\alpha_1 + x_2\alpha_2 + \cdots + x_n\alpha_n\|_1$$

$$\leq |x_1| \|\alpha_1\|_1 + |x_2| \|\alpha_2\|_1 + \cdots + |x_n| \|\alpha_n\|_1$$

$$\leq (|x_1| + |x_2| + \cdots + |x_n|) w = \|\boldsymbol{x}\|_1 w = w.$$

Explicitly,

$$\|\alpha_j\|_1 = \sum_{i=1}^{m} |a_{ij}|.$$

If the r-th column $\|\alpha_r\|_1$ is the largest, we take $x_r = (0, 0 \ldots, 1_{(r)}, 0 \ldots, 0)^T$ so that $\|x_r\| = 1$ and

$$\|Ax_r\|_1 = \sum_{i=1}^m |a_{ir}| = \max_j \|\alpha_j\|_1 = w.$$

In conclusion,

$$\|A\|_1 = \sum_{i=1}^m |a_{ir}| = \max_j \left(\sum_{i=1}^m |a_{ij}| \right), \quad (j = 1, 2, \ldots, n).$$

(ii) By the definition of vector norm $\|x\|_2$, we have known that

$$\|Ax\|_2 = (Ax, Ax)^{1/2} = (x, A^+ Ax)^{1/2}.$$

Because A^+A is positive definite or semidefinite, the n eigenvalues of A^+A are nonnegative. Let y_i be the i-th eigenvector of the eigenvalue λ_i of the matrix $P = A^+A$.

$$Py_i = A^+ Ay_i = \lambda_i y_i.$$

The n column eigenvectors y_i constitute the eigenvector matrix of P, denoted by V. The n eigenvalues constitute the diagonal matrix Λ.

$$PV = V\Lambda.$$

Then, we find

$$\|Ax\|_2 = \max_{\|x\|_2=1} (x, Px)^{1/2} = \max_{\|x\|_2=1} \left(x, V\Lambda V^{-1}x\right)^{1/2}$$

$$\leq \max_{\|x\|_2=1,i} \lambda_i^{1/2} \left(x, VV^{-1}x\right)^{1/2}.$$

We search the largest value of the left hand side. This is simply done by letting x just be the eigenvector corresponding to the largest eigenvalue of P. This gives

$$\|Ax\|_2 = \max_{\|y_i\|_2=1,i} (y_i, Py_i)^{1/2}$$

$$= \max_i (y_i, \lambda_i y_i)^{1/2} = \max_i \left(\lambda_i(A^+A)\right)^{1/2}.$$

Thus, (7.3.4) is proved.

(iii) Let $w = \max_i \left(\sum_{j=1}^{n} |a_{ij}| \right)$, $(i = 1, 2, \ldots, m)$. For any vector $x \in R^n$, we have

$$\|Ax\|_\infty = \max_{1 \le i \le m} \left| \sum_{j=1}^{n} |a_{ij} x_j| \right|$$

$$\le \max_{1 \le i \le m} \left(\sum_{j=1}^{n} |a_{ij}| \right) \|x\|_\infty = w\|x\|_\infty.$$

This equation holds for any $\|x\|_\infty$. Taking the limit $\|x\|_\infty \to 1$ on both sides leads to $\|A\|_\infty \le w$. Assume that the r-th column vector of matrix A has the largest 1-norm:

$$\max_{1 \le i \le m} \sum_{j=1}^{n} |a_{ij}| = \sum_{j=1}^{n} |a_{rj}|.$$

Then we let x be the following vector:

$$x = \left(\frac{a_{r1}^*}{|a_{r1}|}, \frac{a_{r2}^*}{|a_{r2}|}, \ldots, \frac{a_{rn}^*}{|a_{rn}|} \right)^T,$$

where a_{rj}^* is the complex conjugate of a_{rj}. So, $\|x\|_\infty = 1$. Among the n components of the vector Ax, the r-th has the largest absolute value. Therefore,

$$\|Ax\|_\infty = \sum_{j=1}^{n} |a_{rj}| = \max_{1 \le i \le m} \sum_{j=1}^{n} |a_{ij}| = w. \qquad \square$$

The matrix norms $\|A\|_1$, $\|A\|_\infty$ and $\|A\|_2$ thus defined are also called **column-sum norm**, **row-sum norm** and **spectral norm**, respectively. The topics related to the spectral norm will be further discussed in the next subsection.

For matrices $A \in F^{n \times n}, B \in F^{n \times p}$ and vector $y \in R^p$, we have

$$\|ABy\|_2 \le \|A\|_2 \|By\|_2 \le \|A\|_2 \|B\|_2 \|y\|_2.$$

This equation holds for any $\|y\|_2$. Taking $\|y\|_2 \to 1$ limit on both sides leads to

$$\|AB\|_2 = \max_{\|y\|_2 = 1} \|ABy\|_2$$

$$\le \max_{\|y\|_2 = 1} \|A\|_2 \|B\|_2 \|y\|_2 = \|A\|_2 \|B\|_2.$$

Therefore,

$$\|AB\|_2 \leq \|A\|_2 \|B\|_2. \tag{7.3.6}$$

Thus, the matrix norms $\|\cdot\|_2$ are compatible to each other.

Proposition 1. *The induced norm of the unit matrix is always equal to 1:* $\|I\|_\beta = 1$.

Proof.   It follows directly from the definition (7.3.2) that

$$\|I\|_\beta = \max_{\boldsymbol{x} \neq 0} \frac{\|I\boldsymbol{x}\|_\beta}{\|\boldsymbol{x}\|_\beta} = \max_{\boldsymbol{x} \neq 0} \frac{\|\boldsymbol{x}\|_\beta}{\|\boldsymbol{x}\|_\beta} = 1,$$

or

$$\|I\|_\beta = \max_{\|\boldsymbol{x}\|_\beta = 1} \|I\boldsymbol{x}\|_\beta = \max_{\|\boldsymbol{x}\|_\beta = 1} \|\boldsymbol{x}\|_\beta = 1.$$

Please note that the induced norm is merely one among various possible norms. There can be other norms besides the induced norm. The induced norm of the unit matrix is always 1, but this conclusion does not apply to other kinds of norms. For example, one can define the norm of a matrix A to be $\|A\| = 2\|A\|_\beta$, where $\|A\|_\beta$ is the induced norm. This defined norm satisfies the three axioms of norm. But under this definition, the norm of the unit matrix is not 1.   □

2. Frobenius norm of matrix

Beside the three kinds of norms introduced above, Frobenius norm is also a kind of useful norm.

Definition 4. Suppose that a matrix $A \in F^{m \times n}$. Its **Frobenius norm** is defined by

$$\|A\|_{\mathrm{F}} = \left(\sum_{i=1}^{m} \sum_{j=1}^{n} |a_{ij}|^2 \right)^{1/2}. \tag{7.3.7}$$

This norm has some other names: **Euclid norm (E-norm)**, **Schur norm**, **Hilbert-Schmidt norm** and l_2 **norm**.

Equation (7.3.7) can be written as

$$\|A\|_F = \left(\sum_{i=1}^{m} (A^+A)_{ii}\right)^{1/2} = \left(\sum_{j=1}^{n} (AA^+)_{jj}\right)^{1/2}$$

or, simply,

$$\|A\|_F = \left(\text{Tr}(A^+A)\right)^{1/2} = \left(\text{Tr}(AA^+)\right)^{1/2}.$$

As $A = x \in F^{m \times 1}$, $\|A\|_F = \|x\|_2$. Hence, a Frobenius norm is a natural generalization of the 2-norm $\|x\|_2$ of vectors. $\|A\|_F$ can actually be regarded as a norm of a special "vector" A every component of which is a vector. The matrix A can be considered as composed either by row vectors, $A = [A_1, A_2, \ldots, A_n]$, or by column vectors, $A = [B_1, B_2, \ldots, B_n]$. Then, the square of the F-norm of matrix A is the summation of the squares of the 2-norms of either row or column vectors.

$$\|A\|_F^2 = \sum_{i=1}^{m} \|A_i\|_2^2 = \sum_{j=1}^{n} \|B_j\|_2^2.$$

The F-norm defined by (7.3.7) apparently satisfies the three axioms of norm. Now, we prove that the F-norm $\|A\|_F$ satisfies the compatibility.

Consider two matrices A and B. The former is $m \times n$, and can be written in the form of row vectors, while the latter is $n \times k$. Their product $C = AB$ is written in the form of row vectors: $C = [C_1, C_2, \ldots, C_n]$, where the i-row is $C_i = [c_{i1}, c_{i2}, \ldots, c_{ik}]$. The square of the 2-norm of the vector C_i is

$$\|C_i\|_2^2| = \sum_{p=1}^{k} |c_{ip}|^2 = \sum_{p=1}^{k} \left| \sum_{j=1}^{n} a_{ij} b_{jp} \right|^2$$

$$\leq \sum_{p=1}^{k} \sum_{j=1}^{n} |a_{ij}|^2 \sum_{j=1}^{n} |b_{jp}|^2$$

$$= \sum_{j=1}^{n} |a_{ij}|^2 \sum_{p=1}^{k} \sum_{j=1}^{n} |b_{jp}|^2 = \|A_i\|_F^2 \|B\|_F^2,$$

where the Hölder inequality is used. Thus, the square of the F-norm of the matrix C is

$$\|AB\|_{\mathrm{F}}^2\,| = \|C\|_{\mathrm{F}}^2\,| = \sum_{i=1}^{m} \|C_i\|_2^2$$

$$\leq \sum_{i=1}^{m} \|A_i\|_{\mathrm{F}}^2 \|B\|_{\mathrm{F}}^2 = \|A\|_{\mathrm{F}}^2 \|B\|_{\mathrm{F}}^2. \qquad (7.3.8)$$

By (7.3.1), it is seen that the Frobenius norm is of compatibility. It can be verified that $\|A\|_{\mathrm{F}}$ is compatible with $\|x\|_2$.

Theorem 2. *Let $A \in C^{m \times n}$, and U and V are unitary matrices of order m and n, respectively. Then,*

$$\|UAV\|_{\mathrm{F}} = \|A\|_{\mathrm{F}}. \qquad (7.3.9)$$

Proof. Let $A = [A_1, A_2, \ldots, A_n]$, where A_i is the i-th column of A. Then,

$$\|A\|_{\mathrm{F}}^2 = \|A_1\|_2^2 + \|A_2\|_2^2 + \cdots + \|A_n\|_2^2.$$

It follows from $\|UA_i\|_2^2 = \left\|A_i^+ U^+ U A_i\right\|_2 = \|A_i\|_2^2$ that $\|UA\|_{\mathrm{F}}^2 = \|A\|_{\mathrm{F}}^2$. On the other hand, $\|A\|_{\mathrm{F}}^2 = \|A^+\|_{\mathrm{F}}^2$. So,

$$\|UAV\|_{\mathrm{F}}^2 = \left\|V^+ A^+ U^+\right\|_{\mathrm{F}}^2 = \left\|A^+ U^+\right\|_{\mathrm{F}}^2 = \|UA\|_{\mathrm{F}}^2 = \|A\|_{\mathrm{F}}^2. \qquad \square$$

Similarly to vector norms, matrix norms also possess equivalence relation.

Theorem 3. *Let $\|\cdot\|_a$ and $\|\cdot\|_b$ be two kinds of matrix norms. There exist necessarily positive numbers c_1 and c_2, such that for any matrix A,*

$$c_1 \|A\|_b \leq \|A\|_a \leq c_2 \|A\|_b.$$

We point out that although Frobenius norm is a natural generalization of the 2-norm, it is not derived from a vector norm, and so it does not belong to induced norms.

7.3.2. *Spectral norm and spectral radius of matrices*

1. Spectral norm of matrices

The induced matrix norm $\|A\|_2$ defined in the last subsection is also called the spectral norm, which is of very useful application is matrix analysis and system theory. By (7.3.2), the spectral norm is

$$\|A\|_2 = \max_{x \neq 0} \frac{\|Ax\|_2}{\|x\|_2}.$$

Its geometric significance is the largest ratio of the length of the transformed vector $\|Ax\|_2$ to that of the untransformed vector $\|x\|_2$.

In the following, we discuss the properties of the spectral norm.

Theorem 4. *For any $A \in C^{m \times n}$, the following conclusions hold.*

(i) $\displaystyle\max_{\|x\|_2 = \|y\|_2 = 1} |y^+ Ax| = \|A\|_2,\ x \in F^n,\ y \in F^m$.

(ii) $\left\|A^+\right\|_2 = \|A\|_2$.

(iii) $\left\|A^+ A\right\|_2 = \|A\|_2^2$.

Proof. (i) For $\|x\|_2 = \|y\|_2 = 1$, we have

$$\max_{\|x\|_2 = \|y\|_2 = 1} |y^+ Ax| \leq \max_{\|x\|_2 = \|y\|_2 = 1} \|y\|_2 \|Ax\|_2$$

$$= \max_{\|x\|_2 = 1} \|Ax\|_2 = \|A\|_2.$$

Here the Schwartz inequality is used. Now we search the largest value of the left hand side. To do so, we assume $\|x\|_2 = 1$ and let $y = Ax/\|Ax\|_2$. Subsequently, $\|y\|_2 = 1$ and $\|Ax\|_2 - \|A\|_2 \neq 0$. It then follows that

$$\max_{\|x\|_2 = \|y\|_2 = 1} |y^+ Ax| = \frac{\|Ax\|_2^2}{\|Ax\|_2} = \|Ax\|_2 = \|A\|_2. \qquad (7.3.10)$$

(ii) $\displaystyle \|A\|_2 = \max_{\|x\|_2 = \|y\|_2 = 1} |y^+ Ax| = \max_{\|x\|_2 = \|y\|_2 = 1} |x^+ A^+ y| = \left\|A^+\right\|_2,$

$$(7.3.11)$$

where property (i) is used.

(iii) By the compatibility (7.3.6), $\|A^+A\|_2 \leq \|A^+\|_2\|A\|_2$, and by $\|A^+\|_2 = \|A\|_2$ it is known that $\|A^+A\|_2 \leq \|A\|_2^2$. On the other hand,

$$\max_{\|\boldsymbol{x}\|_2 \neq 0} \frac{\|A^+A\boldsymbol{x}\|_2}{\|\boldsymbol{x}\|_2} = \|A^+A\|_2 \geq \max_{\|\boldsymbol{x}\|_2 = 1} \frac{|\boldsymbol{x}^+A^+A\boldsymbol{x}|}{\|\boldsymbol{x}\|_2}$$

$$= \max_{\|\boldsymbol{x}\|_2 = 1} \|A\boldsymbol{x}\|_2^2 = \|A\|_2^2.$$

The inequality stems from that the constraint $\|\boldsymbol{x}\| \neq 0$ on its left hand side is weaker than the constraint $\|\boldsymbol{x}\| = 1$ on the right hand side.

In fact, for $\|\boldsymbol{x}\|_2 = 1$ and the definition $\max_{\|\boldsymbol{x}\|_2=1} \|A\boldsymbol{x}\|_2 = \|A\|_2$, it follows from (i) that

$$\|A^+A\|_2 = \max_{\|\boldsymbol{x}\|_2 = 1} |\boldsymbol{x}^+A^+A\boldsymbol{x}|$$

$$= \max_{\|\boldsymbol{x}\|_2 = 1} \|A\boldsymbol{x}\|_2^2 = \|A\|_2^2. \tag{7.3.12}$$

$\square$

Theorem 5. *Suppose that* $A \in C^{m\times n}$, $U \in C^{m\times m}$, $V \in C^{n\times n}$, $U^+U = I_m$ *and* $V^+V = I_n$. *Then,*

$$\|UAV\|_2 = \|A\|_2. \tag{7.3.13}$$

Proof. Let $\boldsymbol{v} = V\boldsymbol{x}$ and $\boldsymbol{u} = U^+\boldsymbol{y}$. Then $\|\boldsymbol{x}\|_2 = 1$ when and only when $\|\boldsymbol{v}\|_2 = 1$, because $\|\boldsymbol{v}\|_2^2 = \|V\boldsymbol{x}\|_2^2 = |\boldsymbol{x}^+V^+V\boldsymbol{x}| = |\boldsymbol{x}^+\boldsymbol{x}| = \|\boldsymbol{x}\|_2^2$. For the same reason, $\|\boldsymbol{y}\|_2 = 1$ when and only when $\|\boldsymbol{u}\|_2 = 1$. By property (i), we have

$$\|A\|_2 = \max_{\|\boldsymbol{x}\|_2=\|\boldsymbol{y}\|_2=1} |\boldsymbol{y}^+A\boldsymbol{x}|$$

$$= \max_{\|\boldsymbol{v}\|_2=\|\boldsymbol{u}\|_2=1} |\boldsymbol{u}^+UAV\boldsymbol{v}| = \|UAV\|_2.$$

$\square$

Equations (7.3.13) and (7.3.9) show the common feature of the spectral norm and Frobenius norm of matrix.

Theorem 6. *If* $\|A\|_2 < 1$, *then* $I - A$ *is nonsingular and*

$$\left\|(I-A)^{-1}\right\|_2 \leq (1 - \|A\|_2)^{-1}. \tag{7.3.14}$$

Proof. For any nonzero vector x, we have

$$\|(I - A)x\|_2 = \|x - Ax\|_2 \geq \|x\|_2 - \|Ax\|_2$$

$$\geq \|x\|_2 - \|A\|_2 \|x\|_2 = (1 - \|A\|_2) \|x\|_2 > 0.$$

So, if $x \neq 0$, $(I-A)x \neq 0$. This means that the equation $(I-A)x = 0$ has no nonzero solution, i.e., the matrix $I - A$ is nonsingular.

Since $I - A$ is nonsingular,

$$I = (I - A)(I - A)^{-1} = (I - A)^{-1} - A(I - A)^{-1}.$$

Thus, $(I - A)^{-1} = I + A(I - A)^{-1}$. It follows that

$$\left\|(I - A)^{-1}\right\|_2 \leq \|I\|_2 + \|A\|_2 \left\|(I - A)^{-1}\right\|_2$$

$$= 1 + \|A\|_2 \left\|(I - A)^{-1}\right\|_2.$$

This leads to

$$\left\|(I - A)^{-1}\right\|_2 \leq \frac{1}{1 - \|A\|_2}.$$

$\square$

2. Spectral radius of a matrix

We introduce the concept of spectral radius.

Definition 5. Suppose that $A \in F^{n \times n}$ and its eigenvalues are $\lambda_1, \lambda_2, \ldots, \lambda_n$. The number

$$\rho(A) = \max_i (|\lambda_i|, i = 1, 2, \ldots, n)$$

is called the **spectral radius** of A.

The geometric significance of the spectral radius $\rho(A)$ is the minimum radius of the circle centered on the origin and containing all the eigenvalues of A in the complex λ plane.

Theorem 7. *For any matrix $A \in F^{n \times n}$, it follows always that*

$$\rho(A) \leq \|A\|_2. \tag{7.3.15}$$

Proof. Let $\lambda_1, \lambda_2, \ldots, \lambda_n$ be eigenvalues of A, and the k-th one λ_k has the largest absolute value, $|\lambda_k| = \rho(A)$. The eigenvector corresponding to λ_k is $\boldsymbol{x}_k$. We have

$$\|A\|_2 \, \|\boldsymbol{x}_k\|_2 \geq \|A\boldsymbol{x}_k\|_2 = |\lambda_k| \, \|\boldsymbol{x}_k\|_2 = \rho(A) \, \|\boldsymbol{x}_k\|_2 \,.$$

This, divided by $\|\boldsymbol{x}_k\|_2$ on both sides, leads to $\rho(A) \leq \|A\|_2$.   □

This theorem tells us that the absolute value of any eigenvalue of a matrix is less than and equal to the spectral norm of the matrix. Please distinguish (7.3.15) from (7.3.4). The latter means the square root of the largest eigenvalue of matrix A^+A.

Particularly, if A is a normal matrix, $A^+A = AA^+$, the following corollaries stand.

Corollary 1. *If $A \in F^{n \times n}$ and A is a normal matrix, then*

$$\|A\|_2 = \rho(A). \tag{7.3.16}$$

That is to say, the maximum absolute value of the eigenvalues of a normal matrix is just its spectral norm.

Proof.   Because A is a normal matrix, there exists a matrix U with $U^+U = I$ such that

$$U^+AU = \text{diag}\,[\lambda_1, \lambda_2, \ldots, \lambda_n]\,.$$

So, it follows from (7.3.13) that

$$\|A\|_2 = \|\text{diag}[\lambda_1, \lambda_2, \ldots, \lambda_n]\|_2\,.$$

Assume that the normalized eigenvector of λ_j is $\boldsymbol{x} = (x_1, x_2, \ldots, x_n)$:

$$A\boldsymbol{x} = \lambda_j \boldsymbol{x}.$$

Because the $\boldsymbol{x}$ is normalized, $|\boldsymbol{x}|^2 = \sum_{i=1}^{n} |x_i|^2 = 1$. We obtain

$$\|A\|_2 = \max_{\|\boldsymbol{x}\|_2=1} \|A\boldsymbol{x}\|_2 = \max_j \left[\sum_{i=1}^{n} |\lambda_j x_i|^2 \right]^{1/2}$$

$$= \max_j \left[|\lambda_j|^2 \sum_{i=1}^{n} |x_i|^2 \right]^{1/2} = \max_j |\lambda_j| = \rho(A).$$

The last equal sign is just the definition of spectral norm, so that (7.3.16) is proved.   □

7.4. Operator Norms

7.4.1. *Operator norms*

1. Bounded linear operators

We have defined the concept of operators in Section 2.2, where Definitions 1 and 2 stand for any set and Definition 3 stands in inner product spaces. If the words "inner product spaces" are replaced by "normed spaces" in Definition 3, then the concept of linear transformation in normed spaces is defined.

The concept of vector norms is in fact a nonnegative real number defined in some way to characterize the dimension of a vector, and either is that of matrix norms. For any operator T, we also want to characterize its dimension by a nonnegative real number. The concept of **operator norms** is thus introduced. However, it is not easy to define an operator norm. In the following, we merely define the concept of norms for bounded linear operators. To do so, the concept of bounded linear operators must be defined.

Definition 1. Suppose that V and U are two normed linear spaces on the same domain K, and T is an operator from V to U. The operator T is said to be a **linear operator** or **linear transformation** from V to U, if T satisfies the following two conditions.

(i) For all $f, g \in V, T(f + g) = Tf + Tg$. $\qquad$ (7.4.1)

(ii) For all $f \in V$ and all scalars $\alpha \in K, T(\alpha f) = \alpha Tf$. $\qquad$ (7.4.2)

An operator T is said to be a **continuous linear operator** or **continuous linear transformation** if it satisfies not only the conditions (i) and (ii) above, but also the following (iii).

(iii) For $f, g \in V$ and each $\varepsilon > 0$, there exists $\delta > 0$ such that $\|Tf - Tg\| < \delta$, as $\|f - g\| < \varepsilon$. $\qquad$ (7.4.3)

The condition (iii) characterizes the continuity of the action of the operator T, simply called the continuity of operator T. An operator T is said to be a **bounded linear operator** or **bounded linear**

transformation if it satisfies not only the conditions (i) and (ii) above, but also the following (iv).

(iv) For all $f \in V$, there exists a positive constant M such that

$$\|Tf\| \leq M\|f\|. \qquad (7.4.4)$$

The condition (iv) characterizes the boundedness of the action of the operator T, simply called as the boundedness of operator T.

An analogy between the continuity of linear operators and that of functions is helpful. The operator T acts on a vector f. If the vector is regarded as a variable in a function and the operator is regarded as the function, the action of the operator on the vector is written in the form of $Tf = T(f)$. The continuity of a function requires that as the variable varies sufficiently little, the function does so either. A continuous operator is indeed of such a property. Please note that here when we say that the function varies sufficiently little, we means actually the norm of Tf varies sufficiently little.

Theorem 1. *Suppose that V and U are two normed spaces. A linear operator T from V to U is a continuous one if and only if there exists a positive constant M such that for all $f \in V$, $\|Tf\| \leq M\|f\|$.*

This theorem shows that for a linear operator from one normed space to another, conditions (7.4.3) and (7.4.4) are sufficient and necessary to each other, or they are equivalent. Thus, this theorem can be restated as: a linear operator T is continuous if and only if it is bounded. Hereafter, we simply refer to a continuous operator or bounded operator.

Example 1. A linear integral operator K is defined by

$$Kf \equiv \int_a^b k(x,y)f(y)\mathrm{d}y. \qquad (7.4.5)$$

If the f and k are bounded functions on closed interval $[a, b]$, the operator K is a **continuous linear integral operator** or **bounded linear integral operator**.

Let $B(V, U)$ be a set consisting of all the continuous linear operators from V to U. For any two elements $S, T \in B(V, U)$ and scalar α in number domain K, let $S + T$ be a linear operator defined by $(S + T)f = Sf + Tf$, and αS be a linear operator defined by $(\alpha S)f = \alpha Sf$. Then, $B(V, U)$ is a linear space on K with respect to the defined addition and scalar multiplication. That is to say, a set of bounded linear transformations constitute a linear space, in which all the elements are continuous linear operators. Note the discrepancy between such a space and those composed of vectors but not operators. If norms can be defined for all the elements of this space, this is a normed space; if inner products can be defined, this is an inner product space.

2. The norms of bounded linear operators

Now we try to define a norm for bounded linear operators.

Definition 2. Let V be a normed vector space, and T be a bounded transformation acting on V. Suppose that for all $f \in V$, there stands

$$\|Tf\| \leq M\|f\|. \tag{7.4.6}$$

The number M may depend on f. Among all possible M's the smallest one is denoted as T_M and is defined as the norm of the operator T: $\|T\| = T_M$. Please note that (7.4.6) means

$$\|T\| = T_M \leq M. \tag{7.4.7}$$

By this definition,

$$\|Tf\| \leq \|T\|\,\|f\|. \tag{7.4.8}$$

It is easy to prove that the norm thus defined satisfies the three axioms of norms. For instance,

$$\|(T_1 + T_2)f\| = \|T_1 f + T_2 f\| \leq \|T_1 f\| + \|T_2 f\|$$
$$\leq (\|T_1\| + \|T_2\|)\,\|f\|.$$

This stands for all f. The norm of the operator $(T_1 + T_2)$ is $\|T_1 + T_2\| = T_M$ and is the smallest M among those satisfying $\|(T_1 + T_2)f\| \leq M\|f\|$. Therefore, $T_M \leq \|T_1\| + \|T_2\|$, or

$$\|T_1 + T_2\| \leq \|T_1\| + \|T_2\|.$$

This is the triangle inequality. The other two axioms are also satisfied.

From (7.4.8), we further obtain an inequality concerning the norm of product of two operators. After an operator T_1 acts on a function f, the resultant $T_1 f$ is a new function. Then let an operator T_2 act on it. Applying (7.4.6) repeatedly, we get

$$\|T_2 T_1 f\| \leq \|T_2\| \cdot \|T_1 f\| \leq \|T_2\| \cdot \|T_1\| \|f\|. \qquad (7.4.9\text{a})$$

This means that the norm of the product $T_2 T_1$ is less than or equal to $\|T_2\| \cdot \|T_1\|$, i.e.,

$$\|T_2 T_1\| \leq \|T_2\| \cdot \|T_1\|. \qquad (7.4.9\text{b})$$

The operator norms obey compatibility. Furthermore,

$$\|T^n\| \leq \|T\|^n. \qquad (7.4.10)$$

The definition (7.4.6) itself does not tell how to define the norm of an operator T. Nevertheless, if for a vector f, $\|Tf\| = m \|f\|$, then $\|T\| \geq m$, because $\|Tf\| \leq \|T\| \|f\|$. This is helpful for one to define the operator norms.

The definition of the operator norms can be employed to explain the continuity of linear operators. $\|f - g\| \to 0$,

$$\|Kf - Kg\| = \|K(f-g)\| \leq \|K\| \|f-g\| \leq M \|f-g\| \to 0.$$

3. The completeness of normed linear transformation spaces

A space $B(V,U)$ consists of all the bounded linear transformations from normed linear spaces V to U. After the norms of the bounded linear operators in this space are defined, this space is a normed one. The following theorem addresses the completeness of this space.

Theorem 2. *If V and U are Banach spaces, and a space $B(V,U)$ consists of all the bounded linear transformations from V to U, then $B(V,U)$ is a Banach space with respect to the norm in Definition 2.*

This conclusion can be briefly addressed as follows: the set consisting of all the bounded linear transformations in a Banach space is also a Banach space.

Definition 3. Let V and U be Banach spaces. Suppose that there exists a bounded linear operator T from V to U such that for $f \in V$

and $u \in U$, $Tf = u$ and $\|u\| = \|Tf\| = \|f\|$. Then the two Banach spaces are said to be **isomorphic**. If V and U are isomorphic, it is denoted by $V \cong U$.

Theorem 3. *If*

$$\frac{1}{p} + \frac{1}{q} = 1, \quad p \geq 1, \quad (p = 1 \text{ means } q = +\infty), \tag{7.4.11}$$

then

$$(L_q[a, b])^* \cong L_q[a, b]. \tag{7.4.12}$$

This means that under the condition (7.4.11), the dual space of p-integrable and q-integrable spaces are isomorphic. The concept of dual space will be seen below.

Now we define the concept of the convergence in Banach space.

Definition 4. Let V be a Banach space and $\{f_n\}$ be a sequence in V. $\{f_n\}$ is said to be **strongly convergent** to f, if there exists an $f \in V$ such that

$$\lim_{n \to \infty} \|f_n - f\| = 0. \tag{7.4.13}$$

$\{f_n\}$ is said to be **weakly convergent** to f, if there exists an $f \in V$ such that for every $g \in V^*$,

$$\lim_{n \to \infty} g(f_n) = g(f). \tag{7.4.14}$$

Equations (7.4.13) and (7.4.14) can be compared with (7.4.3). Equation (7.4.13) corresponds to the continuity of the variable of a function, and (7.4.14) corresponds to the continuity of a function. By the theory of function continuity, the following theorem is easily understood.

Theorem 4. *The strong convergence contains the weak convergence, but not vice versa.*

Further let us define the convergences of bounded linear operators.

Definition 5. Let V be a Banach space and $\{T_n\}$ be a sequence in $B(V, V)$. $\{T_n\}$ is said to be **uniformly convergent** or **convergent**

in norm to T, if there exists a $T \in B(V, V)$ such that

$$\lim_{n \to \infty} \|T_n - T\| = 0, \qquad (7.4.15a)$$

or in short,

$$\|T_n - T\| \to 0. \qquad (7.4.15b)$$

$\{T_n\}$ is said to be **strongly convergent** to T, if there exists a $T \in B(V, V)$ such that for every $f \in V$, the sequence $\{T_n f\}$ strongly converges to $T f$ in V, i.e.,

$$\lim_{n \to \infty} \|T_n f - T f\| = 0, \qquad (7.4.16a)$$

or in short,

$$\|T_n f\| \to \|T f\|. \qquad (7.4.16b)$$

$\{T_n\}$ is said to be **weakly convergent** to T, if the sequence $\{T_n f\}$ weakly converges to $T f$ in V.

Note that (7.4.15) is independent of f. That is to say this equation holds for any f. In this sense we say that $\{T_n\}$ converges by norm. If the vector f is regarded as a variable of a function and the operator regarded as the function, $T f = T(f)$, then (7.4.15) means that no matter what the "variable f" is, the operator sequence $\{T_n\}$ always converges in norm to T. This is similar to the concept of uniform convergence introduced in Subsection 2.3.1. Equation (7.4.16) stresses the effect of the action of the operator on an f, which is similar to the concept of pointwise convergence introduced in Subsection 2.3.1. The concept of the convergence in norm and strong convergence can be regarded as the generalization of those of uniform convergence and pointwise convergence into Banach spaces. The convergences of function sequences are judged by absolute values, while the convergences in a Banach space consisting of operator elements are judged by norms.

The uniform convergence is a requirement stronger than the pointwise convergence. Similarly, the convergence in norm is a requirement stronger than strong convergence.

Theorem 5. *For an operator sequence from X to X, uniform convergence contains strong convergence, and strong convergence contains weak convergence, but not vice versa.*

In Subsection 7.4.3 below, we will show by the example of projection operator that the convergence by norm is a requirement stronger than strong convergence.

7.4.2. *Adjoint operators*

When an operator acts on a preimage, the resultant image is called the value of the operator. The operator's value is an element in a certain space.

Definition 6. When the value of an operator is a real or complex number, the operator is called a **functional**. If T is a bounded linear transformation from Banach space V to U, where U is a real or complex number space, then U is called a **continuous linear functional** or **bounded linear functional** on V. All of the continuous linear functionals on V constitute a Banach space $B(V, U)$, and is called the **dual space** or **conjugate space** of V, denoted by V^* or V'.

It is possible to understand the dual spaces by looking at several examples: for an n-dimensional Euclid space (n-dimensional real vector space) E, if a column vector $y \in E$, then $x = y^{\mathrm{T}}$ such that all the row vectors transposed from all y constitute an n-dimensional real vector space E^* which is the dual space of E; for an n-dimensional complex vector space U, if a column vector $y \in U$, then $x = y^{+}$ such that all the row vectors transposed from all y constitute an n-dimensional complex vector space U^* which is the dual space of U; for an $m \times n$-dimensional real matrix space $R_{m \times n}$, if a matrix $A \in R_{m \times n}$, then $B = A^{\mathrm{T}}$ such that all the matrices transposed from all A constitute a space $R^*_{m \times n}$ which is the dual space of $R_{n \times m}$; for an $m \times n$-dimensional complex matrix space $C_{m \times n}$, if a matrix $A \in C_{m \times n}$, then $B = A^{+}$ such that all the matrices transposed and conjugated from all A constitute a space $C^*_{m \times n}$ which is the dual space of $C_{m \times n}$. In going from a space to its dual space, we actually make a manipulation, or linear transformation, to every element in the space.

It is easily verified that these examples of spaces and corresponding dual ones observe Theorem 3 when their induced matrix norms are defined.

Having defined the conception of dual space, we can now understand inner product defined by (g, f): it is a scalar composed by a vector f in space V and a vector g in V's dual space V^* according to a specified manner.

In Chapter 2, the conception of adjoint operator was defined based on the conception of inner product. We are going to extend the definition in Banach space (note that the inner product is not necessarily defined in a Banach space).

By Definition 4, the bounded linear transformations from Banach spaces V to U are called bounded linear functionals on V. The action of a bounded linear operator T on an element f is denoted as $T(f)$, and its effect is denoted as $F(f)$. The difference between $T(f)$ and $F(f)$ is as follows: when we emphasize the action of the operator, we write in the form of $T(f)$ which is an operator; when we emphasize the effect of the operator, we write in the form of $F(f)$ which is equivalent to a function of the f. The f in $F(f)$ can be regarded as a variable. In a Hilbert space, the element f itself can be a function. Therefore, $F(f)$ is of the meaning of a functional defined in Chapter one.

When we say the action of a bounded linear operator T generates a functional, the concrete form of the functional has not been defined. The inner product can be a kind of functional. This form has been employed in Chapter 5, see (5.1.6).

Theorem 6. *For every continuous linear functional F on a normed space X, there exists a unique $g_T \in X$ such that for every $f \in X$, there exists $F(f) = (f, g_T)$ and $\|F\| = \|g_T\|$.*

Please note that here (f, g_T) represents a functional, and the inner product is a special case. Because the form of the inner product is relatively simple, this form is often employed in examples.

This theorem shows that on a normed space, each continuous linear functional F of element f is just that formed by f and a specific element g_T in this space, and it is so for every f. Furthermore, the norm of F equals to the norm of g_T.

Definition 7. Let X be a Hilbert space and T be a continuous linear operator from X to X. For a fixed $g \in V$ and any $f \in X$, a functional

$F_g(f)$ composed by Tf and g is made and is denoted as

$$F_g(f) = (Tf, g).$$

The $F_g(f)$ thus defined is a continuous linear functional on X. By Theorem 6, there exists $u \in X$ such that for all $f \in X$, $F_g(f) = (f, u)$. Now let $u = T^*g$. Then T^* is also a continuous linear operator from X to X. T^* is called an **adjoint operator** or **conjugate operator** of T. The basic relation between T and T^* is that for any $f, g \in X$,

$$(Tf, g) = (f, T^*g). \tag{7.4.17}$$

The concept of the conjugate space by Definition 6 is helpful for us to understand the conjugate operator. Equation (7.4.17) shows that if a vector f belongs to space V, then the vector g belongs to the dual space V^*. The two functionals are equal: one is composed by a vector f acted on by a linear operator T and g; the other is composed by f and g acted on by the adjoint operator T^*. So, if the operator T acts on vectors in space V, then the adjoint operator T^* acts on vectors in the dual space V^*. A dual space is also called a conjugate space, so that an adjoint operators are also a conjugate operator.

Example 2. In an n-dimensional Euclid space E, its elements x are column vectors. All the row vectors x^{T} constitute the dual space E^*. Then (7.4.17) in the present case becomes

$$(Bx)^{\mathrm{T}}y = x^{\mathrm{T}}B^{\mathrm{T}}y. \tag{7.4.18}$$

Example 3. In an n-dimensional unitary space U, its elements x are column vectors. All the row vectors x^+ constitute the dual space U^*. Then (7.4.17) in the present case becomes

$$(Ax)^+y = x^+A^+y. \tag{7.4.19}$$

Examples 2 and 3 are the same as Examples 2 and 3 in Subsection 2.2.1, but here the viewpoint of conjugate spaces is involved.

Theorem 7. *Let X be a Hilbert space and T is a continuous linear operator from X to X. Then, $\|T^*\| = \|T\|$.*

Proof. The inner product of a vector with itself is defined as the square of the vector's norm. Then, according to norm's compatibility, we have

$$\|Tx\|^2 = (Tx, Tx) = (T^*Tx, x)$$

$$\leq \|T^*Tx\| \, \|x\| \leq \|T^*\| \, \|Tx\| \, \|x\| \, ,$$

where the first inequality sign comes from Schwartz inequality. Thus, we get $\|Tx\| \leq \|T^*\| \, \|x\|$. On the other hand, by (7.4.8), $\|Tx\| \leq \|T\| \, \|x\|$ so that $\|T^*\| \geq \|T\|$. Exchanging T and T^* immediately leads to $\|T^*\| \leq \|T\|$. Therefore, it is necessary that $\|T^*\| = \|T\|$.

This theorem shows that the norm of a continuous linear operator T is always equal to that of its adjoint. For instance, the norm of a unitary matrix is equal to that of its transpose, and the norm of a unitary matrix is equal to that of its transpose and complex conjugate matrix.   □

Definition 8. Let X be a Hilbert space and T is a continuous linear operator from X to X. T is called a **norm-reservation operator** or **norm-reservation transformation** if for any $x \in X$,

$$\|Tx\| = \|x\| \, . \tag{7.4.20}$$

Apparently, the concept of the norm-reservation transformation is a generalization of that of isometric transformation defined in Subsection 2.2.2. Although (7.4.20) and (2.2.31) are of the same form, the former is understood in the sense of norm. In short, a norm-reservation transformation retains the norm of a vector.

The definition (7.4.20) is equivalent to the following two:

$$T^*T = TT^* = I \tag{7.4.21}$$

and

$$(Tf, Tg) = (f, g). \tag{7.4.22}$$

For example, the inner product of a vector and itself is defined as the square of the norm of this vector. Then,

$$\|Tx\|^2 = (Tx, Tx) = (T^*Tx, x) = \|x\|^2 = (x, x).$$

This leads to (7.4.21). On the other hand, by the definition of adjoint operators, $(Tf, Tg) = (T^*Tf, g) = (f, g)$ which is (7.4.22). Any two of (7.4.20)–(7.4.22) are sufficient and necessary conditions of each other. The norm-reserving operator is also called the unitary operator.

The conceptions of inverse operators, normal operators, self-adjoint operators etc. are the same those in Chapter 2, as long as the inner products there are replaced by functionals here.

By (7.4.21), it is seen that a norm-reserving operator has necessarily an inverse which is just its adjoint and is also norm-reserving.

Theorem 8. *Let T be a continuous linear operator from a Hilbert space X to itself. The sufficient and necessary conditions for T to be a normal one are that $\|T^*x\| = \|Tx\|$ for any $x \in X$.*

7.4.3. *Projection operators*

Definition 9. Given an orthonormalized vector set $\{\phi_i\}$ in V and an operator P_n, the effect of P_n acting on any vector f in V is

$$P_n f \equiv \sum_{i=1}^{n} (\phi_i, f)\phi_i. \qquad (7.4.23)$$

Then P_n is called a **projection operator** which projects f onto a subspace of V composed by the first n vectors of $\{\phi_i\}$.

Definition 10. An operator K is said to be **idempotent** if

$$K^2 = K. \qquad (7.4.24)$$

In the following we discuss some properties of projection operators.

Theorem 9. *Suppose that T is a continuous linear operator from Hilbert space X to itself. The sufficient and necessary conditions for T to be a self-adjoint one are that it is idempotent: $T^2 = TT = T$.*

This theorem reveals two properties of a projection operator P_n.

(i) P_n is a Hermitian operator. By the definition (7.4.23), it is easily seen that for all $f, g \in H$, $(f, P_n g) = (P_n f, g)$. Thus P_n is Hermitian.

(ii) P_n is idempotent.

Proposition 1. *The norm of projection operator P_n is 1.*

Proof. We first show that the norm of P_n satisfies

$$\|P_n\| \leq 1. \tag{7.4.25}$$

To do so, we inspect

$$\|P_n f\| = (P_n f, P_n f)^{1/2} = (f, P_n f)^{1/2}$$

$$= \left[\sum_{i=1}^{n} (f, \phi_i)(\phi_i, f) \right]^{1/2} \leq \|f\|. \tag{7.4.26}$$

At the last step Bessel inequality is used. Hence, $\|P_n f\| \leq \|f\|$ for all f, so that $\|P_n\| \leq 1$.

In the same way it is shown for operator $(I - P_n)$ that

$$\|I - P_n\| \leq 1. \tag{7.4.27}$$

Next we show that the inequality signs in (7.4.25) and (7.4.27) can be removed. There is at least one f which satisfies $\|P_n f\| = \|f\|$. In fact, when $f = \phi_1, \phi_2, \ldots, \phi_n$, we obtain from (7.4.23) that $\|P_n f\| = \|f\|$. For example, as $f = \phi_i$, $1 \leq i \leq n$,

$$P_n f = \sum_{j=1}^{n} (\phi_j, \phi_i)\phi_j = \sum_{j=1}^{n} \delta_{ij}\phi_j = \phi_i = f.$$

Meanwhile, $\|f\| = \|P_n f\| \leq \|P_n\| \|f\|$. This results in $\|P_n\| \geq 1$. Combined with (7.4.25) we get

$$\|P_n\| = 1. \tag{7.4.28}$$

Similarly,

$$\|I - P_n\| = 1. \tag{7.4.29}$$

In summary, the norm of the projection operator is 1. $\qquad\square$

As $n \to \infty$, the inequality sign in (7.4.26) can be changed to be an equality sign, and the equation is just the completeness relation. The completeness relation can be expressed by projection operator:

$$\lim_{n \to \infty} (f, P_n f) = \sum_{i=1}^{\infty} (f, \phi_n)(\phi_n, f) = (f, f). \tag{7.4.30}$$

Proposition 2. *A projection operator sequence* $\{P_n\}$ *strongly converges to unit operator* I.

Proof. By the definition (7.4.23) and the idempotent of the projection operator, we have

$$\|P_n f - If\| = (P_n f - f, P_n f - f)^{1/2}$$
$$= [(P_n f, P_n f) - (P_n f, f) - (f, P_n f) + (f, f)]^{1/2}$$
$$= [(f, P_n f) - 2(f, P_n f) + (f, f)]^{1/2}$$
$$= [(f, f) - (f, P_n f)]^{1/2}.$$

Equation (7.4.30) means that

$$\lim_{n \to \infty} \|P_n f - If\| = 0.$$

By comparison with (7.4.16), it is known that the sequence $\{P_n\}$ strongly converges to I. □

Another way of the proof is as follows. Because (7.4.23) stands for all $f \in V$ and

$$\|P_n f\| = (P_n f, P_n f)^{1/2} = (f, P_n^2 f)^{1/2} = (f, P_n f)^{1/2},$$

$\lim_{n \to \infty} \|P_n f\| = (f, f)^{1/2} = \|If\|$, and as a result P_n approaches I. Then, following (7.4.16) it is known that $\{P_n\}$ strongly converges to I.

Proposition 3. *A projection operator sequence $\{P_n\}$ does not converge in norm to I.*

Proof. If the sequence $\{P_n\}$ converges in norm to I, then, for any given ε, there exists an N such that as $n \geq N$, $\|(P_n - I)f\| \leq \varepsilon \|f\|$ for all $f \in V$ and the N is independent of f. However, the fact is that for any given N, there always are some functions $h \in V$, e.g., $h = \phi_{N+1}$ which makes $(P_N - I)h = (P_N - I)\phi_{N+1} = -\phi_{N+1}$. In this case, $\|(P_N - I)h\| = \|h\|$, so that the sequence $\{P_n\}$ does not converge in norm to I. □

The above two propositions say that the norm of P_n is 1, but the sequence $\{P_n\}$ does not converge in norm to I. It seems that the two conclusions are contradictory. Since the norm of the operator itself is 1, why does not it converge in norm to the unit operator?

As a matter of fact, they are not contradictory. To explain this we recall the triangle inequality of norms:

$$\|P_n\| - \|I\| \le \|P_n - I\|.$$

That the left hand side approaches zero cannot derive that the right hand side goes to zero. In fact, by (7.4.29), the right hand side is 1. The key of this problem is that convergence in norm to I means that $\|P_n - I\| \to 0$ but not $\|P_n\| \to \|I\|$.

This discloses that convergence in norm is a requirement stronger than strong convergence.

Suppose that in a space X, a projection operator, denoted by P_{X_1}, projects a vector in X into its subspace X_1 and another one, denoted by P_{X_2}, projects a vector in X into its subspace X_2.

Definition 11. Let P_{X_1} and P_{X_2} be two projection operators in a space. They are said to be **orthogonal** if $P_{X_1} P_{X_2} = \theta$. P_{X_2} is said to be a **part** of P_{X_1} if $P_{X_1} P_{X_2} = P_{X_2}$.

Here the θ represents either zero vector or zero operator. The action of zero operator on any vector generates a zero vector. Because the projection operator is self-adjoint, $P_{X_1} P_{X_2} = \theta$ leads to $P_{X_1} P_{X_2} = (P_{X_2} P_{X_1})^\dagger = \theta$.

$$P_{X_1} P_{X_2} = P_{X_2} \text{ leads to } P_{X_2} P_{X_1} = P_{X_2}.$$

The sufficient and necessary conditions for the addition, product and subtraction between the two projection operators P_{X_1} and P_{X_2} to be still a projection operator are addressed as follows.

Theorem 10. (i) *The sufficient and necessary conditions for the addition of P_{X_1} and P_{X_2} to be still a projection operator are that P_{X_1} and P_{X_2} are orthogonal to each other. The addition operator*

$$P_{X_1} + P_{X_2} = P_{X_1 + X_2}$$

projects a vector in space X into its subspace which is the direct sum of subspaces X_1 and X_2.

(ii) *The sufficient and necessary conditions for the product of P_{X_1} and P_{X_2} to be still a projection operator are that P_{X_1} and P_{X_2} are* *exchangeable to each other. The product operator*

$$P_{X_1} P_{X_2} = P_{X_1 \cap X_2}$$

projects a vector in space X *into its subspace which is the intersection of subspaces* X_1 *and* X_2.

(iii) *The sufficient and necessary conditions for the subtraction of* P_{X_1} *and* P_{X_2} $P_{X_1} - P_{X_2}$ *to be still a projection operator are that* P_{X_2} *is a part of* P_{X_1}. *The operator* $P_{X_1} - P_{X_2}$ *projects a vector in space* X *into its subspace which is in the subspace* X_1 *but does not include subspace* X_2.

Exercises

1. Show that (7.1.7) and (7.1.8) hold, i.e., starting from the right hand sides, the left hand sides can be derived. (Hint: the two equations reflect $(f, f) = \|f\|^2$.

2. Show that the norm defined by (7.1.22) satisfies the three axioms of norm.

3. Another way of proving Hölder inequality in the integral form.

   (1) Show that as $0 < \alpha < 1$, the function $f(z) = z^\alpha - \alpha z - \beta$ takes its maximum at $z = 1$, and the maximum is zero as $\beta = 1 - \alpha$. Therefore, as $\alpha < 1$ and $\beta = 1 - \alpha$, $z^\alpha \leq \alpha z + \beta$.

   (2) By variable transformation $z = x/y$, show that as $\alpha \leq 1$ and $\beta = 1 - \alpha$, $x^\alpha y^\beta \leq \alpha x + \beta y$,

   (3) Suppose that $|f(x)|^p$ and $|g(x)|^q$ are both integrable, where $1/p + 1/q = 1$. By the conclusion in (2), show that $|fg| \leq \frac{1}{p}|f|^p + \frac{1}{q}|g|^q$, and so $|fg|$ is also integrable.

   (4) Next, under the conditions that $|f(x)|^p$ and $|g(x)|^q$ are integrable, show that

$$\left| \int f(x)g(x)\mathrm{d}x \right| \leq \left[\int |f(x)|^p \mathrm{d}x \right]^{1/p} \left[\int |g(x)|^q \mathrm{d}x \right]^{1/q}.$$

4. Another way of proving Minkowski inequality in the integral form.

   (1) Prove that $|f + g|^p \leq 2^p \left[|f|^p + |g|^p \right]$. Thus, if f and g are p-integrable, $f + g$ is either.

   (2) Prove that for $p > 1$ and $1/p + 1/q = 1$,

   $$\int |f(x) + g(x)|^p \mathrm{d}x \leq \int |f(x)||f(x) + g(x)|^{p-1} \mathrm{d}x$$

   $$+ \int |g(x)||f(x) + g(x)|^{p-1} \mathrm{d}x.$$

   If both functions f and g are p-integrable and belong to L_p, then the function $|f + g|^{p-1}$ is q-integrable and belongs to L_q.

   (3) Apply Hölder inequality to both sides of the inequality in (2) so as to prove Minkowski inequality.

5. Hölder inequality with weights.

   (1) In Subsection 2.1.2, the concept of inner product with weight was defined. Now, if in an n-dimensional vector space there are a group of given positive numbers $\gamma_i, (i = 1, 2, \ldots, n)$ as weight. Consider to properly modify (7.1.14) and redefine u and v so as to show Hölder inequality with weights:

   $$\sum_{k=1}^{n} \gamma_k a_k b_k < \left(\sum_{k=1}^{n} \gamma_k a_k^p \right)^{1/p} \left(\sum_{k=1}^{n} \gamma_k b_k^q \right)^{1/q}.$$

   (2) In Example 14 in Subsection 2.1.2, the concept of inner product with weight in integral form was defined. Now, if in a function space there is a weight function $\rho(x) \geq 0$. Show Hölder inequality with weight in integral form:

   $$\int f(x)g(x)\rho(x)\mathrm{d}x$$

   $$\leq \left(\int f^p(x)\rho(x)\mathrm{d}x \right)^{1/p} \left(\int g^q(x)\rho(x)\mathrm{d}x \right)^{1/q}.$$

6. Prove Minkowski inequality with weights

$$\left(\sum_{i=1}^{n}|a_i + b_i|^p \gamma_i\right)^{1/p} \le \left(\sum_{i=1}^{n}\gamma_i\,|a_i|^p\right)^{1/p} + \left(\sum_{i=1}^{n}\gamma_i\,|b_i|^p\right)^{1/p}$$

and its integral form

$$\left[\int |f(x) + g(x)|^p \rho(x)\mathrm{d}x\right]^{1/p}$$

$$\le \left[\int |f(x)|^p \rho(x)\mathrm{d}x\right]^{1/p} + \left[\int |g(x)|^p \rho(x)\mathrm{d}x\right]^{1/p}$$

7. Show that (1) $\|x - y\| \ge \|x\| - \|y\|$ and (2) $\|A - B\| \ge \|A\| - \|B\|$.

8. Suppose that $A = (a_{ij})_{n\times n}$ is a positive definite symmetric matrix. Show that no matter what numbers $x_1, x_2, \ldots, x_n$ and $y_1, y_2, \ldots, y_n$ are, there always is the following inequality:

$$\left(\sum a_{ij}x_iy_j\right)^2 \le \left(\sum a_{ij}x_ix_j\right)\left(\sum a_{ij}y_iy_j\right).$$

9. For vectors $x = (x_1, x_2, \ldots, x_n)$ with n components, prove the following equivalence relations. (1) $\|x\|_\infty \le \|x\|_1 \le n\|x\|_\infty$. (2) $\frac{1}{\sqrt{n}}\|x\|_2 \le \|x\|_\infty \le n\|x\|_2$. (3) $\|x\|_2 \le \|x\|_1 \le \sqrt{n}\|x\|_2$.

10. For $f(x) = (x - 1/2)^3$, $x \in [0, 1]$, find $\|f\|_1, \|f\|_2, \|f\|_\infty$.

11. Suppose that $A = (a_{ij})_{n\times n}$ and $\|A\|$ is its norm induced from vector a norm. Show that $\|A\| \ge |a_{ij}|$.

12. If $\|A\|$ is an induced matrix norm and $\det A \ne 0$. Show that (1) $\|A^{-1}\| \ge \|A\|^{-1}$ and (2) $\|A^{-1}\|^{-1} = \min_{x \ne 0} \frac{\|Ax\|}{\|x\|}$.

13. Verify that $\|A\|_F$ and $\|x\|_2$ are compatible.

14. Show that $\|A\|_2 \le \|A\|_F$.

15. Evaluate the row-sum norm, column-sum norm, spectral norm and F-norm of the following matrices.
(1) $A = \begin{pmatrix} 3 & -1 & 1 \\ 1 & 1 & 1 \\ 2 & 1 & -1 \end{pmatrix}$. (2) $A = \begin{pmatrix} 0 & a \\ -a & 0 \end{pmatrix}$, where a is a real number.

16. Find $\rho(A)$ of $A = \begin{pmatrix} 1 & 2 & 0 \\ -2 & 1 & 0 \\ 0 & 0 & 2 \end{pmatrix}$.

17. Prove the continuity and linearity of bounded linear transformations in a Banach space.

    (1) Let F be a bounded linear functional in Banach space B. Assume that in B there is a sequence $\{f_n\}$ which converges in norm to $f \in B$. Show that $F(f_n) \to F(f)$, i.e., F is continuous.

    (2) Assume that $f \in L_p$ and $1/p + 1/q = 1$. Then, the functional

    $$F(g) = (f, g) = \int f^*(x)g(x)\mathrm{d}x, \ g \in L_q$$

    defined on L_q is a bounded linear functional.

18. The norm of bounded linear functionals has been defined in Subsection 7.4.1.

    (1) Show that for a functional F defined by $F(g) = (f, g) = \int f^*(x)g(x)\mathrm{d}x$ on L_q, where $f \in L_p$ and $1/p + 1/q = 1$, there exists $||F|| \le ||f||$ where

    $$||f|| = \left[\int |f(x)|^p \mathrm{d}x\right]^{1/p}.$$

    (2) Show that if $f \in L_p$, then $g = \frac{1}{f^*}|f|^{p/q+1}$ belongs to $L_q(1/p + 1/q = 1)$. Show that (i) $|F(g)| = |(f, g)| = ||f||^p$ and (ii) $||f|| \, ||g|| = ||f||^p$. Therefore, $|F(g)| = ||f|| \, ||g||$ and consequently $||F|| \ge ||f||$.
    Combination of (1) and (2) gives $||F|| = ||f||$.

19. By the definition (7.4.23), show that projection operators are self-adjoint and idempotent.

20. Prove Theorem 10 in Subsection 7.4.3. Show that if $P_{X_2}P_{X_1} = P_{X_1}P_{X_2}$, then $P_{X_1} + P_{X_2} - P_{X_1}P_{X_2}$ is a projection operator. What is the subspace into which this operator projects the vectors in X?

Chapter 8

Integral Equations

8.1. Fundamental Theory of Integral Equations

8.1.1. *Definition and classification of integral equations*

1. Definition and classification of integral equations

The definition and classification of integral equations are as follows.

Definition 1. Equations associated with the integrals of unknown functions are called **integral equations**. An integral equation is called a **linear integral equation** if the highest power of the unknown function is one, otherwise called a **nonlinear integral equation**.

Hereafter in this chapter, the unknown function is denoted by f and other functions are known, unless specified.

Example 1. The following equations belong to integral equations.

$$\int_0^1 f(y)\cos y \, dy = 1. \tag{8.1.1}$$

$$\int_0^y f(x)x^4 dx + \sin x = f(y). \tag{8.1.2}$$

$$\int_1^y f^2(x)e^x dx = f(y). \tag{8.1.3}$$

Equations (8.1.1) and (8.1.2) are linear integral equations, and (8.1.3) is a nonlinear one. Comparatively, linear integral equations are easier to solve.

The classification of integral equations are as follows.

Definition 2. Integral equations in the forms of

$$\int_a^b k(x,y)f(y)\mathrm{d}y + g(x) = 0, \qquad (8.1.4)$$

$$\lambda \int_a^b k(x,y)f(y)\mathrm{d}y + g(x) = f(x) \qquad (8.1.5)$$

and

$$\lambda \int_a^b k(x,y)f(y)\mathrm{d}y + g(x) = a(x)f(x) \qquad (8.1.6)$$

are called **Fredholm integral equations**. Equations (8.1.4), (8.1.5) and (8.1.6) are respectively called **Fredholm integral equations of the first, second and third kind**. Obviously, all these Fredholm integral equations are linear ones.

The first kind of these equations are in general not easy to solve. However there are some exceptions. One is the well-known Fourier transformation of the unknown function:

$$g(x) = \frac{1}{\sqrt{2\pi}} \int_{-\infty}^{\infty} e^{\mathrm{i}xy} f(y)\mathrm{d}y. \qquad (8.1.7)$$

Its inverse transformation gives the unknown function:

$$f(y) = \frac{1}{\sqrt{2\pi}} \int_{-\infty}^{\infty} e^{-\mathrm{i}yx} g(x)\mathrm{d}x. \qquad (8.1.8)$$

Other examples are the Laplace transformation and its inverse, Hankel transformation and its inverse, Mellin transformation and its inverse, and so on.

Definition 3. Integral equations in the forms of

$$\int_a^x k(x,y)f(y)\mathrm{d}y + g(x) = 0, \tag{8.1.9}$$

$$\lambda \int_a^x k(x,y)f(y)\mathrm{d}y + g(x) = f(x) \tag{8.1.10}$$

and

$$\lambda \int_a^x k(x,y)f(y)\mathrm{d}y + g(x) = q(x)f(x) \tag{8.1.11}$$

are called respectively **Volterra integral equations of the first, second and third kinds.**

All these Volterra integral equations are linear ones. In general, nonlinear Volterra integral equations are of the following form:

$$f(x) = g(x) + \int_a^x h(x,y,f(y))\mathrm{d}y. \tag{8.1.12}$$

Definition 4. In the above linear integral equations, $k(x,y)$ is called the kernel of integral equations, also called **basic kernel** or **integral kernel**. In the first and second kinds of equations, the function $g(x)$ is called the **free term** of the equations. The equations are called **inhomogeneous** if $g(x)$ is not identical to zero, and called **homogeneous** if $g(x)$ is identical to zero.

In the first kind of Volterra integral equations (8.1.9), if $k(x,x) \neq 0$ and $k(x,y)$ and $g(x)$ are continuous, the equations can be converted to the second kind of equations in the form of (8.1.10). To show this explicitly, we take derivative with respect to x on both sides of (8.1.9), and then divide by $k(x,x)$ to get

$$f(x) + \int_a^x \frac{k_x(x,y)}{k(x,x)} f(y)\mathrm{d}y = \frac{y'(x)}{k(x,x)}.$$

This is just the form of (8.1.10), where the subscript x of $k_x(x,y)$ means the partial derivative with respect to x.

In this chapter we will mainly study the second kind of Fredholm integral equations in the form of (8.1.5). The resolutions of homogeneous and inhomogeneous integral equations will be concerned. The second kind of Volterra integral equations in the form of (8.1.10) will also be discussed. Section 8.3 discusses the resolution of (8.1.12).

Integral equations have many practical applications.

2. Classification of kernels

Some conceptions are defined to describe the properties of kernels.

Definition 5. A $k(x, y)$ is called a **continuous kernel** if it is continuous on $a \le x$, $y \le b$. A kernel $k(x, y)$ is called a **square integrable kernel**, or in short L_2 **kernel**, if it is square integrable:

$$\int_a^b dx \rho(x) \int_a^b dy \, |k(x, y)|^2 \rho(y) < \infty. \qquad (8.1.13)$$

Here the weight function in each integration has been explicitly written, see (2.1.24).

Hereafter in this chapter, all the kernels are L_2 ones unless specified.

Definition 6. A kernel $k(x, y)$ is called a **Hermitian kernel** or **conjugate kernel** if it satisfies $k(x, y) = k^*(y, x)$. It is also denoted as $k^*(y, x) = k^+(x, y)$. If the kernel is real, then $k(x, y) = k(y, x)$, and is called a **symmetric kernel**.

Definition 7. A kernel $k(x, y)$ is called a **degenerate kernel** or **separable kernel** if it can be written in the form of

$$k(x, y) = \sum_{i=1}^n \varphi_i(x) \chi_i^*(y). \qquad (8.1.14)$$

In this form, n is called the **rank** of the kernel $k(x, y)$. This kind of kernels are also called finite rank kernels. As $n = 1$, (8.1.4) is simplified to be

$$k(x, y) = \varphi(x) \chi^*(y). \qquad (8.1.15)$$

Hereafter, the case of $n \ne 1$ in (8.1.14) is called a **finite rank kernel**, and the case of (8.1.15) is called a **separable kernel**. Both are collectively called **degenerate kernels**.

Definition 8. If a kernel is of the form of $k(x,y) = k(x-y)$, then the corresponding integral equation is called a **convolution type one.**

When the interval $[a,b]$ covers the whole real axis or some special regions, the **integral equations of convolution type** can be solved by means of the integral transformation method.

8.1.2. *Relations between integral equations and differential equations*

Some differential equations can be converted to be integral equations, and some integral equations can be converted to be differential equations.

1. Relations between the initial value problems of differential equations and integral equations

Consider the following initial value problem of differential equation of second order.

$$\begin{cases} y'' + p(x)y' + q(x)y = f(x), \\ y(a) = \alpha, \quad y'(a) = \beta, \end{cases} \tag{8.1.16}$$

where $p(x), q(x), f(x)$ are continuous function on $[a,b]$ and $p(x)$ has continuous derivative on $[a,b]$.

Integrating the equation from a to x and using the initial conditions, we get

$$y'(x) - \beta = -p(x)y(x) + p(a)\alpha$$
$$- \int_a^x [q(t) - p'(t)]y(t)\mathrm{d}t + \int_a^x f(t)\mathrm{d}t,$$

where integration by parts is carried out with respect to the term containing y'. The second integration gives

$$\int_a^x \mathrm{d}u \int_a^u f(t)\mathrm{d}t = \left[u \int_a^u f(t)\mathrm{d}t \right]_a^x - \int_a^x u\mathrm{d}u \frac{\mathrm{d}}{\mathrm{d}u} \int_a^u f(t)\mathrm{d}t$$
$$= \int_a^x (x-t)f(t)\mathrm{d}t.$$

Thus, we have

$$y(x) = g(x) + \int_a^x k(x,t)y(t)dt, \qquad (8.1.17)$$

where

$$k(x,t) = (t-x)[q(t) - p'(t)] - p(t) \qquad (8.1.18a)$$

and

$$g(x) = \int_a^x (x-t)f(t)dt + [p(a)\alpha + \beta](x-a) + \alpha. \qquad (8.1.18b)$$

Equation (8.1.17) is of the same form as (8.1.10). Conversely, taking derivatives of (8.1.17) twice, we will achieve (8.1.16). Therefore, the initial problem (8.1.16) is equivalent to the second kind of Volterra equation (8.1.10).

2. Relations between the boundary value problems of differential equations and integral equations

Consider the following two-point boundary value problem of a differential equation of second order.

$$\begin{cases} y'' + p(x)y' + q(x)y = f(x), \\ y(a) = \alpha, \quad y(b) = \beta. \end{cases} \qquad (8.1.19)$$

Under certain conditions, the boundary value problem (8.1.19) is equivalent to a second kind of Fredholm integral equation.

Example 2. Consider the boundary value problems of differential equations:

$$\frac{d^2 f}{dx^2} + \lambda f = 0, \quad f(0) = 0, \quad f(1) = 0, \qquad (8.1.20)$$

where λ is a parameter. This differential equation can be converted to be the following integral equation.

$$f(x) = \lambda \int_0^1 k(x,u)f(u)du, \qquad (8.1.21a)$$

where the kernel is

$$k(x,u) = u(1-x)\theta(x-u) + x(1-u)\theta(u-x),$$

$$0 \le x, \quad u \le 1. \qquad (8.1.21b)$$

The integral equation (8.1.21) is entirely equivalent to the boundary value problem (8.1.20). To confirm this, let us take derivative of the kernel $k(x, u)$ with respect to x two times, and recall the properties of the delta function: $\delta(x)$ is an even function and $\delta'(x)$ is an odd one. So,

$$\frac{\partial^2}{\partial x^2} k(x, u) = -2\delta(x - u) + \delta(x - u) = -\delta(x - u). \qquad (8.1.22)$$

This is substituted into (8.1.21a) to meet (8.1.20). In the present case, $k(x, y)$ is actually the Green's function satisfying (8.1.22).

The kernel (8.1.21b) is a separable one and is a symmetric and square integrable one. Please note the form of (8.1.21b), which reveals that (8.1.21a) is in fact a Volterra equation.

In (8.1.20), the first derivative of y does not appear. The general form of this kind of equation is

$$\frac{d^2 f}{dx^2} = \varphi(x, f(x)), \qquad a \le x \le b \qquad (8.1.23)$$

This equation is equivalent to the following Volterra integral equation:

$$f(x) = A + Bx + \int_a^x (x - u)\varphi\left(u, f(u)\right) du \qquad (8.1.24)$$

where coefficients A and B are to be determined by boundary or initial conditions, see Exercise 2 in this chapter.

3. Volterra integral equations are converted to be differential equations

Taking derivative of Volterra equations (8.1.9) (8.1.12) with respect to x, we can convert these equations to be differential equations. Letting $x = a$ in the original equations gives the initial conditions. When the kernel is in some specific forms, this solving process is convenient.

Example 3. Solve the nonlinear integral equation

$$f(x) = \lambda \int_0^x [1 + f^2(y)] dy.$$

Solution. Taking derivative of both sides with respect to x leads to

$$f'(x) = \lambda[1 + f^2(x)].$$

The initial value is $f(0) = 0$. Thus the solution is

$$f(x) = \tan(\lambda x).$$

8.1.3.   *Theory of homogeneous integral equations*

Similar to the case of differential equations, one first tries to find the solutions of homogeneous equations, and then the solutions of the inhomogeneous equations.

1. Homogeneous Fredholm integral equations

For the second kind of homogeneous Fredholm integral equations, some conclusions have been obtained. Particularly, in the case of Hermitian kernels, the theory is clear and useful.

In order to show the features of the integral equations more explicitly, we rewrite the second kind of homogeneous Fredholm integral equation in the following form

$$f(x) = \lambda \int_a^b dy k(x,y) f(y) \rho(y) = \lambda(k^*, f). \qquad (8.1.25)$$

Here the integral is put down as a form of inner product, and the function $\rho(y)$ is the weight. The weight can be regarded as a factor separated from the kernel, and it is not changed on the integral interval.

Definition 9. In (8.1.25), if the equation has nonzero solutions as $\lambda = \lambda_0$, then λ_0 is called the **eigenvalue of the equation**, also called the **eigenvalue belonging to the kernel** $k(x,y)$. The equation

$$f(x) = \lambda_0 \int_a^b dy k(x,y) f(y) \rho(y) = \lambda_0(k^*, f)$$

is called the **eigenvalue equation**, and all of its nonzero solutions called the **eigenfunctions** belong to eigenvalue λ_0.

Hereafter, the eigenvalues referred are those of homogeneous equation (8.1.25), and belong to the kernel $k(x,y)$.

In the following, when we mention kernels, eigenvalues and eigenfunctions, they are all nonzero. The variables of kernels and other functions are all on interval $[a, b]$. For example, the variables in $k(x, y)$ are $a \leq x, y \leq b$.

Definition 10. If $\lambda = \lambda_i$ is an eigenvalue of $k(x, y)$, and there are m linearly independent eigenfunctions $f_{i,1}(x), f_{i,2}(x), \ldots, f_{i,m}(x)$ belonging to λ_i, then λ_i is said to be **m-degenerate**, and m is called the **rank** of λ_i, also called the **degeneracy** of λ_i.

Theorem 1. *Within any finite domain of the λ plane, there are a finite number of eigenvalues of kernel $k(x, y)$.*

Theorem 2. *Every eigenvalue has at least one eigenfunction belonging to it. There are a finite number of linearly independent eigenfunctions belonging to an eigenvalue.*

Proof. Suppose that λ_i is an eigenvalue of $k(x, y)$, and all the linearly independent eigenfunctions $f_{i,1}(x), f_{i,2}(x), \ldots, f_{i,m}(x)$ belonging to λ_i have been normalized. The eigenvalue equation of the j-th eigenvalue is

$$f_{i,j}(x) = \lambda_i \int_a^b dy k(x, y) f_{i,j}(y) \rho(y) = \lambda_i (k^*, f_{i,j}).$$

Let

$$a_j = \frac{f_{i,j}(x)}{\lambda_i} = \int_a^b dy k(x, y) f_{i,j}(y) \rho(y).$$

This equation can understood as follows: when x is fixed, $k(x, y)$ is regarded as a function of y, and is expanded by the eigenfunctions $f_{i,j}(x)$ of λ_i. Then the expansions coefficients are a_j. By Bessel inequality (2.1.15),

$$\sum_{j=1}^{m} \left| \frac{f_{i,j}(x)}{\lambda_i} \right|^2 \leq \int_a^b dy \, |k(x, y)|^2 \, \rho(y). \qquad (8.1.26)$$

Assume that the eigenfunctions have been normalized. Taking integration on both sides results in

$$\frac{m}{|\lambda_i|^2} \leq \int_a^b dx \rho(x) \int_a^b dy \, |k(x, y)|^2 \, \rho(y). \qquad (8.1.27)$$

Since the $k(x, y)$ is an L_2 kernel, the integration on the right hand side is finite. Therefore, the rank m is finite. $\qquad\square$

Theorem 3. *If λ_0 is an eigenvalue of $k(x, y)$, then its complex conjugate λ_0^* is an eigenvalue belonging to $k^*(y, x)$. That is to say, λ_0^* satisfies equation*

$$f(x) = \lambda_0^* \int_a^b dy k^*(y, x) f(y) \rho(x) = \lambda_0^*(k, f).$$

In addition, the number of the linearly independent eigenfunctions belonging to λ_0^* is the same as that of λ_0.

The above three theorems were proposed by Fredholm, and are also called the first, second and third Fredholm theorems. In these three theorem, the $k(x, y)$ is an L_2 one.

The following theorems involve Hermitian kernels.

Theorem 4. *If the $k(x, y)$ is an L_2 one and is Hermitian, this kernel has at least one eigenvalue.*

This theorem does not stand for non-Hermitian kernels.

Theorem 5. *If the kernel of (8.1.25) is a Hermitian one, then (i) its eigenvalues are real, and (ii) the eigenfunctions belonging to different eigenvalues are orthogonal to each other.*

Proof. (i) Let λ_i be an eigenvalue and $f_i(x)$ be its eigenfunction. Then, the eigenvalue equation is

$$f_i(x) = \lambda_i \int_a^b dy k(x, y) f_i(y) \rho(y) = \lambda_i(k^*, f_i). \qquad (8.1.28)$$

Multiplying $f_i^*(x) \rho(x)$ and then taking integration on both sides give

$$\int_a^b |f_i(x)|^2 \rho(x) dx = \lambda_i \int_a^b dx \int_a^b dy k(x, y) f_i(y) \rho(y) f_i^*(x) \rho(x).$$

Taking complex conjugate and noticing that the kernel is Hermitian, we have

$$\int_a^b |f_i(x)|^2 \rho(x) dx = \lambda_i^* \int_a^b dx \int_a^b dy k^*(x, y) f_i^*(y) \rho(y) f_i(x) \rho(x)$$

$$= \lambda_i^* \int_a^b dx \int_a^b dy k(x, y) f_i^*(x) \rho(x) f_i(y) \rho(y).$$

Subtraction of these two equations leads to

$$(\lambda_i - \lambda_i^*) \int_a^b dx \int_a^b dy k(x, y) f_i^*(x) f_i(y) \rho(x) \rho(y) = 0.$$

Because the integration is not identical to zero, it must be

$$\lambda_i - \lambda_i^* = 0.$$

In conclusion, the eigenvalue is real.

(ii) The eigenvalue equation of the eigenfunction $f_i(x)$ is (8.1.28). For another eigenvalue λ_j, the eigenvalue equation is

$$f_j = \lambda_j(k^*, f_j). \tag{8.1.29}$$

Equation (8.1.28) leads to

$$(f_i, f_j) = \lambda_i((k^*, f_i), f_j) \Rightarrow \lambda_j(f_i, f_j) = \lambda_i \lambda_j((k^*, f_i), f_j)$$

and (8.1.29) leads to

$$(f_i, f_j) = \lambda_j(f_i, (k^*, f_j)) \Rightarrow \lambda_i(f_i, f_j) = \lambda_i \lambda_j(f_i, (k^*, f_j)).$$

The right hand sides of these two equations are the same. This can be verified by explicitly writing the integrations:

$$((k^*, f_i), f_j) = \int_a^b dx f_j(x) \rho(x) \left[\int_a^b dy k(x, y) f_i(y) \rho(y) \right]^*$$

$$= \int_a^b dy f_i^*(y) \rho(y) \int_a^b dx k^*(y, x) f_j(x) \rho(x)$$

$$= (f_i, (k^*, f_j)),$$

where we have used the fact that the kernel is Hermitian. Thus we obtain $(\lambda_i - \lambda_j)(f_i, f_j) = 0$. Since $\lambda_i \neq \lambda_j$, then $(f_i, f_j) = 0$. The conclusion is that the eigenfunctions belonging to different eigenvalues are orthogonal to each other. □

This theorem is in fact specific of the corollary of Theorem 7 in Subsection 2.2.2 in the case of integral operators.

The ranks of different eigenvalues may be different. Rank depends on the kernel. If there are m linear independent eigenfunctions $f_1(x), f_2(x), \ldots, f_m(x)$ belonging to eigenvalue λ_0, then any of its eigenfunction can be written as the linear combination of $f_1(x), f_2(x), \ldots, f_m(x)$. These m functions can be orthonormalized by means of the Gram-Schmidt method introduced in Subsection 2.1.2.

All of the eigenvalues of a kernel $k(x, y)$ can be written in order of their increasing absolute values:

$$\lambda_1, \lambda_2, \ldots, \lambda_k, \ldots \quad (|\lambda_1| \le |\lambda_2| \le \cdots |\lambda_k| \le \cdots). \quad (8.1.30)$$

This sequence is denoted in short as $\{\lambda_i\}$. The rank of an eigenvalue is the times it appears in the $\{\lambda_i\}$. All the eigenfunctions are written in the order following that of their eigenvalues:

$$\{f_i(x)\}, \quad i = 1, 2, \ldots. \quad (8.1.31)$$

We assume that all the eigenfunctions have been normalized, and the independent linear eigenfunctions of one eigenvalue have be orthogonalized to each other.

Definition 11. The sequences $\{\lambda_i\}$ (8.1.30) and $\{f_i(x)\}$ (8.1.31) are called the **eigenvalue sequence** and **eigenfunction sequence** of the kernel $k(x, y)$, respectively, or in short, $\{\lambda_i\}$ and $\{f_i(x)\}$ are called the **eigen-system** of $k(x, y)$.

By the same procedure as used in proving Theorem 2, we can show that if $\{\lambda_i\}$ and $\{f_i(x)\}$ are the eigenvalue sequence and eigenfunction sequence of the kernel $k(x, y)$, then

$$\sum_{i=1}^{\infty} \left| \frac{f_i(x)}{\lambda_i} \right|^2 \le \int_a^b dy \, |k(x, y)|^2 \, \rho(y). \quad (8.1.32)$$

The weighted integral on both sides give

$$\sum_{i=1}^{\infty} \frac{1}{|\lambda_i|^2} \le \int_a^b dx \rho(x) \int_a^b dy \, |k(x, y)|^2 \, \rho(y). \quad (8.1.33)$$

As a matter of fact, the left hand side of (8.1.26) is merely a part of that of (8.1.32), and the left hand side of (8.1.27) is merely a part of that of (8.1.33). Therefore, (8.1.32) and (8.1.33) naturally contain (8.1.26) and (8.1.27).

Theorem 6 (Expansion of kernel in terms of eigenfunctions). *Suppose that a kernel $k(x, y)$ is an L_2 one and is Hermitian, $\{\lambda_i\}$ and $\{\psi_i(x)\}$ are its eigenvalue sequence and eigenfunction sequence.*

Then, the kernel can be expanded by

$$k(x,y) = \sum_{i=1}^{\infty} \frac{1}{\lambda_i} \psi_i(x) \psi_i^*(y). \qquad (8.1.34)$$

This expansion converges in the mean-square sense to $k(x,y)$.

Proof. If λ_i is an eigenvalue of a kernel $k(x,y)$ and $\psi_i(x)$ is its corresponding eigenfunction, then (8.1.28) stands. Now we expand the $k(x,y)$ by its eigenfunctions:

$$k(x,y) = \sum_{i=1}^{\infty} c_i(x) \psi_i^*(y).$$

The expansion coefficients are evaluated by

$$c_i(x) = \int_a^b k(x,y)\psi_i(y)\rho(x)\mathrm{d}y = \frac{\psi_i(x)}{\lambda_i}.$$

Thus, (8.1.34) is proved. □

Equation (8.1.34) can be compared to the expansion of Green's function (6.1.14).

From (8.1.34) it is seen that if the kernel is Hermitian, the eigenvalues are real.

Theorem 7. *If a kernel is not a degenerate one, it has infinite number of eigenvalues.*

Theorem 8. *The sufficient and necessary conditions for a kernel to be a degenerate one are that it has finite number of eigenvalues.*

If a kernel is a degenerate one, (8.1.34) can be rewritten as

$$k(x,y) = \sum_{i=1}^{m} \frac{1}{\lambda_i} \psi_i(x) \psi_j^*(y). \qquad (8.1.35)$$

Here the summation on the right hand side covers finite terms.

2. Homogeneous Volterra integral equations

The second and third kinds of homogeneous Volterra integral equations can be written in the following form:

$$f(x) = \lambda \int_a^x k(x,y)f(y)\rho(y)\mathrm{d}y. \qquad (8.1.36)$$

This equation has no nonzero solutions for any λ values. Thus, homogeneous Volterra equations have no eigenvalues. This conclusion will be naturally drawn in the end of subsection 8.2.2.

8.2. Iteration Technique for Linear Integral Equations

8.2.1. *The second kind of Fredholm integral equations*

1. Expressing the equations by operators

Equation (8.1.5) is recast by resorting to an operator into a form as follows:

$$f = g + \lambda K f. \tag{8.2.1}$$

Here K is an operator which is defined by

$$K f \equiv \int_a^b k(x,y) f(y) \rho(y) \mathrm{d}y \equiv (k^*, f). \tag{8.2.2}$$

In Example 1 in Subsection 7.4.1 we have known that it is a linear operator. It is Hermitian when the kernel $k(x,y)$ is Hermitian.

The eigenvalue λ_0 of the kernel $k(x,y)$ can also be said to be the eigenvalue of the operator K, which highlights the characteristic of the eigenvalue.

Equation (8.2.1) is rearranged to be

$$(I - \lambda K)f = g. \tag{8.2.3}$$

Here I is also an operator, and its effect is that

$$If = (I, f). \tag{8.2.4}$$

It does not substantially change the function $f(y)$. In this sense, I is a unit operator. The kernel of I is $k(x,y) = \delta(x-y)/\rho(y)$.

If the inverse operator $(I - \lambda K)^{-1}$ can be evaluated, the solution f of (8.2.1) will be

$$f = (I - \lambda K)^{-1} g. \tag{8.2.5}$$

2. Neumann series and its convergence conditions

Although (8.2.5) is formally written, this form brings some merits. We have learnt in Chapter 7 that an operator can be characterized by its norm. If λK is a "small" operator in some sense, then the following expansion is valid:

$$(I - \lambda K)^{-1} = I + \lambda K + \lambda^2 K^2 + \cdots . \tag{8.2.6}$$

It is required that an action of K on any element in V yields another element in V and the operator K satisfies

$$K^2 = KK, \quad K^3 = KK^2. \tag{8.2.7}$$

So, when the series

$$f = g + \lambda K g + \lambda^2 K^2 g + \lambda^3 K^3 g + \cdots \tag{8.2.8}$$

converges, it is the solution of (8.2.1). Following the definition (8.2.2), (8.2.8) is recast to be

$$f(x) = g(x) + \lambda(k^*, g) + \lambda^2(k^*, (k^*, g)) + \cdots , \tag{8.2.9}$$

or

$$f(x) = g_0(x) + \lambda g_1(x) + \lambda^2 g_2(x) + \lambda^3 g_3(x) + \cdots , \tag{8.2.10}$$

where

$$g_0(x) = g(x) \tag{8.2.11}$$

and

$$g_n(x) = (k^*, g_{n-1}) = K g_{n-1}. \tag{8.2.12}$$

Please note that although the series (8.2.8) seems obtained from the expansion (8.2.6), it is actually generated from the successive iteration of (8.2.1). Therefore, this method is called an iteration one, but in the following, we will focus on whether the series (8.2.8) is convergent.

Definition 1. Let $k_1(x, x_1) = k(x, x_1)$,

$$k_n(x, x_1) = (k, k_{n-1}), \quad (n = 2, 3, \ldots). \tag{8.2.13}$$

The kernels thus defined are called the **iteration kernels**, and $k_n(x, x_1)$ is called the **n-iteration kernel** of $k(x, x_1)$. Correspondingly, $k(x, x_1)$ is called a **basic kernel**.

If $k(x, y)$ is Hermitian, its n-iteration kernel $k_n(x, y)$ is also Hermitian. By means of the iteration kernels, (8.2.12) can be expressed as

$$g_n(x) = (k_n^*, g). \qquad (8.2.14)$$

We denote

$$R(x, y; \lambda) = \sum_{n=0}^{\infty} k_{n+1}(x, y)\lambda^n. \qquad (8.2.15)$$

Then, (8.1.5) is rewritten as

$$f(x) = g(x) + \lambda(R, g). \qquad (8.2.16)$$

Definition 2. $\sum_{n=0}^{\infty} k_{n+1}(x, y)\lambda^n$ in (8.2.15) is called a **Neumann series**, and the summation $R(x, y; \lambda)$ is called the **resolvent kernel** of (8.1.5).

The resolvent kernel satisfies equations

$$R(x, y; \lambda) = k(x, y) + \lambda \int_a^b dx_1 k(x, x_1) R(x_1, y; \lambda)\rho(x_1) \qquad (8.2.17a)$$

and

$$R(x, y; \lambda) = k(x, y) + \lambda \int_a^b dx_1 k(x_1, y) R(x, x_1; \lambda)\rho(x_1). \qquad (8.2.17b)$$

If the resolvent kernel exists, the solution of (8.1.5) is given by (8.2.16).

The Neumann series was named by mathematicians, while physicists usually call it **Born series**. This is because the physicist Born arrived at this series when he was studying the scattering of a particle with a high energy by a potential in quantum mechanics.

Obviously the condition of having a solution is that the series (8.2.8) must be convergent. Now we investigate its convergence.

Two conditions are postulated. One is that $|k(x, y)|$ is bounded on $x, y \in [a, b]$:

$$\max_{x, y \in [a, b]} |k(x, y)|\rho(y) = M. \qquad (8.2.18)$$

The other is that

$$\int_a^b |g(x)| dx = C \qquad (8.2.19)$$

is finite.

If the conditions (8.2.18) and (8.2.19) are satisfied, it can be proven that with an additional condition (8.2.24) below, $\lim_{N\to\infty} \sum_{n=0}^{N} g_n(x)\lambda^n$ uniformly converges to the function $f(x)$. We now show by reduction *ad absurdum* that the summation of the absolute value of all terms of $\{g_n(x)\lambda^n\}$ is less than a geometric series. That is to say, as $n \geq 1$, there is

$$|g_n(x)| \leq CM[M(b-a)]^{n-1}. \tag{8.2.20}$$

Proof. As $n = 1$,

$$|g_1(x)| \leq \int_a^b |k(x,y)g(y)|\rho(y)\mathrm{d}y$$

$$\leq M \int_a^b |g(y)|\mathrm{d}y \leq MC. \tag{8.2.21}$$

Assume that (8.2.20) stands for $n = m - 1$. Then, for $n = m$, we have

$$|g_m(x)| \leq \int_a^b |k(x,y)g_{m-1}(y)|\rho(y)\mathrm{d}y$$

$$\leq MC[M(b-a)]^{m-2} \int_a^b |k(x,y)|\rho(y)\mathrm{d}y$$

$$\leq MC[M(b-a)]^{m-1}. \tag{8.2.22}$$

This is just (8.2.20). It follows that

$$\left| f(x) - \sum_{n=0}^{N} g_n(x)\lambda^n \right| \leq |g_{N+1}(x)\lambda^{N+1}| + |g_{N+2}(x)\lambda^{N+2}| + \cdots$$

$$\leq MC\lambda \sum_{m=N}^{\infty} [M(b-a)]^m \lambda^m$$

$$\leq MC[|\lambda| M(b-a)]^N \sum_{m=0}^{\infty} [|\lambda| M(b-a)]^m.$$

$$\tag{8.2.23}$$

If

$$|\lambda| M(b-a) < 1, \tag{8.2.24}$$

then

$$\left| f(x) - \sum_{n=0}^{N} g_n(x)\lambda^n \right| \le \frac{MC[|\lambda| M(b-a)]^N}{1 - |\lambda| M(b-a)}. \qquad (8.2.25)$$

So, as $N \to \infty$, $\sum_{n=0}^{N} g_n(x)\lambda^n$ uniformly converges to $f(x)$. The convergence speed depends on the value of $\lambda M(b-a)$. The constant C given by (8.2.19) merely guarantees a finite number on the right hand side of (8.2.25), but does not influence the convergence speed.

If for all y, $k(x,y)$ is continuous with respect to x, and $g(x)$ is continuous, then $g_n(x)$, where every term is given by (8.2.12), is also continuous, and either is (8.2.16). Hence, it is obtained that

$$(I - \lambda K)f = f - \lambda K f$$

$$= \sum_{n=0}^{\infty} g_n(x)\lambda^n - \lambda K \sum_{n=0}^{\infty} g_n(x)\lambda^n$$

$$= g_0 + \lambda g_1 + \lambda^2 g_2 + \cdots - \lambda K \left(g_0 + \lambda g_1 + \lambda^2 g_2 + \cdots \right).$$
$$(8.2.26)$$

Combination of (8.2.8)–(8.2.11) results in

$$(I - \lambda K)f = g_0 = g. \qquad (8.2.27)$$

Thus, the Neumann series gives the solution. Because the series $\sum_{n=0}^{N} g_n(x)\lambda^n$ gradually approximates the solution $f(x)$ with N increasing, this method is called the **successive approximation method**.

In summary, there are three conditions (8.2.18), (8.2.19) and (8.2.24) for a Neumann series to converge to the solution. Sometimes, however, the last condition (8.2.24), i.e., $\lambda M < (b-a)^{-1}$, is not necessary. In some cases, this condition can be cancelled, but the Neumann series still converges to the solution. For example, in the case that the kernel is the form of $k(x,y) = \xi(x)\eta(y)$ and happens to be $\int_a^b \xi(x)\eta(x)\rho(x)\mathrm{d}x = 0$, we will obtain from (8.2.9) that

$$f(x) = g(x) + \lambda \int_a^b \mathrm{d}x' \xi(x)\eta(x')g(x')\rho(x')$$

$$+ \lambda^2 \int_a^b \mathrm{d}x' \int_a^b \mathrm{d}x'' \xi(x)\eta(x')\xi(x')\rho(x')\eta(x'')g(x'')\rho(x'') + \cdots$$

$$= g(x) + \lambda \int_a^b \xi(x)\eta(x')g(x')\rho(x')\mathrm{d}x',$$

where $g_2(x) = 0$. This equation shows that for any value of $\lambda M(b-a)$, the Neumann series is always convergent.

3. Solution under bounded linear transformation

We are going to present a more rigorous evaluation of the above convergence by calculating the operator's norm.

Consider that K is a bounded linear transformation in a Banach space and functions f and g are in space V. There is an operator equation

$$f = g + \lambda K f, \quad g \in V. \tag{8.2.28}$$

The operator K here may not be limited to an integral operator. We have known in Chapter 7 that it was possible to characterize an operator by evaluating its norm. If the norm of the operator K in (8.2.28) satisfies the condition

$$\|\lambda K\| < 1, \tag{8.2.29}$$

we are able to construct a series

$$f_n = \sum_{m=0}^n \lambda^m K^m g \tag{8.2.30}$$

to find the solution of (8.2.8).

Theorem 1. *Under the condition* (8.2.29), *the equation* (8.2.28) *has one and only one solution* $f = \lim_{n \to \infty} \sum_{m=0}^n \lambda^m K^m g$.

Proof.   First, we prove that $\lim_{n\to\infty} \sum_{m=0}^{n} \lambda^m K^m g$ is the solution. Before doing so, we show that the $\{f_n\}$ constructed by (8.2.30) is a Cauchy sequence. As $\nu > \mu$,

$$\|f_\nu - f_\mu\| = \left\| \sum_{m=\mu+1}^{\nu} \lambda^m K^m g \right\| \leq \sum_{m=\mu+1}^{\nu} \|\lambda^m K^m\| \|g\|$$

$$= \|g\| \, \|\lambda K\|^{\mu+1} \sum_{m=0}^{\nu-\mu-1} \|\lambda K\|^m$$

$$\leq \|g\| \, \|\lambda K\|^{\mu+1} \sum_{m=0}^{\infty} \|\lambda K\|^m$$

$$= \|g\| \, \|\lambda K\|^{\mu+1} \left(1 - \|\lambda K\|\right)^{-1}. \tag{8.2.31}$$

Since $(1 - \|\lambda K\|)^{-1}$ is finite, as long as μ is large enough, $\|f_\nu - f_\mu\|$ can be smaller than any given number. Therefore, $\{f_n\}$ is a Cauchy sequence.

On the other hand, space V is complete, so that $f_n \to f$. By the continuity of bounded operators, $K f_n \to K f$. Each term $K f_n$ can be evaluated as

$$\lambda K f_n = \lambda K \sum_{m=0}^{n} \lambda^m K^m g = \sum_{m=1}^{n+1} \lambda^m K^m g$$

$$= \sum_{m=0}^{n+1} \lambda^m K^m g - g = f_{n+1} - g.$$

Therefore, $(f_{n+1} - g) \to \lambda K f$, and its limit is $f - g = \lambda K f$. This demonstrates that f is the solution of (8.2.28).

Next, we show the uniqueness of the solution. Suppose that there are two different solutions f_1 and f_2. It follows that

$$\|f_1 - f_2\| = \|\lambda K(f_1 - f_2)\| \leq \|\lambda K\| \cdot \|f_1 - f_2\|.$$

Since $\|\lambda K\| < 1$, it must be that $\|f_1 - f_2\| = 0$. By (8.2.30), f_1 is identical to f_2. The conclusion is that the solution is unique.   □

This theorem indicates that when $\|\lambda K\| < 1$ the series (8.2.30) is convergent, but does not address that as $\|\lambda K\| > 1$ it is divergent. Hence, $\|\lambda K\| < 1$ is a sufficient but not a necessary condition.

If $\|\lambda K\|$ is sufficiently small, taking the first few terms in (8.2.30) will be enough to obtain a satisfactory result.

We apply the above conclusion to square integral space, i.e., $k(x,y)$ is an L_2 kernel.

$$\int \rho(x)\mathrm{d}x \int \rho(y)\mathrm{d}y\, |k(x,y)|^2 < \infty. \qquad (8.2.32)$$

Assume that function f is in the square integral space L_2, $f \in L_2$. It follows

$$\int |f(y)|^2 \rho(y)\mathrm{d}y < \infty. \qquad (8.2.33)$$

Now we show that the linear integral operator K defined by (8.2.2) transforms an element in L_2 to an element in L_2, as long as the conditions (8.2.32) and (8.2.33) are satisfied. By Schwartz inequality (2.1.8), we know that

$$|(k,f)|^2 \le (k,k)(f,f).$$

This means that

$$|Kf|^2 = |(k,f)|^2 \le \int |k(x,y)|^2 \rho(y)\mathrm{d}y \int |f(y)|^2 \rho(y)\mathrm{d}y.$$

We multiply on both sides by $\rho(x)$ and take integration with respect to x.

$$\int |Kf|^2 \rho(x)\mathrm{d}x \le \iint |k(x,y)|^2 \rho(y)\mathrm{d}y\rho(x)\mathrm{d}x \int |f(y)|^2 \rho(y)\mathrm{d}y. \qquad (8.2.34)$$

By the conditions (8.2.32) and (8.2.33), the function Kf is also square integrable, i.e., $Kf \in L_2$.

The norm of the operator K can be defined as follows. We adopt the Hölder norm for a function f by (7.2.3). Its 2-norm is

$$\|f\| = \left(\int |f(y)|^2 \rho(y)\mathrm{d}y\right)^{1/2}.$$

By this definition, the norm of the function Kf, observing (8.2.34), is

$$\|Kf\|^2 = \int \mathrm{d}x\, |Kf|^2 \rho(x)$$

$$\le \int \rho(x)\mathrm{d}x \int \rho(y)\mathrm{d}y\, |k(x,y)|^2 \|f\|^2. \qquad (8.2.35)$$

Recall the definition of the norm of an operator (7.4.6). It is seen that (8.2.35) gives the upper limit of K's norm:

$$\|K\| \leq \left[\int \rho(x)\mathrm{d}x \int \rho(y)\mathrm{d}y \, |k(x,y)|^2 \right]^{1/2}. \tag{8.2.36}$$

Thus, to meet (8.2.29), it is required that

$$|\lambda|^2 \int \rho(x)\mathrm{d}x \int \rho(y)\mathrm{d}y \, |k(x,y)|^2 < 1. \tag{8.2.37}$$

This is the condition under which a Neumann series converges and a series expansion method can be used to solve the integral equation (8.2.1).

4. Scattering of a microscopic particle by a potential

Here we present an example to show how to evaluate precisely the norm of an integral operator in order to solve an integral equation.

In quantum mechanics, the wave function of a microscopic particle obeys Schrödinger equation:

$$-\frac{\hbar^2}{2m}\nabla_r^2 \psi(\boldsymbol{r}) + V(\boldsymbol{r})\psi(\boldsymbol{r}) = E\psi(\boldsymbol{r}). \tag{8.2.38a}$$

It is rearranged to become

$$\left(\nabla_r^2 + \frac{2m}{\hbar^2}E \right) \psi(\boldsymbol{r}) = \frac{2m}{\hbar^2}V(\boldsymbol{r})\psi(\boldsymbol{r}). \tag{8.2.38b}$$

The corresponding homogeneous equation is

$$\left(\nabla_r^2 + \frac{2m}{\hbar^2}E \right) \varphi(\boldsymbol{r}) = 0. \tag{8.2.38c}$$

If a Green's function satisfies the equation

$$\left(\nabla_r^2 + \frac{2m}{\hbar^2}E \right) G(\boldsymbol{r},\boldsymbol{r}') = \delta(\boldsymbol{r}-\boldsymbol{r}'), \tag{8.2.38d}$$

then the solution of (8.2.38b) can be expressed by

$$\psi(\boldsymbol{r}) = \varphi(\boldsymbol{r}) + \int G_0(\boldsymbol{r},\boldsymbol{r}',E)V(\boldsymbol{r}')\psi(\boldsymbol{r}')\mathrm{d}\boldsymbol{r}'. \tag{8.2.38e}$$

The integral equation (8.2.38e) is completely equivalent to (8.2.38b), both describing a particle moving in a potential $V(\boldsymbol{r})$ in three-dimensional space. The solution of (8.2.38d) has been given by (6.2.15):

$$G_0(\boldsymbol{r}, \boldsymbol{r}', E) = -\frac{m}{2\pi\hbar^2} \frac{e^{iq|\boldsymbol{r}-\boldsymbol{r}'|}}{|\boldsymbol{r}-\boldsymbol{r}'|}, \tag{8.2.39}$$

where

$$q = \sqrt{2mE/\hbar^2}. \tag{8.2.40}$$

The function $\varphi(\boldsymbol{r})$ is the solution of (8.2.38c) and represents the motion of the particle in a free space. For bound states $(E < 0)$, $\varphi(\boldsymbol{r}) \equiv 0$. Then the q in (8.2.40) takes a positive imaginary part. In the case of scattering $(E \geq 0)$,

$$\varphi(\boldsymbol{r}) = \frac{1}{(2\pi)^{3/2}} e^{i\boldsymbol{q}\cdot\boldsymbol{r}}.$$

Then, (8.2.38e) becomes

$$\psi(\boldsymbol{r}) = \frac{1}{(2\pi)^{3/2}} e^{i\boldsymbol{q}\cdot\boldsymbol{r}} - \int \frac{m}{2\pi\hbar^2} \frac{e^{iq|\boldsymbol{r}-\boldsymbol{r}'|}}{|\boldsymbol{r}-\boldsymbol{r}'|} V(\boldsymbol{r}')\psi(\boldsymbol{r}')\mathrm{d}\boldsymbol{r}'. \tag{8.2.41}$$

In this case, $\mathrm{Im}\, q = 0$. Equation (8.2.41) is called the **Lippmann-Schwinger equation**. It describes the course that a particle takes moving from infinity to the vicinity of the potential and is scattered by the potential.

Equation (8.2.41) is a second kind of Fredholm integral equation with a kernel

$$k(\boldsymbol{r}, \boldsymbol{r}', E) = G_0(\boldsymbol{r}, \boldsymbol{r}', E)V(\boldsymbol{r}'). \tag{8.2.42}$$

Now the integral operator K is defined by

$$K\psi(\boldsymbol{r}) = \int G_0(\boldsymbol{r}, \boldsymbol{r}', E)V(\boldsymbol{r}')\psi(\boldsymbol{r}')\mathrm{d}\boldsymbol{r}'. \tag{8.2.43}$$

We intend to evaluate the upper limit of this operator's norm to find under what conditions (8.2.38e) can be solved by the iteration method.

By the definition of the kernel (8.2.42), the weight $\rho(\boldsymbol{r}) = 1$. Thus it follows from (8.2.36) that the upper limit of the norm of K should be

$$\|K\|^2 \leq \int \mathrm{d}\boldsymbol{r} \int \mathrm{d}\boldsymbol{r}' \left|k(\boldsymbol{r}, \boldsymbol{r}', E)\right|^2$$

$$= \int \mathrm{d}\boldsymbol{r} \int \mathrm{d}\boldsymbol{r}' \left|G_0(\boldsymbol{r}, \boldsymbol{r}', E)V(\boldsymbol{r}')\right|^2. \qquad (8.2.44)$$

We substitute the expression of the Green's function (8.2.39) into (8.2.44) and let $q = \operatorname{Re} q + i\operatorname{Im} q$.

$$\|K\|^2 \leq \left(\frac{m}{2\pi\hbar^2}\right)^2 \int \mathrm{d}\boldsymbol{r} \int \mathrm{d}\boldsymbol{r}' \frac{e^{-2\operatorname{Im} q|\boldsymbol{r}-\boldsymbol{r}'|}}{|\boldsymbol{r}-\boldsymbol{r}'|^2} \left|V(\boldsymbol{r}')\right|^2. \qquad (8.2.45)$$

Note that $\mathrm{d}\boldsymbol{r} \equiv r^2 \sin\theta \mathrm{d}r\mathrm{d}\theta\mathrm{d}\varphi$. Let $\boldsymbol{r} = \boldsymbol{\rho} + \boldsymbol{r}'$. Then the integral becomes

$$\int \mathrm{d}\boldsymbol{\rho} \int \mathrm{d}\boldsymbol{r}' \frac{e^{-2\rho\operatorname{Im} q}}{\rho^2} \left|V(\boldsymbol{r}')\right|^2 = 4\pi \int_0^\infty \mathrm{d}\rho \frac{\rho^2 e^{-2\rho\operatorname{Im} q}}{\rho^2} \int \mathrm{d}\boldsymbol{r} \left|V(\boldsymbol{r})\right|^2$$

$$= 4\pi \frac{1}{2\operatorname{Im} q} \int \mathrm{d}\boldsymbol{r} \left|V(\boldsymbol{r})\right|^2.$$

Thus,

$$\|K\|^2 \leq \frac{m^2}{2\pi\hbar^4\operatorname{Im} q} \int |V(\boldsymbol{r})|^2 \,\mathrm{d}\boldsymbol{r}. \qquad (8.2.46)$$

This integral is finite for $\operatorname{Im} q \neq 0$ which corresponding to the case of $E < 0$. That is to say, for the problems of bound states, (8.2.38e) can be solved by series expansion. In other words, this integral equation can always be solved if $\operatorname{Im} q \neq 0$.

As $\operatorname{Im} q = 0$, in the case of $E > 0$, (8.2.46) indicates that the upper limit of K's norm is infinity. So we do not known whether the integral equation (8.2.38e) can be solved or not. In reality, the scattering problem can be solved, and this is a very significant topic.

Is it possible to lower the upper limit of the operator's norm to a finite value? The answer is yes. Let us recall the definition of the integral operator (8.2.2), where there is a weight $\rho(x)$. When the kernel is defined by (8.2.42), we have defaulted that $\rho(x) = 1$. In this way, we overestimate the upper limit of the operator's norm.

Actually, there is arbitrariness in defining the kernel and weight. We try to define an alternative kernel to obtain a finite upper limit.

By inspection of (8.2.42), we find that a natural way is to select the kernel as

$$k_\rho(r, r', E) = G_0(r, r', E) \qquad (8.2.47)$$

and accordingly, the weight is

$$\rho(r) \equiv V(r), \qquad (8.2.48)$$

the condition being that $V(r)$ does not change its sign in the whole space. Under this definition, (8.2.43) becomes

$$K\psi(r) = \int G_0(r, r', E)\psi(r')\rho(r')dr'. \qquad (8.2.49)$$

It actually remains the same, but its kernel has been changed. Let us evaluate the upper limit of its norm under the new kernel. By (8.2.36),

$$\|K\|^2 \leq \int \rho(r)dr \int \rho(r')dr' \left| G_0(r, r', E) \right|^2$$

$$= \left(\frac{m}{2\pi\hbar^2}\right)^2 \int dr \int dr' V(r)V(r') \frac{e^{-2\mathrm{Im}\,q|r-r'|}}{|r - r'|^2}.$$

$$(8.2.50)$$

The further calculation depends on the expression of the potential. If it is a central potential, $V(r) = V(r)$, then the following argument transformation is helpful: $s = r+r'$, $t = r-r'$, $u = |r - r'|$ and $drdr' = \pi^2(s^2 - t^2)udsdtdu$. After the transformation, (8.2.50) becomes

$$\|K\|^2 \leq \left(\frac{m}{2\hbar^2}\right)^2 \int_0^\infty ds \int_0^s du \int_{-u}^u dt\, (s^2 - t^2) \frac{e^{-2u\mathrm{Im}\,q}}{u} V(r)V(r').$$

$$(8.2.51)$$

For instance, in the case of Yukawa potential $V(r) = g^2 e^{-\mu r}/r$, this is calculated to be

$$\|K\| \leq \left[\frac{2}{1 + 2\mathrm{Im}\,(q/\mu)}\right]^{1/2} \frac{mg^2}{\sqrt{\mu}\hbar^2}. \qquad (8.2.52)$$

This result is always finite even when $\mathrm{Im}\,q = 0$, which can be compared to (8.2.46). When $E > 0$, $\|K\| \leq \sqrt{2}\frac{mg^2}{\mu\hbar^2}$. As long as g is sufficiently small, $\|K\| < 1$ can be reached. So, (8.2.38e) can be solved by the iteration method.

The discrepancy between (8.2.50) and (8.2.45) comes from the different kernels. It is seen that there are more than one possibility of choosing the kernel. In the present case, the definition of the kernel is changed from (8.2.43) to (8.2.49). This results in that the evaluation is changed from (8.2.45) to (8.2.50). The way of redefining the kernel, as well as the weight, is called the **weight function method**.

8.2.2.   *The second kind of Volterra integral equations*

In the Volterra integral equation (8.1.10), the action of the operator K is slightly different from (8.2.2). It is

$$Kf = \int_a^x k(x,y)f(y)\rho(y)\mathrm{d}y,$$

which is still a linear one. For the series expansion to be convergent, the condition (8.2.18) of the kernel $k(x,y)$ should be satisfied, but there is no need of the constraint $\lambda M < (b-a)^{-1}$. Similarly to (8.2.10), the equation is expanded to be

$$f(x) = \tilde{g}_0(x) + \lambda\tilde{g}_1(x) + \lambda^2\tilde{g}_2(x) + \cdots, \tag{8.2.53}$$

where

$$\tilde{g}_0(x) \equiv g(x) \tag{8.2.54}$$

and

$$\tilde{g}_n(x) = \int_a^x k(x,y)\tilde{g}_{n-1}(y)\rho(y)\mathrm{d}y, \quad a \le x \le b. \tag{8.2.55}$$

As $n = 1$,

$$|\tilde{g}_1(x)| \le \int_a^x |k(x,y)\tilde{g}_0(y)|\,\rho(y)\mathrm{d}y$$

$$\le \int_a^x |k(x,y)|\,\rho(y)\,|\tilde{g}_0(y)|\,\mathrm{d}y \le M\int_a^b |\tilde{g}_0(y)|\,\mathrm{d}y = MC.$$

Here the integral upper limit is extended to b and (8.2.18) and (8.2.19) are employed. From $n = 2$, there is no need of extending the upper limit.

$$|\tilde{g}_2(x)| \leq \int_a^x |k(x,y)|\, \rho(y)\, |\tilde{g}_1(y)|\, \mathrm{d}y$$

$$\leq M \int_a^x |\tilde{g}_1(y)|\, \mathrm{d}y \leq M \int_a^x MC\, \mathrm{d}y = MC[M(x-a)].$$

It is easily shown that as $n \geq 2$,

$$|\tilde{g}_n(x)| \leq \int_a^x |k(x,y)|\, \rho(y)\, |\tilde{g}_{n-1}(y)|\, \mathrm{d}y \leq M \int_a^x |\tilde{g}_{n-1}(y)|\, \mathrm{d}y$$

$$\leq M \int_a^x MC\frac{[M(x-a)]^{n-2}}{(n-2)!}\, \mathrm{d}y = MC\frac{[M(x-a)]^{n-1}}{(n-1)!}.$$

$$(8.2.56)$$

So, the Neumann series is always convergent, independent of the value of $\lambda M(b-a)$. That is to say, the Neumann series can always give the solution of the second kind of Volterra integral equations.

As long as the upper integral limit b in (8.2.13) is changed to be x, the iteration kernel of the Volterra integral equation is obtained. The form of the resolvent kernel is still (8.2.15). When the upper integral limit b in (8.2.17) is changed to be x, we achieve the solution expressed by the resolvent kernel.

Equation (8.2.56) shows that Neumann series of the Eq. (8.1.10) is always convergent. This series, however, is generated from (8.2.54). As long as $g(x) = 0$, this series must be zero. Therefore, the corresponding homogeneous equation

$$f(x) = \lambda \int_a^x k(x,y) f(y)\rho(y)\mathrm{d}y$$

does not have a nonzero solution. This conclusion has been given in the end of Subsection 8.1.3: homogeneous Volterra equations did not have eigenvalues.

The conclusions in this subsection can be summarized as the following theorem.

Theorem 2. *Under the conditions* (8.2.18) *and* (8.2.19), *the second kind of Volterra integral equation has a uniformly convergent series in the form of* (8.2.53) *for any λ value. The limit function of the* *series* (8.2.53) *is the solution of the equation, and the solution is unique.*

The uniqueness of the solution is easily proven. Suppose that $f_1(x)$ and $f_2(x)$ are the solutions of the inhomogeneous Volterra equations, then $f_1(x) - f_2(x)$ satisfies the homogeneous Volterra equations, which is zero.

8.3.   Iteration Technique of Inhomogeneous Integral Equations

In this section, we investigate how to solve nonlinear Volterra integral equations in the form of (8.1.12):

$$f(x) = g(x) + \int_a^x h(x, y, f(y))\mathrm{d}y. \tag{8.3.1}$$

8.3.1.   *Iteration procedure*

Let

$$f_0(x) = g(x). \tag{8.3.2}$$

Successive iteration yields

$$f_n(x) = g(x) + \int_a^x h(x, y, f_{n-1}(y))\mathrm{d}y. \tag{8.3.3}$$

If the equation is linear, $h(x, y, f(y)) = k(x, y)f(y)$, (8.3.3) is simplified to be

$$f_n = \sum_{m=0}^n K^m g,$$

which is just the form of (8.2.30).

In general cases, we hope that the iteration procedure (8.3.3) can give the solution. It is necessary to investigate the convergence of the iteration.

The basic assumption is that when

$$|f(y) - g(y)| \leq \Delta, \quad y \in [a, b], \tag{8.3.4}$$

on square domain

$$a \leq x \leq b, \quad a \leq y \leq b, \tag{8.3.5}$$

the integrand in (8.3.1) is bounded:

$$|h(x, y, f(y))| \leq M. \tag{8.3.6}$$

Starting from (8.3.2) we make an iteration:

$$f_1(x) = g(x) + \int_a^x h(x, y, f_0(y)) dy = g(x) + \int_a^x h(x, y, g(y)) dy.$$

Under the condition

$$|f_0(y) - g(y)| = 0 < \Delta, \tag{8.3.7}$$

it is deduced that

$$|f_1(x) - g(x)| = |f_1(x) - f_0(x)|$$
$$\leq \int_a^x |h(x, y, f_0(y))| dy \leq M(x - a). \tag{8.3.8}$$

Next step is to consider if there stands an inequality

$$|f_1(y) - g(y)| \leq \Delta, \quad y \in [a, b]. \tag{8.3.9}$$

From (8.3.8) it is known that as $\Delta = M(y - a)$ or $y = \Delta/M + a$, (8.3.8) stands. As a matter of fact, since the upper integral limit is x, $a \leq y \leq x$, as $x \leq \Delta/M + a$, (8.3.8) still stands:

$$|f_1(x) - f_0(x)| \leq \int_a^{\Delta/M+a} |h(x, y, f_0(y))| dy \leq \Delta. \tag{8.3.10}$$

So, as $y \leq \Delta/M + a$, (8.3.9) also stands.

With the confinement

$$a \leq x \leq (\Delta/M) + a, \quad a \leq y \leq (\Delta/M) + a \tag{8.3.11}$$

(8.3.10) always stands. In the case of $\Delta/M + a \geq b$, (8.3.11) is in agreement to (8.3.5), so that the confinement (8.3.11) is not needed.

Since (8.3.9) is valid, we consider

$$f_2(x) \equiv g(x) + \int_a^x h(x, y, f_1(y)) dy.$$

Now that as $a \leq x \leq (\Delta/M) + a$, $|f_1(x) - g(x)| \leq \Delta$, then

$$|f_2(x) - g(x)| = \int_a^x h(x, y, f_1(y)) dy$$

$$\leq M(x - a) \leq \Delta. \qquad (8.3.12)$$

Under the condition (8.3.6), for the nth iteration (8.3.3), we always get

$$|f_n(x) - g(x)| \leq \int_a^x |h(x, y, f_{n-1}(y))| dy$$

$$\leq M(x - a) \leq \Delta. \qquad (8.3.13)$$

That is to say, the difference of the result of the nth iteration and $g(x)$ is less than a definite value.

However, this does not guarantee the convergence of the iteration series. The convergence requires that as n is sufficiently large, f_n and f_{n-1} ought to be sufficiently close to each other.

8.3.2. *Lipschitz condition*

In order to guarantee the convergence of the iteration series, a so-called **Lipschitz condition** is applied. This condition requires that there is a finite number N such that as long as

$$|\phi(y) - g(y)| \leq \Delta \qquad (8.3.14a)$$

and

$$|\psi(y) - g(y)| \leq \Delta, \qquad (8.3.14b)$$

$$|h(x, y, \phi(y)) - h(x, y, \psi(y))| \leq N |\phi(y) - \psi(y)|. \qquad (8.3.15)$$

The specific of the Lipschitz condition in our present case is that as long as

$$|f_n(y) - g(y)| \leq \Delta \qquad (8.3.16a)$$

and

$$|f_{n-1}(y) - g(y)| \leq \Delta, \qquad (8.3.16b)$$

it must be that

$$|h(x, y, f_n(y)) - h(x, y, f_{n-1}(y))| \leq N |f_n(y) - f_{n-1}(y)|. \qquad (8.3.17)$$

From (8.3.9) and (8.3.7), we have $|f_1(y) - g(y)| \leq \Delta$ and $|f_0(y) - g(y)| = 0 < \Delta$. Then following Lipschitz condition, we get

$$|f_2(x) - f_1(x)| \leq N \int_a^x |f_1(y) - f_0(y)| \, dy.$$

By (8.3.10), $|f_1(y) - f_0(y)| = |f_1(y) - g_0(y)| \leq \Delta$, so that

$$|f_2(x) - f_1(x)| \leq N\Delta(x - a).$$

Further, by (8.3.13), $|f_2(x) - g(x)| \leq \Delta$, so that

$$|f_3(x) - f_2(x)| \leq N \int_a^x |f_2(y) - f_1(y)| \, dy$$

$$\leq N \int_a^x N\Delta(y - a) dy = \frac{1}{2!} \Delta N^2 (x - a)^2.$$

The process goes successively. At the nth step, because $|f_n(x) - g(x)| \leq \Delta$, we have

$$|f_n(x) - f_{n-1}(x)| \leq N \int_a^x |f_{n-1}(y) - f_{n-2}(y)| \, dy$$

$$\leq N \int_a^x \frac{\Delta N^{n-2}(y - a)^{n-2}}{(n-2)!} dy$$

$$= \frac{\Delta N^{n-1}(x - a)^{n-1}}{(n-1)!}.$$

This is rewritten as

$$|f_n(x) - f_{n-1}(x)| \leq \frac{\Delta}{(n-1)!} \left(\frac{\Delta N}{M}\right)^{n-1}. \tag{8.3.18}$$

In summary, for all $x \subset [a, (\Delta/M) \mid a]$, as $n \to \infty$, $|f_n(x) - f_{n-1}(x)|$ uniformly approaches zero.

The last step is to prove that $f_n(x)$ goes to the solution of (8.3.1). The result of the νth step is written as

$$f_\nu(x) = f_0(x) + \sum_{n=1}^{\nu} [f_n(x) - f_{n-1}(x)],$$

where each term is controlled by (8.3.18) as long as $x \in [a, (\Delta/M) + a]$. It is apparently that as $\nu \to \infty$, $f_\nu(x)$ converges to a function $f_\infty(x)$:

$$\lim_{\nu \to \infty} f_\nu(x) = f_\infty(x). \tag{8.3.19}$$

It should be confirmed that $f_\infty(x)$ is just the solution of (8.3.1). By (8.3.3),

$$f_\nu(x) = g(x) + \int_a^x h(x, y, f_{\nu-1}(y))dy$$

$$= g(x) + \int_a^x h(x, y, f(y))dy$$

$$+ \int_a^x [h(x, y, f_{\nu-1}(y)) - h(x, y, f(y))]\, dy.$$

When $\nu \to \infty$, the limit of the left hand side is (8.3.19), so that

$$f_\infty(x) = g(x) + \int_a^x h(x, y, f(y))dy + \lim_{\nu \to \infty} R_\nu,$$

where

$$R_\nu = \int_a^x [h(x, y, f_{\nu-1}(y)) - h(x, y, f(y))]dy.$$

Applying Lipschitz condition, we have

$$|R_\nu| \le N \int_a^x |f_{\nu-1}(y) - f(y)|\, dy$$

$$\le N\Delta \sum_{n=\nu}^\infty \frac{N^{n-1}}{(n-1)!} \int_a^x (y-a)^{n-1}dy$$

$$\le \Delta \sum_{n=\nu}^\infty \frac{N^n(x-a)^n}{n!} \le \Delta \sum_{n=\nu}^\infty \left(\frac{N\Delta}{M}\right)^n \frac{1}{n!}.$$

It goes to zero as $\nu \to \infty$. Thus,

$$f_\infty(x) = g(x) + \int_a^x h(x, y, f(y))dy.$$

This demonstrates that the function obtained by (8.3.19) is indeed the solution of the equation (8.3.1): $f_\infty(x) = f(x)$.

The keys in the above proof process are to confirm if the assumed conditions are satisfied. (1) The iteration conditions are

(8.3.4)–(8.3.6), i.e., $|h(x, y, f(y))| \leq M$ as long as $|f(y) - g(y)| \leq \Delta$. (2) The convergence condition is Lipschitz condition, i.e., $|h(x, y, \phi(y)) - h(x, y, \psi(y))| \leq N |\phi(y) - \psi(y)|$, as long as $|\phi(y) - g(y)| \leq \Delta$ and $|\psi(y) - g(y)| \leq \Delta$. (3) The control parameter Δ, M and N are given. The convergence domain is determined by Δ/M, the convergence speed depends on Δ/M and constant N in Lipschitz condition. It is known by (8.3.18) that the smaller the Δ/M and N, the faster the convergence speed.

8.3.3. *Use of contraction*

Now we repeat the above proof by use of the concept of contraction introduced in Subsection 2.2.1. In this way, the proof process appears quite concise. The right hand side of (8.3.1) is regarded as the result of the action of an operator T on the function f:

$$Tf(x) = g(x) + \int_a^x h(x, y, f(y)) \mathrm{d}y. \qquad (8.3.20)$$

Then, (8.3.1) is recast to be $f(x) = Tf(x)$, which is the form of (2.2.3). It is easily seen that

$$|Tf - Tp| = \left| \int_a^x h(x, y, f(y)) \mathrm{d}y - \int_a^x h(x, y, p(y)) \mathrm{d}y \right|$$

$$\leq \int_a^x \mathrm{d}y \, |h(x, y, f(y)) - h(x, y, p(y))| \,.$$

Applying Lipschitz condition (8.3.15), we have

$$|Tf - Tp| \leq \int_a^x \mathrm{d}y N \, |f - p|$$

$$\leq N(x - a) \max_{a \leq x, y \leq b} |f - p| \,.$$

This inequality holds on $x \in [a, b]$. We define a distance $\rho(f, p) = \max_{a \leq x \leq b} |f - p|$. Then,

$$\rho(Tf, Tp) \leq N(x - a)\rho(f, p).$$

Under the condition $N(x - a) \leq N(b - a) < 1$, the operator T defined by (8.3.20) is a contraction. If this condition is not satisfied, we apply the operator once more, and the distance between the resultant elements are obtained by the Lipschitz condition:

$$\rho(T^2 f, T^2 p) \le \int_a^x dy N \, |Tf - Tp|$$

$$\le N \int_a^x dy N(x - a) \le N^2 \frac{(x - a)^2}{2} \rho(f, p).$$

After m steps,

$$\rho(T^m f, T^m p) \le N^m \frac{(x - a)^m}{m!} \rho(f, p)$$

$$\le N^m \frac{(b - a)^m}{m!} \rho(f, p).$$

It is seen that if m is large enough such that $N^m \frac{(b-a)^m}{m!} < 1$, then the operator T^m is a contraction. In summary, once the Lipschitz condition (8.3.15) is satisfied, the inhomogeneous Volterra integral equation can always be solved by means of the iteration method.

8.3.4.   *Anharmonic vibration of a spring*

Example 1. An anharmonic potential is of the form of

$$V(x) = \frac{1}{2} k x^2 + \frac{1}{3} a x^3,$$

where a is a small positive constant. In this potential, the equation of motion of a mass point m is $m\ddot{x} + kx + ax^2 = 0$, where $\omega_0^2 = k/m$, $\varepsilon = a/m > 0$. The operator $d^2/dt^2 + \omega_0^2$ has a Green's function (6.3.15) under the one-end boundary condition:

$$G(t, t') = \frac{1}{\omega_0} \sin[\omega_0(t - t')]\theta(t - t').$$

The initial conditions, the state of the mass at initial instant $t = 0$, are

$$x(0) = x_0 > 0, \quad \dot{x}(0) = 0.$$

In the absence of the inhomogeneous term, the solution of the homogeneous equation is $x_0 \cos \omega_0 t$. With the inhomogeneous term, the solution is expressed by (6.1.4):

$$x(t) = x_0 \cos \omega_0 t - \int_0^t \frac{1}{\omega_0} \sin \omega_0(t - t')\varepsilon x^2(t')dt'.$$

We now assume that $\Delta = x_0$. Equation (8.3.4) requires that

$$|x(t) - x_0 \cos \omega_0 t| \leq x_0. \qquad (8.3.21)$$

In the present example, $h(t, t', x(t')) = -\frac{\varepsilon}{\omega_0} \sin \omega_0 (t - t') x^2(t')$. Obviously,

$$\left| h(t, t', x(t')) \right| \leq \frac{\varepsilon}{\omega_0} x^2(t'). \qquad (8.3.22)$$

Compared to (8.3.6), it is obtained that $M = 4\varepsilon x_0^2/\omega_0$. On the other hand,

$$\left| h(t, t', x(t')) \right| \leq (\varepsilon/\omega_0) \left[x(t') - x_0 \cos \omega_0 t' + x_0 \cos \omega_0 t' \right]^2$$
$$\leq (\varepsilon/\omega_0) \left[|x(t') - x_0 \cos \omega_0 t'| + x_0 |\cos \omega_0 t'| \right]^2.$$

For any t that satisfies (8.3.21), it must be that $|h(t, t', x(t'))| < 4\varepsilon x_0^2/\omega_0$. Then (8.3.22) leads to

$$\left| h(t, t', x(t')) - h(t, t', \xi(t')) \right| \leq (\varepsilon/\omega_0)|x(t') + \xi(t')| \cdot |x(t') - \xi(t')|.$$

This equation is reformed to the following form:

$$\left| h(t, t', x(t')) - h(t, t', \xi(t')) \right|$$
$$\leq (\varepsilon/\omega_0) \left| x(t') - x_0 \cos \omega_0 t' + \xi(t') - x_0 \cos \omega_0 t' + 2x_0 \cos \omega_0 t' \right|$$
$$\cdot \left| x(t') - \xi(t') \right|.$$

For all t and t' that satisfy $|x(t') - x_0 \cos \omega_0 t'| \leq x_0$ and $|\xi(t') - x_0 \cos \omega_0 t'| \leq x_0$, we have

$$\left| h(t, t', x(t')) - h(t, t', \xi(t')) \right| \leq (4\varepsilon x_0/\omega_0) \left| x(t') - \xi(t') \right|.$$

By comparing Lipschitz condition (8.3.15), $N = 4\varepsilon x_0/\omega_0$. Therefore, we have obtained the three control parameters for the iteration to converge:

$$\Delta = x_0, \quad M = 4\varepsilon x_0^2/\omega_0, \quad N = 4\varepsilon x_0/\omega_0.$$

The iteration can be carried out step by step. Let $x_0(t) = x_0 \cos \omega_0 t$. We have

$$x_1(t) = x_0 \cos \omega_0 t - \frac{\varepsilon}{\omega_0} \int_0^t \sin \omega_0(t - t') x_0^2(t') \mathrm{d}t'. \qquad (8.3.23a)$$

It follows that

$$x_1(t) = x_0 \left[\cos \omega_0 t + \delta \left(\frac{1}{6} \cos 2\omega_0 t + \frac{1}{3} \cos \omega_0 t - \frac{1}{2} \right) \right] \qquad (8.3.23b)$$

and

$$x_2(t) = x_0 \left[\cos \omega_0 t + \delta \left(\frac{1}{6} \cos 2\omega_0 t + \frac{1}{3} \cos \omega_0 t - \frac{1}{2} \right) \right]$$

$$+ x_0 \left[\delta^2 A + \delta^3 B + \omega_0 t \left(\frac{5}{12} \delta^2 \sin \omega_0 t + \frac{5}{36} \delta^3 \sin \omega_0 t \right) \right],$$

$$(8.3.23c)$$

with

$$A = \frac{1}{48} \cos 3\omega_0 t + \frac{1}{9} \cos 2\omega_0 t + \frac{29}{144} \cos \omega_0 t - \frac{1}{3}$$

and

$$B = \frac{1}{1080} \cos 4\omega_0 t + \frac{1}{144} \cos 3\omega_0 t - \frac{1}{27} \cos 2\omega_0 t$$

$$+ \frac{753}{2160} \cos \omega_0 t - \frac{23}{72},$$

and so on.

8.4.   Fredholm Linear Equations with Degenerated Kernels

In this section, we investigate the resolution of the second kind of Fredholm integral equations with a degenerate kernel.

8.4.1.   *Separable kernels*

Separable kernels have been defined by (8.1.15):

$$k(x, y) = \lambda \phi(x) \psi^*(y). \qquad (8.4.1)$$

Here both ϕ and ψ are functions in a Hilbert space, and the $k(x, y)$ is an L_2 kernel. In this section, we will not write the weight function explicitly. The L_2 kernel means, by (8.2.37), that

$$\int |k_s(x,y)|^2 \, \mathrm{d}x\mathrm{d}y < 1 \qquad (8.4.2)$$

Thus, the condition (8.2.29) $\|K\| \leq 1$ is satisfied, and the Neumann series is convergent.

1. Solutions of inhomogeneous equations

For equation

$$f(x) = g(x) + \int k(x,y)f(y)\mathrm{d}y$$

$$= g(x) + \lambda \int \phi(x)\psi^*(y)f(y)\mathrm{d}y, \qquad (8.4.3)$$

its iteration series is

$$f(x) = g(x) + \int \mathrm{d}y_1 k(x,y_1)g(y_1)$$

$$+ \int \mathrm{d}y_1 \int \mathrm{d}y_2 k(x,y_1)k(y_1,y_2)g(y_2) + \cdots .$$

It follows from (8.4.2) that $k(x,y)$ as a function of two variables is square integrable. So, the two functions ϕ and ψ are also square integrable. Now the kernel is separated. Denoting $(\psi,\phi) = \int \mathrm{d}y\psi^*(y)\phi(y)$, we can write

$$f(x) = g(x) + \lambda\phi(x)(\psi,g) + \lambda^2\phi(x)(\psi,\phi)(\psi,g)$$

$$+ \lambda^3\phi(x)(\psi,\phi)^2(\psi,g) + \cdots$$

$$= g(x) + \lambda\phi(x)(\psi,g) \sum_{n=0}^{\infty} \lambda^n(\psi,\phi)^n. \qquad (8.4.4)$$

If

$$|\lambda(\psi,\phi)| < 1, \qquad (8.4.5)$$

the series in (8.4.1) is convergent.

As a matter of fact, (8.4.5) has already been met due to (8.4.2). Equation (8.4.2) indicates that $|\lambda|^2 \int |\phi(x)|^2\mathrm{d}x \int |\psi(y)|^2 \, \mathrm{d}y < 1$. By

Schwartz inequality, $|(\psi, \phi)| \leq [(\phi, \phi)(\psi, \psi)]^{1/2}$, we have

$$|\lambda(\psi, \phi)| \leq [|\lambda|^2 (\phi, \phi)(\psi, \psi)]^{1/2}$$

$$= [|\lambda|^2 \int |\phi(x)|^2 dx \int |\psi(y)|^2 dy]^{1/2} < 1.$$

Since the convergence condition (8.4.5) has been satisfied, the summation in (8.4.4) can be carried out, and the result is

$$f(x) = g(x) + \lambda \frac{(\psi, g)}{1 - \lambda(\psi, \phi)} \phi(x). \tag{8.4.6}$$

This is a solution in a closed form. It is obtained under the condition (8.4.5). This condition can actually be relaxed. Now we rewrite the integral in the form of an inner product:

$$f(x) = g(x) + \lambda \int \phi(x)\psi^*(y)f(y)dy$$

$$= g(x) + \lambda\phi(x)(\psi, f). \tag{8.4.7}$$

Then use ψ to make an inner product on both sides:

$$(\psi, f) = (\psi, g) + \lambda(\psi, \phi)(\psi, f). \tag{8.4.8}$$

When

$$\lambda(\psi, \phi) \neq 1, \tag{8.4.9}$$

we obtain

$$(\psi, f) = \frac{(\psi, g)}{1 - \lambda(\psi, \phi)}. \tag{8.4.10}$$

This is substituted back into (8.4.7) to get (8.4.6). Therefore, the only condition for (8.4.6) is (8.4.9).

When $\lambda(\psi, \phi) = 1$, it is still possible that there is a solution. From (8.4.6) it is known that if

$$(\psi, g) = 0. \tag{8.4.11}$$

the right hand side of (8.4.6) is a zero to zero type, which can be a finite number. The solution is

$$f(x) = g(x) + B\phi(x), \tag{8.4.12}$$

where B is any constant. It is easily confirmed that under the condition (8.4.11), (8.4.12) satisfies (8.4.8).

2. Solutions of homogeneous equations

Equation (8.4.6) does not apply to homogeneous equations. This is because if we let

$$g(x) = 0 \qquad (8.4.13)$$

in (8.4.6), then it will be identical to zero. However, it is possible for a homogeneous equation to have nonzero solutions. Under the condition of (8.4.13), (8.4.8) becomes

$$(\psi, f) = \lambda(\psi, \phi)(\psi, f). \qquad (8.4.14)$$

In this case, as

$$\lambda(\psi, \phi) = 1, \qquad (8.4.15)$$

(ψ, f) can be an arbitrary constant, i.e., it can have nonzero solutions. Then the solutions of (8.4.7) are

$$f(x) = A\phi(x), \qquad (8.4.16)$$

where A is any constant.

Thus, the interesting fact is that (8.4.15) that makes homogeneous equations have nonzero solutions is just the condition that makes inhomogeneous equations have no nonzero solutions. This fact implies that the divergence of the Neumann series is closely related to the solutions of homogeneous equations.

Both (8.4.12) and (8.4.16) are obtained under the condition $\lambda(\psi, \phi) = 1$, and the latter is a special case of the former at $g(x) = 0$. As $\lambda(\psi, \phi) \neq 1$ and $g(x) = 0$, (8.4.6) and (8.4.14) tell that there is no nonzero solution for inhomogeneous and homogeneous equations, respectively.

In summary, the separable kernel always lead to the solutions with simple closed forms, such as (8.4.6), (8.4.12) and (8.4.16). We distinguish them as five cases, which are listed in Table 8.1.

3. Scattering of a microscopic particle by a nonlocal potential

Now we apply the above conclusions to investigate Schrödinger equation in a nonlocal potential. If a particle moves in a local potential, Schrödinger equation is in the form of (8.2.38a). For many-body

Table 8.1. When the kernel is separable, $k(x,y) = \lambda\phi(x)\psi^*(y)$, the conditions for the second kind of Fredholm integral equations to have solutions and the expressions of the solutions. A and B are constants.

Equations	Cases	Conditions	Expressions of solution $f(x)$
Inhomogeneous equations	I	$\lambda(\psi,\phi) \neq 1$	$g(x) + \lambda\dfrac{(\psi,g)\phi(x)}{1-\lambda(\psi,\phi)}$
	II	$\lambda(\psi,\phi) = 1, (\psi,g) \neq 0$	No solution
	III	$\lambda(\psi,\phi) = 1, (\psi,g) = 0$	$g(x) + B\phi(x)$
Homogeneous equations	IV	$\lambda(\psi,\phi) \neq 1$	0
	V	$\lambda(\psi,\phi) = 1$	$A\phi(x)$

problems, particles usually are in a nonlocal potential, and the corresponding Schrödinger equation is of the form of

$$-\frac{\hbar^2}{2m}\nabla^2\psi + \int U(\boldsymbol{r},\boldsymbol{r}')\psi(\boldsymbol{r}')\mathrm{d}\boldsymbol{r}' = E\psi(\boldsymbol{r}). \qquad (8.4.17)$$

Imitating the procedure (8.2.38a)–(8.2.38e), we are able to write the formal solution of (8.4.17) as

$$\psi(\boldsymbol{r}) = \frac{1}{(2\pi)^{3/2}}e^{i\boldsymbol{k}\cdot\boldsymbol{r}}$$
$$-\frac{m}{2\pi\hbar^2}\iint\frac{e^{iq|\boldsymbol{r}-\boldsymbol{r}'|}}{|\boldsymbol{r}-\boldsymbol{r}'|}U(\boldsymbol{r}',\boldsymbol{r}'')\psi(\boldsymbol{r}'')\mathrm{d}\boldsymbol{r}'\mathrm{d}\boldsymbol{r}'', \qquad (8.4.18)$$

Please note that here the wave vectors $\boldsymbol{k}$ and $\boldsymbol{q}$ are respectively obtained from the left hand sides of Eqs. (8.2.38c) and (8.2.38d). Hence, they are the same in magnitude, $k = q$, but may not be the same in direction.

$$|\boldsymbol{k}| = |\boldsymbol{q}| = \left(\frac{2mE}{\hbar^2}\right)^{1/2}. \qquad (8.4.19)$$

Equation (8.4.18) is an integral equation. As a comparison, (8.4.17) is an integral differential equation. These two equations are equivalent to each other. Equation (8.4.18) is substantially the same as (8.2.41), both being Lippmann-Schwinger equation of scattering problem.

We take the Fourier transformation of (8.4.18), as well we all the three factors in the integrand.

The Fourier component of the nonlocal potential is

$$U(\boldsymbol{p}', \boldsymbol{p}'') = \int \int e^{-i\boldsymbol{p}'\cdot\boldsymbol{r}'} e^{-i\boldsymbol{p}''\cdot\boldsymbol{r}''} U(\boldsymbol{r}', \boldsymbol{r}'') d\boldsymbol{r}' d\boldsymbol{r}''. \qquad (8.4.20)$$

Then,

$$\int U(\boldsymbol{r}', \boldsymbol{r}'')\psi(\boldsymbol{r}'')d\boldsymbol{r}''$$

$$= \int \frac{1}{(2\pi)^6} \int \int e^{i\boldsymbol{p}'\cdot\boldsymbol{r}'} e^{i\boldsymbol{p}''\cdot\boldsymbol{r}''} U(\boldsymbol{p}', \boldsymbol{p}'')d\boldsymbol{p}'d\boldsymbol{p}''\psi(\boldsymbol{r}'')d\boldsymbol{r}''$$

$$= \frac{1}{(2\pi)^6} \int \int e^{i\boldsymbol{p}'\cdot\boldsymbol{r}'} U(\boldsymbol{p}', \boldsymbol{p}'')d\boldsymbol{p}'d\boldsymbol{p}''\psi(-\boldsymbol{p}''). \qquad (8.4.21)$$

The Fourier component of Green's function is (6.2.31). Thus, the second term of (8.4.18) becomes

$$-\frac{m}{2\pi\hbar^2} \int \int \frac{e^{iq|\boldsymbol{r}-\boldsymbol{r}'|}}{|\boldsymbol{r}-\boldsymbol{r}'|} U(\boldsymbol{r}', \boldsymbol{r}'')\psi(\boldsymbol{r}'')d\boldsymbol{r}'d\boldsymbol{r}''$$

$$= -\frac{2m}{\hbar^2} \frac{1}{(2\pi)^3} \int \int \frac{e^{i\boldsymbol{q}_1\cdot(\boldsymbol{r}-\boldsymbol{r}')}d\boldsymbol{q}_1}{q_1^2 - q^2 + i\varepsilon}$$

$$\times \frac{1}{(2\pi)^6} \int \int e^{i\boldsymbol{p}'\cdot\boldsymbol{r}'} U(\boldsymbol{p}', \boldsymbol{p}'')d\boldsymbol{p}'d\boldsymbol{p}''\psi(-\boldsymbol{p}'')d\boldsymbol{r}'$$

$$= -\frac{2m}{\hbar^2} \frac{1}{(2\pi)^6} \int \int \frac{e^{i\boldsymbol{q}_1\cdot\boldsymbol{r}}d\boldsymbol{q}_1}{q_1^2 - q^2 + i\varepsilon} U(\boldsymbol{q}_1, -\boldsymbol{p}'')d\boldsymbol{p}''\psi(\boldsymbol{p}'').$$

The integration with respect to coordinates are transformed to that with respect to momentum. So, the Fourier transformation of (8.4.18) is

$$\psi(\boldsymbol{p}) = \frac{\delta(\boldsymbol{k} - \boldsymbol{p})}{(2\pi)^{3/2}} + \frac{2m}{\hbar^2} \frac{1}{(2\pi)^3} \int \frac{U(\boldsymbol{p}, -\boldsymbol{p}_1)}{p^2 - q^2 + i\varepsilon}\psi(\boldsymbol{p}_1)d\boldsymbol{p}_1. \qquad (8.4.22)$$

This is the Lippmann-Schwinger equation in momentum space.

The physical process we are considering is as follows. A particle with wave vector $\boldsymbol{k}$ is incident from infinity to the vicinity of the potential to be scattered, and then goes to infinity. The incident free particle is a plane wave, which is the first term of (8.4.18). The motion toward infinity should behave as a spherical wave. This means that at a far observation point in the p direction, the scattering wave, i.e., the second term of (8.4.18) should be of the form of $\frac{e^{ipr}}{r}$. The total wave function is the superposition of the incident and scattering waves. Hence, (8.4.18) is simplified to be the form of

$$\psi(r) = \frac{1}{(2\pi)^{3/2}}e^{ik\cdot r} + f(k,q)\frac{e^{iqr}}{r}. \qquad (8.4.23)$$

The factor $f(k,p)$ is called scattering amplitude, which depends on the wavevector of the incident and scattering waves. Its square is the scattering probability, an experimentally measurable quantity. We hope that the scattering amplitude can be evaluated. Note that the second term of (8.4.23) has just the form of (6.5.16), which shows that they have the same mathematical origin.

We have written (8.4.23) purely based on physical consideration. It is in fact rigorously derived from (8.4.18). Suppose that the potential takes effect in a finite domain. Then at places far away, the potential goes to zero. In (8.4.18), at the points where $r \leq r'$, we take the following approximation: $|r - r'| \approx 1 - n \cdot r'$, where n is the unit vector in the direction of r. After these approximations, (8.4.18) is simplified to be

$$\psi(r) = \frac{e^{ik\cdot r}}{(2\pi)^{3/2}} - \frac{m}{2\pi\hbar^2}\frac{e^{iqr}}{r}\iint e^{-iq\cdot r'}U(r',r'')\psi(r'')\mathrm{d}r'\mathrm{d}r'',$$

$$(8.4.24)$$

where in the exponential, $qn = q$. This is the scattering wave vector along the direction of r, and it depends on the observing point. After defining

$$f(k,q) = -\frac{m}{2\pi\hbar^2}\iint e^{-iq\cdot r'}U(r',r'')\psi(r'')\mathrm{d}r'\mathrm{d}r'', \qquad (8.4.25)$$

(8.4.24) becomes the form of (8.4.23). Equation (8.4.25) demonstrates that the scattering amplitude $f(k,q)$ is just the Fourier transformation of $-\frac{m}{2\pi\hbar^2}\int U(r',r'')\psi(r'')\mathrm{d}r''$.

In order to evaluate the scattering amplitude $f(\boldsymbol{k}, \boldsymbol{q})$, we have to establish an equation it satisfies. Using the inverse Fourier transformation of (8.4.20), (8.4.25) becomes

$$f(\boldsymbol{k}, \boldsymbol{q}) = -\frac{m}{2\pi\hbar^2} \int \frac{1}{(2\pi)^6} \int$$

$$\times \int e^{-i\boldsymbol{q}\cdot\boldsymbol{r}'} e^{i\boldsymbol{p}'\cdot\boldsymbol{r}'} U(\boldsymbol{p}', -\boldsymbol{p}'') d\boldsymbol{p}' d\boldsymbol{p}'' \psi(\boldsymbol{p}'') d\boldsymbol{r}'$$

$$= -\frac{m}{2\pi\hbar^2} \frac{1}{(2\pi)^3} \int U(\boldsymbol{q}, -\boldsymbol{p}'') \psi(\boldsymbol{p}'') d\boldsymbol{p}''. \qquad (8.4.26)$$

Consequently, (8.4.22) is simplified to be

$$\psi(\boldsymbol{p}) = \frac{\delta(\boldsymbol{k} - \boldsymbol{p})}{(2\pi)^{3/2}} + 4\pi \frac{f(\boldsymbol{k}, \boldsymbol{p})}{p^2 - k^2 + i\varepsilon}, \qquad (8.4.27)$$

where q is replaced by k, since we have mentioned that $k = q$. Multiplying on both sides by $-\frac{m}{2\pi\hbar^2} \frac{1}{(2\pi)^3} U(\boldsymbol{k}, \boldsymbol{p})$ and integrating with respect to $\boldsymbol{p}$, we obtain the equation that $f(\boldsymbol{k}, \boldsymbol{q})$ should satisfy:

$$f(\boldsymbol{k}, \boldsymbol{q}) = -\frac{m}{2\pi\hbar^2} \frac{U(\boldsymbol{q}, -\boldsymbol{k})}{(2\pi)^{3/2}}$$

$$-\frac{2m}{\hbar^2} \frac{1}{(2\pi)^3} \int d\boldsymbol{p} \frac{U(\boldsymbol{q}, -\boldsymbol{p})}{p^2 - k^2 + i\varepsilon} f(\boldsymbol{k}, \boldsymbol{p}). \qquad (8.4.28)$$

For a general nonlocal potential, it is not easy to solve (8.4.28). Here we merely consider a simple case that the nonlocal potential is separable:

$$U(\boldsymbol{r}, \boldsymbol{r}') = -g^2 v(\boldsymbol{r}) v(\boldsymbol{r}'). \qquad (8.4.29)$$

Subsequently, its Fourier component is also separable:

$$U(\boldsymbol{k}, \boldsymbol{p}) = -g^2 v(\boldsymbol{k}) v(\boldsymbol{p}). \qquad (8.4.30)$$

Then (8.4.28) becomes

$$f(\boldsymbol{k}, \boldsymbol{q}) = \frac{m}{2\pi\hbar^2} \frac{v(\boldsymbol{q}) v(-\boldsymbol{k})}{(2\pi)^{3/2}}$$

$$+ \frac{2m}{\hbar^2} \frac{1}{(2\pi)^3} \int v(\boldsymbol{q}) \frac{v(-\boldsymbol{p})}{p^2 - q^2 + i\varepsilon} f(\boldsymbol{k}, \boldsymbol{p}) d\boldsymbol{p}.$$

$$(8.4.31)$$

It is seen that when the potential is separable as in (8.4.29), the integral equation (8.4.18) does not have a separable kernel, but its Fourier transformation (8.4.31) has.

We make the following correspondence:

$$x \to \boldsymbol{q}, \quad y \to \boldsymbol{p}, \qquad (8.4.32\text{a})$$

$$f(x) \to f(\boldsymbol{k}, \boldsymbol{q}), \quad g(x) \to \frac{mg^2}{2\pi\hbar^2} \frac{v(\boldsymbol{q})v(-\boldsymbol{k})}{(2\pi)^{3/2}} \qquad (8.4.32\text{b})$$

and

$$\phi(x) \to v(\boldsymbol{q}), \quad \psi \to \frac{v(-\boldsymbol{p})}{p^2 - k^2 + \mathrm{i}\varepsilon}. \qquad (8.4.32\text{c})$$

Let $\tau = \frac{4\pi mg^2}{\hbar^2}$ and $\lambda = \frac{2mg^2}{\hbar^2} = \frac{\tau}{2\pi}$. Then, (8.4.31) becomes the form of (8.4.3) or (8.4.7). Please note that in the inner product, the weight is $1/(2\pi)^3$. Following (8.4.6), we easily put down the expression of the solution:

$$
\begin{aligned}
f(\boldsymbol{q}) &= \frac{\tau}{8\pi^2(2\pi)^{3/2}} v(\boldsymbol{q})v(-\boldsymbol{k}) \left(1 + \frac{\lambda I}{1 - \lambda I}\right) \\
&= \frac{\tau}{8\pi^2(2\pi)^{3/2}} \frac{v(\boldsymbol{q})v(-\boldsymbol{k})}{1 - \lambda I}
\end{aligned}
\qquad (8.4.33)
$$

where for the sake of convenience, we have defined

$$I = \frac{1}{(2\pi)^3} \int \frac{v(\boldsymbol{p})v(-\boldsymbol{p})\mathrm{d}\boldsymbol{p}}{p^2 - (k - \mathrm{i}\varepsilon)^2}. \qquad (8.4.34)$$

In the form of (8.4.33), the scattering amplitude can be evaluated easily.

Example 1. Suppose that the potential is of the form of

$$v(\boldsymbol{r}) = \mathrm{e}^{-\mu r}/r. \qquad (8.4.35)$$

The examples of such kind of potentials are Yukawa potential in nuclear physics and screen potential of electrons in metals. The Fourier transformation of this potential has been evaluated. Equation (8.4.35) is of the same form of (6.2.11a), Green's function in three-dimensional space, and its Fourier component is (6.2.31). Let

$\lambda = -\mu^2$ in (6.2.11a). Then the Fourier component of (8.4.35) is immediately obtained:

$$v(\boldsymbol{p}) = \frac{4\pi}{p^2 + \mu^2}. \tag{8.4.36}$$

It depends on the dimension of wavevector but not its direction, which indicates that the calculated scattering amplitude does either. Using this potential, (8.4.34) becomes

$$I = \frac{1}{(2\pi)^3} \int_0^\infty \frac{4\pi p^2 dp}{p^2 - (k - i\varepsilon)^2} \left(\frac{4\pi}{p^2 + \mu^2}\right)^2.$$

It is calculated by supplementing semicircle at the upper half plane. The result is

$$I = -\frac{2\pi}{\mu(k + i\mu)^2}.$$

This is in turn substituted into (8.4.33) to get

$$f(\boldsymbol{k}) = \frac{1}{(2\pi)^{3/2}} \frac{2\tau(k + i\mu)}{(k + i\mu)^2 + 2\pi\lambda/\mu} \frac{1}{k - i\mu} \frac{1}{(q^2 + \mu^2)^2}.$$

Please note that here $\boldsymbol{k}$ and $\boldsymbol{q}$ are respectively the wave vectors of the incident and scattering waves. This result shows that the scattering amplitude merely relies on the magnitudes of the wave vectors, independent of their directions. For elastic scattering, they are the same. Then the final result is

$$f(\boldsymbol{k}, \boldsymbol{q}) = f(E) = \frac{1}{(2\pi)^{3/2}} \frac{2\tau}{(k^2 + \mu^2)^2 + \tau(k - i\mu)^2/\mu}.$$

In this case, $f(\boldsymbol{k}, \boldsymbol{p})$ depends on the dimension of the wave vector of the incident or scattering wave, so that in the final expression, it is written in the form depending on energy $f(E)$.

8.4.2. *Kernels with a finite rank*

1. Linear equations

Finite rank kernels have been defined by (8.1.14). The action of an operator K_F with a finite rank kernel on a vector f in a Banach space V is of the following form:

$$K_F f \equiv \sum_{n=1}^{N} \phi_n(\psi_n, f). \qquad (8.4.37)$$

That is to say, the operator projects the function f onto finite number of basis functions. Due to this fact, the action of the operator can be converted to the computation of a matrix with finite ranks.

In principle, the (ψ_n, f) in (8.4.37) may be a general linear transformation, not limited to inner product. Nevertheless, inner products are often encountered. If (ψ_n, f) is just an inner product, (8.4.37) becomes explicitly

$$K_F f = \int dx \sum_{n=1}^{N} \phi_n \psi_n(x) f(x). \qquad (8.4.38)$$

The vector in the dual space of V is denoted by $\tilde{f}$. The adjoint of K_F is denoted by K_F^+. K_F^+ acts on the vector $\tilde{f}$, and the effect is

$$K_F^+ \tilde{f} \equiv \sum_{n=1}^{N} \psi_n(\phi_n, \tilde{f}). \qquad (8.4.39)$$

Suppose that there are linear equations

$$f - \lambda K_F f = g, \qquad (8.4.40)$$

where f and g are vectors in space V.

Definition 1. Linear equation

$$\tilde{h} - \lambda^* K_F^+ \tilde{h} = 0 \qquad (8.4.41)$$

is called conjugate homogeneous equations or adjoint homogeneous equation of (8.4.40), where $\tilde{h}$ is the vector in the dual space of V. The explicit forms of these two equations follow (8.4.37) and (8.4.39):

$$f = g + \lambda \sum_{n=1}^{N} \phi_n(\psi_n, f) \qquad (8.4.42)$$

and

$$\tilde{h} - \lambda^* \sum_{n=1}^{N} \psi_n(\phi_n, \tilde{h}) = 0. \qquad (8.4.43)$$

The following process is analogous to the case of separable kernels. Imitating (8.4.8), we make the inner product of (8.4.42) by ψ_m:

$$(\psi_m, f) = (\psi_m, g) + \lambda \sum_{n=1}^{N} (\psi_m, \phi_n)(\psi_n, f). \tag{8.4.44}$$

Let

$$\alpha_{mn} \equiv (\psi_m, \phi_n), \quad a_n \equiv (\psi_n, f), \quad b_m \equiv (\psi_m, g). \tag{8.4.45}$$

Equation (8.4.44) becomes

$$a_m - \lambda \sum_{n=1}^{N} \alpha_{mn} a_n = b_m. \tag{8.4.46}$$

Similarly, making the inner product of (8.4.43) ϕ_m leads to

$$\tilde{a}_m - \lambda^* \sum_{n=1}^{N} \alpha_{nm}^* \tilde{a}_n = 0, \tag{8.4.47}$$

where

$$\tilde{a}_n \equiv (\phi_n, \tilde{h}). \tag{8.4.48}$$

In the form of matrices, (8.4.46) and (8.4.47) are

$$(I - \lambda M) a = b \tag{8.4.49}$$

and

$$(I - \lambda M)^+ \tilde{a} = 0, \tag{8.4.50}$$

respectively. Thus, Eqs. (8.4.40) and (8.4.41) are transferred to linear algebraic equations (8.4.49) and (8.4.50).

Since (8.4.49) and (8.4.50) have the same forms as (2.2.35) and (2.2.37), the alternative theorem for the solutions of the former can be applied. In the present case, if λ happens to be an eigenvalue of the matrix M, then the condition for (8.4.49) to have solutions is that

$$(b, \tilde{a}) = \sum_{n=1}^{N} b_n^* \tilde{a}_n = 0. \tag{8.4.51}$$

The inhomogeneous term in (8.4.49) should be orthogonal to the solutions $\tilde{a}$ of the homogeneous equations. Accordingly, (8.4.40) has solutions under this condition.

The condition (8.4.51) means that

$$\sum_{n=1}^{N} (g, \psi_n)(\phi_n, \tilde{h}) = \left(g, \sum_{n=1}^{N} \psi_n(\phi_n, \tilde{h}) \right) = 0, \qquad (8.4.52)$$

which, by definition (8.4.39), is

$$(g, K_F^+ \tilde{h}) = 0. \qquad (8.4.53)$$

If $\tilde{h} \neq 0$, then it follows from (8.4.41) that $\lambda^* K_F^+ \tilde{h} = \tilde{h}$. For all $\tilde{h}$ satisfying $\tilde{h} - \lambda^* K_F^+ \tilde{h} = 0$, the equations (8.4.40) have solution as long as $(g, \tilde{h}) = 0$.

Please note that if $(I - \lambda M)^+ \tilde{a} = 0$ has no solutions, which means that $(I - \lambda M)^{-1}$ is nonzero, then equation $(I - \lambda A)a = b$ has a unique solution. Accordingly, if $\tilde{h} - \lambda^* K_F^+ \tilde{h} = 0$ has no solution, then the equation $f - \lambda K_F f = g$ has a unique solution.

These discussions are summarized into the following theorem.

Theorem 1. *If K_F is a finite rank transformation and $\tilde{h}_i$ satisfies adjoint homogeneous equation $\lambda^* K_F^+ \tilde{h}_i = \tilde{h}_i$, equation $f = g + \lambda K_F f$ has solutions when and only when $(g, \tilde{h}_i) = 0$ for all i. If the adjoint homogeneous equation does not have a solution, then the inhomogeneous equation has a unique solution. The similar conclusions apply to operator K_F^+.*

Corollary 1. *If K_F is a finite rank operator, then the equation $\tilde{f} = \tilde{g} + \lambda^* K_F^+ \tilde{f}$ has solutions when and only when $(\tilde{g}, h_i) = 0$ for all the solutions h_i of the homogeneous equation $\lambda K_F h_i = h_i$, and the adjoint inhomogeneous equation has a unique solution.*

Corollary 2. *Equations $\lambda K_F h_i = h_i$ and $\lambda^* K_F^+ \tilde{h}_i = \tilde{h}_i$ have the same number of linearly independent solutions.*

Theorem 1 and its two corollaries are called the **Fredholm alternative theorem**. "Alternative" means that one can have one of two possibilities: either inhomogeneous equation $(I - \lambda K)f = g$ has a unique solution, or adjoint homogeneous equation has at least one solution. By the way, this theorem is for the case of finite rank kernels. For general kernels, Fredholm theorem has a more general conclusion. The statement is as follows.

Suppose that λ_0 is an eigenvalue of a kernel $k(x, y)$. The sufficient and necessary conditions for the inhomogeneous equation

$$\lambda_0 \int_a^b k(x, y) f(y) \rho(y) dy + g(x) = f(x)$$

to have solutions are that the free term $g(x)$ is orthogonal to all the eigenfunctions of the adjoint homogeneous equation

$$\lambda_0^* \int_a^b k^*(y, x) f(y) \rho(y) dy = f(x).$$

This theorem does not require that the kernels are degenerate.

In the following, the procedure of resolution in the case of finite rank kernels is presented.

2. Resolvent operator

If $f - \lambda K_F f = g$ has only one solution, then by $(I - \lambda M)a = b$ we have $a = (I - \lambda M)^{-1}b$ or

$$a_n \equiv (\psi_n, f) \equiv \sum_{m=1}^{N} [(I - \lambda M)^{-1}]_{nm} b_m$$

$$= \sum_{m=1}^{N} [(I - \lambda M)^{-1}]_{nm} (\psi_m, g).$$

This is substituted into (8.4.42) to get

$$f = g + \lambda \sum_{n=1}^{N} \sum_{m=1}^{N} \phi_n [(I - \lambda M)^{-1}]_{nm} (\psi_m, g). \tag{8.4.54}$$

This is the required solution when $\lambda^* K_F^+ \tilde{h} = \tilde{h}$ has no solution, or equivalently, when $(I - \lambda M)^+ \tilde{a} = 0$ has no solution.

We define a new operator R_{K_F}:

$$R_{K_F} f \equiv \sum_{n=1}^{N} \sum_{m=1}^{N} [(I - \lambda M)^{-1}]_{mn} \phi_m (\psi_n, f). \tag{8.4.55}$$

Thus, (8.4.54) can be written in a concise form:

$$f = g + \lambda R_{K_F} g = (I + \lambda R_{K_F})g. \tag{8.4.56}$$

The operator R_{K_F} is called the **resolvent operator** corresponding to K_F. By the definition (8.4.55), as long as $(I - \lambda M)^{-1}$ exists, the resolvent operator does so. The determinant $\det(I - \lambda A)$ is an N-order polynomial of λ, so that has N roots among which some may be degenerate. R_{K_F} is analytical in the whole complex λ plane except the finite number of the roots of $\det(I - \lambda A) = 0$. Therefore, R_{K_F} is a **meromorphic function**. The definition of a meromorphic function is that it is analytic everywhere except finite number of poles. At the poles of R_{K_F}, equation $\lambda K_F h = h$ has solutions, or equivalently, $\lambda^* K_F^+ \tilde{h} = \tilde{h}$ has solutions.

3. Solving procedure

Now we are at the stage to put down the solving procedure of (8.4.40) with a finite rank kernel. For the sake of clarity, some equations above are repeated in the following.

Inhomogeneous equation (8.4.42) and its adjoint homogeneous equation (8.4.43) are as follows.

$$f = g + \lambda \sum_{n=1}^{N} \phi_n (\psi_n, f). \tag{8.4.57}$$

$$\tilde{h} - \lambda^* \sum_{n=1}^{N} \psi_n (\phi_n, \tilde{h}) = 0. \tag{8.4.58}$$

Denote that

$$\alpha_{mn} \equiv (\psi_m, \phi_n), \quad a_n \equiv (\psi_n, f),$$
$$b_m \equiv (\psi_m, g), \quad \tilde{a}_n \equiv (\phi_n, \tilde{h}). \tag{8.4.59}$$

The α_{mn} and b_m can be calculated from given functions, in which a_n and $\tilde{a}_n$ involve the unknown function f and are to be determined. In fact, $a_n \equiv (\psi_n, f)$ and $\tilde{a}_n \equiv (\phi_n, \tilde{h})$ can be solved from equations (8.4.49) and (8.4.50):

$$(I - \lambda M)a = b \tag{8.4.60}$$

and

$$(I - \lambda M)^+ \tilde{a} = 0, \tag{8.4.61}$$

where $(M)_{mn} = \alpha_{mn}$. In the real number space, (8.4.61) is simplified to be

$$(I - \lambda M)a = 0. \tag{8.4.62}$$

(1) Compute α_{mn} and b_m from (8.4.47) and (8.4.48), and calculate the eigenvalues λ_i, $(i = 1, 2, \ldots, N_\beta \leq N)$ of $\det(I - \lambda M) = 0$, some of which may be degenerate.

It should be noted that the eigenvalues here are from equation $\det(I - \lambda M) = 0$. This should be distinguished from those of linear algebraic equations where the eigenvalues of a matrix M are computed from $\det(\lambda I - M) = 0$.

(2) For $\lambda \neq \lambda_i$, $(i = 1, 2, \ldots, N_\beta)$, the inhomogeneous equation (8.4.60) has only one solution, but its adjoint homogeneous equation (8.4.61) does not have nonzero solutions.

When the calculated a is substituted into (8.4.57), one can solve the solution f. This case corresponds to Case I in Table 8.1. For homogeneous equations $g = 0$, the solution has to be zero, which correspond to Case IV in Table 8.1.

If the iteration technique introduced in Subsection 8.2.1 is used, the upper limit of $|\lambda|$ should be evaluated. There are two ways to do so.

One is to use conditions (8.2.18) and (8.2.24):

$$\max_{x,y \in [a,b]} |k(x, y)| \rho(y) = M \qquad (8.4.63a)$$

and

$$|\lambda| < 1/M(b - a). \qquad (8.4.63b)$$

The other is the condition (8.2.37):

$$|\lambda|^2 \int \rho(x)\mathrm{d}x \int \rho(y)\mathrm{d}y\, |k(x, y)|^2 < 1. \qquad (8.4.64)$$

One of (8.4.63) and (8.4.64) being satisfied, the Neumann series method introduced in Subsection 8.2.1 can be applied. Usually, the upper limit of $|\lambda|$ evaluated by (8.4.64) is higher than that by (8.4.63).

The iteration method cannot be applied to two cases: one is the homogeneous equations with $g = 0$ and the other is the inhomogeneous equations with $g \neq 0$ but λ being just eigenvalues.

(3) The case that λ is one of the eigenvalues.

Suppose that the ith eigenvalue λ_i is k_i-degenerate so that has k_i linearly independent eigenvectors, denoted by $\phi_{i,j}$, $(j = 1, 2, \ldots, k_i)$. For each λ_i, the number of $\tilde{a}$ solved by (8.4.61) is k_i. Each $\tilde{a}$ is substituted into (8.4.58) to find $\tilde{h}$ which in fact corresponds to the eigenvector $\phi_{i,j}$ of λ_i.

In the case of inhomogeneous equation $g = 0$, the transpose and complex conjugate of $\phi_{i,j}$ is the solution of the homogeneous equation $f - \lambda K_F f = 0$. This corresponds to Case V in Table 8.1.

In the case of inhomogeneous equation $g \neq 0$, the condition $(g, \phi_{i,j}) = 0$ in Fredholm alternative theorem is employed. For each λ_i, we calculate $(g, \phi_{i,j})$, $(j = 1, 2, \ldots, k_i)$. Two cases are distinguished as follows.

One is that there is at least one of $\phi_{i,j}$ such that $(g, \phi_{i,j}) \neq 0$. Then (8.4.57) has no solution corresponding to eigenvalue λ_i. This corresponds to Case II in Table 8.1.

The other is that for $\phi_{i,j}$, $(g, \phi_{i,j}) = 0$, $(j = 1, 2, \ldots, k_i)$ hold. Then the solutions corresponding to λ_I are $f(x) = g(x) + \sum_{j=1}^{k_i} A_j \phi_{i,j}(x)$, where A_j, $(j = 1, 2, \ldots, k_i)$ are arbitrary constants. This corresponds to Case III in Table 8.1.

Example 2. Solve the integral equation

$$f(x) = x + \lambda \int_0^\pi \frac{\sin(x + x')}{\pi} f(x') dx'. \qquad (8.4.65)$$

Solution. We apply the standard procedure presented above. The equation is expanded into the following form:

$$f(x) = x + \lambda \frac{\sin x}{\sqrt{\pi}} \int_0^\pi \frac{\cos x'}{\sqrt{\pi}} f(x') dx'$$

$$+ \lambda \frac{\cos x}{\sqrt{\pi}} \int_0^\pi \frac{\sin x'}{\sqrt{\pi}} f(x') dx'. \qquad (8.4.66a)$$

Thus it is obvious that it is of a finite rank kernel. Compared to (8.4.57), we have $g(x) = x$, $N = 2$ and

$$\{\phi_n\} = \frac{1}{\sqrt{\pi}}\{\sin x, \cos x\}, \quad \{\psi_n\} = \frac{1}{\sqrt{\pi}}\{\cos x, \sin x\}.$$

(1) We compute the quantities defined by (8.4.59).

$$a = \begin{pmatrix} a_1 \\ a_2 \end{pmatrix} = \begin{pmatrix} (\psi_1, f) \\ (\psi_2, f) \end{pmatrix} = \frac{1}{\sqrt{\pi}} \int_0^\pi f(x') \begin{pmatrix} \cos x' \\ \sin x' \end{pmatrix} dx'.$$

$$b = \begin{pmatrix} b_1 \\ b_2 \end{pmatrix} = \begin{pmatrix} (\psi_1, g) \\ (\psi_2, g) \end{pmatrix} = \frac{1}{\sqrt{\pi}} \begin{pmatrix} -2 \\ \pi \end{pmatrix}.$$

$$M = \begin{pmatrix} (\psi_1, \phi_1) & (\psi_1, \phi_2) \\ (\psi_2, \phi_1) & (\psi_2, \phi_2) \end{pmatrix} = \frac{1}{2} \begin{pmatrix} 0 & 1 \\ 1 & 0 \end{pmatrix}.$$

Hence, (8.4.66a) becomes

$$f(x) = x + a_1 \lambda \frac{\sin x}{\sqrt{\pi}} + a_2 \lambda \frac{\cos x}{\sqrt{\pi}}. \tag{8.4.66b}$$

Making inner products of this equations by ψ_1 and ψ_2, respectively, we obtain

$$a_1 = b_1 + \frac{\lambda}{2} a_2, \quad a_2 = b_2 + \frac{\lambda}{2} a_1.$$

The matrix in (8.4.60) is now

$$I - \lambda M = \begin{pmatrix} 1 & -\lambda/2 \\ -\lambda/2 & 1 \end{pmatrix}.$$

From its determinant, the eigenvalues are computed: $\lambda = \pm 2$.

(2) $\lambda \neq \pm 2$

(i) By $a = (I - \lambda M)^{-1} b$, it is solved that

$$a_1 = \frac{2}{\sqrt{\pi}} \frac{\pi \lambda - 4}{4 - \lambda^2}, \quad a_2 = \frac{4}{\sqrt{\pi}} \frac{\pi - \lambda}{4 - \lambda^2}.$$

They are substituted into (8.4.66b) to obtain the solution

$$f(x) = x + 2\lambda \frac{\pi \lambda - 4}{\pi (4 - \lambda^2)} \sin x + 4\lambda \frac{\pi - \lambda}{\pi (4 - \lambda^2)} \cos x. \tag{8.4.67}$$

(ii) Iteration method

Making iteration of (8.4.65) repeatedly, we get Neumann series as follows:

$$f(x) = x + \lambda \int_0^\pi dx' k(x, x') x' + \lambda^2 \int_0^\pi dx'$$

$$\times \int_0^\pi dx'' k(x, x') k(x', x'') x'' + \cdots, \tag{8.4.68}$$

where the kernel is $k(x, x') = \frac{1}{\pi} \sin(x + x')$. It is necessary to find the convergence criterion of this series.

On one hand, by (8.4.63), it is must be $|k(x, x')| \leq 1/\pi$. So, as $|\lambda| < 1$ the series can be used to find the solution.

On the other hand, by (8.4.64), we obtain

$$\frac{|\lambda|}{\pi} \left[\int_0^\pi \int_0^\pi \sin^2(x + x') dx dx' \right]^{1/2} = \frac{|\lambda|}{\sqrt{2}} < 1. \qquad (8.4.69)$$

So, as $|\lambda| < \sqrt{2}$, the Neumann series converges.

It is seen that the upper limit of $|\lambda|$ obtained by (8.4.64) is larger than that by (8.4.63).

The iteration results are as follows.

$$f_0(x) = x, \qquad (8.4.70a)$$

$$f_1(x) = x + \lambda \left(\cos x - \frac{2}{\pi} \sin x \right), \qquad (8.4.70b)$$

$$f_2(x) = x + \lambda \left(\cos x - \frac{2}{\pi} \sin x \right) + \lambda^2 \left(\frac{1}{2} \sin x - \frac{1}{\pi} \cos x \right),$$

$$(8.4.70c)$$

$$\cdots\cdots .$$

Figure 8.1 plots the three iteration results as $\lambda = 1$ and the rigorous solution. It is seen that as $\lambda = 1$, $f_2(x)$ has been a quite good approximation of the rigorous result.

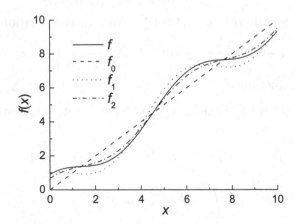

Fig. 8.1.   The rigorous solution $f(x)$ of (8.4.65) and its first three iteration results $f_0(x), f_1(x), f_2(x)$ as $\lambda = 1$.

It should be noted that (8.4.67) is the rigorous solution valid for any λ value other than $\lambda \neq \pm 2$, while (8.4.68) or (8.4.70) is the approximate solution valid for $|\lambda| < \sqrt{2}$.

(3) Homogeneous equation when $\lambda = \pm 2$

In this case, (8.4.66b) becomes

$$f_i(x) = a_1 \lambda_i \frac{\sin x}{\sqrt{\pi}} + a_2 \lambda_i \frac{\cos x}{\sqrt{\pi}}, \qquad (8.4.71)$$

where $\lambda = \lambda_i$ is an eigenvalue of matrix M.

As $\lambda = \pm 2$, $\det(I - \lambda M) = 0$. The degeneracy of each eigenvalue is 1. From homogeneous linear algebraic equations

$$\begin{pmatrix} 1 & -\lambda_i/2 \\ -\lambda_i/2 & 1 \end{pmatrix} \begin{pmatrix} a_{i,1} \\ a_{i,2} \end{pmatrix} = \begin{pmatrix} 0 \\ 0 \end{pmatrix},$$

the $a_{i,1}$ and $a_{i,2}$ are solved for $\lambda_1 = 2$ and $\lambda_2 = -2$. Then they are substituted into (8.4.71) to get two eigenfunctions:

$$f_+(x) = A_+(\sin x + \cos x) = \sqrt{\frac{2}{\pi}} \sin\left(x + \frac{\pi}{4}\right) \qquad (8.4.72a)$$

and

$$f_-(x) = A_-(\sin x - \cos x) = \sqrt{\frac{2}{\pi}} \sin\left(x - \frac{\pi}{4}\right). \qquad (8.4.72b)$$

At the last steps, the coefficients have been normalized. The solutions of the homogeneous equation

$$f(x) = \pm 2 \int_0^\pi \frac{\sin(x + x')}{\pi} f(x')\mathrm{d}x' \qquad (8.4.73)$$

are the two functions of (8.4.72).

What about the solution of the inhomogeneous equation (8.4.65) as $\lambda = \pm 2$? The equation is

$$f(x) = x \pm 2 \int_0^\pi \frac{\sin(x + x')}{\pi} f(x')\mathrm{d}x'$$

$$= x \pm \lambda_i \left(a_1 \frac{\sin x}{\sqrt{\pi}} + a_2 \frac{\cos x}{\sqrt{\pi}}\right). \qquad (8.4.74)$$

The free term $g(x) = x \neq 0$. It is easily calculated that $(g, f_+) \neq 0$ and $(g, f_-) \neq 0$. The conclusion is that (8.4.74) does not has solutions as $\lambda = \pm 2$.

In summary, we have mainly introduced the solving procedures in two cases. One is the Neumann series generated from iteration method when the norm of the linear integral operator K is sufficiently small. The other is to apply the Fredholm alternative theorem when the kernel is a finite rank one. In the cases of degenerate kernels, there exist solutions in closed forms except Case II in Table 8.1.

8.4.3. *Expansion of kernel in terms of eigenfunctions*

In general, the kernel is not degenerate. However, if the eigenvalues and eigenfunctions of $k(x, y)$ are known, $k(x, y)$ can be expanded by them:

$$k(x, y) = \sum_{i=1}^{\infty} \frac{f_i(y) f_i^*(x)}{\lambda_i}. \qquad (8.4.75)$$

This has been proved by Theorem 6 in Subsection 8.1.3.

In the form of (8.4.75), $k(x, y)$ appears to be a degenerate one, and is expanded by its eigenfunctions. This is distinguished from a degenerate kernel discussed above which is expanded by a function set instead of its eigenfunction set.

For the first kind of Fredholm integral equations

$$g(x) = \int_a^b k(x, y) f(y) \mathrm{d}y, \qquad (8.4.76)$$

we have the following theorem.

Theorem 2 (Schmidt-Picard Theorem). *Suppose that $k(x, y)$ is Hermitian and $\{\lambda_i\}$ and $\{f_i(x)\}$ are its eigen-system. The sufficient and necessary conditions for Eq. (8.4.76) to have solutions are that the following series*

$$\sum_{i=1}^{\infty} |\lambda_i g_i|^2 \quad \text{with } g_i = (f_i, g)$$

is convergent. If $\{f_i(x)\}$ is complete, Eq. (8.4.76) has a unique solution

$$f(x) = \sum_{i=1}^{\infty} \lambda_i g_i f_i(x). \tag{8.4.77}$$

Equation (8.4.77) is easily proved. Substitution of (8.4.75) into (8.4.76) yields

$$g(x) = \int_a^b \sum_{i=1}^{\infty} \frac{f_i(y) f_i^*(x)}{\lambda_i} f(y) \mathrm{d}y.$$

Multiplying on both sides by $f_k(x)$ and integrating with respect to x lead to

$$g_k = \int_a^b f_k^*(x) g(x) \mathrm{d}x = \int_a^b \frac{f_k^*(y)}{\lambda_k} f(y) \mathrm{d}y, \tag{8.4.78}$$

where the orthonormalization of the eigenfunctions are utilized. Since $\{f_i(x)\}$ is complete, assuming $f(x) = \sum_{i=1}^{\infty} c_i f_i(x)$ in (8.4.78) gives

$$g_k = \int_a^b \frac{f_k^*(y)}{\lambda_k} \sum_{j=1}^{\infty} c_j f_j(y) \mathrm{d}y = \frac{c_k}{\lambda_k}. \tag{8.4.79}$$

So,

$$c_k = \lambda_k g_k. \tag{8.4.80}$$

Equation (8.4.77) is achieved.

It should be noted that if the eigenfunction set of $k(x, y)$ is complete, the unique solution of the equation (8.4.76) is acquired. If it is not, then the solution of the equation is expanded by eigenfunctions that are not complete. The other functions in this complete set are orthogonal to the kernel. For example, suppose that a solution $f(x)$ of (8.4.75) has been known. If a function $\phi(x)$ is orthogonal to the kernel $k(x, y)$, i.e., $\int_a^b k(x, y) \phi(y) \mathrm{d}y = 0$, then $f(x) + \phi(x)$ is also a solution of this equation. This is a feature of the first kind of Fredholm integral equation. The second kind of Fredholm equations do not have such a feature. The following is an example.

Example 3. Find the solutions of the equation

$$\int_{-\pi}^{\pi} \sin(x + y) f(y) \mathrm{d}y = 3 \sin x + 2 \cos x.$$

Solution. The kernel is expanded to be

$$k(x, y) = \sin(x + y) = \sin x \cos y + \cos x \sin y.$$

Compared to (8.4.76), we get $N = 2$ and

$$\{\phi_n\} = \{\sin x, \cos x\}, \quad \{\psi_n\} = \{\cos x, \sin x\}.$$

The matrix is

$$M = \begin{pmatrix} (\psi_1, \phi_1) & (\psi_1, \phi_2) \\ (\psi_2, \phi_1) & (\psi_2, \phi_2) \end{pmatrix} = \int_{-\pi}^{\pi} dx \begin{pmatrix} \cos x \sin x & \cos^2 x \\ \sin^2 x & \sin x \cos x \end{pmatrix}$$

$$= \pi \begin{pmatrix} 0 & 1 \\ 1 & 0 \end{pmatrix}.$$

Its eigenvalue equation is $\det(I - \lambda M) = 1 - \lambda^2 \pi^2$. The eigenvalues are $\lambda = \pm 1/\pi$. The kernel in this example is of the same form of that of Example 2, but they have different eigenvalues. Even if the forms of the kernels are the same, different integral intervals will generate different eigenvalues. The eigenvectors are $\begin{pmatrix} 1 \\ 1 \end{pmatrix}$ and $\begin{pmatrix} 1 \\ -1 \end{pmatrix}$. The eigenfunctions are $\{\sin x + \cos x, \sin x - \cos x\}$. Their normalizations are $\{\psi_1, \psi_2\} = \frac{1}{\sqrt{2\pi}}\{\sin x + \cos x, \sin x - \cos x\}$.

The kernel is expanded by the eigenfunctions:

$$k(x, y) = \frac{1}{2}(\sin x + \cos x)(\sin y + \cos y)$$

$$- \frac{1}{2}(\sin x - \cos x)(\sin y - \cos y).$$

It is evaluated by (8.4.78) that

$$g_1 = \frac{1}{2\pi} \int_{-\pi}^{\pi} (\sin y + \cos y)(3 \sin y + 2 \cos y) dy = \frac{5}{2}$$

and

$$g_2 = \frac{1}{2\pi} \int_{-\pi}^{\pi} (\sin y - \cos y)(3 \sin y + 2 \cos y) dy = \frac{1}{2}.$$

Assume that one solution is the linear combination of the eigenfunctions:

$$f_0(x) = c_1 \psi_1 + c_2 \psi_2.$$

The coefficients can be computed by (8.4.80): $c_1 = \frac{5}{2\pi}, c_2 = -\frac{1}{2\pi}$.
Hence, one solution is obtained:

$$f_0(x) = \frac{5}{2\pi}(\sin x + \cos x) - \frac{1}{2\pi}(\sin x - \cos x) = \frac{2}{\pi}\sin x + \frac{3}{\pi}\cos x.$$

Now the eigenfunction set $\{\psi_1, \psi_2\}$ of the kernel is not a complete one. $f_0(x)$ is not expanded by a complete set. The complete set should be $\{\sin nx \pm \cos nx\}$. Therefore, the function

$$\psi(x) = a_0 + \sum_{j=2}^{\infty}(a_n \sin nx + b_n \cos nx)$$

is orthogonal to the kernel:

$$\int_{-\pi}^{\pi} \sin(x+y)\psi(y)dy = 0.$$

Here the coefficients $a_0, a_n, b_n, (n = 2, 3, \ldots)$ are arbitrary constants. Thus, the solutions of the equation are

$$f(x) = \frac{2}{\pi}\sin x + \frac{3}{\pi}\cos x + \psi(x).$$

There are infinite number of solutions as the coefficients $a_0, a_n, b_n, (n = 2, 3, \ldots)$ vary.

8.5. Integral Equations of Convolution Type

8.5.1. *Fredholm integral equations of convolution type*

For Fredholm integral equations of convolution type, they can be solved by means of the Fourier transformation when the interval is the whole real axis.

First of all, let us briefly recall the formalism of the Fourier transformation. Its definition is

$$Q(q) = \int_{-\infty}^{\infty} f(x)e^{-iqx}dq \tag{8.5.1}$$

and its inverse is

$$f(x) = \frac{1}{2\pi} \int_{-\infty}^{\infty} Q(q)e^{iqx}dq. \tag{8.5.2}$$

If expressed by a linear integral operator F, they appear as

$$Q = F[f] \tag{8.5.3a}$$

and

$$f = F^{-1}[Q]. \tag{8.5.3b}$$

Suppose that a function $f(x)$ has the following convolution form:

$$f(x) = \int_{-\infty}^{\infty} f_1(x - x_1) f_2(x_1) dx_1, \tag{8.5.4}$$

denoted in short by $f = f_1 * f_2$. The convolution has a property that $f_1 * f_2 = f_2 * f_1$.

The Fourier transformation of the convolution is

$$F[f_1 * f_2] = F[f_1] \cdot F[f_2]. \tag{8.5.5}$$

This equation is called the convolution theorem. The Fourier transformations and their inverses of some functions can be found in specialized tables.

1. The second kind of Fredholm integral equations of convolution type

The equation is

$$\int_{-\infty}^{\infty} k(x - y) f(y) dy + g(x) = f(x). \tag{8.5.6}$$

Taking the Fourier transformations on both sides and denoting

$$Q = F[f], \quad G = F[g], \quad K = F[k], \tag{8.5.7}$$

we achieve

$$K(q)Q(q) + G(q) = Q(q). \tag{8.5.8}$$

If $1 - K(q) \neq 0$, we have

$$Q(q) = \frac{G(q)}{1 - K(q)}. \tag{8.5.9}$$

Its inverse Fourier transformation gives the solution:.

$$f(x) = F^{-1}[Q(k)] = \frac{1}{2\pi} \int_{-\infty}^{\infty} \frac{G(k)}{1 - K(k)} e^{ikx} dk. \tag{8.5.10}$$

2. The first kind of Fredholm integral equations of convolution type

The equation is

$$\int_{-\infty}^{\infty} k(x-y)f(y)\mathrm{d}y = g(x). \tag{8.5.11}$$

Taking the Fourier transformations on both sides, we obtain

$$K(q)Q(q) = G(q). \tag{8.5.12}$$

If $K(q) \neq 0$, we get

$$Q(q) = \frac{G(q)}{K(q)}. \tag{8.5.13}$$

Its inverse Fourier transformation gives the solution

$$f(x) = F^{-1}[Q(q)] = \frac{1}{2\pi} \int_{-\infty}^{\infty} \frac{G(q)}{K(q)} e^{iqx} \mathrm{d}q. \tag{8.5.14}$$

In addition, when the kernel is just the sine or cosine functions and the interval $[a, b]$ happens to be the positive half axis, the Fourier sine or cosine transformations can be applied.

Example 1. The kernel in (8.5.11) happens to be Fermi distribution function

$$k(x) = \frac{1}{e^x + 1}. \tag{8.5.15}$$

Find the expression of the function $f(x)$.

Solution. This is the first kind of Fredholm integral equations of convolution type. The first step is to evaluate the Fourier transformation of the kernel.

$$K(\omega) = \int_{-\infty}^{\infty} \frac{e^{ix\omega}}{e^x + 1} \mathrm{d}x. \tag{8.5.16}$$

Supplement an infinite semicircle path at the upper half plane to constitute a closed integral contour. There are infinite number of simple poles positioned at $x = (2n+1)\pi i$ in this half plane. Hence,

$$K(\omega) = 2\pi i \sum_{n=0}^{\infty} e^{i\omega(2n+1)\pi i} = 2\pi i \frac{e^{-\omega\pi}}{1 - e^{-\omega 2\pi}} = \frac{\pi i}{\sinh \omega \pi}. \tag{8.5.17}$$

Then the solution of the equation is calculated by (8.5.14):

$$f(x) = \frac{1}{2\pi} \int_{-\infty}^{\infty} \frac{G(\omega)}{K(\omega)} e^{-ix\omega} d\omega$$

$$= \frac{1}{2\pi} \int_{-\infty}^{\infty} \frac{\sinh\omega\pi}{\pi i} e^{-ix\omega} \int_{-\infty}^{\infty} g(x') e^{ix'\omega} dx' d\omega$$

$$= \frac{1}{2\pi i} \int_{-\infty}^{\infty} g(x') dx' \frac{1}{2\pi} \int_{-\infty}^{\infty} e^{i(x'-x)\omega} \sinh\omega\pi d\omega, \qquad (8.5.18)$$

where $G(\omega)$ is the Fourier component of $g(x)$. The last step is the Fourier transformation of hyperbolic sine function, which can be found in mathematics handbook:

$$\frac{1}{2\pi} \int_{-\infty}^{\infty} e^{i(x'-x)\omega} \sinh\omega\pi d\omega$$

$$= \delta(x' - x - i\pi) - \delta(x' - x + i\pi). \qquad (8.5.19)$$

Thus,

$$f(x) = \frac{1}{2\pi i} [g(x + i\pi) - g(x - i\pi)], \qquad (8.5.20)$$

where (5.1.24) is employed. The final result can be expressed by a more compact form:

$$\frac{1}{2\pi i} [g(x + i\pi) - g(x - i\pi)] = -\frac{1}{\pi} \text{Im} \, g(x - i\pi). \qquad (8.5.21)$$

For a give function g, the solution f can immediately be written.

8.5.2. *Volterra integral equations of convolution type*

For Volterra integral equations of convolution type, when the lower limit of integral $a = 0$, they can be solved by means of the Laplace transformation.

Here we us briefly recall the formalism of the Laplace transformation. Its definition is

$$Q(p) = \int_{0}^{\infty} f(x) e^{-px} dx. \qquad (8.5.22)$$

If expressed by a linear integral operator L, it appears as

$$Q = L[f]. \tag{8.5.23}$$

This is the transformation from the original function f to the image function Q. Its inverse transformation is denoted by

$$f = L^{-1}[Q]. \tag{8.5.24}$$

Suppose that a function $f(x)$ has the following convolution form:

$$f(x) = \int_0^x f_1(x - x_1) f_2(x_1) dx_1 \tag{8.5.25}$$

denoted in short by $f = f_1 * f_2$. The convolution has a property that $f_1 * f_2 = f_2 * f_1$.

The Laplace transformation of the convolution is

$$L[f_1 * f_2] = L[f_1] \cdot L[f_2] \tag{8.5.26}$$

This equation is called convolution theorem. The Laplace transformations and their inverses of some functions can be found in specialized tables.

1. The second kind of Volterra integral equations of convolution type

The equation is

$$\int_0^x k(x - y) f(y) dy + g(x) = f(x) \tag{8.5.27}$$

Taking the Fourier transformations on both sides and denoting

$$Q = L[f], \quad G = L[g], \quad K = L[k], \tag{8.5.28}$$

we achieve

$$K(p)Q(p) + G(p) = Q(p). \tag{8.5.29}$$

If $1 - K(p) \neq 0$, we have

$$Q(p) = \frac{G(p)}{1 - K(p)}. \tag{8.5.30}$$

Its inverse Laplace transformation gives the solution:

$$f(x) = L^{-1}[Q(p)]. \tag{8.5.31}$$

Example 2. Find the solution of the integral equation

$$f(x) = \sin x + 2 \int_0^x \cos(x - y) f(y) dy.$$

Solution. Taking the Laplace transformation according to (8.5.28), we get

$$G(p) = L[\sin x] = \frac{1}{p^2 + 1}, \quad K(p) = L[2\cos(x)] = \frac{2p}{p^2 + 1}.$$

Then it follows from (8.5.30) that

$$Q(p) = \frac{1/(p^2 + 1)}{1 - 2p/(p^2 + 1)} = \frac{1}{(p - 1)^2}.$$

In mathematics handbook we can find the formulas

$$L[x^n] = \frac{n!}{p^{n+1}}, \quad L^{-1}\left[\frac{n!}{p^{n+1}}\right] = x^n,$$

$$L^{-1}[Q(p - \alpha)] = e^{\alpha x} L^{-1}[Q(p)]. \tag{8.5.32}$$

Thus the required solution is

$$f(x) = L^{-1}\left[\frac{1}{(p - 1)^2}\right] = x e^x.$$

2. The first kind of Volterra integral equations of convolution type

The convolution equation is

$$\int_0^x k(x - y) f(y) dy = g(x). \tag{8.5.33}$$

Taking the Laplace transformations on both sides, we obtain

$$K(p)Q(p) = G(p) \tag{8.5.34}$$

If $K(p) \neq 0$, we get

$$Q(p) = \frac{G(p)}{K(p)}. \tag{8.5.35}$$

Its inverse Laplace transformation (8.5.31) gives the solution.

Example 3. Find the solution of the equation

$$x^3 = \int_0^x \left[1 - 4(x-y) + \frac{3}{2}(x-y)^2 \right] f(y) dy.$$

Solution. This is the first kind of Volterra integral equations of convolution type. Taking the Laplace transformation on both sides and making use of (8.5.32), we get

$$G(p) = \frac{3!}{p^4}, \quad K(p) = \frac{1}{p} - \frac{4}{p^2} + \frac{3}{p^3} = \frac{p^2 - 4p + 3}{p^3}.$$

It follows from (8.5.35) that

$$Q(p) = \frac{6}{p^4} \frac{p^3}{p^2 - 4p + 3} = \frac{1}{p-3} - \frac{3}{p-1} + \frac{2}{p}.$$

Referring to (8.5.32), the Laplace inverse transformation gives the solution:

$$f(x) = L^{-1} \left[\frac{1}{p-3} - \frac{3}{p-1} + \frac{2}{p} \right] = e^{3x} - 3e^x + 2.$$

Example 4. Solve the equation

$$x^\alpha = \int_0^x (x-y)^\beta f(y) dy, \quad \alpha \geq 0, \quad \beta > -1.$$

Solution. Taking the Laplace transformation on both sides leads to

$$L[x^\alpha] = L[x^\beta] Q(p).$$

As α is not an integer, there is a pair of Laplace transformation and its inverse:

$$L[x^\alpha] = \frac{\Gamma(\alpha+1)}{p^{\alpha+1}}, \quad L^{-1} \left[\frac{1}{p^\alpha} \right] = \frac{x^{\alpha-1}}{\Gamma(\alpha)}.$$

Thus

$$Q(p) = \frac{\Gamma(\alpha+1)}{\Gamma(\beta+1)p^{\alpha-\beta}}.$$

Its inverse Laplace transformation gives the required solution

$$f(x) = L^{-1}[Q(p)] = \frac{\Gamma(\alpha+1)}{\Gamma(\beta+1)} L^{-1} \left[\frac{1}{p^{\alpha-\beta}} \right]$$

$$= \frac{\Gamma(\alpha+1)}{\Gamma(\beta+1)\Gamma(\alpha-\beta)} x^{\alpha-\beta-1}.$$

3. Nonlinear Volterra integral equations of convolution type

The following special convolution integral equation

$$\lambda \int_0^x f(x - y)f(y)\mathrm{d}y + g(x) = f(x) \qquad (8.5.36)$$

can also be solved by the Laplace transformation. Although this equation is nonlinear, the integral happens to be the form of convolution. Taking the Laplace transformation on both sides leads to

$$\lambda Q(p)Q(p) + G(p) = Q(p),$$

or

$$\lambda Q(p)Q(p) - Q(p) + G(p) = 0. \qquad (8.5.37)$$

It follows that

$$Q(p) = \frac{1}{2\lambda}\left[1 \pm \sqrt{1 - 4\lambda G(p)}\right]. \qquad (8.5.38)$$

Its inverse Laplace transformation $f(x) = L^{-1}[Q(p)]$ gives the solution of the original equation.

Example 5. Solve the equation

$$\lambda \int_0^x f(x - y)f(y)\mathrm{d}y = \frac{x^3}{6}.$$

Solution. The result of the Laplace transformation is

$$\lambda Q^2(p) = \frac{1}{p^4},$$

or

$$Q(p) = \pm\frac{1}{\sqrt{\lambda}p^2}.$$

Its inverse transformation gives the solution

$$f(x) = \pm\frac{x}{\sqrt{\lambda}}.$$

8.6.   Integral Equations with Polynomials

8.6.1.   *Fredholm integral equations with polynomials*

In Fredholm integral equations, if the kernel $k(x, y)$ and free term $g(x)$ are both the polynomials of x, the unknown function can be assumed to be a polynomial of x. Explicitly, suppose that the second kind of Fredholm integral equations are of the form of

$$\lambda \int_a^b \sum_{i=1}^n \varphi_i(y) x^i f^k(y) \mathrm{d}y + \sum_{i=1}^m a_j x^j = f(x), \quad m \geq n. \qquad (8.6.1)$$

We can assume

$$f(x) = \sum_{i=1}^m b_i x^l.$$

After this is substituted into (8.6.1), comparing coefficients on the two sides can give the values of all b_i. This is a special case of nonlinear integral equations. As $k = 1$, it goes back to linear equations.

Example 1. Solve

$$60 \int_0^1 xy f^2(y) \mathrm{d}y + 1 + 20x - x^2 = f(x).$$

Solution. By inspection of the equation, we can set $f(x) = a + bx + cx^2$. This is substituted into the equation. By comparing the coefficients on both sides, we obtain

$$a = 1, \quad c = -1$$

and

$$b - 60 \left(\frac{1}{2} + \frac{2}{3}b + \frac{1}{4}b^2 - \frac{1}{2} - \frac{2}{5}b + \frac{1}{6} \right) = 20, \quad b = \frac{1}{2}(-1 \pm 7\mathrm{i}).$$

Therefore, the required solution is

$$f(x) = 1 + \frac{1}{2}(-1 \pm 7\mathrm{i})x - x^2.$$

Equation (8.6.1) can be further extended to the following form:

$$\lambda \int_a^b \sum_{i=1}^n \varphi_i(y) x^i f^k(y) \mathrm{d}y + \sum_{i=1}^m a_j x^j = f^l(x). \qquad (8.6.2)$$

As long as the parameter m, n, k, l are appropriate, the similar method can be applied.

Example 2. Solve

$$\lambda \int_0^1 xy f(y) \mathrm{d}y + 1 + x^2 = f^2(x).$$

Solution. Let $f(x)$ be the form of $f(x) = a + bx$ and be substituted into the equation. By comparing the coefficients, we get

$$a^2 = 1, \quad b^2 = 1, \quad \lambda = \frac{2ab}{a/2 + b/3}.$$

Four pairs of eigenvalues a and b and their corresponding solutions are obtained, see Table 8.2.

Table 8.2.   The parameter values of the four solutions and corresponding eigenvalues in Example 2.

a	b	λ	$f(x)$
1	1	12/5	$1 + x$
1	-1	12	$1 - x$
-1	1	-12	$-1 + x$
-1	-1	-12/5	$-1 - x$

8.6.2.   *Generating function method*

We have learnt in Chapter 3 that the solutions $\{Q_n(x)\}$ of an ordinary differential equations of second kind are a complete function set on $[a, b]$ and orthogonal to each other with weight $\rho(x)$. Tables 3.1 and 3.2 list the series expressions and corresponding weights in various cases. These solutions have their generating functions, which are listed in Tables 3.5 and 3.6. In the following, we assume that $\{Q_n(x)\}$ have been normalized. Please note the interval that each differential equation applies.

We consider such a case that the interval $[a, b]$ of a first kind of Fredholm integral equation

$$\int_a^b k(x, y) f(y) \mathrm{d}y = g(x) \tag{8.6.3}$$

happens to be the same as that of an ordinary differential equation of second order and the kernel $k(x, y)$ happens to be the generating function $G(x, y)$ of the solutions of this differential equation multiplied by the corresponding weight $\rho(y)$:

$$k(x, y) = G(x, y)\rho(y). \qquad (8.6.4)$$

Then this kind of equations can be solved by the **generating function method**.

By definition, a generating function $G(x, y)$ is expanded by

$$G(x, y) = \sum_{n=0}^{\infty} c_n Q_n(y) x^n. \qquad (8.6.5)$$

Here the coefficients c_n are introduced so as to include both the ordinary and exponential generating functions in Tables 3.5 and 3.6, and even other cases. Since the set $\{Q_n(x)\}$ is complete, the unknown $f(y)$ can be expanded by it:

$$f(y) = \sum_{k=0}^{\infty} a_k Q_k(y). \qquad (8.6.6)$$

After the equations (8.6.4)–(8.6.6) are substituted into (8.6.3), we get

$$\int_a^b \sum_{n=0}^{\infty} c_n Q_n(y) x^n \rho(y) \sum_{k=0}^{\infty} a_k Q_k^*(y) dy$$

$$= \sum_{n=0}^{\infty} c_n a_n N_n x^n = g(x), \qquad (8.6.7)$$

where

$$N_n = \int_a^b Q_n(y) Q_n^*(y) \rho(y) dy. \qquad (8.6.8)$$

Make Taylor expansion for the function $g(x)$:

$$g(x) = \sum_{n=0}^{\infty} \frac{1}{n!} g^{(n)}(0) x^n. \qquad (8.6.9)$$

Comparing the coefficients of the terms with the same powers, we obtain

$$a_n = \frac{1}{n! c_n N_n} g^{(n)}(0). \tag{8.6.10}$$

Example 3. Solve equation.

$$\int_{-1}^{1} \frac{f(y)}{\sqrt{1 - 2xy + x^2}} dy = x + 1.$$

Solution. The kernel is the generating function of Legendre polynomials $\{P_n(x)\}$ with corresponding weight being 1, and the interval is just that $\{P_n(x)\}$ apply. It has been known that

$$N_n = \int_{-1}^{1} P_n(y) P_n^*(y) dy = \frac{2}{2n + 1}.$$

Then it follows from (8.6.7) that

$$\sum_{n=0}^{\infty} \frac{2a_n}{2n + 1} x^n = x + 1.$$

In this example, $g(x)$ has been naturally a polynomial. By comparison of coefficients of the terms with the same powers, we get $a_0 = \frac{1}{2}$, $a_1 = \frac{3}{2}$, $a_n = 0$, $(n \geq 2)$. At last, (8.6.6) gives the solution:

$$f(x) = \frac{1}{2} P_0(x) + \frac{3}{2} P_1(x) = \frac{1}{2} + \frac{3}{2} x.$$

Example 4. Solve the equation

$$\int_{0}^{\infty} \frac{1}{1 - x} \exp\left(-\frac{xy}{1 - x}\right) e^{-y} f(y) dy = 1 - x, \quad |x| < 1.$$

Solution. The kernel contains the generating function of Laguerre polynomials and corresponding weight. It is known that

$$N_n = \int_{-1}^{1} e^{-y} L_n(y) L_n^*(y) dy = (n!)^2.$$

The expansion of the generating function is

$$\frac{1}{1 - t} \exp\left(-\frac{xt}{1 - t}\right) = \sum_{n=0}^{\infty} \frac{L_n(x)}{n!} t^n.$$

It follows from (8.6.7) that,

$$\sum_{n=0}^{\infty} \frac{1}{n!} a_n (n!)^2 x^n = 1 - x.$$

Comparison of the coefficients results in $a_0 = 1$, $a_1 = -1$, $a_n = 0$, $(n \geq 2)$. The required solution is

$$f(x) = L_0(x) - L_1(x) = 1 - (1 - x) = x.$$

Exercises

1. Show that (8.1.21) satisfies the boundary conditions of (8.1.20). The derivative two times of (8.1.21) will results in (8.1.20). On the contrary, how can one obtain (8.1.21) from (8.1.20)?

2. Volterra integral equation (8.1.24) is equivalent to the differential equation of second order (8.1.23). The coefficients A and B are determined by boundary or initial conditions. Determine A and B in the following three cases.

   (1) $f(a) = \alpha$, $f(b) = \beta$.

   (2) $f(a) = \alpha$, $f'(a) = \beta$.

   (3) $f(b) = \alpha$, $f'(b) = \beta$.

3. Prove (8.2.56). Show that as $N \to \infty$, $\sum_{n=0}^{N} \lambda^n \tilde{g}_n(x)$ uniformly converges to the solution $f(x)$ of (8.1.10).

4. Verify the following solutions of Volterra equations.

   (1) $f(x) = 1 - x$ is the solution of $\int_0^x dy e^{x-y} f(y) = x$.

   (2) $f(x) = \frac{1}{\pi\sqrt{x}}$ is the solution of $\int_0^x \frac{f(y)}{\sqrt{x-y}} dy = 1$.

5. Convert the following problems of differential equations to corresponding integral equations.

   (1) $y'' + y = \cos x$, $y(0) = 0$, $y'(0) = 1$

   (2) $y'' + (1 + x^2)y = \cos x$, $y(0) = 0$, $y'(0) = 2$.

   (3) $y'' + 4y = \varphi(x)$, $0 < x < \pi/2$, $y(0) = 0$, $y(\pi/2) = 0$.

6. Convert the following integral equations to the problems of differential equations and then find their solutions.

   (1) $x - \int_0^x dy\, e^{x-y} f(y) = f(x)$.

   (2) $\int_0^x dy\, e^{x-y} f(y) = x$.

   (3) $1 + 2 \int_0^x \frac{2y+1}{(2x+1)^2} f(y) dy = f(x)$.

7. Show that if $k(x, y)$ is Hermitian, then its n-iteration kernel $k_n(x, y)$ is also Hermitian.

8. Solve the following integral equations by means of iteration method.

   (1) $f(x) = 1 + \lambda \int_0^x dy f(y)$. (2) $f(x) = 1 + \lambda \int_0^\infty dy e^{-(x+y)} f(y)$.
   Determine the λ value at which iteration series converges in each case.

9. Solve the following integral equations by means of the iteration method.

   (1) $f(x) = \frac{5}{6}x + \frac{1}{2}\int_0^1 dy f(y)xy$.

   (2) $f(x) = e^x - \frac{1}{2}e + \frac{1}{2} + \frac{1}{2}\int_0^1 dy f(y)$.

   (3) $f(x) = 1 + \int_0^x dy f(y)(y - x)$.

   (4) $f(x) = x + \int_0^x dy f(y)(y - x)$.

10. Solve the following integral equations.

    (1) $f(x) = \sin x - \frac{1}{4}x + \frac{1}{4}\int_0^{\pi/2} dy f(y)xy$.

    (2) $f(x) = 1 - 2x - 4x^2 + \int_0^x dy f(y)[3 + 6(x - y) - 4(x - y)^2]$.

11. Assume that the potential $V(r)$ in (8.2.38) is nonnegative. Multiply $[V(r)]^{1/2}$ on both sides of (8.2.38e). Show that if we define

$$\tilde{\psi}(r) = [V(r)]^{1/2}\psi(r), \tilde{\varphi}(r) = [V(r)]^{1/2}\varphi(r)$$

and $\tilde{k}(r, r', E) = [V(r)]^{1/2}G_0(r, r', E)[V(r')]^{1/2}$, then,

$$\tilde{\psi}(r) = \tilde{\varphi}(r) + \int \tilde{k}(r, r', E)\tilde{\psi}(r')dr'.$$

The linear integral operator with kernel $\tilde{k}(r, r', E)$ is denoted as $\tilde{K}$. Show that the norm of $\tilde{K}$ is the same as that of (8.2.49) obtained by the weight function method. Show that if we apply the iteration technique to the equation of $\tilde{\psi}(r)$, $[V(r)]^{1/2}$ does not appear in iteration series, i.e., the expression of solution $\tilde{\psi}(r)$ is the usual Born series besides a $[V(r)]^{1/2}$ factor.

12. When $V(r) = V_0 e^{-\mu r}$, evaluate the norm of the Lippmann-Schwinger equation. At what V_0 value will the Born series converge?

13. Suppose that there is a one-dimensional Schrödinger equation with a delta potential: $-\frac{\hbar^2}{2m}\frac{\mathrm{d}^2}{\mathrm{d}x^2}\psi(x) - V_0\delta(x)\psi(x) = E\psi(x)$.

    (1) In the case of $V_0 > 0$, show that no matter how larger V_0 is, there is only one bound state, and determine the energy and wavefunction of this state.

    (2) In the case of $V_0 < 0$, a particle with a positive energy E is incident from infinity. Evaluate the transmitted and reflected waves.

14. Show, by means of the weight function method, that in one-dimensional case, as long as $\frac{1}{\hbar v}\int_{-\infty}^{\infty} V(x)\mathrm{d}x < 1$ where $V(x)$ is the potential, the iteration series of the Lippmann-Schwinger equation converges.

15. Show that if $V(r)$ changes its sign as r varies, (8.2.48) should be changed to be $\rho(r) = |V(r)|$ in the weight function method. Evaluate the norm of the Lippmann-Schwinger integral operator under such a weight function.

16. (1) For a one-dimensional Lippmann-Schwinger equation, define $\psi(x) = e^{\mathrm{i}kx}\phi(x)$ and then show that $\phi(x)$ satisfies equation

$$\phi(x) = 1 - \frac{\mathrm{i}m}{\hbar^2 k}\int_{-\infty}^{x} V(x')\phi(x')\mathrm{d}x' - \frac{\mathrm{i}m}{\hbar^2 k}e^{-2\mathrm{i}kx}$$

$$\times \int_{x}^{\infty} e^{2\mathrm{i}kx'}V(x')\phi(x')\mathrm{d}x'.$$

    (2) Show that as k is very large, or the incident energy is very large, the last term of this equation can be neglected.

(3) After the last term is neglected, the solution will be changed from $\phi(x)$ to $\phi_E(x)$. Show that $\phi_E(x)$ satisfies the equation

$$\frac{d}{dx}\phi_E(x) = -\frac{im}{\hbar^2 k}V(x)\phi_E(x)$$

and its boundary condition is $\phi_E(-\infty) = 1$. Prove the solution $\phi_E(x) = \exp\left\{-\frac{im}{\hbar^2 k}\int_{-\infty}^{x} V(x')dx'\right\}$. This approximation is called eikonal approximation. This approximation turns out helpful in analyzing high energy scattering problems when extended to three-dimensional space.

(4) When $V(x) = \begin{cases} V_0, & |x| \leq a/2 \\ 0, & |x| > a/2 \end{cases}$, show that the last term in (1) is indeed very small as $k \to \infty$. Use $\phi_E(x)$ in (3) instead of $\phi(x)$ to do the integration in (1), and the integration is simple. Show that this term is indeed small as $E \to V_0$.

17. Program to compute $x_0(t), x_1(t), x_2(t)$ expressed by (8.3.23a), (8.3.23b), (8.3.23c), plot the curves.

18. If in a nonlinear Volterra integral equation, the upper integral limit x is replaced by b, it is turned to a nonlinear Fredholm equation. In Subsection 8.3.3, we have made use of the concept of contraction to address the condition for the iteration of the nonlinear Volterra equation. Please in the same way address the condition for the iteration of the nonlinear Fredholm equation.

19. Show that under the condition (8.4.11), (8.4.12) is the solution of (8.4.3).

20. Consider a damping oscillator satisfying the equation

$$\frac{d^2}{dt^2}x(t) + 2\gamma\frac{d}{dt}x(t) + \omega_0^2 x(t) = 0.$$

(1) Show that if $x(0) = x_0$ and $\dot{x}(0) = 0$, $x(t)$ satisfies integral equation

$$x(t) = x_0 \cos\omega_0 t + \frac{2\gamma}{\omega_0}x_0 \sin\omega_0 t - 2\gamma$$

$$\times \int_0^t dt' \cos[\omega_0(t-t')]x(t').$$

(2) Iterate this equation few times, and show that the results are the same as the expansion of the rigorous solution to the same order.

21. Consider a separable potential with the form $-g^2 v(r)v(r')$ where $v(r) = e^{-\mu r}/r$. Find the unique bound solution of Schrödinger equation. Show that when and only when $g^2 > \hbar^2 \mu^3/4\pi m$ where m is the mass of the bound particle, the bound state exists. Meanwhile, show that the energy of the bound state is $E_B = -\frac{1}{2m}\left[\left(\frac{4\pi m g^2}{\mu \hbar^2}\right)^{1/2} - \mu\right]$. What is the wavefunction of this bound state?

22. Consider the quantity $f^s(E) = \frac{2\tau}{(q_i^2+\mu^2)^2+(q_i-i\mu)^2/\mu}$, where $\tau = 4\pi m g^2/\hbar^2$ and $q_i = \sqrt{2mE/\hbar^2}$. Show that if E is a complex number, then $f^s(E)$ is a singly-valued function on two leaves of Riemann surface cut by positive real E axis. Show that $f^s(E)$ as a function of E has a second order singularity at $E = -\mu^2 \hbar^2/2m$ on the first leaf and a simple pole at $E = -(\sqrt{\tau/\mu}+\mu)^2\hbar^2/2m$ on the second leaf. In addition, it has a singularity representing a bound state which is on either the first or second leaf. If $\sqrt{\tau/\mu} < \mu$, the bound state is at $E = -(\sqrt{\tau/\mu} - \mu)^2\hbar^2/2m$ on the second leaf. If $\sqrt{\tau/\mu} > \mu$, the bound state is at $E = -(\mu - \sqrt{\tau/\mu})^2\hbar^2/2m$ on the first leaf. This E is just the energy of the bound state determined in the last exercise. Thus we see that as g^2 decreases such that it is below the critical value of generating a bound state, the effect of the bound state does not inconceivably disappear, but the singularity corresponding to the bound state moves from the first leaf (physical leaf) to the second leaf.

23. Prove (8.4.69).

24. Find solutions of the following integral equations with separable kernels.

(1) $f(x) = 2x - \pi + 4\int_0^{\pi/2} \sin^2 x f(y)dy$.
(2) $f(x) = 1 + \lambda \int_0^1 \cos(q \ln y) f(y)dy$.
(3) $f(x) = \sin x + \lambda \int_0^{\pi/2} \sin x \cos y f(y)dy$.

(4) $f(x) = \frac{1}{\sqrt{1-x^2}} + \lambda \int_0^1 \cos^{-1} x f(y) dy.$

(5) $f(x) = x + \lambda \int_0^{2\pi} (\sin x)|\pi - y| f(y) dy.$

25. Numerically compute up to $f_2(x)$ of (8.4.70) at $\lambda = -3, 1.8$, and plot the curves. Do you think if the iteration converges at these two λ values?

26. Solve the integral equation $f(x) = \sin x + \cos x + \lambda \int_0^\pi \frac{\sin(x+x')}{\pi} f(x') dx'.$

27. Solve the following equations.

(1) $f(x) = x + \lambda \int_0^\infty e^{-(x+y)} f(y) dy.$

(2) $f(x) = x + \frac{\lambda}{\pi} \int_0^{\pi/2} dy \cos(x+y) f(y).$

28. Consider the integral equation

$$f(x) = x + (\lambda/\pi) \int_0^\pi x' \sin(x+x') f(x') dx', \quad x \in [0, \pi].$$

(1) Solve it by means of iteration technique. To do so, the upper limit of $|\lambda|$ should be evaluated which guarantees the convergence of the iteration series.

   (i) Evaluate the upper limit of $|\lambda|$ by the conditions $\max_{x,y\in[a,b]} |k(x,y)| = M$ and $\lambda < 1/M(b-a).$

   (ii) Evaluate the upper limit of $|\lambda|$ by calculation of the norm of the integral operator. Note that choose appropriate kernel and weight so as to get the upper limit as large as possible.

   (iii) Iterate to the second order terms.

(2) Find the rigorous solution. Take into account both cases that λ is and is not eigenvalues. What about the case of homogeneous equation?

29. Consider integral equation $f(x) = 1 + 4 \int_0^{1/2} dy \sqrt{1-xy} f(y).$ Take finite rank approximation of the kernel as follows.

(1) $\sqrt{1-xy} \approx 1$ and calculate up to $f_1(x).$

(2) $\sqrt{1-xy} \approx 1 - \frac{1}{2}xy$ and calculate up to $f_2(x).$

(3) $\sqrt{1 - xy} \approx 1 - \frac{1}{2}xy - \frac{1}{8}x^2y^2$ and calculate up to $f_3(x)$.
Compare $f_1(x)$, $f_2(x)$, $f_3(x)$.

30. Operators K_F and R_{K_F} are defined by $K_F f \equiv \sum_{n=1}^{N} \phi_n(\psi_n, f)$
and
$R_{K_F} f \equiv \sum_{n=1}^{N} \sum_{m=1}^{N} [(I - \lambda A)^{-1}]_{nm} \phi_n(\psi_m, f)$. Show that $(1 - \lambda K_F)(1 + \lambda R_{K_F}) = I$.

31. Three operator K, K_N and P_N are defined by $Kf \equiv \int_a^b k(x, y) f(y) dy$, $K_N f \equiv \sum_{m,n=1}^{N} (\phi_n, f)(\phi_n \phi_m^*, k)(\phi_m, f)$ and $P_N f \equiv \sum_{n=1}^{N} (\phi_n, f) \phi_n$. Show that $P_N K P_N = K_N$.

32. Prove (8.5.19) and (8.5.21).

33. Solve the following second kind of Volterra integral equations by means of the Laplace transformation.

(1) $xe^{2x} - \int_0^x e^{2(x-y)} f(y) dy = f(x)$.

(2) $1 + x + \int_0^x e^{-n(x-y)} f(y) dy = f(x)$.

(3) $\sin x - \int_0^x \sinh(x - y) f(y) dy = f(x)$.

(4) $1 + \int_0^x \sin(x - y) \cos(x - y) f(y) dy = f(x)$.

(5) $1 - 2x - 4x^2 + \int_0^x \left[3 + 6(x - y) - 4(x - y)^2 \right] f(y) dy = f(x)$.

34. Solve the following first kind of Volterra integral equations by means of the Laplace transformation.

(1) $\int_0^x e^{-(x-y)} f(y) dy = e^{-x} + x - 1$.

(2) $\int_0^x \cos(x - y) f(y) dy = x \sin x$.

(3) $\int_0^x (x - y) f(y) dy = \cosh x - 1$.

(4) $\int_0^x \sqrt{x - y} f(y) dy = x^2 \sqrt{x}$.

(5) $\int_0^x (x - y) \sin(x - y) f(y) dy = \sin^2 x$.

35. Show that for any $g(x)$, the solution of the integral equation

$$f(x) = \int_0^x \sin(x - y) f(y) dy + g(x)$$

is

$$f(x) = \int_0^x (x - y)g(y)\mathrm{d}y + g(x).$$

36. Show that the integral equation

$$f(x) = a + bx + \int_0^x \mathrm{d}y f(y)[c + d(x - y)]$$

has a solution $f(x) = Ae^{\alpha x} + Be^{\beta x}$, where A, B, α, β depend on a, b, c, d.

37. Solve the following Volterra integral equations of convolution type.

(1) $\int_0^x f(x - y)f(y)\mathrm{d}y = A^2 x^\alpha$.

(2) $\int_0^x f(x - y)f(y)\mathrm{d}y = A^2 e^{\beta x}$.

(3) $\int_0^x f(x - y)f(y)\mathrm{d}y = A^2 x^\alpha e^{\beta x}$.

(4) $\int_0^x f(x - y)f(y)\mathrm{d}y = A \sin \alpha x$.

(5) $\int_0^x f(x - y)f(y)\mathrm{d}y = Ae^{\beta x} \sin \alpha x$.

38. Consider the nonlinear integral equation

$$2f(x) = \int_0^x f(x - y)f(y)\mathrm{d}y + g(x).$$

Show that the Laplace transformation of the solution of this equation is $Q(p) = \frac{G(p)}{1+\sqrt{1-G(p)}}$, where $Q(p) = [f(x)], G(p) = L[g(x)]$.

(1) Show that as $g(x) = \sin x$, the solution is $f(x) = J_1(x)$.
(2) What is the solution as $g(x) = -\sinh x$?

39. Solve the following integral equations of polynomial type.

(1) $\int_{-1}^1 (xy + x^2 y^2)f^2(y)\mathrm{d}y = f(x)$.

(2) $\int_{-1}^1 x^2 y^2 f^3(y)\mathrm{d}y = f(x)$.

40. Solve the following integral equations by means of generating function method.

(1) $\int_{-0}^{\infty} \frac{1}{1-x} \exp\left(-\frac{xy}{1-x}\right) e^{-y} f(y) dy = 2 - x^2.$

(2) $\int_{-1}^{1} \frac{f(y)}{\sqrt{1-2xy+x^2}} dy = 2x^3 - 2x.$

(3) $\int_{-1}^{1} \frac{f(y)}{\sqrt{1-2xy+x^2}} dy = \frac{1}{1-x}.$

Chapter 9

Application of Number Theory in Inverse Problems in Physics

9.1. Chen-Möbius Transformation

9.1.1. *Introduction*

1. The statement of inverse problems in physics

The fundamental laws in physics are expressed in the forms of equations in mathematics, such as Newton equation of motion in classical physics, Schrödinger equation in quantum physics, and so on. The equations present the relationship between physical quantities and govern the motion of systems. In principle, among certain physical quantities, when given some of them, others can be obtained. For examples, when given the force to a mass point, the orbit of the mass point and its position as a function of time can be calculated. The orbit is an experimentally measurable quantity. In quantum mechanics, from a given potential, the scattering amplitude of a particle scattered by this potential can be calculated. The scattering amplitude is also an experimentally measurable quantity.

The process of treating a problem is that starting from some **intrinsic physical quantities** of a system, some experimentally measurable quantities are derived observing fundamental physical laws. Then, the calculated results are compared with the measured ones. In this way, whether the theory is correct or not is tested. An example is the computation of the specific heat of a crystal. According to the theory of crystal lattice vibration, the specific heat of a crystal can be evaluated from phonon density of states. The converse work, computation of the phonon density of states from the specific heat, is very difficult to do. However, the specific heat can be measured in experiments easily. Therefore, it is desirable to explore an inverse transformation, or in short inversion, in order to calculate the density of states from the specific heat. This example indicates that one often needs to deal with reverse problems.

Usually, starting from a set of physical quantities A, another set of physical quantities B can be calculated according to physical laws. The reverse work, calculation of A from B, is called an **inverse problem** or a **reverse problem**. This is a general statement of an inverse problem. In each specialized field, the inverse problem has its own concrete statement.

In a dynamic system, it is to find the force acting on a mass point based on its law of motion. A famous example is the movement of planets. Kepler summarized the three laws of planetary motion. Based on the laws, Newton deduced that the planets should suffer an attractive force that was inverse to the square of the distance between a planet and the sun. This demonstrates that the very first problem dealt with in mechanics is just an inverse problem. In a quantum system, usually the known quantity is the scattering amplitude of a particle measured experimentally, from which the potential distribution that the particle suffers is derived. The inverse problem has significant application in practice.

So, the inverse problems come from the requirements in reality. Physical quantities A are usually intrinsic ones, and are difficult to detect in experiments. But physical quantities B can be measured experimentally.

In mathematical form, the relationship between the quantities A and B appear as transformation and inverse transformation, which may be either simple or complicated. The simplest one is the well-known Fourier transformation. Through a pair of formulas of known transformation and its inverse, it is possible to evaluate one set of quantities from another.

The inverse problem is to explore the mathematical formula required by the inverse transformation. In the case of the specific heat of a crystal, it is to find an explicit expression to calculated density of states by means of specific heat. For various inverse problems, corresponding methods have been proposed. An important means is the transformations and their inverse in mathematics, such as the Fourier transformation, Laplace transformation, Hankel transformation, Mellin transformation, and so on. Solving integral equations introduced in Chapter 8 is also one of the methods dealing with inverse problems. Nevertheless, physical formulas do not always happen by chance in the forms of the known mathematical transformations. Therefore, it is desired to explore new transformations.

Chen had a breakthrough in inverse problems. He opened a new path: he found a mathematical means in number theory, and successfully applied number theory to solve inverse problems in physics. His main idea was to reform the Möbius inverse in number theory so as to make it applicable in the case of continuous variables.

2. The significance of integers in physics

In classical physics, physical quantities are usually continuous. However, in quantum physics, the integers play an important role. The features of discrete values become remarkable. The followings are some examples.

The energies of a microscopic particle in a stable state are discrete. That is to say, there is an interval between adjacent energy levels. The form of existence and motion of a particle obeys Schrödinger equation, which is a differential equation of second order. We have learnt in Chapter 3 that the eigenvalues of such equations are discrete, and constitute a monotonically rising sequence. The energy of the particle must take one of the eigenvalues.

A stable system of a particle in a bound state is of discrete energy eigenvalues. For a system consisting of a great amount of microscopic particles, the allowed energies are also discrete. For example, the energy distribution of electrons in a solid state appears as energy bands. In each energy band, the energies are substantially discrete, but they are so dense that the band can be treated as if a continuous spectrum. A continuous spectrum can be regarded as the limit case of a discrete one when the discrete energies are sufficiently dense.

The discreteness of eigenenergies causes that the absorption and emission of energies of a system should have a least quantum. The electromagnetic field consists of photons, which shows that the energy distribution in the field is essentially discontinuous, although it behaves more like a wave when its wavelength is comparatively longer.

Quantum information and quantum communication are fields rapidly developed recently. They are to implement the storage, transformation and transmission of information by making use of microscopic states and their variation. In the theory of the quantum information and quantum communication, the knowledge of number theory are employed, such as congruence needed in the quantum Fourier transformation, finding prime factors of numbers with hundreds of digits needed in quantum encoding, and so on.

These examples indicate that to describe the physical quantities of microscopic systems, it is necessary resorting to the knowledge of number theory. Since number theory mainly investigates the functions of natural numbers, no one thought that it would be useful in physics. Chen's work revealed that number theory had significant applications in the fields of physics, material science, and so on.

9.1.2.   *Möbius transformation*

1. Basic concepts and theorems in number theory

We first introduce some basic concepts and theorems in number theory.

Definition 1. A function $f(n)$ is called a **number-theoretic function** if it has a definite value for any positive integer n.

For any positive integer, there is a fundamental theorem as follows.

Theorem 1 (Fundamental theorem of arithmetic). *Any positive integer can be expressed as the products of powers of some prime* *numbers:*

$$n = p_1^{r_1} p_2^{r_2} p_3^{r_3} \cdots\cdots p_s^{r_s} = \prod_{i=1}^{s} p_i^{r_i}, \qquad (9.1.1)$$

and the expression is unique. Every factor in (9.1.1) is a prime number and they are different from each other. The powers are natural numbers.

Denote (m, n) as the **greatest common divisor** of integers m and n, and $[m, n]$ as the **least common multiple** of them. It must be that $[m, n](m, n) = mn$.

Definition 2. A number-theoretic function $f(n)$ is called a **completely multiplicative function** if for any positive integers m and n, there stands $f(m)f(n) = f(mn)$. A function $f(n)$ is called a **multiplicative function** or **relatively multiplicative function** if there stands $f(m)f(n) = f(mn)$ for $(m, n) = 1$, m and n have no common divisor other than 1, or they are coprime positive integers.

Multiplicative functions are of the following properties.

(i) If $f(n)$ is a nonzero multiplicative function, then $f(1) = 1$.
(ii) If $f(n)$ is a multiplicative function, then $f([m, n])f((m, n)) = f(m)f(n)$.
(iii) If $g(n)$ and $h(n)$ are multiplicative functions, then their product $g(n)h(n)$ is still a multiplicative function, and $f(n) = \sum_{d/n} g(d)h(\frac{n}{d}) = \sum_{d/n} g(\frac{n}{d})h(d)$ is also a multiplicative one, where d/n means that the summation over all possible integer factors d of n, including 1 and n.

A typical completely multiplicative function is $f(n) = n^a$, where a is any complex number.

It is easy to prove the formula

$$\sum_{d/n} f(d) = \sum_{(n/d)/n} f(n/d). \qquad (9.1.2)$$

It is obvious that d involves all integer factors of n, and so does n/d.

Table 9.1.   Some values of the Möbius function.

n	1	2	3	4	5	6	7	8	9	10
$\mu(n)$	1	-1	-1	0	-1	1	-1	0	0	1
n	11	12	13	14	15	16	17	18	19	20
$\mu(n)$	-1	0	-1	1	1	0	-1	0	-1	0
n	21	22	23	24	25	26	27	28	29	30
$\mu(n)$	1	1	-1	0	0	1	0	0	-1	-1

Definition 3. A particular number-theoretic function, called the **Möbius function**, is defined as follows.

$$\mu(n) = \begin{cases} 1, & n = 1, \\ (-1)^r, & n \text{ are the products of } r \text{ different prime factors,} \\ 0, & \text{other } n. \end{cases}$$

$$(9.1.3)$$

Table 9.1 lists some values of the Möbius function.

Theorem 2. *The Möbius function is a multiplicative function.*

Proof.   When $(m, n) = 1$, m and n can be respectively expressed by

$$n = p_1^{r_1} p_2^{r_2} p_3^{r_3} \cdots \cdots p_s^{r_s}, \quad m = q_1^{u_1} q_2^{u_2} q_3^{u_3} \cdots \cdots q_t^{u_t}, \qquad (9.1.4a)$$

where all p and q are different. In this case, if one of the powers of factors in n is larger than 1, e.g., $r_i > 1$, we have $\mu(n) = \mu(p_1^{r_1} p_2^{r_2} p_3^{r_3} \cdots \cdots p_s^{r_s}) = 0$. So,

$$\mu(n)\mu(m) = 0 = \mu(p_1^{r_1} p_2^{r_2} p_3^{r_3} \cdots \cdots p_s^{r_s} m).$$

If all the powers in (9.1.4a) are 1, i.e.,

$$n = p_1 p_2 p_3 \cdots \cdots p_s, \quad m = q_1 q_2 q_3 \cdots \cdots q_t, \qquad (9.1.4b)$$

then

$$\mu(n)\mu(m) = (-1)^s (-1)^t = (-1)^{s+t} = \mu(nm).$$

The conclusion is that as $(m, n) = 1$ we always get

$$\mu(n)\mu(m) = \mu(nm). \qquad (9.1.5)$$

If one p_i and one q_j in (9.1.4b) are the same, e.g., $p_1 = q_1$, then $\mu(n) = (-1)^s$ and $\mu(m) = (-1)^t$. However, $\mu(nm) = \mu(p_1^2 p_2 p_3 \cdots p_s q_2 q_3 \cdots q_t) = 0$, which will results in $\mu(n)\mu(m) \neq \mu(nm)$. The conclusion is that $\mu(n)$ is not a completely multiplicative function.

□

Following the sum rule,

$$\sum_{n/k} \delta_{n,1} = 1. \tag{9.1.6}$$

Theorem 3.

$$\sum_{n/k} \mu(n) = \delta_{k,1} = \begin{cases} 1, & k = 1, \\ 0, & k \text{ is other natural numbers except } 1. \end{cases}$$

$$\tag{9.1.7}$$

Proof. As $k = 1$, there is only one term $k = 1$ in (9.1.7). By definition (9.1.3), the result is 1. As $k \neq 1$, any positive integer can be expressed in the form of (9.1.1), which says that k contains s different prime factors $p_1, p_2, \ldots, p_s$. The summation in (9.1.7) covers the terms containing different prime factors:

$$\sum_{n/k} \mu(n) = \mu(1) + \sum_{i=1}^{s} \mu(p_i) + \sum_{i,j=1,i\neq j}^{s} \mu(p_i p_j)$$

$$+ \cdots + \mu(p_1 p_2 \cdots p_s). \tag{9.1.8}$$

Other terms, if any, include at least a factor with power greater than 1, e.g., p_i^2. Such terms are necessarily zero. On the right hand side of (9.1.8), the first summation contains s terms, each being the value -1. The second summation contains C_s^2 terms, which is the number of combination of taking 2 different numbers from s prime numbers, each being value 1. The next term contains C_s^3 terms, and so on. The final result should be

$$\sum_{n/k} \mu(n) = 1 + (-1)C_s^1 + (-1)^2 C_s^2 + \cdots + (-1)^s C_s^s$$

$$= \sum_{i=0}^{s} (-1)^i C_s^i = (1 - 1)^s = 0.$$

□

2. Möbius transformation

Theorem 4. *Suppose that $F(n)$ and $f(n)$ are number-theoretic functions. If*

$$F(n) = \sum_{d/n} f(d) = \sum_{(n/k)/n} f(d = n/k) = \sum_{k/n} f(n/k), \qquad (9.1.9)$$

then

$$f(n) = \sum_{d/n} \mu(d) F(n/d), \qquad (9.1.10)$$

*and vice versa. $F(n)$ is called the **Möbius transformation** of $f(n)$, and $f(n)$ is the **inverse Möbius transformation** of $F(n)$.*

Proof. We derive (9.1.10) from (9.1.9). Starting from the right hand side of (9.1.10), by (9.1.9), we gain

$$\sum_{d/n} \mu(d) F(n/d) = \sum_{d/n} \mu(d) \sum_{k/\frac{n}{d}} f(k) = \sum_{d/n} \mu(d) \sum_{kd/n} f(k)$$

$$= \sum_{k/n} f(k) \sum_{kd/n} \mu(d) = \sum_{k/n} f(k) \delta_{n/k,1} = f(n).$$

This is just the left hand side of (9.1.10). Conversely, we can derive (9.1.9) from (9.1.10). Starting from the right hand side of (9.1.9), by (9.1.10), we gain

$$\sum_{d/n} f(d) = \sum_{d/n} f\left(\frac{n}{d}\right) = \sum_{d/n} \sum_{k/\frac{n}{d}} \mu(k) F\left(\frac{n}{dk}\right) = \sum_{d/n} \sum_{kd/n} \mu(k) F\left(\frac{n}{dk}\right)$$

$$= \sum_{k/n} F(k) \sum_{kd/n} \mu\left(\frac{n}{dk}\right) = \sum_{k/n} F(k) \delta_{n/k,1} = F(n).$$

This is just the left hand side of (9.1.9).   □

It is seen that when proving (9.1.10) from (9.1.9), (9.1.7) is needed. Once Theorem 4 is proved, it is seen that Theorem 3 is just a particular case of Theorem 4. That is to say, (9.1.7) shows that the function $\delta_{k,1}$ is just the inverse transformation of function 1, and its transformation is (9.1.6).

The Möbius transformation is a fundamental and important transformation. The following two theorems are just its generalization.

Theorem 5. *Suppose that $F(n)$, $f(n)$ and $r(n)$ are number-theoretic functions, and $r(n)$ is a completely multiplicative one. If*

$$F(n) = \sum_{d/n} r(d) f\left(\frac{n}{d}\right), \tag{9.1.11}$$

then

$$f(n) = \sum_{d/n} r(d)\mu(d)F(n/d), \tag{9.1.12}$$

and vice versa. Equations (9.1.11) and (9.1.12) are called generalized Möbius transformation and inverse transformation.

Proof.   We merely prove that (9.1.12) can be derived from (9.1.11).

$$\sum_{d/n} r(d)\mu(d)F(n/d) = \sum_{d/n} r(d)\mu(d) \sum_{k/\frac{n}{d}} r(k) f\left(\frac{n}{kd}\right)$$

$$= \sum_{kd/n} r(kd)\mu(d) f\left(\frac{n}{kd}\right)$$

$$= \sum_{m/n, d/m} r(m)\mu(d) f\left(\frac{n}{m}\right)$$

$$= \sum_{m/n} r(m) f\left(\frac{n}{m}\right) \sum_{d/m} \mu(d)$$

$$= \sum_{m/n} r(m) f\left(\frac{n}{m}\right) \delta_{m,1} = r(1)f(n) = f(n).$$

$$\tag{9.1.13}$$

$\square$

Equation (9.1.13) requires that

$$\sum_{m/n, d/m} r(m)\mu(d) f\left(\frac{n}{m}\right) = f(n).$$

That is to say,

$$\sum_{d/m} r(m)\mu(d) = \delta_{m,1}. \tag{9.1.14}$$

It is easily seen that all the completely multiplicative functions meet (9.1.14). This equation is obvious, since

$$\sum_{d/m} r(m)\mu(d) = r(m)\sum_{d/m}\mu(d) = r(m)\delta_{m,1} = r(1)\delta_{m,1} = \delta_{m,1}.$$

Theorem 6. *Suppose that $F(n)$, $f(n)$ and $s(n)$ are number-theoretic functions. If*

$$F(n) = \sum_{d/n} s(d)f\left(\frac{n}{d}\right), \tag{9.1.15}$$

then

$$f(n) = \sum_{d/n} J(d)F(n/d), \tag{9.1.16}$$

where

$$\sum_{d/n} J(d)s\left(\frac{n}{d}\right) = \delta_{n,1}. \tag{9.1.17}$$

Proof. Substituting (9.1.15) into (9.1.16), we get

$$\sum_{d/n} J(d)F(n/d) = \sum_{d/n} J(d)\sum_{k/\frac{n}{d}} s(k)f\left(\frac{n}{kd}\right)$$

$$= \sum_{d/n}\sum_{kd/n} J(d)s(k)f\left(\frac{n}{kd}\right)$$

$$= \sum_{m/n} f\left(\frac{n}{m}\right)\delta_{m,1} = f(n).$$

It is also easy to obtain (9.1.15) from (9.1.16). $\square$

The discrepancy of Theorem 5 and Theorem 6 is that the function $r(n)$ is a completely multiplicative function, while the function $s(n)$ in the latter is not. Comparison of (9.1.17) and (9.1.14) shows that if $s(n)$ is a completely multiplicative function, then $J(d) = \mu(d)$.

9.1.3. Chen-Möbius transformation

1. Chen-Möbius transformation

We are going to reform the Möbius transformation. The key is that conceptually, n is not considered as a natural number, but is understood as n intervals. Based on this consideration, number-theoretic functions $F(n)$ and $f(n)$ are replaced by ordinary functions $A(\omega)$ and $B(\omega)$, where the variable ω is continuous. Nevertheless, we consider that ω is divided into n intervals. Then we let $n \rightarrow \infty$ so as to implement the transition from discrete variable to continuous variable: $F(n) \rightarrow A(\omega)$ and $f(n/d) \rightarrow B(\omega/d)$. Because n goes to infinity, any natural number is its factor. Thus, the summation over d/n in (9.1.9) should be replaced by the summation over all the natural numbers: from 1 to ∞. That is to say,

$$A(\omega) = \sum_{n=1}^{\infty} B(\omega/n). \qquad (9.1.18)$$

Its inverse is

$$B(\omega) = \sum_{n=1}^{\infty} \mu(n) A(\omega/n). \qquad (9.1.19)$$

These two equations are the **Chen-Möbius inverse transformation**.

The following is the proof given by Chen. If (9.1.18) stands, then,

$$\sum_{n=1}^{\infty} B(\omega/n) = \sum_{n=1}^{\infty} \sum_{m=1}^{\infty} \mu(m) A(\omega/mn)$$

$$= \sum_{k=1}^{\infty} \sum_{m/k} \mu(m) A(\omega/k) = \sum_{k=1}^{\infty} \delta_{k,1} A(\omega/k) = A(\omega).$$

This is just (9.1.18). If (9.1.18) stands, then,

$$\sum_{n=1}^{\infty} \mu(n)A(\omega/n) = \sum_{n=1}^{\infty} \mu(n) \sum_{m=1}^{\infty} B(\omega/mn)$$

$$= \sum_{k=1}^{\infty} \sum_{n/k} \mu(n)B(\omega/k) = \sum_{k=1}^{\infty} \delta_{k,1} B(\omega/k) = B(\omega).$$

This is just (9.1.19).

After Chen's primary generalization of the Möbius transformation, Hughes *et al.* suggested that this generalization could be proved in a broader sense by means of the Mellin transformation and the Riemann-ζ function. Chen's work soon attracted mathematicians' attention.

Möbius transformation has some generalizations. Similarly, Chen-Möbius transformation has also some generalizations. One is that

$$A(\omega) = \sum_{n=1}^{\infty} B(n\omega) \tag{9.1.20}$$

and its inverse

$$B(\omega) = \sum_{n=1}^{\infty} \mu(n)A(n\omega). \tag{9.1.21}$$

On the right hand sides of formulas (9.1.18) and (9.1.19), the argument is divided by n, while on those of (9.1.20) and (9.1.21), the argument is multiplied by n. Both cases hold.

Here we merely address the transformations (9.1.20) and (9.1.21). In Subsection 9.3.1, we will derive them by resorting to enlargement operator.

For the time being, we prove them in view of number theory. If (9.1.21) holds,

$$\sum_{n=1}^{\infty} B(n\omega) = \sum_{n=1}^{\infty} \sum_{m=1}^{\infty} \mu(m)A(mn\omega)$$

$$= \sum_{k=1}^{\infty} \sum_{m/k} \mu(m)A(k\omega) = \sum_{k=1}^{\infty} \delta_{k,1} A(k\omega) = A(\omega).$$

This is just (9.1.20). If (9.1.20) holds,

$$\sum_{n=1}^{\infty} \mu(n)A(n\omega) = \sum_{n=1}^{\infty} \mu(n) \sum_{m=1}^{\infty} B(mn\omega)$$

$$= \sum_{k=1}^{\infty}\sum_{n/k} \mu(n)B(k\omega) = \sum_{k=1}^{\infty} \delta_{k,1}B(k\omega) = B(\omega).$$

This is just (9.1.21).

In putting down (9.1.18)–(9.1.21), we have assumed that all the summations were convergent. Hereafter in this chapter, we always postulate that all the summations are convergent.

Following appendix 9A, it can be tested that a more generalized transformation is as follows: $A(x) = \sum_{n=1}^{\infty} B(n^{\alpha}x)$ and $B(x) = \sum_{n=1}^{\infty} \mu(n)A(n^{\alpha}x)$, where α is any complex number. In view of this transformation, (9.1.18)–(9.1.21) are just particular cases of them when $\alpha = -1$ and $\alpha = 1$. Some pairs of functions A and B were listed in Millane R P, Möbius transform pairs. *J. Math. Phys.*, 1993, **34**(2): 875.

The Chen-Möbius inverse transformation has found its extensive applications in various fields such as quantum field theory, condensed matter physics, material design theory, remote sensing and so on. Among them, the first application was done by Chen. Hereafter in this chapter, all the basic formulas were first obtained by Chen.

2. The inverse problem of blackbody radiation

Here is a primary application of the Chen-Möbius inverse transformation.

When receiving the light coming from a remote star, we can measure the total energy of unit frequency interval at a certain frequency, i.e., radiant emittance $W(\nu)$, a function of frequency. By means of statistical mechanics, its expression is as follows:

$$W(\nu) = \frac{2h\nu^3}{c^2} \int_0^{\infty} \frac{a(T)}{e^{h\nu/k_{\mathrm{B}}T} - 1} \mathrm{d}T, \qquad (9.1.22)$$

where $a(T)$ is the energy density varying with temperature. We hope to evaluate conversely the energy density $a(T)$ of the blackbody by means of $W(\nu)$. The process is as follows. Let $u = \frac{h}{k_B T}$. Then,

$$
W(\nu) = \frac{2h\nu^3}{c^2} \int_\infty^0 \frac{a(h/k_B u)}{e^{u\nu} - 1} d\frac{h}{k_B u}
$$

$$
= \frac{2h^2\nu^3}{k_B c^2} \sum_{n=1}^\infty \int_0^\infty \frac{a\,(h/k_B u)}{u^2} e^{-nu\nu} du
$$

$$
= \frac{2h^2\nu^3}{k_B c^2} \int_0^\infty \frac{1}{x} \sum_{n=1}^\infty \frac{na(nh/k_B x)}{x} e^{-x\nu} dx
$$

$$
= \frac{2h^2\nu^3}{k_B c^2} \int_0^\infty \frac{1}{x} G(x) e^{-x\nu} dx, \tag{9.1.23}
$$

where

$$
\sum_{n=1}^\infty \frac{na(nh/k_B x)}{x} = G(x). \tag{9.1.24}
$$

Comparing (9.1.24) and (9.1.18), we make correspondence

$$
G(\omega) \rightarrow A(\omega) \quad \text{and} \quad \frac{a(h/k_B x)}{x} \rightarrow B(\omega). \tag{9.1.25}
$$

So, the inverse transformation of (9.1.24) should be

$$
a(h/k_B x) = x \sum_{n=1}^\infty \mu(n) G\left(\frac{x}{n}\right), \tag{9.1.26}
$$

and consequently, the inverse of (9.1.23) is

$$
G(x) = xL^{-1}\left[\frac{k_B c^2}{2h^2\nu^3} W(\nu)\right] = \frac{k_B c^2}{2h^2} xL^{-1}\left[\frac{W(\nu)}{\nu^3}\right]. \tag{9.1.27}
$$

This equation is substituted into (9.1.26) to get

$$
a(T) = \frac{k_B c^2}{2h^2} T^2 \sum_{n=1}^\infty \frac{\mu(n)}{n} L^{-1}\left[\frac{W(\nu)}{\nu^3}\right]. \tag{9.1.28}
$$

On the right hand side, after the Laplace transformation is made, the argument x should be replaced back to temperature. In this way, we have implemented the required invers transformation. The key step is to make use of the Chen-Möbius transformation. Equation (9.1.28) enables one to evaluate $a(T)$ in terms of experimentally measurable $W(\nu)$, so as to test the correctness of theoretical models.

9.2. Inverse Problem in Phonon Density of States in Crystals

9.2.1. *Inversion formula*

According to the theory of vibration of crystal lattices, the **specific heat** of a crystal lattice as a function of temperature T is expressed by

$$C_\mathrm{V}(T) = \frac{\partial E(T)}{\partial T} = r k_\mathrm{B} \int_0^\infty \left(\frac{h\nu}{k_\mathrm{B}T}\right)^2 \frac{\mathrm{e}^{h\nu/k_\mathrm{B}T}}{\left(\mathrm{e}^{h\nu/k_\mathrm{B}T}-1\right)^2} g(\nu)\mathrm{d}\nu,$$

(9.2.1)

where r is atom number in a crystal cell, h and k_B are respectively Planck and Boltzmann constants and $g(\nu)$ as a function of ν is density of state of phonons. The integration in (9.2.1) covers all possible vibration frequencies of the crystal. The total number of the vibration modes in a crystal is $3N$, so that there is a normalization condition as follows:

$$\int_0^\infty g(\nu)\mathrm{d}\nu = 3N.$$

(9.2.2a)

As a matter of fact, the frequency spectrum of phonons has an upper limit, denoted as ν_m. The upper limit usually relies on temperature. So, (9.2.2a) can be written as

$$\int_0^{\nu_m(T)} g(\nu)\mathrm{d}\nu = 3N.$$

(9.2.2b)

The phonon density of states $g(\nu)$ is an intrinsic physical quantity of a crystal lattice system, from which the specific heat of the lattice can be calculated. However, this quantity as a function of frequency in fact is not known beforehand, while the specific heat $C_\mathrm{V}(T)$ can easily be measured in experiments. Therefore, what one needs is to evaluate $g(\nu)$ in terms of $C_V(T)$ obtained from experiments. Then, the resultant $g(\nu)$ can be utilized to calculate the quantities of the crystal lattice such as its total energy, equation of state, and so on.

In solid state physics, a Debye temperature was defined which is proportional to the upper limit of vibration frequency:

$$\Theta_D(T) = h\nu_m(T)/k_B. \tag{9.2.3}$$

This is a characteristic quantity expressing the upper limit of frequency but in the dimension of temperature. Usually, the Debye temperature is also a function of temperature.

We are going to derive reversely the expression of $g(\nu)$ in terms of $C_V(T)$. Let $u = h/k_B T$. Then, (9.2.1) becomes

$$C_V(T) = rk_B \int_0^\infty \frac{(u\nu)^2 e^{-u\nu}}{(1 - e^{-u\nu})^2} g(\nu) d\nu.$$

Since $u\nu > 1$, $e^{-u\nu} < 1$ can be expanded by the Taylor series:

$$C_V(h/k_B u) = rk_B \sum_{n=1}^\infty \int_0^\infty n\, (u\nu)^2 e^{-nu\nu} g(\nu) d\nu. \tag{9.2.4}$$

Let $\omega = n\nu$. Then,

$$C_V(h/k_B u) = rk_B \sum_{n=1}^\infty \int_0^\infty \left(u\frac{\omega}{n}\right)^2 e^{-u\omega} g\left(\frac{\omega}{n}\right) d\omega$$

$$= rk_B u^2 \int_0^\infty e^{-u\omega} \sum_{n=1}^\infty \left(\frac{\omega}{n}\right)^2 g\left(\frac{\omega}{n}\right) d\omega$$

$$= rk_B u^2 L\left[G(\omega)\right], \tag{9.2.5}$$

where $L[\;]$ represents the Laplace transformation. $L[f(\omega)] = \int_0^\infty e^{-u\omega} f(\omega) d\omega = F(u)$. Its inverse is $L^{-1}[F(u)] = f(\omega)$. Note that in (9.2.5) the definition of the function G is

$$G(\omega) = \sum_{n=1}^\infty \left(\frac{\omega}{n}\right)^2 g\left(\frac{\omega}{n}\right). \tag{9.2.6}$$

Comparing (9.2.7) and (9.1.18), we make correspondence

$$G(\omega) \to A(\omega) \quad \text{and} \quad \omega^2 g(\omega) \to B(\omega). \tag{9.2.7}$$

It is seen that the inverse of (9.2.7) is

$$g(\omega) = \frac{1}{\omega^2} \sum_{n=1}^{\infty} \mu(n) G\left(\frac{\omega}{n}\right). \tag{9.2.8}$$

Substitution of (9.2.8) into (9.2.5) brings

$$g(\nu) = \frac{1}{r k_B \nu^2} \sum_{n=1}^{\infty} \mu(n) L_n^{-1} \left[\frac{C_V(h/k_B u)}{u^2} \right], \tag{9.2.9}$$

where it is denoted that

$$L_n^{-1}[F(u)] = f\left(\frac{\omega}{n}\right). \tag{9.2.10}$$

After the inverse Laplace transformation, ω should be replaced back by ω/n.

In derivation above, (9.1.18) is used. We can also alternatively employ (9.1.20). To do so, we recast (9.2.4) to be

$$\frac{C_V(h/k_B u)}{r k_B u} = \sum_{n=1}^{\infty} nu \int_0^{\infty} e^{-nu\nu} \nu^2 g(\nu) d\nu$$

$$= \sum_{n=1}^{\infty} nu L_n[\nu^2 g(\nu)]. \tag{9.2.11}$$

Comparison to (9.1.20) shows that the correspondences we should make now are $A(u) \to \frac{C_V(h/k_B u)}{r k_B u}$ and $B(nu) \to nu L_n \left[\nu^2 g(\nu)\right]$. Then the inverse transformation (9.1.21) is used to get

$$uL[\nu^2 g(\nu)] = \sum_{n=1}^{\infty} \mu(n) \frac{C_V(h/k_B nu)}{r k_B nu}. \tag{9.2.12}$$

This equation is divided by u, and then the inverse Laplace transformation is made:

$$g(\nu) = \frac{1}{rk_B\nu^2} \sum_{n=1}^{\infty} \mu(n)L^{-1}\left[\frac{C_V(h/k_Bnu)}{nu^2}\right]. \qquad (9.2.13)$$

Equations (9.2.13) and (9.2.9) have slight difference in forms, but they are essentially the same.

Thus, we make use of Chen-Möbius inverse transformation to smartly obtain the expression of $g(\nu)$ in terms of $C_V(T)$. As long as the expression of $C_V(T)$ as a function of temperature is known, the phonon density of states as a function of frequency can be evaluated. Either (9.2.9) or (9.2.13) can be applied. It depends on which one is more convenient for a specific problem. These two formulas are valid for the whole temperature range. However, the inverse Laplace transformations cannot be done because the explicit form of $C_V(T)$ is not known. In low and high temperatures, $C_V(T)$ has well-known approximate expressions. In the following, we discuss low- and high-temperature approximations, respectively.

9.2.2. *Low-temperature approximation*

1. The inverse transformation at low temperature

When temperature is very low, $T \to 0$, it has been concluded from experiments that $C_V(T)$ has the form of

$$C_V(T) = a_3 T^3 + a_5 T^5 + a_7 T^7 + \cdots = \sum_{n=1}^{\infty} a_{2n+1}T^{2n+1}. \qquad (9.2.14)$$

That is to say, the expansion of the specific heat merely contains the odd powers of temperature, with the first term being cubic. It is assumed that the coefficients a_{2n+1} have been determined from experiments. Equation (9.2.14) is rewritten in the following form:

$$C_V\left(\frac{h}{k_Bu}\right) = \sum_{n=1}^{\infty} a_{2n+1}\left(\frac{h}{k_Bu}\right)^{2n+1}. \qquad (9.2.15)$$

This is substituted into (9.2.9) to get

$$g(\nu) = \frac{1}{rk_B\nu} \sum_{m=1}^{\infty} \frac{a_{2m+1}}{(2m+2)!}\left(\frac{h\nu}{k_B}\right)^{2m+1} \sum_{n=1}^{\infty} \frac{\mu(n)}{n^{2m+2}}.$$

In Appendix 9A, the following two equations are given:

$$\sum_{n=1}^{\infty} \frac{\mu(n)}{n^{2m}} = \frac{1}{\zeta(2m)} \tag{9.2.16}$$

and

$$\zeta(2m) = (-1)^{m+1} \frac{(2\pi)^{2m}}{2(2m)!} B_{2m},$$

where $\zeta(x)$ is the **Riemann-ζ function**. Thus, we obtain

$$g(\nu) = \frac{2}{rk_{\mathrm{B}}^2 \nu} \sum_{m=1} a_{2m+1} \frac{(-1)^{m+1}(h\nu/k_{\mathrm{B}})^{2m+1}}{(2\pi)^{2m+2} B_{2m+2}} = \sum_{m=1} A_{2m} \nu^{2m}. \tag{9.2.17}$$

It is seen that the expansion of the phonon density of states merely contains the even powers of frequency, with the first term being square.

By the normalization condition (9.2.2), the upper limit of frequency ν_m can be evaluated in principle. For example, if only the lowest term is taken,

$$g(\nu) = -\frac{2}{rk_{\mathrm{B}}^2 \nu} a_3 \frac{(h\nu/k_{\mathrm{B}})^3}{(2\pi)^4 B_4} = \frac{15a_3 h^3}{4\pi^4 rk_{\mathrm{B}}^5} \nu^2, \tag{9.2.18}$$

then by (9.2.2) we gain

$$\frac{5a_3 h^3}{4\pi^4 k_{\mathrm{B}}^4} \nu_m^3 = 3Nr = 3r \frac{R}{k_{\mathrm{B}}}, \tag{9.2.19}$$

where R is the gas constant. Thus, $\nu_m = \left(\frac{12\pi Rr}{5a_3}\right)^{1/3} \frac{\pi k_{\mathrm{B}}}{h}$. Subsequently, the Debye temperature is

$$\Theta_{\mathrm{D}} = \left(\frac{12\pi Rr}{5a_3}\right)^{1/3} \pi. \tag{9.2.20}$$

In (9.2.14), all the coefficients are independent of temperature. In doing the inverse Laplace transformation, integration with respect to temperature has been carried out. In this way, the obtained Debye temperature is independent of temperature.

2. Variation of Debye temperature with temperature

It is known from solid state physics that when the specific heat of a crystal is proportional to the cubic of temperature, the Debye temperature in fact relies on temperature. Accordingly, the specific heat should be rewritten in the form of

$$C_V(T) = b_3(T)T^3, \tag{9.2.21}$$

where the coefficient varies with temperature. However, from experiments, the specific heat is in the form of (9.2.14). Hence, it is fitted by (9.2.21). One simple case is that the lowest three terms are kept:

$$C_V(T) = a_3 T^3 + a_5 T^5 + a_7 T^7. \tag{9.2.22}$$

When written in the form of (9.2.21), it should be that

$$b_3(T) = a_3 + a_5 T^2 + a_7 T^4. \tag{9.2.23}$$

In doing the inverse Laplace transformation, $b_3(T)$ is considered a constant. Hence, the obtained frequency spectrum through the inverse is that

$$
g(\nu) = \frac{1}{r k_B^2 \nu^2} \sum_{n=1}^{\infty} \mu(n) L_n^{-1} \left[\frac{1}{u^2} b_3(T) \left(\frac{h}{k_B u} \right)^3 \right]
$$
$$
= \frac{b_3(T)}{r k_B^2 \nu^2} \left(\frac{h}{k_B} \right)^3 \sum_{n=1}^{\infty} \mu(n) L_n^{-1} \left[\frac{1}{u^5} \right] = \frac{b_3(T) h^3}{r k_B^5} \frac{\nu^2}{4!} \sum_{n=1}^{\infty} \frac{\mu(n)}{n^4}.
$$

By means of formulas in appendix 9A, it is calculated that

$$g(\nu) = \frac{15 b_3(T) h^3}{4\pi^4 r k_B^5} \nu^2. \tag{9.2.24}$$

This result shows that as long as the specific heat is proportional to the cubic of temperature, in the form of (9.2.21), the frequency spectrum is proportional to the square of frequency.

Equation (9.2.24) is actually the same as (9.2.18) which is the first term of (9.2.17). One merely needs to replace a_3 in (9.2.18) by $b_3(T)$ in (9.2.23). The Debye temperature is

$$\Theta_{\mathrm{D}}(T) = \left[\frac{12\pi R r}{5(a_3 + a_5 T^2 + a_7 T^4)} \right]^{1/3} \pi, \tag{9.2.25}$$

which reflects the dependence on temperature. As the temperature goes to zero, the Debye the temperature approaches a constant, which is just (9.2.20).

It is seen that by means of the Chen-Möbius inverse transformation, a concise method is achieved to calculate the Debye temperature from the experimental specific heat.

9.2.3. *High-temperature approximation*

1. The inverse transformation at high temperature

As the temperature goes to infinity, the specific heat will approach a constant. The dependence of the specific heat on temperature is as follows.

$$C_{\mathrm{V}}(T) = b_0 - b_2 T^{-2} + b_4 T^{-4} - b_6 T^{-6} + \cdots, \tag{9.2.26}$$

or

$$C_{\mathrm{V}}\left(\frac{h}{k_{\mathrm{B}} n u}\right) = \sum_{m=0} b_{2m} \left(\frac{k_{\mathrm{B}} n u}{h}\right)^{2m}. \tag{9.2.27}$$

When this is substituted into (9.2.13), we get

$$g(\nu) = \frac{1}{r k_{\mathrm{B}} \nu^2} \sum_{n=1}^{\infty} \mu(n) L^{-1} \left[\frac{1}{n u^2} \sum_{m=0} b_{2m} \left(\frac{k_{\mathrm{B}} n u}{h}\right)^{2m} \right].$$

Here the following inverse Laplace transformations are needed:

$$L^{-1}[u^{-2}] = \nu, \quad L^{-1}[u^m] = \delta^{(m)}(\nu), \quad m \geq 0.$$

So,

$$g(\nu) = \frac{b_0}{r k_{\mathrm{B}} \nu^2} \frac{\nu}{\zeta(1)} + \frac{1}{r k_{\mathrm{B}} \nu^2} \sum_{m=1} b_{2m} \left(\frac{k_{\mathrm{B}}}{h}\right)^{2m} \frac{\delta^{(2m-2)}(\nu)}{\zeta(1-2m)}, \tag{9.2.28}$$

where (9.2.16) and (9A.5) in appendix 9A are used. Because $\zeta(1) \to \infty$, the first term in (9.2.28) is zero. As a result,

$$g(\nu) = \frac{1}{rk_B\nu^2} \sum_{m=1} \frac{b_{2m}}{\zeta(1-2m)} \left(\frac{k_B}{h}\right)^{2m} \delta^{(2m-2)}(\nu). \qquad (9.2.29)$$

The conclusion is that the constant term in (9.2.26) does not contribute to the frequency spectrum of phonons. Therefore, the discussion below will not concern the constant term in (9.2.26).

If only one term containing b_2 in (9.2.26) remained, it will lead to an unreasonable result. This is because (9.2.29) will give

$$g(\nu) = \frac{b_1}{rk_B\nu^2\zeta(-1)} \left(\frac{k_B}{h}\right)^2 \delta(\nu)$$

which requires that the frequency has to be zero and the value of $g(0)$ is infinite.

2. Einstein unimodal approximation

Let us assume that the coefficients in (9.2.26) happen to be

$$b_{2m} = b \left(\frac{h}{k_B}\nu_E\right)^{2m} \frac{\zeta(1-2m)}{(2m-2)!}. \qquad (9.2.30)$$

They are substituted into (9.2.29) to produce

$$g(\nu) = \frac{b}{rk_B\nu^2} \left(\frac{h}{k_B}\nu_E\right)^2 \sum_{m=1} \frac{\nu_E^{2m-2}}{(2m-2)!} \delta^{(2m-2)}(\nu)$$

$$= \frac{bh^2\nu_E^2}{rk_B^3\nu^2} \sum_{m=0} \frac{\nu_E^{2m}}{(2m)!} \delta^{(2m)}(\nu) = \frac{bh^2}{rk_B^3} \delta(\nu - \nu_E). \qquad (9.2.31)$$

This result shows that the vibration has only one frequency, which is the Einstein unimodal approximation. This is sometimes a good approximation for the optical branch of lattice vibration.

In reality, it is of course impossible to obtain infinite terms in (9.2.26). We reasonably assume that the first two terms are known:

$$C_V(T) = -b_2T^{-2} + b_4T^{-4}.$$

Then, by (9.2.29) we gain

$$g(\nu) = \frac{1}{rk_{\mathrm{B}}\nu^2} \left[-\frac{b_2}{\zeta(-1)} \left(\frac{k_{\mathrm{B}}}{h}\right)^2 \delta(\nu) + \frac{b_4}{\zeta(-3)} \left(\frac{k_{\mathrm{B}}}{h}\right)^4 \delta^{(2)}(\nu) \right]$$

$$= -\frac{b_2}{rk_{\mathrm{B}}\nu^2\zeta(-1)} \left[\delta(\nu) - \frac{\zeta(-1)}{\zeta(-3)} \frac{b_4}{b_2} \left(\frac{k_{\mathrm{B}}}{h}\right)^2 \delta^{(2)}(\nu) \right],$$

where $\zeta(-1) = -\frac{1}{12}$ and $\zeta(-3) = \frac{1}{120}$. Let

$$\nu_{\mathrm{E}} = \frac{k_{\mathrm{B}}}{h} \sqrt{\frac{20 b_4}{b_2}}. \tag{9.2.32}$$

Then, the spectrum becomes

$$g(\nu) = \frac{12 b_2}{rk_{\mathrm{B}}\nu^2} \left[\delta(\nu) + \nu_{\mathrm{E}}^2 \delta^{(2)}(\nu) \right]$$

$$\approx \frac{12 b_2}{rk_{\mathrm{B}}\nu^2} \delta(\nu - \nu_{\mathrm{E}}) = \frac{12 b_2}{rk_{\mathrm{B}}\nu_{\mathrm{E}}^2} \delta(\nu - \nu_{\mathrm{E}}). \tag{9.2.33}$$

Here an approximation is made so as to retrieve the Einstein unimodal approximation. The Einstein frequency is expressed by (9.2.32) and can be evaluated by experimentally measured coefficients.

3. Bimodal approximation

Suppose that the first four terms have been known:

$$C_{\mathrm{V}}(T) = -b_2 T^{-2} + b_4 T^{-4} + b_6 T^{-6} + b_8 T^{-8}.$$

Then, similarly to (9.2.33), we will find that the frequency spectrum contains two isolated peaks. The derivation process is as follows.

In (9.2.29), the first four terms remained.

$$g(\nu) = \frac{1}{rk_{\mathrm{B}}\nu^2} \left[-\frac{b_2}{\zeta(-1)} \left(\frac{k_{\mathrm{B}}}{h}\right)^2 \delta(\nu) + \frac{b_4}{\zeta(-3)} \left(\frac{k_{\mathrm{B}}}{h}\right)^4 \delta^{(2)}(\nu) \right.$$

$$\left. -\frac{b_6}{\zeta(-5)} \left(\frac{k_{\mathrm{B}}}{h}\right)^6 \delta^{(4)}(\nu) - \frac{b_8}{\zeta(-7)} \left(\frac{k_{\mathrm{B}}}{h}\right)^8 \delta^{(6)}(\nu) \right]$$

$$= \frac{12b_2}{k\nu^2} \left(\frac{k_B}{h}\right)^2 \left\{ B_1 \left[\delta(\nu) + \frac{1}{2}\nu_1^2\delta^{(2)}(\nu) \right.\right.$$

$$\left. + \frac{1}{4!}\nu_1^4\delta^{(4)}(\nu) + \frac{1}{6!}\nu_1^6\delta^{(6)}(\nu)\right]$$

$$\left. + B_2 \left[\delta(\nu) + \frac{1}{2}\nu_2^2\delta^{(2)}(\nu) + \frac{1}{4!}\nu_2^4\delta^{(4)}(\nu) + \frac{1}{6!}\nu_2^6\delta^{(6)}(\nu)\right] \right\}$$

$$\approx \frac{12b_2}{k\nu^2} \left(\frac{k_B}{h}\right)^2 [B_1\delta(\nu - \nu_1) + B_2\delta(\nu - \nu_2)].$$

Indeed, the spectrum contains two peaks under the approximation. The two frequencies are ν_1 and ν_2, and their strengths are proportional to B_1 and B_2. The four values are determine by the following equations. $B_1 + B_2 = 1$, $B_1\frac{1}{2}\nu_1^2 + B_2\frac{1}{2}\nu_2^2 = 10\frac{b_4}{b_2}\left(\frac{k_B}{h}\right)^2$, $B_1\frac{1}{4!}\nu_1^4 + B_2\frac{1}{4!}\nu_2^4 = 21\frac{b_6}{b_2}\left(\frac{k_B}{h}\right)^4$ and $B_1\frac{1}{6!}\nu_1^6 + B_2\frac{1}{6!}\nu_2^6 = 20\frac{b_8}{b_2}\left(\frac{k_B}{h}\right)^6$. Numerical computation is needed.

9.3. Inverse Problem in the Interaction Potential between Atoms

A crystal consists of atoms at lattice sites. There are interactions between the atoms. These interactions cause the vibration of the crystal lattice. The total energy of the lattice is the summation of all the interaction energies between atoms. The simplest consideration is that the interaction between two atoms is independent of any other atom. This is a kind of classical model: every atom is regarded as a geometric point and the effect of electronic cloud on the interaction is neglected. Such a kind of interatomic potential is called **pair potential** between atoms, which is a function of distance r between the two atoms, denoted by $\Phi(r)$. We consider ideal crystals without defects. A crystal lattice is of translational invariance. The total energy of a crystal can be evaluated by first-principle calculation. The corresponding inverse problem is to find the pair potentials by means of the calculated total energy. The Chen-Möbius transformation proves valuable.

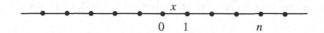

Fig. 9.1. A one-dimensional Bravais lattice.

In this section, the simplest crystal lattices are considered: every crystal cell contains only one atom, so-called Bravais lattices.

9.3.1. *One-dimensional case*

The atom arrangement in a one-dimensional Bravais lattice is sketched in Fig. 9.1.

1. Transformation between total energy and pair potential

Let the distance between the nearest neighbor atoms be x. The distance of any two atoms is the integer times of x. The pair potential denoted by Φ, the total energy of the lattice can be expressed by

$$E(x) = \frac{1}{2} \sum_{n \neq 0} \Phi(nx) = \sum_{n=1}^{\infty} \Phi(nx). \qquad (9.3.1)$$

Since it is a Bravais lattice, what we need to calculate is merely the interaction energy between one atom and all others. It should be pointed out that this calculated energy, denoted by E_1, is different from the average interaction energy between one atom and all the others, denoted by E_2. E_1 is just two times of E_2.

$E(x)$ is a function of the nearest neighbor distance x. What we want to do is to express conversely $\Phi(x)$ in terms of $E(x)$.

We define an **enlargement operator** T_n. Its effect is to enlarge the argument by n times: $x \rightarrow nx$.

$$T_n \Phi(x) = \Phi(nx). \qquad (9.3.2)$$

One property of this operator is that if two operators act successively, the result will be

$$T_m T_n = T_{mn}. \qquad (9.3.3)$$

This property shows that the enlargement operator on one-dimensional periodical lattices is a completely multiplicative function.

By means of the enlargement operator, (9.3.1) is recast to be

$$E(x) = \sum_{n=1}^{\infty} T_n \Phi(x). \qquad (9.3.4)$$

Now $\sum_{n=1}^{\infty} T_n$ is an operator acting on the function $\Phi(x)$. Formally, we can always apply an inverse operator on both sides of (9.3.4) to get

$$\Phi(x) = \left[\sum_{n=1}^{\infty} T_n \right]^{-1} E(x). \qquad (9.3.5)$$

Note that the n in T_n is an integer. We recall the first theorem in this chapter, (9.1.1). A result derived from this theorem is that the summation of all the positive integers can be written as follows:

$$\sum_{n=1}^{\infty} n = \sum_{n=1}^{\infty} \prod_{i=1}^{s} p_i^{r_i}$$

$$= \left(1 + p_1 + p_1^2 + p_1^3 + \cdots \right) \left(1 + p_2 + p_2^2 + p_2^3 + \cdots \right)$$

$$\times \left(1 + p_3 + p_3^2 + p_3^3 + \cdots \right) \cdots \cdots. \qquad (9.3.6a)$$

The terms on the left and right hand sides are one-to-one correspondent. This equation is called the Euler formula. In each parentheses, the summation can be formally carried out:

$$\sum_{n=1}^{\infty} n = \prod_{i=1}^{\infty} \left(1 + p_i + p_i^2 + p_i^3 + \cdots \right) = \prod_{i=1}^{\infty} \frac{1}{1 - p_i}. \qquad (9.3.6b)$$

Its inverse is

$$\left[\sum_{n=1}^{\infty} n \right]^{-1} = \left[\prod_{i=1}^{\infty} \frac{1}{1 - p_i} \right]^{-1} = \prod_{i=1}^{\infty} (1 - p_i)$$

$$= (1 - p_1)(1 - p_2)(1 - p_3) \cdots$$

$$= 1 - p_1 - p_2 \cdots + p_1 p_2 + p_1 p_3 + \cdots$$

$$= 1 + \sum_{\{p_1 p_2 \cdots p_s\}} (-1)^s p_1 p_2 \cdots p_s. \qquad (9.3.7a)$$

In each term on the right hand side, every factor p_i in $p_1 p_2 \cdots p_s$ is different from the others. Each term is negative (positive) if it contains an odd (even) number of factors. If the power of one factor p_i is greater than 1, this term should be zero. Thus, (9.3.7a) can be recast to

$$\left[\sum_{n=1}^{\infty} n \right]^{-1} = 1 + \sum_{\{p_1 p_2 \cdots p_s\}} (-1)^s p_1 p_2 \cdots p_s = \sum_{n=1}^{\infty} \mu(n) n, \quad (9.3.7b)$$

where $\mu(n)$ is the Möbius function. The summation over n in (9.3.7b) can cover all positive integers. When n includes a prime factor with its power greater than 1, the Möbius function guarantees this term to be zero. Therefore, (9.3.7b) and (9.3.7a) are exactly the same.

It should be noted that the operator $[\sum_{n=1}^{\infty} T_n]^{-1}$ on the right hand side of (9.3.5) is of the same form of $[\sum_{n=1}^{\infty} n]^{-1}$ on the right hand side of (9.3.7). Hence, imitating the latter, the former can be written in the following form:

$$\left[\sum_{n=1}^{\infty} T_n \right]^{-1} = \left[\prod_{i=1}^{\infty} (1 + T_{p_i} + T_{p_i}^2 + T_{p_i}^3 + \cdots) \right]^{-1}$$

$$= \left[\prod_{p_i=1}^{\infty} \frac{1}{1 - T_{p_i}} \right]^{-1} = \prod_{p_i=1}^{\infty} (1 - T_{p_i})$$

$$= 1 - T_{p_1} - T_{p_2} - T_{p_3} - \cdots + T_{p_1} T_{p_2} + T_{p_1} T_{p_3} + \cdots$$

$$= 1 + \sum_{\{p_1 p_2 \cdots p_s\}} (-1)^s T_{p_1 p_2 \cdots p_s} = \sum_{n=1}^{\infty} \mu(n) T_n. \quad (9.3.8)$$

One should distinguish between (9.3.7) and (9.3.8). Firstly, the former involves purely the arithmetic of natural numbers, where p_i^2 means the product of two p_i, while the latter concerns the operator of enlargement of argument by integer times, where $T_{p_i}^2$ represents successive operations of T_{p_i} two times. Secondly, the operator T_n in (9.3.8) means the operation will be executable physically, and the final effect of a series of successive of operations is still an executable operation. In this sense, we think the summation $\sum_{n=1}^{\infty} T_n$ is convergent. The inverse of an executable operation is still an executable one. Therefore, both sides of (9.3.8) are convergent. This convergence is guaranteed physically and resorts to the form of the formulas in number theory. As for (9.3.6), intuition tells us that the summation cannot be convergent. However, we have to go beyond the apparent divergence. When the left hand side of (9.3.6a) is expanded, any finite terms equal to those on the left. What is more, the intuitive divergence does not necessarily mean that there is no result. There is a commentary on this point in the end of Appendix 9A. In fact, (9.3.6) is just the expression of the Riemann-ζ function with its argument being -1, $\zeta(-1)$, see (9A.1). Equation (9.3.7b) is just the special case of (9A.7) where $x = -1$.

With the help of (9.3.8), the result of (9.3.5) becomes

$$\Phi(x) = \sum_{n=1}^{\infty} \mu(n) T_n E(x) = \sum_{n=1}^{\infty} \mu(n) E(nx). \qquad (9.3.9)$$

In this way we have successfully obtained the inverse transformation of (9.3.1). If the total energy $E(x)$ as a function of distance x can be calculated in some way, the pair potential $\Phi(x)$ between atoms can be computed by means of (9.3.9).

The transformation (9.3.1) and its inverse (9.3.9) are just that of (9.1.20) and its inverse (9.1.21). Hence, we have proved the Chen-Möbius transformation by means of the enlargement operator.

2. A test of Chen-Möbius transformation

In the derivation above, we have resorted to infinite sums and products in number theory. Their convergences in mathematics need to be studied carefully, but the corresponding physical operations are definitely convergent. Anyhow, we have obtained the transformation (9.3.1) and its inverse (9.3.9) not through rigorous derivation. It is thought that a test to the formulas should be helpful. The test is done as follows. It is noticed that when the argument x of $\Phi(x)$ is expanded to its m times, $\Phi(x) \rightarrow \Phi(mx)$, then the total energy $E(x)$ becomes $E(mx)$. Accordingly, we can write the following two matrices.

$$
\begin{Bmatrix}
1 & 1 & 1 & 1 & 1 & 1 & \dots \\
0 & 1 & 0 & 1 & 0 & 1 & \dots \\
0 & 0 & 1 & 0 & 0 & 1 & \dots \\
0 & 0 & 0 & 1 & 0 & 0 & \dots \\
0 & 0 & 0 & 0 & 1 & 0 & \dots \\
0 & 0 & 0 & 0 & 0 & 1 & \dots \\
\vdots & \vdots & \vdots & \vdots & \vdots & \vdots & \ddots
\end{Bmatrix}
\begin{pmatrix}
\Phi(x) \\
\Phi(2x) \\
\Phi(3x) \\
\Phi(4x) \\
\Phi(5x) \\
\Phi(6x) \\
\vdots
\end{pmatrix}
=
\begin{pmatrix}
E(x) \\
E(2x) \\
E(3x) \\
E(4x) \\
E(5x) \\
E(6x) \\
\vdots
\end{pmatrix}
\qquad (9.3.10a)
$$

and

$$
\begin{Bmatrix}
\mu(1) & \mu(2) & \mu(3) & \mu(4) & \mu(5) & \mu(6) & \dots \\
0 & \mu(1) & 0 & \mu(2) & 0 & \mu(3) & \dots \\
0 & 0 & \mu(1) & 0 & 0 & \mu(2) & \dots \\
0 & 0 & 0 & \mu(1) & 0 & 0 & \dots \\
0 & 0 & 0 & 0 & \mu(1) & 0 & \dots \\
0 & 0 & 0 & 0 & 0 & \mu(1) & \dots \\
\vdots & \vdots & \vdots & \vdots & \vdots & \vdots & \ddots
\end{Bmatrix}
\begin{pmatrix}
E(x) \\
E(2x) \\
E(3x) \\
E(4x) \\
E(5x) \\
E(6x) \\
\vdots
\end{pmatrix}
=
\begin{pmatrix}
\Phi(x) \\
\Phi(2x) \\
\Phi(3x) \\
\Phi(4x) \\
\Phi(5x) \\
\Phi(6x) \\
\vdots
\end{pmatrix} .
$$

$$(9.3.10b)$$

After defining two column vectors $\Phi = (\Phi(x), \Phi(2x), \Phi(3x), \dots)^{\mathrm{T}}$ and $E = (E(x), E(2x), E(3x), \dots)^{\mathrm{T}}$, (9.3.10a) and (9.3.10b) are written in a concise way:

$$U\Phi = E \qquad (9.3.11a)$$

and

$$VE = \Phi. \qquad (9.3.11b)$$

Obviously, matrices U and V are inverses of each other.

$$UV = VU = I. \qquad (9.3.12)$$

The elements of (9.3.12) are

$$\mu(1) = 1, \quad \mu(1) + \mu(2) = 0, \quad \mu(1) + \mu(3) = 0,$$
$$\mu(1) + \mu(2) + \mu(4) = 0, \dots.$$

These are just the cases of (9.1.7) where $k = 1, 2, 3, 4, \dots$.

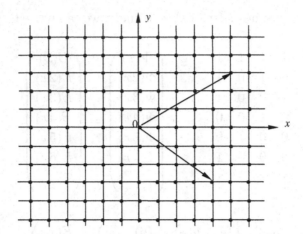

Fig. 9.2.   A two-dimensional square Bravais lattice.

9.3.2.   *Two-dimensional case*

1. Transformation between total energy and pair potential

The atom arrangement in a two-dimensional square Bravais lattice is depicted in Fig. 9.2.

Now there are two coordinates, more complex than the one-dimensional case. For simplicity, a square lattice is considered: the distances between the nearest atoms in the x and y directions are the same, denoted by x. Even in this simplest case, the distance between any two atoms may not be just the integer times of the nearest distance.

We use a pair of indices (i, j) to label every lattice site. Then to label all the sites in a plane we need a two-dimensional array. A two-dimensional array is a countable set. Thus, it is possible to rearrange the array in some way so as to use only one index n to label every site in the two-dimensional lattice. The distance between any two atoms is denoted by $b(n)x$. Apparently, there should be

$$b(n) = \sqrt{i^2 + j^2}, \tag{9.3.13}$$

where i and j are integers.

Because it is a Bravais lattice, we merely need to calculate the interaction energy between one atom and all others. The atom is positioned at the origin. So, the total energy can be written in the form of

$$E(x) = \sum_{n=1}^{\infty} r(n)\Phi(b(n)x), \qquad (9.3.14)$$

where n is the index of the nth nearest neighbor with its distance being $b(n)x$. The number of atoms with the same distance is denoted by $r(n)$ and is called the coordination number of the nth nearest neighbors. The summation in (9.3.14) covers all possible atoms.

The transformation formulas in the one-dimensional case is not valid in the present case. Fortunately, we have the following Euler theorem. Before giving the theorem, we introduce the concept of multiplication closure.

Definition 1. Suppose that in a set an operation between elements called multiplication is defined. If the multiplication of two element in this set is still an element in this set, then this set is said to be **closed** with respect to the multiplication, and also said to be of **multiplication closure**.

In the present two-dimensional lattice case, the elements in the set are the distances $b(n)$ between atoms in the two-dimensional lattice. Arithmetic product is defined as multiplication.

Theorem 1 (Euler Theorem). $b(n)$ *is closed with respect to multiplication*:

$$b(m)b(n) = b(k). \qquad (9.3.15)$$

Proof.

$$(i_1^2 + j_1^2)(i_2^2 + j_2^2)$$
$$= i_1^2 i_2^2 + i_1^2 j_2^2 + j_1^2 i_2^2 + j_1^2 j_2^2 + 2i_1 i_2 j_1 j_2 - 2i_1 i_2 j_1 j_2$$
$$= (i_1 i_2 + j_1 j_2)^2 + (i_1 j_2 - i_2 j_1)^2. \qquad \square$$

Equation (9.3.15) shows that the multiplication of any two elements in this set is still an element in the set. Therefore, (9.3.15) reflects the multiplication closure. Physically, in the two-dimensional lattice, if there is an atom B away from atom A by a distance $b(n)x$, and there is an atom C away from B by distance $b(m)x$, then there must be an atom away from A by distance $b(m)b(n)x$, which is also a site in the lattice, but not necessarily C. For examples, if $(i_1, j_1) = (1, 1)$ and $(i_2, j_2) = (1, 0)$, then $(i_1^2 + j_1^2)(i_2^2 + j_2^2) = 2 = 1^2 + 1^2$.

Making use of the Euler theorem, we are able to put down the inverse transformation of (9.3.14). We assume that the inverse exists, as in the one-dimensional case. By comparison of (9.3.1) and (9.3.14), it is seen that the right hand side of the latter has an additional factor $r(n)$. We assume that the inverse of (9.3.14) has the same form as (9.3.1), but the factor function before $E(x)$ may be different from a Möbius function. The factor is denoted by $K(n)$, which is to be determined through derivation. Thus, we put down the inverse transformation of (9.3.14) in the following form:

$$\Phi(x) = \sum_{n=1}^{\infty} K(n)E(b(n)x), \qquad (9.3.16)$$

where the summation covers the range the same as that in (9.3.14).

Now let us determine the function $K(n)$ in (9.3.16). Substitution of (9.3.14) into (9.3.16) leads to

$$\Phi(x) = \sum_{n=1}^{\infty} K(n)E(b(n)x) = \sum_{n=1}^{\infty} K(n) \sum_{m=1}^{\infty} r(m)\Phi(b(m)b(n)x)$$

$$= \sum_{k=1,\, m=1, b(m)b(n)=b(k)}^{\infty} \quad K(n)r(m)\Phi(b(k)x)$$

$$= \sum_{k=1}^{\infty} \left\{ \sum_{b(n)|b(k)}^{\infty} K(n)r\left(b^{-1}\left[\frac{b(k)}{b(n)}\right]\right) \right\} \Phi(b(k)x)$$

$$= \sum_{k=1}^{\infty} \delta_{k1}\Phi(b(k)x). \qquad (9.3.17)$$

Euler theorem shows that as long as $b(n)$ and $b(m)$ are known, $b(k) = b(n)b(m)$ is necessary in this lattice, and the corresponding $r(k)$ can be found. Conversely, if $b(k)$ and $b(n)$ are sites in the lattice,

$b(m) = b(k)/b(n)$ is also one. That is to say, in (9.3.17),

$$b^{-1}\left[\frac{b(k)}{b(n)}\right] = m \qquad (9.3.18)$$

must be an integer. However, $b^{-1}[b(n)/b(m)]$ may not be an integer in the case that at least one of $b(n)$ and $b(m)$ is not in the lattice. Then such terms can be ruled out in (9.3.17), or the corresponding coordination number r is set to zero. This is the meaning of $\sum_{b(n)|b(k)}^{\infty}$ in (9.3.17).

From (9.3.17), it should be that

$$\sum_{\substack{b(n)|b(k)}}^{\infty} K(n)r\left(b^{-1}\left[\frac{b(k)}{b(n)}\right]\right) = \delta_{k1}. \qquad (9.3.19)$$

This is the equation to calculate $K(n)$ by given coordination numbers $r(m)$.

Equation (9.3.19) can be regarded as a development of the Möbius formula $\sum_{n/k}\mu(n) = \delta_{k1}$ which is the simplest case of all coordination numbers being 1. In the one-dimensional case, all of the coordination numbers on the right hand side of (9.3.1) are 1, so that the coefficients on the right hand side of (9.3.9) are just the Möbius function. In the two-dimensional case, the coefficients $K(n)$ calculated from the coordination numbers $r(m)$ by (9.3.19) are not the Möbius function.

2. Computation of coefficients

Let us calculate the coefficients $K(n)$ needed in (9.3.16). To do so, we first calculate the coordination numbers $r(n)$. The square distances $b^2(n)$ and $r(n)$ up to the 20th nearest neighbors are listed in Table 9.2.

Because all $b^2(n)$ are integers, we let $l = b^2(n)$ and write $r(n) = r[b^2(n)] = r(l)$, where $r(l)$ is expressed as follows.

$$r(l) = \begin{cases} 4, & l = i^2 \text{ or } 2i^2, \ m \neq 0, \\ 8, & l = i^2 + j^2 \text{ and } 0 \neq |i| \neq |j| \neq 0, \\ 12, & l = i_1^2 + j_1^2 = i_2^2 + j_2^2, \ i_1 + ij_1 \neq i_2 + ij_2, \\ 0, & l \neq i^2 + j^2. \end{cases} \qquad (9.3.20)$$

Mathematics for Physicists

Table 9.2.   The square distances $b^2(n)$ and coordination $r(n)$ up to the 20th nearest neighbors.

n	1	2	3	4	5	6	7	8	9	10
$b^2(n)$	1	2	4	5	8	9	10	13	16	17
$r(n)$	4	4	4	8	4	4	8	8	4	8
n	11	12	13	14	15	16	17	18	19	20
$b^2(n)$	18	20	25	26	29	34	36	37	40	41
$r(n)$	4	8	12	8	8	8	4	8	8	8

The condition for the coordination number to be 12 is that the integer l can be written as the square sum of two pairs of integers: $k = i_1^2 + j_1^2 = i_2^2 + j_2^2$, and (i_1, j_1) are not identical to (i_2, j_2). For examples, $(3, 4)$ and $(5, 0)$, $(5, 5)$ and $(1, 7)$, $(5, 12)$ and $(13, 0)$, and so on.

Now $K(n)$'s can be calculated one by one. By (9.3.15), $b(k) = b(m)b(n)$. Once the integers k and n are given, $b(m)$ is calculated by (9.3.15) and then the corresponding m and $r(m)$ are found in Table 9.2.

As $k = 1$, $b(k) = b(1) = 1$. So, $n = 1$ and $r[b^{-1}(1/b(1))] = r(1) = 4$. By $K(1)r(1) = 1$, we get $K(1) = 1/4$.

As $k = 2$, $b(k) = b(2) = \sqrt{2}$.

$$n = 1, r[b^{-1}(\sqrt{2}/b(1))] = r[b^{-1}(\sqrt{2}/1)]$$

$$= r[b^{-1}(\sqrt{2})] = r(2) = 4$$

$$n = 2, r[b^{-1}(\sqrt{2}/b(2))] = r[b^{-1}(\sqrt{2}/\sqrt{2})]$$

$$= r[b^{-1}(1)] = r(1) = 4$$

From $K(1)r(2) + K(2)r(1) = 0$, we solve $K(2) = -K(1) = -1/4$.

As $k = 3$, $b(k) = b(3) = 2$. $n = 1$ and $2 = b(1)b(3)$ so that $m = 3$ and $r(3) = 4$; $n = 2$ and $2 = b(2)b(2)$, so that $m = 2$ and $r(2) = 4$; $n = 1$ and $b(3) = b(3)b(1)$, so that $m = 1$ and $r(1) = 4$.

$$n = 1, r[b^{-1}(2/b(1))]$$

$$= r[b^{-1}(2/1)] = r[b^{-1}(2)] = r(3) = 4.$$

$$n = 2, r[b^{-1}(2/b(2))] = r[b^{-1}(2/\sqrt{2})]$$
$$= r[b^{-1}(\sqrt{2})] = r(2) = 4.$$
$$n = 3, r[b^{-1}(2/b(3))]$$
$$= r[b^{-1}(2/2)] = r[b^{-1}(1)] = r(1) = 4.$$

From $K(1)r(3) + K(2)r(2) + K(3)r(1) = 0$, we solve $K(3) = 0$.
As $k = 4$, $b(k) = b(4) = \sqrt{5}$.

$$n = 1, \ r[b^{-1}(\sqrt{5}/b(1))] = r[b^{-1}(\sqrt{5}/1)]$$
$$= r[b^{-1}(\sqrt{5})] = r(4) = 8.$$
$$n = 2, \ r[b^{-1}(\sqrt{5}/b(2))]$$
$$= r[b^{-1}(\sqrt{5}/\sqrt{2})] = 0.$$
$$n = 3, \ r[b^{-1}(\sqrt{5}/b(3))]$$
$$= r[b^{-1}(\sqrt{5}/2)] = 0.$$
$$n = 4, \ r[b^{-1}(\sqrt{5}/b(4))] = r[b^{-1}(\sqrt{5}/\sqrt{5})]$$
$$= r[b^{-1}(1)] = r(1) = 4.$$

So, $K(1)r(4) + K(4)r(1) = 0$. It is solved that $K(4) = -\frac{K(1)r(4)}{r(1)} = -\frac{1}{8}$.

The process goes successively in this way. Actually, a code is made to let the computer do the work.

3. The introduction of virtual lattice sites

In the end, we mention by the way that in Table 9.2, not all of the positive integers are included when $b^2(n)$ are listed. We can also artificially list $b^2(n)$ of all positive integers into the table. Those that actually do not exist are called **virtual sites** or **false sites**, which will not bring substantial influence as along as their coordination numbers are set to zero. In this way, Table 9.2 is extended to Table 9.3. At first glance, the sites listed in the latter are much more than those in the former. Nevertheless, the coordination numbers of all the virtual sites are zero. Therefore, both tables reflect the same real two-dimensional crystal lattice.

Table 9.3. The square distances $b^2(n)$ and $r(n)$ up to 20th nearest neighbors. The virtual sites with zero coordination numbers are added artificially.

n	1	2	3	4	5	6	7	8	9	10
$b^2(n)$	1	2	3	4	5	6	7	8	9	10
$r(n)$	4	4	0	4	8	0	0	4	4	8
n	11	12	13	14	15	16	17	18	19	20
$b^2(n)$	11	12	13	14	15	16	17	18	19	20
$r(n)$	0	0	8	0	0	4	8	4	0	8

9.3.3. *Three-dimensional case*

1. Transformation between total energy and pair potential

For simplicity, a cubic lattice is considered. That is to say, the distances between the nearest atoms in the three Cartesian coordinates are the same, denoted by x. Similar to the two-dimensional case, the total energy can be written in the form of

$$E(x) = \sum_{n=1}^{\infty} r_0(n)\Phi(b_0(n)x),\qquad(9.3.21)$$

where n is the index labeling the nth nearest neighbors with distance to the origin being $b_0(n)x$ and coordination number being $r_0(n)$. The summation in (9.3.21) covers all possible atoms. These features are the same as those in the two-dimensional case.

It should be pointed out that it is possible to rearrange the array in some way so as to use only one index n to label every site in the two-dimensional lattice. For a three-dimensional lattice, it is impossible to do so. However, in (9.3.21) we have actually used only one index n to label all the lattice sites, just as in the two-dimensional case. The reason is as follows.

Actually, numerical computation is inevitable. In counting the sites in the lattice, we will not really sum all the sites until infinity. The total interaction energy of one atom and all others is finite, so that the summation is definitely convergence. Therefore, one merely needs to sum until a distance far enough. That is to say, the atom cluster considered is finitely large, which contains a finite number of atoms. For finite atoms, it is always possible to arrange them in such a way that only one index n labels all the sites. Formally in (9.3.21), the upper limit of the sum is written as infinity.

The Euler theorem in the two-dimensional case does not apply to the three-dimensional case. When $b(n)x$ and $b(m)x$ are known sites and $[b(n)]^2$ and $[b(m)]^2$ are respectively the square sum of three integers, the product $[b(m)b(n)]^2$ is not necessarily the square sum of three integers. This indicates that in the three-dimensional crystal lattice, the distances between atoms have no multiplication closure as in two-dimensional case. Still, we are able to mimic the two-dimensional case and treat the lattice as follows.

(1) If the product $[b(m)b(n)]^2$ happen to be the square sum of three integers, this is a real site with distance $b(m)b(n) = b(k)$ and its coordination number is $r(k)$.

(2) If the product $[b(m)b(n)]^2$ is not the square sum of three integers, we assume that there still exists the "site" with distance $b(m)b(n) = b(k)$, but with its coordination number being zero, $r(k) = 0$. This does not bring any effect. The additional "sites" are called "virtual sites". They are similar to those with $r(k) = 0$ in Table 9.3.

With the virtual sites included, (9.3.21) is rewritten in the following form:

$$E(x) = \sum_{n=1}^{\infty} r_0(n)\Phi(b_0(n)x) = \sum_{n=1}^{\infty} r(n)\Phi(b(n)x), \qquad (9.3.22)$$

where the summation covers all the real distances $b_0(n)x$, plus the virtual sites produced by $b_0(m)b_0(n) = b(k)$. The corresponding coordination numbers are as follows.

If $b(n) \in b_0(n)$, $r(n) = r_0(n)$.
If $b(n) \notin b_0(n)$, $r(n) = 0$.
Or

$$r(n) = \begin{cases} r_0\left(b_0^{-1}[b(n)]\right), & \text{if } b(n) \in \{b_0(n)\}, \\ 0, & \text{if } b(n) \notin \{b_0(n)\}. \end{cases} \qquad (9.3.23)$$

After introducing the virtual sites, the distances in the lattice are of multiplication closure as in (9.3.15). Now we write down the inverse of (9.3.22) following (9.3.16):

$$\Phi(x) = \sum_{n=1}^{\infty} K(n)E(b(n)x), \qquad (9.3.24)$$

where the function $K(n)$ is derived following the process as in (9.3.17), and it should meet the condition formally the same as (9.3.19):

$$\sum_{b(d)|b(n)}^{\infty} K(n)r\left(b^{-1}\left[\frac{b(n)}{b(d)}\right]\right) = \delta_{n1}. \qquad (9.3.25)$$

Please note that $r(n)$ appearing in (9.3.25) may be zero, as stipulated by (9.3.23).

Following the two-dimensional case, the procedure of calculating $K(n)$ is as follows.

(1) The distances $b_0(n)$ and coordination numbers $r_0(n)$ of the nth nearest neighbors are computed. These are real sites in the lattice.
(2) All possible $b_0(m)b_0(n) = b(k)$ are produced so as to get $b(k)$. If $b^2(k)$ is not a square sum of three integers, it is a virtual site and its coordination number is zero, $r(k) = 0$.
(3) $K(n)$'s are calculated by means of (9.3.25). Finally, the pair potential can be computed in terms of (9.3.24). As a comparison, the step (2) was lacking in the two-dimensional case.

2. Possible forms of pair potentials

The pair potential between two atoms is always finite. Intuitively, as the distance between the two atoms becomes infinite, the pair potential should go to zero.

The above formulas for inversion of pair potential assumes that the interaction between atoms does not have directionality, i.e., isotropic, which is coincide to the metal bonds in crystals.

It is desired that the numerical results obtained by the inversion can be fitted by a proper mathematical expression.

In solid state physics, the simplest mathematical form describing the interaction between atoms is the so-called Lennard-Jones potential:

$$\Phi(x) = -\frac{A}{x^6} + \frac{B}{x^{12}}. \tag{9.3.26}$$

This form possesses merely two parameters and so is too simple. When the real interaction is complicated, this form is not sufficient to fit the numerical results.

Another form is the so-called Morse potential:

$$\Phi(x) = D_0[e^{-\gamma(x/R_0-1)} - 2e^{-\gamma(x/R_0-1)/2}]. \tag{9.3.27}$$

This form possesses three parameters, and can reflect some details of interactions, so that it is usually used.

There is one potential named as the Rahman-Stillinger-Lemberg (RSL2) potential which is of the following form:

$$\Phi(r) = D_0 e^{y(1-r/R_0)} + \frac{a_1}{1 + e^{b_1(r-c_1)}}$$

$$+ \frac{a_2}{1 + e^{b_2(r-c_2)}} + \frac{a_3}{1 + e^{b_3(r-c_3)}}. \tag{9.3.28}$$

This form includes more parameters, so that reflects more details of interactions. When the crystal structures or components are complicated, the RSL2 has to be used. For atoms at two sides of an interfacial, the pair potential is definitely complicated, such that the RSL2 form is inevitable.

A pair potential obtained in a crystal may be used in other crystals. That is to say, the pair potential is transplantable.

By means of the first-principle calculation, the total energy in one crystal cell can be calculated. Then, the pair potential as a function of distance between atoms can be evaluated in terms of the Chen-Möbius transformation. The mathematical form of the pair potential may be chosen as (9.3.27) or (9.3.28) or others. The relative parameters are calculated and stored in a potential database. These parameters can be employed to calculate the potentials in other crystals. A great amount of work in this field has been done and an abundant database has been established by Chen's group.

A final point should be stressed. The pair potential between atoms in a crystal is different from that in the free space. In a crystal, the pair potential has involved the influence of background atoms. Even if the two atoms are the same in the crystal and in free space, the pair potentials are different. The potentials inversed from a crystal may be applied in other crystals, but cannot be applied in the free space.

3. Some remarks

Here we present some remarks about the potential inversion in crystals.

Firstly, Chen smartly introduced a new method and concept. The new method is the application of number-theoretical formulas including transformation of number-theoretic functions, the inversion of the multiplicative function and Euler theorem representing multiplication closure. The new concepts appear in two aspects. One is to turn the discrete arguments in a number-theoretic function to continuous ones so as to apply the functions to physical systems easily. The other is to introduce the concept of virtual sites. This concept guarantees that the multiplication closure is met, and the pair potential inversion in the two- and three-dimensional cases can be done. Without the virtual sites, the multiplication closure could not be met for the distances between atoms in two- and three dimensional crystals.

Secondly, we have merely discussed the cases of Bravais lattices, i.e., there is only one atom in a crystal cell. These are the simplest cases. For complex lattices such as body-centered and face-centered cubics, the idea of the inversion is the same. For multi-component crystals, the cases are complicated. For instance, in the case of Fe_3Al, the procedure is as follows: from Fe and Al elemental crystals, Fe-Fe and Al-Al pair potentials are obtained, respectively; then, in the Fe_3Al alloy, all the possible pair potentials are written, when Fe-Fe and Al-Al pair potentials are subtracted, the Fe-Al potentials are left. We do not intend to introduce the details here because the derivation is cumbersome.

For some crystals such as NaCl, the existing crystals are not enough to provide cohesive energy that is required for the inversion of potentials. In this case, virtual crystals are necessary. This is a new technique introduced by Chen.

Thirdly, the Chen-Möbius transformation is of universal application. That is to say, it can be applied to different kinds of compounds and crystals, e.g., the transplantability mentioned above. Particularly, it can be used to inverse pair potentials in crystals containing rare earth elements. Besides the Chen-Möbius transformation, there are other inversion methods to produce the pair potentials in solids. They are usually difficult to deal with crystals containing rare earth elements.

Fourthly, experience shows that once a pair potential making a crystal stable is obtained, applying external forces can compress and deform the crystal to some extent. Then if the forces are withdrawn, the crystal can retrieve the stable state. This shows the stability of the inversed pair potentials. If the displacement of atoms and distortion of the crystal are beyond some extent, the crystal structure will change. Thus, structural phase transitions of crystals can be studied.

Finally, we have mentioned that in the course of inversion, atoms are regarded as geometric points, and their atomic structures, especially the distributions of valence electrons, are not considered. This is the limitation of this method. Nevertheless, the pair potentials are obtained by the inversion of the total energies of crystals. The total energies are in turn calculated by means of first-principle calculation, which is based on quantum mechanics. Therefore, in calculation of the total energies, the factors arising from electronic structures have been partly considered, and thus the inversed pair potentials have partly reflected the influences of electronic structures. The obtained pair potentials can be used to investigate the mechanical properties. This method assumes that the pair potentials are isotropic, so that it cannot be applied to study the properties of crystals when electronic structures play a role.

9.4.   Additive Möbius Inversion and Its Applications

The inverse transformation introduced in the previous two sections can be called the multiplicative Möbius transformation. In the two sections the multiplicative Möbius transformation has been mainly applied to the inverse problems in Bose systems and to the pair potentials between atoms in materials. Besides this multiplicative Möbius transformation, there is a kind of additive Möbius transformation. The additive Möbius transformations, although not obtained from himself, were the derivatives of Möbius' primary idea, so that had his name. It was hardly thought that the additive Möbius transformation would have its applications. Therefore, even in mathematical handbooks, this kind of transformation was seldom present. Chen smartly explored its application in inverse problems in physics. It turns out that this kind of transformation can have wide applications. In this chapter, we merely present a few simple examples. In this section, we are going to introduce the additive Möbius transformation and its applications in inverse problems in Fermion systems and in discrete signal systems. In the next section, its applications in the inverse problems in pair potentials between interfacial atoms will be introduced.

9.4.1. *Additive Möbius inversion of functions and its applications*

1. Additive Möbius transformation of function sequences

Additive Möbius transformation of function sequences means a pair of transformation and its inverse as follows.

Theorem 1. *If*

$$P_m(x) = \sum_{n=0}^{\infty} a_n Q_{n+m}(x), \qquad (9.4.1)$$

then

$$Q_m(x) = \sum_{n=0}^{\infty} b_n P_{n+m}(x), \qquad (9.4.2)$$

and vice versa. Here $a_n \neq 0$, *and the series* $\{a_n\}$ *and* $\{b_n\}$ *satisfy the relation*

$$\sum_{m=0}^{n} a_m b_{n-m} = \delta_{n,0}. \qquad (9.4.3)$$

Proof. Beginning from the right hand side of (9.4.2) and making use of (9.4.1), we have

$$\sum_{n=0}^{\infty} b_n P_{n+m}(x) = \sum_{n=0}^{\infty} b_n \sum_{k=0}^{\infty} a_k Q_{n+m+k}(x)$$

$$= \sum_{s=0}^{\infty} \sum_{k=0}^{s} a_k b_{s-k} Q_{m+s}(x)$$

$$= \sum_{s=0}^{\infty} \delta_{s,0} Q_{m+s}(x) = Q_m(x). \qquad (9.4.4)$$

Thus, the left hand side of (9.4.2) is reached. Similarly, beginning with the right hand side of (9.4.1) and making use of (9.4.2), we get

$$\sum_{n=0}^{\infty} a_n Q_{n+m}(x) = \sum_{n=0}^{\infty} a_n \sum_{k=0}^{\infty} b_k P_{n+m+k}(x)$$

$$= \sum_{s=0}^{\infty} \sum_{n=0}^{s} a_n b_{s-n} P_{m+s}(x)$$

$$= \sum_{s=0}^{\infty} \delta_{s,0} P_{m+s}(x) = P_m(x). \qquad (9.4.5)$$

Thus, the left hand side of (9.4.1) is reached. Here we have assumed that the double sums in (9.4.4) and (9.4.5) were absolutely convergent. $\qquad \square$

Equations (9.4.1) and (9.4.2) are defined as **additive Möbius transformation of function sequences** and its inverse. The condition is that (9.4.3) is satisfied.

2. Application in Fermion systems

Suppose that in the following convolution integral

$$P(x) = \int_{-\infty}^{\infty} Q(y) \Phi(y - x) \mathrm{d}y, \qquad (9.4.6)$$

the kernel is the Fermi distribution function

$$\Phi(x) = \frac{1}{e^x + 1} \tag{9.4.7}$$

and the function $Q(y)$ can have derivatives taken arbitrary times.

The problem is now to solve function $Q(y)$ under the kernel (9.4.7) provided that $P(x)$ is given. Equation (9.4.6) is the first kind of Fredholm integral equation. We employ the additive Möbius transformation of function sequences.

Because $Q(y)$ can have derivatives taken arbitrary times, we are able to expand it by the Taylor series:

$$Q(y) = \sum_{n=0}^{\infty} \frac{Q^{(n)}(x)}{n!}(y - x)^n. \tag{9.4.8}$$

This is substituted into (9.4.6) to give

$$P(x) = \int_{-\infty}^{\infty} \sum_{n=0}^{\infty} \frac{Q^{(n)}(x)}{n!}(y - x)^n \Phi(y - x)dy$$

$$= \sum_{n=0}^{\infty} \frac{Q^{(n)}(x)}{n!} \int_{-\infty}^{\infty} t^n \Phi(t)dt.$$

We denote

$$a_n = \frac{1}{n!} \int_{-\infty}^{\infty} t^n \Phi(t)dt. \tag{9.4.9}$$

Then,

$$P(x) = \sum_{n=0}^{\infty} a_n Q^{(n)}(x). \tag{9.4.10}$$

The premise for this equation to hold is that the integral (9.4.9) must be convergent. However, under the kernel (9.4.7), the integral equation is

$$P(x) = \int_{-\infty}^{\infty} \frac{Q(y)}{e^{y-x} + 1} dy. \tag{9.4.11}$$

The integral $\int_{-\infty}^{\infty} \frac{t^n}{e^t+1} dt$ is divergent. Actually, as $t \to -\infty$, the integrand itself is divergent. To achieve a convergent integral, the integrand itself must be convergent rapidly enough as $t \to \pm\infty$. To this aim, we take derivative of (9.4.6) with respect to x once.

$$P_1(x) = P'(x) = \int_{-\infty}^{\infty} \frac{e^{y-x}}{(e^{y-x} + 1)^2} Q(y) dy. \tag{9.4.12}$$

This integral has a kernel

$$\Phi(x) = \frac{e^x}{(e^x + 1)^2}, \tag{9.4.13}$$

which goes to zero rapidly enough as $t \to \pm\infty$.

When the process (9.4.8)–(9.4.10) is applied to (9.4.12), we obtain

$$P_1(x) = \sum_{n=0}^{\infty} a_n Q^{(n)}(x), \tag{9.4.14}$$

where

$$a_n = \frac{1}{n!} \int_{-\infty}^{\infty} \frac{t^n e^t}{(e^t + 1)^2} dt = \frac{1}{n!} \int_{-\infty}^{\infty} \frac{t^n}{(e^{t/2} + e^{-t/2})^2} dt. \tag{9.4.15}$$

This integral is finite.

Since the kernel (9.4.13) can have derivatives taken arbitrary times, the $P_1(x)$ is assumed also so. Taking derivative of (9.4.14) m times, we gain

$$P_1^{(m)}(x) = \sum_{n=0}^{\infty} a_n Q^{(n+m)}(x). \tag{9.4.16}$$

This equation has the same form as (9.4.1). So, the additive Möbius transformation of functions can be applied.

From (9.4.2), we immediately get

$$Q^{(m)}(x) = \sum_{n=0}^{\infty} b_n P_1^{(m+n)}(x). \tag{9.4.17}$$

In particular, as $m = 0$, we have

$$Q(x) = \sum_{n=0}^{\infty} b_n P_1^{(n)}(x). \tag{9.4.18}$$

Thus the expression of the required function $Q(x)$ is explicitly given. The next task is to calculate coefficients a_n by (9.4.15), and then b_n by (9.4.3).

It is easily seen from (9.4.15) that as n is an odd integer, the integrand is an odd function, so that

$$a_{2n+1} = 0, \quad n \text{ is a natural number.} \qquad (9.4.19)$$

We calculate a_{2n}.

$$a_0 = 2 \int_0^\infty \frac{e^t}{(e^t + 1)^2} dt = 1,$$

$$a_{2n} = \frac{2}{(2n)!} \int_0^\infty \frac{t^{2n} e^t}{(e^t + 1)^2} dt = \frac{2}{(2n)!} c_{2n}.$$

The integral is

$$c_k = \int_0^\infty \frac{t^k e^t}{(e^t + 1)^2} dt = \int_0^\infty t^k \sum_{m=1}^\infty (-1)^{m+1} m e^{-mt} dt$$

$$= (-1)^{k+1} k! \sum_{m=1}^\infty \frac{(-1)^m}{m^k} = (-1)^{k+1} (1 - 2^{1-k}) \zeta(k) k!.$$

The last step is left to readers as Exercise 5 in this chapter. Hence, we have

$$a_{2n} = \frac{2}{(2n)!} (1 - 2^{1-2n}) \zeta(2n)(2n)! = 2(1 - 2^{1-2n}) \zeta(2n)$$

$$= 2(1 - 2^{1-2n})(-1)^{n+1} \frac{(2\pi)^{2n}}{2(2n)!} B_{2n} = (-1)^{n+1} \frac{(2^{2n} - 2)\pi^{2n}}{(2n)!} B_{2n},$$

$$(9.4.20)$$

where B_{2n}'s are Bernoulli's numbers, see Appendix 9A. In the following we put down the first five a_{2n}.

$$a_2 = \frac{\pi^2}{6}, \quad a_4 = \frac{7\pi^4}{360}, \quad a_6 = \frac{31\pi^6}{42 \times 360},$$

$$a_8 = \frac{127\pi^8}{15 \times 8!}, \quad a_{10} = \frac{511\pi^{10}}{66 \times 9!}. \qquad (9.4.21)$$

Based on formulas in Appendix 9B, it can be evaluated from (9.4.3) that

$$b_0 = 1, \quad b_2 = -\frac{\pi^2}{3!}, \quad b_4 = \frac{\pi^4}{5!}, \quad b_6 = -\frac{\pi^6}{7!},$$

$$b_8 = \frac{\pi^8}{9!}, \quad b_{10} = -\frac{\pi^{10}}{11!}, \dots.$$

As a matter of fact, in the present case, it happens that the coefficient b_{2n} can be evaluated uniformly. We now do the work. Let us multiply both sides of (9.4.15) by u^n and then take summation over n with the confinement $0 < u < 1$. The result is

$$\sum_{n=0}^{\infty} a_n u^n = \sum_{n=0}^{\infty} \frac{1}{n!} \int_{-\infty}^{\infty} \frac{u^n t^n e^t}{(e^t + 1)^2} dt$$

$$= \int_{-\infty}^{\infty} \frac{e^{ut} e^t}{(e^t + 1)^2} dt = \int_{0}^{\infty} \frac{z^u}{(z + 1)^2} dz.$$

In the last step, argument replacement $z = e^t$ is made. Now integration by parts gives

$$\sum_{n=0}^{\infty} a_n u^n = \frac{z^u}{(z + 1)^2} \bigg|_0^{\infty} + u \int_{0}^{\infty} \frac{z^{u-1}}{z + 1} dz = \frac{\pi u}{\sin \pi u}, \qquad (9.4.22)$$

where the definite integral formula is used:

$$\int_{0}^{\infty} \frac{z^{u-1}}{z + 1} dz = \frac{\pi}{\sin \pi u}. \qquad (9.4.23)$$

Equation (9.4.22) reveals that $\frac{\pi u}{\sin \pi u}$ is just the generating function of series $\{a_n\}$. Consider the following product:

$$\sum_{n=0}^{\infty} a_n u^n \sum_{m=0}^{\infty} b_m u^m = \sum_{k=0}^{\infty} \sum_{n=0}^{k} a_n u^n b_{k-n} u^{k-n}$$

$$= \sum_{k=0}^{\infty} u^k \sum_{n=0}^{k} a_n b_{k-n} = \sum_{n=0}^{\infty} u^n \delta_{n,0} = u^0 = 1,$$

where (9.4.3) is used. This equation means that the generating function of $\{b_n\}$ is just the inverse of that of $\{a_n\}$:

$$\sum_{n=0}^{\infty} b_n u^n = \frac{1}{\sum_{n=0}^{\infty} a_n u^n} = \frac{\sin \pi u}{\pi u}.$$

The right hand side is expanded by the Taylor series:

$$\frac{\sin \pi u}{\pi u} = \sum_{n=0}^{\infty} \frac{(-1)^n \pi^{2n}}{(2n + 1)!} u^{2n}.$$

Therefore, we easily obtain

$$b_{2n} = \frac{(-1)^n \pi^{2n}}{(2n+1)!}, \quad b_{2n+1} = 0, \quad (n = 0, 1, 2, \ldots). \tag{9.4.24}$$

The series $\{b_n\}$ is put into (9.4.18) to get

$$Q(x) = \sum_{n=0}^{\infty} \frac{(-1)^n \pi^{2n}}{(2n+1)!} P_1^{(2n)}(x)$$

$$= \sum_{n=0}^{\infty} \frac{(-1)^n \pi^{2n}}{(2n+1)!} P^{(2n+1)}(x). \tag{9.4.25}$$

It is easily seen that the right hand side is just the imaginary part of the following function:

$$P(x - i\pi) = \sum_{n=0}^{\infty} \frac{(-i\pi)^n}{n!} P^{(n)}(x)$$

$$= \sum_{n=0}^{\infty} \frac{(-1)^n \pi^{2n}}{(2n)!} P^{(2n)}(x) - i \sum_{n=0}^{\infty} \frac{(-1)^n \pi^{2n+1}}{(2n+1)!} P^{(2n+1)}(x).$$

Therefore, we express the solution of the integral equation (9.4.11) in a concise manner:

$$Q(x) = -\frac{1}{\pi} \mathrm{Im} P(x - i\pi). \tag{9.4.26}$$

This integral equation has been solved in Example 1 in Subsection 8.5.1. Here we apply additive Möbius transformation of function sequences to gain the same result. This inversion method turns out to be a proper one, and can be applied to inverse problems.

3. Density of states in solids

In solid state physics, the Fermi energy E_F of an electron system varies with temperature T. The electron number within a unit volume is expressed by the following integral:

$$n(E_\mathrm{F}, T) = \int_0^{\infty} \frac{\rho(E, T)}{e^{(E - E_\mathrm{F})/k_\mathrm{B}T} + 1} dE, \tag{9.4.27a}$$

where $\rho(E)$ is the density of states of electrons. When E_F is much greater than $k_\mathrm{B}T$, the lower limit of the integral can be extended to infinity without causing significant error:

$$n(E_\mathrm{F}, T) = \int_{-\infty}^{\infty} \frac{\rho(E, T)}{e^{(E - E_\mathrm{F})/k_\mathrm{B}T} + 1} \mathrm{d}E. \tag{9.4.27b}$$

Let

$$y = \frac{E}{k_\mathrm{B}T}, \quad x = \frac{E_\mathrm{F}}{k_\mathrm{B}T}, \quad Q(y) = k_\mathrm{B}T\rho(k_\mathrm{B}Ty). \tag{9.4.28}$$

Then the integral becomes the form of (9.4.11).

If the number density of electrons as a function of temperature is measured experimentally, the density of states can be evaluated by (9.4.25):

$$k_\mathrm{B}T\rho(k_\mathrm{B}Tx, T) = \sum_{m=0}^{\infty} \frac{(-1)^m \pi^{2m}}{(2m+1)!} \frac{\partial^{2m+1}}{\partial x^{2m+1}} n(x, T).$$

Replacement back of the argument from (9.4.28) brings

$$\rho(E_\mathrm{F}, T) = \sum_{m=0}^{\infty} \frac{(-1)^m (\pi k_\mathrm{B}T)^{2m}}{(2m+1)!} \frac{\partial^{2m+1}}{\partial E_\mathrm{F}^{2m+1}} n(E_\mathrm{F}, T). \tag{9.4.29}$$

In reality, it is impossible to take infinite terms. Usually, the first two terms are taken.

$$\rho(E_\mathrm{F}, T) = \frac{\partial}{\partial E_\mathrm{F}} n(E_\mathrm{F}, T) - \frac{\pi^2}{6} (k_\mathrm{B}T)^2 \frac{\partial^3}{\partial E_\mathrm{F}^3} n(E_\mathrm{F}, T).$$

This is the commonly used expression of electronic density of states as a function of Fermi energy and temperature.

4. The expression of the Dirac delta function by the Fermi distribution function

If in (9.4.11), $Q(y) = \delta(y - x_0)$, then $P(x) = \frac{1}{e^{x_0 - x} + 1}$. Equation (9.4.25) manifests as an expression of the Dirac delta function by the Fermi distribution function:

$$\delta(x - x_0) = \sum_{m=0}^{\infty} \frac{(-1)^m \pi^{2m}}{(2m+1)!} \frac{\partial^{2m+1}}{\partial x^{2m+1}} \frac{1}{e^{x_0 - x} + 1}. \tag{9.4.30}$$

9.4.2. *Additive Möbius inversion of series and its applications*

1. Additive Möbius inversion of series

Additive Möbius inversion of series means a pair of transformation and its inverse as follows.

Theorem 2. *If*

$$g_n = \sum_{m=0}^{n} a_m f_{n-m}, \tag{9.4.31}$$

then,

$$f_n = \sum_{m=0}^{n} b_m g_{n-m}, \tag{9.4.32}$$

and vice versa. Here $a_n \neq 0$, *and the series* $\{a_n\}$ *and* $\{b_n\}$ *satisfy the relation*

$$\sum_{m=0}^{n} a_m b_{n-m} = \delta_{n,0}. \tag{9.4.33}$$

Proof. Beginning from the right hand side of (9.4.32) and making use of (9.4.31), we have

$$\sum_{m=0}^{n} b_m g_{n-m} = \sum_{m=0}^{n} b_m \sum_{k=0}^{n-m} a_k f_{n-m-k}$$

$$= \sum_{s=0}^{n} \sum_{m=0}^{s} a_m b_{s-m} f_{n-s} = \sum_{s=0}^{\infty} \delta_{s,0} f_{n-s} = f_n.$$

$$\tag{9.4.34}$$

Thus, the left hand side of (9.4.32) is reached. Similarly, beginning with the right hand side of (9.4.31) and making use of (9.4.32), we get

$$\sum_{m=0}^{n} a_m f_{n-m} = \sum_{m=0}^{n} a_m \sum_{k=0}^{n-m} b_k g_{n-m-k}$$

$$= \sum_{s=0}^{\infty} \sum_{m=0}^{s} a_m b_{s-m} g_{n-s} = \sum_{s=0}^{\infty} \delta_{s,0} g_{n-s} = g_n.$$

$$\tag{9.4.35}$$

Thus, the left hand side of (9.4.31) is reached.   □

Equations (9.4.31) and (9.4.32) are defined as **additive Möbius transformation of series** and its inverse. The condition is that (9.4.33) is satisfied.

2. Applications to discrete signal systems

Suppose that there is a series of input signals $\{x_n\}$ and output signals $\{y_n\}$ with

$$x_n = y_n = 0 \quad \text{when} \quad n < 0. \tag{9.4.36}$$

The relationship between the two series is

$$\sum_{m=0}^{n} a_m y_{n-m} = \sum_{m=0}^{n} b_m x_{n-m}, \tag{9.4.37}$$

where the coefficients $\{a_n\}$ and $\{b_n\}$ are determined by the intrinsic properties of the system and have been given. In (9.4.37), the output signals appear as an implicit function. Our aim is to turn $\{y_n\}$ to an explicit function expressed by the input signals $\{x_n\}$.

Because $\{a_n\}$ and $\{b_n\}$ are known, by use of the formulas in appendix 9B, it is easy to calculate series $\{c_n\}$ through the reciprocal relation

$$\sum_{m=0}^{n} a_m c_{n-m} = \delta_{n,0}. \tag{9.4.38}$$

The result of (9.4.37) is denoted by w_n:

$$w_n = \sum_{m=0}^{n} a_m y_{n-m} = \sum_{m=0}^{n} b_m x_{n-m}.$$

Now we apply the additive Möbius transformation of series. According to (9.4.32), we have

$$y_n = \sum_{m=0}^{n} c_m w_{n-m} = \sum_{m=0}^{n} c_m \sum_{k=0}^{n-m} b_k x_{n-m-k} = \sum_{s=0}^{n} \sum_{k=0}^{s} c_{s-k} b_k x_{n-s}.$$

Let

$$h_s = \sum_{k=0}^{s} c_{s-k} b_k. \tag{9.4.39}$$

then,

$$y_n = \sum_{s=0}^{n} h_s x_{n-s}. \qquad (9.4.40)$$

This is the required explicit expression.

The following example shows an application.

Example 1. The relationship between the input and output signals is

$$y_n + 0.2y_{n-1} - 0.24y_{n-2} = x_n + x_{n-1}, \qquad (9.4.41)$$

and

$$x_n = \begin{cases} 1, & n \geq 0. \\ 0, & n < 0. \end{cases}$$

Find the first few values of the series $\{y_n\}$.

Solution. Comparison of (9.4.41) and (9.4.37) gives

$$a_0 = 1, \quad a_1 = 0.2, \quad a_2 = -0.24, \quad a_m = 0, \quad m > 2$$

and

$$b_0 = 1, \quad b_1 = 1, \quad b_m = 0, \quad m > 1.$$

The first step is to calculate $\{c_n\}$ following (9.4.38), and the results are listed in the second row in the following table. The second step is to calculate $\{h_n\}$ following (9.4.39), see the third row in the following table. The third step is to calculate $\{y_n\}$ following (9.4.40), see the fourth row in the table.

n	0	1	2	3	4	...
c_n	1	0.2	0.28	-0.104	0.88	...
h_n	1	0.8	0.08	0.176	-0.016	...
y_n	1	1.8	1.88	2.056	2.04	...

Conversely, one is able to derive the input signals by known input signals. The procedure is as follows. Through the reciprocal relation

$$\sum_{m=0}^{n} b_m d_{n-m} = \delta_{n,0}, \qquad (9.4.42)$$

the series $\{d_n\}$ can be obtained. Then following (9.4.32), we calculate

$$x_n = \sum_{m=0}^{n} d_m w_{n-m} = \sum_{m=0}^{n} d_m \sum_{k=0}^{n-m} a_k y_{n-m-k}$$

$$= \sum_{s=0}^{n} \sum_{k=0}^{s} d_{s-k} a_k y_{n-s}.$$

Let

$$g_s = \sum_{k=0}^{s} d_{s-k} a_k. \qquad (9.4.43)$$

Then,

$$x_n = \sum_{s=0}^{n} h_s y_{n-s}. \qquad (9.4.44)$$

In this way, the input signals are expressed explicitly in terms of the output signals.

It is seen that it is very convenient to treat discrete signals by means of the additive Möbius transformation of series.

9.5. Inverse Problem in Crystal Surface Relaxation and Interfacial Potentials

The additive Möbius transformation enables one to investigate the pair potentials between atoms on surfaces or interface. In this section, only elementary substances are involved in deriving formulas.

9.5.1. *Pair potentials between an isolated atom and atoms in a semi-infinite crystal*

Consider a semi-infinite crystal. It is infinite in the xy plane, and has lattice sites at $z > 0$ along the z direction. This semi-infinite crystal is called a substrate. An atom below its surface is called an isolated atom. The interaction between the isolated atom and those in the substrate is certainly different from those inside a crystal. We take the (001) surface of a simple cubic lattice as an example to show how to inverse the pair potential between the isolated and substrate atoms.

Suppose that there is a simple cubic crystal with the distance between nearest neighbors being a and its surface normal being along the (001) direction. It is vacuum at $z < 0$. An isolated atom is positioned at the origin of xy plane but away from the plane by distance z. Then the total energy of the interaction between this isolated atom and the atoms in the whole substrate is expressed by

$$E(z) = \sum_{k=0}^{\infty} \sum_{i,j=-\infty}^{\infty} \Phi(\sqrt{(z + ka)^2 + (i^2 + j^2)a^2}). \qquad (9.5.1)$$

It is a function of the distance between the isolated atom and the surface. The function Φ represents the pair potential between the isolated atom and a substrate atom, which is a function of the distance between them. Suppose that the total energy on the left hand side of (9.5.1) can be evaluated in some way, such as the first-principle method. Please note that the summation in (9.5.1) is different from those in (9.3.1), (9.3.14) and (9.3.21), so that the inversion formulas in Section 9.3 are not applicable. In the following, we apply the additive Möbius transformation of function sequences, i.e., Theorem 1 in Subsection 9.4.1, to inverse the pair potential Φ.

Let

$$H(z) = \sum_{i,j=-\infty}^{\infty} \Phi(\sqrt{z^2 + (i^2 + j^2)a^2}) \qquad (9.5.2)$$

represent the total interaction energy between the isolated atom and all atoms in an xy plane with distance z between them. Then, (9.5.1) is recast to be

$$E(z) = \sum_{k=0}^{\infty} H(z + ka). \qquad (9.5.3)$$

The meaning of this equation is that the interaction energy between the isolated atom and the whole substrate is the sum of those between the former and every plane in the substrate. We proceed to solve $H(z)$ from (9.5.3), and then obtain $\Phi(r)$ by (9.5.2).

It follows from (9.5.3) that

$$E(z) - E(z + a) = \sum_{k=0}^{\infty} H(z + ka)$$

$$- \sum_{k=0}^{\infty} H(z + (k+1)a) = H(z). \quad (9.5.4)$$

In order to solve (9.5.2), we turn it to be

$$H(z) = \sum_{l=0}^{\infty} h(l)\Phi(\sqrt{z^2 + la^2}). \quad (9.5.5)$$

The coefficients $h(l)$ are as follows.

$$h(l) = \begin{cases} 1, & l = 0. \\ 4, & l = i^2 \text{ or } 2i^2, \text{ and } m \neq 0. \\ 8, & l = i^2 + j^2 \text{ and } 0 \neq |i| \neq |j| \neq 0. \\ 12, & l = i_1^2 + j_1^2 = i_2^2 + j_2^2, i_1 + ij_1 \neq i_2 + ij_2. \\ 0, & l \neq i^2 + j^2. \end{cases} \quad (9.5.6)$$

They are actually the same as those in (9.3.20) except the first row.

Comparison of (9.5.5) and (9.4.1) reveals that the former is a special case of the latter where $m = 0$. Therefore, the inverse transformation should be

$$\Phi(z) = \sum_{n=0}^{\infty} g(n)H(\sqrt{z^2 + na^2}), \quad (9.5.7)$$

where

$$\sum_{k=0}^{m} h(k)g(m - k) = \delta_{m,0}. \quad (9.5.8)$$

It is easy to prove (9.5.7) by means of (9.5.8):

$$\sum_{n=0}^{\infty} g(n)H(\sqrt{z^2 + na^2}) = \sum_{n=0}^{\infty} g(k) \sum_{k=0}^{\infty} h(k)\Phi(\sqrt{z^2 + na^2 + ka^2})$$

$$= \sum_{m=0}^{\infty} \left(\sum_{k=0}^{m} h(k)g(m - k) \right) \Phi(\sqrt{z^2 + ma^2})$$

$$= \sum_{m=0}^{\infty} \delta_{m,0}\Phi(\sqrt{z^2 + ma^2}) = \Phi(z).$$

Substitution of (9.5.4) into (9.5.7) leads to

$$\Phi(z) = \sum_{n=0}^{\infty} g(n)[E(\sqrt{z^2 + na^2}) - E(\sqrt{z^2 + na^2} + a)]. \qquad (9.5.9)$$

Thus, $\Phi(z)$ is obtained by the inversion.

The pair potential thus obtained can be employed in molecular dynamics to evaluate the behavior of an isolated atom near the surface of a crystal. For example, numerical results showed that when a Pd atom from far away was incident to the surface of the MgO crystal, its behavior was closely related to its energy. When the energy was low, the Pd atom would stay at the surface near an O atom; when the energy was raised to some extent, it would bounce back from the surface and goes to infinity; when the energy was further raised, it would enter into the crystal, and would replace a Mg atom in the second plane beneath the surface, crowding out the Mg atom out of the surface. When a cluster consisting of five Pd atoms was incident in a speed to the MgO surface, the cluster would not hold but spread due to interactions between the cluster and the substrate, and the spread atoms attached on the surface. However, if the cluster was an icosahedral consisting of thirteen Pb and impinged the surface with the same speed, it will scroll on the surface keeping the cluster as a whole. It was impossible to predict such kinds of microscopic behaviors by calculation before the pair potentials could be inversed. These examples demonstrated the significance of pair potential inversion.

9.5.2. *Relaxation of atoms at a crystal surface*

The environment of an atom in the surface is quite different from that in the crystal body. The former has less nearest neighbor atoms than the latter. The discrepancy of environments makes the surface atom relaxtion: the distance between the surface and second atom layer is different from that in bulk. Such distance relaxation can be observed in experiments. Accordingly, the pair potential between a surface atom and those in other atom layer is different from that in the bulk. Consequently, the inversion method of this kind of pair potential should be different from that in the bulk. Once the pair potential is obtained, the surface relaxation can be evaluated to be compared with experiments.

A two-dimensional atom layer is called a monolayer. Now the semi-infinite crystal except the surface monolayer is called substrate, and the surface monolayer is called a cover layer. We are going to find the pair potential between one atom in the cover monolayer and that in the substrate in the case of a (001) surface of a simple cubic lattice.

The (001) surface of a simple cubic lattice is depicted in Fig. 9.3. The nearest neighbor distance is a. The distance between the cover and the second monolayer is z. In the cover monolayer, there is only one atom in a cell. Because of the periodicity in the xy plane, we merely need to take one atom in the cover monolayer to calculate its interaction energy with the whole substrate. The position of the atom is set at the origin of the xy plane. The total energy is

$$E(z) = \sum_{k=0}^{\infty} \sum_{i,j=-\infty}^{\infty} \Phi(\sqrt{(z+ka)^2 + (i^2+j^2)a^2}). \qquad (9.5.10)$$

It is a function of the distance z between the cover and the second monolayer. Φ is the required pair potential between an atom in the cover monolayer and that in the substrate.

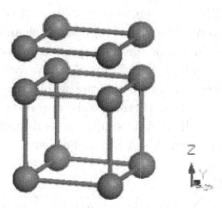

Fig. 9.3. The (001) surface of a simple cubic lattice. The lines are just to connect atoms with no specific meanings. The distance between the surface and the second atom layer is z.

It is seen that the form of (9.5.10) is the same as that of (9.5.1). Both equations represent the total interaction energy between an atom outside of the substrate and an atom in the substrate, but it is by no means that the two energies are the same. In (9.5.1), the atom outside of the substrate is isolated, while in (9.5.10), the atom outside of the substrate is in a two-dimensional lattice plane. Therefore, the two total energies are not the same, and consequently, the inversed pair potentials are different.

Nevertheless, since the forms of (9.5.10) and (9.5.1) are identical, the inversion formulas are also formally the same: (9.5.2)–(9.5.9). Please note that although the inversion formulas are the same, the two total energies calculated by the first-principle are not the same, and subsequently, the inversed pair potentials are different.

The inversed pair potential can be used to determine the relaxation of the surface monolayer.

9.5.3. *Inverse problem of interfacial potentials*

In this subsection, we are going to consider the case of an interface between two elementary substances. The interface is constituted by two semi-infinite crystals, see Fig. 9.4. For simplicity, it is assumed that both lattices are simple cubic with lattice constant being a and the interface is their (001) planes. The distance between the two surfaces is z. In Fig. 9.4, the upper white lattice represents element A and the lower black one represents element B. A crystal cell in A is the one in a monolayer perpendicular to the z axis and extending to infinity along the z direction. We calculate the interaction energy between this cell and the whole semi-infinite crystal B. In the case of Fig. 9.4, there is only one atom in each cell in a monolayer, and a cell in A is simply a semi-infinite atom chain. The total energy between this A atom chain and the substance B is expressed by

$$E(z) = \sum_{l_+,l_-=0}^{\infty} \sum_{i,j=-\infty}^{\infty} \Phi(\sqrt{(z+l_+a+l_-a)^2 + i^2a^2 + j^2a^2}).$$

$$(9.5.11)$$

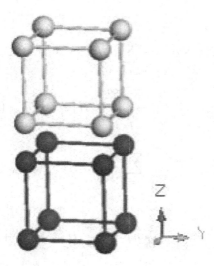

Fig. 9.4. An interface between two simple cubic lattices. The upper white lattice represents element A and the lower black one represents element B.

Here l_+ and l_- mean summation along the z direction within A and B, respectively, and z is the distance between the two surface monolayers of A and B. The total energy is a function of z. Φ represents the pair potential between an atom in A and that in B, and is a function of the distance between the two atoms. Compared with (9.5.1), (9.5.11) merely has an additional summation along the positive z direction. Therefore, the inversion process of the pair potential Φ from the total energy is almost the same as (9.5.1)–(9.5.9).

Let

$$H(z) = \sum_{i,j=-\infty}^{\infty} \Phi(\sqrt{z^2 + (i^2 + j^2)a^2}). \qquad (9.5.12)$$

This is the interaction energy between one atom in surface monolayer of A and the whole surface monolayer of B. In terms of (9.5.12), the total energy $E(z)$ can be recast to the following concise form:

$$E(z) = \sum_{l_+, l_-=0}^{\infty} H(z + l_+ a + l_- a). \qquad (9.5.13)$$

It follows that

$$E(z) - E(z+a) = \sum_{l_+,l_-=0}^{\infty} H(z + l_+a + l_-a)$$

$$- \sum_{l_+,l_-=0}^{\infty} H(z + l_+a + (l_- + 1)a)$$

$$= \sum_{l_+=0}^{\infty} H(z + l_+a) \qquad (9.5.14)$$

and

$$E(z+a) - E((z+a)+a) = \sum_{l_+=0}^{\infty} H(z + (l_+ + 1)a). \qquad (9.5.15)$$

Subtraction between (9.5.14) and (9.5.15) results in

$$H(z) = E(z) - 2E(z+a) + E(z+2a). \qquad (9.5.16)$$

Our task now is to gain the pair potential Φ from (9.5.12). Because (9.5.12) is of the same form as that of (9.5.2), the inversion process (9.5.5)–(9.5.8) can be copied. After inversion, we achieve

$$\Phi(z) = \sum_{n=0}^{\infty} g(n) H(\sqrt{z^2 + na^2}), \qquad (9.5.17)$$

where the coefficients $g(n)$ are evaluated through (9.5.6) and (9.5.8). At last, substitution of (9.5.16) into (9.5.17) leads to

$$\Phi(z) = \sum_{n=0}^{\infty} g(n)[E(\sqrt{z^2 + na^2}) - 2E(\sqrt{z^2 + na^2} + a)$$

$$+ E(\sqrt{z^2 + na^2} + 2a)]. \qquad (9.5.18)$$

Thus, the inversion of the pair potential between interfacial atoms is accomplished.

In summary, the key is to conceive a smart way which can make use of the additive Möbius transformation. As long as this is done, the inversion process is not difficult.

In the cases of interfaces between compounds and elementary substances, between compounds, and so on, the process is almost the same in principle, although consideration of pair potentials between different atoms will make the formalism tedious.

Once the interfacial pair potentials are obtained, the actual configurations atoms near the interface can be numerically calculated by means of molecular dynamics. Furthermore, the defects such as dislocations near the interface and their movement can be investigated.

9.6. Construction of Biorthogonal Complete Function Sets

In Chapter 3, we have introduced the knowledges about the complete eigenfunction sets of an ordinary differential equation and its adjoint equation.

Let $L(x)$ be a differential operator of second order. Suppose that there is following boundary value problem:

$$[\lambda - L(x)]u(x) = 0; \quad a \leq x \leq b, \tag{9.6.1a}$$

$$B(u) = 0, \tag{9.6.1b}$$

where the boundary conditions are homogeneous and satisfies the conditions in Theorem 3 in Subsection 3.6.3. It is assumed that the eigenvalues $\{\lambda_n\}$ and eigenfunction set $\{u_n(x)\}$ of (9.6.1) have been solved. The set $\{u_n(x)\}$ is complete.

The adjoint boundary value problem of (9.6.1) is as follows:

$$[\mu - L^\dagger(x)]v(x) = 0; \quad a \leq x \leq b, \tag{9.6.2a}$$

$$B^\dagger(v) = 0. \tag{9.6.2b}$$

It is assumed that the eigenvalues $\{\mu_n\}$ and eigenfunction set $\{v_n(x)\}$ of (9.6.2) have also been solved. Apparently, the set $\{v_n(x)\}$ is also complete. Every function in the sets $\{u_n(x)\}$ and $\{v_n(x)\}$ has been normalized.

By Theorem 8 in Subsection 2.2.2, there is a biorthogonality between the sets $\{u_n(x)\}$ and $\{v_n(x)\}$:

$$(u_m, v_n) = \int_a^b \rho(x)u_m^*(x)v_n(x)\mathrm{d}x = \delta_{mn}, \tag{9.6.3}$$

where $\rho(x)$ is the weight function of operator

If the problem (9.6.1) is self-adjoint, the sets $\{u_n(x)\}$ and $\{v_n(x)\}$ are actually the same.

Based on the above pair of sets and the biorthogonality between them, one is able to construct another pair of complete sets with biorthogonality as follows.

Suppose that $h(n)$ and $h^{-1}(n)$ are a pair of number-theoretic functions satisfying the following reciprocal relationship:

$$\sum_{n|k} h^{-1}(n) h\left(\frac{k}{n}\right) = \delta_{k,1}. \qquad (9.6.4)$$

Now we construct two new sets of functions. One is that

$$U_k(x) = \sum_{n=1}^{\infty} h(n) u_{nk}(x). \qquad (9.6.5)$$

Here, since n takes all the positive integers, there are infinite terms in every $U_n(x)$. The $\{U_n(x)\}$ is called the derived function set of $\{u_n(x)\}$.

Another one is that

$$V_l(x) = \sum_{n/l} h^{-1}\left(\frac{l}{n}\right) v_n(x). \qquad (9.6.6)$$

Obviously, there are finite terms in every $V_l(x)$. The $\{V_l(x)\}$ is called the derived function set of $\{v_n(x)\}$, and also called the adjoint set of $\{U_n(x)\}$. Equation (9.6.6) is called the adjoint transformation of (9.6.5).

By the transformations (9.6.5) and (9.6.6), even if the sets $\{u_n(x)\}$ and $\{v_n(x)\}$ are the same, the derived sets $\{U_n(x)\}$ and $\{V_l(x)\}$ are not so.

There is also a biorthogonality between the derived sets. This is easily verified.

$$(U_k, V_l) = \left(\sum_{m=1}^{\infty} h(m) u_{mk}, \sum_{n/l} h^{-1}\left(\frac{l}{n}\right) v_n\right)$$

$$= \sum_{m=1}^{\infty} h(m) \sum_{n/l} h^{-1}\left(\frac{l}{n}\right) (u_{mk}, v_n)$$

$$= \sum_{n/l} h\left(\frac{n}{k}\right) h^{-1}\left(\frac{l}{n}\right)$$

$$= \sum_{p/\frac{l}{k}} h(p) h^{-1}\left(\frac{l/k}{p}\right) = \delta_{l/k,1} = \delta_{k,l}. \qquad (9.6.7)$$

Here the weight in the inner product is exactly the same as that in (9.6.3). It is obvious that the biorthogonality between the derived sets is originated from that between $\{u_n(x)\}$ and $\{v_n(x)\}$.

Now, let us inspect the inverse transformations. This inverse of (9.6.5) is

$$u_l(x) = \sum_{n=1}^{\infty} h^{-1}(n) U_{nl}(x). \qquad (9.6.8)$$

This can be easily verified by substitution of (9.6.5) into the right hand side of (9.6.8). The inverse of (9.6.6) is

$$v_n(x) = \sum_{m/n} h\left(\frac{n}{m}\right) V_m(x). \qquad (9.6.9)$$

We further show that the two derived sets are complete. That is to say, any function, as long as it can be expanded by $\{u_n(x)\}$ ($\{v_n(x)\}$), it can also be expanded by $\{U_n(x)\}$ ($\{V_l(x)\}$).

$$f(x) = \sum_{n=1}^{\infty} a_n u_n(x) = \sum_{n=1}^{\infty} a_n \sum_{m=1}^{\infty} h^{-1}(m) U_{mn}(x)$$

$$= \sum_{k=1}^{\infty} \sum_{n/k} a_n h^{-1}\left(\frac{k}{n}\right) U_k(x) = \sum_{k=1}^{\infty} A_k U_k(x). \qquad (9.6.10)$$

The relationship between the coefficients $\{a_n\}$ and $\{A_m\}$ is

$$A_m = \sum_{n/k} h^{-1}\left(\frac{k}{n}\right) a_n. \qquad (9.6.11)$$

It is of the same form of (9.6.6), the adjoint transformation. On the other hand,

$$f(x) = \sum_{n=1}^{\infty} b_n v_n(x) = \sum_{n=1}^{\infty} b_n \sum_{m/n} h\left(\frac{n}{m}\right) V_m(x)$$

$$= \sum_{m=1}^{\infty} \sum_{p=1}^{\infty} b_{pm} h\left(\frac{pm}{m}\right) V_m(x) = \sum_{m=1}^{\infty} B_m V_m(x) \qquad (9.6.12)$$

The relationship between the coefficients $\{b_n\}$ and $\{B_m\}$ is

$$B_m = \sum_{p=1}^{\infty} h(p)b_{pm}, \qquad (9.6.13)$$

which is of the same form of (9.6.5).

In conclusion, based on the complete eigenfunction sets of the boundary value problem (9.6.1) and its adjoint (9.6.2), we are able to construct a pair of new sets by means of (9.6.5) and (9.6.6). The new set $\{u_n(x)\}$ and $\{v_n(x)\}$ are complete, and have biorthogonality as shown by (9.6.7). Please note that each set itself is not of orthogonality, i.e., $(U_n, U_m) \neq 0$ and $(V_n, V_m) \neq 0$ even if $\{u_n(x)\}$ and $\{v_n(x)\}$ are. The constructed sets $\{U_n(x)\}$ and $\{V_l(x)\}$ have their applications. Readers can see examples in Chen's work Böbius Inversion in Physics.

In fact, the transformations (9.6.5) and (9.6.6) can be used once more to generate a pair of new sets complete functions from $\{U_n(x)\}$ and $\{V_l(x)\}$, say $U_{k,1}(x) = \sum_{n=1}^{\infty} h(n)U_{nk}(x)$ and $V_{l,1}(x) = \sum_{n/l} h^{-1}\left(\frac{l}{n}\right)V_n(x)$. The new sets $\{U_n(x)\}$ and $\{V_l(x)\}$ again observe the biorthogonality $(U_{k,1}, V_{l,1}) = \delta_{k,l}$. Therefore, the transformations (9.6.5) and (9.6.6) can be employed repeatedly.

Exercises

1. Please give some multiplicative and completely multiplicative functions.

2. Under the Einstein approximation of phonon frequency spectrum, $g(\nu) \propto \delta(\nu - \nu_E)$, show that the specific heat is of the form of (9.2.26) and the coefficients are expressed by (9.2.30).
   Under high temperature, make bimodal approximation, $g(\nu) = A_1\delta(\nu - \nu_1) + A_2\delta(\nu - \nu_2)$ and find the expression of coefficients in high temperature expansion.

3. (1) Under high temperature, if the experimental result of specific heat has the first three instead of four terms, the bimodal approximation can also be made, but there are three equations to determine four parameters and the result will be

not unique. In this case, an artificial condition is applied, such as letting $B_1 = B_2$. Then the three parameters can be determined. Please derive their expressions.

(2) Show that under high temperature, if the experimental specific heat has only the first two terms, it is impossible to achieve bimodal structure of frequency spectrum.

4. Show that the general form of Euler theorem: for any integers i and j and natural number m and n, there stands $(i_1^{2n} + j_1^{2n})(i_2^{2m} + j_2^{2m}) = k^2 + l^2$. What are the expressions of numbers k and l?

5. Show that $-\sum_{m=1}^{\infty} \frac{(-1)^m}{m^k} = (1 - 2^{1-k})\zeta(k)$.

6. Integrate $\int_0^{\infty} \frac{x^s e^x}{(e^x+1)^2} dx$ and compute the results when $s = 2, 2.5,$ 3, 3.5 and 4.
(Hint: $\sum_{n=1}^{\infty} \frac{(-1)^{n+1}}{n^s} = (1 - \frac{1}{2^{s-1}})\zeta(s)$, where $\zeta(s) = \sum_{n=1}^{\infty} \frac{1}{n^s}$ is Riemann-ζ function.)

7. Prove (9.4.23): $\int_0^{\infty} \frac{z^{u-1}}{z+1} dz = \frac{u}{\sin \pi u}$.

8. For given $y_n + 0.2y_{n-1} - 0.24y_{n-2} - x_n + x_{n-1}$ and $y_n = \begin{cases} 1, & n \geq 0 \\ 0, & n < 0 \end{cases}$, evaluate the first five numbers of $\{x_n\}$.

9. Following the way in Subsection 9.5.1, derive the inversion formulas of interfacial pair potentials of the following crystal surfaces.

(1) (001) surface of a face-centered cubic.
(2) (001) surface of a body-centered cubic.
(3) (011) surface of a face-centered cubic.
(4) (011) surface of a body-centered cubic.
(5) (0001) surface of a hexagonal crystal.
(6) (111) surface of a face-centered cubic.
(7) (111) surface of a body-centered cubic.

10. Following the way in Subsection 9.5.3, derive the inversion formulas of interfacial pair potentials of the following interfaces between elementary substances. The lattice constants of the two

elementary substances are assumed the same.

(1) Face-centered cubic (001)/simple cubic (001) interface.
(2) Face-centered cubic (001)/body-centered cubic (001) interface.
(3) Face-centered cubic (001)/face-centered cubic (001) interface.
(4) Body-centered cubic (001)/simple cubic (001) interface.
(5) Body-centered cubic (001)/body-centered cubic (001) interface.

11. Derive the inversion formulas of interfacial pair potentials of the following compound/elementary substance interfaces, where CsCl and NaCl merely represent the two configurations of AB compounds. The lattice constants in the xy planes of the two substances are assumed the same.

(1) CsCl(001)/simple cubic (001) interface.
(2) CsCl(001)/face-centered cubic (001) interface.
(3) CsCl(001)/body-centered cubic (001) interface.
(4) NaCl(001)/simple cubic (001) interface.
(5) NaCl(001)/face-centered cubic (001) interface.
(6) NaCl(001)/body-centered cubic (001) interface.

12. Derive the inversion formulas of interfacial pair potentials of the following compound/compound interfaces, where CsCl and NaCl merely represent the two configurations of AB compounds. The lattice constants of the two substances are assumed the same.

(1) CsCl(001)/CsCl(001) interface.
(2) CsCl(001)/NaCl(001) interface.

Appendix 9A. Some Values of Riemann ζ Function

Riemann-ζ function is defined by

$$\zeta(x) = \sum_{n=1}^{\infty} \frac{1}{n^x}. \tag{9A.1}$$

Here we merely discuss the case where the argument x is real.

Obviously, as $x > 1$, the summation in (A.1) is absolutely convergent, and the larger the x, the smaller the value of the function. Below listed are some values of the function.

$$\zeta\left(\frac{3}{2}\right) \approx 2.612, \quad \zeta(2) = \frac{\pi^2}{6}, \quad \zeta\left(\frac{5}{2}\right) \approx 1.341, \quad \zeta(3) \approx 1.202,$$

$$\zeta(4) = \frac{\pi^4}{90}, \quad \zeta(6) = \frac{\pi^6}{945}. \tag{9A.2}$$

When x is a positive even integer, i.e., $x = 2m$, the result of the summation of the Riemann-ζ function is as follows:

$$\zeta(2m) = (-1)^{m+1}\frac{(2\pi)^{2m}}{2(2m)!}B_{2m}, \tag{9A.3}$$

where B_{2m} is the Bernoulli number. The generating function of Bernoulli numbers is

$$\frac{x}{e^x - 1} = \sum_{n=1}^{\infty} \frac{B_n}{n!}x^n.$$

The Bernoulli numbers with odd indices except 1 are zero:

$$B_{2k+1} = 0, \quad (k = 1, 2, 3, \ldots).$$

Some Bernoulli numbers are listed below.

$$B_0 = 1, \quad B_1 = -\frac{1}{2}, \quad B_2 = \frac{1}{6}, \quad B_4 = -\frac{1}{30},$$

$$B_6 = \frac{1}{42}, \quad B_8 = -\frac{1}{30}, \quad B_{10} = \frac{5}{66}. \tag{9A.4}$$

When $x = 2, 4, 6$, the values $\zeta(2), \zeta(4), \zeta(6)$ have been listed in (A.2).

When $x \leq 1$, the summation in (A.1) should be divergent in arithmetic sense. However, we inspect the summation in the sense of operation. Then the summation is considered convergent, and carried out by choosing appropriate convergent path to obtain

$$\zeta(1 - 2m) = -\frac{B_{2m}}{2m}, \quad m \geq 1. \tag{9A.5}$$

Some results are given below.

$$\zeta(1) = \infty, \quad \zeta(0) = -\frac{1}{2}, \quad \zeta(-1) = -\frac{1}{12}, \quad \zeta(-3) = \frac{1}{120},$$

$$\zeta(-5) = -\frac{1}{252}, \quad \zeta(-7) = \frac{1}{240}, \quad \zeta(-9) = -\frac{1}{132}. \tag{9A.6}$$

Some results of Riemann-ζ function can be seen in literature.

The argument of Riemann-ζ is not confined to positive integers, which does not agree with the definition of number-theoretic functions. Whatever it is, it is always closely related to number-theoretic functions. Its relation to Möbius function is as follows.

The inverse of a Riemann-ζ function is

$$\frac{1}{\zeta(x)} = \sum_{n=1}^{\infty} \frac{\mu(n)}{n^x}, \tag{9A.7}$$

where $\mu(n)$ is the Möbius function. This equation shows the relationship between Riemann-ζ function and Möbius function. It is verified as follows.

$$\zeta(x) \sum_{n=1}^{\infty} \frac{\mu(n)}{n^x} = \sum_{m=1}^{\infty} \frac{1}{m^x} \sum_{n=1}^{\infty} \frac{\mu(n)}{n^x} = \sum_{m,n=1}^{\infty} \frac{\mu(n)}{(mn)^x}$$

$$= \sum_{k=1}^{\infty} \frac{1}{k^x} \sum_{n/k}^{\infty} \mu(n) = \sum_{k=1}^{\infty} \frac{1}{k^x} \delta_{k,1} = 1. \tag{9A.8}$$

Equation (9A.8) is merely a verification, not a rigorous proof. We now derive it from the left to right hand sides. We evaluate sum $\sum_{n=1}^{\infty} n^x$, where x is real.

$$\sum_{n=1}^{\infty} n^x = \sum_{n=1}^{\infty} \prod_{i=1}^{s} (p_i^{r_i})^x = \sum_{n=1}^{\infty} \prod_{i=1}^{s} (p_i^x)^{r_i}.$$

Compared to (9.3.6), replacement of n and p_i in (9.3.6) by n^x and p_i^x, respectively, will give the result.

$$\sum_{n=1}^{\infty} n^x = (1 + p_1^x + p_1^{2x} + p_1^{3x} + \cdots)(1 + p_2^x + p_2^{2x} + p_2^{3x} + \cdots)$$

$$\times (1 + p_3^x + p_3^{2x} + p_3^{3x} + \cdots) \cdots$$

$$= \prod_{i=1}^{\infty} (1 + p_i^x + (p_i^x)^2 + (p_i^x)^3 + \cdots) = \prod_{i=1}^{\infty} \frac{1}{1 - p_i^x}.$$

Its inverse is

$$\left[\sum_{n=1}^{\infty} n^x \right]^{-1} = \prod_{i=1}^{\infty} (1 - p_i^x) = (1 - p_1^x)(1 - p_2^x)(1 - p_3^x) \cdots$$

$$= 1 - p_1^x - p_2^x \cdots + (p_1 p_2)^x + (p_1 p_3)^x + \cdots$$

$$= 1 + \sum_{\{p_1 p_2 \cdots p_s\}} (-1)^s (p_1 p_2 \cdots p_s)^x$$

$$= \sum_{n=1}^{\infty} \mu(n) n^x.$$

When x is negative, it is written in the form of (9A.7). This ends the general proof of (9A.7).

Finally, we point out that (9A.7) can be extended to

$$\frac{1}{\zeta(\alpha)} = \sum_{n=1}^{\infty} \frac{\mu(n)}{n^\alpha},$$

where α is complex.

In the end of this appendix, we like to present some commentary about the results in (9A.6). Intuitively, such summations should not be convergent. It seems that these results should not be put down. The question is raised: does not the intuitively divergent sums necessarily have results? The answer is no. In mathematics, convergence depends on the path that is taken in convergent sums. Often, an intuitive and convenient path is taken to get a convergent result. For an intuitively divergent sum, if we select an appropriate path, a finite result may be obtained. Such kind of sums has its application in physics. As a matter of fact, in Subsection 9.2.3, with these results seemingly unreasonable in mathematics, the reasonable results in physics are acquired. These sums have necessarily profound origins, which are to be explored.

Appendix 9B. Calculation of Reciprocal Coefficients

The reciprocal coefficients mean that two series a_n and b_n obey the following equation:

$$\sum_{m=0}^{n} a_m b_{n-m} = \delta_{n,0}. \tag{9B.1}$$

According to this equation, one is able to evaluate one of the series from the other. They are explicitly write in the order of

$n = 0, 1, 2, \ldots$:

$$a_0 b_0 = 1, \quad a_0 b_1 + a_1 b_0 = 0,$$
$$a_0 b_2 + a_1 b_1 + a_2 b_0 = 0, \ldots . \tag{9B.2}$$

In matrix form, let

$$A = \begin{Bmatrix} a_0 & a_1 & a_2 & a_3 & \cdots \\ 0 & a_0 & a_1 & a_2 & \cdots \\ 0 & 0 & a_0 & a_1 & \cdots \\ 0 & 0 & 0 & a_0 & \cdots \\ \vdots & \vdots & \vdots & \vdots & \ddots \end{Bmatrix}, \quad B = \begin{Bmatrix} b_0 & b_1 & b_2 & b_3 & \cdots \\ 0 & b_0 & b_1 & b_2 & \cdots \\ 0 & 0 & b_0 & b_1 & \cdots \\ 0 & 0 & 0 & b_0 & \cdots \\ \vdots & \vdots & \vdots & \vdots & \ddots \end{Bmatrix}. \tag{9B.3}$$

It is obvious that

$$AB = BA = I. \tag{9B.4}$$

In the matrix form, it is

$$\begin{Bmatrix} a_0 & a_1 & a_2 & a_3 & \cdots \\ 0 & a_0 & a_1 & a_2 & \cdots \\ 0 & 0 & a_0 & a_1 & \cdots \\ 0 & 0 & 0 & a_0 & \cdots \\ \vdots & \vdots & \vdots & \vdots & \ddots \end{Bmatrix} \begin{Bmatrix} b_0 & b_1 & b_2 & b_3 & \cdots \\ 0 & b_0 & b_1 & b_2 & \cdots \\ 0 & 0 & b_0 & b_1 & \cdots \\ 0 & 0 & 0 & b_0 & \cdots \\ \vdots & \vdots & \vdots & \vdots & \ddots \end{Bmatrix} = \begin{Bmatrix} 1 & 0 & 0 & 0 & \cdots \\ 0 & 1 & 0 & 0 & \cdots \\ 0 & 0 & 1 & 0 & \cdots \\ 0 & 0 & 0 & 1 & \cdots \\ \vdots & \vdots & \vdots & \vdots & \ddots \end{Bmatrix}. \tag{9B.5}$$

The matrices in (9.3.10) are a particular case of (9B.3) that satisfy (9B.5).

We solve b_n one by one:

$$b_0 = \frac{1}{a_0}, \quad b_1 = -\frac{a_1}{a_0^2},$$

$$b_2 = -\frac{1}{a_0}(a_1 b_1 + a_2 b_0) = -\frac{1}{a_0}\left(-\frac{a_1^2}{a_0^2} + \frac{a_2}{a_0}\right), \ldots \tag{9B.6}$$

As $n \geq 1$, because

$$a_0 b_n + \sum_{m=1}^{n} a_m b_{n-m} = 0,$$

we get

$$b_n = -\frac{1}{a_0} \sum_{m=1}^{n} a_m b_{n-m}. \tag{9B.7}$$

Chapter 10

Fundamental Equations in Spaces with Arbitrary Dimensions

In reality, the world we are living in is a three-dimensional space. In spite of that, problems in higher-dimensional spaces have to be studied in physics. For example, the motion in phase space of N microscopic particles is investigated, where we inevitably need to establish coordinates in higher-dimensional spaces and to calculate volume in these spaces. It is well-known that in special relativity, the four-dimensional time-space is necessary to correctly describe objects' movement when their speed is close to that of light. On the other hand, through the investigation of problems in higher-dimensional spaces, one recognizes physical properties of matter in lower-dimensional spaces. For example, quasi-crystals or quasi-periodical materials lack periodicity, so that they are difficult to treat in mathematics. However, they can be regarded as the projections of structures in higher-dimensional spaces into lower-dimensional ones, e.g., one-, two- and three-dimensional quasi-periodical structures can be regarded as the projections from two-, four- and six-dimensional periodical structures. There have been sophisticated approaches to study these periodical structures. The conception of the projection from the higher- to lower dimensional spaces helps one investigate quasi-periodical structures conveniently. One success of this projection theory is that the positions of atoms in the quasi-periodical

structures can be accurately determined theoretically to compare
with those detected by X ray diffraction.

In this chapter, we will introduce some basic knowledge of equations in higher-dimensional spaces. In the end, an example is given
to try to explain the geometric meaning of mass in a space from
the view of higher-dimensional space. Hereafter, when we mention a
higher-dimensional space, we mean that its dimension is larger than
three, unless specified.

10.1.   Euclid Spaces with Arbitrary Dimensions

10.1.1.   *Cartesian coordinate system and spherical coordinates*

1.   Orthogonal curvilinear coordinate systems in N-dimensional spaces

In general, a curvilinear coordinate system in an N-dimensional space
contains N coordinates q_i, $(i = 1, 2, 3, \ldots, N)$. The **line element** in
each dimension is proportional to the differential of the coordinate:

$$\mathrm{d}l_i = h_i \mathrm{d}q_i \quad (i = 1, 2, 3, \ldots, N), \qquad (10.1.1)$$

where h_i is called the **coordinate weight** or **scalar factor** of
the ith coordinate. The **square of the line element** in this
space is

$$\mathrm{d}l^2 = \sum_{i=1}^{N} h_i^2 \mathrm{d}q_i^2. \qquad (10.1.2)$$

The volume element is expressed by

$$\mathrm{d}V = \prod_{i=1}^{N} h_i \mathrm{d}q_i. \qquad (10.1.3)$$

In the first five sections of this chapter, we merely consider the case
where the weighs are positive: $h_i > 0$, $(i = 1, 2, 3, \ldots, N)$. Table 10.1
lists the Cartesian, cylindrical, spherical and parabolic coordinates
and their weights in three-dimensional space.

Table 10.1. Four kinds of coordinates and their weight in the three-dimensional space.

Coordinates	Cartesian	Cylindrical	Spherical	Parabolic
q_1	x	r	r	ξ
q_2	y	θ	θ	η
q_3	z	z	φ	φ
h_1	1	1	1	$\sqrt{\xi^2 + \eta^2}$
h_2	1	r	r	$\sqrt{\xi^2 + \eta^2}$
h_3	1	1	$r\sin\theta$	$\xi\eta$

A position vector r in the N-dimensional space is expressed by

$$r = x_1 e_1 + x_2 e_2 + \cdots + x_N e_N = \sum_{i=1}^{N} x_i e_i, \qquad (10.1.4)$$

where e_i is the unit vector in the ith coordinate axis. When we say orthogonal curvilinear coordinate systems, we mean that

$$e_i \cdot e_j = \delta_{ij}. \qquad (10.1.5)$$

Equations (10.1.2) and (10.1.3) are valid for orthogonal curvilinear coordinate systems.

The coordinate components of r are given by set of numbers $\{x_i\}$. The square of the length of the position vector is

$$r^2 = |r|^2 = r \cdot r = \sum_{i=1}^{N} x_i^2. \qquad (10.1.6)$$

Equations (10.1.4)–(10.1.6) have been in fact included in Table 2.2 in Subsection 2.3.3.

2. Cartesian coordinate system

It is seen from Table 10.1 that in general, every weight is a function of the coordinates: $h_i = h_i(q_1, q_2, \ldots, q_N)$. If all the weights are independent of coordinates, this coordinate system is called a Cartesian system. Apparently, in a this case, every h_i can be set to be 1. Therefore, in a Cartesian system, all the weights are

$$h_i = 1, \quad (i = 1, 2, 3, \ldots, N). \tag{10.1.7}$$

Subsequently, (10.1.2) and (10.1.3) are simplified to be

$$dl^2 = \sum_{i=1}^{N} dx_i^2 \tag{10.1.8}$$

and

$$d^N r = \prod_{i=1}^{N} dx_i \tag{10.1.9}$$

respectively. The volume of an N-dimensional "cubic" with its edge length being a is $V = a^N$. The Cartesian coordinate system in N-dimensional space is a natural extension of those in two- and three-dimensional spaces.

3. Spherical coordinate system

Now we inspect the spherical coordinate system. Similar to the cases in two- and three-dimensional spaces, there must be one coordinate r representing the length of the position vector and $N-1$ coordinates representing angles. The coordinates in the Cartesian system are expressed by those in the spherical system in the following way:

$$x_1 = r\cos\theta_1,$$
$$x_2 = r\sin\theta_1\cos\theta_2,$$
$$x_3 = r\sin\theta_1\sin\theta_2\cos\theta_3,$$
$$\cdots,$$
$$x_{N-1} = r\sin\theta_1\cdots\sin\theta_{N-2}\cos\theta_{N-1},$$
$$x_N = r\sin\theta_1\cdots\sin\theta_{N-2}\sin\theta_{N-1}. \tag{10.1.10}$$

The ranges of the spherical coordinates are as follows:

$$r > 0; \quad 0 \le \theta_i \le \pi, \quad 1 \le i \le N-2; \quad 0 \le \theta_{N-1} \le 2\pi; \quad N \ge 3. \tag{10.1.11}$$

Note that the last angle ranges $[0, 2\pi]$, while the remaining $N - 2$ angles range $[0, \pi]$.

Let us consider the coordinate weights. We take differential of (10.1.10) and then substitute the results into (10.1.8) to achieve the form of (10.1.2). Thus, the weights are obtained:

$$(h_1, h_2, h_3, h_4, \ldots, h_{N-1}, h_{N-2})$$

$$= (1, r, r \sin \theta_1, r \sin \theta_1 \sin \theta_2, \ldots r \sin \theta_1 \sin \theta_2 \cdots \sin \theta_{N-3},$$

$$\times r \sin \theta_1 \sin \theta_2 \cdots \sin \theta_{N-2}). \tag{10.1.12}$$

The Jacobi determinant of the Cartesian and spherical coordinates is

$$J = \frac{\partial(x_1, x_2, \ldots, x_N)}{\partial(r, \theta_1, \ldots, \theta_{N-1})} = r^{N-1} \sin^{N-2} \theta_1 \sin^{N-3} \theta_2 \cdots \sin \theta_{N-2}$$

$$= \prod_{i=1}^{N} h_i. \tag{10.1.13}$$

Substituting the weights in (10.1.12) into (10.1.3), we gain

$$dV = r^{N-1} \sin^{N-2} \theta_1 \sin^{N-3} \theta_2 \cdots$$

$$\times \sin \theta_{N-2} dr d\theta_1 d\theta_2 \cdots d\theta_{N-2} d\theta_{N-1}. \tag{10.1.14}$$

We calculate the volume of an N-dimensional "ball", or hypersphere, with radius R. Before doing so, we evaluate the following integration:

$$I_k = \int_0^\pi \sin^k \theta d\theta = \frac{k-1}{k} I_{k-2}. \tag{10.1.15}$$

It is calculated that $I_k = \frac{(k-1)!!}{k!!} \pi$ as k is even, and $I_k = 2 \frac{(k-1)!!}{k!!}$ as k is odd.

The integration of (10.1.14) in the N-dimensional space results in

$$V = \int_0^R r^{N-1} dr \int_0^\pi \sin^{N-2} \theta_1 d\theta_1 \int_0^\pi \sin^{N-3} \theta_2 d\theta_2 \cdots$$

$$\int_0^\pi \sin \theta_{N-2} d\theta_{n-2} \int_0^{2\pi} d\theta_{N-1}$$

$$= 2\pi \frac{R^N}{N} \prod_{i=1}^{N-2} \int_0^\pi \sin^{N-1-i} \theta_i d\theta_i. \tag{10.1.16}$$

Making use of (10.1.16), we have a uniform result

$$\prod_{i=1}^{n-2} \int_0^\pi \sin^{N-1-i} \theta_i \mathrm{d}\theta_i = \frac{\pi^{N/2-1}}{\Gamma(N/2)}, \tag{10.1.17}$$

which is valid for whatever N being odd and even. So, the volume of the supersphere with radius R is

$$V = \frac{2\pi^{N/2} R^N}{N\Gamma(N/2)}. \tag{10.1.18}$$

We turn to calculate the area S of the hypersurface of the hypersphere. This is done simply by removing the radial integration in (10.1.16). Thus, the surface area is

$$S = \frac{2\pi^{N/2} R^{N-1}}{\Gamma(N/2)}. \tag{10.1.19}$$

This is just the derivative of (10.1.18) with respect to the radius R. By the way, the area element of the surface of an N-dimensional sphere with radius r is

$$\mathrm{d}S = r^{N-1} \sin^{N-2} \theta_1 \sin^{N-3} \theta_2 \cdots \sin \theta_{N-2} \mathrm{d}r \mathrm{d}\theta_1 \mathrm{d}\theta_2 \cdots \mathrm{d}\theta_{N-2} \mathrm{d}\theta_{N-1}$$

$$= \prod_{i=1}^{N-2} h_i \mathrm{d}\theta_i. \tag{10.1.20}$$

10.1.2. *Gradient, divergence and Laplace operator*

1. Gradient of a scalar field

Gradient of a scalar field is the differential of a scalar field $\psi(\boldsymbol{r})$, which results in a vector field. Each component of the vector field is defined by

$$\nabla_i \psi = \lim_{dl_i \to 0} \frac{\psi(q_i + \mathrm{d}q_i) - \psi(q_i)}{\mathrm{d}l_i} = \frac{1}{h_i} \frac{\partial \psi}{\partial q_i}, \tag{10.1.21}$$

and the entire vector is

$$\nabla \psi = \sum_{i=1}^N \frac{1}{h_i} \frac{\partial \psi}{\partial q_i} \boldsymbol{e}_i. \tag{10.1.22}$$

Its geometric meaning is the same as that in three-dimensional space: at one space point, the direction of $\nabla \psi$ of the scalar field ψ is that along which the change rate of the ψ at this point is the largest, and is the normal of the isosurface; the magnitude of $\nabla \psi$ is the change rate along this direction.

By this definition, in Cartesian coordinate system, with the coordinate weights (10.1.7), the expression of the gradient is

$$\nabla \psi = \sum_{i=1}^{N} \frac{\partial \psi}{\partial x_i} e_i. \tag{10.1.23}$$

In the spherical coordinate system, with the weight (10.1.12), the gradient is

$$\nabla \psi = \frac{\partial \psi}{\partial r} e_r + \frac{1}{r} \frac{\partial \psi}{\partial \theta_1} e_{\theta 1} + \frac{1}{r \sin \theta_1} \frac{\partial \psi}{\partial \theta_2} e_{\theta 2} + \frac{1}{r \sin \theta_1 \sin \theta_2} \frac{\partial \psi}{\partial \theta_3} e_{\theta 3}$$

$$+ \cdots + \frac{1}{r \sin \theta_1 \sin \theta_2 \cdots \sin \theta_{N-3}} \frac{\partial \psi}{\partial \theta_{N-2}} e_{\theta N-2}$$

$$+ \frac{1}{r \sin \theta_1 \sin \theta_2 \cdots \sin \theta_{N-2}} \frac{\partial \psi}{\partial \theta_{N-1}} e_{\theta N-1}. \tag{10.1.24}$$

2. Divergence of a vector field

The divergence of a vector field is the differential of a vector field $W(r)$, which results in a scalar field. Its definition is

$$\nabla \cdot W = \lim_{\Delta V \to 0} \frac{1}{\Delta V} \int_{\sigma} W \cdot ds. \tag{10.1.25a}$$

We take three-dimensional space as an example. $W(r)$ has three components, $W = (W_1, W_2, W_3)$. Taking integration over two infinitesimal areas through $q_1 = a_1$ and $q_1 = a_1 + \delta q_1$, respectively, and then subtracting them, we have

$$\int_{a_2}^{a_2+\delta q_2} dq_2 \int_{a_3}^{a_3+\delta q_3} dq_3 h_2(a_1 + \delta q_1, q_2, q_3)$$

$$\times h_3(a_1 + \delta q_1, a_2, a_3) W_1(a_1 + \delta q_1, q_2, q_3)$$

$$- \int_{a_2}^{a_2+\delta q_2} dq_2 \int_{a_3}^{a_3+\delta q_3}$$

$$\times \, dq_3 h_2(a_1, q_2, q_3) h_3(a_1, a_2, a_3) W_1(a_1, q_2, q_3)$$

$$= \left[\frac{\partial}{\partial q_1} (W_1 h_2 h_3) \right]_{q_i=a_i} \delta q_1 \delta q_2 \delta q_3.$$

By $dV = h_1 h_2 h_3 \delta q_1 \delta q_2 \delta q_3$, we further obtain that

$$\nabla \cdot W = \frac{1}{h_1 h_2 h_3} \left[\frac{\partial(h_2 h_3 W_1)}{\partial q_1} + \frac{\partial(h_3 h_1 W_2)}{\partial q_2} + \frac{\partial(h_1 h_2 W_3)}{\partial q_3} \right].$$

In general, the gradient of $W(r)$ in N-dimensional space is expressed by

$$\nabla \cdot W = \frac{1}{\prod_{i=1}^{N} h_i} \sum_{j=1}^{N} \frac{\partial}{\partial q_j} \left(\frac{W_j}{h_j} \prod_{l=1}^{N} h_l \right). \tag{10.1.25b}$$

We emphasize that every weight h_i may be a function of coordinates, so that they cannot be taken out of the differential symbols.

In Cartesian system, with the weight (10.1.7), the gradient has the simplest form:

$$\nabla \cdot W = \sum_{j=1}^{N} \frac{\partial W_j}{\partial x_j}. \tag{10.1.26}$$

Substituting the weights (10.1.12) into (10.1.25) one will acquire the expression of gradient in the spherical coordinate system.

3. Laplace operator

Laplace operator is defined by $\Delta = \nabla^2 = \nabla \cdot \nabla$. When it acts on a scalar field ψ, the effect is $\Delta \psi = \nabla^2 \psi = \nabla \cdot \nabla \psi$. The expression of $\nabla \psi$ has been given by (10.1.23), and the effect of $\nabla \cdot$ by (10.1.26). So, the combination of these two equations gives

$$\nabla^2 \psi = \frac{1}{\prod_{i=1}^{N} h_i} \sum_{j=1}^{N} \frac{\partial}{\partial q_j} \left[\left(\frac{1}{h_j^2} \prod_{l=1}^{N} h_l \right) \frac{\partial \psi}{\partial q_j} \right]. \tag{10.1.27}$$

In a Cartesian system, with the weight (10.1.7), the Laplace operator has the simplest form:

$$\nabla^2 = \sum_{i=1}^{n} \frac{\partial^2}{\partial x_i^2}. \tag{10.1.28}$$

In a spherical system, with the weights (10.1.12), the expression is

$$\nabla^2 \psi = \frac{1}{r^{N-1}} \frac{\partial}{\partial r} \left(r^{N-1} \frac{\partial \psi}{\partial r} \right) + \frac{1}{r^2 \sin^{N-2} \theta_1} \frac{\partial}{\partial \theta_1} \left(\sin^{N-2} \theta_1 \frac{\partial \psi}{\partial \theta_1} \right)$$

$$+ \frac{1}{(r \sin \theta_1)^2 \sin^{N-3} \theta_2} \frac{\partial}{\partial \theta_2} \left(\sin^{N-3} \theta_2 \frac{\partial \psi}{\partial \theta_2} \right)$$

$$+ \frac{1}{(r \sin \theta_1 \sin \theta_2)^2 \sin^{N-4} \theta_3} \frac{\partial}{\partial \theta_3} \left(\sin^{N-3} \theta_3 \frac{\partial \psi}{\partial \theta_3} \right) + \cdots$$

$$+ \frac{1}{(r \sin \theta_1 \cdots \sin \theta_{N-3})^2 \sin \theta_{N-2}} \frac{\partial}{\partial \theta_{N-2}} \left(\sin \theta_{N-2} \frac{\partial \psi}{\partial \theta_{N-2}} \right)$$

$$+ \frac{1}{(r \sin \theta_1 \cdots \sin \theta_{N-2})^2} \frac{\partial^2 \psi}{\partial \theta_{N-1}^2}. \tag{10.1.29}$$

In order to emphasize that the parts of the radial and angular coordinates are separated, we rewrite (10.1.29) into the following form:

$$\nabla^2 \psi = \frac{1}{r^{N-1}} \frac{\partial}{\partial r} \left(r^{N-1} \frac{\partial \psi}{\partial r} \right) + \frac{1}{r^2} L_N^2(\theta) \psi, \tag{10.1.30}$$

where $L_N^2(\theta)$ is the angular part in (10.1.29):

$$L_N^2(\theta) = \frac{1}{\sin^{N-2} \theta_1} \frac{\partial}{\partial \theta_1} \left(\sin^{N-2} \theta_1 \frac{\partial}{\partial \theta_1} \right)$$

$$+ \frac{1}{(\sin \theta_1)^2 \sin^{N-3} \theta_2} \frac{\partial}{\partial \theta_2} \left(\sin^{N-3} \theta_2 \frac{\partial}{\partial \theta_2} \right)$$

$$+ \frac{1}{(\sin \theta_1 \sin \theta_2)^2 \sin^{N-4} \theta_3} \frac{\partial}{\partial \theta_3} \left(\sin^{N-3} \theta_3 \frac{\partial}{\partial \theta_3} \right) + \cdots$$

$$+ \frac{1}{(\sin \theta_1 \cdots \sin \theta_{N-3})^2 \sin \theta_{N-2}} \frac{\partial}{\partial \theta_{N-2}} \left(\sin \theta_{N-2} \frac{\partial}{\partial \theta_{N-2}} \right)$$

$$+ \frac{1}{(\sin \theta_1 \cdots \sin \theta_{N-2})^2} \frac{\partial^2}{\partial \theta_{N-1}^2}. \tag{10.1.31}$$

We call to attention that this is the square of the **total angular momentum**. The total angular momentum may have projection in some subspaces.

4. Curl of a vector field in a three-dimensional space

At the end of this section, we mention that in a three-dimensional space, it is possible defining a curl operator. **The curl of a vector field** is the differential of a vector field $W(r)$, which results in a vector field. Its definition is as follows.

$$\nabla \times V = \frac{1}{h_2 h_3} \left[\partial_2 (h_3 V_3) - \partial_3 (h_2 V_2) \right] e_1$$

$$+ \frac{1}{h_3 h_1} \left[\partial_3 (h_1 V_1) - \partial_1 (h_3 V_3) \right] e_2$$

$$+ \frac{1}{h_1 h_2} \left[\partial_1 (h_2 V_2) - \partial_2 (h_1 V_1) \right] e_3$$

$$= \frac{1}{h_1 h_2 h_3} \begin{vmatrix} h_1 e_1 & h_2 e_2 & h_3 e_3 \\ \partial_1 & \partial_2 & \partial_3 \\ h_1 V_1 & h_2 V_2 & h_3 V_3 \end{vmatrix} \tag{10.1.32}$$

The expressions are easily written by use of the coordinate weights listed in Table 10.1.

10.2. Green's Functions of the Laplace Equation and Helmholtz Equation

In this section, we derive the Green's functions of the Laplace equation and Helmholtz equation, the boundary conditions being that the Green's functions are finite at infinity.

10.2.1. *Green's function of the Laplace equation*

The equation that the Green's function satisfies is

$$\nabla_r^2 G(r, r') = \delta(r - r'). \tag{10.2.1}$$

Infinite vacuum is of uniformity, which includes translational invariance and isotropy. Therefore,

$$G(r, r') = G(r - r') = G(|r - r'|) = G(\rho), \tag{10.2.2}$$

where

$$\rho = \left(\sum_{i=1}^{N} (x_i - x_i')^2 \right)^{1/2}. \tag{10.2.3}$$

This feature has been discussed in the end of Subsection 6.2.3 for one-, two- and three-dimensional spaces.

In the spherical system, it follows from (10.1.10) that

$$|\boldsymbol{r} - \boldsymbol{r}'|^2 = r^2 + r'^2 - 2rr' \left(\cos\theta_1 \cos\theta_1' + \sin\theta_1 \cos\theta_2 \sin\theta_1' \cos\theta_2' \right.$$
$$+ \sin\theta_1 \sin\theta_2 \cos\theta_3 \sin\theta_1' \sin\theta_2' \cos\theta_3'$$
$$+ \cdots + \sin\theta_1 \cdots \sin\theta_{n-2} \sin\theta_1' \cdots \sin\theta_{n-2}'$$
$$\left. \times \cos(\theta_{n-1} - \theta_{n-1}') \right)$$
$$= r^2 + r'^2 - 2rr' \cos\gamma,$$

where γ is the angle between position vectors $\boldsymbol{r}$ and $\boldsymbol{r}'$, with its geometric meaning the same as in two- and three-dimensional spaces.

The translational invariance is manifested by that Green's function is a function of the difference of the two position vectors, and the isotropy is manifested by that Green's function is a function of its argument, independent of the directions. Due to these reasons, it is more convenient to solve Green's function in the spherical system. Because of (10.2.2), it is actually $\nabla_r^2 = \nabla_\rho^2$ in (10.2.1). Thus, (10.2.1) is simplified to be

$$\nabla_\rho^2 G(\rho) = \delta(\rho). \tag{10.2.4}$$

The form of (10.1.31) is used. Since the Green's function is independent of angles, (10.2.4) becomes

$$\frac{1}{\rho^{N-1}} \frac{d}{d\rho} \left[\rho^{N-1} \frac{d}{d\rho} G(\rho) \right] = \delta(\rho). \tag{10.2.5}$$

We apply the piecewise expression method introduced in Subsection 6.1.3. Now that $\rho \geq 0$, we discuss the resolution when $\rho > 0$ and $\rho = 0$. As $\rho > 0$, (10.2.5) is simplified to be

$$\frac{1}{\rho^{N-1}} \frac{d}{d\rho} \left[\rho^{N-1} \frac{d}{d\rho} G(\rho) \right] = 0, \quad N \geq 1. \tag{10.2.6}$$

The equation in this form is valid for $N \geq 1$. As $N \geq 3$, the solution can be expressed by

$$G(\rho) = \frac{C}{\rho^{N-2}}, \quad N \geq 3, \tag{10.2.7}$$

where C is a constant to be determined. Readers are suggested to give the solutions of (10.2.6) when $N = 1$ and 2. The constant C is determined by the condition at $\rho = 0$. On both sides of (10.2.5), we take integration over an infinitesimal sphere with radius γ centered on $\rho = 0$, and then let $\gamma \to 0$. On the right hand side of (10.2.5), the integration is apparently 1. On the left hand side, the result is

$$\int_0^\gamma \rho^{N-1} d\rho \prod_{i=1}^{N-2} \int_0^\pi \sin^{N-1-i} \theta_i d\theta_i$$

$$\times \int_0^{2\pi} d\theta_{N-1} \frac{1}{\rho^{N-1}} \frac{d}{d\rho} \left[\rho^{N-1} \frac{d}{d\rho} G(\rho) \right]$$

$$= \gamma^{N-1} \left[\frac{d}{d\rho} \frac{C}{\rho^{N-2}} \right]_{\gamma=\rho} \frac{2\pi^{N/2}}{\Gamma(N/2)} = -C(N-2) \frac{2\pi^{N/2}}{\Gamma(N/2)} = 1, \tag{10.2.8}$$

where (10.1.17) is employed. Thus we obtain

$$C = -\frac{\Gamma(N/2)}{2\pi^{N/2}(N-2)}. \tag{10.2.9}$$

Subsequently, the required Green's function is

$$G(\rho) = -\frac{\Gamma(N/2)}{2\pi^{N/2}(N-2)\rho^{N-2}}. \tag{10.2.10}$$

This is the expression in the spherical coordinate system. Substituting (10.2.3) into (10.2.10) gives the expression in Cartesian system.

It is easy to verify that in the three-dimensional space, it goes back to (6.2.13). In four-dimensional space, $N = 4$, we have

$$G(\rho) = -\frac{\Gamma(2)}{2\pi^2 2\rho^2} = -\frac{1}{4\pi^2 |\boldsymbol{r} - \boldsymbol{r}'|^2}.$$

10.2.2.   *Green's function of the Helmholtz equation*

The equation that the Green's function satisfies is

$$(\nabla_r^2 + k^2)G(r, r') = \delta(r - r').$$ (10.2.11)

Compared to (10.2.1), a constant on the left hand side is added. The region of the equation is still an infinite vacuum. Therefore, the Green's function is still of the feature of (10.2.2). Accordingly, (10.2.11) is simplified to be

$$(\nabla_\rho^2 + k^2)G(\rho) = \delta(\rho).$$ (10.2.12)

The Laplace operator taking the form of (10.1.30), (10.2.12) becomes

$$\frac{1}{\rho^{N-1}}\frac{d}{d\rho}\left[\rho^{N-1}\frac{d}{d\rho}G(\rho)\right] + k^2 G(\rho) = \delta(\rho).$$ (10.2.13)

Again, we use the method as the last subsection. As $\rho > 0$, (10.2.13) is

$$\left(\rho^2\frac{d^2}{d\rho^2} + (N-1)\rho\frac{d}{d\rho} + k^2\rho^2\right)G(\rho) = 0.$$ (10.2.14)

This equation is close to a Bessel equation. In Subsection 4.5.2 we have known that if the coefficient of the first derivative term is 2, a transformation $G(\rho) = u(\rho)/\sqrt{\rho}$ could turn a Bessel equation to be that satisfied by $u(\rho)$. Now in (10.2.14), the coefficient of that term is $N-1$, the necessary transformation is

$$G(\rho) = \frac{1}{R^{N/2-1}}Z_{N/2-1}(k\rho).$$ (10.2.15)

Thus, (10.2.14) turns to that satisfied by $Z_{N/2-1}(k\rho)$, where Z_ν represents a cylindrical function of ν order. The boundary condition is that at infinity the solution of the equation should be finite. Therefore, Z_ν is selected as a Hankel function. The asymptotic behavior of Hankel functions as $\nu \neq 0$ were shown by (4.5.10). The outward wave solution is

$$G(\rho) = \frac{A}{\rho^{N/2-1}}H_{N/2-1}^{(1)}(k\rho).$$ (10.2.16)

The constant A is determined by the condition at $\rho = 0$. On both sides of (10.2.13), we take integration over an infinitesimal sphere with radius γ centered on $\rho = 0$, and then let $\gamma \to 0$. On the right hand side of (10.2.13), the integration is apparently 1. On the left hand side, the result is

$$\frac{2\pi^{N/2}}{\Gamma(N/2)} \int_0^\gamma \rho^{N-1} d\rho \left[\frac{1}{\rho^{N-1}} \frac{d}{d\rho} \left(\rho^{N-1} \frac{d}{d\rho} \right) + k^2 \right]$$

$$\frac{A}{\rho^{N/2-1}} H_{N/2-1}^{(1)}(k\rho) = 1, \tag{10.2.17}$$

where the angular integrations have been carried out. For radial integration, since γ is infinitely small, we use the asymptotic form (4.5.10c) to simplify (10.2.17) to be

$$-i\frac{4\pi^{N/2-1}A}{N-2} \left(\frac{2}{k}\right)^{N/2-1}$$

$$\times \left[\int_0^\gamma \frac{d}{d\rho} \left(\rho^{N-1} \frac{d}{d\rho} \frac{1}{\rho^{N-2}} \right) d\rho + k^2 \int_0^\gamma \rho d\rho \right] = 1. \tag{10.2.18}$$

There remain two integrals. The first is

$$\int_0^\gamma \frac{d}{d\rho} \left(\rho^{N-1} \frac{d}{d\rho} \frac{1}{\rho^{N-2}} \right) d\rho = \left[\rho^{N-1} \frac{d}{d\rho} \frac{1}{\rho^{N-2}} \right]_{\rho=\gamma}$$

$$= -(N-2). \tag{10.2.19}$$

The second is zero when $\gamma \to 0$. As a result, the constant A is

$$A = -i\frac{1}{4} \left(\frac{k}{2\pi}\right)^{N/2-1}. \tag{10.2.20}$$

When it is substituted into (10.2.16), we get

$$G(\rho) = -i\frac{1}{4} \left(\frac{k}{2\pi\rho}\right)^{N/2-1} H_{N/2-1}^{(1)}(k\rho). \tag{10.2.21}$$

This is the Green's function of a Helmholtz equation in infinite space. By use of (10.2.18), it is seen that as $k = 0$, (10.2.21) degrades to the solution of Laplace equation (10.2.10).

When $N = 1$, 2 and 3, (10.2.21) goes back to (6.2.27), (6.2.21) and (6.2.11), respectively. Here we put down the expression when $N = 4$:

$$G(\rho) = -i\frac{1}{4}\left(\frac{k}{2\pi\rho}\right)^{4/2-1} H^{(1)}_{4/2-1}(k\rho) = -i\frac{k}{8\pi\rho}H^{(1)}_1(k|\boldsymbol{r}-\boldsymbol{r}'|).$$

10.3.   Radial Equations under Central Potentials

10.3.1.   *Radial equation under a central potential in multidimensional spaces*

Suppose that the Schrödinger equation in higher-dimensional spaces is still of the form

$$(-\nabla^2 + V(\boldsymbol{r}))\psi(\boldsymbol{r}) = E\psi(\boldsymbol{r}). \tag{10.3.1}$$

The left hand side is kinetic energy plus potential energy $V(\boldsymbol{r})$. We assume a central potential $V(r)$, i.e., it is independent of the direction of the position vector $\boldsymbol{r}$. Similar to the three-dimensional case, E is called energy, and the solution $\psi(\boldsymbol{r})$ is called a bound state when $E < 0$ or the energy is less than the height of the potential.

Under the spherical coordinates, the Laplace operator is (10.1.30). In this form, (10.3.1) becomes

$$\left[-\frac{1}{r^{N-1}}\frac{\partial}{\partial r}\left(r^{N-1}\frac{\partial}{\partial r}\right) - \frac{1}{r^2}L_N^2(\theta) + V(r)\right]\psi(\boldsymbol{r}) = E\psi(\boldsymbol{r}). \tag{10.3.2}$$

We use separation of variables. Let the wavefunction be expressed by the product of radial and angular factors:

$$\psi(\boldsymbol{r}) = R(r)\zeta(\theta). \tag{10.3.3}$$

Thus, two equations are acquired.

$$L_N^2(\theta)\zeta(\theta) + \lambda\zeta(\theta) = 0. \tag{10.3.4}$$

$$\frac{1}{R(r)}\frac{1}{r^{N-3}}\frac{\partial}{\partial r}\left(r^{N-1}\frac{\partial}{\partial r}\right)R(r) - V(r)r^2 + Er^2 = \lambda. \tag{10.3.5}$$

Equation (10.3.4) is the eigenvalue equation of angular momentum, which will be discussed in the next section. Here we just mention that in N-dimensional space, the eigenvalue λ takes

$$\lambda = l(l + N - 2), \quad l = 0, 1, 2, 3, \ldots. \tag{10.3.6}$$

Equation (10.3.5) is a radial eigenvalue equation. Its solutions are to be discussed in this section. When (10.3.6) is substituted in, (10.3.5) becomes

$$\left[\frac{d^2}{dr^2} + \frac{N-1}{r}\frac{d}{dr} - V(r) + E - \frac{l(l+N-2)}{r^2}\right] R(r) = 0. \quad (10.3.7)$$

This equation is equivalent to a one-dimensional movement of a particle in such a potential: as $r \le 0$, the potential is infinitely high, and as $r > 0$ there may be a potential barrier.

The following transformation is made to eliminate the first derivative term:

$$R = \frac{u}{r^{(N-1)/2}}. \quad (10.3.8)$$

This transformation makes (10.3.7) become

$$u'' - \frac{\alpha}{r^2} + (E - V(r))u = 0, \quad (10.3.9)$$

where

$$\alpha = l(l+N-2) + \frac{(N-3)(N-1)}{4} \quad (10.3.10a)$$

and

$$4\alpha = 4l^2 + 4l(N-2) + (N-2)^2 - 1$$
$$= (2l + N - 2)^2 - 1. \quad (10.3.10b)$$

Equation (10.3.9) is very convenient to solve under some potential forms. For the sake of convenience, hereafter the function u is also called a radial function, which becomes the true radial function when divided by $r^{(N-1)/2}$. In the following, we solve the radial equation for some specific forms of potentials $V(r)$.

We solve the bound states unless the potential is zero. The wavefunction is to be normalized:

$$\int_0^\infty r^{N-1}|R|^2 dr = 1. \quad (10.3.11)$$

This integration means that the boundary conditions should be

$$[r^{(N-1)/2}R]_{r\to\infty} \to 0 \quad (10.3.12a)$$

and

$$[r^{(N-1)/2}R]_{r\to 0} < \infty. \tag{10.3.12b}$$

10.3.2. Helmholtz equation

In this case, $V(r) = 0$. Let $E = k^2$. Because there is no potential, the energy $E > 0$. Accordingly, (10.3.9) appears as

$$u'' + \left(k^2 - \frac{\alpha}{r^2}\right)u = 0. \tag{10.3.13}$$

The transformation

$$u(r) = r^{1/2}p(r) \tag{10.3.14}$$

turns (10.3.13) to be the following form:

$$r^2 p'' + r p' + \left(k^2 r^2 - \left(l + \frac{N}{2} - 1\right)^2\right)p = 0, \tag{10.3.15}$$

where (10.3.10) are employed. The is a Bessel equation of $l + N/2 - 1$ order. Its solution is $p(r) = J_{l+N/2-1}(kr)$. Subsequently,

$$u(r) = r^{1/2}J_{l+N/2-1}(kr). \tag{10.3.16}$$

The radial function follows (10.3.16) and (10.3.8) to be

$$R(r) = \frac{J_{l+N/2-1}(kr)}{r^{N/2-1}}. \tag{10.3.17}$$

This form shows that in odd- (even-) integer-dimensional spaces, the solutions are Bessel functions of half-integer (integer) orders. It has been known by (4.2.15b) that when $v > 0$ and $r \to 0$, $J_v(r) \to \frac{1}{\Gamma(v+1)}(\frac{r}{2})^v$. In this case, the wavefunction is zero at the origin. When $r \to \infty$, (10.3.17) approaches to zero. So, the boundary conditions (10.3.12) are met.

The other linearly independent solution of (10.3.13) is a Bessel function of the second kind, which does not meet the condition that it should be finite as $r \to 0$.

10.3.3. *Infinitely deep spherical potential*

The potential is

$$V(r) = \begin{cases} 0, & r < a, \\ \infty, & r > a. \end{cases} \qquad (10.3.18)$$

The potential barrier is infinitely high. Let $E = k^2$. Although the energy is positive, it is always less than the barrier, so that the solutions are always bound states. As $r < a$, a the equation goes back to (10.3.13), and accordingly, the solution is (10.3.17):

$$R(r) = \frac{C}{r^{N/2-1}} J_{l+N/2-1}(kr) \qquad (10.3.19)$$

as $r > a$. Because of the infinite barrier, the solution must be zero. At $r = a$, the function should also be zero. This requires that

$$J_{l+N/2-1}(k_j a) = 0, \quad j = 1, 2, 3, \dots. \qquad (10.3.20)$$

These equations determine the eigenvalues k_j. The solutions are

$$R_j(r) = \frac{C_j}{r^{N/2-1}} J_{l+N/2-1}(k_j r). \qquad (10.3.21)$$

The constant C_j is evaluated by (10.3.11).

10.3.4. *Finitely deep spherical potential*

The potential is

$$V(r) = \begin{cases} 0, & r < a, \\ V_0, & r > a. \end{cases} \qquad (10.3.22)$$

The only difference of this potential and (10.3.18) is that the barrier height is finite. Since the potential is discontinuous at $r = a$, it is required to solve (10.3.9) in the cases of $r < a$ and $r > a$ separately.

When $r < a$, the equation and its solution are the same as (10.3.13) and (10.3.19), respectively, in the last subsection. So, the solution is

$$R(r) = \frac{C}{r^{N/2-1}} J_{l+N/2-1}(kr), \quad r < a. \qquad (10.3.23)$$

When $r > a$, (10.3.9) becomes

$$u'' + \left(k^2 - V_0 - \frac{\alpha}{r^2} \right) u = 0. \qquad (10.3.24)$$

The cases of $E > V_0$ and $E < V_0$ are discussed separately.

(1) $E > V_0$

Let

$$E - V_0 = q^2. \tag{10.3.25}$$

The replacement of k by q turns (10.3.13) to (10.2.34). Then the same replacement turns (10.3.17) to the solution of (10.3.24):

$$R(r) = \frac{1}{r^{N/2-1}} \left[A J_{l+N/2-1}(qr) + B Y_{l+N/2-1}(qr) \right]$$

$$= \frac{A H^{(1)}_{l+N/2-1}(qr)}{r^{N/2-1}}, \quad r > a. \tag{10.3.26}$$

In this expression, r does not go to zero, so that the general solution includes the Bessel function of the second kind. We consider the solution to be an outward decaying wave as $r \to \infty$. Therefore, eventually Hankel function of the first kind is artificially selected.

At $r = a$, the radial function and its derivative should be continuous.

$$C_j J_{l+n/2-1}(k_j a) = A_j H^{(1)}_{l+N/2-1}(q_j a), \quad j = 1, 2, 3, \ldots. \tag{10.3.27}$$

$$C_j k_j \left[\frac{\mathrm{d}}{\mathrm{d}r} \frac{J_{l+N/2-1}(k_j r)}{r^{N/2-1}} \right]_{r=a}$$

$$= A_j \left[\frac{\mathrm{d}}{\mathrm{d}r} \frac{H^{(1)}_{l+N/2-1}(q_j r)}{r^{N/2-1}} \right]_{r=a}, \quad j = 1, 2, 3, \ldots. \tag{10.3.28}$$

These two equations plus the normalization condition (10.3.11) determine three parameters. They are C_j and A_j, and eigenvalue q_j,

$$q_j = \sqrt{k_j^2 - V_0}. \tag{10.3.29}$$

(2) $E < V_0$

This corresponds to bound states. Let

$$V_0 - E = \kappa^2. \tag{10.3.30}$$

Then (10.3.24) becomes

$$u'' + \left(-\kappa^2 - \frac{\alpha}{r^2} \right) u = 0. \tag{10.3.31}$$

Since in (10.3.13) the replacement of k by q leads to the solution (10.3.26), the replacement of q by $i\kappa$ in (10.3.26) gives the solution of (10.3.31):

$$R(r) = \frac{AH^{(1)}_{l+N/2-1}(i\kappa r)}{r^{N/2-1}}, \quad r > a. \tag{10.3.32}$$

This is a modified Hankel function of the first kind. As a matter of fact, (10.3.31) is just a modified Bessel equation.

At $r = a$, the radial function and its derivative should be continuous.

$$
\begin{aligned}
& C_j J_{l+N/2-1}(k_j a) \\
& \quad = A_j H^{(1)}_{m+N/2-1}(i\kappa_j a), \quad j = 1, 2, 3, \ldots.
\end{aligned}
\tag{10.3.33}
$$

$$
\begin{aligned}
& C_j k_j \left[\frac{d}{dr} \frac{J_{l+N/2-1}(k_j r)}{r^{N/2-1}} \right]_{r=a} \\
& \quad = A_j \left[\frac{d}{dr} \frac{H^{(1)}_{l+N/2-1}(i\kappa_j r)}{r^{N/2-1}} \right]_{r=a}, \quad j = 1, 2, 3, \ldots.
\end{aligned}
\tag{10.3.34}
$$

These two equations plus the normalization condition (10.3.11) determine three parameters. They are C_j and A_j, and eigenvalue κ_j,

$$\kappa_j = \sqrt{V_0 - k_j^2}. \tag{10.3.35}$$

10.3.5. *Coulomb potential*

The potential is inversely proportional to radial distance:

$$V(r) = -\frac{\chi}{r}. \tag{10.3.36}$$

Then (10.3.9) becomes that

$$u'' + \left(E + \frac{\chi}{r} - \frac{\alpha}{r^2} \right) u = 0. \tag{10.3.37}$$

This radial equation is more complicate than those above, for it includes an additional term inversely proportional to the argument. The solution is not a Bessel function any more. Nevertheless,

(10.3.37) is of the same form of the radial equation of a Hydrogen atom, and so the resolution is also the same.

Suppose that $E < 0$. Let

$$E = -\frac{\alpha^2}{4}, \quad n = \frac{\chi}{\alpha}, \quad \rho = \alpha r \tag{10.3.38}$$

and

$$u(\rho) = e^{-\rho/2} F(\rho). \tag{10.3.39}$$

Then, we get the equation that $F(\rho)$ satisfies:

$$F'' - F' + \left(\frac{n}{\rho} - \frac{\alpha}{\rho^2}\right) F = 0. \tag{10.3.40}$$

It has a first kind of singularity at $\rho = 0$. The two indices are reckoned from its indicial equation. We simply follow the theory presented in Subsection 3.5.2. In the present case, we have $c_0 = 0, d_0 = -\alpha$, so that the indicial equation is

$$s(s-1) - \alpha = 0. \tag{10.3.41}$$

Its two roots are

$$s_1 = l + \frac{N-1}{2}, \quad s_2 = -\left(l + \frac{N-3}{2}\right). \tag{10.3.42}$$

One is positive and the other is negative. To meet the condition that at $r = 0$ the wavefunction must be finite, the positive index has to be taken. Let

$$F(\rho) = \rho^{l+(N-1)/2} L(\rho). \tag{10.3.43}$$

Then $L(\rho)$ satisfies the following equation:

$$\rho L'' + (2l + N - 1 - \rho)L' + \left[n - \left(l + \frac{N-1}{2}\right)\right] L = 0, \tag{10.3.44}$$

which is a **confluent hypergeometric equation**. It is concisely written in the form of

$$\rho L'' + (\mu + 1 - \rho)L' + n_r L = 0, \tag{10.3.45}$$

where

$$\mu = 2l + N - 2, \quad n_r = n - \frac{N-1}{2} - l. \tag{10.3.46}$$

Note that μ is always a positive integer. The solutions of (10.3.45) are confluent hypergeometric functions in the form of

$$F(\mu, n_r, \rho) = \sum_{k=0}^{\infty} \frac{(\mu)_k}{k!(n_r)_k}\rho^k, \tag{10.3.47}$$

where Gauss symbols are used, see (3.3.16).

Since we are considering bound states, then as $r \to \infty$, the wavefunction should go to zero. Only when n_r is a positive integer can this condition be met. In this case, (10.3.45) turns to be a generalized Laguerre equation, see (3.4.1). One of its solutions is a generalized Laguerre polynomial, denoted as

$$L_{n_r}^{\mu}(\rho) = \sum_{k=0}^{\infty} \frac{(\mu)_k}{k!(n_r)_k}\rho^k. \tag{10.3.48}$$

We let

$$n = n_r + \frac{N-1}{2} + l = \frac{\chi}{\alpha} = \frac{\chi}{2\sqrt{-E}}. \tag{10.3.49}$$

Then,

$$E = -\frac{\chi^2}{4\left(n_r + l + (N-1)/2\right)^2} = -\frac{\chi^2}{4n^2}. \tag{10.3.50}$$

Since n_r is a positive integer, it is seen from (10.3.49) that for an odd- (even-) integer-dimensional space, n is a positive integer (half odd integer). Obviously, n has a lower bound:

$$n \geq \frac{N-1}{2}.$$

For example, in five-dimensional space, the quantum number $n = 2$, 3, and so forth. In four-dimensional space, $n = 3/2, 5/3$, and so on.

The radial wavefunction is obtained through (10.3.39) and (10.3.43):

$$u(r) = Cr^{l+(N-1)/2}\mathrm{e}^{-\alpha r/2}L_{n-l-(N-1)/2}^{2l+N-2}(\alpha r). \tag{10.3.51}$$

Then by (10.3.8), we have

$$R(r) = Cr^l\mathrm{e}^{-\alpha r/2}L_{n_r}^{\mu}(\alpha r). \tag{10.3.52}$$

10.3.6. *Harmonic potential*

The harmonic potential is proportional to the square of the radial distance:

$$V(r) = \chi^2 r^2. \tag{10.3.53}$$

Equation (10.3.9) becomes

$$u'' + \left(E - \chi^2 r^2 - \frac{\alpha}{r^2}\right) u = 0. \tag{10.3.54}$$

This equation is formally the same as that of a harmonic oscillator in three-dimensional space. The transformation

$$u(r) = e^{-\chi r^2/2} F(r) \tag{10.3.55}$$

yields the equation that $F(r)$ satisfies:

$$F'' - 2\chi r F' + \left(F - \chi - \frac{\alpha}{r^2}\right) F = 0. \tag{10.3.56}$$

It has a first kind of singularity at $r = 0$. Its A_0 matrix is the same as (10.3.41) and correspondingly, the positive eigenvalue of (10.3.42) should be chosen in order to acquire the physically significant solution. Let

$$F(r) = r^{l+(N-1)/2} f(r) \tag{10.3.57a}$$

and

$$\xi = \chi r^2. \tag{10.3.57b}$$

Then we further obtain the equation satisfied by the function $f(\xi)$:

$$\xi f'' + \left(\frac{2l + N}{2} - \xi\right) f' + \left(-\frac{2l + N}{4} + \frac{E}{4\chi}\right) f = 0. \tag{10.3.58}$$

This equation is formally the same as (10.3.45), and is also a confluent hypergeometric equation. Denoting

$$n_r = \frac{E}{4\chi} - \frac{2l + N}{4}, \quad \mu = l + \frac{N}{2} - 1 \tag{10.3.59}$$

and observing the discussion of (10.3.45)–(10.3.48), we know that the bound states require that n_r must be positive integers. The solution of (10.3.58) is then the following generalized Laguerre polynomial:

$$L_{n_r}^{\mu}(\rho) = \sum_{k=0}^{\infty} \frac{(\mu)_k}{k!(n_r)_k} \rho^k. \tag{10.3.60}$$

It is seen from (10.3.59) that for an even- (odd-) integer-dimensional space, μ is a positive integer (half odd integer).

The energy takes discrete values, which are determined by (10.3.59):

$$E = (4n + 2l + N)\chi. \qquad (10.3.61)$$

If we denote $\chi = \frac{\mu\omega}{\hbar}$ and $E = \frac{2\mu}{\hbar^2}\varepsilon$, then the energy is expressed by

$$\varepsilon = \left(2n + l + \frac{N}{2}\right)\hbar\omega. \qquad (10.3.62)$$

The zero-point energy is $\varepsilon_0 = N\hbar\omega/2$. The higher the dimension of a space, the greater the zero-point energy.

Combining (10.3.55), (10.3.57) and (10.3.60), we obtain the radial function:

$$R(r) = Cr^l e^{-\chi r^2/2} L_{n_r}^{l+N/2-1}(\chi r^2). \qquad (10.3.63)$$

10.3.7.   *Molecular potential with both negative powers*

The potential is

$$V(r) = \frac{A}{r^2} - \frac{B}{r}, \quad A > 0, \quad B > 0. \qquad (10.3.64)$$

The feature of this potential is that as $r \to 0$, it is infinitely repulsive, and as $r \to \infty$ it goes to zero. Between the two limits, there is a minimum. Thus, there can exist bound states.

For the present potential, (10.3.9) becomes

$$u'' + \left(E + \frac{B}{r} - \frac{\alpha + A}{r^2}\right)u = 0. \qquad (10.3.65)$$

It can be compared to that of Hydrogen atom (10.3.37). Let

$$\beta = \alpha + A \qquad (10.3.66)$$

and

$$4\beta + 1 = (2h + N - 2)^2. \qquad (10.3.67)$$

Then

$$u'' + \left(E + \frac{B}{r} - \frac{\beta}{r^2}\right)u = 0. \qquad (10.3.68)$$

This equation is of the same form of (10.3.37). As long as χ is replaced by B and α replaced by β, (10.3.37) becomes (10.3.68). In (10.3.50), l is replaced by h which may not be a positive integer. Observing (10.3.38), we have

$$E = -\frac{\beta^2}{4}, \quad n = \frac{B}{\beta}, \quad \rho = \beta r. \tag{10.3.69}$$

The following procedure is the same as that in Subsection 10.3.5. The bound state solution is still of the form of (10.3.52):

$$R(r) = Cr^h e^{-\beta r/2} L_{n_r}^{\mu}(\beta r). \tag{10.3.70}$$

The condition for the equation to have a solution is

$$n = n_r + \frac{N-1}{2} + h = \frac{B}{\beta} = \frac{B}{2\sqrt{-E}}. \tag{10.3.71}$$

Thus, the energy is

$$E = -\frac{B^2}{4(n_r + h + (N-1)/2)^2} = -\frac{B^2}{4n^2}. \tag{10.3.72}$$

Please note that h is not an integer. Therefore, n is neither an integer nor a half integer, but its increments should be integers.

By (10.3.66) and (10.3.67), the magnitude of h is

$$h = \frac{1}{2}\left(\sqrt{4(\alpha+A)+1} - N + 2\right)$$
$$= \frac{1}{2}\left(\sqrt{(2l+N-2)^2 + 4A} - N + 2\right). \tag{10.3.73}$$

10.3.8. *Molecular potential with positive and negative powers*

The potential is

$$V(r) = \frac{A}{r^2} + Br^2, \quad A > 0, \quad B > 0. \tag{10.3.74}$$

It is infinitely repulsive when both $r \to 0$ and $r \to \infty$. Therefore, a minimum exists at a finite distance, a potential well. Bound states can be found.

Under the potential (10.3.74), (10.3.9) becomes

$$u'' + \left(E - Br^2 - \frac{\alpha + A}{r^2}\right)u = 0. \qquad (10.3.75)$$

This is of the same form as that of harmonic oscillator (10.3.54), we make the replacement

$$B \to \chi^2, \quad \alpha + A \to \alpha \qquad (10.3.76)$$

We observe the procedure of Subsection 10.3.6 to write the bound state solution:

$$R(r) = Cr^h e^{-\sqrt{B}r^2/2} L_{n_r}^{h+N/2-1}(\sqrt{B}r^2), \qquad (10.3.77)$$

where n_r and h are defined by (10.3.71) and (10.3.73), respectively. The expression of the energy, following (10.3.61), reads

$$E = (4n + 2h + N)\sqrt{B}. \qquad (10.3.78)$$

10.3.9.   *Attractive potential with exponential decay*

The potential decays with radial distance exponentially:

$$V(r) = -Ce^{-r/a}. \qquad (10.3.79)$$

The radial equation (10.3.9) then becomes

$$u'' + \left(E + Ce^{-r/a} - \frac{\alpha}{r^2}\right)u = 0. \qquad (10.3.80)$$

We merely consider its solutions when $\alpha = 0$. Then, observing (10.3.10), the magnitude of l is determined by $4\alpha = (2l + N - 2)^2 - 1 = 0$. In three-dimensional space, $l = 0$. In higher-dimensional spaces, l cannot be positive. In this case, defining a new variable $\xi = e^{-r/2a}$ is helpful to find the solutions.

10.3.10.   *Conditions that the radial equation has analytical solutions*

Up to now we have solved the eigenvalues and eigenfunctions of the radial equation under some particular potentials. In this subsection, we simply assume the central potential without giving any specific form, and discuss under what conditions the analytical solutions of the radial equation can be gained. To this aim, we postulate that the radial is of the following form:

$$u(r) = C z^{(a+1)/2} e^{-z/2} L_n^a(z) / \sqrt{z'}, \qquad (10.3.81)$$

where $z = z(r)$ is a function of z and $L_n^a(z)$ is an associated Laguerre function. Taking derivative of $u(r)$ with respect to r twice, we get

$$u'' + u \left[-\frac{1}{4} z'^2 + \frac{n+a+1}{2} \frac{z'^2}{z} - \frac{a^2-1}{4} \left(\frac{z'}{z} \right)^2 - \frac{3z''^2}{4z'^2} + \frac{z'''}{2z'} \right] = 0, \qquad (10.3.82)$$

where we have employed an associated Laguerre equation

$$z L_n^{a''} + (a+1-z) L_n^{a'} + n L_n^a = 0. \qquad (10.3.83)$$

The radial function $u(r)$ has to satisfy (10.3.82). That is to say, this equation should be actually (10.3.9). Comparison of the two equations indicates that there should be a constant term in (10.3.83), which represents energy E, and a $1/r^2$ term, which represents angular momentum. If the latter is absent, the angular quantum number $l = 0$. The remaining terms in (10.3.82) related to r are of the meaning of potential. Under which potential it is possible to obtain analytical solutions depends on the form of the function $z(r)$. In the following, we discuss two forms.

(1) Suppose that

$$z(r) = b r^\mu. \qquad (10.3.84)$$

Then, (10.3.83) becomes

$$u'' + u \left[-\frac{1}{4} b^2 \mu^2 r^{2\mu-2} + \frac{n+a+1}{2} b^2 \mu^2 r^{\mu-2} - \frac{a^2 \mu^2 - 1}{4r^2} \right] = 0. \qquad (10.3.85)$$

In this case, the $1/r^2$ term appears automatically. The constant term will appear if $\mu = 1$ or 2.

When $\mu = 1$, the potential is inversely proportional to r. Taking $\mu = 1$ in (10.3.84) and (10.3.81), we get the radial function:

$$u(r) = C r^{(a+1)/2} e^{-br/2} L_n^a(br).$$

This is just the case of Coulomb potential, see (10.3.51), and equation (10.3.85) degrades to (10.3.37).

When $\mu = 2$, the potential is proportional to r^2. Taking $\mu = 2$ in (10.3.85) and (10.3.81), we get the radial function:

$$u(r) = Cr^{a+1/2}e^{-br^2/2}L_n^a(br^2).$$

This is just the case of harmonic potential, see (10.3.63), and equation (10.3.85) degrades to (10.3.54).

(2) Suppose that

$$z(r) = be^{-r/d}. \tag{10.3.86}$$

Then, (10.3.82) becomes

$$u'' + u\left(-\frac{b}{4d^2}e^{-2r/d} + \frac{2n+a+1}{2}\frac{b}{d^2}e^{-r/d} - \frac{a^2}{4d^2}\right) = 0. \tag{10.3.87}$$

In this case, the $1/r^2$ is absent, which corresponds to that of $\alpha = 0$ in (10.3.9). We have known from Subsection 10.3.9 that it applies to the case of the angular momentum $l = 0$ in three-dimensional space. In (10.3.87), a constant term appears automatically. However, the constant a needs to be modified such that $(2n + a + 1)/2$ is independent of both n and a. Then the constant term in (10.3.87) will involve quantum number n. The two terms other than the constant term in (10.3.87) reflect potential. Apparently, the potential can be written in the form of

$$V(r) = De^{2(r_0-r)/d} - 2De^{(r_0-r)/d}. \tag{10.3.88}$$

This form of potential is called Morse potential. It is a kind of molecular potential. We have mentioned it in Chapter 9 when we discussed the pair potential between atoms, see (9.3.27).

The radial function is

$$u(r) = Cbe^{-ar/2d}\exp(-be^{-r/d}/2)L_n^a(be^{-r/d}). \tag{10.3.89}$$

10.4. Solutions of Angular Equations

In the last section, we solved the radial equations under various potentials. In this section, we turn to solve the angular equation

(10.3.4).

$$L_N^2(\theta)\zeta(\theta) + \lambda\zeta(\theta) = 0. \qquad (10.4.1)$$

The operator $L_N^2(\theta)$ is called the square of total angular momentum.

For two- and three-dimensional spaces, the eigenvalues and eigenfunctions of the angular equations have been clearly presented in mathematical physics and quantum physics textbooks. In part 3 of Subsection 3.4.2 of this book, some main formulas were given. Let us now have a brief retrospect of them under the uniform expression of the square of angular momentum operator (10.1.31).

For two-dimensional space, $N = 2$, there is one angular variable, which is denoted as φ. From (10.1.32) it is seen that

$$L_2^2(\varphi) = \frac{\partial^2}{\partial\varphi^2}. \qquad (10.4.2)$$

The range of φ is $[0, 2\pi]$. This range determines that the periodical boundary condition should be adopted. Then, (10.4.1) goes back to (3.4.44), and the eigenvalues are thus determined. The square of the angular momentum takes discrete values: $L_2^2 = m^2$, $m = 0, 1, 2, \ldots$.

For three-dimensional space, $N = 3$, there are two angular variables, denoted as φ and θ, and their ranges are $[0, 2\pi]$ and $[0, \pi]$, respectively. From (10.1.32) we have

$$L_3^2(\theta, \varphi) = \frac{1}{\sin\theta}\frac{\partial}{\partial\theta}\left(\sin\theta\frac{\partial}{\partial\theta}\right) + \frac{1}{\sin^2\theta}\frac{\partial^2}{\partial\varphi^2}. \qquad (10.4.3)$$

So, (10.4.1) turns to be (3.4.43). After separation of variables, one will obtain two equations (3.4.44) and (3.4.45). The equation with respect to θ is an associated Legendre equation (3.4.46). Its eigenvalues are determined by (3.4.47b). The possible values of the square of the total angular momentum are $L_3^2 = l(l+1); l = 0, 1, 2, \ldots$. The equation with respect to φ is (3.4.44), but its eigenvalues m are restrained by the other equation. As soon as l is fixed, the eigenvalues have to take $m = 0, \pm1, \pm2, \ldots, \pm l$. This is the relationship between two quantum numbers l and m. The latter embodies the projection of the total angular momentum on the subspace labeled by angle φ.

Having retrospected the knowledges of angular momenta in two- and three-dimensional spaces, we are ready to discuss higher-dimensional cases.

10.4.1.   *Four-dimensional space*

1. Angular equation with three angular variables

When $N = 4$, there are three angular variables. The volume of a hypersphere with radius R in N-dimensional space has been evaluated by (10.1.18). Its explicit form in four-dimensional space is

$$\int_0^R dr \int_0^\pi r d\theta_1 \int_0^\pi r s_1 d\theta_2 \int_0^{2\pi} r s_1 s_2 d\theta_3 = \frac{\pi^2 R^4}{2}. \qquad (10.4.4)$$

The square of the total angular momentum operator is written observing (10.1.31) as

$$L_4^2(\theta) = \frac{1}{\sin^2\theta_1} \frac{\partial}{\partial\theta_1} \left(\sin^2\theta_1 \frac{\partial}{\partial\theta_1} \right) + \frac{1}{\sin^2\theta_1 \sin\theta_2} \frac{\partial}{\partial\theta_2} \left(\sin\theta_2 \frac{\partial}{\partial\theta_2} \right)$$

$$+ \frac{1}{\sin^2\theta_1 \sin^2\theta_2} \frac{\partial^2}{\partial\theta_3^2}. \qquad (10.4.5)$$

To simplify denotations, hereafter the following shorthands are adopted:

$$s_i = \sin\theta_i, \quad c_i = \cos\theta_i. \qquad (10.4.6)$$

For instances, $s_1 = \sin\theta_1$, $s_1^2 = \sin^2\theta_1$, $c_1 = \cos\theta_1$, and so on. The angular equation (10.4.1) becomes

$$\left[\frac{1}{s_1^2} \frac{\partial}{\partial\theta_1} \left(s_1^2 \frac{\partial}{\partial\theta_1} \right) + \frac{1}{s_1^2 s_2} \frac{\partial}{\partial\theta_2} \left(s_2 \frac{\partial}{\partial\theta_2} \right) + \frac{1}{s_1^2 s_2^2} \frac{\partial^2}{\partial\theta_3^2} \right] \zeta(\theta_1, \theta_2, \theta_3)$$

$$= -\lambda_1 \zeta(\theta_1, \theta_2, \theta_3). \qquad (10.4.7)$$

It is solved by separation of variables. Let

$$\zeta(\theta_1, \theta_2, \theta_3) = \zeta_1(\theta_1)\zeta_2(\theta_2)\zeta_3(\theta_3). \qquad (10.4.8)$$

Then the equation with respect to θ_1 is

$$\frac{\partial}{\partial\theta_1} \left(s_1^2 \frac{\partial}{\partial\theta_1} \right) \zeta_1(\theta_1) + (\lambda_1 s_1^2 - \lambda_2)\zeta_1(\theta_1) = 0 \qquad (10.4.9)$$

and that with respect to θ_2 and θ_3 is

$$\frac{1}{\zeta_2(\theta_2)s_2}\frac{\partial}{\partial\theta_2}\left(s_2\frac{\partial}{\partial\theta_2}\right)\zeta_2(\theta_2) + \frac{1}{\zeta_3(\theta_3)s_2^2}\frac{\partial^2}{\partial\theta_3^2}\zeta_3(\theta_3) + \lambda_2 = 0.$$
(10.4.10a)

The ranges of θ_2 and θ_3 are $[0, \pi]$ and $[0, 2\pi]$, respectively, see (10.1.11), which are the same as those of angulars (θ, φ) in Subsection 3.4.2. As a matter of fact, (10.4.10a) is identical to (3.4.43) under the same boundary conditions. Therefore the discussion of the former simply copies that of the latter. The equation of θ_3 is (3.4.44) and its eigenvalues are m^2. The equation of θ_2 is (3.4.45), i.e., an associated Legendre equation (3.4.46), which is reput down here:

$$(1 - x^2)\frac{\mathrm{d}^2 p}{\mathrm{d}x^2} - 2x\frac{\mathrm{d}p}{\mathrm{d}x} + \left(\lambda_2 - \frac{m^2}{1 - x^2}\right)p = 0. \qquad (10.4.10b)$$

Its eigenvalues are still (3.4.47b):

$$\lambda_2 = l(l+1), \ (l = 0, 1, 2, \ldots), \quad m = 0, \pm 1, \pm 2, \ldots, \pm l. \qquad (10.4.11)$$

Now we look back (10.4.9). After transformation

$$\cos\theta_1 = x,$$

(10.4.9) is turned to be

$$(1 - x^2)y''(x) - 3xy'(x) + \left(\lambda_1 - \frac{l(l+1)}{1 - x^2}\right)y(x) = 0, \qquad (10.4.12)$$

where the expression of λ_2 (10.4.11) has been substituted into. It is quite similar to an associated Legendre equation (10.4.10b). The minor differences are that the coefficient of the first derivative term is 3 instead of 2 and the numerator in the last term is $l(l+1)$ instead of m^2. In order to solve this equation, we investigate the general case.

2. Associated Gegenbauer equation

First of all, we put down **Gegenbauer equation**:

$$(1 - x^2)y'' - 2(\alpha + 1)xy' + \beta(\beta + 2\alpha + 1)y = 0. \qquad (10.4.13)$$

The two parameters α and β can be complex numbers. The interval of x is $[-1, 1]$. Its solutions are called **Gegenbauer functions**. Here we merely consider $\beta = n$ a natural number:

$$(1 - x^2)y'' - 2(\alpha + 1)xy' + n(n + 2\alpha + 1)y = 0,$$

$$n = 0, 1, 2, 3, \ldots, \qquad\qquad (10.4.14)$$

which has been actually listed in Table 3.1. We have been familiar with its two particular cases.

One particular case is $\alpha = 0$. Then it becomes a Legendre equation, see Table 3.1 and (3.4.24). The other is $\alpha = -1/2$. Then it is Chebyshev equation, see Table 3.1 and (3.4.49). As an extension of the latter, when $\alpha = m - 1/2$, where m is positive integers, it becomes associated Chebyshev equation, see (3.4.65).

If α is any complex number other than negative integers, the polynomial solutions are **ultraspheric polynomials**, or **Gegenbauer polynomials** with index α introduced in Subsection 3.3.1, denoted as $G_n^\alpha(x)$ or $G_n(x, \alpha)$. Their series expression was listed in Table 3.2, differential expression in Table 3.3 and generating function in Table 3.5. The details of a Gegenbauer equation and Gegenbauer polynomials can be found in literature.

Now we take derivative k times of the Gegenbauer equation (10.4.14) with respect to x. Then let

$$w(x) = (1 - x^2)^{k/2} y^{(k)}(x). \qquad\qquad (10.4.15)$$

As a result, the function $w(x)$ satisfies the following equation:

$$(1 - x^2)w^{(2)} - 2(\alpha + 1)xw^{(1)}$$

$$+ n(n + 2\alpha + 1)w - \frac{k(k + 2\alpha)}{1 - x^2}w = 0, \qquad (10.4.16)$$

which is called an **associated Gegenbauer equation**. The process of obtaining (10.4.16) from (10.4.14) is actually the same as that of obtaining an associated Legendre equation (3.4.34) from a Legendre equation (3.4.24) in Subsection 3.4.2.

The solutions of associated Gegenbauer equation are denoted as $Q_n^{(k)}(x, \alpha)$. Since the solutions of (10.4.14) are Gegenbauer polynomials $G_n^\alpha(x)$, then by (10.4.15), the expression of $Q_n^{(k)}(x, \alpha)$ is

$$w(x) = Q_n^{(k)}(x, \alpha) = (1 - x^2)^{k/2} \frac{\mathrm{d}^k}{\mathrm{d}x^k} G_n^\alpha(x).$$

Making use of the differential expression of $G_n^\alpha(x)$ listed in Table 3.3, we get

$$Q_n^{(k)}(x,\alpha) = (1-x^2)^{k/2} \frac{d^k}{dx^k} \frac{1}{(1-x^2)^\alpha} \frac{d^n}{dx^n} (1-x^2)^{n+\alpha},$$

which is called an **associated Gegenbauer polynomial**. Here the normalization coefficient is omitted.

On the other hand, if we take derivative of (10.4.14) k times and then let $z(x) = y^{(k)}(x)$ instead of (10.4.15), we will gain

$$(1-x^2)z' - 2(\alpha + k + 1)xz'$$

$$+ (n-k)(n-k+2(\alpha+k)+1)z = 0. \quad (10.4.17a)$$

Since the solutions of (10.4.14) are $G_n(x,\alpha)$, those of (10.4.17a) are $G_{n-k}(x,\alpha+k)$. That is to say: $G_n^{(k)}(x,\alpha) = G_{n-k}(x,\alpha+k)$, which can also be derived from the generating function of Gegenbauer polynomials. Apparently, the derivative of a Gegenbauer polynomial $G_n(x,\alpha)$ k times is still a Gegenbauer polynomial, with the highest power being lowered from n to $n-k$. Consequently, k is a positive integer not greater than n. Thus the solutions of (10.4.17a) are

$$w(x) = (1-x^2)^{k/2} G_{n-k}(x,\alpha+k). \quad (10.4.17b)$$

As $\alpha = 0$, they go back to associated Legendre polynomials.

3. Angular momentum in four-dimensional space

Having the knowledges of Gegenbauer equation, we are able to easily obtain the eigenvalues and eigenfunctions of (10.4.12). We take $\alpha = 1/2$ in (10.4.16) and compare it with (10.4.12). Then we immediately know that the eigenvalues of (10.4.9) should take

$$\lambda_1 = n(n+2), \quad n = 0,1,2,3,\ldots;$$

$$\lambda_2 = k(k+1), \quad k = 0,1,2,\ldots,n, \quad (10.4.18a)$$

and the eigenfunction is

$$\zeta_1(\theta_1) = Q_n^{(k)}\left(\cos\theta_1, \frac{3}{2}\right). \quad (10.4.18b)$$

It seems that the expression of λ_2 is the same as (10.4.11). However, we have pointed out below (10.4.17a) that k has an upper bound which is n.

Now we put the three eigenvalues together.

$$\lambda_1 = n(n+2), \quad n = 0, 1, 2, 3, \dots;$$

$$\lambda_2 = k(k+1), \quad k = 0, 1, 2, \dots, n;$$

$$\lambda_3 = m, \quad m = 0, \pm 1, \pm 2, \dots, \pm k. \qquad (10.4.19)$$

Referring to the physical significance of the angular momenta in three-dimensional space, we give the explanation of the angular momenta (10.4.19) in four-dimensional space as follows. In angular space $(\theta_1, \theta_2, \theta_3)$, the square of the total angular momentum takes the values $n(n+2), n = 0, 1, 2, 3, \dots$; the total angular momentum has a projection in (θ_2, θ_3) subspace, the square of which should take values $k(k+1), k = 0, 1, 2, \dots, n$; this projection has further a projection component in θ_3 subspace with its value being $0, \pm 1, \pm 2, \dots, \pm k$.

We are going to write the total eigenfunction following (10.4.8). Because (10.4.10b) is identical to (3.4.43), the function $\zeta_2(\theta_2)\zeta_3(\theta_3)$ is just spherical function (3.4.48). Combined with (10.4.18b), the solution is

$$\zeta_{nkm}(\theta_1, \theta_2, \theta_3) = Q_n^{(k)}\left(\cos\theta_1, \frac{1}{2}\right) P_k^m(\cos\theta_2)\, e^{im\theta_3}, \qquad (10.4.20a)$$

where the subscripts label the angular momenta. The Legendre function can be written as an associated Gegenbauer polynomial with index $\alpha = 0$:

$$P_k^m(\cos\phi) = Q_k^{(m)}(\cos\phi, 0).$$

In four-dimensional space, any function of three angles can be expanded as the linear combination of eigenfunctions (10.4.20a):

$$\psi(\theta_1, \theta_2, \theta_3) = \sum_{n=0}^{\infty}\sum_{k=0}^{n}\sum_{m=-k}^{k} A_{nkm}\zeta_{nkm}(\theta_1, \theta_2, \theta_3). \qquad (10.4.20b)$$

Finally, let count from (10.4.19) the number of the possible states when n is fixed. Apparently, for a given k, m can take $2k+1$ values. Thus, the total state number is

$$M = \sum_{k=0}^{n}(2k+1) = (n+1)^2. \qquad (10.4.21)$$

10.4.2.   *Five-dimensional space*

In five-dimensional space, the square of the total angular momentum operator is

$$L_5^2(\theta) = \frac{1}{s_1^3} \frac{\partial}{\partial \theta_1} \left(s_1^3 \frac{\partial}{\partial \theta_1} \right) + \frac{1}{s_1^2 s_2^2} \frac{\partial}{\partial \theta_2} \left(s_2^2 \frac{\partial}{\partial \theta_2} \right)$$
$$+ \frac{1}{s_1^2 s_2^2 s_3} \frac{\partial}{\partial \theta_3} \left(s_3 \frac{\partial}{\partial \theta_3} \right) + \frac{1}{s_1^2 s_2^2 s_3^2} \frac{\partial^2}{\partial \theta_4^2}. \quad (10.4.22)$$

Its eigenvalue equation is

$$L_5^2(\theta)\zeta(\theta_1, \theta_2, \theta_3, \theta_4) = -\lambda_1 \zeta(\theta_1, \theta_2, \theta_3, \theta_4). \quad (10.4.23)$$

Let $\zeta(\theta_1, \theta_2, \theta_3, \theta_4) = \zeta_1(\theta_1)\xi(\theta_2, \theta_3, \theta_4)$. Then the two factors satisfy

$$\frac{\partial}{\partial \theta_1} \left(s_1^3 \frac{\partial}{\partial \theta_1} \right) \zeta_1(\theta_1) + (\lambda_1 s_1^2 - \lambda_2)s_1 \zeta_1(\theta_1) = 0 \quad (10.4.24)$$

and

$$\left[\frac{1}{s_2^2} \frac{\partial}{\partial \theta_2} \left(s_2^2 \frac{\partial}{\partial \theta_2} \right) + \frac{1}{s_2^2 s_3} \frac{\partial}{\partial \theta_3} \left(s_3 \frac{\partial}{\partial \theta_3} \right) + \frac{1}{s_2^2 s_3^2} \frac{\partial^2}{\partial \theta_4^2} \right]$$
$$\times \xi(\theta_2, \theta_3, \theta_4) + \lambda_2 \xi(\theta_2, \theta_3, \theta_4) = 0, \quad (10.4.25)$$

respectively.

Equation (10.4.25) is identical to (10.4.7), and has been discussed clearly in the last subsection. The eigenvalues are those in (10.4.19). The λ_2 is actually λ_1 in (10.4.7) and in (10.4.18). We put down

$$\lambda_2 = l_2(l_2 + 2), \quad l_2 = 0, 1, 2, \ldots. \quad (10.4.26)$$

For equation (10.4.24), the transformation $\cos \theta_1 = x$ turns it to be

$$(1 - x^2)y''(x) - 4xy'(x) + \left(\lambda_1 - \frac{l_2(l_2 + 2)}{1 - x^2} \right) y = 0. \quad (10.4.27)$$

Compared to (10.4.16), it is seen that in (10.4.27) $\alpha = 1$. So, the solution of (10.4.24) is

$$\zeta_1(\theta_1) = Q_n^{(k)}(\cos \theta_1, 1). \quad (10.4.28)$$

Accordingly, the eigenvalues should take $\lambda_1 = l_1(l_1 + 3)$. Together with (10.4.19), all the eigenvalues are as follows:

$$\lambda_1 = l_1(l_1 + 3), \quad l_1 = 0, 1, 2, \ldots;$$
$$\lambda_2 = l_2(l_2 + 2), \quad l_2 = 0, 1, 2, 3, \ldots, l_1;$$

$$\lambda_3 = l_3(l_3 + 1), \quad l_3 = 0, 1, 2, \ldots, l_2;$$

$$\lambda_4 = l_4, \qquad\qquad l_4 = 0, \pm1, \pm2, \ldots, \pm l_3. \qquad (10.4.29)$$

Similar to the case of four-dimensional space, we give the explanation of the angular momenta (10.4.29) in five-dimensional space as follows. In angular space $(\theta_1, \theta_2, \theta_3, \theta_4)$, the square of the total angular momentum takes the values $l_1(l_1 + 2), l_1 = 0, 1, 2, 3, \ldots$; the total angular momentum has a projection in $(\theta_2, \theta_3, \theta_4)$ subspace, the square of which should take values $l_2(l_2 + 1), l_2 = 0, 1, 2, \ldots, l_1$; this projection has further a projection component in (θ_3, θ_4) subspace, the square of which should take values $l_3(l_3+1), l_3 = 0, 1, 2, \ldots, l_2$; at last, the angular momentum in the (θ_3, θ_4) subspaces has a projection in θ_4 subspace with its value being $0, \pm1, \pm2, \ldots, \pm l_3$.

The total eigenfunction is the product of (10.4.28) and (10.4.20):

$$\zeta_{l_1 l_2 l_3 l_4}(\theta_1, \theta_2, \theta_3, \theta_4)$$

$$= Q_{l_1}^{(l_2)}(\cos\theta_1, 1)\, Q_{l_2}^{(l_3)}\left(\cos\theta_2, \frac{1}{2}\right) Q_{l_3}^{(l_4)}(\cos\theta_3, 0)\, e^{il_4\theta_4}.$$

$$(10.4.30)$$

In five-dimensional space, any function of four angles can be expanded as the linear combination of eigenfunctions (10.4.30):

$$\psi(\theta_1, \theta_2, \theta_3, \theta_4) = \sum_{l_1=0}^{\infty} \sum_{l_2=0}^{l_1} \sum_{l_3=0}^{l_2} \sum_{l_4=-l_3}^{l_3} A_{l_1 l_2 l_3 l_4} \zeta_{l_1 l_2 l_3 l_4}(\theta_1, \theta_2, \theta_3, \theta_4).$$

10.4.3.   *N-dimensional space*

In the case of six-dimensional space, the eigenvalue equation of the square of the total angular momentum is

$$L_6^2(\theta)\zeta(\theta_1, \theta_2, \theta_3, \theta_4, \theta_5) = -\lambda_1\zeta(\theta_1, \theta_2, \theta_3, \theta_4, \theta_5). \qquad (10.4.31)$$

The solving process is similar to those in four- and five- dimensional spaces. After separation of variables $\zeta(\theta_1, \theta_2, \theta_3, \theta_4, \theta_5) = \zeta_1(\theta_1)\xi(\theta_2, \theta_3, \theta_4, \theta_5)$, one obtains the equation that $\xi(\theta_2, \theta_3, \theta_4, \theta_5)$ satisfies, which is identical to (10.4.23). The equation satisfied by

$\zeta_1(\theta_1)$ can be, after the transformation $\cos\theta_1 = x$, converted to the form of

$$(1 - x^2)y''(x) - 5xy'(x) + \left(\lambda_1 - \frac{l_2(l_2 + 3)}{1 - x^2}\right)y = 0. \qquad (10.4.32)$$

This is the associated Gegenbauer equation with index $\alpha = 3/2$. Its eigenvalue obviously takes $\lambda_1 = l_1(l_1 + 4)$, and its eigenfunction is

$$\zeta_1(\theta_1) = Q_{l_1}^{(l_2)}\left(\cos\theta_1, \frac{3}{2}\right). \qquad (10.4.33)$$

Together with (10.4.29), all the eigenvalues are as follows:

$$\begin{aligned}
\lambda_1 &= l_1(l_1 + 4), & l_1 &= 0, 1, 2, \ldots; \\
\lambda_2 &= l_2(l_2 + 3), & l_2 &= 0, 1, 2, 3, \ldots, l_1; \\
\lambda_3 &= l_3(l_3 + 2), & l_3 &= 0, 1, 2, \ldots, l_2; \\
\lambda_4 &= l_4(l_4 + 1), & l_4 &= 0, 1, 2, \ldots, l_3; \\
\lambda_5 &= l_5, & l_5 &= 0, \pm1, \pm2, \ldots, \pm l_4.
\end{aligned} \qquad (10.4.34)$$

The significance of these quantum numbers can be explained similarly to those in five-dimensional space.

From the discussions above, the conclusion in N-dimensional space is easily induced. In N-dimensional space, there are $N - 1$ angles, the ranges of which have been stipulated in (10.1.11). The eigenvalue equation of the square of total angular momentum is

$$L_N^2(\theta)\zeta(\theta_1, \theta_2, \ldots, \theta_N) = -\lambda_1\zeta(\theta_1, \theta_2, \ldots, \theta_N). \qquad (10.4.35)$$

After separation of variables $\zeta(\theta_1, \theta_2, \ldots, \theta_N) = \zeta_1(\theta_1)\xi(\theta_2, \theta_3, \ldots, \theta_N)$, the eigenequation that $\xi(\theta_2, \theta_3, \ldots, \theta_N)$ satisfies is just that of $L_{N-1}^2(\theta)$ in $N - 1$-dimensional space, and $\zeta_1(\theta_1)$ satisfies

$$\frac{\partial}{\partial\theta_1}\left(s_1^{N-2}\frac{\partial}{\partial\theta_1}\right)\zeta_1(\theta_1)$$

$$+ \left(\lambda_1 s_1^2 - \lambda_2\right)s_1^{N-4}\zeta_1(\theta_1) = 0, \qquad (10.4.36)$$

where $\lambda_2 = l_2(l_2 + N - 3)$ are the eigenvalues of $L_{N-1}^2(\theta)$ in $N - 1$-dimensional space. Through transformation $\cos\theta_1 = x$, (10.4.36) turns to be

$$(1 - x^2)y''(x) - (N - 1)xy'(x)$$

$$+ \left(\lambda_1 - \frac{l_2(l_2 + N - 3)}{1 - x^2}\right) y = 0, \qquad (10.4.37)$$

which is an associated Gegenbauer equation of index $\alpha = (N - 3)/2$. Its eigenvalue takes $\lambda_1 = l_1(l_1 + N - 2)$, and the eigenfunction is

$$\zeta_1(\theta_1) = Q_{l_1}^{(l_2)}\left(\cos\theta_1, \frac{N - 3}{2}\right). \qquad (10.4.38)$$

All of the eigenvalues of (10.4.35) are listed as follows.

$$\lambda_1 = l_1(l_1 + N - 2), \quad l_1 = 0, 1, 2, \dots;$$

$$\lambda_2 = l_2(l_2 + N - 3), \quad l_2 = 0, 1, 2, 3, \dots, l_1;$$

$$\lambda_3 = l_3(l_3 + N - 4), \quad l_3 = 0, 1, 2, \dots, l_2;$$

$$\cdots\cdots$$

$$\lambda_{N-2} = l_{N-2}(l_{N-2} + 1), \quad l_{N-2} = 0, 1, 2, \dots, l_{N-3};$$

$$\lambda_{N-1} = l_{N-1}^2, \quad l_{N-1} = 0, \pm 1, \pm 2, \dots, \pm l_{N-2}. \qquad (10.4.39)$$

The total angular momentum projects to the subspaces successively in a way similar to that in five-dimensional space. The total eigenfunction of (10.4.35) is

$$\zeta_{l_1 l_2 \cdots l_{N-1}}(\theta_1, \theta_2, \dots, \theta_N)$$

$$= Q_{l_1}^{(l_2)}\left(\cos\theta_1, \frac{N-3}{2}\right) Q_{l_2}^{(l_3)}\left(\cos\theta_2, \frac{N-4}{2}\right) \cdots$$

$$\times Q_{l_{N-3}}^{(l_{N-2})}\left(\cos\theta_{N-3}, \frac{1}{2}\right) Q_{l_{N-2}}^{(l_{N-1})}(\cos\theta_{N-2}, 0)e^{il_{N-1}\theta_{N-1}},$$

where $N \geq 3$.

At last, we would like to point out the physical meaning in another viewpoint. If in (10.1.30) the radial variable of ψ is fixed to be R, then Laplace equation is simplified to be

$$\nabla^2\varphi = \frac{1}{R^2}L_N^2(\theta)\varphi = 0, \qquad (10.4.40)$$

which is Laplace equation on the hypersurface of an N-dimensional hypersphere. Although all the variables of this equation are angular ones, it can be regarded as a Laplace equation in an $N-1$-dimensional space. This $N-1$-dimensional space is a hypersurface of an N-dimensional hypersphere. In the following, it is called $N-1$-dimensional angular space. If a constant is added to the operator, (10.4.40) becomes the Helmholtz equation on the hypersurface:

$$\left[\frac{1}{R^2}L_N^2(\theta) + A\right]\varphi = 0. \tag{10.4.41}$$

The physical meaning of the constant A in this equation is the square of the total angular momentum in the N-dimensional space. It is in fact generated in the course of the separation of the radial and angular variables in the Laplace equation of the N-dimensional space.

Since the radius R has been fixed, in the Cartesian coordinates,

$$R^2 = \sum_{i=1}^{N} x_i^2 \tag{10.4.42}$$

is fixed. Note that in this expression, there are still N variables, but their variation is under the constraint of (10.4.42). Therefore, the $N-1$-dimensional subspace of an N-dimensional space has $N-1$ coordinates in one system, but may still have N coordinates in another system with some constraints.

For example, in the three-dimensional space we are familiar with, the Laplace equation is (3.4.40). After separation of radial and angular variables, the equation on the surface of a sphere with radius R can be obtained, (3.4.43), which is just the case of $N=3$ in (10.4.37). The surface of a sphere with radius R is a two-dimensional subspace. In the sphere system, there are two angular variables. Nevertheless, in the Cartesian system,

$$R^2 = x^2 + y^2 + z^2, \tag{10.4.43}$$

which manifests three variables, although their variations are under the constraint (10.4.43).

Therefore, the square of the total angular momentum operator in the N-dimensional space is also regarded as the Laplace operator of the $N-1$-dimensional hypersurface, or the $N-1$-dimensional angular space.

That the square of the total angular momentum operator is inspected in this viewpoint is helpful for us to recognize some equations in non-Euclid spaces.

10.5.　Pseudo Spherical Coordinates

10.5.1.　*Pseudo coordinates in four-dimensional space*

In Subsection 10.1.1, we have given for an N-dimensional space the transformations between Cartesian and spherical coordinates in (10.1.10) and the ranges of the ranges of the angular coordinates in (10.1.11). In fact, for certain dimensions of spaces, there can be other spherical coordinates.

In four-dimensional space, we define another spherical coordination, in which there are still one radial coordinate and three angles (ϕ_1, ϕ_2, ϕ_3). The transformations between them and Cartesian coordinates are as follows:

$$x_1 = rs_1c_2, \quad x_2 = rs_1s_2, \quad x_3 = rc_1c_3, \quad x_4 = rc_1s_3. \quad (10.5.1)$$

Here the denotations of (10.4.6) are adopted, with the angles θ 's there replaced by ϕ's here, e.g., $s_i = \sin\phi_i$, $c_i = \cos\phi_i$. The square of line element in this coordinate system is

$$\begin{aligned} dl^2 &= dx_1^2 + dx_2^2 + dx_3^2 + dx_4^2 \\ &= dr^2 + r^2 d\phi_1^2 + r^2 s_1^2 d\phi_2^2 + r^2 c_1^2 d\phi_3^2. \end{aligned} \quad (10.5.2)$$

This expression is different from that of the usual spherical coordinate system. The coordinates defined by (10.5.1) are tentatively called a **pseudo spherical system**.

Comparison of (10.5.2) with (10.1.2) demonstrates that the expression of the line element square does not contain crosses between coordinates. So, this system is still an orthogonal curvilinear one. Meanwhile, the scalar factors of the four coordinates $(r, \phi_1, \phi_2, \phi_3)$ are $(h_1, h_2, h_3, h_4) = (1, r, rs_1, rc_1)$, respectively. The volume element, following (10.1.3), reads

$$dV = r^3 s_1 c_1 dr d\phi_1 d\phi_2 d\phi_3. \quad (10.5.3)$$

Let us determine the ranges of the three angles. To do so we evaluate the volume of a hypersphere with radius R. Since there are one radial and three angular coordinates, the volume of the hypersphere should be the same as that evaluated by (10.4.4). By the volume element in (10.5.3), we get

$$\int_0^R dr \int_0^{\pi/2} r d\phi_1 \int_0^{2\pi} r s_1 d\phi_2 \int_0^{2\pi} r c_1 d\phi_3$$

$$= \frac{R^4}{4} 2\pi \cdot 2\pi \int_0^{\pi/2} s_1 c_1 d\phi_1 = \frac{\pi^2 R^4}{2}. \tag{10.5.4}$$

Because the volume of the sphere has been fixed, the ranges of the angles are properly chosen such that the integral result is correct. The ranges are as follows:

$$\phi_1 : [0, \pi/2]; \quad \phi_2 : [0, 2\pi]; \quad \phi_3 : [0, 2\pi] \tag{10.5.5}$$

It is seen that the range of one angle is doubled and another is narrowed to a half. The author guesses that the total angles are conserved. The conservation of total angles means that the product of all the angles are constant in an N-dimensional space: in an N-dimensional spherical coordinate system, the ranges of the angles are shown by (10.1.11): one angle is π and $N - 2$ angles are π. Their total product is $2\pi^{N-1}$. As $N = 4$, it is $2\pi^3$.

We put down the gradient, divergence and Laplace operators in the pseudo spherical system.

Observing (10.1.22), the gradient operator is

$$\nabla \psi = \frac{\partial \psi}{\partial r} e_r + \frac{1}{r} \frac{\partial \psi}{\partial \phi_1} e_{\phi 1} + \frac{1}{r s_1} \frac{\partial \psi}{\partial \phi_2} e_{\phi 2} + \frac{1}{r c_1} \frac{\partial \psi}{\partial \phi_3} e_{\phi 3}. \tag{10.5.6}$$

Following (10.1.26), the divergence operator is

$$\nabla \cdot W = \frac{1}{r^3} \frac{\partial}{\partial r} \left(r^3 V_1 \right) + \frac{1}{r s_1 c_1} \frac{\partial}{\partial \phi_1} \left(s_1 c_1 V_2 \right)$$

$$+ \frac{1}{r s_1} \frac{\partial V_3}{\partial \phi_2} + \frac{1}{r c_1} \frac{\partial V_4}{\partial \phi_3}. \tag{10.5.7}$$

Following (10.1.28), the Laplace operator is

$$\nabla^2 = \frac{1}{r^3}\frac{\partial}{\partial r}\left(r^3\frac{\partial}{\partial r}\right) + \frac{1}{r^2 s_1 c_1}\frac{\partial}{\partial \phi_1}\left(s_1 c_1 \frac{\partial}{\partial \phi_1}\right)$$

$$+ \frac{1}{r^2 s_1^2}\frac{\partial^2}{\partial \phi_2^2} + \frac{1}{r^2 c_1^2}\frac{\partial^2}{\partial \phi_3^2}. \qquad (10.5.8)$$

10.5.2.   *Solutions of Laplace equation*

Let us solve the Laplace equation in the pseudo spherical system.

$$\left[\frac{1}{r^3}\frac{\partial}{\partial r}\left(r^3\frac{\partial}{\partial r}\right) + \frac{1}{r^2 s_1 c_1}\frac{\partial}{\partial \phi_1}\left(s_1 c_1 \frac{\partial}{\partial \phi_1}\right)\right.$$

$$\left. + \frac{1}{r^2 s_1^2}\frac{\partial^2}{\partial \phi_2^2} + \frac{1}{r^2 c_1^2}\frac{\partial^2}{\partial \phi_3^2}\right] \psi(\boldsymbol{r}) = 0. \qquad (10.5.9a)$$

The separation of variables

$$\psi(\boldsymbol{r}) = R(r)\zeta(\phi) \qquad (10.5.9b)$$

separates (10.5.9a) into two: radial and angular equations. The former is identical to that in the spherical coordinate system. The angular equation is

$$\left[\frac{1}{s_1 c_1}\frac{\partial}{\partial \phi_1}\left(s_1 c_1 \frac{\partial}{\partial \phi_1}\right) + \frac{1}{s_1^2}\frac{\partial^2}{\partial \phi_2^2} + \frac{1}{c_1^2}\frac{\partial^2}{\partial \phi_3^2}\right]\zeta(\phi) = -\lambda_1\zeta(\phi).$$

$$(10.5.10)$$

Further separation of variables $\zeta(\phi) = \zeta_1(\phi_1)\zeta_2(\phi_2)\zeta_3(\phi_3)$ turns it to three equations. Those of ϕ_2 and ϕ_3 are as follows:

$$\zeta_2''(\phi_2) + l_2^2 \zeta_2(\phi_2) = 0 \qquad (10.5.11a)$$

and

$$\zeta_3''(\phi_3) + l_3^2 \zeta_3(\phi_3) = 0. \qquad (10.5.11b)$$

The range of ϕ_2 and ϕ_3 have been given in (10.5.5). Under the periodical boundary conditions, the eigenfunctions of the equations are

$$\zeta_2(\phi_2) = \exp(il_2\phi_2), \quad \zeta_3(\phi_3) = \exp(il_3\phi_3), \qquad (10.5.12a)$$

and eigenvalues are

$$l_2, l_3 = 0, \pm 1, \pm 2, \ldots. \qquad (10.5.12b)$$

The equation of ϕ_1 is

$$\zeta_1''(\phi_1) + \frac{c_1^2 - s_1^2}{s_1 c_1}\zeta_1'(\phi_1) + \left(\lambda_1 - \frac{l_2^2}{c_1^2} - \frac{l_3^2}{s_1^2}\right)\zeta_1(\phi_1) = 0. \quad (10.5.13)$$

In order to find its solutions, we make the following transformation:

$$\zeta_1 = \cos^{l_2}\phi_1 \sin^{l_3}\phi_1 f(\eta), \quad \eta = \sin^2\phi_1. \quad (10.5.14)$$

This transformation leads to the equation that $f(\eta)$ must satisfy:

$$\eta(1 - \eta)f''(\eta) + [l_3 + 1 - (l_2 + l_3 + 2)\eta]\,f'(\eta)$$
$$+ \frac{1}{4}[\lambda_1 - (l_2 + l_3 + 1)^2 + 1]f(\eta) = 0. \quad (10.5.15)$$

Compared to the hypergeometric equation (10A.1) in Appendix 10A,

$$\eta(1 - \eta)F''(\eta) + [c - (a + b + 1)\eta]F'(\eta) - abF(\eta) = 0, \quad (10.5.16)$$

the relations between coefficients are

$$c = l_3 + 1, \quad a + b = l_2 + l_3 + 1,$$
$$ab = -\frac{1}{4}\left[\lambda_1 - (l_2 + l_3 + 1)^2 + 1\right],$$
$$(10.5.17a)$$

from which it is solved that

$$a = \frac{1}{2}\left(l_2 + l_3 + 1 + \sqrt{1 + \lambda_1}\right),$$
$$b = \frac{1}{2}\left(l_2 + l_3 + 1 - \sqrt{1 + \lambda_1}\right). \quad (10.5.17b)$$

The two linearly independent solutions of (10.5.16) are given by (10A.2):

$$f_1(z) = F(a, b, |c|, \eta) \quad (10.5.18a)$$

and

$$f_2(z) = \eta^{-l_3} F\left(a - l_3, b - l_3, |2 - c|, \eta\right), \quad (10.5.18b)$$

where F is the hypergeometric function:

$$F(a, b, |c|z) = 1 + \frac{ab}{c}z + \frac{a(a+1)b(b+1)}{2!c(c+1)}z^2$$

$$+ \cdots + \frac{(a)_n(b)_n}{n!(c)_n}z^n + \cdots . \qquad (10.5.19)$$

The range of the angle ϕ_1 is $[0, \pi/2]$, see (10.5.5). Consequently, the range of the variable η is $[0, 1]$, see (10.5.14).

The necessary condition for the solution to be finite at $\eta = 1$ is that

$$a = -n_r, \qquad (10.5.20)$$

where n_r takes natural numbers, because this condition guarantees that there are finite terms in (10.5.18a). Inspection of the special solution (10.5.18b) shows that it does not meet the necessary condition. This is because, when l_3 is positive, the factor η^{-l_3} makes it divergent at $\eta = 0$; when l_3 is negative, $a - l_3 = |l_3| - n_r$ can be positive (the position of $a - l_3$ in (10.5.18b) is the same as that of a in (10.5.18a)). Thus, (10.5.18b) is abandoned, and only (10.5.18a) is retained in the general solution.

Substitution of (10.5.20) into (10.5.18) gives the value of λ_1:

$$\lambda_1 = (l_2 + l_3 + 2n_r)(l_2 + l_3 + 2n_r + 2). \qquad (10.5.21)$$

Let

$$l_1 = l_2 + l_3 + 2n_r. \qquad (10.5.22)$$

The solution of (10.5.15) is

$$f(\eta) = F(-n_r, b, |c|, \eta)$$

$$= F\left(-\frac{1}{2}(l_1 - l_2 - l_3), \frac{1}{2}(l_1 + l_2 + l_3) + 1, |l_3 + 1|, \eta\right).$$

$$(10.5.23)$$

Observing (10.5.21), (10.5.22) and (10.5.12b), all the eigenvalues are as follows:

$$\lambda_1 = l_1(l_1 + 2), \quad l_1 = 0, 1, 2, \ldots; \qquad (10.5.24a)$$

$$l_2, l_3 = 0, \pm 1, \pm 2, \ldots, \pm l_1. \qquad (10.5.24b)$$

The square of the total angular momentum is the same as that in a spherical system, see the first line of (10.4.19). The total angular momentum can have projections into subspaces of angles ϕ_2 and ϕ_3 simultaneously. The two subspaces have equal weights. As a comparison, in a spherical coordinate system, the total angular momentum can have its projections into subspace (ϕ_2, ϕ_3), and then the projections in this subspace can have projections into the further subspace of ϕ_3. Each time the projections are into the next subspace. Equation (10.5.23) indicates that the magnitudes of l_2 and l_3 are bounded and the difference between l_1 and $l_2 + l_3$ is an even integer.

We count the number of the possible total states when l_1 is fixed. Equation (10.5.24b) tells us that l_2 and l_3 cover the same range of eigenvalues. It seems that both can take $2l_1 + 1$ values. However, the difference between l_1 and $l_2 + l_3$ should be an even integer. This means that l_2 and l_3 can respectively take integers from $-l_1$ to l_1, but with increment being 2, i.e., each takes l_1+1 values. Thus the number of total states is $(l_1 + 1)^2$, exactly the same as (10.4.21).

The total eigenfunction is written by the combination of (10.5.12a) and (10.5.14):

$$\zeta_{l_1 l_2 l_3}(\phi_1, \phi_2, \phi_3)$$

$$= F\left(-\frac{1}{2}(l_1 - l_2 - l_3), \frac{1}{2}(l_1 + l_2 + l_3) + 1, |l_3 + 1| \sin^2 \phi_1\right)$$

$$\times \cos^{l_2} \phi_1 \sin^{l_3} \phi_1 \exp(il_2\phi_2) \exp(il_3\phi_3). \qquad (10.5.25)$$

According to group theory, the angular momentum in four-dimensional space can be regarded as the direct product of those in two three-dimensional spaces. This feature implies that there are two subspaces with equal weight, so that is probably a reflection of the pseudo spherical coordinate system. The feature cannot be explicitly reflected in the usual spherical coordinate system, since there is no equal weight subspaces.

10.5.3. *Five- and six-dimensional spaces*

In the cases of five- and six-dimensional spaces, we merely present the coordinate transformations and the final conclusions. How to establish and to solve Laplace equations are left to the readers.

1. Five-dimensional space

The transformations between Cartesian coordinates $(x_1, x_2, x_3, x_4, x_5)$ and pseudo spherical coordinates $(r, \phi_1, \phi_2, \phi_3, \phi_4)$ are as follows:

$$x_1 = rc_1, \quad x_2 = rs_1c_2c_3, \quad x_3 = rs_1c_2s_3,$$

$$x_4 = rs_1s_2c_4, \quad x_5 = rs_1s_2s_4. \tag{10.5.26}$$

Putting down the square of the line element following (10.1.2), we obtain the coordinate weight:

$$(h_1, h_2, h_3, h_4, h_5) = (1, r, rs_1, rs_1c_2, rs_1s_2). \tag{10.5.27}$$

The ranges of the four angles are as follows:

$$\phi_1 : [0, \pi]; \quad \phi_2 : [0, \pi/2]; \quad \phi_3 : [0, 2\pi]; \quad \phi_4 : [0, 2\pi]. \tag{10.5.28}$$

With these two equations, we are able to write the Jacobi determinant between Cartesian and pseudo spherical coordinates by (10.1.13), and then calculate the volume of a hypersphere with radius R. The expressions of the gradient, divergence and Laplace operators can be written down in terms of formulas in Subsection 10.1.2.

We solve Laplace equation. After separation of radial and angular coordinates, we find that the equations of radial and ϕ_1 coordinates are exactly the same as those of radial and θ_1 coordinates in five-dimensional spherical system. Therefore, the square of the total angular momentum is $l_1 = l_2(l_2 + 1)$. The equation with respect to (ϕ_2, ϕ_3, ϕ_4) is the same as that in four-dimensional pseudo spherical system (10.5.10). Thus, we know that in four-dimensional $(\phi_1, \phi_2, \phi_3, \phi_4)$ angular space, there is a three-dimensional (ϕ_2, ϕ_3, ϕ_4) angular subspace. The total angular momentum has its projection into this angular subspace, following the rule the same as that in five-dimensional spherical system. Then in the (ϕ_2, ϕ_3, ϕ_4) angular subspace, the rules of the angular projections follow those in four-dimensional pseudo spherical system, i.e., (10.5.24).

2. Six-dimensional space

The transformations between Cartesian coordinates $(x_1, x_2, x_3, x_4, x_5, x_6)$ and pseudo spherical coordinates $(r, \phi_1, \phi_2, \phi_3, \phi_4, \phi_5)$ are as follows:

$$x_1 = rc_1, \quad x_2 = rs_1c_2, \quad x_3 = rs_1s_2c_3c_4, \quad x_4 = rs_1s_2c_3s_4,$$

$$x_5 = rs_1s_2s_3c_5, \quad x_6 = rs_1s_2s_3s_5. \qquad (10.5.29)$$

Putting down the square of the line element following (10.1.2), we obtain the coordinate weight:

$$(h_1, h_2, h_3, h_4, h_5, h_6) = (1, r, rs_1, rs_1s_2, rs_1s_2c_3, rs_1s_2s_3) \quad (10.5.30)$$

The ranges of the four angles are as follows:

$$\phi_1 : [0, \pi]; \quad \phi_2 : [0, \pi]; \quad \phi_3 : [0, \pi/2];$$

$$\phi_4 : [0, 2\pi]; \quad \phi_5 : [0, 2\pi] \qquad (10.5.31)$$

With these two equations, we are able to write the Jacobi determinant between Cartesian and pseudo spherical coordinates by (10.1.13), and then calculate the volume of a hypersphere with radius R. The expressions of the gradient, divergence and Laplace operators can be written down in terms of formulas in Subsection 10.1.2.

We solve Laplace equation. After separation of radial and angular coordinates, we find that the equations of radial, ϕ_1 and ϕ_2 coordinates are exactly the same as those in six-dimensional spherical system. Therefore, the square of the total angular momentum is λ_1 in (10.4.34), and the next quantum number is λ_2 in (10.4.34). The equation with respect to (ϕ_3, ϕ_4, ϕ_5) is the same as that in four-dimensional pseudo spherical system (10.5.10). Thus, we know that in five-dimensional $(\phi_1, \phi_2, \phi_3, \phi_4, \phi_5)$ angular space, there is a four-dimensional $(\phi_2, \phi_3, \phi_4, \phi_5)$ angular subspace, which in turn has a three-dimensional (ϕ_3, ϕ_4, ϕ_5) angular subspace.

The square of the total angular momentum takes value λ_1 in (10.4.34). The total angular momentum has its projections firstly into the $(\phi_2, \phi_3, \phi_4, \phi_5)$ angular subspace, following the rule the same as that in six-dimensional spherical system. The square of the projection component takes the value λ_2 in (10.4.34), and it in turn has projections into the (ϕ_2, ϕ_3, ϕ_4) angular subspace, following the rule the same as that in five-dimensional spherical system. Finally in the (ϕ_3, ϕ_4, ϕ_5) angular subspace, the rules of the angular projections follow those in four-dimensional pseudo spherical system, i.e., (10.5.24).

10.6.   Non-Euclidean Space

10.6.1.   *Metric tensor*

Under orthogonal curvilinear systems, the square of the line element was expressed by (10.1.2), in which there was no cross terms between different coordinates. In general, this is not always so, and the square of the line element should be written in the following form:

$$dl^2 = \sum_{i,j=1}^{N} g_{ij} dq_i dq_j. \qquad (10.6.1)$$

The matrix g in this equation is called a **metric tensor**, or in short, a **metric** in the N-dimensional space. In orthogonal curvilinear systems, the metrics are just diagonal:

$$g_{ij} = g_i \delta_{ij}. \qquad (10.6.2)$$

In the following, we discuss the cases of orthogonal curvilinear systems, so that the metrics are diagonal.

The square of the line element expressed by (10.1.2) is a more special case: each diagonal matrix element is just the square of the coordinate weight,

$$g_{ij} = h_i^2 \delta_{ij}. \qquad (10.6.3)$$

Obviously, the metric is of N order in an N-dimensional space.

Here we present some examples of metrics. Since they are all diagonal, only diagonal elements are written.

In two-dimensional space, Cartesian coordinates are (x, y), and the corresponding metric is

$$g_{11} = g_{22} = 1. \qquad (10.6.4)$$

The polar coordinates are (r, θ) and the corresponding metric is as follows:

$$g_{11} = 1, \quad g_{22} = r. \qquad (10.6.5)$$

In three-dimensional space, Cartesian coordinates are (x, y, z), and the corresponding metric is

$$g_{11} = g_{22} = g_{33} = 1. \qquad (10.6.6)$$

The cylindrical coordinates are (r, θ, z) and the corresponding metric is as follows:

$$g_{11} = g_{33} = 1, \quad g_{22} = r. \tag{10.6.7}$$

The spherical coordinates are (r, θ, φ) and the corresponding metric is as follows:

$$g_{11} = 1, \quad g_{22} = r, \quad g_{33} = r \sin \theta. \tag{10.6.8}$$

In four-dimensional space, Cartesian coordinates are (x_1, x_2, x_3, x_4), and the corresponding metric is

$$g_{11} = g_{22} = g_{33} = g_{44} = 1. \tag{10.6.9}$$

The spherical coordinates are $(r, \theta_1, \theta_2, \theta_3)$ and the corresponding metric is as follows:

$$g_{11} = 1, \quad g_{22} = r, \quad g_{33} = r \sin \theta_1, \quad g_{44} = r \sin \theta_1 \sin \theta_2 \tag{10.6.10}$$

The pseudo spherical coordinates are $(r, \phi_1, \phi_2, \phi_3)$ and the corresponding metric is as follows:

$$g_{11} = 1, \quad g_{22} = r, \quad g_{33} = r \sin \theta_1, \quad g_{44} = r \sin \theta_1. \tag{10.6.11}$$

In special relativity, space and time cannot be completely separated, and they compose a four-dimensional vector. This four-dimensional vector reflects the movement of a particle in different uniform reference systems related by Lorentz transformation. The coordinates in this four-dimensional spacetime are $(x, y, z, \mathrm{i}ct)$, where time is multiplied by light speed to become the fourth coordinate in this space. In this form, the fourth coordinate is explicitly different from the other three. However, we have known that the coordinates are converted through Lorentz transformation. The transformation makes the mutual conversion among the four coordinates, so that one is unable to tell which coordinate is a purely time or space one. Because of this reason, we prefer to put down the vector in the form of (x_1, x_2, x_3, x_4), stressing that they are equivalent. The square of the line element is selected as:

$$\mathrm{d}l^2 = \mathrm{d}x_1^2 + \mathrm{d}x_2^2 + \mathrm{d}x_3^2 - \mathrm{d}x_4^2. \tag{10.6.12}$$

The minus sign in the last term comes from the square of the imaginary number i multiplied to ct. Thus we obtain the metric in this four-dimensional space:

$$g_{11} = g_{22} = g_{33} = 1, \quad g_{44} = -1. \tag{10.6.13}$$

According to special relativity, the square of the line element is invariant under the transformation of coordinates.

In fact, in all spaces, the square of the line element dl^2 is an invariant. That is why the metric is defined through dl^2. An invariant in a space is a very important quantity, which may reflect some basic characteristics in this space.

The examples above show that in one space, different coordinates can be established and the metric of the space relies on the coordinates, e.g., (10.6.6)–(10.6.8) in three–dimensional space and (10.6.9)–(10.6.11) in four-dimensional space.

In general, the elements of a metric depends on coordinates. Therefore, the metric is also called a **metric tensor field**, or in short, a **metric field**.

If the metric of a space can be transformed to be a unit one by a certain coordinate transformation, and the ranges of all the transformed coordinates are $[-\infty, \infty]$, then this space is called **Euclid space**. In above examples, the metrics from (10.6.4) to (10.6.11) reflect Euclid spaces. The four-dimensional special relativistic space is not a Euclid one. In discussing (10.4.41), we have mentioned the hypersphere surface with fixed radius, which is called angular space and is neither a non-Euclid one.

If all the elements of a diagonal metric are positive, the metric can always be transformed to be a unit one. A feature of a Euclid space is that all the element of its diagonal metric are positive.

If all the elements in a metric change their sign, the properties of the space are unaltered. Therefore, the two cases of a metric before and after it is multiplied by -1 is are thought the same.

If all the elements of a diagonal metric are positive, it is called a **Riemann metric**. If among the elements, one is positive and others are positive, the metric is called a **Lorentz metric**. The four-dimensional space of special relativity has a Lorentz metric. This metric is also called a **Minkowski metric**. A space described by Minkowski metric is called a **Minkowski space**.

If not all the elements in a metric are of the same sign, the space is non-Euclid. For instance, the four-dimensional space of special relativity is a Minkowski space.

In (10.1.2), the square of the line element is defined. This definition is valid for Euclid space. In the case of non-Euclid space, elements of a diagonal metric may have different signs. Therefore, the definition (10.1.2) ought to be extended to the following form:

$$dl^2 = \sum_{i=1}^{N} \eta_i h_i^2 dq_i^2, \tag{10.6.14}$$

where η_i is either $+1$ or -1. For examples, in (10.6.4)–(10.6.11), $\eta_i = 1$. In (10.6.13), $\eta_1 = \eta_2 = \eta_3 = 1, \eta_4 = -1$.

The coefficients $\{\eta_i\}$ in (10.6.14) are called **metric weights**. Metric weight can be either $+1$ or -1. In all of Euclid spaces, metric weight are all $+1$, which is more clearly seen in Cartesian coordinates. Therefore, the metric weight are also the weight of the dot product of two vectors. That is to say, if there are two vectors x and y, in Cartesian coordinates they are expressed by $(x_1, x_2, \ldots, x_N)$ and $(y_1, y_2, \ldots, y_N)$, then the dot product of them is defined by

$$\boldsymbol{x} \cdot \boldsymbol{y} = \sum_{i=1}^{N} \eta_i x_i y_i, \tag{10.6.15}$$

which is the same as (2.1.5). In this sense, (10.6.14) actually means the dot product of a vector with components $h_i dq_i$ with itself.

Let us recall the definitions of the gradient, divergence and Laplace operators. The definition of the gradient operator (10.1.22) remains unchanged. However, the divergence defined by (10.1.25) appears as the dot product of two vectors, so that the weights $\{\eta_i\}$ should be included. Thus the divergence is reformed to be

$$\nabla \cdot \boldsymbol{W} = \frac{1}{\prod_{i=1}^{N} h_i} \sum_{j=1}^{N} \eta_j \frac{\partial}{\partial q_j} \left(\frac{W_j}{h_j} \prod_{l=1}^{N} h_l \right). \tag{10.6.16}$$

Laplace operator $\Delta = \nabla^2 = \nabla \cdot \nabla$ also appears as the dot product of two vectors, and the weight $\{\eta_i\}$ are also required. So, (10.1.27) is reformed to be

$$\nabla^2 = \frac{1}{\prod_{i=1}^{N} h_i} \sum_{j=1}^{N} \eta_j \frac{\partial}{\partial q_j} \left[\left(\frac{1}{h_j^2} \prod_{l=1}^{N} h_l \right) \frac{\partial}{\partial q_j} \right]. \tag{10.6.17}$$

Mathematics for Physicists

The Laplace equation is

$$\nabla^2 \psi = 0. \tag{10.6.18}$$

The Helmholtz equation is

$$\nabla^2 \psi + k^2 \psi = 0. \tag{10.6.19}$$

In a Euclid space, the square of the line element defined by (10.1.2) is not negative. If the square of the line element is zero, then the variation of every coordinate is zero, i.e., the coordinate are not changed. By contrast, in a non-Euclid space, the elements of the diagonal metric can be negative. Consequently, the square of the line element defined by (10.6.14) can be negative or zero. In the case of

$$dl^2 = \sum_{i=1}^{N} \eta_i h_i^2 dq_i^2 = 0, \tag{10.6.20}$$

Every coordinate in (10.6.14) still can vary but under the constraint $dl^2 = 0$. This is somewhat similar to the case encountered when discussing (10.4.40). Equation (10.6.20) actually defines an $N-1$-dimensional subspace.

In the following, we take a five-dimensional Minkowski space as an example to show how to solve a Laplace equation in a non-Euclid space.

10.6.2. Five-dimensional Minkowski space and four-dimensional de Sitter space

1. Five-dimensional Minkowski space

The coordinates are $(x_0, x_1, x_2, x_3, x_4)$, and the square of the line element is

$$ds^2 = dx_0^2 - dx_1^2 - dx_2^2 - dx_3^2 - dx_4^2. \tag{10.6.21}$$

Following the convention in literature, the coordinate indices are labeled by 0 to 4 instead of the usual 1 to 5. It is seen from (10.6.21) that the metric in this space is

$$g = \text{diag}(1, -1, -1, -1, -1). \tag{10.6.22}$$

The coordinates $(x_0, x_1, x_2, x_3, x_4)$ are still called Cartesian coordinate system. This is because all the coordinate weight are independent of coordinates, and their magnitudes are 1 each with a sign factor. Of course, they are essentially different from Cartesian coordinates in a Euclid space where all the metric weights are $+1$.

Now we define a "pseudo spherical coordinate system" in this space, denoted by $(r, \phi_0, \phi_1, \phi_2, \phi_3)$. The relations between them and those of Cartesian coordinates are as follows:

$$x_0 = r \sinh \phi_0,$$

$$x_1 = r \cosh \phi_0 \cos \phi_1 \cos \phi_2,$$

$$x_2 = r \cosh \phi_0 \cos \phi_1 \sin \phi_2$$

$$x_3 = r \cosh \phi_0 \sin \theta_1 \cos \phi_3,$$

$$x_4 = r \cosh \phi_0 \sin \phi_1 \sin \phi_3. \tag{10.6.23}$$

Substitution of (10.6.23) into (10.6.21) gives the expression of $\mathrm{d}s^2$ by the pseudo spherical coordinates:

$$\mathrm{d}s^2 = -\mathrm{d}r^2 + r^2 \mathrm{d}\phi_0^2 - r^2 \cosh^2 \phi_0 \mathrm{d}\phi_1^2$$
$$- r^2 \cosh^2 \phi_0 \cos^2 \phi_1 \mathrm{d}\phi_2^2 - r^2 \cosh^2 \phi_0 \sin^2 \phi_1 \mathrm{d}\phi_3^2. \tag{10.6.24}$$

So, in the pseudo spherical coordinate system the metric is

$$g = \mathrm{diag}\left(1, -r^2, -r^2 \cosh^2 \phi_0, -r^2 \cosh^2 \phi_0 \cos^2 \phi_1, \right.$$
$$\left. -r^2 \cosh^2 \phi_0 \sin^2 \phi_1\right) \tag{10.6.25}$$

and the coordinate weight are

$$(h_0, h_1, h_2, h_3, h_4) = (1, r, r \cosh \phi_0 \cos \phi_1, r \cosh \phi_0 \sin \phi_1). \tag{10.6.26}$$

It is interesting to compare (10.6.24) and (10.6.26) with (10.5.26) and (10.5.27), the latter being the transformations between Cartesian and pseudo spherical coordinates in a five-dimensional Euclid space. There is discrepancy in angles. The ranges of the angles (ϕ_1, ϕ_2, ϕ_3) in (10.6.23) are as follows:

$$\phi_1 : [0, \pi/2]; \quad \phi_2 : [0, 2\pi]; \quad \phi_3 : [0, 2\pi]. \tag{10.6.27}$$

They are the same as those of pseudo spherical coordinates in a Euclid space, see (10.5.5) or (ϕ_2, ϕ_3, ϕ_4) in (10.5.28). The "angle" ϕ_0

in (10.6.23) takes its value in $[-\infty, \infty]$. So, it is not a true angle, and can be called a "pseudo angle".

2. Laplace equation

Observing (10.6.17), we put down the Laplace equation in the pseudo spherical coordinates:

$$\nabla^2 \psi = -\frac{1}{r^4}\frac{\partial}{\partial r}\left(r^4\frac{\partial}{\partial r}\right)\psi + \frac{1}{r^2\cosh^3\phi_0}\frac{\partial}{\partial\phi_0}\left(\cosh^3\phi_0\frac{\partial}{\partial\phi_0}\right)\psi$$

$$-\frac{1}{r^2\cosh^2\phi_0\sin\phi_1\cos\phi_1}\frac{\partial}{\partial\phi_1}\left(\sin\phi_1\cos\phi_1\frac{\partial}{\partial\phi_1}\right)\psi$$

$$-\frac{1}{r^2\cosh^2\phi_0\cos^2\phi_1}\frac{\partial^2\psi}{\partial\phi_2^2} - \frac{1}{r^2\cosh^2\phi_0\sin^2\phi_1}\frac{\partial^2\psi}{\partial\phi_3^2} = 0.$$

$$(10.6.28)$$

Compared to (10.1.29), the coordinate r retains the implication of radial one. By separation of variables,

$$\psi(\boldsymbol{r}) = R(r)\zeta(\phi_0,\phi_1,\phi_2,\phi_3), \qquad (10.6.29)$$

the radial and angular equations are obtained as follows:

$$\frac{1}{r^2}\frac{\partial}{\partial r}\left(r^4\frac{\partial}{\partial r}\right)R(r) - \lambda R(r) = 0 \qquad (10.6.30)$$

and

$$\left[\frac{1}{\cosh^3\phi_0}\frac{\partial}{\partial\phi_0}\left(\cosh^3\phi_0\frac{\partial}{\partial\phi_0}\right)\right.$$

$$-\frac{1}{\cosh^2\phi_0\sin\phi_1\cos\phi_1}\frac{\partial}{\partial\phi_1}\left(\sin\phi_1\cos\phi_1\frac{\partial}{\partial\phi_1}\right)$$

$$\left.-\frac{1}{\cosh^2\phi_0\cos^2\phi_1}\frac{\partial^2}{\partial\phi_2^2} - \frac{1}{\cosh^2\phi_0\sin^2\phi_1}\frac{\partial^2}{\partial\phi_3^2}\right]\zeta + \lambda\zeta = 0.$$

$$(10.6.31)$$

The angular equation is of specific physical meaning to be discussed below.

3. Four-dimensional de Sitter space

The angular equation (10.6.31) in five-dimensional Minkowski space is just the Helmholtz equation in its four-dimensional angular subspace after the radial coordinate r is separated. A similar discussion was encountered when (10.4.41) was concerned. In fact, if a hypersurface with fixed "radius" ρ

$$-x_0^2 + x_1^2 + x_2^2 + x_3^2 + x_4^2 = \rho^2 \tag{10.6.32}$$

is chosen, then the square of the line element in this subspace is the result of letting $\mathrm{d}r^2 = 0$ in (10.6.24) (since $r = \rho$ is invariant):

$$\mathrm{d}s^2 = \left(-\mathrm{d}\phi_0^2 + \cosh^2\phi_0\mathrm{d}\phi_1^2 + \cosh^2\phi_0\cos^2\phi_1\mathrm{d}\phi_2^2\right.$$
$$\left. + \cosh^2\phi_0\sin^2\phi_1\mathrm{d}\phi_3^2\right)\rho^2. \tag{10.6.33}$$

In the five-dimensional Minkowski space, the four-dimensional subspace defined by (10.6.32) is called a **de Sitter space** or the **de Sitter spacetime**. Because the solution with the highest symmetry of Einstein cosmological equation is closely related to de Sitter spacetime, this space is of important significance for modern cosmology.

In the de Sitter spacetime, the four coordinates $(\phi_0, \phi_1, \phi_2, \phi_3)$ are "angular ones" in Minkowski space. The square of the line element is (10.6.33). Referring to (10.6.14), one immediately knows that the coordinates are

$$(h_0, h_1, h_2, h_3)$$
$$= (\rho, \rho\cosh\phi_0, \rho\cosh\phi_0\cos\phi_1, \rho\cosh\phi_0\sin\phi_1) \tag{10.6.34}$$

and the metric is

$$g = \mathrm{diag}\left(\rho^2, -\rho^2\cosh^2\phi_0, -\rho^2\cosh^2\phi_0\cos^2\phi_1,\right.$$
$$\left. -\rho^2\cosh^2\phi_0\sin^2\phi_1\right). \tag{10.6.35}$$

Following the definition (10.6.17), we put down The Helmholtz equation in this space

$$\nabla^2\zeta + \gamma\zeta = 0, \tag{10.6.36}$$

where the Laplace operator is

$$\nabla^2 = \frac{1}{\rho^2\cosh^3\phi_0}\frac{\partial}{\partial\phi_0}\left(\cosh^3\phi_0\frac{\partial}{\partial\phi_0}\right)$$
$$-\frac{1}{\rho^2\cosh^2\phi_0\sin\phi_1\cos\phi_1}\frac{\partial}{\partial\phi_1}\left(\sin\phi_1\cos\phi_1\frac{\partial}{\partial\phi_1}\right)$$

$$-\frac{1}{\rho^2 \cosh^2 \phi_0 \cos^2 \phi_1} \frac{\partial^2}{\partial \phi_2^2} - \frac{1}{\rho^2 \cosh^2 \phi_0 \sin^2 \phi_1} \frac{\partial^2}{\partial \phi_3^2}.$$

$$(10.6.37)$$

The Helmholtz equation in this space is in fact (10.6.31). The constant λ, apart from a constant ρ^2, in these two equations is of the meaning of the "square of the total angular momentum" in the five-dimensional Minkowski space. This is similar to the discussion below (10.4.41).

In order to solve (10.6.37), we use separation of variables:

$$\zeta(\phi_0, \phi_1, \phi_2, \phi_3) = \zeta_0(\phi_0)\zeta_1(\phi_1)\zeta_2(\phi_2)\zeta_3(\phi_3). \qquad (10.6.38)$$

The equation of function $\zeta_1(\phi_1)\zeta_2(\phi_2)\zeta_3(\phi_3)$ is

$$\left[\frac{1}{\sin\phi_1 \cos\phi_1} \frac{\partial}{\partial\phi_1} \left(\sin\phi_1 \cos\phi_1 \frac{\partial}{\partial\phi_1} \right) \right.$$

$$\left. + \frac{1}{\cos^2\phi_1} \frac{\partial^2}{\partial\phi_2^2} + \frac{1}{\sin^2\phi_1} \frac{\partial^2}{\partial\phi_3^2} \right] \zeta_1\zeta_2\zeta_3 + \lambda_1\zeta_1\zeta_2\zeta_3 = 0.$$

$$(10.6.39)$$

This equation is exactly the same as (10.5.10), the latter being the angular equation under pseudo coordinate system in four-dimensional Euclid space. Therefore, the eigenvalues can be written following (10.5.22) and (10.5.24):

$$\lambda_1 = l_1(l_1 + 2); \qquad (10.6.40a)$$

$$l_2, l_3 = 0, \pm 1, \pm 2, \ldots, \pm l_1; \qquad (10.6.40b)$$

$$l_1 = l_2 + l_3 + 2n_r. \qquad (10.6.40c)$$

In this three-dimensional angular subspace, the projections of angular momentum observe the discussion below (10.5.24). The eigenfunction follows (10.5.25) to be

$$\zeta_{l_1 l_2 l_3}(\phi_1, \phi_2, \phi_3)$$

$$= F\left(-\frac{1}{2}(l_1 - l_2 - l_3), \frac{1}{2}(l_1 + l_2 + l_3) + 1, |l_3 + 1| \sin^2\phi_1 \right)$$

$$\times \cos^{l_2}\phi_1 \sin^{l_3}\phi_1 \exp(i l_2 \phi_2) \exp(i l_3 \phi_3). \qquad (10.6.41)$$

By separation of variables (10.6.38), we obtain the equation with respect to ϕ_0 as follows:

$$\frac{1}{\cosh^3 \phi_0} \frac{\partial}{\partial \phi_0} \left(\cosh^3 \phi_0 \frac{\partial}{\partial \phi_0} \right) \zeta_0(\phi_0)$$

$$+ \left[\frac{l_1(l_1 + 2)}{\cosh^2 \phi_0} - \gamma \rho^2 \right] \zeta_0(\phi_0) = 0. \qquad (10.6.42)$$

Let us recall equation (10.5.9a). When it is treated by separation of variables, the obtained angular equation was (10.5.10), but the radial equation was not put down. In fact, that radial equation is quite similar to (10.6.42), and it was indeed easily converted to (10.6.42) as long as the argument r there is replaced by $\cosh \phi_0$ and the derivative with respect to r is replaced by ϕ_0. These minor discrepancy arises from the different coordinate weight of the two systems. Equation (10.6.42) is still called a "radial equation". Its resolution is as follows. Let

$$\zeta_0(\phi_0) = \cosh^{l_1} \phi_0 \xi(x) \qquad (10.6.43a)$$

and

$$x = -\sinh^2 \phi_0. \qquad (10.6.43b)$$

Then, $\xi(x)$ satisfies the equation

$$x(1 - x)\xi'' + \left[\frac{1}{2} - \left(l_1 + \frac{5}{2} \right) x \right] \xi'$$

$$- \frac{1}{4} \left[l_1(l_1 + 3) - \gamma \rho^2 \right] \xi = 0. \qquad (10.6.44)$$

Compared to (10A.1) in Appendix 10A, it shows that

$$c = \frac{1}{2}, \quad a + b = l_1 + \frac{3}{2}, \quad ab = \frac{1}{4} \left[l_1(l_1 + 3) - \gamma \rho^2 \right], \qquad (10.6.45)$$

from which two parameters a and b are solved:

$$a = \frac{1}{2} \left[l_1 + \frac{3}{2} \pm \left(\frac{9}{4} + \lambda \rho^2 \right)^{1/2} \right],$$

$$b = \frac{1}{2} \left[l_1 + \frac{3}{2} \mp \left(\frac{9}{4} + \lambda \rho^2 \right)^{1/2} \right]. \qquad (10.6.46)$$

The two linearly independent solutions of (10.6.45) are

$$\xi_1(x) = F\left(\frac{1}{2}\left[l_1 + \frac{3}{2} \pm \left(\frac{9}{4} + \lambda\rho^2\right)^{1/2}\right],\right.$$

$$\left.\frac{1}{2}\left[l_1 + \frac{3}{2} \mp \left(\frac{9}{4} + \gamma\rho^2\right)^{1/2}\right], \frac{1}{2}, x\right) \qquad (10.6.47a)$$

and

$$\xi_2(x) = \sqrt{x}F\left(\frac{1}{2}\left[l_1 + \frac{5}{2} \pm \left(\frac{9}{4} + \gamma\rho^2\right)^{1/2}\right],\right.$$

$$\left.\frac{1}{2}\left[l_1 + \frac{5}{2} \mp \left(\frac{9}{4} + \gamma\rho^2\right)^{1/2}\right], \frac{3}{2}, x\right). \qquad (10.6.47b)$$

According to the discussion in Appendix 10A, in order for the solution to be finite, it must be that

$$a = -n_s, \qquad (10.6.48)$$

where n_s takes natural numbers. Combined with (10.6.46), the constant $\gamma\rho^2$ in (10.6.42) should take the following values:

$$\gamma\rho^2 = (2n_s - l_1)(2n_s - l_1 - 3). \qquad (10.6.49)$$

Let

$$l_0 = 2n_s - l_1 - 3. \qquad (10.6.50)$$

Equation (10.6.49) is recast to be

$$\gamma\rho^2 = l_0(l_0 + 3). \qquad (10.6.51)$$

The general solution of (10.6.42) is the linear combination of (10.6.47a) and (10.6.47b). Note that the range of x, by (10.6.43), is $[-\infty, 0]$. Equation (10.6.43) has been continuated from $[0,1]$ to the whole complex plane except positive half real axis.

The eigenfunction $\zeta(-\sin^2\phi_0, \lambda)$ of (10.6.37) should be the combination of (10.6.41) and (10.6.43):

$$\zeta(\phi_0, \phi_1, \phi_2, \phi_3)$$

$$= \cosh^{l_1}\phi_0 \xi(-\sin^2\phi_0, \gamma)$$

$$\times F\left(-\frac{1}{2}(l_1 - l_2 - l_3), \frac{1}{2}(l_1 + l_2 + l_3) + 1, |l_3 + 1|\sin^2\phi_1\right)$$

$$\times \cos^{l_2}\phi_1 \sin^{l_3}\phi_1 \exp(il_2\phi_2)\exp(il_3\phi_3). \qquad (10.6.52)$$

Before further discussion, let us recall the concept of angular momentum discussed previously. In the expression of the Laplace operator (10.1.30), if the radial coordinate r is fixed, the remaining part is the square of the total angular momentum operator. Or, as has been pointed out in Section 10.4, the Laplace equation in N-dimensional Euclid space can be separated to be radial and angular equations. The angular one is the eigenvalue equation of the square of the total angular momentum, (10.4.1). Its eigenvalue has definite physical meaning: the square of the total angular momentum.

Now the Laplace equation in five-dimensional Minkowski space (10.6.28) can also be separated by (10.6.29) to radial and "angular" equations. The angular one is (10.6.31). As an analogy, it can be regarded as the eigenequation of the square of the "total angular momentum" of the "angular" subspace, and accordingly, its eigenvalue is regarded as the square of the "total angular momentum". For explicitly, the "angular" here is called "pseudo angular" such as pseudo angular momentum and so on.

In four-dimensional Euclid space, the three angles can be either spherical or pseudo spherical coordinates, see Subsections 10.4.1 and 10.5.1, respectively. In de Sitter space, apart from ϕ_0, the remaining three angles are just pseudo angular coordinates, see (10.6.27). This shows us that it is possible that these three angles can be described by usual spherical ones. That is to say, it is possible to replace the transformations (10.6.23) by the following ones.

$$x_0 = r \sinh \theta_0, \quad x_1 = r \cosh \theta_0 \cos \theta_1,$$

$$x_2 = r \cosh \theta_0 \sin \theta_1 \cos \theta_2,$$

$$x_3 = r \cosh \theta_0 \sin \theta_1 \sin \theta_2 \cos \theta_3,$$

$$x_4 = r \cosh \theta_0 \sin \theta_1 \sin \theta_2 \sin \theta_3, \tag{10.6.53}$$

where the ranges of the three angles are as follows:

$$\theta_1 : [0, \pi]; \quad \theta_2 : [0, \pi]; \quad \theta_3 : [0, 2\pi]. \tag{10.6.54}$$

The verification is left to the readers.

4. The geometric meaning of the mass in de Sitter space

In quantum mechanics, the equation of motion of a free particle is a Schrödinger equation. In relativistic quantum mechanics, a spinless free particle follows Klein-Gordon equation:

$$\left(-\frac{1}{c^2}\frac{\partial^2}{\partial t^2} + \nabla^2\right)\psi = m^2c^2\psi. \qquad (10.6.55)$$

Here we have let Planck constant $\hbar = 1$. The parameter m in this equation is of explicit physical meaning: the mass square of the particle. The operator ∇^2 is the Laplace operator in three-dimensional space. The parameter c is light speed in vacuum, which is also a universal constant factor to remedy dimensions. It has been mentioned that time and space should be regarded as a whole, and the four-dimensional coordinates (x, y, z, ct) should be written in a general form (x_1, x_2, x_3, x_4). The square of the line element is (10.6.12). The metric weights of this four-dimensional spacetime are three negatives and one positive as follows:

$$\eta_1 = \eta_2 = \eta_3 = -1, \quad \eta_4 = 1. \qquad (10.6.56)$$

The left hand side of (10.6.55) is actually the Laplace operator in four-dimensional spacetime with metric (10.6.56). Because in (10.6.55), the derivative with respect to time has a metric weight 1 and others are -1, the metric weight being 1 can be called the coordinate being called "time-like" and those being -1 called "space-like". These names arise from the three-dimensional space and one-dimensional time of the world we are living in.

Now we turn to inspect de Sitter space, the metric weights of which are also three negatives and one positive. Therefore, in this four-dimensional spacetime, there are three space-like and one time-like coordinates. We mimic (10.6.55) to write down the **Klein-Gordon equation** in de Sitter space:

$$\nabla^2\psi + m^2\psi = 0, \qquad (10.6.57)$$

where the Laplace operator is (10.6.36). In comparison with (10.6.55), the second term on the left hand side is of the physical meaning the mass square of a free particle in de Sitter four-dimensional spacetime. Here a possible universal constant factor is neglected, which is not necessarily light speed, and should be determined by dimensions. Then one will immediately see that (10.6.57) is just (10.6.37). That is to say, the square of the pseudo angular momentum in de Sitter spacetime is proportional to the minus of the mass square of a free particle in this space:

$$\lambda \propto -m^2. \tag{10.6.58}$$

The eigenvalue solved from (10.6.44) is shown by (10.6.51). We achieve

$$m \propto i\sqrt{l_0(l_0 + 3)}. \tag{10.6.59}$$

One sees two features of the mass of a free particle in de Sitter space. One is that the mass is a purely imaginary number, which arises from the property of de Sitter spacetime. The other is that it takes discrete values. This is because the Klein-Gordon equation in de Sitter space is the "angular part" of the Laplace equation in Minkowski space with one dimension higher. By the experiences, when solving the angular equation, the eigenvalues are always discrete.

The result (10.6.58) has been obtained by means of group theory. In this book we retrieve the result by analytically solving the equation. This way helps us to understand the physical reason of this result clearly. We can say that the square of the mass in de Sitter space is just the minus of the square of the pseudo angular momentum in five-dimensional Minkowski space. In summary, in such a space, mass is originated from pseudo angular momentum. This causes the mass to take discrete numbers, but not continuous. We emphasize that here we are discussing the origin of mass in the aspect of geometry, but not the true physical origin. The real origin of mass must come from the elementary particle theory. Here we mainly provide a cognition in viewpoint of the property of space.

Since quantum numbers are contained in the expression of the mass, a question is raised that how to count the degeneracy. Let us do this work. Note that (10.6.59) is a simplified form. The explicit form is obtained when (10.6.50) is substituted into it:

$$m \propto i\sqrt{(2n_s - l_1)(2n_s - l_1 - 3)}. \tag{10.6.60}$$

In the expression, n_s is a basic quantum number, because it determines the physically significant solution of (10.6.44). Because $l_1 \geq 0$, it is required that $2n_s \geq 3$. Thus n_s should take integers greater than 1: $n_s = 2, 3, 4, \ldots$.

Because of (10.6.40b), for a given l_1, the degeneracy is $(l_1 + 1)^2$, see discussion above (10.5.25). Therefore, the total degeneracy when n_s is fixed is as follows:

$$\sum_{l_1=0}^{2n_s-3} (l_1 + 1)^2 = \frac{1}{3}(2n_s - 3)(n_s - 1)(4n_s - 5). \qquad (10.6.61)$$

The degeneracy we are discussing can be viewed in two aspects. On one hand, we fix the values of n_s and l_1. So, the mass is fixed according to (10.6.60), but from (10.6.40) it is seen that l_2 and l_3 can still take different values, and the total degeneracy is $(l_1 + 1)^2$. When mass is fixed, a particle can be in different states. These states arise from the eigenvalue equation of angular subspace, i.e., the equation of the square of total angular momentum, (10.6.39). This is similar to the case in a Euclid space where an angular equation can be separated from a Helmholtz equation. The angular momentum can take discrete values and can be along different directions in the space, while mass remain unchanged. On the other hand, when the quantum number n_s is fixed, l_1 can take $2n_s - 2$ numbers: $l_1 = 0, 1, 2, \ldots, 2n_s - 3$. Accordingly, the mass can take $2n_s - 2$ values. This can be understood as follows. By separation of variables (10.6.38), (10.6.42) is obtained which plays a role of radial equation, and its solutions are radial wavefunctions each belonging to a certain quantum number n_s. Under one radial wavefunction, a particle can have different mass. In another word, particles with different masses can have the same wavefunction. As n_s is fixed, different masses are due to different l_1 values which label the pseudo angular momentum. It is seen from (10.6.60) that the greater the mass, the smaller the magnitude of the pseudo angular momentum and subsequently the smaller the angular degeneracy. When the radial wavefunction is fixed, the mass of a particle is still closely related to its pseudo angular momentum. This case has not been encountered yet in Euclid spaces.

The present case shows that to understand a physical quantity in a space, we start from the space with one dimension higher. This viewpoint may be very helpful. At last, we mention that Dirac first studied the fundamental equations of an electron in de Sitter space and happened to acquire (10.6.59) without any discussion.

In this chapter, we inspect the basic equations in higher-dimensional spaces, mainly Laplace and Helmholtz equations.

10.6.3. *Maxwell equations in de Sitter spacetime*

At the end of this section, we would like to discuss possible Maxwell equations in de Sitter spacetime. In our spacetime, there can be electromagnetic fields. They follow Maxwell equations, which were expressed in Subsection 1.7.1. The de Sitter spacetime is also a four-dimensional one. As an analogy, we guess that there can also be "electromagnetic fields" in that spacetime and that they follow the corresponding "Maxwell equations". Then what form should the equations be? Now we try to derive them. To do so, we make use of the variational principle introduced in Chapter 1, and follow the procedure in Subsection 1.7.1. A Lagrangian is suggested and then Euler-Lagrange equations (1.7.3) are applied to derive the required equations.

It is suggested the Lagrangian is of the form of (1.7.5), the same as that in our spacetime:

$$L = -\frac{1}{4\mu}F_{\mu\nu}F^{\mu\nu} + J^{\mu}A_{\mu}. \tag{10.6.62}$$

Here, μ in the denominator is regarded as "permeability" in de Sitter spacetime. Please note that it is not to be confused with that in the subscripts and superscripts. J and A are four-dimensional electric current density vector and potential vector, respectively. The function F is the electromagnetic tensor, the definition of which follows (1.7.6):

$$F_{\mu\nu} = \frac{\partial A_{\nu}}{\partial x^{\mu}} - \frac{\partial A_{\mu}}{\partial x^{\nu}} = \partial_{\mu}A_{\nu} - \partial_{\nu}A_{\mu}. \tag{10.6.63}$$

We have used the shorthand $\frac{\partial}{\partial x^{\mu}} = \partial_{\mu}$.

In our spacetime, four-dimensional vectors $\boldsymbol{J}$, $\boldsymbol{A}$ and coordinates $\boldsymbol{x}$ were written in (1.7.7), where they were all of the form of three-dimensional space plus one-dimensional time. That means that space and time can be separated. In de Sitter spacetime, however, it is not so. Consequently, the combination of (10.6.62) and (10.6.33) will not lead to the form of (1.7.8). In fact, in our spacetime, electric field and magnetic field can be distinguished easily, while in de Sitter spacetime, how to distinguish them is to be studied.

In this subsection, Einstein sum rule is adopted: that in one term, two letters in indices are the same means that summation should be taken with respect to this index. In de Sitter spacetime, we have ordered the four indexes as 0, 1, 2, 3. Therefore, the summation is from 0 to 3. Nevertheless, if three letters are the same, then there is no need of summation. This case means that the letters have taken a definite value.

In (10.6.62) and (10.6.33), some indices appear as superscripts, which concerns the conventions in tensor theory. We do not introduce the conventions, but merely give the quantities needed here. The metric of de Sitter spacetime was given by (10.6.35):

$$(g_{\mu\nu}) = \mathrm{diag}\left(\rho^2, -\rho^2 \cosh^2 \phi_0, -\rho^2 \cosh^2 \phi_0 \cos^2 \phi_1, \right.$$
$$\left. -\rho^2 \cosh^2 \phi_0 \sin^2 \phi_1\right). \tag{10.6.64}$$

Its inverse is written as $(g^{\mu\nu})$. Therefore,

$$(g^{\mu\nu}) = \mathrm{diag}\left(\frac{\cosh^2 \phi_0}{\rho^2 \cosh^2 \phi_0}, -\frac{1}{\rho^2 \cosh^2 \phi_0}, \right.$$
$$\left. -\frac{\cos^{-2} \phi_1}{\rho^2 \cosh^2 \phi_0}, -\frac{\sin^{-2} \phi_1}{\rho^2 \cosh^2 \phi_0}\right). \tag{10.6.65}$$

By the way, the determinant of the metric is

$$g = \det(g_{\mu\nu}) = -\rho^8 \cosh^6 \phi_0 \sin^2 \phi_1 \cos^2 \phi_1. \tag{10.6.66a}$$

The following quantity is often used:

$$\sqrt{-g} = \rho^4 \cosh^3 \phi_0 \sin \phi_1 \cos \phi_1. \tag{10.6.66b}$$

The subscripts in $F_{\lambda\tau}$ can be raised to be superscripts by multiplying the metric:

$$F^{\mu\nu} = g^{\mu\lambda} g^{\nu\tau} F_{\lambda\tau}. \tag{10.6.67}$$

After applying this equation, the Lagrangian becomes

$$L = -\frac{1}{4\mu} g^{\mu\lambda} g^{\nu\tau} F_{\lambda\tau} F_{\mu\nu} + J^\mu A_\mu.$$ (10.6.68)

In some cases, multiplying a factor $\sqrt{-g}$ will greatly reduce the derivation, but does not reduce understanding of the physical meaning. So, we define a new Lagrangian as follows:

$$L' = \sqrt{-g} L.$$ (10.6.69)

In the following, we drop the prime of the new Lagrangian. Thus, it is expressed by

$$L = -\frac{1}{4\mu} \sqrt{-g} g^{\mu\lambda} g^{\nu\tau} F_{\lambda\tau} F_{\mu\nu} + \sqrt{-g} J^\mu A_\mu.$$ (10.6.70)

When (10.6.63) is substituted in, it becomes

$$L = -\frac{1}{2\mu} \sqrt{-g} g^{\mu\lambda} g^{\nu\tau} \partial_\mu A_\nu \left(\partial_\lambda A_\tau - \partial_\tau A_\lambda \right) + \sqrt{-g} J^\mu A_\mu.$$ (10.6.71)

Please note that by exchanging indices, we have merged $\partial_\mu A_\nu - \partial_\nu A_\mu$ into one term. Nevertheless, another factor $\partial_\lambda A_\tau - \partial_\tau A_\lambda$ cannot be done so.

In de Sitter spacetime, the four-dimensional coordinates are

$$\{x^\mu\} = \{\phi_0, \phi_1, \phi_2, \phi_3\}.$$

The derivatives with respect to the coordinates are

$$\left\{ \frac{\partial}{\partial x^\mu} \right\} = \left\{ \frac{\partial}{\partial \phi_0}, \frac{\partial}{\partial \phi_1}, \frac{\partial}{\partial \phi_2}, \frac{\partial}{\partial \phi_3} \right\}.$$ (10.6.72)

With these notations, Euler-Lagrange equations (1.7.3) or (1.3.16) are written as

$$\frac{\partial L}{\partial A_\alpha} - \frac{\partial}{\partial \phi_\mu} \frac{\partial L}{\partial (\partial_\mu A_\alpha)} = 0, \quad (\alpha = 0, 1, 2, 3).$$ (10.6.73)

Having the preparation above, we are now at the stage to derive the equations that the four-dimensional vector should obey by substitution of (10.6.71) into (10.6.73). The first term in (10.6.73) is easily obtained:

$$\frac{\partial L}{\partial A_\alpha} = \sqrt{-g} J^\alpha.$$ (10.6.74)

The second term in (10.6.73) is complicated. It is processed step by step. It is sure that as $\alpha = \mu$ this term is zero. Because $(g^{\mu\nu})$ is a diagonal matrix, the summation over λ and τ in (10.6.71) can be carried out to get

$$g^{\mu\lambda}g^{\nu\tau}\partial_\mu A_\nu(\partial_\lambda A_\tau - \partial_\tau A_\lambda)$$

$$= \sum_{\mu,\nu=0}^{3} g^{\mu\mu}g^{\nu\nu}\partial_\mu A_\nu \left(\partial_\mu A_\nu - \partial_\nu A_\mu\right). \qquad (10.6.75)$$

Then the following derivative is easily computed.

$$\frac{\partial}{\partial(\partial_\beta A_\alpha)} \sum_{\mu,\nu=0}^{3} g^{\mu\mu}g^{\nu\nu}\partial_\mu A_\nu(\partial_\mu A_\nu - \partial_\nu A_\mu)$$

$$= \sum_{\mu,\nu=0}^{3} \left[\left(\frac{\partial}{\partial(\partial_\beta A_\alpha)} g^{\mu\mu}g^{\nu\nu}\partial_\mu A_\nu\right)(\partial_\mu A_\nu - \partial_\nu A_\mu) \right.$$

$$\left. + \partial_\mu A_\nu \frac{\partial}{\partial(\partial_\beta A_\alpha)}(g^{\mu\mu}g^{\nu\nu}\partial_\mu A_\nu - g^{\mu\mu}g^{\nu\nu}\partial_\nu A_\mu) \right]$$

$$= 2g^{\beta\beta}g^{\alpha\alpha}(\partial_\beta A_\alpha - \partial_\alpha A_\beta). \qquad (10.6.76)$$

This implies that

$$\frac{\partial L}{\partial(\partial_\beta A_\alpha)} = -\frac{1}{\mu}\sqrt{-g}g^{\beta\beta}g^{\alpha\alpha}(\partial_\beta A_\alpha - \partial_\alpha A_\beta). \qquad (10.6.77)$$

In the following, we assume that μ is independent of coordinates. Let us put down the equation for $\alpha = 0$. Combination of (10.6.74) and (10.6.77) gives

$$-\frac{\partial}{\partial\phi_1}\left[\sqrt{-g}g^{11}g^{00}(\partial_1 A_0 - \partial_0 A_1)\right]$$

$$-\frac{\partial}{\partial\phi_2}\left[\sqrt{-g}g^{22}g^{00}(\partial_2 A_0 - \partial_0 A_2)\right]$$

$$-\frac{\partial}{\partial\phi_3}\left[\sqrt{-g}g^{33}g^{00}(\partial_3 A_0 - \partial_0 A_3)\right]$$

$$= -\mu\sqrt{-g}J^0. \qquad (10.6.78)$$

Inspection of (10.6.65) tells us that $(g^{\mu\nu})$ is independent of coordinates ϕ_2 and ϕ_2. Therefore, (10.6.78) is simplified to be

$$\frac{1}{\sqrt{-g}}\frac{\partial}{\partial\phi_1}\left(\sqrt{-g}g^{11}g^{00}\right)(\partial_1 A_0 - \partial_0 A_1) + g^{11}g^{00}\partial_1(\partial_1 A_0 - \partial_0 A_1)$$

$$+ g^{22}g^{00}\partial_2(\partial_2 A_0 - \partial_{20}A_2) + g^{33}g^{00}\partial_3(\partial_{33}A_0 - \partial_0 A_3) = \mu J^0.$$

$$(10.6.79)$$

Now we substitute the matrix elements in (10.6.65) into (10.6.67), so as to gain the required equation:

$$2\cot 2\phi_1(\partial_1 A_0 - \partial_0 A_1) + \partial_1(\partial_1 A_0 - \partial_0 A_1)$$

$$+ \sec^2\phi_1\partial_2(\partial_2 A_0 - \partial_0 A_2) + \csc^2\phi_1\partial_3(\partial_3 A_0 - \partial_0 A_3)$$

$$= -\mu\rho^4\cosh^2\phi_0 J^0. \qquad (10.6.80a)$$

In the same procedure, the other three equations are easily obtained. They are listed as follows.

$$\sinh\phi_0\cosh\phi_0(\partial_0 A_1 - \partial_1 A_0) + \cosh^2\phi_0\partial_0(\partial_0 A_1 - \partial_1 A_0)$$

$$- \sec^2\phi_1\partial_2(\partial_2 A_1 - \partial_1 A_2) - \csc^2\phi_1\partial_3(\partial_3 A_1 - \partial_1 A_3)$$

$$= -\mu\rho^4\cosh^4\phi_0 J^1. \qquad (10.6.80b)$$

$$\sinh\phi_0\cosh\phi_0(\partial_0 A_2 - \partial_2 A_0) + \cosh^2\phi_0\partial_0(\partial_0 A_2 - \partial_2 A_0)$$

$$- \sec\phi_1\csc\phi_1(\partial_1 A_2 - \partial_2 A_1) - \partial_1(\partial_1 A_2 - \partial_2 A_1)$$

$$- \csc^2\phi_1\partial_3(\partial_3 A_2 - \partial_2 A_3) = -\mu\rho^4\cosh^4\phi_0\cos^2\phi_1 J^2.$$

$$(10.6.80c)$$

$$\sinh\phi_0\cosh\phi_0(\partial_0 A_3 - \partial_3 A_0) + \cosh^2\phi_0\partial_0(\partial_0 A_3 - \partial_3 A_0)$$

$$+ \sec\phi_1\csc\phi_1(\partial_1 A_3 - \partial_3 A_1) - \partial_1(\partial_1 A_3 - \partial_3 A_1)$$

$$- \sec^2\phi_1\partial_2(\partial_2 A_3 - \partial_3 A_2) = -\mu\rho^4\cosh^4\phi_0\sin^2\phi_1 J^3.$$

$$(10.6.80d)$$

These four equations are the required Maxwell equations in de Sitter spacetime. In the course of derivation, we have assumed that μ is independent of coordinates. Otherwise the terms containing the derivatives of μ with respect to coordinates should be added.

In a vacuum, $\mu = \mu_0$ is a constant. Physical quantities in vacuum should obey the symmetry that de Sitter spacetime possesses. From this aspect, the potential vector was explicitly derived as follows.

$$A_0 = 0, \tag{10.6.81a}$$

$$A_1 = -C\varphi_{N+1\lambda_1} \cosh\phi_0 \cos^{-1}\phi_1 e^{-i\phi_3}, \tag{10.6.81b}$$

$$A_2 = -iC\varphi_{N+1\lambda_1} \cosh\phi_0 \sin\phi_1 e^{-i\phi_3}, \tag{10.6.81c}$$

$$A_3 = iC\varphi_{N+1\lambda_1} \cosh\phi_0 \sin\phi_1 e^{-i\phi_3} = -A_2, \tag{10.6.81d}$$

where

$$\varphi_{N\lambda_1} = (\varphi_{\lambda_1})^N = (\cosh\phi_0 \cos\phi_1 e^{-i\phi_2})^N. \tag{10.6.82}$$

The vector potential certainly observes the equation (10.6.80). When (10.6.81) is substituted into (10.6.80), it is obtained that

$$J^0 = 0, \tag{10.6.83a}$$

$$J^i = g^{ii}\frac{(n+2)(n+3)}{\mu\rho^2}A_i, \quad i = 1,2,3, \tag{10.6.83b}$$

Here n takes positive integers. The factor $(n+2)(n+3)$ arises from the fundamental properties of de Sitter spacetime.

In (10.6.83), the current is linearly proportional to the potential vector. This is not occasional. In our space, the electric current is also linearly proportional to vector potential in vacuum. We present the derivation in the following.

When a particle with mass m and charge q is moving with velocity $\boldsymbol{v}$ in an electric field $\boldsymbol{E}$ and magnetic field $\boldsymbol{B}$, it feels a so-called Lorentz force exerted by the fields:

$$\boldsymbol{F} = q(\boldsymbol{E} + \boldsymbol{v} \times \boldsymbol{B}). \tag{10.6.84}$$

If there are n identical charges in unit volume, then electric current is defined by

$$\boldsymbol{j} = nq\boldsymbol{v}. \tag{10.6.85}$$

Suppose that the magnetic field is zero or just parallel to the velocity $\boldsymbol{v}$. Taking derivative of (10.6.85) with respect to time and applying

Newton's second law yield

$$\frac{\mathrm{d}}{\mathrm{d}t}\boldsymbol{j} = nq\frac{\mathrm{d}}{\mathrm{d}t}\boldsymbol{v} = \frac{nq}{m}\boldsymbol{F} = \frac{nq^2}{m}\boldsymbol{E}. \tag{10.6.86}$$

The integration of this equation results in

$$\boldsymbol{j} = -\frac{nq^2}{m}\boldsymbol{A}. \tag{10.6.87}$$

It shows that the electric current is linearly proportional to the vector potential, and the coefficient is proportional to square of the charge and inverse to the mass.

When in a media, the interaction between the charged particle and the matter has to be taken into account. A simplest example is the movement of electron in a metal crystal. Every electron in moving collides atoms at the crystal lattice sites. Thus, different electrons have different instant velocities. Nevertheless, because of the force exerted by the electric field, the electron system as a whole has a net velocity along the field direction, called drift velocity. In this case, the electric current is still defined by (10.6.85) but with $\boldsymbol{v}$ being the drift velocity. It was derived that

$$\boldsymbol{j} = \sigma\boldsymbol{E}. \tag{10.6.88}$$

Thus, the electric current is linearly electric field instead of vector potential. The coefficient is called conductivity, which reflects the factor in the crystal that hinders the movement of electrons. There is resistance in a metal, while in vacuum, the resistance is zero.

In a superconductor, there is no resistance. On this point, a superconductor and vacuum are of the same feature. Consequently, (10.6.87) stands for superconductors either. Historically, people first studied the electron movement in a metal to obtain (10.6.88). Later, when superconductivity was discovered, (10.6.87) was realized.

At first glance, (10.6.83) and (10.6.87) seem to be the same form. We make some discussion about them.

Firstly, (10.6.83) tells us that three component of current are linearly proportional to potential vector in de Sitter spacetime. For the ϕ_0 component, $J^0 = 0$ since $A_0 = 0$. What is the relation between A_0 and J^0 if one of them is not zero? We are unable to answer. Similarly, (10.6.87) is valid for three space component in our space but not the time component. For the time component, if charge density ρ is zero in the whole vacuum, then the electrostatic potential φ is zero either. The relation between them is not linear if one of them is nonzero. By Coulomb's law, if there is a point charge in space, then there is an electric potential in every point in the space, the value of which is inverse to the square of the distance between this point and the charge. In our spacetime, time coordinate is definitely different from space coordinates. It seems that it is so either in de Sitter spacetime.

Secondly, (10.6.81) is not obtained by solving Maxwell equations (10.6.80), but was obtained based on the symmetry that de Sitter spacetime should have. Equation (10.6.87) was derived by use of two experimental laws in our space. One is that the force exerted by an electric field on a charge is linearly proportional to both the charge and the field, (10.6.84). The other is Newton's second law in deriving (10.6.86). Inspired by this fact, we guess (10.6.83) may imply a possible "Newton's second law" in de Sitter spacetime in some form.

Thirdly, in (10.6.87), permeability μ does not appear. This is because there is no magnetic field but electric field in the vacuum. In our space, electric field and magnetic field can exist independently. In one region, there can be either electrostatic field or static magnetic field. In (10.6.83), permeability μ appears in the relation between current and vector potential. This indicates that in de Sitter spacetime, electric field and magnetic field are closely related to each other. There will be no cases where only electric field or magnetic field exists.

Fourthly, in (10.6.87), the coefficient is proportional to the square of electron charge and inverse to the mass. One may reasonably postulate that the coefficient in (10.6.83) is also proportional to the square of electron charge and inverse to the mass. It was discussed in the last subsection that in de Sitter spacetime mass took discrete values. It is naturally conjectured that the charge also take discrete values.

At last, we guess that when there is a media in de Sitter space, an equation similar to (10.6.88) is possible. Correspondingly, there will probably be a concept about conductivity or resistance.

Exercises

1. The solution of the radial equation separated from Helmholtz equation is (10.3.17). The function

$$R_{N,l}(z) = \frac{Z_{l+N/2-1}(z)}{z^{N/2-1}} \tag{1}$$

is tentatively called spherical Bessel function of l order in N-dimensional space, where the numerator Z is cylindrical function introduced in Chapter 4. In the case of three-dimensional space $N = 3$, $R_{N,l}(z) = R_{3,l}(z)$ is just spherical Bessel function, see (4.5.13). As has been shown in Subsection 4.5.2, spherical Bessel functions with integer orders in three-dimensional space $R_{3,l}(z)$ can be expressed by elementary functions. For other N, can the spherical Bessel functions $R_{N,l}(z)$ defined by (1) be expressed by elementary functions? Is it possible to derive the recursion formulas of $R_{N,l}(z)$?

2. Show that when $N = 1, 2, 3$, the solution (10.2.21) goes back to (6.2.27b), (6.2.21b) and (6.2.11b), respectively, in Chapter 6.

3. Find the minimum position r_0 of the potential (10.3.64) and the potential at this position $V(r_0)$. Is $V(r_0)$ positive or negative? Plot the potential curve when $A = 1$ and $B = 2$.

4. Find the minimum position r_0 of the potential (10.3.74) and the potential at this position $V(r_0)$. Is $V(r_0)$ positive or negative? Plot the potential curve when $A = 1$ and $B = 1$.

5. Solve equation (10.3.80) when $\alpha = 0$.

6. Derive (10.3.82) from (10.3.81).

7. If the potential in (10.3.87) is of the form of (10.3.88), what is the expression of D? What is the expression a in (10.3.87)? Here r_0 is assumed to be known.

8. Show that taking derivative of (10.4.14) k times and making transformation (10.4.15) will lead to (10.4.16).

9. It has been known that one solution of associated Gegenbauer equation (10.4.16) is associated Gegenbauer polynomials
   (10.4.17b). Try to find another linearly independent solution.
   Try to derive the recursion formulas of associated Gegenbauer
   polynomials.

10. In spherical coordinate system of four-dimensional space, when
    l_1 is fixed, the number of states is expressed by (10.4.21). Find
    the number of state in six-dimensional space when l_1 in (10.4.34)
    is fixed. Find the number of state in five-dimensional space when
    l_1 in (10.4.29) is fixed.

11. Derive (10.5.15) from (10.5.13) through transformation (10.5.14).

12. In three-dimensional space, two-particle Green's function satis-
    fies the following Helmholtz equation:

    $$\left(\nabla_{r1}^2 + \nabla_{r2}^2 + k^2\right) G(r_1, r'_1; r_2, r'_2) = \delta\left(r_1 - r'_1\right) \delta\left(r_2 - r'_2\right).$$

    Let the angle between $\boldsymbol{R} = \boldsymbol{r}_1 + \boldsymbol{r}_2$ and the z axis be 2α. After
    separation of variables, one can obtain the equation that $\rho(\alpha)$ as
    a function of α satisfies:

    $$\frac{1}{\sin^2 \alpha \cos^2 \alpha} \frac{d}{d\alpha} \left[\sin^2 \alpha \cos^2 \alpha \frac{d}{d\alpha} \rho(\alpha)\right]$$
    $$+ \left[\lambda - \frac{l_1(l_1 + 1)}{\cos^2 \alpha} - \frac{l_2(l_2 + 1)}{\sin^2 \alpha}\right] \rho(\alpha) = 0.$$

    Use transformation (10.5.14) to solve this equation and find
    expressions of eigenfunctions and its eigenvalues λ.

13. In five-dimensional space, the pseudo spherical coordinates are
    defined by (10.5.26). Show that the coordinate coefficients are
    expressed by (10.5.27). Calculate the volume of a hypersphere
    with radius R. Write the Jacobi determinant between Cartesian
    and pseudo spherical coordinates observing (10.1.13). Put down
    the expressions of gradient, divergence and Laplace operators. By
    means of separation of variables, solve Laplace equation, and find
    the eigenfunctions and eigenvalues with respect to every angular
    coordinate.

14. In six-dimensional space, the pseudo spherical coordinates are defined by (10.5.29). Show that the coordinate coefficients are expressed by (10.5.27). Calculate the volume of a hypersphere with radius R. Write the Jacobi determinant between Cartesian and pseudo spherical coordinates observing (10.1.13). Put down the expressions of gradient, divergence and Laplace operators. By means of separation of variables, solve Laplace equation, and find the eigenfunctions and eigenvalues with respect to every angular coordinate.

15. We have discussed the spherical systems from four- to six-dimensional spaces. Actually, there is only one pseudo spherical system defined by (10.5.1), which is in four-dimensional space. In five- and six-dimensional spaces, there is respectively only one angular subspace having this pseudo spherical system. Is it possible that there exist other pseudo spherical coordinates in five- and six-dimensional spaces? Is it possible that there exist other pseudo spherical coordinates in higher-dimensional spaces?

16. Prove (10.6.24).

17. Verify (10.6.21) in terms of (10.6.53).

18. Substitute (10.6.63) into (10.6.70) to obtain (10.6.71). Note that between two factors $\frac{\partial A_\nu}{\partial x^\mu} - \frac{\partial A_\mu}{\partial x^\nu}$ and $\frac{\partial A_\tau}{\partial x^\lambda} - \frac{\partial A_\lambda}{\partial x^\tau}$, only one can be reduced.

19. When μ is independent of coordinate, we have derived Eq. (10.6.80a). Please derive Eqs. (10.6.80b), (10.6.80c) and (10.6.80d).

Appendix 10A. Hypergeometric Equation and Hypergeometric Functions

The ordinary differential equation of second order

$$z(1-z)w''(z) + (c - (a+b+1)z)w'(z) - abw(z) = 0 \qquad (10A.1)$$

is called a **hypergeometric equation**. It has three parameters a, b and c. Its two linearly independent solutions are

$$w_1(z) = F(a, b, |c|, z) \qquad (10A.2a)$$

and

$$w_2(z) = z^{1-c} F\left(1 + a - c, 1 + b - c, |2 - c|, z\right). \qquad (10A.2b)$$

The function F can be expanded by Taylor series near $z = 0$:

$$F(a, b, cz) = 1 + \frac{ab}{c} z + \frac{a(a+1)b(b+1)}{2!c(c+1)} z^2$$

$$+ \cdots + \frac{(a)_n(b)_n}{n!(c)_n} z^n + \cdots . \qquad (10A.3)$$

This expansion is called a **hypergeometric series**. Its radius of convergence is 1. It is analytical within the circle $|z| < 1$. When analytically continuated to the whole complex plane, it is called a **hypergeometric function**.

The general solution of (10A.1) is the linear combination of (10A.2a) and (10A.2b).

Some properties of hypergeometric functions are listed below.

From (10A.3) it is seen that at $z = 0$,

$$F(a, b, c, 0) = 1.$$

Equation (10A.1) is invariant under exchange of a and b. Its solutions necessarily have this property:

$$F(a, b, c, z) = F(b, a, c, z)$$

In general, the series (10A.3) contains infinite terms. Therefore, even for finite z value, it may be divergent. When a is a negative integer,

$$a = -n \qquad (10A.4)$$

There are $n + 1$ terms in $F(-n, b, c, z)$:

$$F(-n, b, c, z) = 1 - \frac{nb}{c} z + \frac{n(n+1)b(b+1)}{2!c(c+1)} z^2$$

$$- \cdots + (-1)^n \frac{(b)_n}{(c)_n} z^n. \qquad (10A.5)$$

This guarantees that for any finite z, the value of the function is finite. In this case, this series is called **Jacobi series**, which is just Jacobi polynomials mentioned in Chapter 3. In general, Jacobi polynomials are defined by

$$J_n^{(r,s)}(z) = F(-n, n + r, s, z).\qquad\text{(10A.6)}$$

Some formulas concerning Jacobi polynomials were listed in Tables 3.1, 3.2, 3.3, 3.6, 3.7.

References

1. Al-Gwaiz MA, Sturm-Liouville Theory and its Application. London: Springer-Verlag London Limited, 2008.
2. Andrews GE, Askey R, Roy R. Special function. Cambridge: Cambridge University Press, 2000.
3. Andrews LC, Special Functions for Engineers and Applied Mathematicians. New York: Mechillan Publishing Company, Adivision of Macmillan, Inc., 1985.
4. Byron FW, Fuller RW. Mathematics of Classical and Quantum Physics, Vol. 1 & 2, New York: Addison-Wesley, 1969.
5. Chen NX, Möbius Inversion in Physics. Singapore: World Scientific, 2010.
6. Coddington EA and Levinson N, Theory of Differential Equations. New York: McGraw-Hill Book Company, Inc., 1955.
7. Dwight HB, Tables of Integrals and Other Mathematical Data. 4th ed. Chap. 12. New York: The MacMillan Company, 1964.
8. Hanson GW and Yakolev AB, Operator Theory for Electromagnetics-An Introduction, Berlin, Springer-Verlag, 2002
9. Hassani S, Mathematical Physics, A Modern Introduction to Its Foundations. New York: Springer-Verlag, 1999.
10. Harrington RF, Time-Harmonic Electromagnetic Fields. New York: McGraw-Hill Book Co., 1961.
11. Landau LD and Lifshitz EM, The classical Theory of Fields. New York: Pergmon Press, 1976.
12. Rivlin TJ. The Chebyshev Polynomials. New York: John Wiley & Sons, Inc., 1974.
13. Schrödinger E, Annalen der Physik, 1926, 79(4); Schrödinger E. Collected Papers on Wave Mechanics (Translated from the second German edition) London and Glasgow: Blackie & Son Limited,1928.
14. Stakgold I, Green's Functions and Boundary Value Problems. New York: John Wiley & Sons, 1979.

15. Szekeres PA, Course in Modern Mathematical Physics. Cambridge: Cambridge University Press, 2004.
16. Wang HY. Green's Functions in Condensed Matter Physics. Beijing: Alpha Science International ltd. and Science Press, 2012.
17. Watson GN. A treatise on the theory of bessel functions, Cambridge: The University Press, 1952.

Answers of Selected Exercises

Chapter 1

2. (1) $\delta J = \int_a^b \dfrac{y'}{\sqrt{1+y'^2}} \delta y' \mathrm{d}x.$

(2) $\Delta J = \int_{x_0}^{x_1} [(x^2 + 2y + y')\delta y + y\delta y'] \mathrm{d}x$

$\qquad + \int_{x_0}^{x_1} [\delta y \delta y' + (\delta y)^2] \mathrm{d}x.$

$\qquad \delta J = \int_{x_0}^{x_1} [(x^2 + 2y + y')\delta y + y\delta y'] \mathrm{d}x.$

3. (1) $y = \dfrac{1}{2C_2}(x + C_0)^2 + C_2.$ (2) $y = \sinh(C_3 x + C_1).$

4. (1) The extremal curve is $y = -\frac{1}{4}x^2 + 1$. The increment $\Delta J = \delta J + \int_0^2 (\delta y'^2)\mathrm{d}x$. Since $\int_0^2 (\delta y'^2)\mathrm{d}x \geq 0$, $\Delta J \geq 0$. The extremal curve makes the functional take a minimum.

   (2) The extremal curve is $y = x^3 + 2x + 1$. The increment $\Delta J = \delta J + \int_0^1 (\delta y \delta y' + \delta y'^2)\mathrm{d}x$.

6. (1) $y = C_1 e^{-2\mathrm{i}x} + C_2 e^{2\mathrm{i}x} + C_3 e^{-2x} + C_4 e^{2x}.$

   (2) $y = \dfrac{1}{7!}x^7 + \dfrac{1}{5!}C_1 x^5 + \dfrac{1}{4!}C_2 x^4 + \dfrac{1}{6}C_3 x^3 + \dfrac{1}{2}C_4 x^2 + C_5 x + C_6.$

7. (1) $y = (C_1 x + C_2)e^{ix} + (C_3 x + C_4)e^{-ix}$, $z = 2y + y''$.
   (2) $y = C_1 x + C_2$, $z = C_3 x + C_4$.

8. (1) $\dfrac{\partial^2 z}{\partial x^2} - \dfrac{\partial^2 z}{\partial y^2} = 0$. (2) $\dfrac{\partial^2 u}{\partial x^2} + \dfrac{\partial^2 u}{\partial y^2} + \dfrac{\partial^2 u}{\partial z^2} = f(x, y, z)$.

10. (1) $y = \pm 2 \sin k\pi x$. (2) $\lambda = -\dfrac{10}{11}$, $y = -\dfrac{5}{2}x^2 + \dfrac{7}{2}x$, $z = x$.

11. $z = a\varphi + b$

12. $p(x)y'' + p'(x)y' - (q(x) + \lambda r(x))y = 0$; $y(0) = 0$, $y(x_1) = 0$.

13. $y = 0$.

15. $y = C_1 \cosh\left(\dfrac{x - x_0}{C_1} + \cosh^{-1}\dfrac{y_0}{C_1}\right)$.

16. (1) $y = \dfrac{3}{3 + e^2}e^x + \dfrac{e^2}{3 + e^2}e^{-x}$. (2) $y = \dfrac{6e}{3e^2 + 1}e^x + \dfrac{2e}{3e^2 + 1}e^{-x}$.
    (3) $y = 0$.

18.
$$\frac{d}{dt}\boldsymbol{p} = e(\boldsymbol{E} + \boldsymbol{v} \times \mathbf{B}), \quad \text{where}$$

$$\boldsymbol{p} = \frac{\boldsymbol{v}}{\sqrt{c^2 - v^2}}, \boldsymbol{E} = -e\nabla\varphi - \frac{\partial}{\partial t}\mathbf{A}.$$

19. $-i\hbar\frac{\partial}{\partial t}\psi^* = -\frac{\hbar^2}{2m}\nabla^2\psi^* + U(r, t)\psi^*$ and $i\hbar\frac{\partial}{\partial t}\psi = -\frac{\hbar^2}{2m}\nabla^2\psi + U(r, t)\psi$. These are Schrödinger equation and its complex conjugate.

Chapter 2

10. The set of orthonomalized basis is
$$A = \left\{ \frac{1}{\sqrt{2\pi}}, \frac{\sin x}{\sqrt{\pi}}, \frac{\sin 2x}{\sqrt{\pi}}, \ldots, \frac{\sin nx}{\sqrt{\pi}} \right\}.$$

11. The orthonomalized basis is as follows:
$$e_1 = \frac{1}{\sqrt{2}}, \quad e_2 = \sqrt{\frac{3}{2}}x, \quad e_3 = \sqrt{\frac{45}{8}}\left(x^2 - \frac{1}{3}\right),$$

$$e_4 = \sqrt{\frac{175}{8}}\left(x^3 - \frac{3}{5}x\right), \quad e_5 = \frac{105}{8\sqrt{2}}\left(x^4 - \frac{6}{7}x^2 + \frac{3}{35}\right),$$

$$e_6 = \frac{63}{8}\sqrt{\frac{11}{2}}\left(x^5 - \frac{10}{9}x^3 + \frac{5}{21}x\right),$$

$$e_7 = \frac{231}{16}\sqrt{\frac{13}{2}}\left(x^6 - \frac{15}{11}x^4 + \frac{5}{11}x^2 - \frac{5}{231}\right).$$

They are the first seven normalized Legendre polynomials.

15. $L^\dagger v(x) = \left(-\dfrac{d}{dx} + 1\right)v(x), 0 < x < 1; v(1) - \alpha v(0) = 0.$

18. The adjoint boundary conditions are $c_2^* v'(1) - c_1^* v(1) = 0, v'(1) - \gamma^* v'(0) = 0, v'(1) - v(1) + v(0) = 0.$

19. The conditions for the differential operator of the third order are $r_3 = -r_3^*, r_2 = (r_2 - 3r_3')^*, r_1 = (-3r_3'' + 2r_2' - r_1)^*, r_0 = (-r_3''' + r_2'' - r_1' + r_0)^*.$

Chapter 3

13. $f_1(x) = \displaystyle\sum_{m=0}^{\infty} \frac{(-1)^m (2m)!(4m+3)}{2^{2m+1}(m!)^2(m+1)} P_{2m+1}(x).$

$f_2(x) = \dfrac{1}{2} + \displaystyle\sum_{m=1}^{\infty} \frac{(-1)^{m+1}(2m-2)!(4m+1)}{2^{2m}(m+1)!(m-1)!} P_{2m}(x).$

14. $f(x) = \dfrac{1}{2}(1-\alpha) + \dfrac{1}{2}\displaystyle\sum_{n=1}^{\infty}[P_{n-1}(\alpha) - P_{n+1}(\alpha)]P_n(x).$

16. (2) $Q_0(x) = \dfrac{1}{2}\ln\left(\dfrac{1+x}{1-x}\right). Q_1(x) = -1 + \dfrac{x}{2}\ln\left(\dfrac{1+x}{1-x}\right).$

$Q_2(x) = \dfrac{1}{2}(3x^2 - 1)\ln\left(\dfrac{1+x}{1-x}\right) - \dfrac{3}{2}x.$

(3) $Q_0(\cos\theta) = \ln\cot\dfrac{\theta}{2}. Q_1(\cos\theta) = \cos\theta\ln\cot\dfrac{\theta}{2} - 1.$

$Q_2(\cos\theta) = \dfrac{1}{2}(3\cos^2\theta - 1)\ln\cot\dfrac{\theta}{2} - \dfrac{3}{2}\cos\theta.$

(4) $P_1^1(x) = (1 - x^2)^{1/2}, \quad P_2^1(x) = 3(1 - x^2)^{1/2}x,$

$P_2^2(x) = 3(1 - x^2), \quad P_3^1(x) = \frac{3}{2}(1 - x^2)^{1/2}(1 - 5x),$

$P_3^2(x) = 15(1 - x^2)x, \quad P_3^3(x) = 15(1 - x^2)^{3/2}.$

$Q_1^1(x) = (1 - x^2)^{1/2}\left[\frac{1}{2}\ln\left(\frac{1+x}{1-x}\right) + \frac{x}{1-x^2}\right].$

$Q_2^1(x) = (1 - x^2)^{1/2}\left[\frac{3}{2}x\ln\left(\frac{1+x}{1-x}\right) + \frac{3x^2 - 2}{1-x^2}\right].$

$Q_2^2(x) = (1 - x^2)^{1/2}\left[\frac{3}{2}\ln\left(\frac{1+x}{1-x}\right) + \frac{5x - 3x^2}{(1-x^2)^2}\right].$

22. Let $x = \sqrt{2}u/\sqrt{B_1}$. Then $\dfrac{d^2y}{du^2} - 2u\dfrac{dy}{du} + \lambda\dfrac{2}{B_1}y = 0.$

34. The transformed equation is $(p^2 - 1)f''(p) + 3pf'(p) - [(2S+1)^2 - 1]f(p) = 0$. By comparison with (3.4.65) and (3.4.67), it is seen that only when $2S = n$ can the equation have solutions. That is to say, S must be positive integers or half-integers. The required solution is

$$f(u) = R\frac{[V(u) + w(u)]^{n+1} - [V(u) + w(u)]^{-n-1}}{2^n[(Q+R)^{n+1} - (Q-R)^{n+1}]w(u)},$$

where $w(u) = \sqrt{[V(u)]^2 - 4(Q^2 - R^2)}.$"

35. (1) The first kind of singularities: $z = 1, z = \infty$. The second kind of singularities: $z = 0$. (2) The first kind of singularities: $z = 0, z = -1/3, z = \infty$.

37. (i) When $\lambda_1 - \lambda_2 = \sqrt{1 + 4\alpha} \neq n$, or $\alpha \neq \frac{1}{4}(n^2 - 1)$, $w_1(z) = z^{\lambda_1}, w_2(z) = z^{\lambda_2}$.

(ii) When $\lambda_1 - \lambda_2 = \sqrt{1 + 4\alpha} = 0$, or $\alpha = -1/4, \lambda_1 = \lambda_2 = 1/2$, $w_1(z) = z^{1/2}, w_2(z) = z^{1/2}\ln z$.

(iii) When $\lambda_1 - \lambda_2 = \sqrt{1 + 4\alpha} = 2n+1$ are odd integers, $w_1(z) = z^{n+1}, w_2(z) = z^{-n}$.

(iv) When $\lambda_1 - \lambda_2 = \sqrt{1 + 4\alpha} = 2n$ are even integers, $w_1(z) = z^{n+1/2}, w_2(z) = z^{-n+1/2}$.

The solutions in (iii) are rational functions. The conclusion is that when $\alpha = n(n+1)$, where n are integer numbers, the solutions are rational ones.

38. (1) $w_1(z) = \sum\limits_{k=0}^{\infty} C_{4k} z^{4k}$, $w_2(z) = \sum\limits_{k=0}^{\infty} C_{4k+1} z^{4k+1}$.

(2) $w_1(z) = \sum\limits_{l=0}^{\infty} C_{2l} z^{2l}$, $w_2(z) = \sum\limits_{l=0}^{\infty} C_{2l+1} z^{2l+1}$.

43. For the boundary problem

$$u'(x) = f(x), \quad 0 < x < 1; \quad u(0) = 0, \quad u(1) = 0, \qquad (1)$$

its adjoint boundary problem is

$$v'(x) = 0, 0 < x < 1. \qquad (2)$$

There is no boundary condition for (2). This is because the original boundary problem (1) has one more boundary condition. The solution of (2) is $v(x) = 1$. Hence the compatibility condition is $\int_0^1 f(x)\mathrm{d}x = 0$. We choose a function $f(x) = x - 1/2$ which satisfies the compatibility condition. Accordingly, the solution of (1) is $u(x) = x^2/2 - x/2$.

45. The boundary value problem is not a self-adjoint one unless α_1 is a complex conjugate of β_2 and α_2 is a complex conjugate of β_1.

Chapter 4

4. By (4.2.19),

$$\int \frac{\mathrm{d}z}{z[J_\nu(z)]^2} = -\frac{\pi}{2\sin\nu\pi} \frac{J_{-\nu}(z)}{J_\nu(z)} + C.$$

By (4.2.20),

$$\int \frac{\mathrm{d}z}{z[J_\nu(z)]^2} = \frac{\pi}{2} \frac{Y_\nu(z)}{J_\nu(z)} + C.$$

5. (1) $\dfrac{\pi}{2} \ln \dfrac{Y_\nu(z)}{J_\nu(z)} + C.$ (2) $-\dfrac{\pi}{2} \dfrac{J_\nu(z)}{Y_\nu(z)} + C.$ (3) $\dfrac{\pi}{2} \tan^{-1} \dfrac{Y_\nu(z)}{J_\nu(z)} + C.$

10. When $m = 2k$ is even,

$$z^m = z^{2k} = 2^{2k+1} \sum_{n=0}^{\infty} J_{2n}(z) \frac{(n+k-1)!}{(n-1)!} \prod_{i=0}^{k-1} (n-i),$$

where all the terms with $n \le k-1$ vanish. When $m = 2k+1$ is odd,

$$z^m = z^{2k+1} = 2^{2k+1} \sum_{n=0}^{\infty} J_{2n+1}(z)(2n+1) \prod_{i=1}^{k} (n+i) \prod_{i=1}^{k} (n-i+1).$$

35. Make transformation $w(z) = J_\nu(z)/z^{(n-1)/2}$, where $\nu = [(n-1)^2/4 + \gamma]^{1/2}$. Here n and γ can be any real numbers.

36. The resultant equation is $u^2 w''(u) + u w'(u) + \frac{4a}{m^2} u^2 w(u) = 0$. The general solution of the original equation is $w(z) = A J_0(\frac{2}{m}\sqrt{a}e^{mz/2}) + B Y_0(\frac{2}{m}\sqrt{a}e^{mz/2})$, where A and B are arbitrary constants.

39. $f(x) = \dfrac{1}{2} \displaystyle\sum_{n=1}^{\infty} \dfrac{J_1(\mu_n)}{[J_1(2\mu_n)]^2 k_n} J_0(k_n x).$

40. $x = \displaystyle\sum_{n=1}^{\infty} \dfrac{2}{k_n J_2(k_n)} J_1(k_n x).$

41. $x^3 = \displaystyle\sum_{n=1}^{\infty} \dfrac{2(8-k_m^2)}{k_n^3 J_1'(k_n)} J_1(k_n x).$

42. $1 - x^2 = \displaystyle\sum_{n=1}^{\infty} \dfrac{8}{k_n^3 J_1(k_n)} J_0(k_n x).$

44. $\displaystyle\sum_{n=1}^{\infty} \dfrac{1}{k_n^2} = \dfrac{1}{4(\nu+1)}$ where $J_\nu(k_n) = 0$, $(n = 1,2,3,\ldots; \nu > -1)$.

45. $u_0 - 2u_0 \displaystyle\sum_{n=1}^{\infty} \dfrac{1}{k_n J_1(k_n)} J_0(k_n r/R) \exp\left(-\dfrac{k_n^2 a^2}{R^2}t\right).$

46. $\displaystyle\sum_{m,n=1}^{\infty} \sin m\varphi J_m(\lambda_{m,n} r/R)(A_{m,n} e^{i\lambda_{m,n}at} + B_{m,n} e^{-i\lambda_{m,n}at}),$

where $\lambda_{m,n}$'s meet $J_m(\lambda_{m,n}) = 0$, and $A_{m,n}$ and $B_{m,n}$ are constants to be determined by the initial conditions.

47. $\dfrac{4Lq}{\pi^2} \displaystyle\sum_{n=0}^{\infty} \dfrac{I_n(n\pi r/L)}{n^2 I'_n(n\pi r_0/L)} \sin \dfrac{n\pi}{L} z,$

where $I_n(x)$ is a modified Bessel function with integer order n.

48. $\dfrac{2}{R^2} \displaystyle\sum_{n=0}^{\infty} \dfrac{J_0(k_n r)}{(1 + k^2/k_n^2) J_0^2(k_n R)} \dfrac{\sinh k_n z}{\sinh k_n h} \displaystyle\int_0^R r f(r) J_0(k_n r) \mathrm{d}r,$

where k_n is a positive root of $k J_0(Rx) - x J_1(Rx) = 0$.

49. $\displaystyle\sum_{n=0}^{\infty} J_0(k_n r)(C_n \cos q_n t + D_n \sin q_n t) e^{-ht},$

where k_n is a positive root of $J_0(lx) = 0$, $q_n = \sqrt{k_n^2 a^2 - h^2}$, $C_n = \dfrac{2}{J_1^2(k_n l) l^2} \displaystyle\int_0^l r\varphi(r) J_0(k_n r) \mathrm{d}r$ and $D_n = hC_n/q_n$.

50. $\displaystyle\sum_{n=0}^{\infty} A_n I_0 \left(\dfrac{n\pi r}{h} \right) \sin \dfrac{n\pi z}{h},$

where $A_n = \dfrac{2}{h I_0(n\pi a/h)} \displaystyle\int_0^h f(x) \sin \dfrac{n\pi z}{h} \mathrm{d}z$. Specially, when $f(z) = f_0$ is a constant, $A_n = \dfrac{2f_0[1-(-1)^n]}{n\pi I_0(n\pi a/h)}$.

Chapter 5

7. (1) 8. (2) $-1/\pi$. (3) $2/\pi$.
8. $\delta(x - 1) - \delta(x + 1)$.
9. (2) $\dfrac{\mathrm{d}}{\mathrm{d}x^2} \delta(x^2 - 1) = \frac{1}{4}[\delta'(x + 1) - \delta'(x - 1)]$, which represents two dipoles with the opposite signs and same strength $1/4$, at positions $x+1$ and $x-1$, respectively. It can be regarded as a quadrupole.
11. $L[\delta(x - y)] = e^{py}$, $L[\delta^{(n)}(x)] = p^n$. $L^{-1}[u; u \to v] = \delta^{(1)}(v)$, $L^{-1}[u^m; u \to v] = \delta^{(m)}(v)$.

18. $\dfrac{\pi}{2}[\delta(a + b) + \delta(a - b)]$.

19. (1) Let $I_n = \int_{-\infty}^{\infty} e^{at} \sin bt \, \delta^{(n)}(t) dt$. $I_0 = 0$. $I_1 = -b$. $I_2 = 2ab$.

(2) Let $J_n = \int_{-\infty}^{\infty} (\cos t + \sin t) \delta^{(n)} (t^3 + t^2 + t) dt$ and $t_{1,2} = \dfrac{-1 \pm i\sqrt{3}}{2}$.

$$J_0 = 1 - \frac{i}{\sqrt{3}} (t_2(\cos t_1 + \sin t_1) - t_1(\cos t_2 + \sin t_2)).$$

$$J_1 = -1 - \frac{i}{\sqrt{3}} (t_2(\sin t_1 - \cos t_1) - t_1(\sin t_2 - \cos t_2)).$$

23. $\delta(\boldsymbol{r} - \boldsymbol{r}_0) = \dfrac{1}{2\pi r^2 \sin\theta} \delta(r - r_0)\delta(\theta - \theta_0) = \dfrac{1}{2\pi r^2 \sin\theta}\delta(r - r_0)\delta(\theta)$.

24. (1) $\delta(\boldsymbol{r} - \boldsymbol{r}_0) = \dfrac{1}{2\pi r}\delta(r)\delta(z - z_0)$. (2) When $z_0 = 0$, $\delta(\boldsymbol{r} - \boldsymbol{r}_0) = \dfrac{1}{2\pi r}\delta(r)\delta(z)$.

26. $\varphi(x) = ix(e^{\cos x - i \sin x} - 1)$.

Chapter 6

3. $G(x, x') = \left(1 - \dfrac{\sin kx'}{k}\right) e^{ikx}\theta(x - x')$

$\qquad + \left(-\dfrac{\sin kx}{k}e^{ikx'} + e^{ikx}\right)\theta(x' - x)$

4. (1) Piecewise expression method:

$$G(x, x') = \frac{\cos\lambda(\pi - x')}{\lambda\sin\lambda\pi}\cos\lambda x + \frac{\sin\lambda(x - x')}{\lambda}\theta(x - x').$$

Eigenfunction method:

(i) As λ is not an integer,

$$G(x, x'; \lambda) = \frac{1}{\lambda^2\pi} + \frac{2}{\pi}\sum_{n=1}^{\infty}\frac{\cos nx' \cos nx}{\lambda^2 - n^2}.$$

(ii) As λ is an integer m,

$$G(x, x'; m) = \frac{2}{\pi} \cos mx' \cos mx + \frac{1}{m^2\pi}$$

$$+ \frac{2}{\pi} \sum_{n=1, n \neq m}^{\infty} \frac{\cos nx' \cos nx}{m^2 - n^2}, \quad m \neq 0;$$

$$G(x, x'; 0) = \frac{1}{\pi} - \frac{2}{\pi} \sum_{n=1}^{\infty} \frac{\cos nx' \cos nx}{n^2}.$$

(2) Piecewise expression method:

$$G(x, x') = \left[-\frac{1}{4} + \frac{1}{2}(x' - x) \right] \theta(x' - x)$$

$$+ \left[-\frac{1}{4} + \frac{1}{2}(x - x') \right] \theta(x - x')$$

Eigenfunction method:

$$G(x, x'; 0) = -\sum_{n=1}^{\infty} \frac{\varphi_n(x')\varphi_n(x)}{(2n+1)^2},$$

where $\varphi_n(x) = A \sin \lambda_n \pi x + B \cos \lambda_n \pi x$, $\lambda_n = (2n+1)\pi$, $\frac{1}{2}(A^2 + B^2) = 1$.

(3) Piecewise expression method:

$$G(x, x') = \frac{i}{2(e^i - 1)}$$

$$\times [(e^{i(1+x-x')} + e^{-i(x-x')})\theta(x' - x) + (e^{i(x-x')}$$

$$+ e^{-i(1-x+x')})\theta(x - x')].$$

Eigenfunction method:

$$G(x, x'; 1) = -\sum_{n=1}^{\infty} \frac{\varphi_n(x')\varphi_n(x)}{1 - 4n^2\pi^2},$$

where $\varphi_n(x) = A_n \sin 2n\pi x + B_n \cos 2n\pi x$, $\frac{1}{2}(A^2 + B^2) = 1$.

(4) Piecewise expression method:

$$G(x, x') = -\frac{1}{2k(k-1)}[(ke^{-kx'}e^{kx} - e^{kx'}e^{-kx})\theta(x' - x)$$

$$+ (ke^{-kx'}e^{kx} + e^{kx'}e^{-kx})\theta(x - x')].$$

Eigenfunction method:

(i) As $k^2 \neq k_n^2$, $G(x, x'; k^2) = \sum_{n=1}^{\infty} \dfrac{\varphi_n(x')\varphi_n(x)}{k^2 - k_n^2}$.

(ii) As

$$k^2 = k_m^2,$$

$$G(x, x'; k^2) = \varphi_m(x')\varphi_m(x) + \sum_{n=1}^{\infty} \frac{\varphi_n(x')\varphi_n(x)}{k_m^2 - k_n^2};$$

$$\varphi_n(x) = Ae^{k_n x} + Be^{-k_n x}, \, k_0 = \pm 1, \, k_n = in\pi,$$

$$n = 1, 2, \ldots, |A|^2 + |B|^2 = 1.$$

(5) $G(x, x') = \dfrac{x' - x'^2}{2}x - \dfrac{1 + x'}{2}x^2 + \left(\dfrac{x'^2}{2} + \dfrac{1}{2}x^2\right)\theta(x - x')$

5. (1) $G(x, x') = -\dfrac{1}{2}e^{-|x-x'|}$.

(2) $G(x, x') = -\dfrac{1}{2(e^{2a} - e^{-2a})}\{[(e^{2a}e^{-x'} - e^{x'})e^x$

$\qquad + (e^{-2a}e^{x'} - e^{-x'})e^{-x}]\theta(x' - x)$

$\qquad - [(e^{x'} - e^{-2a}e^{-x'})e^x + (e^{-x'} - e^{2a}e^{x'})e^{-x}]$

$\qquad \times \theta(x - x')\}$

9. $G(x, x'; \lambda) = \dfrac{\sin \lambda x \cos \lambda(\pi - x')}{2\lambda \cos^2(\lambda\pi/2)} + \dfrac{\sin \lambda(x' - x)}{\lambda}\theta(x - x')$.

The eigenvalues are $\lambda = 2k + 1$, where the Green's function is of singularities of second order.

10. (2) $G(R) = -\dfrac{\Gamma(2)}{2\pi^2 2R^2} = -\dfrac{1}{4\pi^2|\boldsymbol{r} - \boldsymbol{r}'|^2}$.

11. $G(x, x'; z) = \dfrac{1}{2i\sqrt{z}}\left(e^{i\sqrt{z}|x-x'|} + e^{i\sqrt{z}|x|}e^{i\sqrt{z}|x'|}\dfrac{V_0}{2i\sqrt{z} - V_0}\right)$.

As $V_0 \to \infty$, $G(x, x'; z) = \dfrac{1}{2i\sqrt{z}}(e^{i\sqrt{z}|x-x'|} - e^{i\sqrt{z}|x|}e^{i\sqrt{z}|x'|})$.

12.

$$G_\pm(x, x'; z) = \frac{i}{2\hbar c} \left[\begin{pmatrix} \zeta(z) & \theta(x - x') \\ \theta(x - x') & 1/\zeta(z) \end{pmatrix} e^{ik|x-x'|} \right.$$

$$\left. + \begin{pmatrix} \zeta(z) & \theta(x + x') \\ \theta(x + x') & 1/\zeta(z) \end{pmatrix} e^{\pm ik|x+x'|} \right],$$

where $\zeta(z) = \dfrac{z + mc^2}{\sqrt{z^2 - m^2 c^4}}$ and $\theta(x - x')$ is the step function.

13. Under the condition $\int_a^b dx \rho(x) g^*(x, x_1; \lambda_m^*) \varphi_m(x) = c\varphi_m(x_1)$, where c is a real number, the modified Green's function is of the property $g^*(x_1, x_2; \lambda_m^*) - g(x_2, x_1; \lambda_m) = 0$.

18. The angular Green's function should observe the equations

$$\left(\frac{1}{r^2} \frac{\partial^2}{\partial^2 \theta} + k^2 - k_{m,n}^2 + \frac{m^2}{r^2} \right) g(\theta, \theta') = \delta(\theta - \theta').$$

Since the argument r is concerned in this equation, the piecewise expression method cannot be used along the angular direction.

20. $G(\boldsymbol{r}, \boldsymbol{r}') = \displaystyle\sum_{l=0}^{\infty} \sum_{m=-l}^{l} \sin\theta Y_l^{m*}(\theta', \varphi') Y_l^m(\theta, \varphi) \frac{k^2}{j_l(ka)}$

$$\times \{ [j_l(kr')y_l(ka) - y_l(kr')j_l(ka)]j_l(kr)\theta(r' - r)$$

$$+ [y_l(ka)j_l(kr) - j_l(ka)y_l(kr)]j_l(kr')\theta(r - r') \}$$

21. $G(\boldsymbol{r}, \boldsymbol{r}') = \dfrac{4}{ab} \displaystyle\sum_{n,m=1}^{\infty} \dfrac{\cos k_x x' \cos k_y y' \cos k_x x \cos k_y y}{k^2 - (k_x^2 + k_y^2)}.$

22. Eigenfunction method:

$$G(\boldsymbol{r}, \boldsymbol{r}') = \frac{1}{2\pi} \sum_{n=1}^{\infty} \sum_{m=-\infty}^{\infty} \frac{\varphi_{n,m}(r')\varphi_{n,m}(r)}{k^2 - k_{n,m}^2} e^{im(\theta - \theta')},$$

where $\varphi_{n,m}(r) = AJ_m(k_{n,m}r) + BY_m(k_{n,m}r)$
Piecewise expression method:

$$G(\boldsymbol{r}, \boldsymbol{r}') = \frac{1}{2\pi} \sum_{m=-\infty}^{\infty} g_m(r, r') e^{im(\theta - \theta')},$$

where

$$g_m(r, r') = \frac{U_m(b, r)U_m(a, r')}{U_m(a, b)} - U_m(r, r')\theta(r' - r),$$

where $U_m(x, y) = J_m(kx)Y_m(ky) - J_m(ky)Y_m(kx)$ and k not an eigenvalue.

23. $$G(r, r') = \frac{i\pi k}{2(\pi - \alpha)} \sum_v \cos\frac{n\pi(\theta' - \alpha)}{2(\pi - \alpha)} \cos\frac{n\pi(\theta - \alpha)}{2(\pi - \alpha)}$$

$$\times \begin{cases} H_v^{(2)}(kr')J_v(kr), r' < r \\ J_v(kr')H_v^{(2)}(kr), r > r' \end{cases}$$

24. $q(t) = EC(1 - e^{-t/RC})$.

25. $$G(x, x'; \lambda) = \frac{ie^{i\lambda(x'-x)}}{1 - e^{i\lambda}}\theta(x' - x) + \frac{ie^{i\lambda(1+x'-x)}}{1 - e^{i\lambda}}\theta(x - x').$$

26. $$G_\omega^p(t - t') = \frac{e^{-i\omega(t-t'-T/2)}}{2i\sin(\omega T/2)}\theta(t - t') + \frac{e^{-i\omega(t-t'+T/2)}}{2i\sin(\omega T/2)}\theta(t' - t).$$

Although the boundary condition was regarded as periodic one, the equation itself is not, and thus the solution is not of periodicity: $G_\omega^p(t + T) \neq G_\omega^p(t)$.

$$G_\omega^a(t - t') = \frac{e^{-i\omega(t'-t-T/2)}}{2\cos(\omega T/2)}\theta(t - t') - \frac{e^{-i\omega(t'-t+T/2)}}{2\cos(\omega T/2)}\theta(t' - t).$$

There is no properties such as $G_\omega^a(t) = -G_\omega^a(t + T)$.

Chapter 7

10. $\|f\|_1 = \dfrac{1}{32}$. $\|f\|_2 = \dfrac{1}{8\sqrt{7}}$. $\|f\|_\infty = \dfrac{1}{8}$.

15. (1) $\|A\|_\infty = 5$. $\|A\|_1 = 6$. $\|A\|_2 = 3.7888019$. $\|A\|_F = 2\sqrt{5}$.
    (2) $\|A\|_\infty = |a|$. $\|A\|_1 = |a|$. $\|A\|_2 = |a|$. $\|A\|_F = \sqrt{2}|a|$.

16. $\rho(A) = \sqrt{5}$.

Chapter 8

2. (1) $B = \dfrac{1}{b-a}[\beta - \alpha - \displaystyle\int_a^b (b-u)\varphi(u, f(u))\mathrm{d}u]$,

$A = \alpha - \dfrac{a}{b-a}[\beta - \alpha - \displaystyle\int_a^b (b-u)\varphi(u, f(u))\mathrm{d}u]$.

(2) $A = \alpha - \beta a$, $B = \beta$.

(3) $A = \alpha - b\beta + \displaystyle\int_a^b u\varphi(u, f(u))\mathrm{d}u$, $B = \beta - \displaystyle\int_a^b \varphi(u, f(u))\mathrm{d}u$.

5. (1) $f(x) = x - \cos x + 1 - \displaystyle\int_0^x (x-u)f(u)\mathrm{d}u$.

(2) $f(x) = 2x - \cos x + 1 - \displaystyle\int_0^x (x-u)(1+u^2)f(u)\mathrm{d}u$.

(3) $f(x) = -x\displaystyle\int_0^{\pi/2}(1 - \dfrac{2}{\pi}u)(\varphi(u) - 4f(u))\mathrm{d}u$

$+ \displaystyle\int_0^x (x-u)(\varphi(u) - 4f(u))\mathrm{d}u$.

6. (1) $f(x) = x - \dfrac{1}{2}x^2$. (2) $f(x) = 1 - x$. (3) $f(x) = \dfrac{4x+1}{2x+1}$.

8. (1) $f(x) = e^{\lambda x}$. For any real λ the iteration series converges.
(2) $f(x) = 1 + \dfrac{2\lambda}{2-\lambda}e^{-x}$. For any real λ except 2 the iteration series converges.

9. (1) $f(x) = x$. (2) $f(x) = e^x$. (3) $f(x) = \cos x$. (4) $f(x) = \sin x$.

10. (1) $f(x) = \sin x$. (2) $f(x) = e^x$.

12. If the kernel is chosen as (8.2.42) and the weight is 1, the condition that the Born series converges is that $|V_0| < \dfrac{2\mu\hbar^2\sqrt{\mu\mathrm{Im}q}}{m}$, from which the series does not converge when $\mathrm{Im}q = 0$. If the kernel is chosen as (8.2.47) and the weight is (8.2.48), i.e., by the weight function method, the condition becomes,

$$\dfrac{1}{2w}\left(\dfrac{m}{\hbar^2}\right)^2 V_0^2\left[\dfrac{-8}{(w-\mu)^3} - \dfrac{6}{\mu^2} + \dfrac{1}{w(w-\mu)^2}\right.$$

$$\left. + \dfrac{1}{w^2}\left(\dfrac{1}{(w-\mu)} - \dfrac{1}{\mu}\right)\right] \leq 1$$

as $w = 2\mathrm{Im}\,q \neq 0$, and $V_0 \leq \dfrac{\sqrt{3}\hbar^2\mu^2}{2\sqrt{2m}}$ as $\mathrm{Im}\,q = 0$.

13. The wave function is $\psi(x) = \varphi(x) + \frac{i\alpha\varphi(0)}{1-i\alpha}e^{ik|x|}$ where $k = \sqrt{2mE/\hbar^2}$ and $\alpha = mV_0/\sqrt{2mE}\hbar$.

(1) $V_0 > 0$. If $E < 0$ and $\varphi(x)$ is a bounded state. The second term $\frac{i\alpha\varphi(0)}{1-i\alpha}e^{ik|x|}$ is a bounded state, the resonant energy of which is $E = -mV_0^2/2\hbar^2$.

(2) $V_0 < 0$. If $E > 0$ and $\varphi(x) = \frac{1}{\sqrt{2\pi}}e^{ikx}$, then $\psi(x) = \frac{1}{\sqrt{2\pi}}e^{ikx} + \frac{1}{\sqrt{2\pi}}\frac{i\alpha}{1-i\alpha}e^{ik|x|}$. The reflection and transmission coefficients are $r = \frac{\alpha^2}{1+\alpha^2}$ and $t = \frac{1}{1+\alpha^2}$, respectively.

24. (1) $f(x) = \dfrac{\pi^2}{\pi - 1}\sin^2 x + 2x - \pi$.

(2) $f(x) = 1 + \lambda I = 1 + \dfrac{\lambda}{1+q^2-\lambda} = \dfrac{1+q^2}{1+q^2-\lambda}$. This is a constant, and the condition is $\lambda \neq 1 + q^2$.

(3) $f(x) = \dfrac{2}{2 - \lambda}\sin x$, $\lambda \neq 2$.

(4) $f(x) = \dfrac{1}{\sqrt{1 - x^2}} + \dfrac{\lambda}{1 - \lambda}\dfrac{\pi}{2}\cos^{-1} x$, $\lambda \neq 1$.

(5) $f(x) = x + \lambda I \sin x = x + \lambda\pi^3 \sin x$.

26. $f(x) = \dfrac{\sin x + \cos x}{1 - \lambda/2}$, $\lambda \neq 2$.

27. (1) $f(x) = x + \dfrac{2\lambda}{2 - \lambda}e^{-x}$, $\lambda \neq 2$.

(2) $f(x) = x + \dfrac{\lambda}{\pi}\dfrac{1}{1 - \lambda^2/16 + \lambda^2/4\pi^2}$

$$\times \left\{\left[\left(1+\frac{\lambda}{4}\right)\left(\frac{\pi}{2}-1\right) - \frac{\lambda}{2\pi}\right]\cos x \right.$$
$$\left. - \left[\frac{\lambda}{2\pi}\left(\frac{\pi}{2}-1\right) + 1 - \frac{\lambda}{4}\right]\sin x\right\}, \quad \lambda \neq \pm\frac{4\pi}{\sqrt{\pi^2 - 4}}$$

28. (1) (i) By $|k(x,y)| \leq 1$, the condition for the iteration series to converge is $\lambda < 1/\pi$. (ii) By (8.2.42) and the weight is 1,

the condition is $|\lambda| < \frac{\sqrt{6}}{\pi}$. By the weight function method, the condition becomes $|\lambda| < 2\sqrt{\frac{2}{\pi^2-1}}$. (iii) To the first order, the solution is $f_1(x) = x + 2\lambda \sin x + \lambda\frac{\pi^2-4}{\pi}\cos x$.

(2) $f(x) = x + a_1\dfrac{\lambda}{\pi}\sin x + a_2\dfrac{\lambda}{\pi}\cos x,$

where

$$\lambda_1 = \frac{4}{1+\pi}, \lambda_2 = \frac{4}{1-\pi}$$

and

$$\begin{pmatrix} a_1 \\ a_2 \end{pmatrix} = \frac{\pi}{(1-\lambda/2)^2 - (\lambda\pi/2)^2}\begin{pmatrix} -2(4-\lambda) - \lambda\pi^2 \\ -2\lambda\pi + \pi(4-\lambda) \end{pmatrix}.$$

33. (1) $f(x) = e^x - e^{2x}$.

(2) $f(x) = 1 + \dfrac{1}{n-1} - \dfrac{1}{(n-1)^2} + \dfrac{n}{n-1}x + \dfrac{-n+2}{(n-1)^2}e^{-n+1}$.

(3) $f(x) = 2\sin x - x$.

(4) $f(x) = \dfrac{4}{3} - \dfrac{1}{3}\cos\sqrt{3}x$.

(5) $f(x) = e^x$.

34. (1) $f(x) = x$. (2) $f(x) = -2\sin x$. (3) $f(x) = \cosh x$.

(4) $f(x) = \dfrac{15}{4}x$. (5) $f(x) = \dfrac{1}{4}\left(-\dfrac{1}{x} - x + \dfrac{9}{2}\sin 2x\right)$.

36. $A = \dfrac{b+a\alpha}{(c^2-4d)^{1/2}}$, $B = -\dfrac{b+a\beta}{(c^2-4d)^{1/2}}$, $\alpha = \dfrac{1}{2}[c + (c^2-4d)^{1/2}]$,

$\beta = \dfrac{1}{2}[c - (c^2-4d)^{1/2}]$.

37. (1) $f(x) = \pm A\dfrac{\Gamma^{1/2}(\alpha+1)}{\Gamma(\alpha/2+1/2)}x^{(\alpha-1)/2}$. (2) $f(x) = \pm A\dfrac{e^{\beta x}}{\sqrt{\pi x}}$.

(3) $f(x) = \pm A\Gamma^{1/2}(\alpha+1)e^{\beta x}\dfrac{x^{(\alpha-1)/2}}{\Gamma((\alpha+1)/2)}$.

(4) $f(x) = \pm A\sqrt{\alpha}J_0(\alpha x)$. (5) $f(x) = \pm A\sqrt{\alpha}e^{\beta x}J_0(\alpha x)$

38. (2) $f(x) = -I_1(x)$.

39. (1) $f(x) = \pm\dfrac{15}{4\sqrt{7}}x + \dfrac{5}{4}x^2$. (2) $f(x) = \pm\dfrac{3}{\sqrt{2}}x^2$.

40. (1) $f(x) = 1 + 2x - \dfrac{1}{2}x^2$. (2) $f(x) = \dfrac{5}{2}(7x^3 - 3x)$.

   (3) $f(x) = \displaystyle\sum_{n=0}^{\infty} \dfrac{2n+1}{2}P_n(x)$.

Author Index

Subject Index

Printed in the United States
By Bookmasters